Elementary Algebra
Graphs & Authentic Applications

Jay Lehmann

College of San Mateo

To Keri, who has made this journey
worth every heart-felt step.

PEARSON

Prentice
Hall

Upper Saddle River, New Jersey 07458

Library of Congress Cataloging-in-Publication Data

Lehmann, Jay.
   Elementary algebra : graphs and authentic applications / Jay Lehmann.
      p. cm.
   Includes bibliographical references and index.
   ISBN 0-13-220164-X (alk. paper)
1. Algebra--Textbooks.   I. Title.
   QA152.3.L45 2008
   512.9--dc22

                                                    2007006833

Executive Editor: *Paul Murphy*
Vice President and Editorial Director, Mathematics: *Christine Hoag*
Project Manager: *Mary Beckwith*
Senior Development Editor: *Karen Karlin*
Media Project Manager, Developmental Mathematics: *Audra J. Walsh*
Senior Managing Editor: *Linda Mihatov Behrens*
Associate Managing Editor: *Bayani Mendoza de Leon*
Manufacturing Buyer: *Maura Zaldivar*
Associate Director of Operations: *Alexis Heydt-Long*
Director of Marketing: *Patrice Jones*
Executive Marketing Manager: *Kate Valentine*
Marketing Assistant: *Jennifer de Leeuwerk*
Editorial Assistant/Print Supplements Editor: *Georgina Brown*
Art Director/Interior Designer: *Maureen Eide*
Cover Designers: *Juan R. López/Kristine Carney*
Art Editor: *Thomas Benfatti*
Creative Director: *Juan R. López*
Manager, Cover Visual Research & Permissions: *Karen Sanatar*
Production Management/Composition: *ICC Macmillan, Inc.*
Art Studio: *Scientific Illustrators, Laserwords*
Cover Image: *Aaron Graubart/Stone/Getty Images, Inc.*

© 2008 by Pearson Education, Inc.
Pearson Prentice Hall
Pearson Education, Inc.
Upper Saddle River, New Jersey 07458

All rights reserved. No part of this book may be reproduced, in any form
or by any other means, without permission in writing from the publisher.

Pearson Prentice Hall™ is a trademark of Pearson Education, Inc.

Printed in the United States of America

10 9 8 7 6 5 4 3 2 1

ISBN-13: 978-0-13-220164-3
ISBN-10:   0-13-220164-X

Pearson Education Ltd., *London*
Pearson Education Australia Pty. Limited, *Sydney*
Pearson Education Singapore, Pte. Ltd.
Pearson Education North Asia Ltd., *Hong Kong*
Pearson Education Canada Ltd., *Toronto*
Pearson Educación de Mexico, S.A. de C.V.
Pearson Education—Japan, *Tokyo*
Pearson Education Malaysia, Pte. Ltd

Contents

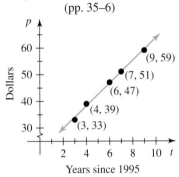

Average Ticket Prices for the Top-50-Grossing Concert Tours (pp. 35–6)

Presidential Election Voter Turnout (p. 85)

Year	Percent of Eligible Voters Who Voted
1980	59.2
1984	59.9
1988	57.4
1992	61.9
1996	54.2
2000	54.7
2004	60.7

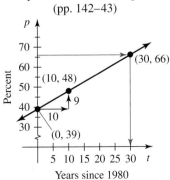

People Who Use Tax Preparers (pp. 142–43)

College Freshmen Whose Average Grade in High School Was an A (p. 207)	
Year	Percent
1990	29.4
1995	36.1
1998	39.8
2001	44.1
2004	47.5

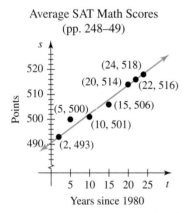

Average SAT Math Scores (pp. 248–49)

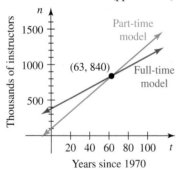

Numbers of Part-Time and Full-Time Instructors (pp. 287–88)

Average Times Spent Driving
Each Day (pp. 347–48)

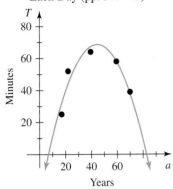

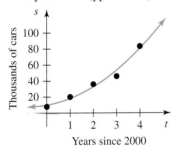

Sales of Gasoline–Electric
Hybrid Cars (pp. 484–85)

Total Revenues from
Snack-Vending Machines
(pp. 441–42)

Year	Revenue (billions of dollars)
1997	22.1
1998	23.3
1999	24.5
2000	25.6
2001	24.3
2002	23.1

Households That Pay Bills
Online (pp. 531–32)

Year	Households That Pay Bills Online (millions)	Total Number of Households (millions)
1998	1.7	102.5
1999	4.1	103.9
2000	5.2	104.7
2001	7.3	108.2
2002	9.9	109.3

Adults Who Watch Cable
Television (pp. 581–82)

11 MORE RADICAL EXPRESSIONS AND EQUATIONS 566

A USING A TI-83 OR TI-84 GRAPHING CALCULATOR 590

Preface

"The question of common sense is always, 'What is it good for?'—a question which would abolish the rose and be answered triumphantly by the cabbage."
—James Russell Lowell

These words seem to suggest that poet and editor James Russell Lowell (1819–1891) took elementary algebra. How many times have your students asked, "What is it good for?" After years of responding "You'll find out in the next course," I began a quest to develop a more satisfying and substantial response to my students' query.

APPROACH AND ORGANIZATION

Curve-Fitting Approach Although there are many ways to center an elementary algebra course around authentic applications, I chose a curve-fitting approach for several reasons. A curve-fitting approach

- allows great flexibility in choosing interesting, authentic, current situations to model.
- emphasizes key concepts and skills in a natural, substantial way.
- deepens students' understanding of equations in two variables, because it requires students to describe these equations graphically, numerically, symbolically, and verbally.
- unifies the many diverse topics of a typical elementary algebra course.

To fit a curve to data, students learn the following four-step modeling process:

1. Examine the data set to determine which type of model, if any, to use.
2. Find an equation of the model.
3. Verify that the model fits the data.
4. Use the model to make estimates and predictions.

This four-step process weaves together topics that are crucial to the course. Students must notice numerical patterns from data displayed in tables, recognize graphical patterns in scattergrams, find equations of models, graph models, and solve equations.

Not only does curve fitting foster cohesiveness within chapters, but it also creates a parallel theme for each set of chapters that introduces and discusses a new type of model. This structure enhances students' abilities to observe similarities and differences among fundamental models such as linear models, quadratic models, rational models, and radical models.

Unique Organization Many college students who take elementary algebra had significant difficulties with the equivalent course in high school. These students face a greater challenge in the college course, because they must complete the course in one semester, rather than in two. Instead of presenting the material in the "same old way," this text provides a unique organization that will better aid students in succeeding.

The text uses modeling to provide the "big picture" before going into details. For example, Chapter 1 gives an overview of linear modeling, which is the main theme of Chapters 1–6, and Section 7.2 provides an overview of quadratic modeling, which is a major focus of Chapters 7–9. Using modeling to provide the big picture not only is good pedagogy, but also sets the tone that this course will be different, interesting, alive, and relevant, inviting students' creativity into the classroom.

The organization of the text is also unique in that students perform graphing (in Chapter 3) before they manipulate expressions and equations (in most of Chapters 4–11). That way, students can visualize the mathematics early on, making them better able to solve equations in one variable graphically. This early-graphing organization also buys students a bit more time to find their "sea legs" before moving on to the more challenging manipulation work.

Several sections discuss how to use graphs and tables to solve equations in one variable.

HOMEWORK SETS

The homework sets have been carefully structured so that the exercises are well paired and progress gradually in difficulty. Students receive ample opportunities to master both procedures and concepts. In addition, two special types of exercises are included to help students succeed in elementary algebra and prepare them for their next course:

Related Review Exercises   These exercises relate current concepts to previously learned concepts. The exercises occur near the end of all Homework sections in Chapters 5–11 and serve as constant reinforcement, helping students draw connections between previously learned and new material.

Expressions, Equations, and Graphs   These exercises are designed to help students gain a solid understanding of those core concepts, including how to distinguish among them. The exercises are included at the end of all Homework sections in Chapters 5–11, but their foundation is laid in Section 4.5, which is devoted to making such distinctions.

BUILDING STUDY SKILLS

The following features have been included throughout the text to help students improve their study skills, to motivate students, and to provide just-in-time support.

Tips for Success   Many sections close with tips for success in mathematics. These tips are intended to help students succeed in the course. A complete listing is included in the index.

Warnings   These are discussions (flagged by the margin entry "WARNING") that address students' common misunderstandings about key concepts and help the students avoid such misunderstandings.

Chapter Opener   Each chapter begins with a description of an authentic situation that can be modeled by the concepts discussed in the chapter.

TECHNOLOGY, EXPLORATIONS, AND LABS

Technology   The text assumes that students have access to technology, such as the TI-83 or TI-84 graphing calculator. Technology of this sort allows students to create scattergrams and check the fit of a model quickly and accurately. It also empowers students to verify their results from Homework exercises and efficiently explore mathematical concepts in the Group Explorations.

The text supports instructors in holding students accountable for all aspects of the course without the aid of technology, including finding equations of linear models. (Linear regression equations are included in the Answers section, because it can be difficult or impossible to anticipate which points a student will choose in trying to find a reasonable equation.)

Appendix A: Using a TI-83 or TI-84 Graphing Calculator   Appendix A contains step-by-step instructions for using the TI-83 and TI-84 graphing calculators. A subset of this appendix can serve as a tutorial early in the course. In addition, when the text requires a new calculator skill, students are referred to the appropriate section in Appendix A.

Group Explorations All sections of the text contain one or two explorations that support student investigation of a concept. Instructors use explorations as collaborative activities during class time or as part of homework assignments. Some explorations lead students to think about concepts introduced in the current section. Other, "Looking Ahead," explorations are directed-discovery activities that introduce key concepts to be discussed in the section that follows. The explorations empower students to become active explorers of mathematics and can open the door to the wonder and beauty of the subject.

"Analyze This" Labs Laboratory assignments have been included at the end of most chapters, to increase students' understanding of concepts and the scientific method. These labs reinforce the idea that mathematics is useful. They are also an excellent avenue for more in-depth writing assignments.

Some of the labs are about global warming and have been written at a higher reading level than the rest of the text in order to give students a sense of what it is like to perform research. Students will find that by carefully reading (and possibly rereading) the background information, they can comprehend the information and apply concepts they have learned in the course to make estimates and predictions about this compelling, current, and authentic situation.

RESOURCES FOR INSTRUCTORS

Instructor's Resource Manual This manual contains suggestions for pacing the course and creating homework assignments. It discusses how to incorporate technology and how to structure lab and project assignments. The manual also contains section-by-section suggestions for presenting lectures and for undertaking the explorations in the text.

Instructor's Solutions Manual This manual includes complete solutions to the even-numbered exercises in the homework sections of the text.

MyMathLab® Instructor Version MyMathLab is a series of text-specific, easily customizable online courses for Prentice Hall textbooks in mathematics and statistics. Powered by CourseCompass™ (Pearson Education's online teaching and learning environment) and MathXL® (our own online homework, tutorial, and assessment system), MyMathLab gives you the tools you need to deliver all or a portion of your course online, whether your students are in a lab setting or working from home. MyMathLab provides a rich and flexible set of course materials, featuring free-response exercises that are algorithmically generated for unlimited practice and mastery. Students can also use online tools, such as animations and a multimedia textbook, to independently improve their understanding and performance. Instructors can use MyMathLab's homework and test managers to select and assign online exercises correlated directly with the textbook, and they can also create and assign their own online exercises and import TestGen tests for added flexibility. MyMathLab's online gradebook—designed specifically for mathematics and statistics—automatically tracks students' homework and test results and gives the instructor control over how to calculate final grades. Instructors can also add offline (paper-and-pencil) grades to the gradebook. MyMathLab is available to qualified adopters. For more information, visit our website at www.mymathlab.com, or contact your Prentice Hall sales representative.

NEW MathXL® Instructor Version MathXL® is a powerful online homework, tutorial, and assessment system that accompanies Prentice Hall textbooks in mathematics or statistics. With MathXL, instructors can create, edit, and assign online homework and tests by using algorithmically generated exercises correlated with the textbook at the objective level. Instructors can also create and assign their own online exercises and import TestGen tests for added flexibility. All student work is tracked in MathXL's online gradebook. Students can take chapter tests in MathXL and receive personalized study plans based on their test results. The study plan diagnoses weaknesses and links students directly to tutorial exercises for the objectives they need to study and retest. MathXL is available to qualified adopters. For more information, visit our website at www.mathxl.com, or contact your Prentice Hall sales representative.

NEW InterAct Math Tutorial Website: www.interactmath.com   Get practice and tutorial help online! This interactive tutorial website provides algorithmically generated practice exercises that correlate directly with the exercises in the textbook. Students can retry an exercise as many times as they like, with new values each time, for unlimited practice and mastery. Every exercise is accompanied by an interactive guided solution that provides helpful feedback for incorrect answers, and students can also view a worked-out sample problem that steps them through an exercise similar to the one they're working on.

TestGen   TestGen enables instructors to build, edit, print, and administer tests by using a computerized bank of questions developed to cover all the objectives of the text. TestGen is algorithmically based, allowing instructors to create multiple, but equivalent, versions of the same question or test with the click of a button. Instructors can also modify test bank questions or add new questions. Tests can be printed or administered online. The software is available on a dual-platform Windows/Macintosh CD-ROM.

RESOURCES FOR STUDENTS

Student Solutions Manual   This manual contains the complete solutions to the odd-numbered exercises in the Homework sections of the text.

MathXL® Tutorials on CD   This interactive tutorial CD-ROM provides algorithmically generated practice exercises that are correlated with the exercises in the textbook at the objective level. Every practice exercise is accompanied by an example and a guided solution designed to involve students in the solution process. The software provides helpful feedback for incorrect answers and can generate printed summaries of students' progress.

GETTING IN TOUCH

I would love to hear from you and would greatly appreciate receiving your comments or questions regarding this text. If you have any questions, please ask them, and I will respond.

Thank you for your interest in preserving the rose.

Jay Lehmann
MathnerdJay@aol.com

To the Student

You are about to embark on an exciting journey. In this course, you will not only learn more about algebra but also find out how to apply algebra to describe and make predications about authentic situations. This text contains data that describe hundreds of situations. Most of the data have been collected from recent newspapers and Internet postings, so the information is current and of interest to the general public. I hope that includes you.

Working with authentic data will make mathematics more meaningful. While working with data about authentic situations, you will learn the meaning of mathematical concepts. As a result, the concepts will be easier to learn, because they will be connected to familiar contexts. And you will see that almost any situation can be viewed mathematically. That vision will help you understand the situation and make estimates and predictions.

Many of the problems you will explore in this course involve data collected in a scientific experiment, survey, or census. The practical way to deal with such data sets is to use technology. So, a graphing calculator or computer system is required.

Hands-on explorations are rewarding and fun. This text contains explorations with step-by-step instructions that will lead you to *discover* concepts, rather than hear or read about them. Because discovering a concept is exciting, it is more likely to leave a lasting impression on you. Also, as you progress through the explorations, your ability to make intuitive leaps will improve, as will your confidence in doing mathematics. Over the years, students have remarked to me time and time again that they never dreamed that learning math could be so much fun.

This text contains special features to help you succeed. Many sections contain a Tips for Success feature. These tips are meant to inspire you to try new strategies to help you succeed in this course and future courses. Some tips might remind you of strategies you have used successfully in the past but have forgotten. If you browse through all of the tips early in the course, you can take advantage of as many of them as you wish. Then, as you progress through the text, you'll be reminded of your favorite strategies. A complete listing of Tips for Success is included in the index.

Other special features that are designed to support you in the course include warnings (flagged by the margin entry "WARNING"), which can help you avoid common misunderstandings; Key Points summaries, which can help you review and retain concepts and skills addressed in the chapter you have just studied; Related Review exercises, which can help you understand current concepts in the context of previously learned concepts; and Expressions, Equations, and Graphs exercises, which can help you understand and distinguish among these three core concepts.

Feel free to contact me. It is my pleasure to read and respond to e-mails from students who are using my text. If you have any questions or comments about the text, feel free to contact me.

Jay Lehmann
MathnerdJay@aol.com

ACKNOWLEDGMENTS

Writing a modeling textbook is an endurance run that I could not have completed without the dedicated assistance of many people. First, I am greatly indebted to Keri, my wife, who yet again served as an irreplaceable sounding board for the multitude of decisions that went into creating this text. In particular, I credit her internal divining rod in selecting captivating data from a mound of data sets I have collected. And thanks to Dylan, our eight-year-old artist, who did rough drafts of many of the cartoons in the text.

I have received much support from the following instructors in my district: Robert Biagini-Komas, Ken Brown, Cheryl Gregory, Bob Hasson, and Rick Hough. Over the years, they have given much sound advice in responding to my countless e-mail inquiries.

I acknowledge several people at Prentice Hall. I am very grateful to Editorial Director Christine Hoag, who has shared in my vision of this text and has made significant investments in making that vision happen. The text has been greatly enhanced through the support of Executive Editor Paul Murphy, who has made a multitude of contributions, including assembling an incredible team to develop and produce the text. The team includes Project Manager Mary Beckwith, who has orchestrated the many aspects of this project, leading to a significantly better book. And I am deeply grateful to Senior Development Editor Karen Karlin, who has given incredible support in clarifying the text, updating hundreds of data sets, and responding to literally thousands of my queries.

I also thank these reviewers, whose thoughtful, detailed comments helped me sculpt this text into its current form:

Scott Adamson, *Chandler-Gilbert Community College*
Thomas Adamson, *Phoenix College*
Mona Baarson, *Jackson Community College*
Sam Bazzi, *Henry Ford Community College*
Joel Berman, *Valencia Community College—East*
Laurie Burton, *Western Oregon University*
James Cohen, *Los Medanos College*
Cathy Gardner, *Grand Valley State University*
Kathryn Gundersen, *Three Rivers Community College*
Tracey Hoy, *College of Lake County*
Charles Klein, *De Anza College*
Diane Mathios, *De Anza College*
Jim Matovina, *Community College of Southern Nevada*
Jane Mays, *Grand Valley State University*
Tim Merzenich, *Chemeketa Community College*
Jason Miner, *Santa Barbara City College*
Camille Moreno, *Cosumnes River College*
Ellen Musen, *Brookdale Community College*
Charlie Naffziger, *Central Oregon Community College*
Chris Nord, *Chemeketa Community College*
Ellen Rebold, *Brookdale Community College*
James Ryan, *State Center Community College District, Clovis*
David Shellabarger, *Lane Community College*
Laura Smallwood, *Chandler-Gilbert Community College*
Janet Teeguarden, *Ivey Tech State College*
Lenove Vest, *Lower Columbia College*
Karen Wiechelman, *University of Louisiana at Lafayette*

Index of Applications

Introduction to Modeling

Shoot for the moon. Even if you miss, you'll land among the stars.

—Les Brown

Table 1 Average Ticket Prices for Top-50-Grossing Concert Tours

Year	Average Ticket Price (dollars)
1998	33
1999	39
2001	47
2002	51
2004	59

Source: *Pollstar*

Think about the last concert you attended. What was the ticket price? Was it worth it? The average ticket price for the top-50-grossing concert tours has increased greatly (see Table 1). In Example 2 of Section 1.4, we will predict when the average ticket price will be $80.

In this course, we will discuss how to describe the relationship between two quantities that occur in an authentic situation. For example, we will describe how the average ticket price for the top-50-grossing concert tours has changed over time. In Chapters 1–6, we will focus on how to use (straight) lines to describe authentic situations. In Chapters 7–11, we will discuss other types of *curves* that can be used to describe authentic situations.

1.1 VARIABLES AND CONSTANTS

Objectives

▹ Know the meaning of *variable* and *constant*.

▹ Know the meaning of *counting numbers, integers, rational numbers, irrational numbers, real numbers, positive numbers,* and *negative numbers*.

▹ Use a number line to describe numbers.

▹ Graph data.

▹ Find the average (or mean) of a group of numbers.

▹ Know how to describe a concept or procedure.

In this section, we work with *variables* and *constants,* two extremely important building blocks of algebra. We also discuss various types of numbers and how to describe numbers visually.

Variables

In arithmetic, we work with numbers. In algebra, we work with *variables* as well as numbers.

DEFINITION Variable

A **variable** is a symbol which represents a quantity that can vary.

For example, we can define h to be the height (in feet) of a specific child. Height is a quantity that varies: As time passes, the child's height will increase. So, h is a variable. When we say $h = 4$, we mean that the child's height is 4 feet.

We will discuss other roles of a variable in Sections 2.1 and 4.3.

Example 1   Using a Variable to Represent a Quantity

1. Let s be a car's speed (in miles per hour). What is the meaning of $s = 60$?
2. Let n be the number of people (in millions) who work from home at least once a week during normal business hours. For the year 2000, $n = 23$. What does that mean in this situation?
3. Let t be the number of years since 2005. What is the meaning of $t = 4$?

Solution

1. The speed of the car is 60 miles per hour.
2. In 2000, 23 million people worked from home at least once a week during normal business hours.
3. $2005 + 4 = 2009$; so, $t = 4$ represents the year 2009.   ■

There are many benefits to using variables. For example, in Problem 2 of Example 1, we found that the simple equation "$n = 23$" means the same thing as the wordy sentence "23 million people worked from home at least once a week during normal business hours." Variables can help us describe some situations with a small amount of writing.

In Problem 3 of Example 1, we described the year 2009 by using $t = 4$. So, our definition of t allows us to use smaller numbers to describe various years—an approach that will be helpful throughout the course.

We will see other benefits of variables as we proceed through the course.

Example 2   Using a Variable to Represent a Quantity

Choose a symbol to represent the given quantity. Explain why the symbol is a variable. Give two numbers that the variable can represent and two numbers that it cannot represent.

1. the weight (in pounds) of a baby at birth
2. the number of people who live in a two-bedroom house

Solution

1. Let w be the weight (in pounds) of a baby at birth. The weight of a baby at birth can vary, so w is a variable. For example, w can represent the numbers 6 and 8, because babies can weigh 6 or 8 pounds at birth. The variable w does not represent 0 or 300, because babies cannot weigh 0 or 300 pounds at birth!
2. Let n be the number of people who live in a two-bedroom house. The number of people who live in a two-bedroom house can vary, so n is a variable. For example, n can represent the numbers 2 and 3, because 2 or 3 people can live in a two-bedroom house. The variable n cannot represent the numbers 5000 or $\frac{1}{2}$, because 5000 people cannot live in a two-bedroom house and half of a person doesn't make sense.   ■

In Problem 1 of Example 2, we stated that the units of w are pounds. Without stating the units of w, "$w = 10$" could mean that the baby's weight was 10 ounces, 10 pounds, or 10 tons! In defining a variable, it is important to describe the variable's units.

Constants

A variable is a symbol that represents a quantity that can vary. When we use a symbol to represent a quantity that does *not* vary, we call that symbol a *constant*. So, 2, 0, 4.8, and π are constants. The constant π is approximately equal to 3.14.

DEFINITION   Constant

A **constant** is a symbol which represents a specific number (a quantity that does *not* vary).

---

1 inch

1 inch

Figure 1 One square inch

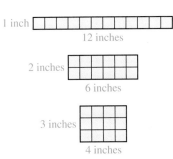

1 inch

12 inches

2 inches

6 inches

3 inches

4 inches

Figure 2 Three possible rectangles of area 12 square inches

In the next example, we will compare the meanings of a variable and a constant while we consider the widths, lengths, and areas of some rectangles. The **area** (in square inches) of a flat surface is the number of square inches that it takes to cover the surface (see Fig. 1). The area of a rectangle is equal to the rectangle's length times its width.

Example 3   Comparing Constants and Variables

A rectangle has an area of 12 square inches. Let W be the width (in inches), L be the length (in inches), and A be the area (in square inches).

1. Sketch three possible rectangles of area 12 square inches.
2. Which of the symbols W, L, and A are variables? Explain.
3. Which of the symbols W, L, and A are constants? Explain.

Solution

1. We sketch three rectangles for which the width times the length is equal to 12 square inches (see Fig. 2).
2. The symbols W and L are variables, since they represent quantities that vary.
3. The symbol A is a constant, because in this problem the area does not vary—the area is always 12 square inches.   ■

Counting Numbers

When we describe people, it often helps to describe them in terms of certain categories, such as gender, ethnicity, and employment. In mathematics, it helps to describe numbers in terms of categories, too. We begin by describing the *counting numbers,* which are the numbers 1, 2, 3, 4, 5, and so on.

DEFINITION   Counting numbers (natural numbers)

The **counting numbers,** or **natural numbers,** are the numbers

$$1, 2, 3, 4, 5, \ldots$$

The three dots mean that the pattern of the numbers shown continues without ending. In this case, the pattern continues with 6, 7, 8, and so on. When a list of numbers goes on forever, we say that there are an *infinite* number of numbers.

Integers

Next, we describe the *integers,* which include the counting numbers and other numbers.

DEFINITION   Integers

The **integers** are the numbers

$$\ldots, -3, -2, -1, 0, 1, 2, 3, \ldots$$

The three dots on both sides mean that the pattern of the numbers shown continues without ending in both directions. In this case, the pattern continues with $-4, -5, -6$, and so on, and with 4, 5, 6, and so on.

The **positive integers** are the numbers 1, 2, 3, ... The **negative integers** are the numbers $-1, -2, -3, \ldots$. The integer 0 is neither positive nor negative. So, the integers consist of the counting numbers (which are positive integers), the negative integers, and 0.

The Number Line

1 unit   1 unit

−3 −2 −1 0 1 2 3

Figure 3 The number line

We can visualize numbers on a *number line* (see Fig. 3).

Each point (location) on the number line represents a number. The numbers increase from left to right. We refer to the distance between two consecutive integers on the number line as 1 *unit* (see Fig. 3).

Example 4   Graphing Integers on a Number Line

Draw dots on a number line to represent the integers between -2 and 3, inclusive.

Solution

The integers between -2 and 3, inclusive, are $-2, -1, 0, 1, 2,$ and 3. "Inclusive" means to include the first and last numbers, which in this case are -2 and 3. We sketch a number line and draw dots at the appropriate locations for the numbers $-2, -1, 0, 1, 2,$ and 3 (see Fig. 4).

Figure 4 Graphing the numbers $-2, -1, 0, 1, 2,$ and 3

When we draw dots on a number line, we say that we are "plotting points" or "graphing numbers."

In Example 4, we worked with the integers between -2 and 3, inclusive: $-2, -1, 0, 1, 2,$ and 3. Here are the integers between -2 and 3: $-1, 0, 1,$ and 2. We did not **WARNING**   include -2 or 3, because the word "inclusive" was not used. When working with such problems, it is important to check whether the word "inclusive" is used.

Rational Numbers

For a fraction $\frac{n}{d}$, we call n the **numerator** and d the **denominator.** The dash between the numerator and the denominator is the **fraction bar:**

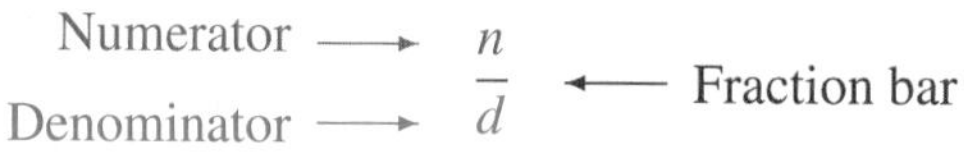

$$\text{Numerator} \longrightarrow \quad \frac{n}{d} \quad \longleftarrow \text{Fraction bar}$$
$$\text{Denominator} \longrightarrow$$

A fraction can be used to describe a part of a whole. For example, consider the meaning of $\frac{5}{8}$ of a pizza. If we divide the pizza into 8 slices of equal area, 5 of the slices make up $\frac{5}{8}$ of the pizza (see Fig. 5).

The number $\frac{5}{8}$ is called a *rational number*.

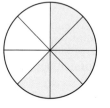

Figure 5 $\frac{5}{8}$ of a pizza

DEFINITION   Rational numbers

The **rational numbers** are the numbers that can be written in the form $\frac{n}{d}$, where n and d are integers and d is nonzero.

We specify that d is nonzero because, as we shall see later, division by zero does not make sense.

Here are some examples of rational numbers:

$$\frac{3}{7} \qquad \frac{-2}{5} \qquad 4 = \frac{4}{1}$$

Rational numbers include all the integers, since any integer n can be written as $\frac{n}{1}$.

Irrational Numbers

There are numbers represented on the number line that are *not* rational. These numbers are called **irrational numbers.** An irrational number *cannot* be written in the form $\frac{n}{d}$, where n and d are integers and d is nonzero. The number $\sqrt{2}$ is the number greater than zero that we multiply by itself to get 2. The number $\sqrt{2}$ is an irrational number. Here are some more examples of irrational numbers:

$$\pi \qquad \sqrt{3} \qquad \sqrt{5}$$

We know that $\sqrt{9} = 3 = \dfrac{3}{1}$, because $3 \times 3 = 9$. So, $\sqrt{9}$ is rational (not irrational).

Decimals

Any rational number or irrational number can be written as a decimal number.

A rational number can be written as a decimal number that either terminates or repeats:

$$\frac{3}{4} = \underbrace{0.75}_{\text{terminates}} \qquad \frac{3}{11} = 0.\underbrace{27272727\ldots}_{\text{repeats}}$$

We can use an overbar to write the repeating decimal $0.272727\ldots = 0.\overline{27}$.

An irrational number can be written as a decimal number that neither terminates nor repeats. It is impossible to write all the digits of an irrational number, but we can approximate the number by rounding. For example, earlier we *approximated* π by rounding to the second decimal place: $\pi \approx 3.14$.

Real Numbers

Recall that each point on the number line represents a number. We call all of the numbers represented by all of the points on the number line the *real numbers*.

> **DEFINITION** Real numbers
>
> The **real numbers** are all of the numbers represented on the number line.

The real numbers are made up of the rational numbers and the irrational numbers. Here are some real numbers:

$$-1.8 \qquad -1 \qquad -\frac{7}{10} \qquad 0 \qquad 0.4 \qquad \frac{6}{5} \qquad \pi$$

We graph these real numbers in Fig. 6.

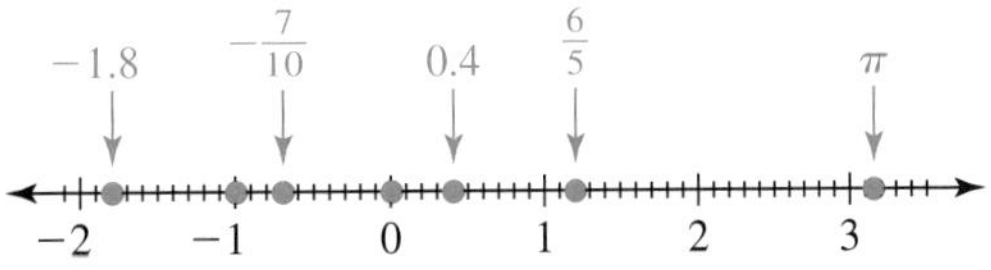

Figure 6 Graphing the real numbers -1.8, -1, $-\dfrac{7}{10}$, 0, 0.4, $\dfrac{6}{5}$, and π

We use an arrow in labeling each point that does not fall on a labeled tick mark.

Example 5 Graphing Real Numbers on a Number Line

Graph the number on a number line.

1. $-\dfrac{7}{4}$ **2.** 2.3

Solution

1. We draw a number line so that the distance between tick marks is $\dfrac{1}{4}$ unit (see Fig. 7).

To graph $-\dfrac{7}{4}$, we draw a dot at the seventh tick mark to the left of 0.

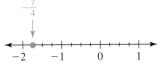

Figure 7 Graphing the number $-\dfrac{7}{4}$

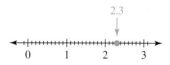

Figure 8 Graphing the number 2.3

2. We draw a number line so that the distance between tick marks is $0.1 = \dfrac{1}{10}$ unit (see Fig. 8). To graph 2.3, we draw a dot at the third tick mark to the right of 2. ∎

Figure 9 illustrates how the various types of numbers we have discussed so far are related. In particular, it shows that every counting number is an integer, every integer is a rational number, and every rational number is a real number. It also shows that irrational numbers are the real numbers that are not rational.

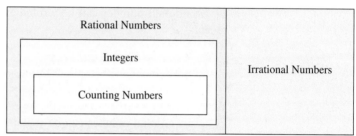

All numbers shown here are real numbers.

Figure 9 The real numbers

Example 6   Identifying Types of Numbers

Consider the following numbers:

$$-8 \qquad -2.56 \qquad 0 \qquad \frac{3}{5} \qquad 2 \qquad \sqrt{13} \qquad 98$$

Which of these numbers are the given type of number?

1. Counting numbers   3. Rational numbers   5. Real numbers
2. Integers   4. Irrational numbers

Solution

1. The counting numbers are 2 and 98.
2. The integers are $-8, 0, 2,$ and 98.
3. The rational numbers are the numbers $-8, -2.56, 0, \dfrac{3}{5}, 2,$ and 98.
4. The irrational number is $\sqrt{13}$.
5. The real numbers are all seven of the numbers. ∎

Graphing Data

Data are values of quantities that describe authentic situations. We often can get a better sense of data by graphing than by just looking at the data values.

Example 7   Graphing Data

Over the course of a semester, a student takes five quizzes. Here are the points he earned on the quizzes, in chronological order: 0, 4, 7, 9, 10. Let q be the number of points earned by the student on a quiz.

1. Graph the student's scores on a number line.
2. Did the quiz scores increase, decrease, stay approximately constant, or none of these?
3. Did the *increases* in the quiz scores increase, decrease, stay approximately constant, or none of these?

Solution

1. We sketch a number line and write "q" to the right of the number line and the units "Points" underneath the number line (see Fig. 10). Then we graph the numbers 0, 4, 7, 9, and 10.

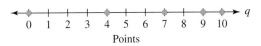

Figure 10 Graphing the quiz scores

2. From the opening paragraph, we know that the quiz scores increased. (From the graph alone, we cannot tell that the quiz scores increased, because the order of the quizzes is not indicated.)

3. As we look from left to right at the points plotted on the graph, we see that the distance between adjacent points decreases. This means that the increases in the quiz scores decreased. That is, the jump from 0 to 4 is greater than the jump from 4 to 7, and so on. ■

It is often helpful to use a single number to represent a group of numbers. One such number is the *average* (or *mean*).

DEFINITION Average, mean

To find the **average** (or **mean**) of a group of numbers, we divide the sum of the numbers by the number of numbers in the group.

For example, to find the average of the quiz scores included in Example 7, we first add the scores: $0 + 4 + 7 + 9 + 10 = 30$. We then divide the total, 30, by the number of quiz scores, 5:

$$30 \div 5 = 6 \text{ points}$$

So, the average quiz score is 6 points.

In general, the average of a group of numbers estimates the center of the numbers graphed on a number line. Figure 11 illustrates this concept for the quiz data.

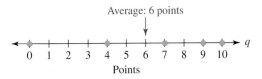

Figure 11 The average quiz score, 6 points, estimates the center of the graphed quiz data

WARNING Always include the units of an average. For example, we say that the average quiz score is 6 *points,* not 6.

Averages can be especially helpful in comparing two data sets. For instance, a student whose average quiz score is 9 points would have performed much better, in general, than a student whose average quiz score is 3 points.

Example 8 Graphing Data and the Mean

The numbers (in thousands) of breast enlargements or breast reconstruction operations in the United States for the years 2000, 2001, 2002, and 2003 are 250, 265, 285, and 300, respectively (Source: *American Society of Plastic Surgeons*). Let n be the number (in thousands) of such operations in a given year.

1. Graph the data. Find the average of the data values and indicate it on the graph.

2. Did the number of operations per year increase, decrease, stay approximately constant, or none of these from 2000 to 2003, inclusive? Explain.

3. Did the *increases* in the number of operations per year increase, decrease, stay approximately constant, or none of these from 2000 to 2003, inclusive? Explain.

Solution

1. We sketch a number line and write "n" to the right of the number line and the units "Thousands of operations" underneath the number line (see Fig. 12). Since the data values are between 250 and 300, inclusive, we write the numbers 250, 260, 270, 280, 290, and 300 equally spaced on the number line. Then we graph the numbers 250, 265, 285, and 300.

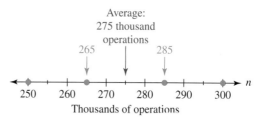

Figure 12 Graphing the data

To find the average, we divide the sum of the data values, $250 + 265 + 285 + 300 = 1100$, by 4: $1100 \div 4 = 275$ thousand operations. We indicate the average, 275 thousand operations, in Fig. 12.

2. From the opening paragraph, we know that the number of operations per year is increasing. (From the graph alone, we cannot tell that the number of operations is increasing, because the years are not indicated.)

3. By reading the opening paragraph of this example and viewing the almost equal spacing between dots on the graph, we see that the increases in the number of operations per year were approximately constant.   ■

In Fig. 12, we wrote the numbers 250, 260, 270, 280, 290, and 300 on the number line. **When we write numbers on a number line, they should increase by a fixed amount and be equally spaced.**

There are two limitations with graphing the data in Example 8 on a *single* number line. One limitation is that the years are not indicated. The other limitation is that it is not clear what to do when the number of operations for two of the years are the same. We will address both limitations in Section 1.2.

Positive and Negative Numbers

The **negative numbers** are the real numbers less than 0, and the **positive numbers** are the real numbers greater than 0 (see Fig. 13).

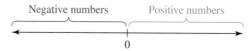

Figure 13 The location of the negative numbers and the positive numbers on the number line

Some examples of negative numbers are -13, -5.2, $-\dfrac{3}{4}$, and $-\sqrt{2}$. Some examples of positive numbers are 13, 5.2, $\dfrac{3}{4}$, and π. As we discussed earlier, the number 0 is neither positive nor negative.

The negative numbers include the negative integers, and the positive numbers include the positive integers.

We say that the *sign* of a negative number is negative and that the *sign* of a positive number is positive.

To include zero, we define the *nonnegative* numbers as the positive numbers together with 0. Likewise, we define the *nonpositive* numbers as the negative numbers together with 0.

Example 9   Graphing a Negative Quantity

A person bounces several checks and, as a result, is charged service fees. If b is the balance (in dollars) of the checking account, what value of b represents the fact that the person owes \$50? Graph the number on a number line.

Solution

Since the person *owes* money, the value of b is negative: $b = -50$. We graph -50 on a number line in Fig. 14.

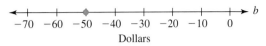

Dollars

Figure 14 Graphing the number $b = -50$ ■

Describing a Concept or Procedure

In some homework exercises, you will be asked to describe, in general, a concept or procedure.

GUIDELINES ON WRITING A GOOD RESPONSE

- Create an example that illustrates the concept or outlines the procedure. Looking at examples or exercises may jump-start you into creating your own example.
- Using complete sentences and correct terminology, describe the key ideas or steps of your example. You can review the text for ideas, but write your description in your own words.
- Describe also the concept or the procedure in general without referring to your example. It may help to reflect on several examples and what they all have in common.
- In some cases, it will be helpful to point out the similarities and the differences between the concept or the procedure and other concepts or procedures.
- Describe the benefits of knowing the concept or the procedure.
- If you have described the steps in a procedure, explain why it's permissible to follow these steps.
- Clarify any common misunderstandings about the concept, or discuss how to avoid making common mistakes when following the procedure.

Example 10   Responding to a General Question about a Concept

Describe the meaning of *variable*.

Solution

Let t be the number of hours that a person works in a department store. The symbol t is an example of a variable, because the value of t can vary. In general, a variable is a symbol that stands for an amount that can vary. A symbol that stands for an amount that does *not* vary is called a constant.

There are many benefits to using variables. We can use a variable to concisely describe a quantity; using the earlier definition of t, we see that the equation $t = 8$ means that a person works in a department store for 8 hours. By using a variable, we can also use smaller numbers to describe various years.

In defining a variable, it is important to describe its units.   ■

group exploration

Reasonable values of a variable

1. Let u be the number of units (credits or hours) a student is currently taking at your college. You may replace the word "units" with "credits" or "hours" if appropriate.
   a. Which of the following values of u are reasonable in this situation? Explain.
      i. $u = 15$ iii. $u = 200$ v. $u = 15.1$
      ii. $u = -5$ iv. $u = 15.5$ vi. $u = 0$
   b. Describe all of the real numbers that are reasonable values of u. Use a number line, a list of numbers, words, or some other way to describe these numbers.

2. A few months ago, a person bought a Porsche 911 Carrera Turbo for $180,000. It has a 16.9-gallon fuel tank. Let g be the amount of gasoline (in gallons) that is in the tank.
   a. Which of the following values of g are reasonable in this situation? Explain.
        i. $g = 7$ **iii.** $g = -4$ **v.** $g = 0$
        ii. $g = 18$ **iv.** $g = 16.9$ **vi.** $g = 10.392$
   b. Describe all of the real numbers that are reasonable values of g.

3. The legal capacity of a club is 180 people. Let n be the number of people who are at the club. You may assume that the number of people in the club never exceeds the legal limit. Describe all of the reasonable values of n.

TIPS FOR SUCCESS: Take Notes

It is always a good idea to take notes during classroom activities. Not only will you have something to refer to later when doing the homework, but also you will have something to help you prepare for tests. In addition, taking notes makes you become even more involved with the material, which means that you will probably increase both your understanding and your retention of it.

HOMEWORK 1.1

FOR EXTRA HELP ▶

Student Solutions Manual   PH Math/Tutor Center   MathXL®   MyMathLab

---

Respond to the questions in Exercises 1–12 by using complete sentences.

1. Let n be the number (in thousands) of fans who attend a Coldplay rock concert. What does $n = 25$ mean in this situation?

2. Let n be the number of home runs hit by Barry Bonds in a season. For 2003, the value of n is 45. What does $n = 45$ mean in this situation?

3. Let n be the number (in millions) of Americans who use a cell phone. For 2003, the value of n is 159 (Source: *Cellular Telecommunications Industry Association*). What does $n = 159$ mean in this situation?

4. Let p be the percentage of children ages 9–13 who participate in organized physical activity. The value of p is about 39 for 2002 (Source: *Infoplease*). What does $p = 39$ mean in this situation?

5. Let s be the annual iPod® sales (in millions). The value of s is 4.4 for 2004 (Source: *iPodlounge*). What does $s = 4.4$ mean in this situation?

6. Let p be the percentage of American workers who are in a union. For 2004, the value of p is 13 (Source: *U.S. Census Bureau*). What does $p = 13$ mean in this situation?

7. Let p be a company's annual profit (in thousands of dollars). What does $p = -45$ mean in this situation?

8. Let T be the temperature (in degrees Fahrenheit). What does $T = -10$ mean in this situation?

9. Let t be the number of years since 2000. What does $t = 9$ mean in this situation?

10. Let t be the number of years since 1995. What does $t = 7$ mean in this situation?

11. Let t be the number of years since 2005. What does $t = -3$ mean in this situation?

12. Let t be the number of years since 2000. What does $t = -2$ mean in this situation?

For Exercises 13–20, choose a variable name for the given quantity. Give two numbers that the variable can represent and two numbers that it cannot represent.

13. The height (in inches) of a person

14. The amount of time (in hours) that a student prepares for an exam

15. The price (in dollars) of an audio CD

16. The number of students enrolled in an algebra class

17. The total time (in hours) that a person works in a week

18. The temperature (in degrees Fahrenheit) in an oven

19. The annual salary (in thousands of dollars) of a person

20. The value (in thousands of dollars) of a new home

21. A rectangle has an area of 24 square inches. Let W be the width (in inches), L be the length (in inches), and A be the area (in square inches).
    a. Sketch three possible rectangles of area 24 square inches.
    b. Which of the symbols W, L, and A are variables? Explain.
    c. Which of the symbols W, L, and A are constants? Explain.

22. A rectangle has an area of 36 square feet. Let W be the width (in feet), L be the length (in feet), and A be the area (in square feet).
    a. Sketch three possible rectangles of area 36 square feet.
    b. Which of the symbols W, L, and A are variables? Explain.
    c. Which of the symbols W, L, and A are constants? Explain.

23. The *perimeter* of a rectangle is the sum of the lengths of all four sides. A rectangle has a perimeter of 20 feet. Let W be the width, L be the length, and P be the perimeter, all with units in feet.
  a. Sketch three possible rectangles of perimeter 20 feet.
  b. Which of the symbols W, L, and P are variables? Explain.
  c. Which of the symbols W, L, and P are constants? Explain.

24. The *perimeter* of a rectangle is the sum of the lengths of all four sides. A rectangle has a perimeter of 16 inches. Let W be the width, L be the length, and P be the perimeter, all with units in inches.
  a. Sketch three possible rectangles of perimeter 16 inches.
  b. Which of the symbols W, L, and P are variables? Explain.
  c. Which of the symbols W, L, and P are constants? Explain.

25. The length of a rectangle is 3 inches more than the width. Let W be the width (in inches), L be the length (in inches), and A be the area (in square inches).
  a. Sketch three possible rectangles of length 3 inches more than the width.
  b. Which of the symbols W, L, and A are variables? Explain.
  c. Which of the symbols W, L, and A are constants? Explain.

26. The length of a rectangle is twice the width. Let W be the width, L be the length, and P be the perimeter, all with units in inches. [**Hint:** *Twice* means to multiply by 2.]
  a. Sketch three possible rectangles in which the length is twice the width.
  b. Which of the symbols W, L, and P are variables? Explain.
  c. Which of the symbols W, L, and P are constants? Explain.

27. The width of a rectangle is 2 yards. Let W be the width, L be the length, and P be the perimeter, all with units in yards.
  a. Sketch three possible rectangles of width 2 yards.
  b. Which of the symbols W, L, and P are variables? Explain.
  c. Which of the symbols W, L, and P are constants? Explain.

28. The width of a rectangle is 5 centimeters. Let W be the width (in centimeters), L be the length (in centimeters), and A be the area (in square centimeters).
  a. Sketch three possible rectangles of width 5 centimeters.
  b. Which of the symbols W, L, and A are variables? Explain.
  c. Which of the symbols W, L, and A are constants? Explain.

Graph all of the given numbers on one number line.

29. $5, -2, 0, -3, 4, -1$

30. $-4, 1, -6, 2, 7, -3$

31. $-\dfrac{2}{3}, -1, \dfrac{7}{3}, 1, -\dfrac{5}{3}, 2$

32. $\dfrac{1}{4}, 0, -2, -\dfrac{5}{4}, \dfrac{9}{4}, 1$

33. $2.8, 3.6, 0.4, 2.5, 1.1, 1.8$ **34.** $3, 1.5, 2.3, 0.9, 2.7, 3.4$

35. $-2, 3.1, 1.2, -1.8, 0.5, 1$ **36.** $1, 0.2, -2.4, -0.7, 1.9, -1$

Graph the numbers on a number line.

37. Counting numbers between 3 and 8

38. Counting numbers between 1 and 5

39. Integers between -2 and 2, inclusive

40. Integers between -4 and 4, inclusive

41. Integers between -1 and 4, inclusive

42. Integers between -6 and 3, inclusive

43. Negative integers between -4 and 4

44. Positive integers between -4 and 4

For Exercises 45–50, consider the following numbers:

$$-9.7 \qquad -4 \qquad 0 \qquad \frac{3}{5} \qquad \sqrt{7} \qquad 3 \qquad \pi \qquad 356$$

Which of these numbers are the given type of number?

45. Counting numbers   **46.** Integers

47. Negative integers   **48.** Rational numbers

49. Irrational numbers   **50.** Real numbers

Give three examples of the following types of numbers.

51. Negative integers

52. Positive integers

53. Negative integers less than -7

54. Negative integers greater than -5

55. Integers that are not counting numbers

56. Rational numbers that are not integers

57. Rational numbers between 1 and 2

58. Irrational numbers between 1 and 10

59. Real numbers between -3 and -2

60. Real numbers that are not rational numbers

Use points on a number line to describe the given values of a variable. Find the average of the values and indicate it on the number line.

61. A student goes to a college for six semesters. Here are the numbers of units (credits or hours) taken per semester: 10, 12, 6, 9, 15, 14. Let u be the number of units taken in one semester.

62. During the summer, a student visits a music website five times. Here are the numbers of songs downloaded: 2, 0, 1, 5, 4. Let n be the number of songs downloaded in one visit to the website.

63. The percentages of disposable personal annual income spent on food for various years are 16%, 14%, 13%, 11%, and 10%. Let p be the percentage of disposable personal annual income spent on food.

64. The percentages of airline flights that are on time for various years are 79%, 82%, 75%, 77%, and 76%. Let p be the percentage of flights in a year that are on time.

For Exercises 65–68, use points on a number line to describe the given values of a variable.

65. The average annual lost time (in hours) due to traffic congestion on highways for various years is 22, 30, 24, 27, and 32. Let L be the average annual lost time (in hours) due to traffic congestion.

66. The U.S. annual per person consumption of sports drinks (in gallons) for various years is 1.9, 2.5, 2.1, 2.3, and 2.2. Let c be the per person consumption (in gallons per year) of sports drinks in a year.

67. The low temperatures (in degrees Fahrenheit) for three days in December in Chicago are 5°F above zero, 4°F below zero, and 6°F below zero. Let F be the low temperature (in degrees Fahrenheit) for one day.

68. Here are a company's annual profits and losses for various years: loss of $5 million, profit of $3 million, and loss of $8 million. Let p be the company's annual profit (in millions of dollars).

69. The number of low-carbohydrate (low-carb) ice cream products in the years 2000, 2001, 2002, 2003, and 2004 is 0, 2, 9, 19, and 62, respectively (Source: *Productscan Online*). Let n be the number of low-carb ice cream products.
    a. Use points on a number line to describe the given values of n. Find the average of the values and indicate it on the number line.
    b. Did the number of low-carb ice cream products increase, decrease, stay approximately constant, or none of these between 2000 and 2004, inclusive? Explain.
    c. Did the annual *increases* in the number of low-carb ice cream products increase, decrease, stay approximately constant, or none of these between 2000 and 2004, inclusive? Explain.

70. The sales (in millions) of digital cameras in the years 2000, 2001, 2002, 2003, and 2004 are 4.5, 7.0, 9.4, 13.0, and 18.2, respectively (Source: *Photo Marketing Association International*). Let s be the digital camera sales (in millions) in a year.
    a. Use points on a number line to describe the given values of s. Find the average of the values and indicate it on the number line.
    b. Did the annual digital camera sales increase, decrease, stay approximately constant, or none of these from 2000 to 2004, inclusive? Explain.
    c. Did the *increases* in the annual digital camera sales increase, decrease, stay approximately constant, or none of these from 2000 to 2004, inclusive? Explain.

71. The number of complaints (in thousands) of identity theft to the Federal Trade Commission (FTC) in the years 2001, 2002, 2003, and 2004 is 86, 162, 215, and 247, respectively (Source: *FTC*). Let n be the number of complaints (in thousands) in a year.
    a. Use points on a number line to describe the given values of n. Find the average of the values and indicate it on the number line.
    b. Did the number of complaints per year increase, decrease, stay approximately constant, or none of these between 2001 and 2004, inclusive? Explain.
    c. Did the *increases* in the number of complaints per year increase, decrease, stay approximately constant, or none of these between 2001 and 2004, inclusive? Explain.

72. The annual total profits (in billions of dollars) of credit card companies in the years 1999, 2000, 2001, 2002, 2003, and 2004 are 14, 20, 24, 27, 30, and 32, respectively (Sources: *CardWeb.com; CardData*). Let p be the total profit (in billions of dollars) of credit card companies in a year.
    a. Use points on a number line to describe the given values of p. Find the average of the values and indicate it on the number line.
    b. Did the annual total profit increase, decrease, stay approximately constant, or none of these from 1999 to 2004, inclusive? Explain.
    c. Did the *increases* in the annual total profit increase, decrease, stay approximately constant, or none of these from 1999 to 2004, inclusive? Explain.

73. Let T be the temperature in degrees Fahrenheit.
    a. What value of T represents the temperature that is $5°F$ below zero?
    b. A student says that T represents only positive numbers and zero, because there is no negative sign. Is the student correct? Explain.

74. A student says that the integers between 2 and 5 are the numbers 2, 3, 4, and 5. Is the student correct? Explain.

75. a. Find the average of each pair of numbers. Then plot the two given numbers and their average on a number line.
       i. 7, 9   **ii.** 1, 5   **iii.** 2, 8
    b. What patterns do you notice in your work in part (a)?
    c. How many real numbers are there between 0 and 1? Explain. [**Hint:** Use the concept of average to help you show that you can keep finding more numbers.]

76. We can describe how far apart two numbers are on the number line. For example, the numbers 3 and 7 are 4 units apart. How far apart are two consecutive integers on the number line? How far apart are two consecutive even integers? How far apart are two consecutive odd integers?

77. List the various types of numbers discussed in this section and describe the meanings of each type. (See page 9 for guidelines on writing a good response.)

78. Describe how to graph a negative quantity. (See page 9 for guidelines on writing a good response.)

1.2 SCATTERGRAMS

Objectives

▸ Know the meaning of *ordered pair, coordinate,* and *coordinate system.*
▸ Create scattergrams.
▸ Know the meaning of *independent variable* and *dependent variable.*
▸ Read bar graphs.
▸ Plot points on a coordinate system.

In Section 1.1, we used a single number line to describe values of a quantity. In this section, we will use a pair of number lines to describe two quantities that are related. For instance, in Example 3 we will compare the average price of a Super Bowl ticket with the Super Bowl number.

When One Number Line Is Not Enough

A number line is convenient for displaying values of a variable. However, there are limitations to using a *single* number line. For example, suppose a student earns the following points (listed in chronological order) from taking five quizzes: 7, 6, 9, 8, 9. We let q stand for the number of points earned by the student on a quiz. We use a number line to graph the scores in Fig. 15.

Figure 15 Graphing the scores

There are two limitations with using one number line to graph these scores. First, the graph does not show that there were two scores of 9 points each. Second, the graph does not show which was the first score, the second score, and so on.

The Coordinate System

There is a way to address both limitations. To begin, we let n be the quiz number. For the first quiz, $n = 1$. For the second quiz, $n = 2$, and so on. Next, we organize the values of the variables n and q (the quiz score) in Table 2.

The "1" and "7" in the first row of Table 2 indicate that when $n = 1$, $q = 7$. This means that the student's score on the first quiz was 7 points. If we agree to write the quiz number first and the quiz score second, we can use the ordered pair $(1, 7)$ to mean that when $n = 1$, $q = 7$. An **ordered pair** is a pair of numbers (written in parentheses and separated by a comma) for which the order of the numbers is meaningful. We call each of the numbers in an ordered pair a **coordinate.** For $(1, 7)$ in this situation, we call 1 the *n-coordinate* and 7 the *q-coordinate*.

The ordered pair $(2, 6)$ indicates that when $n = 2$, $q = 6$. This means that the student's score on the second quiz was 6 points, which agrees with the second row of Table 2.

We call pairs of numbers such as $(3, 9)$ ordered pairs, because the order in which the numbers appear matters: The ordered pair $(3, 9)$ means that the student's score on the third quiz was 9 points, whereas the ordered pair $(9, 3)$ means that the student's score on the ninth quiz was 3 points.

We graph the ordered pairs by using *two* number lines, which are called **axes** (singular: **axis**). To start, we draw a horizontal number line called the *n*-axis and a vertical number line called the *q*-axis (see Fig. 16). We refer to such a pair of axes as a **coordinate system.** The **origin** is the intersection point of the axes. The axes divide the coordinate system into four regions called **quadrants,** which we call Quadrants I, II, III, and IV. The quadrants do not include the axes.

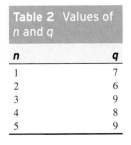

Table 2 Values of n and q

n	q
1	7
2	6
3	9
4	8
5	9

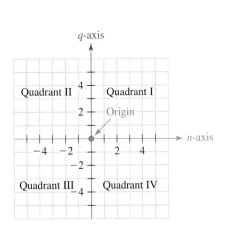

Figure 16 Coordinate system

Scattergrams

Next, we plot the ordered pair (3, 9) shown in the third row of Table 2. To do so, we start at the origin, look 3 units to the right and 9 units up, and then draw a dot (see Fig. 17). In Fig. 18 we plot all the ordered pairs listed in Table 2.

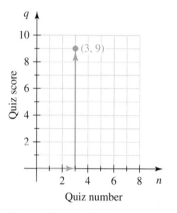

Figure 17 Plot (3, 9)

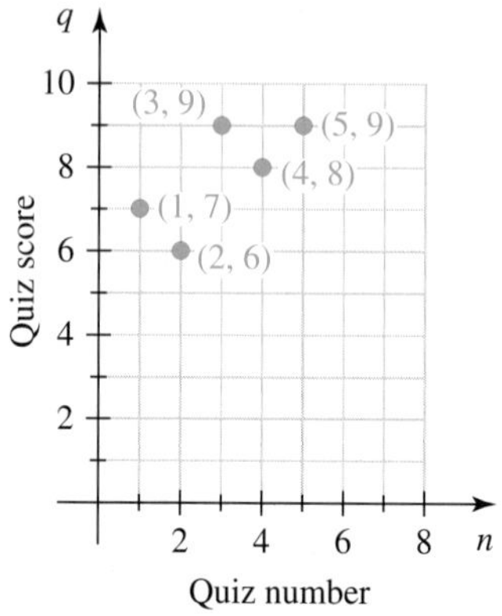

Figure 18 Plot the ordered pairs from Table 2

Note that we have addressed both of the limitations of using just one number line. The coordinate system in Fig. 18 shows that there are two scores of 9, and it also shows which score is from the first quiz, the second quiz, and so on.

As we look at the plotted points in Fig. 18 from left to right, the points, in general, go upward. This means that the quiz scores, in general, are increasing.

A graph of plotted ordered pairs, such as the graph in Fig. 18, is called a **scattergram.**

Independent and Dependent Variables

The student's score q on a quiz depends on the quiz number n. For example, if a certain quiz is more difficult than the others, the student may earn a lower score on that quiz than on the other quizzes. Because q depends on n, we call q the *dependent variable.*

The quiz number does not depend on the student's score, however—n is independent of q. We call n the *independent variable.*

DEFINITION   Identifying independent and dependent variables

Assume that an authentic situation can be described by using the variables t and p, and assume that p depends on t. Then

- We call t the **independent variable.**
- We call p the **dependent variable.**

Example 1   Identifying Independent and Dependent Variables

For each situation, identify the independent variable and the dependent variable.

1. Let L be the loudness (in decibels) of the sound produced by a foghorn that is d miles away from you.
2. A car is traveling at speed s (in mph) on a dry asphalt road, and the brakes are suddenly applied. Let d be the stopping distance (in feet).

Solution

1. The farther you are from the foghorn, the softer the sound you will hear. So, the loudness L depends on the distance d. Thus, L is the dependent variable and d is the independent variable. (Notice that the distance does *not* depend on the loudness.)

2. The greater the traveling speed, the greater the stopping distance will be. So, the stopping distance d depends on the traveling speed s. Thus, d is the dependent variable and s is the independent variable. (Notice that the traveling speed does *not* depend on the stopping distance.) ∎

For an ordered pair (a, b), we write the value of the independent variable in the first (left) position and the value of the dependent variable in the second (right) position. Note that we did just that for the quiz score application. That is, we listed values of the independent variable n in the first position and values of the dependent variable q in the second position. For example, the ordered pair (4, 8) means that when $n = 4$, $q = 8$. In other words, the student's score on the fourth quiz is 8 points.

Example 2   Determining the Meaning of an Ordered Pair

1. Let n be the total number of Quiznos Sub® restaurants at t years since 2000. What does the ordered pair (4, 3500) mean in this situation?
2. Let p be a runner's pulse rate (in beats per minute) when his speed is s miles per hour. What does the ordered pair (10, 160) mean in this situation?

Solution

1. The total number of restaurants depends on the year. So, n is the dependent variable and t is the independent variable. The ordered pair (4, 3500) means that $t = 4$ and $n = 3500$. There were 3500 restaurants in $2000 + 4 = 2004$.
2. The runner's pulse rate depends on his speed. So, p is the dependent variable and s is the independent variable. The ordered pair (10, 160) means that $s = 10$ and $p = 160$. When the runner's speed is 10 miles per hour, his pulse rate is 160 beats per minute. ∎

Table 3 Values of n and q

n	q
1	7
2	6
3	9
4	8
5	9

For tables of ordered pairs, we list the values of the independent variable in the first (left) column and the values of the dependent variable in the second (right) column. For example, in Table 3 the values of n are in the first column and the values of q are in the second column.

For coordinate systems, we describe the values of the independent variable with the horizontal axis and the values of the dependent variable with the vertical axis. For example, in Fig. 18 the horizontal axis is the n-axis and the vertical axis is the q-axis.

Columns of Tables and Axes of Coordinate Systems

Assume that an authentic situation can be described by using two variables. Then

- For tables, the values of the independent variable are listed in the first column and the values of the dependent variable are listed in the second column (see Table 4).
- For coordinate systems, the values of the independent variable are described by the horizontal axis and the values of the dependent variable are described by the vertical axis (see Fig. 19).

Table 4 Position of the Variables

Independent Variable	Dependent Variable
*	*
*	*
*	*

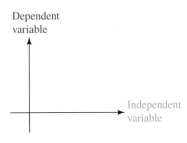

Figure 19 Positions of the variables

More about Scattergrams

Example 3   Creating a Scattergram

Table 5 Average Ticket Prices for the Super Bowl	
Super Bowl Number	**Average Ticket Price (dollars)**
1 (I)	12
5 (V)	15
10 (X)	20
15 (XV)	40
20 (XX)	75
25 (XXV)	150
30 (XXX)	250
35 (XXXV)	325
40 (XL)	550

Source: *NFL.com*

The average ticket prices for the Super Bowl are shown in Table 5 for various years. Let p be the average ticket price (in dollars) and n be the Super Bowl number.

1. Draw a scattergram of the data.
2. For the Super Bowls described in Table 5, which Super Bowl had the highest average ticket price? What was that price?
3. Describe any patterns you see in the prices.

Solution

1. A scattergram of the data is shown in Fig. 20. It makes sense to think of p as the dependent variable, because the average ticket price depends on the Super Bowl number (and not the other way around). So, we let the vertical axis be the p-axis. Note that we write the variable names "n" and "p" and the units "Super Bowl number" and "Dollars" on the appropriate axes.

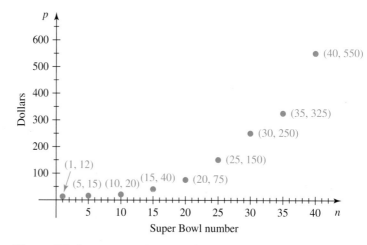

Figure 20 Scattergram of average Super Bowl ticket prices

Recall from Section 1.1 that when we write numbers on an axis, they should increase by a fixed amount and be equally spaced. Since the Super Bowl numbers are between 1 and 40, inclusive, we write the numbers 5, 10, 15, ..., 40 equally spaced on the n-axis. Since the prices shown in Table 5 are between $12 and $550, inclusive, we write the numbers 100, 200, 300, 400, 500, and 600 on the p-axis.

   The ordered pair (1, 12) indicates that the average ticket price was $12 for Super Bowl I.

2. From Table 5 and the scattergram in Fig. 20, we see that the highest average ticket price was $550 in Super Bowl XL.
3. From Table 5 and the scattergram in Fig. 20, we see that average ticket prices have increased. We also see that from Super Bowl I to Super Bowl XXX, inclusive, the *increases* in average ticket prices have increased. That is, as we look from left to right at the first seven points plotted on the coordinate system, we see that the vertical distance between adjacent points increases. ■

In Example 3, creating a scattergram helped us notice some patterns. Such observations will help us make predictions about authentic situations later in this course.

Example 4   Creating a Scattergram with Age Groups

A householder is the person in whose name a house, condominium, or apartment is owned or rented. The percentages of householders who own a home are listed in Table 6 for various age groups.

Table 6 Percentages of Householders Who Own a Home		
Age Group (years)	Age Used to Represent Age Group (years)	Percent
15–24	19.5	18
25–34	29.5	46
35–44	39.5	66
45–54	49.5	75
55–64	59.5	80
65–74	69.5	81
75–84	79.5	77

Source: *U.S. Census Bureau*

Let p be the percentage of householders who own a home when they are at age a years.

1. Draw a scattergram of the data.
2. Describe any patterns in the percentages of householders who own a home.

Solution

1. A look at the first row of Table 6 suggests that we use $a = 19.5$ to represent the age group from 15 years to 24 years. The age 19.5 years is the average of the ages 15 years and 24 years. (Try it.) Likewise, we will use 29.5 to represent the age group from 25 years to 34 years and so on.

   A scattergram of the data is shown in Fig. 21. It makes sense to think of p as the dependent variable, because the percentage of householders depends on the age group (and not the other way around). So, we let the vertical axis be the p-axis. Note that we write the variable names "a" and "p" and the units "Age in years" and "Percent" on the appropriate axes.

   The ordered pair (19.5, 18) indicates that, for the age group from 15 years to 24 years, 18% of householders own a home.

   Since the ages used to represent age groups are between 19.5 years and 79.5 years, inclusive, we write the numbers 10, 20, 30, ..., 80 equally spaced on the a-axis. Since the percents shown in Table 6 are between 18% and 81%, inclusive, we write the numbers 20, 40, 60, 80 and 100 equally spaced on the p-axis.

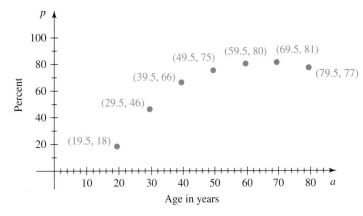

Figure 21 Scattergram of homeowner data

2. From Table 6 and the scattergram in Fig. 21, we see that until about age 70, the percentage of householders who own a home increases as their age increases. After about age 70, the percentage of householders who own a home decreases as their age increases. We also see that until about age 70, the increase in the percentage is decreasing: After every 10 years, the percentage of householders who own a home increases, but by less than it did in the previous 10 years. ∎

In Example 5, we will define a variable to represent time.

Example 5 Defining a Variable for Time

Let t be the number of years since 1990. Find the values of t that represent the years 1990, 1996, 1999, 2000, and 2003.

Solution

We can represent 1990 by $t = 0$, because 1990 is 0 years after 1990. We can represent 1996 by $t = 6$, because 1996 is 6 years after 1990. We list the value of t for each of the years 1990, 1996, 1999, 2000, and 2003 in Table 7.

Table 7 Values of t		
Year	Years since 1990 t	
1990	0	because $1990 - 1990 = 0$
1996	6	$1996 - 1990 = 6$
1999	9	$1999 - 1990 = 9$
2000	10	$2000 - 1990 = 10$
2003	13	$2003 - 1990 = 13$

The values of t in Table 7 are much smaller numbers than the years they represent. When working with authentic situations, we will often perform calculations that involve years. Using definitions similar to the one in Example 5 will enable us to perform those calculations with smaller numbers. It is also easier to label the axes of a coordinate system with smaller numbers.

Example 6 Creating a Scattergram with Zigzag Lines on an Axis

A *life expectancy* is a prediction of how long a person will live. Table 8 shows life expectancies at birth for Americans in various years.

Let L be the life expectancy at birth (in years) for an American born t years after 1980.

1. Create a scattergram of the data.
2. Describe any patterns in life expectancies of Americans.

Table 8 Life Expectancies	
Year of Birth	Life Expectancy (years)
1980	73.7
1985	74.7
1990	75.4
1995	75.8
2000	76.9
2003	77.6

Source: *U.S. Census Bureau*

Table 9 Values of t and L	
t (years since 1980)	L (years of life)
0	73.7
5	74.7
10	75.4
15	75.8
20	76.9
23	77.6

Solution

1. First, we list the values of t and L in Table 9. For example, $t = 0$ represents 1980, because 1980 is 0 years after 1980. Also, $t = 5$ represents 1985, because 1985 is 5 years after 1980.

   A scattergram of the data is shown in Fig. 22. Since L is the dependent variable, we let the vertical axis be the L-axis. We use zigzag lines on the L-axis to indicate that the part of the axis between 0 and about 71 is not displayed. This is done so that we can show a clear view of the data points without having to make the coordinate system exceedingly tall.

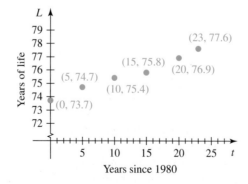

Figure 22 Life expectancy scattergram

2. From Table 9 and Fig. 22, we see that the life expectancy of Americans has been increasing fairly steadily since 1980.

Using zigzag lines on the *L*-axis in Fig. 22 helped us have a clear view of the data points. To see how poor a view we have without the zigzag lines, see Fig. 23.

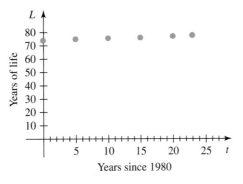

Figure 23 Poor view of life expectancy scattergram (Fig. 22), without zigzag lines

Without the zigzag lines, it is very difficult to plot the data points at the correct height. Also, in Section 1.3 we will begin to estimate coordinates of points by reading graphs, and it would be very difficult to do so accurately by using the graph in Fig. 23. In fact, it is hard to tell that the heights of the points increase from left to right.

Bar Graphs

A **bar graph** is a diagram with two axes that we can use to compare measurements of two or more items (see Fig. 24). Along one axis we list the items, and along the other axis we mark tick marks, write numbers, and write the units of the measurements. We use a bar to indicate the measurement of each item.

Example 7   Reading a Bar Graph

The revenues (in millions of dollars) of some Broadway-based movie musicals are illustrated in the bar graph in Fig. 24 (Source: *Box Office Mojo*).

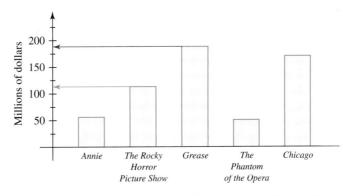

Figure 24 Bar graph of lifetime revenues of some Broadway-based movie musicals

1. Estimate the revenue of *The Rocky Horror Picture Show*.
2. Which Broadway-based movie musical listed in the bar graph has had the largest revenue? What is that revenue?

Solution

1. We look at the top of the bar for *The Rocky Horror Picture Show* and then look to the left at the vertical axis. It appears that the revenue is about $113 million.
2. We identify the tallest bar, which is the one for *Grease*. We look at the top of the bar and then look to the left at the vertical axis. It appears that the revenue is about $188 million. ∎

Plotting Points on a Coordinate System

When we plot points that are not being used to describe authentic situations, we call the horizontal axis the *x-axis* and the vertical axis the *y-axis*. Then *x* is the independent variable and y is the dependent variable. The ordered pair $(6, 3)$ means that $x = 6$ and $y = 3$. So, the x-coordinate is 6 and the y-coordinate is 3.

Example 8   Plotting Points

Plot the points $(3, 4)$, $(-5, -3)$, $(-4, 2)$, and $(5, -4)$ on a coordinate system.

Solution

We plot the ordered pairs $(3, 4)$ and $(-5, -3)$ in Fig. 25, and we plot the ordered pairs $(-4, 2)$ and $(5, -4)$ in Fig. 26.

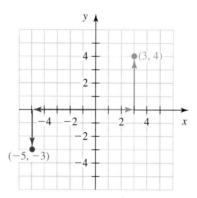

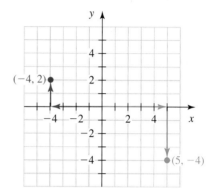

Figure 25 Plotting the ordered pairs $(3, 4)$ and $(-5, -3)$

Figure 26 Plotting the ordered pairs $(-4, 2)$ and $(5, -4)$

group exploration

Looking ahead: Linear modeling

An airplane is beginning to descend. Let a be the altitude (in thousands of feet) of the airplane at t minutes since it began its descent. Some pairs of values of t and a are shown in Table 10.

1. Create a scattergram of the data.
2. Draw a line that contains the points in your scattergram. We call the line a *linear model*.
3. Use your line to estimate the altitude of the airplane 8 minutes after it began its descent. [**Hint:** On the line, locate the point whose t coordinate is 8.]
4. Use your line to estimate when the airplane reached an altitude of 10 thousand feet.
5. What was the altitude of the airplane when it began its descent?
6. Use your line to estimate when the airplane reached the ground.

Table 10 Altitudes of an Airplane

Time (minutes) t	Altitude (thousands of feet) a
0	24
5	20
10	16
15	12
20	8

TIPS FOR SUCCESS: Study Time

For each hour of class time, study for at least two hours outside class. If your math background is weak, you may need to spend more time studying.

   One way to study is to do what you are doing now: Read the text. Class time is a great opportunity to be introduced to new concepts and to see how they fit together with previously learned ones. However, there is usually not enough time to address details as well as a textbook can. In this way, a textbook can serve as a supplement to what you learn in class.

HOMEWORK 1.2 FOR EXTRA HELP ▶

Student Solutions Manual   PH Math/Tutor Center   Math XL MathXL®   MyMathLab MyMathLab

Plot the given points in a coordinate system.

1. (5, 1)   **2.** (2, 3)   **3.** (4, −2)   **4.** (3, −4)

5. (−5, 4)   **6.** (−1, 3)   **7.** (−3, −6)   **8.** (−5, −2)

9. (0, 2)   **10.** (0, −4)   **11.** (−3, 0)   **12.** (1, 0)

13. (2.5, −4.5)   **14.** (−3.5, 1.5)

15. (−1.3, −3.9)   **16.** (−2.4, −4.1)

17. What is the x-coordinate of the ordered pair (2, −4)?

18. What is the y-coordinate of the ordered pair (2, −4)?

For Exercises 19–28, identify the independent variable and the dependent variable.

19. Let n be the number of hours that a student studies for a quiz, and let s be the student's score (in points) on the quiz.

20. Let t be the number of years a person has worked for a company, and let s be the person's salary (in dollars).

21. Let h be the height (in inches) of a girl, and let a be the age (in years) of the girl.

22. Let p be the percentage of colleges that would accept a student whose grade point average (GPA) is g points.

23. Let T be the tuition (in dollars) for enrolling in c credits (units or hours) of classes.

24. Let p be the percentage of men at age a years who have gray hair.

25. Let A be the floor area (in square feet) of a classroom, and let n be the number of students who can comfortably fit into the classroom.

26. A person cooks a potato in an oven for an hour and then removes the potato and allows it to cool. Let t be the number of minutes since the potato was removed from the oven, and let F be the temperature (in degrees Fahrenheit) of the potato.

27. Let t be the number of seconds after a baseball is hit upward, and let h be the baseball's height (in feet).

28. Let p be the percentage of people at age a years who own a computer.

For Exercises 29–36, describe what the given ordered pair represents.

29. Let n be the average number of magazine subscriptions sold per week by a telemarketer who works t hours per week. What does the ordered pair (32, 43) mean in this situation?

30. Let c be the total cost (in dollars) of buying n pens. What does the ordered pair (5, 10) mean in this situation?

31. Let p be the percentage of Americans at age A years who say they volunteer. What does the ordered pair (21, 38) mean in this situation?

32. Let p be the percentage of Americans who have ever purchased a movie on pay-per-view at t years since 1995. What does the ordered pair (6, 30) mean in this situation?

33. Let b be the amount of defense spending (in billions of dollars) at t years since 2000. What does the ordered pair (2, 328) mean in this situation?

34. Let a be the total amount of money spent on ads (in billions of dollars) in the United States at t years since 2000. What does the ordered pair (1, 106.6) mean in this situation?

35. Let p be the number of travelers (in millions) who booked trips online at t years since 2005. What does the ordered pair (−2, 42) mean in this situation?

36. Let p be the percentage of Americans who have confidence in executives running major corporations at t years since 2005. What does the ordered pair (−1, 12) mean in this situation?

37. Create a scattergram of the ordered pairs listed in Table 11.

Table 11 Some Ordered Pairs	
x	**y**
2	5
7	9
11	10
14	9
16	5

38. Create a scattergram of the ordered pairs listed in Table 12.

Table 12 Some Ordered Pairs	
x	**y**
−5	2
−4	3
−1	5
4	9
11	15

39. Find the coordinates of points A, B, C, D, E, and F shown in Fig. 27.

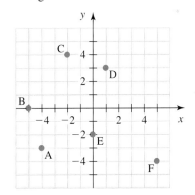

Figure 27 Exercise 39

40. Find the coordinates of points A, B, C, D, E, and F shown in Fig. 28.

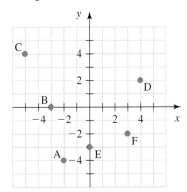

Figure 28 Exercise 40

41. The number of pages in each of the books in the Harry Potter series is listed in Table 13.

Table 13 Numbers of Pages in the Books of the Harry Potter Series	
Book Number	**Number of Pages**
1	309
2	341
3	435
4	734
5	870
6	652

Let p be the number of pages and b be the corresponding book number.
a. Create a scattergram of the data.
b. Which book has the greatest number of pages?
c. From which book to the next did the number of pages increase the most? Explain how you can tell this by inspecting your scattergram.

42. The average life-spans of various denominations of bills are shown in Table 14. For example, the average life-span of a $5 bill is 2 years before it is taken out of circulation due to wear and tear.

Table 14 Life-spans of Denominations of Bills	
Value of Bill (dollars)	**Life-span (years)**
1	1.5
5	2
10	3
20	4
50	9

Source: *Federal Reserve System*

Let L be the life-span (in years) of a bill that is worth d dollars.
a. Create a scattergram of the data.
b. Explain why it makes sense that the average life-span of a $50 bill is greater than the average life-span of a $1 bill.
c. Each year, many more $1 bills are printed than $50 bills. Give at least two reasons that this makes sense.

43. The average times that mothers spend doing paid work per week are shown in Table 15 for various years.

Table 15 Average Time Mothers Spend Doing Paid Work	
Year	**Average Time (hours)**
1965	9
1975	16
1985	21
1995	26
2003	22

Source: *Bureau of Labor Statistics*

a. Let a be the average time (in hours) that mothers spend doing paid work per week at t years since 1960. For example, $t = 0$ represents 1960, and $t = 5$ represents 1965. Create a scattergram of the data.

b. For which of the years shown in Table 15 was the average time that mothers spent doing paid work the greatest? What was that average time?
c. For which of the years shown in Table 15 was the average time that mothers spent doing paid work the least? What was that average time?

44. The numbers of books published about cats are shown in Table 16 for various years.

Table 16 Numbers of Books Published About Cats	
Year	**Number of Books**
1999	138
2000	98
2001	92
2002	73
2003	73
2004	120

Source: *Books in Print, 2005*

a. Let n be the number of books published about cats in the year that is t years since 1990. For example, $t = 0$ represents 1990, and $t = 9$ represents 1999. Create a scattergram of the data.
b. In what year from 1999 to 2004 was there the greatest number of books published about cats? How many were published?
c. In what year from 1999 to 2004 was there the least number of books published about cats? How many were published?

45. The numbers of new products containing the artificial sweetener Splenda® are shown in Table 17 for various years.

Table 17 Numbers of New Products Containing Splenda	
Year	**Number of New Products Containing Splenda**
2000	183
2001	261
2002	365
2003	561
2004	1330

Source: *Productscan Online*

Let n be the number of new products containing Splenda in the year that is t years since 2000.
a. Create a scattergram of the data.
b. Did the number of new products that contain Splenda increase, decrease, stay approximately constant, or none of these? Explain.
c. Did the annual *increase* in the number of new products that contain Splenda increase, decrease, stay approximately constant, or none of these? Explain.

46. The amounts of money that Americans spend annually on clothing for infants, toddlers, and preschoolers are shown in Table 18 for various years. Let s be the annual amount (in billions of dollars) spent by Americans on clothing for infants, toddlers, and preschoolers in the year that is t years since 2000.
a. Create a scattergram of the data.
b. Did the annual amount spent by Americans on clothing for infants, toddlers, and preschoolers increase, decrease, stay approximately constant, or none of these?

Table 18 Monies Spent on Clothing for Infants, Toddlers, and Preschoolers

Year	Money Spent (billions of dollars)
2000	13.2
2001	13.9
2002	14.7
2003	15.4
2004	16.1

Source: *Packaged Facts*

c. Did the *increase* in the annual amount spent by Americans on clothing for infants, toddlers, and preschoolers increase, decrease, stay approximately constant, or none of these?

47. The numbers of automobile accidents per 1000 licensed drivers are shown in Table 19 for various age groups.

Table 19 Automobile Accidents

Age Group (years)	Age Used to Represent Age Group (years)	Accident Rate (number of accidents per 1000 licensed drivers)
16	16	190.3
17	17	163.2
18	18	142.9
19	19	127.8
20–29	24.5	91.4
30–39	34.5	54.7
40–49	44.5	43.9
50–59	54.5	36.4
60–69	64.5	31.3
over 69	75	32.1

Source: *National Highway Traffic Safety Administration*

Let r be the automobile accident rate (number of accidents per 1000 licensed drivers) for licensed drivers at age a years.
a. Create a scattergram of the data.
b. Which age group shown in Table 19 has the lowest accident rate?
c. Which age group shown in Table 19 has the highest accident rate?
d. Between what two consecutive drivers' ages does there seem to be the greatest change in the accident rate? Explain why we can't be sure that this is true, because of the way the data are described in Table 19.
e. Many states put limits on teenage driving. For example, some states do not allow 16-year-old drivers to drive at night. Some states require parental supervision at all times. Why do you think that these regulations were adopted?

48. The percentages of Americans of various age groups who are ordering more takeout food than they did two years ago are shown in Table 20. Let p be the percentage of Americans at age a years who are ordering more takeout food than they did two years ago.
a. Create a scattergram of the data.
b. Which of the points in your scattergram is highest? What does that mean in this situation?
c. Which of the points in your scattergram is lowest? What does that mean in this situation?
d. Do the heights of the points in your scattergram increase or decrease from left to right? What does that mean in this situation?

Table 20 Percentages of Americans Who Are Ordering More Takeout Food than They Did Two Years Ago

Age Group (years)	Age Used to Represent Age Group (years)	Percent
18–24	21.0	34
25–34	29.5	31
35–44	39.5	27
45–54	49.5	17
55–64	59.5	15
over 64	70.0	7

Source: *National Restaurant Association Survey*

49. Several inventions are listed in Table 21, along with the years they were invented and how long it took for one-quarter of the U.S. population to use them ("mass use").

Table 21 Number of Years until Inventions Reached Mass Use

Invention	Year Invented	Years until Mass Use
Electricity	1873	46
Telephone	1876	35
Gasoline-Powered Automobile	1886	55
Radio	1897	31
Television	1923	29
Microwave Oven	1953	36
VCR	1965	13
Personal Computer	1975	16
Mobile Phone	1985	11
CD Player	1985	8
World Wide Web	1991	7
DVD Player	1997	5

Source: *Newsweek*

Let M be the number of years elapsed until an invention reached mass use if it was invented at t years since 1870.
a. Create a scattergram of the data.
b. Compare the time it took to reach mass use for recent inventions versus earlier inventions. In your opinion, why did this happen?
c. Does the datum for the microwave oven fit the pattern you described in part (b)? Explain.
d. For a while after the microwave oven was invented, many people feared that it would cause radiation poisoning, blindness, or impotence. Discuss the impact of these fears in terms of your response to part (c).
e. Explain why the datum for the gasoline-powered automobile does not fit the pattern you described in part (b). Why do you think this happened?

50. The percentages of adults of various age groups who approve of single men raising children on their own are shown in Table 22. Let p be the percentage of adults at age a years who approve of single men raising children on their own.
a. Create a scattergram of the data.
b. Which age group shown in Table 22 has the most faith in single men raising children on their own?
c. Which age group shown in Table 22 has the least faith in single men raising children on their own?

Table 22 Percentages of Adults Who Approve of Single Men Raising Children on Their Own

Age Group (years)	Age Used to Represent Age Group (years)	Percentage
18–34	26.0	81
35–44	39.5	73
45–54	49.5	73
55–64	59.5	66
over 64	70	47

Source: *Taylor Nelson Sofres*

51. The average starting salaries for employees with a bachelor's degree are illustrated in the bar graph in Fig. 29 for various fields of study (Source: *National Association of Colleges and Employers*).

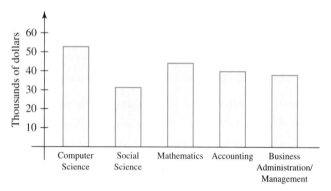

Figure 29 Average starting salaries

a. For which field shown is the average starting salary the highest? What is that salary?

b. For which field shown is the average starting salary the lowest? What is that salary?

c. Estimate the average beginning salary for employees with a mathematics degree.

52. In baseball, a grand slam is a home run with the bases loaded. The top five numbers of career grand slams for major league baseball players are illustrated in Fig. 30 (Source: *MLB.com*).

a. Estimate Robin Ventura's number of career grand slams.

b. Who has the record number of career grand slams? What is that number of grand slams?

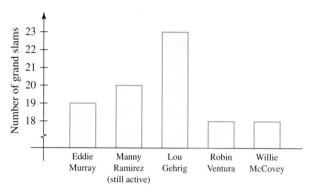

Figure 30 Most career grand slams

c. Of the five baseball players listed in Fig. 30, Manny Ramirez is the only active major league baseball player. Estimate how many grand slams he needs to tie the record.

53. List five ordered pairs whose *x*-coordinate is 3. Then create a scattergram of the ordered pairs. What do you notice about the arrangement of the points in your scattergram? Explain why this makes sense.

54. List five ordered pairs whose *y*-coordinate is 2. Then create a scattergram of the ordered pairs. What do you notice about the arrangement of the points in your scattergram? Explain why this makes sense.

55. The points where the sides of a triangle, rectangle, or any other polygon meet are called *vertices*. A square has vertices at (2, 1) and (2, 5). How many possible positions are there for the other two vertices? Find the coordinates of the vertices for two of these possible positions.

56. The points where the sides of a triangle, rectangle, or any other polygon meet are called *vertices*. A rectangle has vertices at (2, 4) and (7, 4). How many possible positions are there for the other two vertices? Find the coordinates of the vertices for four of these possible positions.

57. Describe the signs of the *x*-coordinate and the *y*-coordinate for a point that lies in the given quadrant.

a. Quadrant I   **c.** Quadrant III

b. Quadrant II   **d.** Quadrant IV

58. Compare a number line with a coordinate system. When is it useful to describe data by a number line? When is it useful to describe data by a coordinate system?

59. Compare the meaning of *dependent variable* with the meaning of *independent variable*.

1.3 EXACT LINEAR RELATIONSHIPS

If we did the things we are capable of, we would astound ourselves.

—*Thomas Edison*

Objectives

▷ Know the meaning of *linearly related, model, linear model, input,* and *output*.

▷ Use a linear model to make estimates and predictions.

▷ Use a scattergram to help decide whether to model a situation with a linear model.

▷ Know the meaning of *intercept*.

▷ Find intercepts of a line and of a linear model.

In this section, we will use a scattergram to help us sketch a line that can be used to describe an authentic situation, such as the descent of a hot-air balloon. We will then use the line to make estimates and predictions about the situation.

Linear Models

Example 1   Using a Line to Describe an Authentic Situation

Table 23 Altitudes of a Balloon	
Time (minutes) t	Altitude (feet) a
0	2200
2	1800
4	1400
6	1000
8	600

A person lowers her hot-air balloon by gradually releasing air from the balloon. Let a be the balloon's altitude (in feet) above the ground after she has released air in the balloon for t minutes. Values of t and a are listed in Table 23.

1. Create a scattergram of the data.
2. Draw the line that contains the points of the scattergram.

Solution

1. We draw a scattergram in Fig. 31.

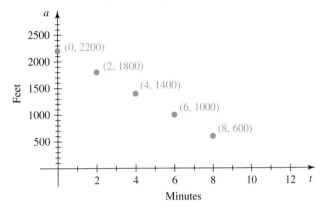

Figure 31 Scattergram of balloon data

2. In mathematics, a "line" means a *straight* line. In Fig. 32, we draw the line that contains the data points shown in Fig. 31.

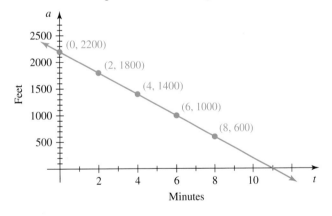

Figure 32 Line that contains the points of the scattergram

The scattergram in Fig. 31 accurately describes the altitude of the balloon at 0, 2, 4, 6, and 8 minutes. But it does not describe the altitude at other times.

If we imagine the line in Fig. 32 to be made up of points, then we can use the line to describe the altitudes at 0, 2, 4, 6, and 8 minutes accurately. We can also use the line to estimate the altitude for other times between 0 and 8 minutes. These results will be good estimates, provided that the altitude of the balloon declined steadily.

In addition, we can use the line to predict altitudes for times a little after 8 minutes. However, these predictions will be accurate only if the altitude of the balloon continued to decline steadily.

If the altitudes of the balloon over a span of time are described accurately by a line, we say that time and altitude (and the variables t and a) are *linearly related* for that span of time.

DEFINITION Linearly related

If two quantities of an authentic situation are described accurately by a line, then the quantities (and the variables representing those quantities) are **linearly related.**

The process of choosing a line to represent the relationship between balloon altitudes and time is an example of *modeling*.

DEFINITION Model

A **model** is a mathematical description of an authentic situation.

We call the line in Fig. 32 a *linear model*. In Chapters 7–11, we will discuss other types of models. The term "model" is being used in much the same way as it is used in "airplane model." Just as an airplane designer can use the behavior of an airplane model in a wind tunnel to predict the behavior of an actual airplane, a linear model can be used to predict what might happen in a situation in which two variables are linearly related.

DEFINITION Linear model

A **linear model** is a line that describes the relationship between two quantities in an authentic situation.

Using a Linear Model to Make Estimates and Predictions

Example 2   Making Estimates and Predictions

1. Use the linear model shown in Fig. 32 to estimate the balloon's altitude when air has been released for 5 minutes.
2. Use the linear model to predict when the balloon's altitude is 400 feet.

Solution

1. To estimate the altitude of the balloon when air has been released for 5 minutes, we locate the point on the linear model where the t-coordinate is 5 (see Fig. 33). The a-coordinate of that point is 1200. So, *according to the model*, the altitude is 1200 feet when air has been released for 5 minutes.

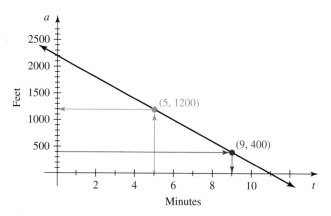

Figure 33 Using the linear model to make an estimate and a prediction

We verify our work by checking that our result is consistent with the values in Table 24. Since the altitude is 1400 feet after 4 minutes and 1000 feet after 6 minutes, it makes sense that the altitude would be between 1000 and 1400 feet when air has been released for 5 minutes, which checks with our result of 1200 feet.

2. To find when the balloon's altitude is 400 feet, we locate the point on the linear model where the a-coordinate is 400 (see Fig. 33). The t-coordinate of that point is 9. So, *according to the linear model,* the altitude is 400 feet when air has been released for 9 minutes.

Table 24 Altitudes of a Balloon

Time (minutes) t	Altitude (feet) a
0	2200
2	1800
4	1400
6	1000
8	600

Again, we verify our work by checking that our result is consistent with the values in Table 24. Since the altitude is 600 feet when air has been released for 8 minutes, it follows that air would have been released for more than 8 minutes for the altitude to be less than 600 feet, which checks with our result of 9 minutes. ■

Input and Output

In Example 2, we found that when the value of the independent variable t is 5, the corresponding value of the dependent variable a is 1200. We say that the *input* 5 leads to the *output* 1200. The blue arrows in Fig. 33 show the action of the input $t = 5$ leading to the output $a = 1200$.

DEFINITION Input, output

An **input** is a permitted value of the *independent* variable that leads to at least one **output**, which is a permitted value of the *dependent* variable.

For a value to be permitted, it must make physical sense and be defined. For instance, in Example 2 the value -50 is not a permitted value of the variable a, because it does not make sense for the balloon's altitude to be -50 feet. Later in the course we will discuss values that are not permitted for mathematical reasons.

Sometimes we will go "backward," from an output back to an input. For instance, in Example 2 we found that the output $a = 400$ originates from the input $t = 9$. The red arrows in Fig. 33 show the action of going backward from the output $a = 400$ to the input $t = 9$.

When to Use a Line to Model Data

Next, we will discuss how to determine whether an authentic situation can be described well by a linear model.

Example 3 Deciding whether to Use a Line to Model Data

Consider the scattergrams of data for situations 1, 2, and 3 shown in Figs. 34, 35, and 36, respectively. For each situation, determine whether a linear model would describe the situation well.

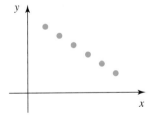

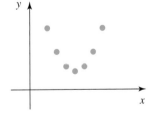

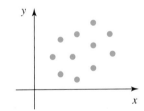

Figure 34 Scattergram for situation 1

Figure 35 Scattergram for situation 2

Figure 36 Scattergram for situation 3

Solution

It appears that the data points for situation 1 lie on a line; a linear model would describe situation 1 well. The data points for situation 2 do not lie close to one line; a linear model would not describe situation 2. (In Chapters 7–9, we will discuss a type of nonlinear model that would describe situation 2 well.) The data points for situation 3 do not lie near a line; a linear model would not describe this situation. ■

We create a scattergram of data to determine whether the data points lie on a line. If the points lie on a line, then we draw the line and use it to make estimates and predictions.

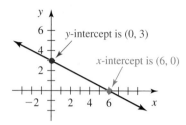

Figure 37 Intercepts of a line

Intercepts of a Line

Consider the line sketched in Fig. 37. The line intersects the x-axis at the point $(6, 0)$. The point $(6, 0)$ is called the *x-intercept*. Also, the line intersects the y-axis at the point $(0, 3)$. The point $(0, 3)$ is called the *y-intercept*.

DEFINITION Intercepts of a line

An **intercept** of a line is any point where the line and an axis (or axes) of a coordinate system intersect. There are two types of intercepts of a line sketched on a coordinate system with an x-axis and a y-axis:

- An **x-intercept** of a line is a point where the line and the x-axis intersect (see Fig. 38). The y-coordinate of an x-intercept is 0.
- A **y-intercept** of a line is a point where the line and the y-axis intersect (see Fig. 38). The x-coordinate of a y-intercept is 0.

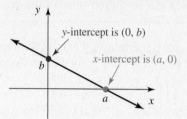

Figure 38 Intercepts of a line

Example 4 Finding Intercepts and Coordinates

Refer to Fig. 39 for the following problems.

1. Find the x-intercept of the line.
2. Find the y-intercept of the line.
3. Find y when $x = -6$.
4. Find x when $y = -3$.

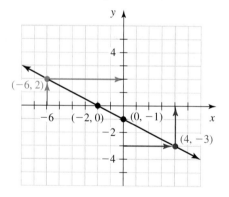

Figure 39 Problems 1–4 of Example 4

Solution

1. The line and the x-axis intersect at $(-2, 0)$. So, the x-intercept is $(-2, 0)$.
2. The line and the y-axis intersect at $(0, -1)$. So, the y-intercept is $(0, -1)$.
3. The blue arrows in Fig. 39 show that the input $x = -6$ leads to the output $y = 2$. So, $y = 2$ when $x = -6$.
4. The red arrows in Fig. 39 show that the output $y = -3$ originates from the input $x = 4$. So, $x = 4$ when $y = -3$. ∎

Intercepts of a Linear Model

Suppose that a linear model describes the relationship between two variables t and p, where p depends on t. Then the t-intercept is a point where the line and the t-axis intersect, and the p-intercept is a point where the line and the p-axis intersect (see Fig. 40).

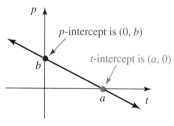

Figure 40 Intercepts of a linear model

Example 5   Finding Intercepts of a Linear Model

1. The linear model from Examples 1 and 2 is shown in Fig. 41. Find the t-intercept of the model. What does it mean in this situation?
2. Find the a-intercept of the model. What does it mean in this situation?

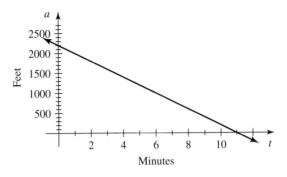

Figure 41 Altitudes of balloon model

Solution

1. The line intersects the t-axis at the point $(11, 0)$. See Fig. 42. So, the t-intercept is $(11, 0)$. This means that when $t = 11$, $a = 0$. The model predicts that the balloon reached the ground 11 minutes after air began to be released from the balloon.

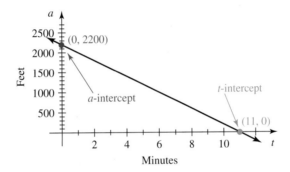

Figure 42 Intercepts of the linear model

2. The line intersects the a-axis at the point $(0, 2200)$. See Fig. 42. So, the a-intercept is $(0, 2200)$. This means that when $t = 0$, $a = 2200$. The model estimates that the balloon's altitude was 2200 feet when air was first released from the balloon. In fact, this estimate is the actual altitude (see Table 24 on p. 26).   ■

Example 6   Using a Linear Model to Make Estimates

An underground rock band manufactures 500 CDs of its original music and tries to sell the CDs at the band's concerts. Let P be the profit (in dollars) from selling a total of n CDs. Some values of n and P are listed in Table 25.

1. Sketch a scattergram of the data. Then draw a reasonable model.
2. Estimate the band's profit from selling all 500 CDs.
3. If the band loses $975, estimate how many CDs will have been sold.
4. Estimate the P-intercept of the model. What does it mean in this situation?
5. Estimate the n-intercept of the model. What does it mean in this situation?

Table 25 Profits from Selling CDs

Sales (number of CDs) n	Profit (dollars) P
50	−1150
100	−800
150	−450
300	600
350	950

Solution

1. The scattergram is shown in Fig. 43 (the black points). Since the points lie on a line, we use the line as a model.
2. The blue arrows show that the input $n = 500$ leads to the output $P = 2000$. So, if the band sells 500 CDs, the profit will be $2000.
3. A negative value of P represents a loss. The red arrows show that the output $P = -975$ originates from the input $n = 75$. So, if the band sells 75 CDs, it will lose $975.

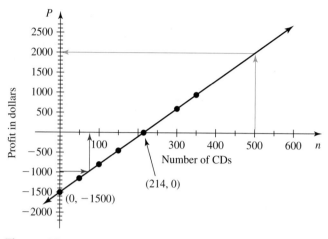

Figure 43 Scattergram and linear model of the CD situation

Melted Zipper members:

Jay   Jim   Steve

4. The linear model and the P-axis intersect at the point $(0, -1500)$. This means that the band will lose \$1500 if no CDs are sold.

5. The linear model and the n-axis intersect at about the point $(214, 0)$. This means that the band will have a profit of about 0 dollars (and hence break even) if 214 CDs are sold. ∎

It is impossible to do a perfect job of sketching models and estimating coordinates of points. Your results for homework exercises will likely be different from the answers provided near the end of this textbook. However, if you do a careful job, your results should be close to those in the book.

group exploration

Looking ahead: Approximately linearly related variables

Most four-year colleges offer an early-decision admissions process, in which students can apply early. By applying early, students can hear back from colleges earlier, which can lessen stress levels for those students who receive acceptance letters. Students who apply via a college's early-acceptance system must commit to enrolling in the college if they are accepted. In a seven-year study, researchers found that a student applicant has a better chance of being accepted to a college through early decision than by regular decision. The number of students who apply to college under early-decision plans has increased greatly since 1997 (see Table 26).

Let n be the number of students (in thousands) who applied to college under early-decision plans at t years since 1995.

1. Create a scattergram of the data.

2. Draw a line that comes close to all of the data points in your scattergram.

3. Use your line to estimate the number of students who applied under early-decision programs in 2004.

4. Use your line to predict when 100 thousand students will apply under early-decision programs.

5. What is the n-intercept of the linear model? What does it mean in this situation?

Table 26 Students Who Applied via Early-Decision Plans

Year	Number of Students (thousands)
1997	42
1998	50
1999	54
2000	63
2001	67
2002	70
2003	79

Source: *The College Board*

TIPS FOR SUCCESS: Get in Touch with Classmates

It is wise to exchange phone numbers and e-mail addresses with some classmates. If any of you has to miss class, then you have someone to contact to find out what you missed and what homework was assigned.

HOMEWORK 1.3 FOR EXTRA HELP ▶

Student Solutions Manual   PH Math/Tutor Center   Math XL MathXL®   MyMathLab MyMathLab

For Exercises 1–6, refer to Fig. 44.

1. Find y when $x = -2$.
2. Find y when $x = 4$.
3. Find x when $y = -2$.
4. Find x when $y = 4$.
5. What is the x-intercept of the line?
6. What is the y-intercept of the line?

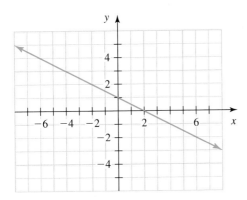

Figure 44 Exercises 1–6

For Exercises 7–12, refer to Fig. 45.

7. Find y when $x = -3$.
8. Find y when $x = 6$.
9. Find x when $y = -3$.
10. Find x when $y = 0$.
11. What is the y-intercept of the line?
12. What is the x-intercept of the line?

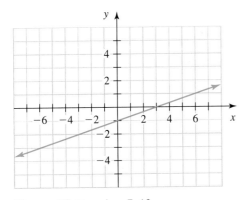

Figure 45 Exercises 7–12

13. Some ordered pairs are listed in Table 27.
  a. Create a scattergram of the points shown in Table 27.
  b. Draw a line that contains the points in your scattergram.
  c. Find y when $x = 3$.
  d. Find x when $y = 6$.

Table 27 Some Ordered Pairs	
x	**y**
2	20
4	16
6	12
8	8
10	4

  e. What is the y-intercept of your line?
  f. What is the x-intercept of your line?

14. Some ordered pairs are listed in Table 28.

Table 28 Some Ordered Pairs	
x	**y**
-6	4
-2	8
2	12
6	16
10	20

  a. Create a scattergram of the points shown in Table 28.
  b. Draw a line that contains the points in your scattergram.
  c. Find y when $x = 4$.
  d. Find x when $y = 17$.
  e. What is the y-intercept of your line?
  f. What is the x-intercept of your line?

15. Water is steadily pumped out of a flooded basement. Let v be the volume of water (in thousands of gallons) that remains in the basement t hours after water began to be pumped. A linear model is shown in Fig. 46.

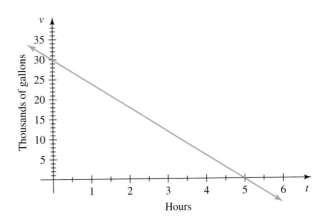

Figure 46 Linear model—Exercise 15

  a. How much water is in the basement after 2 hours of pumping?
  b. After how many hours of pumping will 5 thousand gallons remain in the basement?

c. How much water was in the basement before any water was pumped out?

d. After how many hours of pumping will all the water be pumped out of the basement?

16. Let B be the balance (in dollars) of a student's checking account at t months since the student opened the account. A linear model is shown in Fig. 47.

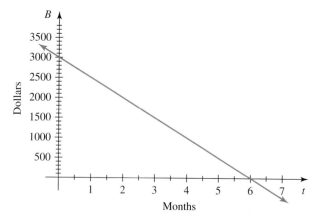

Figure 47 Linear model—Exercise 16

a. What was the balance 3 months after the student opened the account?

b. When was the balance $500?

c. What is the B-intercept of the model? What does it mean in this situation?

d. What is the t-intercept of the model? What does it mean in this situation?

17. A scattergram for a situation is graphed in Fig. 48. Is there a line that is a reasonable model of this situation? If yes, sketch the line. If no, explain why not.

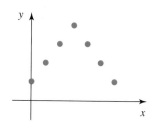

Figure 48 Scattergram—Exercise 17

18. A scattergram for a situation is graphed in Fig. 49. Is there a line that is a reasonable model of this situation? If yes, sketch the line. If no, explain why not.

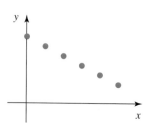

Figure 49 Scattergram—Exercise 18

19. Some ordered pairs are listed in Table 29.

Table 29 Some Ordered Pairs	
x	**y**
0	10
5	5
8	2
10	1
12	2
13	5
14	10

a. Create a scattergram of the data shown in Table 29.

b. Is there a linear relationship between x and y? Explain.

20. Some ordered pairs are listed in Table 30.

Table 30 Some Ordered Pairs	
x	**y**
2	1
3	6
4	9
5	10
6	9
7	6
8	1

a. Create a scattergram of the data shown in Table 30.

b. Is there a linear relationship between x and y? Explain.

21. Let d be the distance traveled (in miles) after a student has driven for t hours (not counting pit stops). Some pairs of values of t and d are shown in Table 31.

Table 31 Times and Distances for a Car	
t (hours)	**d (miles)**
0	0
1	60
2	120
3	180
4	240

a. Create a scattergram of the data. Then draw a linear model.

b. Estimate how far the student has traveled in 2.5 hours.

c. Estimate how long it took the student to travel 210 miles.

22. A student works part time at the college bookstore. Let p be the student's pay (in dollars) for working t hours. Some pairs of values of t and p are shown in Table 32.

Table 32 Pay for Working t Hours	
t (hours)	**p (dollars)**
0	0
5	40
10	80
15	120
20	160

a. Create a scattergram of the data. Then draw a linear model.

b. Estimate the student's pay for working 7 hours.

c. Estimate the number of hours the student must work to earn $96.

23. Let E be a college's enrollment (in thousands of students) at t years since the college began. Some pairs of values of t and E are shown in Table 33.

Table 33 Ages and Enrollments of a College	
t (years)	E (thousands of students)
0	5
1	7
2	9
3	11
4	13

a. Create a scattergram of the data. Then draw a linear model.
b. Predict the enrollment when it has been 6 years since the college opened.
c. Predict when the enrollment will reach 19 thousand students.

24. Let s be a person's salary (in thousands of dollars) after he has worked t years at a company. Some pairs of values of t and s are shown in Table 34.

Table 34 Years Worked and Salary	
t (years)	s (thousands of dollars)
0	20
2	24
4	28
6	32
8	36

a. Create a scattergram of the data. Then draw a linear model.
b. Estimate the person's salary after he has worked 5 years at the company.
c. Estimate when the person's salary will be $34 thousand.
d. What is the s-intercept of the model? What does it mean in this situation?

25. Let v be the value (in dollars) of a company's stock at t years since 2000. Some pairs of values of t and v are shown in Table 35.

Table 35 Values of a Stock	
t (years)	v (dollars)
1	28
2	24
4	16
6	8
7	4

a. Create a scattergram of the data. Then draw a linear model.
b. Estimate when the value of the stock was $12.
c. What is the t-intercept of the model? What does it mean in this situation?
d. What is the v-intercept of the model? What does it mean in this situation?

26. Let p be the profit (in millions of dollars) of a company for the year that is t years since 2000. Some pairs of values of t and p are shown in Table 36.

Table 36 Profits of a Company	
t (years)	p (millions of dollars)
1	20
2	18
3	16
5	12
6	10

a. Create a scattergram of the data. Then draw a linear model.
b. Predict when the profit will be $2 million.
c. What is the p-intercept of the model? What does it mean in this situation?
d. What is the t-intercept of the model? What does it mean in this situation?

27. Let g be the number of gallons of gasoline that remain in a car's gasoline tank after the car has been driven d miles since the tank was filled. Some pairs of values of d and g are shown in Table 37.

Table 37 Miles Traveled and Gallons of Gasoline	
d (miles)	g (gallons)
40	11
80	9
120	7
160	5
200	3
240	1

a. Create a scattergram of the data. Then draw a linear model.
b. Estimate how much gasoline is in the tank after the driver has gone 140 miles since last filling up.
c. Estimate the number of miles driven since the tank was last filled if 2 gallons of gasoline remain in the tank.
d. Find the d-intercept of the model. What does it mean in this situation?
e. Find the g-intercept of the model. What does it mean in this situation?

28. Let v be the value (in thousands of dollars) of a car when it is t years old. Some pairs of values of t and v are listed in Table 38.

Table 38 Ages and Values of a Car	
t (years)	v (thousands of dollars)
1	18
3	14
5	10
7	6
9	2

a. Create a scattergram of the data. Then draw a linear model.
b. Estimate the age of the car when it is worth $4 thousand.
c. Estimate the value of the car when it is 6 years old.
d. What is the v-intercept of the model? What does it mean in this situation?
e. What is the t-intercept of the model? What does it mean in this situation?

29. Let r be the revenue (in millions of dollars) of a company for the year that is t years since 2000. Some pairs of values of t and r are shown in Table 39.

Table 39 Revenues of a Company	
t (years)	r (millions of dollars)
0	8
1	11
3	17
5	23
6	26

a. Create a scattergram of the data. Then draw a linear model.
b. Predict the revenue in 2010.
c. Estimate when the revenue was $14 million.
d. What is the r-intercept of the model? What does it mean in this situation?

30. Let v be the value (in dollars) of a company's stock at t years since 2000. Some pairs of values of t and v are shown in Table 40.

Table 40 Values of a Stock	
t (years)	v (dollars)
0	15
2	19
4	23
5	25
6	27

a. Create a scattergram of the data. Then draw a linear model.
b. Estimate the value of the stock in 2003.
c. Predict when the value of the stock will be $35.
d. What is the v-intercept of the model? What does it mean in this situation?

31. Let a be the altitude (in thousands of feet) of an airplane at t minutes since the airplane began its descent. Some pairs of values of t and a are shown in Table 41.

Table 41 Altitudes of an Airplane	
t (minutes)	a (thousands of feet)
0	36
5	30
10	24
15	18
20	12

a. Create a scattergram of the data. Then draw a linear model.
b. Use your model to estimate the airplane's altitude 12 minutes after it began its descent.
c. Use your model to estimate when the airplane will reach the ground.
d. Assume that your line does a good job of modeling the airplane's descent up until the last 2000 feet, at which point the airplane then descends at a slower rate than before. Is your estimate in part (c) an underestimate or an overestimate? Explain.

32. Let a be the altitude (in feet) of a hot-air balloon after the air in the balloon is released for t minutes. Some pairs of values of t and a are shown in Table 42.

Table 42 Altitudes of a Balloon	
t (minutes)	a (feet)
0	1800
1	1600
3	1200
4	1000
6	600

a. Create a scattergram of the data. Then draw a linear model.
b. Estimate the balloon's altitude after air has been released for 5 minutes.
c. Estimate when the balloon will reach the ground.
d. Assume that your line does a good job of modeling the balloon's descent up until the last 400 feet, at which point the balloon then descends at a faster rate than before. Is your estimate in part (c) an underestimate or an overestimate? Explain.

33. Let p be the percentage of major U.S. firms that perform drug tests on employees and/or job applicants at t years since 1980. Some pairs of values of t and p are shown in Table 43.

Table 43 Percentages of Firms That Perform Drug Tests	
t (years)	p (percent)
7	22
10	51
12	72
15	78
18	74
21	67

Source: *American Management Association*

a. Create a scattergram of the data.
b. Are the variables t and p linearly related? Explain.

34. Let p be the percentage of flights that are delayed at t years since 1995. Some pairs of values of t and p are shown in Table 44.

Table 44 Percentages of Flights That Are Delayed	
t (years)	p (percent)
3	21
4	22
5	24
6	20
7	17
8	16
9	20
10	21

Source: *Bureau of Transportation Statistics*

a. Create a scattergram of the data.
b. Are the variables t and p linearly related? Explain.

35. A student says that the *y*-intercept of the ordered pair $(2, 5)$ is 5. Is the student correct? Explain.

36. A student says that the *x*-intercept of the ordered pair $(-3, 4)$ is -3. Is the student correct? Explain.

37. A student says that the *x*-intercept of a line is $(0, 2)$. Is the student correct? Explain.

38. A student says that the *y*-intercept of a line is $(5, 0)$. Is the student correct? Explain.

39. A student says that the *x*-intercept of a line is 5. Is the student correct? Explain.

40. Are there any lines for which the *x*-intercept is the same point as the *y*-intercept? If yes, sketch such a line, and what is that point? If no, explain why not.

41. Sketch three distinct lines that all have the same *x*-intercept.

42. Sketch three distinct lines that all have the same *y*-intercept.

43. a. Sketch a nonvertical line in a coordinate system. Find any outputs for the given input. State how many outputs there are for that single input.

  i. the input 2   **ii.** the input 4   **iii.** the input -3

  b. For your line, a single input leads to how many outputs? Explain.

  c. For *any* nonvertical line, a single input leads to how many outputs? Explain.

44. In your own words, describe the meaning of *linear model*. (See page 9 for guidelines on writing a good response.)

45. Describe how to find a linear model of a situation and how to use the model to make estimates and predictions. (See page 9 for guidelines on writing a good response.)

1.4 APPROXIMATE LINEAR RELATIONSHIPS

Objectives

▷ Know the meaning of *approximately linearly related*.

▷ Use a linear model to make estimates and predictions.

▷ Find errors in estimations.

▷ Know the meaning of *model breakdown*.

In this section, we will use a line to model a situation in which data points lie close to the line, but not necessarily on the line.

Modeling when Variables Are Approximately Linearly Related

Example 1   Using a Line to Model Data

The average ticket prices for the top-50-grossing concert tours are shown in Table 45 for various years. Let p be the average ticket price (in dollars) at t years since 1995. Sketch a scattergram of the data, and draw a line that comes close to the points of the scattergram.

Table 45 Average Ticket Prices for Top-50-Grossing Concert Tours

Year	Average Ticket Price (dollars)
1998	33
1999	39
2001	47
2002	51
2004	59

Source: *Pollstar*

Solution

First, we list values of t and p in Table 46. For example, $t = 3$ represents 1998, because 1998 is 3 years after 1995; and $t = 4$ represents 1999, because 1999 is 4 years after 1995.

Next, we sketch a scattergram in Fig. 50. It makes sense to think of p as the dependent variable, so we let the vertical axis be the p-axis. Since t is the independent variable, the horizontal axis is the t-axis.

Table 46 Using Values of t to Stand for the Years

Number of Years since 1995 *t*	Average Ticket Price (dollars) *p*
3	33
4	39
6	47
7	51
9	59

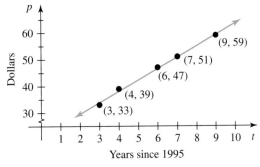

Figure 50 Average ticket price scattergram, and model

Then we sketch a line that comes close to points of the scattergram (see Fig. 50).

The line needn't contain any of the points, but it should come close to all of them. Figure 51 shows that many such lines are possible.

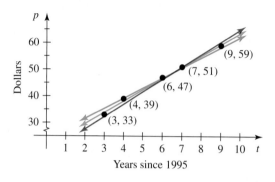

Figure 51 A few of the many reasonable linear models ■

In Fig. 50, we sketched a line that describes the average-ticket-price situation. However, this description is not exact. For example, the line does not describe exactly what happened in the years 1998, 1999, 2001, 2002, or 2004, because the line does not contain any of the data points. However, the line does come quite close to these data points, so it suggests very good approximations for those years.

If the points in a scattergram of data lie close to (or on) a line, we say that the relevant variables are **approximately linearly related.** For the concert tour situation, the variables t and p are approximately linearly related.

Using a Linear Model to Make Estimates and Predictions

Since all of the ticket price data points lie close to our linear model in Fig. 50, it seems reasonable that data points for the years between 1998 and 2004 that are not shown in Table 45 might also lie close to the line. Similarly, it is reasonable that data points for at least a few years before 1998 and for at least a few years after 2004 might also lie near the line.

Example 2 Making Estimates and a Prediction

1. Use the linear model shown in Fig. 50 to estimate the average ticket price in 2000.
2. Use the linear model to estimate the average ticket price in 2003.
3. Use the model to predict in which year the average ticket price will be $80.

Solution

1. The year 2000 corresponds to $t = 5$, because $2000 - 1995 = 5$. The blue arrows in Fig. 52 show that the input $t = 5$ leads to the approximate output $p = 42$. So, according to the model, the average ticket price in 2000 was approximately $42.

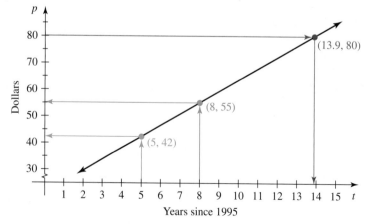

Figure 52 Average-ticket-price model

Table 47 Average Ticket Prices	
Year	**Average Ticket Price (dollars)**
1998	33
1999	39
2001	47
2002	51
2004	59

We verify our work by checking that our result is consistent with the values shown in Table 47. Since the average ticket price was $39 in 1999 and $47 in 2001, it follows that the average ticket price in 2000 probably would be between $39 and $47, which checks with our result of $42.

2. The year 2003 corresponds to $t = 8$, because $2003 - 1995 = 8$. The green arrows in Fig. 52 show that the input $t = 8$ leads to the approximate output $p = 55$. So, according to the model, the approximate average ticket price in 2003 was $55. This result is consistent with the values in Table 47.

3. The red arrows in Fig. 52 show that the output $p = 80$ originates from the approximate input $t = 13.9 \approx 14$. So, according to the linear model, the average ticket price in about $1995 + 14 = 2009$ will be $80.

Since the average ticket price in 2004 was $59 and average ticket prices have been increasing, it follows that the model would predict that the average ticket price sometime *after* 2004 would be $80, which checks with our result of 2009. ∎

We create a scattergram of data to determine whether the relevant variables are approximately linearly related. If so, we draw a line that comes close to the data points and use the line to make estimates and predictions.

WARNING   It is a common error to try to find a line that contains the greatest number of points. However, our goal is to find a line that comes close to *all* of the data points. For example, even though model 1 in Fig. 53 does not contain any of the data points shown, it fits the complete set of data points much better than does model 2, which contains three data points.

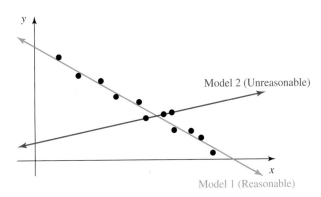

Figure 53 Comparing the fit of two models

Errors in Estimations

WARNING   It is also a common error to confuse the meaning of data points and points that lie on a linear model. Data points are accurate descriptions of an authentic situation. Points on a model may or may not be accurate descriptions.

For example, the data in Table 47 are accurate values for the average ticket prices for the given years. For the linear model in Fig. 50 (p. 35), some points on the line describe the situation well, but some points on the line do not. The advantage of using the linear model is that we can estimate the average ticket price for years other than those in Table 47.

The **error** in an estimate is the amount by which the estimate differs from the actual value. For an overestimate, the error is positive. For an underestimate, the error is negative. If the estimate is equal to the actual value, then the error is 0.

Example 3   Calculating Errors

1. In Example 2, we estimated that the average ticket price was $42 in 2000. The actual average ticket price was $45. Calculate the error in the estimate.
2. In Example 2, we estimated that the average ticket price was $55 in 2003. The actual average ticket price was $52. Calculate the error in the estimate.

Solution

1. Since $45 - 42 = 3$ and we underestimated the actual price, the error is -3 dollars.
2. Since $55 - 52 = 3$ and we overestimated the actual price, the error is 3 dollars. ■

By viewing the average-ticket-price scattergram and model in the same coordinate system, we can see why our estimate of the average ticket price in 2000 is an underestimate. Since the linear model at $(5, 42)$ is *below* the data point $(5, 45)$, the model *underestimates* the average ticket price in 2000 (see Fig. 54).

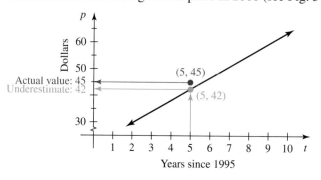

Figure 54 Comparing the data point and the model for 2000

We can also see why our estimate of the average ticket price in 2003 is an overestimate. Since the linear model at $(8, 55)$ is *above* the data point $(8, 52)$, the model *overestimates* the average ticket price in 2003 (see Fig. 55).

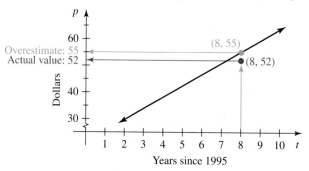

Figure 55 Comparing the data point and the model for 2003

Underestimates and Overestimates

Suppose that an independent variable t and a dependent variable p are approximately linearly related. Then

- If a linear model is below a data point (a, b), the model underestimates the value of p when $t = a$.
- If a linear model is above a data point (a, b), the model overestimates the value of p when $t = a$.

Table 48 Numbers of Viewers of the Miss America Pageant

Year	Number of Viewers (millions)
1990	25.4
1994	20.7
1998	14.3
2002	12.1
2004	9.8

Source: *Nielsen Media Research*

Model Breakdown

If a model gives an estimate that is not reasonably close to the actual value, we say that *model breakdown* has occurred. We will see another type of model breakdown in Example 4.

Example 4   Intercepts of a Linear Model; Model Breakdown

The numbers of television viewers of the Miss America Pageant are shown in Table 48 for various years.

1. Let n be the number of television viewers (in millions) at t years since 1990. Find a linear model that describes the relationship between t and n.
2. Find the n-intercept of the model. What does the n-intercept mean in this situation?

3. In 2004, the ABC network dropped the pageant due to low ratings (only 9.8 million viewers). Suppose that no network will air the show if there are 5 million or fewer viewers. Predict when no network will air the pageant.
4. Find the t-intercept of the model. What does the t-intercept mean in this situation?
5. Use the model to predict the number of viewers in 2017.

Solution

Table 49 Values of t and n for Pageant Data

Number of Years since 1990 t	Number of Viewers (millions) n
0	25.4
4	20.7
8	14.3
12	12.1
14	9.8

1. We describe the data in terms of t and n in Table 49. Next, we sketch a scattergram (see Fig. 56). It appears that t and n are approximately linearly related, so we sketch a line that comes close to the data points.

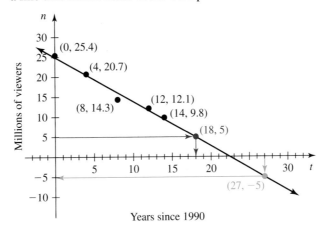

Figure 56 Pageant scattergram and model

2. The n-intercept is $(0, 25)$, or $n = 25$ when $t = 0$. According to the model, there were 25 million viewers in 1990. This estimate is close to the actual 25.4 million viewers in that year.
3. The red arrows in Fig. 56 show that the output $n = 5$ originates from the approximate input $t = 18$. According to the model, there will be 5 million viewers in $1990 + 18 = 2008$. If this viewership trend continues, no networks will air the pageant in 2008 and thereafter.
4. The approximate t-intercept is $(22, 0)$, or $n = 0$ when $t = 22$, which represents the year $1990 + 22 = 2012$. According to the model, if the pageant is on the air in 2012, there will be no viewers. However, one would expect that, at minimum, friends and relatives of the contestants would watch! If we continue to assume that no network will air the pageant if there are 5 million or fewer viewers, then the pageant will no longer be on the air (see Problem 3 of this solution).
5. The year 2017 corresponds to $t = 2017 - 1990 = 27$. To predict the number of viewers, we locate the point on the linear model where the t-coordinate is 27 and see that the n-coordinate is about -5 (see the blue arrows in Fig. 56). So, according to the model, there will be -5 million viewers in 2017. This prediction does not make sense. ■

In Problem 5 in Example 4, we found that when $t = 27$, $n = -5$. The model predicts that there would be -5 million viewers in 2017, which does not make sense. So, $t = 27$ is *not* an input, and $n = -5$ is *not* an output. This is an example of model breakdown.

DEFINITION Model breakdown

When a model yields a prediction that does not make sense or an estimate that is not a good approximation, we say that **model breakdown** has occurred.

If you are using a model to make an estimate or a prediction and model breakdown occurs, you should describe the result, state that model breakdown occurred, and explain why you know it has occurred.

In Section 1.3, we discussed why your estimates and predictions in the homework exercises will likely be different from the answers given near the end of this textbook. Here we note another reason that this will likely happen: If variables are approximately linearly related, there will be many reasonable linear models to choose from. However, if you do a careful job, your results should be close to those in the textbook.

group exploration

Looking ahead: Expressions

1. An instructor adds 5 bonus points to each student's test score. For the given original test score, find the original test score plus 5 points.
   a. 74 points
   b. 88 points
   c. *r* points [**Hint:** Think about how you got your results for parts (a) and (b).]

2. A student gets paid $8 per hour for working at a music store. For the given number of hours worked, find the total amount of money earned (in dollars).
   a. 5 hours
   b. 9 hours
   c. *t* hours

3. In a lottery, some people won $200, which they will share equally. For the given number of people, find each person's share (in dollars).
   a. 4 people
   b. 10 people
   c. *n* people

TIPS FOR SUCCESS: Practice Exams

When studying for an exam (or quiz), try creating your own exam to take for practice. To create your exam, select several homework exercises from each section on which you will be tested. Choose a variety of exercises that address concepts your instructor has emphasized. Your test should include many exercises that are moderately difficult and some that are challenging. Completing such a practice test will help you reflect on important concepts and pin down what types of problems you need to study more.

It is a good idea to work on the practice exam for a predetermined period. Doing so will help you get used to a timed exam, build your confidence, and lower your anxiety about the real exam.

If you are studying with another student, you can each create a test and then take each other's test. Or you can create a test together and each take it separately.

HOMEWORK 1.4

FOR EXTRA HELP ▶

 Student Solutions Manual   PH Math/Tutor Center   Math XL MathXL®   MyMathLab MyMathLab

---

1. Some ordered pairs are listed in Table 50.

   Table 50 Some Ordered Pairs

x	y
1	13
3	9
5	8
7	4
9	2

   a. Create a scattergram of the points shown in Table 50.
   b. Are the variables *x* and *y* linearly related, approximately linearly related, or neither?
   c. Draw a line that comes close to the points in your scattergram.

   d. Which point on your line has *x*-coordinate 8?
   e. Which point on your line has *y*-coordinate 6?
   f. What is the *y*-intercept of your line?
   g. What is the *x*-intercept of your line?

2. Some ordered pairs are listed in Table 51.

   Table 51 Some Ordered Pairs

x	y
−8	−5
−5	−2
−2	0
0	5
3	7

   a. Create a scattergram of the points shown in Table 51.

b. Are the variables x and y linearly related, approximately linearly related, or neither?

c. Draw a line that comes close to the points in your scattergram.

d. Which point on your line has x-coordinate -1?

e. Which point on your line has y-coordinate -3?

f. What is the y-intercept of your line?

g. What is the x-intercept of your line?

3. The number of presidential polls conducted during the first seven months of an election year has increased greatly in the past two decades (see Table 52).

Table 52 Numbers of Presidential Election Polls	
Year	**Number of Polls**
1980	26
1984	42
1988	50
1992	86
1996	99
2000	136

Source: *Roper Center for Public Opinion Research*

a. Let n be the number of presidential polls conducted during the first seven months of the election year that is t years since 1980. For example, $t = 0$ represents 1980, and $t = 4$ represents 1984. Create a scattergram of the data.

b. Draw a line that comes close to the points in your scattergram.

c. Predict in which presidential election there will be about 150 polls in the first seven months of the year. [**Hint:** Round your result to the nearest presidential election year.]

d. Predict the number of polls in the first seven months of the 2008 presidential election year. About how many polls is this per week?

4. The **median** of a group of numbers is the number in the middle (or the average of the two numbers in the middle) when the numbers are listed in order from smallest to largest. The median ages of cars in the United States are shown in Table 53 for various years.

Table 53 Median Ages of Automobiles	
Year	**Median Age (years)**
1970	4.9
1975	5.4
1980	6.0
1985	6.9
1990	6.5
1995	7.7
2000	8.3
2003	8.6

Source: *Bureau of Transportation Statistics*

a. Let a be the median age (in years) of cars at t years since 1970. For example, $t = 0$ represents 1970, and $t = 5$ represents 1975. Create a scattergram of the data.

b. Draw a line that comes close to the points in your scattergram.

c. Use your model to predict the median age of cars in 2011.

d. Use your model to estimate when the median age of cars was 8 years.

5. The total numbers of animal and plant species in the United States that are listed as endangered or threatened are shown in Table 54 for various years.

Table 54 Numbers of Endangered or Threatened Species	
Year	**Number of Species Listed**
1980	281
1985	384
1990	596
1995	962
2000	1244
2005	1264

Source: *U.S. Fish and Wildlife Service*

Let n be the number of species that are listed as endangered or threatened at t years since 1980.

a. Create a scattergram of the data.

b. Are the variables t and n linearly related, approximately linearly related, or neither? Explain.

c. Draw a line that comes close to the points in your scattergram.

d. Estimate when 1000 species were listed.

e. Predict the number of species that will be listed in 2011.

6. Amazon.com® is the largest retailer on the Internet. Its revenues are shown in Table 55 for various years.

Table 55 Amazon.com Revenues	
Year	**Revenue (billions of dollars)**
1997	0.1
1998	0.6
1999	1.6
2000	2.8
2001	3.1
2002	3.9
2003	5.26
2004	6.92

Source: *USA Today*

Let r be Amazon.com's revenue (in billions of dollars) for the year that is t years since 1995.

a. Create a scattergram of the data.

b. Are the variables t and r linearly related, approximately linearly related, or neither? Explain.

c. Draw a line that comes close to the points in your scattergram.

d. Use your model to estimate when Amazon.com's revenue was $5 billion.

e. Use your model to predict Amazon.com's revenue in 2011.

7. If there are too many ticketed passengers for a flight, a person can volunteer to be "bumped" onto another flight. The voluntary bumping rates for large U.S. airlines (number of bumps per 10,000 passengers, January through September) are shown in Table 56 for various years.

Table 56 Voluntary Bumping Rates of U.S. Airlines

Year	Bumping Rate (number of bumps per 10,000 passengers)
2000	20
2001	18
2002	17
2003	15
2004	13
2005	11

Source: *U.S. Department of Transportation*

Let r be the voluntary bumping rate (number of bumps per 10,000 passengers) at t years since 2000.
a. Create a scattergram of the data.
b. Draw a line that comes close to the points in your scattergram.
c. What is the r-intercept of the model? What does it mean in this situation?
d. Predict when the voluntary bumping rate will be 4 bumps per 10,000 passengers.
e. What is the t-intercept of the model? What does it mean in this situation?

8. The percentages of disposable income put aside by Americans for savings are shown in Table 57 for various years. *Disposable income* is income (after taxes) that is available to a person for saving or spending.

Table 57 Percentages of U.S. Disposable Income Put Aside for Savings

Year	Percent
1985	9
1990	8
1995	6
2000	2
2004	2

Source: *U.S. Commerce Department, Bureau of Economic Analysis*

Let p be the percentage of disposable income put aside for savings in the year that is t years since 1985.
a. Create a scattergram of the data.
b. Draw a line that comes close to the points in your scattergram.
c. What is the p-intercept of the model? What does it mean in this situation?
d. What is the t-intercept of the model? What does it mean in this situation?
e. Predict the percentage of disposable income put aside for savings in 2013. [**Hint:** Remember, if you think that model breakdown occurs, say so, say where, and explain why.]

9. The percentages of Americans who went to the movies at least once in the past year are shown in Table 58 for various age groups. Let p be the percentage of Americans at age a years who go to movies.
a. Create a scattergram of the data.
b. Draw a line that comes close to the points in your scattergram.
c. Estimate what percentage of Americans at age 19 go to the movies.

Table 58 Percentages of Americans Who Go to the Movies

Age Group (years)	Age Used to Represent Age Group (years)	Percent
18–24	21.0	88
25–34	29.5	79
35–44	39.5	73
45–54	49.5	65
55–64	59.5	46
65–74	69.5	38
over 74	80	28

Source: *U.S. National Endowment for the Arts*

d. At what age do half of Americans go to the movies?
e. What is the a-intercept of the model? What does it mean in this situation?

10. The death rate from heart disease in the United States has decreased greatly in the past four decades (see Table 59).

Table 59 Death Rates Due to Heart Disease

Year	Death Rate (number of deaths per 100,000 people)
1960	559
1970	493
1980	412
1990	322
2000	258
2003	236

Source: *U.S. Center for Health Statistics*

Let r be the death rate (number of deaths per 100,000 people) from heart disease for the year that is t years since 1960.
a. Create a scattergram of the data.
b. Draw a line that comes close to the points in your scattergram.
c. What is the t-intercept of the model? What does it mean in this situation?
d. Predict the death rate from heart disease in 2010. The U.S. population will be about 310 million in that year. How many Americans will die of heart disease in 2010?

11. The average salaries of public school teachers are shown in Table 60 for various years.

Table 60 Average Salaries of Public School Teachers

Year	Average Salary (in thousands of dollars)
1985	24
1990	31
1995	37
2000	42
2004	46

Source: *National Center for Education Statistics*

Let s be the average salary (in thousands of dollars) at t years since 1980.
a. Create a scattergram of the data.
b. Draw a line that comes close to the points in your scattergram.

c. What is the s-intercept of the model? What does it mean in this situation?

d. Predict the average salary in 2010.

e. Predict when the average salary will reach $50 thousand.

12. Due to an increased use of fax machines, cell phones, pagers, and modems, the number of area codes has increased greatly since 1998 (see Table 61).

Table 61 Numbers of Area Codes	
Year	**Number of Area Codes**
1998	186
1999	204
2000	226
2001	239
2002	262
2003	274

Source: *NeuStar, Inc.*

Let n be the number of area codes at t years since 1995.

a. Create a scattergram of the data.

b. Draw a line that comes close to the points in your scattergram.

c. What is the n-intercept of the model? What does it mean in this situation?

d. Use your model to estimate the number of area codes in 2004.

e. Use your model to predict when there will be 400 area codes.

13. The loudness of sound can be measured by using a *decibel scale*. Some examples of sounds at various sound levels are listed in Table 62.

Table 62 Examples of Sound Levels	
Sound Level (decibels)	**Example**
0	Faintest sound heard by humans
20	Whisper
40	Inside a running car
60	Normal conversation
80	Noisy street corner
100	Soft-rock concert
120	Threshold of pain

Source: Math and Music, *Garland and Kahn, 1995*

The sound level of music from a Pioneer MT-2000® stereo-CD-receiver system is controlled by the system's volume number. The sound levels of music for various volume numbers are shown in Table 63.

Table 63 Sound Levels of Music Played by a Stereo	
Volume Number	**Sound Level (decibels)**
6	60
8	66
10	69
12	74
14	78
16	82
18	86
20	90

Source: *J. Lehmann*

Let S be the sound level (in decibels) for a volume number n.

a. Create a scattergram of the data in Table 63.

b. Draw a line that comes close to the points in your scattergram.

c. Use your model to estimate the sound level when the volume number is 19.

d. Use your model to estimate for what volume number the sound level is comparable to that of a noisy street corner (see Table 62).

14. The percentages of disposable income spent on food by Americans are shown in Table 64 for various years.

Table 64 Percentages of Disposable Income Spent on Food	
Year	**Percent**
1970	14
1975	14
1980	13
1985	12
1990	11
1995	11
2000	10
2003	10

Source: *ERS/USDA*

Let p be the percentage of disposable income spent on food at t years since 1970.

a. Create a scattergram of the data.

b. Draw a line that comes close to the points in your scattergram.

c. Use your model to estimate the percentage of disposable income spent on food in 2011.

d. Describe some possible reasons that the percentage of disposable income spent on food has declined since 1970.

15. Computer virus infection rates are shown in Table 65 for various years.

Table 65 Computer Virus Infection Rates	
Year	**Infection Rate (number of infections per 1000 PCs per month)**
1999	80
2000	91
2001	113
2002	105
2003	108
2004	116

Source: *ICSA Labs*

Let I be the infection rate (number of infections per 1000 PCs per month) at t years since 1995. A scattergram of the data and a linear model are sketched in Fig. 57.

a. Use the linear model to estimate the infection rate in 2001.

b. What was the actual infection rate in 2001?

c. Is your result in part (a) an underestimate or an overestimate? Explain how you can tell this from the graph of the scattergram and the sketch of the model. Calculate the error in the estimate.

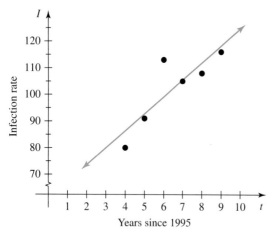

Figure 57 Virus scattergram and model—Exercises 15 and 16

16. Refer to Exercise 15, including Table 65 and Fig. 57.
  a. Use the linear model to estimate the infection rate in 1999.
  b. What was the actual infection rate in 1999?
  c. Is your result in part (a) an underestimate or an overestimate? Explain how you can tell this from the graph of the scattergram and the sketch of the model. Calculate the error in the estimate.

17. Suppose that an independent variable t and a dependent variable p are approximately linearly related. If a data point (c, d) is below a linear model, does the model underestimate or overestimate the value of p when $t = c$? Explain.

18. Suppose that an independent variable t and a dependent variable p are approximately linearly related. If a linear model is below a data point (c, d), does the model underestimate or overestimate the value of p when $t = c$? Explain.

19. When modeling a situation in which the variables are approximately linearly related, different students may all do good work yet not get the same results. Draw a scattergram and at least two reasonable linear models to show how this is possible.

20. Which is more desirable, finding a linear model that contains several, but not all, data points or finding a linear model that does not contain any data points but comes close to all data points? Include in your discussion some sketches of scattergrams and linear models.

21. Compare the meaning of *linearly related* with that of *approximately linearly related*.

22. A student comes up with a shortcut for modeling a situation. Instead of plotting all of the given data points, the student plots only two of the data points and draws a line that contains the two chosen points. Give an example to illustrate what can go wrong with this shortcut.

23. A person collects data by doing research. If the data points lie exactly on a line, will all points on the line describe the situation exactly? Explain.

analyze this

GLOBAL WARMING LAB

Many scientists are greatly concerned that the average temperature of the surface of Earth has increased since 1900 (see Table 66).

Table 66 Average Surface Temperatures of Earth			
Year	**Average Temperature (degrees Fahrenheit)**	**Year**	**Average Temperature (degrees Fahrenheit)**
1900	57.1	1955	57.0
1905	56.7	1960	57.2
1910	56.8	1965	56.9
1915	57.3	1970	57.3
1920	56.9	1975	57.2
1925	56.9	1980	57.7
1930	57.1	1985	57.4
1935	57.1	1990	58.1
1940	57.5	1995	58.0
1945	57.2	2000	57.9
1950	56.9	2003	58.4

Source: *NASA–GISS*

Although it may not seem that the average temperatures shown in Table 66 have increased much, many scientists believe that an increase as small as 3.6°F could be a dangerous climate change.* Global warming would cause the extinction of plants and animals, lead to severe water shortages, create more extreme weather events, increase the number of heat-related illnesses and deaths, and melt glaciers, which would raise ocean levels and, thus, submerge coastlands.

Despite these alarming predictions, not all experts are concerned. Robert Mendelsohn, an environmental economist at Yale University, argues that "global warming will increase agricultural production in the northern half of the United States" and that "the southern half will be able to maintain its current level of production." Even from a global perspective, he believes, the benefits of global warming will offset the damages.[†]

To test such theories, Peter S. Curtis, an ecologist at Ohio State University, and colleagues ran experiments which showed that increased carbon dioxide levels do in fact increase plant growth. However, the nutritional value

---

*"Meeting the Climate Challenge: Recommendations of the International Climate Change Taskforce," The Institute for Public Policy Research/The Center for American Progress/The Australian Institute, January 2005.

[†]*The Impact of Climate Change on the United States Economy,* R. Mendelsohn and J. Neumann (eds.), 1999, Cambridge University Press, Cambridge, UK (reprinted as a paperback in 2004).

of the produce was lower—so much lower that the increase in growth did not make up for the decrease in nutrition.[‡]

Some other theories suggest benefits to global warming. The scientific report *Impacts of a Warming Arctic* points out that it will be easier to extract oil from the Arctic due to less extensive and thinner sea ice. However, the study also says that oil spills are more difficult to clean up in icy seas than in open waters, and many species would suffer from such spills. In addition, potential structural problems such as broken pipelines could mean that the costs outweigh the benefits.[§]

A study done by economist Thomas Gale Moore at Stanford University suggests that a 4.5°F increase in temperature could reduce deaths in the United States by 40,000 per year and that medical costs might be reduced by at least $20 billion annually. Moore also points out that most people prefer warmer climates.[¶]

Most scientists do not share Moore's perspective. On a global scale, they believe that warmer climates could increase the spread of diseases such as malaria and, thus, increase the global death rate and medical costs.

Although there may be some relatively small and short-lived benefits to global warming, the vast majority of scientists agree that global warming is already taking its toll and that further warming would bring catastrophic results.

Glaciologists report that, over the past century, glaciers around the globe have been melting.[‖] Biologists note that many species throughout the world have changed their habitats in search of cooler climates.[**] One species, the golden toad, was not able to migrate and as a result has become extinct due to heat stress.[††]

Looking to the future, an international study, the most comprehensive analysis of its kind, predicts that 15 to 37 percent of all species of plants and animals—well over a million species—will become extinct by 2050.[*] Klaus Toepfer, head of the United Nations Environment Programme (UNEP), said, "If one million species become extinct ... it is not just the plant and animal kingdoms and the beauty of the planet that will suffer. Billions of people, especially in the developing world, will suffer too as they rely on nature for such essential goods and services as food, shelter and medicines."[†]

[‡]"Plant Reproduction under Elevated CO_2 Conditions," L. M. Jablonski, X. Wang, and P. S. Curtis, *New Phytologist* (156):9–26, 2002.

[§]Report of the Arctic Climate Impact Assessment, Cambridge University Press, 2004.

[¶]"In sickness or in health: The Kyoto Protocol versus global warming," Hoover Institution, Stanford University, August 2000.

[‖]National Snow and Ice Data Center, 2003.

[**]T. L. Root et al., "Fingerprints of global warming on wild animals and plants," January 2, 2003, *Nature*, 421:57–60.

[††]J. A. Pounds et al., "Biological response to climate change on a tropical mountain," 1999, *Nature (London)*, 398(6728):611–615.

[*]C. D. Thomas et al., "Extinction risk from climate change," January 8, 2004, *Nature*, 427:145–148.

[†]Reported by UNEP, January 8, 2004; see www.unep.org.

Analyzing the Situation

1. Discuss those theories which describe the benefits of global warming and whether they are likely correct.

2. What are some possible costs of global warming? In your opinion, do the possible costs outweigh the possible benefits? Explain.

3. Let A be the average surface temperature of Earth (in degrees Fahrenheit) at t years since 1900. *Carefully* draw a scattergram of the data in Table 66.

4. On the basis of your scattergram, in what year is it first clear that global warming is occurring? Explain. Also, explain why it makes sense that the first World Climate Conference convened in 1979.

5. Are the variables t and A approximately linearly related for the years 1900–2000? Are the variables approximately linearly related for the years 1965–2000? Explain.

6. Draw a line that comes close to the points in your scattergram for the years 1965–2000.

7. Use your linear model to predict the average global temperature in 2010.

VOLUME LAB

In this lab, you will explore the relationship between the volume of some water in a cylinder and the height of the water. Check with your instructor whether you should collect your own data or use the data listed in Table 67.

Table 67 Heights of Water in a Cylinder with Radius 4.45 Centimeters	
Height (centimeters)	**Volume (ounces)**
0	0
0.9	2
1.9	4
2.9	6
3.8	8
4.8	10
5.7	12

Source: *J. Lehmann*

Materials

You will need the following items:

- A "perfect" cylinder (the diameter of the top should equal the diameter of the base) that can hold at least 8 ounces of water
- At least 8 ounces of water
- A $\frac{1}{4}$-cup measuring cup
- A ruler

Recording of Data

Pour $\frac{1}{4}$ cup (2 ounces) of water into the cylinder, and measure the height of the water, using units of centimeters. Then continue adding $\frac{1}{4}$ cup of water and measuring the height after you have added each $\frac{1}{4}$ cup until there is at least 8 ounces of water in the cylinder. Also, measure the height of the cylinder in units of centimeters.

Analyzing the Data

1. Display your data in a table similar to Table 67. If you are using the data in Table 67, the height of the cylinder is 12 centimeters.

2. Let V be the volume of water (in ounces) in the cylinder when the height is h centimeters. Assume that V is the dependent variable. Draw a scattergram of the data.

3. Draw a line that comes close to the points in your scattergram.

4. What is the V-intercept of your model? What does it mean in this situation?

5. Use the model to estimate the volume of water when the height of the water is 3 centimeters.

6. Use the model to estimate the height of 7 ounces of water in the cylinder.

7. What is the height of the cylinder? Use this height and the model to estimate the maximum amount of water that the cylinder can hold.

8. Indicate on your graph of the model where model breakdown occurs. Also, describe in words when model breakdown occurs.

LINEAR GRAPHING LAB: TOPIC OF YOUR CHOICE

Your objective in this lab is to use a linear model to describe some authentic situation. Choose a situation that has not been discussed in this text. Your first task will be to find some data. Almanacs, newspapers, magazines, and scientific journals are good resources. You may want to try searching on the Internet. Or you can conduct an experiment. Choose something that interests you!

Analyzing the Situation

1. What two quantities did you explore? Define variables for the quantities. Include units in your definitions.

2. Which variable is the dependent variable? Which variable is the independent variable? Explain.

3. Describe how you found your data. If you conducted an experiment, provide a careful description with specific details of how you ran your experiment. If you didn't conduct an experiment, state the source of your data.

4. Include a table of your data.

5. Create a scattergram of your data. (If your data are not approximately linear, find some data that are.)

6. Draw a line that comes close to the points in your scattergram.

7. Choose a value for your independent variable. On the basis of your chosen value for the independent variable, use your model to find a value for your dependent variable. Describe what your result means in the situation you are modeling.

8. Choose a value for your dependent variable. On the basis of your chosen value for the dependent variable, use your model to find a value for your independent variable. Describe what your result means in the situation you are modeling.

9. Comment on your lab experience.
   a. For example, you might address whether this lab was enjoyable, insightful, and so on.
   b. Were you surprised by any of your findings? If so, which ones?
   c. How would you improve your process for this lab if you were to do it again?
   d. How would you improve your process if you had more time and money?

Chapter Summary

Key Points
OF CHAPTER 1

Variable (Section 1.1)	A **variable** is a symbol which represents a quantity that can vary.
Constant (Section 1.1)	A **constant** is a symbol which represents a specific number (a quantity that does *not* vary).
Counting numbers or natural numbers (Section 1.1)	The **counting numbers,** or **natural numbers,** are the numbers 1, 2, 3, 4, 5, ...
Integers (Section 1.1)	The **integers** are the numbers ..., −3, −2, −1, 0, 1, 2, 3, ...
Rational numbers (Section 1.1)	The **rational numbers** are the numbers that can be written in the form $\frac{n}{d}$, where n and d are integers and d is nonzero.
Real numbers (Section 1.1)	The **real numbers** are all the numbers represented on the number line.

Irrational numbers (Section 1.1)	The **irrational numbers** are the real numbers that are not rational.
Average or mean (Section 1.1)	To find the **average** (or **mean**) of a group of numbers, we divide the sum of the numbers by the number of numbers in the group.
Negative numbers and positive numbers (Section 1.1)	The **negative numbers** are the real numbers less than 0, and the **positive numbers** are the real numbers greater than 0.
Scattergram (Section 1.2)	A **scattergram** is a graph of plotted ordered pairs.

Identifying independent and dependent variables (Section 1.2)

Assume that an authentic situation can be described by using the variables t and p, and assume that p depends on t. Then

- We call t the **independent variable.**
- We call p the **dependent variable.**
- For an **ordered pair** (a, b), we write the value of the independent variable in the first (left) position and the value of the dependent variable in the second (right) position.

Columns of tables and axes of coordinate systems (Section 1.2)

Assume that an authentic situation can be described by using two variables. Then

- For tables, the values of the independent variable are listed in the first column and the values of the dependent variable are listed in the second column (see Table 4).
- For coordinate systems, the values of the independent variable are described by the horizontal axis and the values of the dependent variable are described by the vertical axis.

Linearly related (Section 1.3)	If two quantities of an authentic situation are described accurately by a line, then the quantities (and the variables representing those quantities) are **linearly related.**
Model (Section 1.3)	A **model** is a mathematical description of an authentic situation.
Linear model (Section 1.3)	A **linear model** is a line that describes the relationship between two quantities in an authentic situation.
Input and output (Section 1.3)	An **input** is a permitted value of the *independent* variable that leads to at least one **output,** which is a permitted value of the *dependent* variable.
Determine whether data points lie on a line (Section 1.3)	We create a scattergram of data to determine whether the data points lie on a line. If the points lie on a line, then we draw the line and use it to make estimates and predictions.

Intercepts of a line (Section 1.3)

An **intercept** of a line is any point where the line and an axis (or axes) of a coordinate system intersect. There are two types of intercepts of a line sketched on a coordinate system with an x-axis and a y-axis:

- An ***x*-intercept** of a line is a point where the line and the x-axis intersect. The y-coordinate of an x-intercept is 0.
- A ***y*-intercept** of a line is a point where the line and the y-axis intersect. The x-coordinate of a y-intercept is 0.

Approximately linearly related (Section 1.4)	If the points in a scattergram of data lie close to (or on) a line, we say that the relevant variables are **approximately linearly related.**
Determine whether variables are approximately linearly related (Section 1.4)	We create a scattergram of data to determine whether the relevant variables are approximately linearly related. If so, we draw a line that comes close to the data points and use the line to make estimates and predictions.

Underestimates and overestimates (Section 1.4)

Suppose that an independent variable t and a dependent variable p are approximately linearly related. Then

- If a linear model is below a data point (a, b), the model underestimates the value of p when $t = a$.
- If a linear model is above a data point (a, b), the model overestimates the value of p when $t = a$.

Model breakdown (Section 1.4)

When a model yields a prediction that does not make sense or an estimate that is not a good approximation, we say that **model breakdown** has occurred.

CHAPTER 1 REVIEW EXERCISES

1. Let r be the annual DVD revenue (in billions of dollars). For 2003, the value of r is 17.5 (Source: *Adams Media Research*). What does that mean in this situation?

2. Let t be the number of years since 1995. What does $t = 16$ represent?

3. Choose a variable name for the percentage of students who are full-time students at a college. Give two numbers that the variable can represent and two numbers that it cannot represent.

4. A rectangle has a perimeter of 40 inches. Let W be the width, L be the length, and P be the perimeter, all with units in inches.
   a. Sketch three possible rectangles with a perimeter of 40 inches.
   b. Which of the symbols W, L, and P are variables? Explain.
   c. Which of the symbols W, L, and P are constants? Explain.

5. Graph the numbers $-2, -\frac{3}{2}, 0, 1, \frac{5}{2}$, and 3 on a number line.

6. Graph the negative integers between -5 and 5 on a number line.

7. Here are a company's profits and losses for various years: profit of $2 million, loss of $4 million, loss of $1 million, profit of $3 million. Let p be the profit (in millions of dollars). Use points on a number line to describe the profits and losses of the company.

8. Plot the points $(2, 4)$, $(-3, -1)$, $(5, -2)$, and $(-4, 5)$ in a coordinate system.

9. What is the y-coordinate of the ordered pair $(3, -6)$?

10. What is the x-coordinate of the ordered pair $(-4, -7)$?

For Exercises 11 and 12, identify the independent variable and the dependent variable.

11. Let p be the percentage of Americans at age a years who own a home.

12. Let a be the average salary (in dollars) for a person with t years of education.

13. Let n be the total number of U.S. billionaires at t years since 2000. What does the ordered pair $(4, 313)$ mean in this situation?

14. Let n be the number of injuries (in thousands) incurred on amusement park rides at t years since 1990. What does the ordered pair $(10, 10.6)$ mean in this situation?

15. Create a scattergram of the ordered pairs listed in Table 68.

Table 68 Some Ordered Pairs	
x	**y**
-5	-2
-3	4
0	1
2	3
4	-5

16. The average gas mileages of cars are shown in Table 69 for various years.

Table 69 Average Gas Mileages of Cars	
Year	**Average Gas Mileage (miles per gallon)**
1970	14
1980	16
1990	20
2000	22
2003	22

Source: *U.S. Federal Highway Administration*

Let g be the average gas mileage (in miles per gallon) of cars at t years since 1970.
   a. Create a scattergram of the data.
   b. For which of the years shown in Table 69 was the average gas mileage of cars the highest?
   c. For which of the years shown in Table 69 was the average gas mileage of cars the lowest?

17. The countries with the top six percentages of electricity generated by nuclear power are shown in the bar graph in Fig. 58 (Source: *International Atomic Energy Agency*).

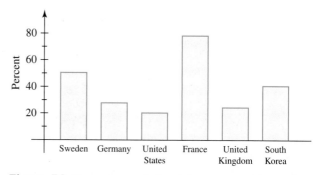

Figure 58 Percentages of electricity generated by nuclear power

   a. Which country generates the largest percentage of its electricity by nuclear power? Estimate that percentage.
   b. Which of the countries included in Fig. 58 generates the smallest percentage of its electricity by nuclear power? Estimate that percentage.
   c. Estimate the percentage of Sweden's electricity that is generated by nuclear power.

For Exercises 18–23, refer to Fig. 59.

18. Find y when $x = -2$.

19. Find y when $x = 6$.

20. Find x when $y = -4$.

21. Find x when $y = 1$.

22. What is the y-intercept of the line?

23. What is the x-intercept of the line?

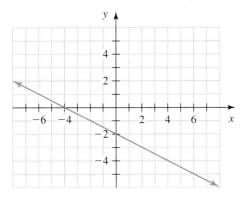

Figure 59 Exercises 18–23

24. Some ordered pairs are listed in Table 70.

Table 70 Some Ordered Pairs	
x	**y**
1	11
2	10
5	7
8	4
9	3

  a. Create a scattergram of the points shown in Table 70.
  b. Draw a line that contains the points in your scattergram.
  c. Find y when $x = 11$.
  d. Find x when $y = 5$.
  e. What is the x-intercept of your line?
  f. What is the y-intercept of your line?

25. Some ordered pairs are listed in Table 71.

Table 71 Some Ordered Pairs	
x	**y**
1	10
2	8
3	6
4	4
5	2
6	0

  a. Create a scattergram of the points shown in Table 71.
  b. Are the variables x and y linearly related, approximately linearly related, or neither?

26. Let p be the profit (in millions of dollars) of a company for the year that is t years since 2000. Some pairs of values of t and p are shown in Table 72.

Table 72 Profits of a Company	
t (years)	**p** (millions of dollars)
1	20
3	16
4	14
6	10
7	8

  a. Create a scattergram of the data. Then draw a linear model.
  b. Predict the profit in 2010.
  c. Estimate when the profit was $18 million.
  d. What is the p-intercept of the model? What does it mean in this situation?
  e. What is the t-intercept of the model? What does it mean in this situation?

27. What is the y-coordinate of an x-intercept of a line?

28. Some ordered pairs are listed in Table 73.

Table 73 Some Ordered Pairs	
x	**y**
1	23
3	17
6	10
8	9
9	4

  a. Create a scattergram of the points shown in Table 73.
  b. Are the variables x and y linearly related, approximately linearly related, or neither?
  c. Draw a line that comes close to the points in your scattergram.
  d. Which point on your line has x-coordinate 5?
  e. Which point on your line has y-coordinate 20?
  f. What is the y-intercept of your line?
  g. What is the x-intercept of your line?

29. The percentages of American adults who are obese are shown in Table 74 for various years.

Table 74 Percentages of American Adults Who Are Obese	
Year	**Percent**
1990	12
1992	13
1994	14
1996	17
1998	18
2000	20
2002	22
2004	25

Source: *National Center for Chronic Disease Prevention and Health Promotion*

Let p be the percentage of American adults who are obese at t years since 1990.
  a. Create a scattergram of the data.
  b. Draw a line that comes close to the points in your scattergram.
  c. Use your model to predict the percentage of American adults who were obese in 2006.
  d. Use your model to predict when 30% of Americans will be obese.

30. Willie Mays, with all-around talent, was one of the greatest baseball players of all time. Mays's statistics on stolen bases from 1956 to 1963 are shown in Table 75. Let n be Willie Mays's number of stolen bases in the year that is t years since 1955.
  a. Create a scattergram of the data.
  b. Draw a line that comes close to the points in your scattergram.

Table 75 Willie Mays: Numbers of Stolen Bases

Year	Number of Stolen Bases
1956	40
1957	38
1958	31
1959	27
1960	25
1961	18
1962	18
1963	8

Source: The Sports Encyclopedia: Baseball 2004, D. S. Neft et al., 2004, St. Martin's Press, NY.

c. Find the n-intercept of the model. What does it mean in this situation?

d. Find the t-intercept of the model. What does it mean in this situation?

e. In 1955, Mays stole 24 bases. Does your linear model underestimate or overestimate his number of stolen bases in that year? Has model breakdown occurred? Explain.

f. In 1971, Mays stole 23 bases. Does your linear model underestimate or overestimate his number of stolen bases in that year? Has model breakdown occurred? Explain.

CHAPTER 1 TEST

1. A rectangle has an area of 36 square feet. Let W be the width (in feet), L be the length (in feet), and A be the area (in square feet).
   a. Sketch three possible rectangles of area 36 square feet.
   b. Which of the symbols W, L, and A are variables? Explain.
   c. Which of the symbols W, L, and A are constants? Explain.

2. Graph the integers between -4 and 2, inclusive, on a number line.

3. The low temperatures (in degrees Fahrenheit) for four days in January in Indianapolis, Indiana, are 5°F below zero, 7°F above zero, 2°F above zero, and 3°F below zero. Let F be the low temperature (in degrees Fahrenheit) for any one day. Use points on a number line to describe the given values of F.

4. The number of electric cars (in thousands) in use in the United States for various years is 4.5, 5.2, 7.0, 8.7, and 10.4. Let n be the number of electric cars (in thousands) in use. Use points on a number line to describe the given values of n. Find the average of the values and indicate it on the number line.

5. Let c be the total cost (in dollars) of n tickets to a hip-hop concert. What is the dependent variable? Explain.

6. Although most ATMs are owned by banks, a growing number of ATMs are privately owned. Let p be the percentage of ATMs that are privately owned at t years since 2000. What does the ordered pair (3, 27) mean in this situation?

7. The percentages of Americans of various age groups who were without health insurance at some point in 2002 or 2003 are shown in Table 76.

Table 76 Percentages of Americans Who Were Uninsured

Age Group (years)	Age Used to Represent Age Group (years)	Percent
0–17	8.5	37
18–24	21	50
25–44	34.5	33
45–54	49.5	21
55–64	59.5	17

Source: The Lewin Group for Families USA

Let p be the percentage of Americans who were without health insurance at age a years.
   a. Create a scattergram of the data.
   b. Which point in your scattergram is highest? What does that mean in this situation?
   c. Which point in your scattergram is lowest? What does that mean in this situation?

For Exercises 8–11, refer to Fig. 60.

8. Find y when $x = -4$.

9. Find x when $y = 1$.

10. What is the y-intercept of the line?

11. What is the x-intercept of the line?

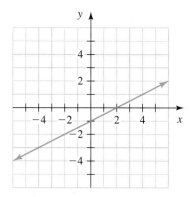

Figure 60 Exercises 8–11

12. Let s be a person's salary (in thousands of dollars) after she has worked t years at a company. Some pairs of values of t and s are shown in Table 77.

Table 77 Years Worked and Salary

t (years)	s (thousands of dollars)
0	21
2	25
3	27
5	31
6	33

a. Create a scattergram of the data. Then draw a linear model.
b. Estimate the person's salary after she has worked 4 years at the company.
c. Estimate when the person's salary will be $35 thousand.
d. What is the *s*-intercept of the model? What does it mean in this situation?

13. Describe, in your own words, the meaning of *linear model*.

14. The number of certified massage therapists and body workers has more than tripled from 1999 to 2004 (see Table 78). Let *n* be the number (in thousands) of certified massage therapists and body workers at *t* years since 1995.
   a. Create a scattergram of the data.
   b. Draw a line that comes close to the points in your scattergram.
   c. Predict the number of massage therapists and body workers in 2010.
   d. Predict when there will be 100 thousand massage therapists and body workers.

Table 78 Certified Massage Therapists and Body Workers

Year	Number of Massage Therapists and Body Workers (thousands)
1999	24
2000	39
2001	41
2002	55
2003	68
2004	81

Source: *National Certification Board for Therapeutic Massage and Bodywork*

15. Suppose that an independent variable *t* and a dependent variable *p* are approximately linearly related. If a data point (c, d) is below a linear model that describes the relationship between *t* and *p*, does the model underestimate or overestimate the value of *p* when $t = c$? Explain.

Chapter 2

Operations and Expressions

I've missed more than 9000 shots in my career. I've lost more than 300 games. Twenty-six times I've been trusted to take the game-winning shot and missed. I've failed over and over and over again in my life. And that is why I succeed.

—*Michael Jordan*

Table 1 Numbers of AIDS Deaths in the United States

Year	Number of AIDS Deaths
1996	37,787
1997	21,923
1998	17,930
1999	16,273
2000	15,245
2001	14,175
2002	16,371
2003	18,017

Source: *Centers for Disease Control and Prevention*

Although the number of AIDS deaths in the United States decreased from 1996 to 2001, it has increased since then (see Table 1). Why do you think the decrease and then the increase occurred? In Example 9 in Section 2.4, we will calculate how the number of deaths has changed in various years.

In this chapter, we will describe authentic quantities by using *expressions*. We will discuss how to perform operations and in which order we should perform them, which will help us use expressions to find values of authentic quantities. We will also discuss how to use subtraction and division to compare quantities pertaining to an authentic situation, such as the AIDS data just described.

2.1 EXPRESSIONS

Objectives

▷ Know the meaning of *expression* and of *evaluate an expression*.
▷ Use expressions to describe authentic quantities.
▷ Evaluate expressions.
▷ Translate English phrases to and from mathematical expressions.
▷ Know two roles of a variable.

In this section, we will work with expressions—a very important concept in algebra.

Expressions

Addition, subtraction, multiplication, and division are examples of *operations*. In arithmetic, you performed operations with numbers. Since variables represent numbers, we can perform operations with variables, too.

Example 1 Using Operations with Variables and Numbers

Each employee at a small company receives a $500 bonus at the end of the year. For each employee's annual salary shown, find the employee's annual salary plus bonus.

1. $28,000   **2.** $32,000   **3.** s dollars

Solution

1. The employee's annual salary plus bonus is $28,000 + 500 = 28,500$ dollars.
2. The employee's annual salary plus bonus is $32,000 + 500 = 32,500$ dollars.
3. In Problems 1 and 2, we added the annual salary and $500, the bonus, to find the results. So, the employee's annual salary plus bonus (in dollars) is $s + 500$. ∎

In Example 1, we took s to be an employee's annual salary and $s + 500$ to be the employee's annual salary plus bonus. We call s and $s + 500$ *expressions*.

> **DEFINITION** Expression
>
> An **expression** is a constant, a variable, or a combination of constants, variables, operation symbols, and grouping symbols, such as parentheses.

Here are some more examples of expressions:
$$t + 6 \qquad \pi \qquad L + W - 9 \qquad y \qquad 4 \qquad 5 \div (x + 2)$$

In Example 1, we used a variable to represent a quantity from an authentic situation. Sometimes we use variables to represent numbers in a math problem that is not being used to describe an authentic situation. In this case, we often use x for the variable. For example, we could let x represent a number. In this case, x could be *any* number.

To avoid confusing the multiplication symbol $\times$ and the variable name x, we use $\cdot$ or no operation symbol to indicate multiplication. For example, each of the following expressions describes multiplying 2 by 3:
$$2 \cdot 3 \qquad 2(3) \qquad (2)3 \qquad (2)(3)$$
And each of the following expressions describes multiplying 2 by k:
$$2 \cdot k \qquad 2k \qquad 2(k) \qquad (2)k \qquad (2)(k)$$

Using Expressions to Describe Authentic Quantities

We can use expressions to describe authentic quantities. In Example 2, we will find such an expression by noticing a pattern as we calculate values of a quantity.

Example 2 Describing a Quantity

A hot-dog stand sells hot dogs for \$2 apiece. Find the total cost of buying the given number of hot dogs.

1. 3 hot dogs   **2.** 5 hot dogs   **3.** 8 hot dogs   **4.** n hot dogs

Solution

1. Three hot dogs cost $3(2) = 6$ dollars.
2. Five hot dogs cost $5(2) = 10$ dollars.
3. Eight hot dogs cost $8(2) = 16$ dollars.
4. In Problems 1–3, we found the total cost by multiplying the number of hot dogs by 2, the cost (in dollars) per hot dog. So, if there are n hot dogs, the total cost (in dollars) is $n(2)$. ∎

In Example 3, we will use a table to help us find an expression that describes an authentic quantity.

Table 2 Driving Times and Distances

Driving Time (hours)	Distance (miles)
1	$75 \cdot 1$
2	$75 \cdot 2$
3	$75 \cdot 3$
4	$75 \cdot 4$
t	$75 \cdot t$

Example 3 Using a Table to Find an Expression

A person drives at a constant speed of 75 miles per hour. Find the distance traveled in 1, 2, 3, and 4 hours of driving at that speed. Show the arithmetic to help you see the pattern. Organize the calculations in a table, and include an expression that stands for the distance traveled in t hours.

Solution

First we create Table 2. From the last row of the table, we see that the expression $75t$ represents the distance traveled (in miles) in t hours. ∎

Evaluating Expressions

In Example 3, we used $75t$ to describe the distance traveled (in miles) in t hours. This means that if the driving time is 5 hours, the distance traveled is $75(5) = 375$ miles. To find the distance, we substituted 5 for t. We say we have *evaluated* the expression $75t$ for $t = 5$.

DEFINITION Evaluate an expression

We **evaluate an expression** by substituting a number for each variable in the expression and then calculating the result. If a variable appears more than once in the expression, the same number is substituted for that variable each time.

When we evaluate an expression, it is good practice to use parentheses each time a number is substituted for a variable. For example, here we evaluate $5x$ for $x = 3$:

$$5(3) = 15$$

This strategy will be especially helpful when we evaluate an expression for a negative number, which we will begin to do in Section 2.3.

Example 4 Evaluating Expressions

1. In Example 1, we used s to represent an employee's annual salary (in dollars) and $s + 500$ to represent the employee's annual salary plus bonus (in dollars). Evaluate $s + 500$ for $s = 40,000$ and describe the meaning of the result.
2. In Example 2, we used n to represent the number of hot dogs bought and $n(2)$ to represent the total cost (in dollars) of n hot dogs. Evaluate $n(2)$ for $n = 4$ and describe the meaning of the result.

Solution

1. We substitute 40,000 for s in $s + 500$:

$$(40,000) + 500 = 40,500$$

So, the annual salary plus bonus is \$40,500.
2. We substitute 4 for n in $n(2)$:

$$(4)(2) = 8$$

So, the total cost of 4 hot dogs is \$8. ∎

Translating English Phrases to and from Expressions

In Example 2, we used English to describe an authentic situation:

"A hot-dog stand sells hot dogs for \$2 apiece."

Then we translated this information into mathematics:

"So, if there are n hot dogs, the total cost (in dollars) is $n(2)$."

In Problem 2 of Example 4, we performed mathematics by evaluating the expression $n(2)$ for $n = 4$ to find the total cost (in dollars) of 4 hot dogs:

$$(4)(2) = 8$$

Finally, we translated the result into English:

"So, the total cost of 4 hot dogs is \$8."

The following process describes the "big picture" of using mathematics to find results for authentic situations.

Using Mathematics to Find Results for Authentic Situations

1. Describe a situation in English.
2. Translate the English description into mathematics.
3. Perform mathematics to get a desired result.
4. Translate the mathematical result into English.

As you read Example 5, try to identify the four steps just described.

Example 5 Finding and Evaluating an Expression

Some students rent a house together. Each roommate pays an equal share of the $1200 monthly rent.

1. Let n be the number of roommates. Use a table to help find an expression that describes each roommate's share (in dollars) of the rent.
2. Evaluate your expression in Problem 1 for $n = 5$. What does the result mean in this situation?

Solution

1. First we create Table 3. We show the arithmetic to help us see the pattern. From the last row of the table, we see that the expression $1200 \div n$ represents each roommate's share of the rent (in dollars).
2. We substitute 5 for n in $1200 \div n$:

$$1200 \div (5) = 240$$

So, each roommate pays $240 per month. ∎

Table 3 Number of Roommates and Share of Rent

Number of Roommates	Share of Rent (dollars)
1	$1200 \div 1$
2	$1200 \div 2$
3	$1200 \div 3$
4	$1200 \div 4$
n	$1200 \div n$

When we translate from English to mathematics, or vice versa, the following definitions are helpful:

DEFINITION Product, factor, and quotient

Let a and b be numbers. Then

- The **product** of a and b is ab. We call a and b **factors** of ab.
- The **quotient** of a and b is $a \div b$, where b is not zero.

For example, since $6 \cdot 3 = 18$, the number 18 is the product of 6 and 3 and the numbers 6 and 3 are factors of 18. The quotient of 6 and 3 is $6 \div 3 = 2$.

Here are some examples of English phrases or sentences and mathematical expressions that have the same meaning:

Operation	English Phrase or Sentence	Mathematical Expression
Addition	A number plus 3	$x + 3$
	The sum of a number and 3	$x + 3$
	The total of a number and 3	$x + 3$
	Add a number and 3.	$x + 3$
	3 more than a number	$x + 3$
	A number increased by 3	$x + 3$
Subtraction	A number minus 3	$x - 3$
	The difference of a number and 3	$x - 3$
	Subtract 3 from a number.	$x - 3$
	3 less than a number	$x - 3$
	A number decreased by 3	$x - 3$
Multiplication	Multiply 3 by a number.	$3x$
	3 times a number	$3x$
	The product of 3 and a number	$3x$
	Twice a number	$2x$
	One-third of a number	$\frac{1}{3}x$
Division	Divide a number by 3.	$x \div 3$
	The quotient of a number and 3	$x \div 3$
	The ratio of a number to 3	$x \div 3$

WARNING To subtract 2 from 5, we write $5 - 2$, not $2 - 5$. Suppose you have $5 and you take $2 from the $5. Then you have $5 - 2 = 3$ dollars left. So, subtracting 2 from 5 is $5 - 2$.

Example 6 Translating from English to Mathematics

Let x be a number.

1. Translate the English phrase "The product of 2 and the number" into an expression.
2. Evaluate your result in Problem 1 for $x = 3$.
3. Evaluate your result in Problem 1 for $x = 7$.

Solution

1. The expression is $2x$.
2. $2(3) = 6$
3. $2(7) = 14$ ■

Roles of a Variable

In Chapter 1, we used a variable to represent a quantity that can vary. In Example 6, we used a variable for another reason: In the expression $2x$, the variable x is used as a *placeholder* for a number to be substituted for x. First, we substituted 3 for x. Then, we substituted 7 for x.

Roles of a Variable
Here are two roles of a variable:*   1. A variable can represent a quantity that can vary.   2. In an expression, a variable is a placeholder for a number.

Sometimes a variable serves both roles. For example, consider the variable n in Example 5. Recall that n represents the number of roommates, which can vary over time or from house to house. The variable n also serves as a placeholder for a number in the expression $1200 \div n$, which describes each roommate's share of the rent (in dollars).

Example 7 Translating from English to Mathematics

Let x be a number. Translate the English phrase or sentence into an expression. Then evaluate the expression for $x = 6$.

1. The quotient of the number and 3
2. Subtract the number from 8.

Solution

1. The expression is $x \div 3$. Next, we evaluate $x \div 3$ for $x = 6$:

$$(6) \div 3 = 2$$

2. The expression is $8 - x$. Next, we evaluate $8 - x$ for $x = 6$:

$$8 - (6) = 2$$ ■

Example 8 Translating from Mathematics to English

Let x be a number. Translate the expression into an English phrase.

1. $6 - x$ 2. $8x$

Solution

1. The difference of 6 and the number
2. The product of 8 and the number ■

---

*We will discuss one more role of a variable in Section 4.3.

Figure 1 The length L and width W of a rectangle

Expressions with More than One Variable

An expression may contain more than one variable. For example, let W be the width (in feet) and let L be the length (in feet) of a rectangle (see Fig. 1).

Recall that the area of a rectangle is equal to the length times the width of the rectangle, so the area (in square feet) is equal to the expression LW. We can evaluate the expression LW for $L = 4$ and $W = 3$:

$$(4)(3) = 12$$

So, a 3-foot by 4-foot rectangle has an area of 12 square feet.

Note the power of algebra in that the expression LW *concisely* tells us how to find the area of *any* rectangle, no matter what its dimensions are.

Example 9   Evaluating an Expression in Two Variables

If it takes a student T minutes to complete a test that has n questions, then $T \div n$ is the average time (in minutes) taken to respond to one question. Evaluate $T \div n$ for $T = 48$ and $n = 16$. What does the result mean in this situation?

Solution

We substitute 48 for T and 16 for n in the expression $T \div n$ and then calculate the result:

$$48 \div 16 = 3$$

If it takes a student 48 minutes to respond to 16 questions, the average response time is 3 minutes per question.   ■

Example 10   Translating from English to Mathematics

Write the phrase as a mathematical expression, and then evaluate the result for $x = 8$ and $y = 4$.

1. The sum of x and y
2. The quotient of x and y

Solution

1. The expression is $x + y$. Next, we evaluate $x + y$ for $x = 8$ and $y = 4$:

$$(8) + (4) = 12$$

2. The expression is $x \div y$. Next, we evaluate $x \div y$ for $x = 8$ and $y = 4$:

$$(8) \div (4) = 2$$ ■

group exploration

Expressions used to describe a quantity

Consider the expression $x + 2$. Suppose that a child has grown 2 inches within the last year. We could define x to be the child's height (in inches) last year, and then $x + 2$ would be the child's current height (in inches).

Describe a situation in which x represents a meaningful quantity and the expression given describes another meaningful quantity.

1. $x + 3$
2. $x - 4$
3. $3x$
4. $x \div 2$

For each of the four expressions, evaluate it for a reasonable value of x and describe the meaning of the result.

TIPS FOR SUCCESS: Make Good Use of This Text

You can get more out of this course by making good use of the text. Before class, consider previewing the material for 10 minutes. You can do this by reading the objectives and the boxed statements. Even if what you read doesn't make much sense to you, previewing will flag key concepts that you can focus on during class time.

After class, read the relevant section(s). When looking at each example, figure out how it goes from one step to the next.

Then begin working on the homework assignment. If you have difficulty with an exercise, locate a similar example to help guide you. You may need to seek outside help for more challenging exercises. If you needed to look at examples or get outside help for a large number of exercises, then it is important that you keep doing additional exercises until you are self-sufficient. After all, unless your instructor allows open-book tests or collaborative tests, you won't be able to read the text or seek help from others during an exam.

HOMEWORK 2.1   FOR EXTRA HELP ▶

Student Solutions Manual   PH Math/Tutor Center   MathXL®   MyMathLab

---

For Exercises 1–12, evaluate the expression for $x = 6$.

1. $x + 2$ **2.** $5 + x$ **3.** $9 - x$ **4.** $x - 4$

5. $7x$ **6.** $x(9)$ **7.** $x \div 3$ **8.** $30 \div x$

9. $x + x$ **10.** $x - x$ **11.** $x \cdot x$ **12.** $x \div x$

13. If a person buys n audio CDs, the total cost is $13n$ dollars. Evaluate $13n$ for $n = 4$. What does your result mean in this situation?

14. If a person weighs w pounds and then loses 7 pounds, the person's new weight is $w - 7$ pounds. Evaluate $w - 7$ for $w = 160$. What does your result mean in this situation?

15. If a student earns a total of T points on five tests, then $T \div 5$ is the student's average test score (in points). If a student earns a total of 440 points on five tests, what is the student's average test score?

16. For the period 1970–2002, if c is the number of two-year colleges in the United States, then $c + 695$ is approximately the number of four-year colleges and universities. There were about 1791 two-year colleges in 2002. Estimate the number of four-year colleges and universities in 2002.

17. Each share of a certain stock is worth $5.
   a. Complete Table 4 to help find an expression that describes the total value (in dollars) of n shares of the stock. Show the arithmetic to help you see a pattern.

Table 4 Number of Shares and Total Value	
Number of Shares	**Total Value (dollars)**
1	
2	
3	
4	
n	

   b. Evaluate the expression you found in part (a) for $n = 7$. What does your result mean in this situation?

18. A pair of socks sells for $3.
   a. Complete Table 5 to help find an expression that describes the total cost (in dollars) of n pairs of socks. Show the arithmetic to help you see a pattern.

Table 5 Number of Pairs and Total Cost	
Number of Pairs	**Total Cost (dollars)**
1	
2	
3	
4	
n	

   b. Evaluate the expression found in part (a) for $n = 9$. What does your result mean in this situation?

19. Each student at a community college pays a student services fee of $12.
   a. Complete Table 6 to help find an expression that describes the total cost (in dollars) of tuition plus the services fee if a student pays t dollars for tuition. Show the arithmetic to help you see a pattern.

Table 6 Tuition and Total Cost	
Tuition (dollars)	**Total Cost (dollars)**
400	
401	
402	
403	
t	

   b. Evaluate the expression you found in part (a) for $t = 417$. What does your result mean in this situation?

20. A person is driving 5 miles per hour over the speed limit.

a. Complete Table 7 to help find an expression that describes the driving speed (in miles per hour) if the speed limit is s miles per hour. Show the arithmetic to help you see a pattern.

Table 7 Speed Limit and Driving Speed

Speed Limit (miles per hour)	Driving Speed (miles per hour)
35	
40	
45	
50	
s	

b. Evaluate the expression you found in part (a) for $s = 65$. What does your result mean in this situation?

21. The enrollment fee for a college student is $87 per hour (unit or credit).

a. Complete Table 8 to help find an expression that describes the total cost (in dollars) of enrolling in n hours of classes. Show the arithmetic to help you see a pattern.

Table 8 Tuition and Total Cost

Number of Hours of Courses	Total Cost (dollars)
1	
2	
3	
4	
n	

b. Evaluate the expression you found in part (a) for $n = 15$. What does your result mean in this situation?

22. The length of a rectangular garden is 20 feet.

a. Complete Table 9 to help find an expression that describes the area (in square feet) of the rectangle if the width is w feet. Show the arithmetic to help you see a pattern.

Table 9 Width and Area

Width (feet)	Area (square feet)
1	
2	
3	
4	
w	

b. Evaluate the expression you found in part (a) for $w = 10$. What does your result mean in this situation?

Let x be a number. Translate the English phrase or sentence into a mathematical expression. Then evaluate the expression for $x = 8$.

23. The number plus 4

24. 8 minus the number

25. The quotient of the number and 2

26. Add 6 and the number.

27. Subtract 5 from the number.

28. 15 more than the number

29. The product of 7 and the number

30. The difference of the number and 7

31. 16 divided by the number   **32.** Multiply the number by 5.

Let x be a number. Translate the expression into an English phrase.

33. $x \div 2$ **34.** $6 \div x$ **35.** $7 - x$

36. $x - 2$ **37.** $x + 5$ **38.** $4 + x$

39. $9x$ **40.** $x(5)$ **41.** $x - 7$

42. $x + 3$ **43.** $x(2)$ **44.** $x \div 5$

Evaluate the expression for $x = 6$ and $y = 3$.

45. $x + y$ **46.** $y + x$ **47.** $x - y$

48. xy **49.** yx **50.** $x \div y$

For Exercises 51–54, translate the phrase into a mathematical expression. Then evaluate the expression for $x = 9$ and $y = 3$.

51. The product of x and y **52.** The sum of x and y

53. The difference of x and y **54.** The quotient of x and y

55. If a car travels at a constant speed of r miles per hour for t hours, it will travel rt miles. Evaluate rt for $r = 62$ and $t = 3$. What does your result mean in this situation?

56. Let b be the balance (in dollars) of a checking account. If a check is written for d dollars, then the new balance (in dollars) is $b - d$. Evaluate $b - d$ for $b = 3758$ and $d = 994$. What does your result mean in this situation?

57. If a car can travel m miles on g gallons of gasoline, then the car's gas mileage is $m \div g$ miles per gallon. Evaluate $m \div g$ for $m = 240$ and $g = 12$. What does your result mean in this situation?

58. If T is the total cost (in dollars) for n students to go on a ski trip, then $T \div n$ is the cost (in dollars) per student. Evaluate $T \div n$ for $T = 9000$ and $n = 20$. What does your result mean in this situation?

59. Let C be the total cost (in dollars) of manufacturing some computers and R be the total revenue (in dollars) from selling the computers. Then $R - C$ is the total profit (in dollars). If the total cost of manufacturing some computers is $315 thousand and the total revenue from selling the computers is $485 thousand, what is the total profit?

60. For the period 1992–1999, if E is the average verbal SAT score (in points) for a certain year, then the average math SAT score (in points) for that year is approximately $E + t$, where t is the number of years since 1992. The average verbal SAT score was 505 points in 1999. Estimate the average math SAT score in 1999.

61. A person gets paid $5t$ dollars for t hours of work.

a. Evaluate $5t$ for $t = 1$, $t = 2$, $t = 3$, and $t = 4$. Describe the meaning of these results.

b. Refer to your results from part (a) to determine how much the person gets paid per hour. Explain.

c. Compare your result from part (b) with the expression $5t$. What do you notice?

62. The total price of n loaves of bread is $3n$ dollars.

a. Evaluate $3n$ for $n = 1$, $n = 2$, $n = 3$, and $n = 4$. Describe the meaning of these results.

b. Refer to your results from part (a) to determine the cost per loaf of bread. Explain.

c. Compare your result from part (b) with the expression $3n$. What do you notice?

63. A person drives $50t$ miles in t hours.
  a. Evaluate $50t$ for $t = 1$, $t = 2$, $t = 3$, and $t = 4$. Describe the meaning of your results.
  b. Refer to your results from part (a) to determine at what speed the person is traveling. Explain.
  c. Compare your result from part (b) with the expression $50t$. What do you notice?

64. An elevator rises $2t$ yards in t seconds.
  a. Evaluate $2t$ for $t = 1$, $t = 2$, $t = 3$, and $t = 4$. Describe the meaning of your results.
  b. Refer to your results from part (a) to determine at what speed the elevator is rising. Explain.

  c. Compare your result from part (b) with the expression $2t$. What do you notice?

65. Compare the meaning of *variable* with the meaning of *expression*. (See page 9 for guidelines on writing a good response.)

66. Give an example of an expression containing a variable, and then evaluate it three times to get three different results.

67. Give an example of a variable that is used to represent a quantity that varies.

68. Give an example of a variable that is used as a placeholder for a number in an expression.

2.2 OPERATIONS WITH FRACTIONS

Objectives

▹ Know the meaning of a fraction.
▹ Know that division by zero is undefined.
▹ Know the rules for $a \cdot 1$, $\dfrac{a}{1}$, and $\dfrac{a}{a}$.
▹ Perform operations with fractions.
▹ Find the prime factorization of a number.
▹ Simplify fractions.

In this section, we perform operations with fractions, which are used in numerous fields, including music, social science, business, computer science, architecture, chemistry, engineering, political science, medicine, and aeronautics.

Meaning of a Fraction

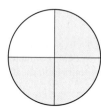

Figure 2 $\dfrac{3}{4}$ of a pizza

A fraction can be used to describe a part of a whole. For example, consider the meaning of $\dfrac{3}{4}$ of a pizza. If we divide the pizza into 4 slices of equal area, 3 of the slices make up $\dfrac{3}{4}$ of the pizza (see Fig. 2).

The fraction $\dfrac{a}{b}$ means $a \div b$. For example, $\dfrac{8}{4} = 8 \div 4 = 2$. So 8 quarters of pizza make 2 pizzas with 4 slices each (see Fig. 3).

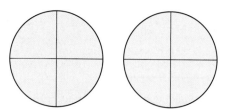

Figure 3 The 8 quarters of pizza make 2 pizzas

Division by Zero

We can think of division in terms of repeated subtraction. For example, $17 \div 5$ is equal to 3 with a remainder of 2 (try it). This means that if we subtract 5 from 17 three times, the result is 2 (the remainder):

$$17 - 5 = 12, \qquad 12 - 5 = 7, \qquad 7 - 5 = 2$$

Note that the remainder, 2, is less than the divisor, 5.

As a matter of fact, the remainder must always be less than the divisor. This rule will help us see that division by 0 is undefined. For example, consider $8 \div 0$. No matter how many times we subtract 0 from 8, the result is always 8:

$$8 - 0 = 8, \qquad 8 - 0 = 8, \qquad 8 - 0 = 8, \text{ and so on}$$

If $8 \div 0$ is defined, the remainder would have to be the repeated result 8. Since the remainder must be less than the divisor, it is implied that 8 is less than 0, which is false. So, $8 \div 0$ is undefined. In fact, any number divided by 0 is undefined.

Division by Zero

The fraction $\dfrac{a}{b}$ is undefined if $b = 0$. Division by 0 is undefined.

For example, $\dfrac{6}{0}$ is undefined. If you use a calculator to divide by zero, the screen will likely display "Error," "ERR:," "E," or "ERR: Divide by 0" to indicate that division by zero is undefined.

WARNING However, the fraction $\dfrac{0}{6}$ *is* defined. In fact, $\dfrac{0}{6} = 0$. For example, if a person eats zero sixths of a pizza, this means that the person didn't eat any pizza.

Rules for $a \cdot 1$, $\dfrac{a}{1}$, and $\dfrac{a}{a}$

The products $4 \cdot 1 = 4$, $5 \cdot 1 = 5$, and $8 \cdot 1 = 8$ suggest the following property:

Multiplying a Number by 1

$$a \cdot 1 = a$$

In words: A number multiplied by 1 is that same number.

When we write statements such as $a \cdot 1 = a$, we mean that if we evaluate $a \cdot 1$ and a for *any* value of a in both expressions, the results will be equal. We say that the expressions $a \cdot 1$ and a are **equivalent expressions.**

The quotients $\dfrac{4}{1} = 4 \div 1 = 4$, $\dfrac{5}{1} = 5 \div 1 = 5$, and $\dfrac{8}{1} = 8 \div 1 = 8$ suggest the following property:

Dividing a Number by 1

$$\frac{a}{1} = a$$

In words: A number divided by 1 is that same number.

Finally, the quotients $\dfrac{4}{4} = 4 \div 4 = 1$, $\dfrac{5}{5} = 5 \div 5 = 1$, and $\dfrac{8}{8} = 8 \div 8 = 1$ suggest the following property:

Dividing a Nonzero Number by Itself

If a is nonzero, then
$$\frac{a}{a} = 1$$

In words: A nonzero number divided by itself is 1.

The properties $a \cdot 1 = a$, $\dfrac{a}{1} = a$, and $\dfrac{a}{a} = 1$ (where a is nonzero) will help us when we work with fractions.

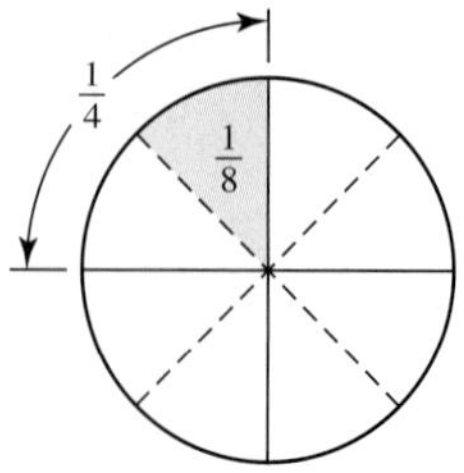

Figure 4 $\frac{1}{2}$ of $\frac{1}{4}$ of a pizza is $\frac{1}{8}$ of a pizza

Multiplication of Fractions

Figure 4 illustrates that $\frac{1}{2}$ of $\frac{1}{4}$ of a pizza is $\frac{1}{8}$ of a pizza. We can calculate this result by finding the product $\frac{1}{2} \cdot \frac{1}{4}$:

$$\frac{1}{2} \cdot \frac{1}{4} = \frac{1 \cdot 1}{2 \cdot 4} = \frac{1}{8}$$

Multiplying Fractions

If b and d are nonzero, then

$$\frac{a}{b} \cdot \frac{c}{d} = \frac{ac}{bd}$$

In words: To multiply two fractions, write the numerators as a product and write the denominators as a product.

Example 1   Finding the Product of Two Fractions

Find the product $\frac{2}{5} \cdot \frac{3}{7}$.

Solution

$$\frac{2}{5} \cdot \frac{3}{7} = \frac{2 \cdot 3}{5 \cdot 7} \qquad \text{Write numerators and denominators as products: } \frac{a}{b} \cdot \frac{c}{d} = \frac{ac}{bd}$$

$$= \frac{6}{35} \qquad \text{Find products.} \qquad ■$$

Prime Factorization

When we work with fractions, it can sometimes help to work with prime numbers.

DEFINITION   Prime number

A **prime number,** or **prime,** is any counting number larger than 1 whose only positive factors are itself and 1.

Here are the first 10 primes:

$$2, 3, 5, 7, 11, 13, 17, 19, 23, 29$$

Sometimes when we work with fractions, it is helpful to write a number as a product of primes. We call this product the **prime factorization** of the number.

Example 2   Writing a Number as a Product of Primes

Write 54 as a product of primes.

Solution

$$54 = \underset{\downarrow\downarrow}{6} \cdot \underset{\downarrow\downarrow}{9} \qquad \text{Write 54 as a product of two numbers.}$$

$$= 2 \cdot 3 \cdot 3 \cdot 3 \qquad \text{Find prime factorizations of 6 and 9.}$$

The prime factorization of 54 is $2 \cdot 3 \cdot 3 \cdot 3$.   ■

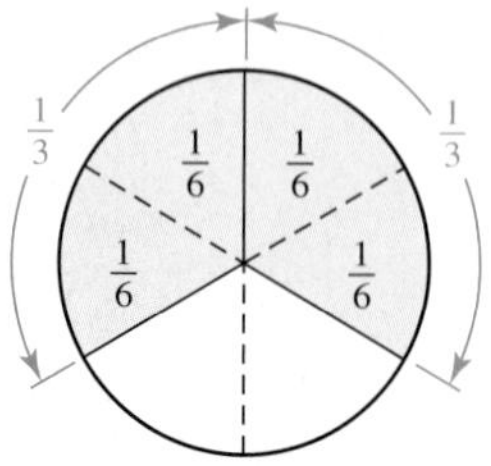

Simplifying Fractions

Figure 5 $\frac{4}{6} = \frac{2}{3}$

Figure 5 illustrates that $\frac{4}{6} = \frac{2}{3}$. We say that $\frac{2}{3}$ is *simplified,* because the numerator and denominator do not have positive factors other than 1 in common. The fraction $\frac{4}{6}$ is not simplified, because the numerator and the denominator have a common factor of 2. To **simplify** a fraction, we write it as an equal fraction in which the numerator and the denominator do not have any common positive factors other than 1.

Example 3   Simplifying a Fraction

Simplify $\dfrac{4}{6}$.

Solution

We begin to simplify $\dfrac{4}{6}$ by finding the prime factorizations of the numerator 4 and the denominator 6:

$$\frac{4}{6} = \frac{2 \cdot 2}{2 \cdot 3} \qquad \text{Find prime factorizations of numerator and denominator.}$$

$$= \frac{2}{2} \cdot \frac{2}{3} \qquad \frac{ab}{cd} = \frac{a}{c} \cdot \frac{b}{d}$$

$$= 1 \cdot \frac{2}{3} \qquad \text{Simplify: } \frac{2}{2} = 1$$

$$= \frac{2}{3} \qquad 1 \cdot a = a$$

Our result matches with what we found in Fig. 5. By performing long division or using a calculator, we can check that both fractions are equal to the repeating decimal $0.\overline{6}$. ∎

Simplifying fractions can make it easier to work out certain problems. Also, if two fractions are simplified, it is easy to tell whether they are equal. **If the result of an exercise is a fraction, simplify it.**

Example 4   Simplifying a Fraction

Simplify $\dfrac{30}{42}$.

Solution

We begin to simplify $\dfrac{30}{42}$ by finding the prime factorizations of the numerator, 30, and the denominator, 42:

$$\frac{30}{42} = \frac{2 \cdot 3 \cdot 5}{2 \cdot 3 \cdot 7} \qquad \text{Find prime factorizations of numerator and denominator.}$$

$$= \frac{2 \cdot 3}{2 \cdot 3} \cdot \frac{5}{7} \qquad \frac{ac}{bd} = \frac{a}{b} \cdot \frac{c}{d}$$

$$= 1 \cdot \frac{5}{7} \qquad \text{Simplify: } \frac{2 \cdot 3}{2 \cdot 3} = 1$$

$$= \frac{5}{7} \qquad 1 \cdot a = a$$ ∎

Simplifying a Fraction

To simplify a fraction,

1. Find the prime factorizations of the numerator and denominator.
2. Find an equal fraction in which the numerator and the denominator do not have common positive factors other than 1 by using the property

$$\frac{ab}{ac} = \frac{a}{a} \cdot \frac{b}{c} = 1 \cdot \frac{b}{c} = \frac{b}{c}$$

where a and c are nonzero.

In Example 5, we multiply two fractions and simplify the result.

Example 5 Finding the Product of Two Fractions

Find the product $\dfrac{8}{9} \cdot \dfrac{15}{4}$.

Solution

$$\dfrac{8}{9} \cdot \dfrac{15}{4} = \dfrac{8 \cdot 15}{9 \cdot 4}$$ Write numerators and denominators as products: $\dfrac{a}{b} \cdot \dfrac{c}{d} = \dfrac{ac}{bd}$

$$= \dfrac{2 \cdot 2 \cdot 2 \cdot 3 \cdot 5}{3 \cdot 3 \cdot 2 \cdot 2}$$ Find prime factorizations.

$$= \dfrac{2 \cdot 5}{3}$$ Simplify: $\dfrac{2 \cdot 2 \cdot 3}{2 \cdot 2 \cdot 3} = 1$

$$= \dfrac{10}{3}$$ Multiply. ∎

Division of Fractions

The **reciprocal** of $\dfrac{a}{b}$ is $\dfrac{b}{a}$. For example, the reciprocal of $\dfrac{3}{8}$ is $\dfrac{8}{3}$. We will need to find the reciprocal of a fraction when we divide two fractions.

Dividing Fractions

If b, c, and d are nonzero, then

$$\dfrac{a}{b} \div \dfrac{c}{d} = \dfrac{a}{b} \cdot \dfrac{d}{c}$$

In words: To divide by a fraction, multiply by its reciprocal.

Example 6 Finding the Quotient of Two Fractions

Find the quotient $\dfrac{3}{4} \div \dfrac{1}{8}$.

Solution

$$\dfrac{3}{4} \div \dfrac{1}{8} = \dfrac{3}{4} \cdot \dfrac{8}{1}$$ Multiply by reciprocal of $\dfrac{1}{8}$, which is $\dfrac{8}{1}$: $\dfrac{a}{b} \div \dfrac{c}{d} = \dfrac{a}{b} \cdot \dfrac{d}{c}$

$$= \dfrac{3 \cdot 8}{4 \cdot 1}$$ Write numerators and denominators as products.

$$= \dfrac{3 \cdot 2 \cdot 2 \cdot 2}{2 \cdot 2 \cdot 1}$$ Find prime factorizations.

$$= \dfrac{3 \cdot 2}{1}$$ Simplify: $\dfrac{2 \cdot 2}{2 \cdot 2} = 1$

$$= 6$$ $\dfrac{a}{1} = a$

Our result makes sense, because $\dfrac{3}{4}$ of a pizza divided into slices, each of size $\dfrac{1}{8}$ of the pizza, gives 6 slices (see Fig. 6).

We can use a graphing calculator to check our work in Example 6 (see Fig. 7). ∎

When you use a calculator to check work with fractions, it is good practice to enclose each fraction in parentheses. You will see the importance of using parentheses when we discuss the order of operations in Section 2.6.

To find the reciprocal of 6, we use the fact $6 = \dfrac{6}{1}$. So, the reciprocal of 6 is $\dfrac{1}{6}$.

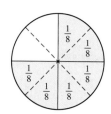

Figure 6 $\dfrac{3}{4}$ of a pizza divided into slices of size $\dfrac{1}{8}$ of the pizza gives 6 slices of pizza

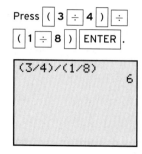

Press (3 ÷ 4) ÷ (1 ÷ 8) ENTER.

(3/4)/(1/8)
            6

Figure 7 Verify the work

Example 7 Evaluating an Expression

Evaluate $\frac{a}{b} \div c$ for $a = 21$, $b = 2$, and $c = 3$.

Solution

We substitute $a = 21$, $b = 2$, and $c = 3$ into the expression $\frac{a}{b} \div c$:

$$\frac{(21)}{(2)} \div (3) = \frac{21}{2} \div \frac{3}{1} \qquad \text{Write 3 as a fraction: } 3 = \frac{3}{1}$$

$$= \frac{21}{2} \cdot \frac{1}{3} \qquad \text{Multiply by reciprocal of } \frac{3}{1}, \text{ which is } \frac{1}{3}: \frac{a}{b} \div \frac{c}{d} = \frac{a}{b} \cdot \frac{d}{c}$$

$$= \frac{21 \cdot 1}{2 \cdot 3} \qquad \text{Write numerators and denominators as products.}$$

$$= \frac{3 \cdot 7 \cdot 1}{2 \cdot 3} \qquad \text{Find prime factorizations.}$$

$$= \frac{7}{2} \qquad \text{Simplify: } \frac{3}{3} = 1$$ ■

In Example 7, the result is $\frac{7}{2}$, which is an improper fraction (that is, the numerator is larger than the denominator). For *nonmodeling* exercises, if a fractional result is in improper form, we will leave it in that form. For *modeling* exercises, if a result is in improper form, we will write it as a mixed number. For example, we say that a car trip takes $3\frac{1}{2}$ hours rather than $\frac{7}{2}$ hours.

Addition of Fractions

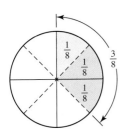

Figure 8 $\frac{1}{8}$ pizza plus $\frac{2}{8}$ pizza is $\frac{3}{8}$ pizza

Figure 8 illustrates that $\frac{1}{8}$ of a pizza plus $\frac{2}{8}$ of a pizza is equal to $\frac{3}{8}$ of a pizza. This illustration suggests that, to find the sum $\frac{1}{8} + \frac{2}{8}$, we add the numerators 1 and 2 and write the result, 3, over the common denominator, 8:

$$\frac{1}{8} + \frac{2}{8} = \frac{1+2}{8}$$

$$= \frac{3}{8}$$

Adding Fractions with the Same Denominator

If b is nonzero, then

$$\frac{a}{b} + \frac{c}{b} = \frac{a+c}{b}$$

In words: To add two fractions with the same denominator, add the numerators and write the result above the common denominator.

Example 8 Adding Fractions with the Same Denominator

Find the sum $\frac{4}{15} + \frac{6}{15}$.

Solution

$$\frac{4}{15} + \frac{6}{15} = \frac{4+6}{15} \qquad \text{Write numerators as a sum and keep common denominator: } \frac{a}{b} + \frac{c}{b} = \frac{a+c}{b}$$

$$= \frac{10}{15} \qquad \text{Find sum.}$$

$$= \frac{2}{3} \qquad \text{Simplify.}$$ ■

Least Common Denominators

To find the sum $\frac{1}{4} + \frac{5}{6}$, in which the denominators of the fractions are different, we find an equal sum of fractions in which the denominators are equal. First, we list the multiples of 4 and the multiples of 6:

$$\textbf{Multiples of 4:}\quad 4, 8, 12, 16, 20, 24, 28, 32, 36, \ldots$$
$$\textbf{Multiples of 6:}\quad 6, 12, 18, 24, 30, 36, 42, 48, 54, \ldots$$

Common multiples of 4 and 6 are

$$12, 24, 36, \ldots$$

Note that 12 is the least (lowest) number in the list. We call it the least common multiple of 4 and 6. The **least common multiple (LCM)** of a group of numbers is the smallest number that is a multiple of *all* of the numbers in the group.

To find the sum $\frac{1}{4} + \frac{5}{6}$, we use the fact $\frac{a}{a} = 1$, where a is nonzero, to write an equal sum of fractions in which each denominator is equal to the LCM, 12:

$$\frac{1}{4} + \frac{5}{6} = \frac{1}{4} \cdot 1 + \frac{5}{6} \cdot 1 \qquad a = a \cdot 1$$

$$= \frac{1}{4} \cdot \frac{3}{3} + \frac{5}{6} \cdot \frac{2}{2} \qquad 1 = \frac{a}{a}$$

$$= \frac{3}{12} + \frac{10}{12} \qquad \text{Multiply numerators and multiply denominators:} \quad \frac{a}{b} \cdot \frac{c}{d} = \frac{ac}{bd}$$

$$= \frac{13}{12} \qquad \text{Add numerators and keep common denominator:} \quad \frac{a}{b} + \frac{c}{b} = \frac{a+c}{b}$$

We also call 12 the least common denominator of $\frac{1}{4}$ and $\frac{5}{6}$. The **least common denominator (LCD)** of a group of fractions is the LCM of the denominators of all of the fractions.

Example 9   Adding Fractions with Different Denominators

Find the sum $\frac{5}{8} + \frac{5}{6}$.

Solution

We list multiples of 8 and multiples of 6:

$$\textbf{Multiples of 8:}\quad 8, 16, 24, 32, 40, 48, \ldots$$
$$\textbf{Multiples of 6:}\quad 6, 12, 18, 24, 30, 36, \ldots$$

The LCD is 24. We write an equal sum of fractions in which each denominator is 24:

$$\frac{5}{8} + \frac{5}{6} = \frac{5}{8} \cdot \frac{3}{3} + \frac{5}{6} \cdot \frac{4}{4} \qquad \text{LCD is 24.}$$

$$= \frac{15}{24} + \frac{20}{24} \qquad \text{Multiply numerators and multiply denominators:} \quad \frac{a}{b} \cdot \frac{c}{d} = \frac{ac}{bd}$$

$$= \frac{35}{24} \qquad \text{Add numerators and keep common denominator:} \quad \frac{a}{b} + \frac{c}{b} = \frac{a+c}{b}$$

We use a graphing calculator to verify the work (see Fig. 9).

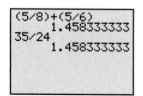

Figure 9 Verify the work

Subtraction of Fractions

The rule for subtracting two fractions with the same denominator is similar to the rule for adding such fractions, except that we subtract the numerators.

Subtracting Fractions with the Same Denominator

If b is nonzero, then

$$\frac{a}{b} - \frac{c}{b} = \frac{a-c}{b}$$

In words: To subtract two fractions with the same denominator, subtract the numerators and write the result above the common denominator.

Example 10 Subtracting Fractions with the Same Denominator

Find the difference $\dfrac{5}{8} - \dfrac{3}{8}$.

Solution

$$\frac{5}{8} - \frac{3}{8} = \frac{5-3}{8}$$ Write numerators as a difference and keep common denominator: $\dfrac{a}{b} - \dfrac{c}{b} = \dfrac{a-c}{b}$

$$= \frac{2}{8}$$ Find difference.

$$= \frac{1}{4}$$ Simplify. ∎

Subtracting fractions with different denominators is similar to adding them. The first step is to rewrite each fraction so that each denominator is the LCD.

Example 11 Subtracting Fractions with Different Denominators

Find the difference $\dfrac{8}{9} - \dfrac{3}{5}$.

Solution

We list the multiples of 9 and the multiples of 5:

Multiples of 9:   9, 18, 27, 36, 45, 54, 63, 72, 81, . . .

Multiples of 5:   5, 10, 15, 20, 25, 30, 35, 40, 45, . . .

The LCD is 45. We now rewrite each fraction with the denominator 45:

$$\frac{8}{9} - \frac{3}{5} = \frac{8}{9} \cdot \frac{5}{5} - \frac{3}{5} \cdot \frac{9}{9}$$ LCD is 45.

$$= \frac{40}{45} - \frac{27}{45}$$ Multiply numerators and multiply denominators: $\dfrac{a}{b} \cdot \dfrac{c}{d} = \dfrac{ac}{bd}$

$$= \frac{13}{45}$$ Subtract numerators and keep common denominator: $\dfrac{a}{b} - \dfrac{c}{b} = \dfrac{a-c}{b}$ ∎

Adding (or Subtracting) Fractions with Different Denominators

To add (or subtract) two fractions with different denominators, use the fact that $\dfrac{a}{a} = 1$, where a is nonzero, to write an equal sum (or difference) of fractions for which each denominator is the LCD.

---

group exploration

Illustrations of simplifying fractions and operations with fractions

Draw a picture of a pizza to show that the true statement makes sense. [**Hint:** See Figs. 4, 5, 6, and 8.]

1. $\dfrac{6}{8} = \dfrac{3}{4}$

2. $\dfrac{5}{8} + \dfrac{2}{8} = \dfrac{7}{8}$

3. $\dfrac{5}{6} - \dfrac{4}{6} = \dfrac{1}{6}$

4. $\dfrac{1}{2} \cdot \dfrac{1}{3} = \dfrac{1}{6}$

5. $\dfrac{2}{3} \div \dfrac{1}{6} = 4$

TIPS FOR SUCCESS: Review Your Notes as Soon as Possible

How often do you get confused by class notes you wrote earlier the same day, even though the class activities made sense to you? If this happens a lot, review your notes as soon after class as possible. Even reviewing your notes for just a few minutes between classes will help. This will increase your likelihood of remembering what you learned in class and will give you the opportunity to add new comments to your notes while the class experience is still fresh in your mind.

HOMEWORK 2.2

FOR EXTRA HELP ▶

Student Solutions Manual · PH Math/Tutor Center · MathXL® · MyMathLab

1. What is the denominator of $\dfrac{3}{7}$?

2. What is the numerator of $\dfrac{2}{5}$?

Write the number as a product of primes.

3. 20 **4.** 18 **5.** 36 **6.** 24

7. 45 **8.** 27 **9.** 78 **10.** 105

Simplify. Then use a calculator to check your work.

11. $\dfrac{6}{8}$ **12.** $\dfrac{10}{14}$ **13.** $\dfrac{3}{12}$ **14.** $\dfrac{7}{28}$

15. $\dfrac{18}{30}$ **16.** $\dfrac{27}{54}$ **17.** $\dfrac{20}{50}$ **18.** $\dfrac{49}{63}$

19. $\dfrac{5}{25}$ **20.** $\dfrac{9}{81}$ **21.** $\dfrac{20}{24}$ **22.** $\dfrac{15}{18}$

Perform the indicated operation. Then use a calculator to check your work.

23. $\dfrac{1}{3} \cdot \dfrac{2}{5}$ **24.** $\dfrac{6}{7} \cdot \dfrac{4}{9}$ **25.** $\dfrac{4}{5} \cdot \dfrac{3}{8}$ **26.** $\dfrac{2}{3} \cdot \dfrac{5}{6}$

27. $\dfrac{5}{21} \cdot 7$ **28.** $\dfrac{5}{12} \cdot 2$ **29.** $\dfrac{5}{8} \div \dfrac{3}{4}$ **30.** $\dfrac{7}{12} \div \dfrac{2}{3}$

31. $\dfrac{8}{9} \div \dfrac{4}{3}$ **32.** $\dfrac{4}{7} \div \dfrac{8}{3}$ **33.** $\dfrac{2}{3} \div 5$ **34.** $\dfrac{4}{9} \div 2$

35. $\dfrac{2}{7} + \dfrac{3}{7}$ **36.** $\dfrac{5}{9} + \dfrac{2}{9}$ **37.** $\dfrac{5}{8} + \dfrac{1}{8}$ **38.** $\dfrac{2}{15} + \dfrac{8}{15}$

39. $\dfrac{4}{5} - \dfrac{3}{5}$ **40.** $\dfrac{5}{7} - \dfrac{2}{7}$ **41.** $\dfrac{11}{12} - \dfrac{7}{12}$ **42.** $\dfrac{13}{18} - \dfrac{9}{18}$

43. $\dfrac{1}{4} + \dfrac{1}{2}$ **44.** $\dfrac{1}{3} + \dfrac{5}{9}$ **45.** $\dfrac{5}{6} + \dfrac{3}{4}$ **46.** $\dfrac{3}{8} + \dfrac{1}{6}$

47. $4 + \dfrac{2}{3}$ **48.** $2 + \dfrac{3}{7}$ **49.** $\dfrac{7}{9} - \dfrac{2}{3}$ **50.** $\dfrac{3}{4} - \dfrac{1}{2}$

51. $\dfrac{5}{9} - \dfrac{2}{7}$ **52.** $\dfrac{5}{6} - \dfrac{4}{7}$ **53.** $3 - \dfrac{4}{5}$ **54.** $1 - \dfrac{9}{7}$

Perform the indicated operation. If the fraction is undefined, say so. Then use a calculator to check your work.

55. $\dfrac{3172}{3172}$ **56.** $\dfrac{62}{62}$ **57.** $\dfrac{599}{1}$ **58.** $\dfrac{215}{1}$

59. $\dfrac{842}{0}$ **60.** $\dfrac{713}{0}$ **61.** $\dfrac{0}{621}$ **62.** $\dfrac{0}{798}$

63. $\dfrac{824}{631} \cdot \dfrac{631}{824}$ **64.** $\dfrac{173}{190} \cdot \dfrac{190}{173}$

65. $\dfrac{544}{293} - \dfrac{544}{293}$ **66.** $\dfrac{345}{917} - \dfrac{345}{917}$

Evaluate the given expression for $w = 4$, $x = 3$, $y = 5$, and $z = 12$.

67. $\dfrac{w}{z}$ **68.** $\dfrac{z}{x}$ **69.** $\dfrac{x}{w} \div \dfrac{y}{z}$

70. $\dfrac{y}{z} \cdot \dfrac{w}{x}$ **71.** $\dfrac{x}{w} - \dfrac{y}{z}$ **72.** $\dfrac{y}{x} + \dfrac{y}{z}$

Use a calculator to compute. Round the result to two decimal places.

73. $\dfrac{19}{97} \cdot \dfrac{65}{74}$ **74.** $\dfrac{67}{71} \cdot \dfrac{381}{399}$

75. $\dfrac{684}{795} \div \dfrac{24}{37}$

76. $\dfrac{149}{215} \div \dfrac{31}{52}$

77. $\dfrac{89}{102} - \dfrac{59}{133}$

78. $\dfrac{614}{701} + \dfrac{391}{400}$

For Exercises 79 and 80, draw a picture of a pizza to show that the true statement makes sense.

79. $\dfrac{2}{8} = \dfrac{1}{4}$

80. $\dfrac{1}{4} + \dfrac{2}{4} = \dfrac{3}{4}$

81. A rectangular plot of land has a length of $\dfrac{2}{5}$ mile and a width of $\dfrac{1}{4}$ mile. What is the area of this plot?

82. A rectangular picture has a width of $\dfrac{2}{3}$ foot and a length of $\dfrac{3}{4}$ foot. What is the perimeter of this picture?

83. For an elementary algebra course, total course points are calculated by adding points earned on homework assignments, quizzes, tests, and the final exam. If the total of scores on tests is worth $\dfrac{1}{2}$ of the course points and the final exam score is worth $\dfrac{1}{4}$ of the course points, what fraction of the course points comes from homework assignments and quizzes?

84. A family spends $\dfrac{1}{3}$ of its income for the mortgage and $\dfrac{1}{6}$ of its income for food. What fraction of its income remains?

For Exercises 85 and 86, let x be a number. Translate the expression into an English phrase.

85. $\dfrac{x}{3}$

86. $\dfrac{5}{x}$

87. Some friends pay a total of $19 for a pizza. Each of the n friends pays an equal share of the cost. Complete Table 10 to help find an expression that describes the cost (in dollars) per person. Show the arithmetic to help you see a pattern.

Table 10 Cost per Person for the Pizza

Number of People	Cost per Person (dollars per person)
2	
3	
4	
5	
n	

88. A tutor charges $45 for a tutoring session that lasts for t hours. Complete Table 11 to help find an expression that describes the cost (in dollars) per hour. Show the arithmetic to help you see a pattern.

Table 11 Cost per Hour for the Session

Total Time (hours)	Cost per Hour (dollars per hour)
2	
3	
4	
5	
t	

89. a. Perform the indicated operation.

i. $\dfrac{5}{6} \cdot \dfrac{2}{3}$ **ii.** $\dfrac{5}{6} \div \dfrac{2}{3}$

iii. $\dfrac{5}{6} + \dfrac{2}{3}$ **iv.** $\dfrac{5}{6} - \dfrac{2}{3}$

b. Compare the methods you used to perform the operations in part (a). Describe how the methods are similar and how they are different.

90. a. Find each product.

i. $\dfrac{2}{3} \cdot \dfrac{3}{2}$ **ii.** $\dfrac{4}{7} \cdot \dfrac{7}{4}$ **iii.** $\dfrac{1}{6} \cdot \dfrac{6}{1}$

b. On the basis of your results from part (a), use words to describe a property of a fraction and its reciprocal. Then describe the property in terms of variables.

91. A student tries to find the product $\dfrac{1}{2} \cdot \dfrac{1}{3}$:

$$\dfrac{1}{2} \cdot \dfrac{1}{3} = \left(\dfrac{1}{2} \cdot \dfrac{3}{3} \right) \cdot \left(\dfrac{1}{3} \cdot \dfrac{2}{2} \right)$$
$$= \dfrac{3}{6} \cdot \dfrac{2}{6}$$
$$= \dfrac{6}{36}$$
$$= \dfrac{1}{6}$$

What would you tell the student?

92. A student tries to find the sum $\dfrac{2}{3} + \dfrac{5}{6}$:

$$\dfrac{2}{3} + \dfrac{5}{6} = \dfrac{2+5}{3+6} = \dfrac{7}{9}$$

Describe any errors. Then find the sum correctly.

93. A student tries to find the product $2 \cdot \dfrac{3}{5}$:

$$2 \cdot \dfrac{3}{5} = \dfrac{6}{10} = \dfrac{3}{5}$$

Describe any errors. Then find the product correctly.

94. A student tries to find the product $3 \cdot \dfrac{7}{2}$:

$$3 \cdot \dfrac{7}{2} = \dfrac{7}{3 \cdot 2} = \dfrac{7}{6}$$

Describe any errors. Then find the product correctly.

95. Greenskeepers have been mowing golf putting surfaces progressively lower over the past half century (see Table 12).

Table 12 Grass Heights on Golf Putting Surfaces

Decade	Year Used to Represent Decade	Grass Height (inches)
1950s	1955	$\dfrac{1}{4}$
1960s	1965	$\dfrac{7}{32}$
1970s	1975	$\dfrac{3}{16}$
1980s	1985	$\dfrac{5}{32}$
1990s	1995	$\dfrac{1}{8}$

Source: *Golf Course Superintendents Association of America*

Let h be the grass height (in inches) at t years since 1950.
   a. Create a scattergram of the data. [**Hint:** Use a common denominator of 32 for grass height fractions.]
   b. Draw a line that contains the data points.
   c. Estimate the grass height in 2005.
   d. Predict when the grass height will be $\dfrac{1}{16}$ inch.
   e. Find the t-intercept. What does it mean in this situation?

96. a. Use a number line to find the distance between 0 and each number.
   i. 4   **ii.** −3   **iii.** −6
   b. *Without* using a number line, describe in general how you can find the distance between 0 and a number on a number line.

97. Explore how to add two negative numbers:
   a. Use a calculator to find each sum of two negative numbers.
      i. $-1 + (-5)$ **ii.** $-6 + (-2)$
      iii. $-3 + (-4)$
   b. What pattern do you notice? If you do not see a pattern, continue finding sums of any two negative numbers until you do.
   c. Without using a calculator, find the sum $-4 + (-5)$. Then use a calculator to check your work.
   d. State a general rule for how to add two negative numbers.

98. Explore how to add two numbers with different signs:
   a. First, consider the case in which the positive number is farther from 0 on the number line than the negative number is. Use a calculator to find the following sums:
      i. $5 + (-2)$ **ii.** $7 + (-1)$ **iii.** $8 + (-3)$
   b. What pattern do you notice in your work from part (a)? If you do not see a pattern, continue finding similar sums until you do.
   c. Now consider the case in which the positive number is closer to 0 on the number line than the negative number is. Use a calculator to find the following sums:
      i. $2 + (-5)$ **ii.** $1 + (-7)$ **iii.** $3 + (-8)$
   d. What pattern do you notice in your work from part (c)?
   e. Now consider the case in which the two numbers are the same distance from 0 on the number line, but on opposite sides of 0. Use a calculator to find the following sums:
      i. $4 + (-4)$ **ii.** $7 + (-7)$ **iii.** $9 + (-9)$
   f. What pattern do you notice in your work from part (e)?
   g. Without using a calculator, find each sum. Then use a calculator to check your work.
      i. $6 + (-4)$ **ii.** $3 + (-7)$ **iii.** $6 + (-6)$
   h. State a general rule for how to add two numbers with different signs.

2.3 ADDING REAL NUMBERS

Objectives

▷ Find the opposite of a number.

▷ Find the opposite of the opposite of a number.

▷ Find the absolute value of a number.

▷ Add real numbers by thinking in terms of money, the number line, and absolute value.

▷ Add real numbers pertaining to authentic situations.

▷ Find an expression to model an authentic quantity.

In this section, our main objective is to add real numbers.

The Opposite of a Number

Note that in Fig. 10 both the numbers −3 and 3 are 3 units from 0 on the number line, but they are on opposite sides of 0. We say that −3 is the *opposite* of 3, that 3 is the *opposite* of −3, and that −3 and 3 are *opposites*.

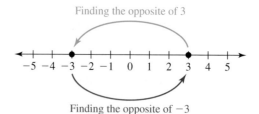

Figure 10 Finding the opposite of 3 and the opposite of −3

Two numbers are called **opposites** of each other if they are the same distance from 0 on the number line, but are on opposite sides of 0. We find the opposite of a number by writing a negative sign in front of the number. For example, the opposite of 3 is -3 (see Fig. 10).

Now consider this true statement:

The opposite of -3 is 3 (see Fig. 10).

In symbols, we write

The opposite of -3 is equal to 3.
$$-\underbrace{}\quad \underbrace{(-3)}\quad \underbrace{=}\quad \underbrace{3}$$

Here are some more examples of finding the opposite of a negative number:

$$-(-4) = 4$$
$$-(-7) = 7$$

We can use a graphing calculator to find $-(-7)$. See Fig. 11. We use the button $\boxed{-}$ for subtraction and the button $\boxed{(-)}$ for negative numbers and for taking opposites.

We can view $-(-7)$ as finding the opposite of -7 or as finding the opposite of the opposite of 7.

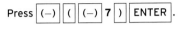

Press $\boxed{(-)}\,\boxed{(}\,\boxed{(-)}\,\boxed{7}\,\boxed{)}\,\boxed{ENTER}$.

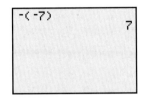

Figure 11 Calculating $-(-7)$

Finding the Opposite of the Opposite of a Number

$$-(-a) = a$$

In words: The opposite of the opposite of a number is equal to that same number.

We use parentheses to separate two opposite symbols or an operation symbol and an opposite symbol.

Example 1 Finding Opposites

Find the opposite.

1. $-(-3)$ **2.** $-(-(-3))$

Solution

1. $-(-3) = 3$ $-(-a) = a$
2. $-(-(-3)) = -(3)$ $-(-a) = a$
$\qquad\qquad\quad = -3$ Write without parentheses.   ∎

Absolute Value

The *absolute value* of a number a, written $|a|$, is the distance that the number a is from 0 on the number line.

DEFINITION Absolute value

The **absolute value** of a number is the distance that the number is from 0 on the number line.

So $|-3| = 3$, because -3 is a distance of 3 units from 0, and $|3| = 3$, because 3 is a distance of 3 units from 0 (see Fig. 12).

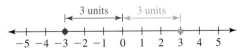

Figure 12 Both -3 and 3 are a distance of 3 units from 0.

Press 2nd 0 ENTER
(−) 3) ENTER .

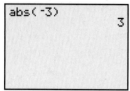

Figure 13 Calculating |−3|

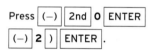

Figure 15 |−2| = 2

Press (−) 2nd 0 ENTER
(−) 2) ENTER .

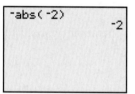

Figure 16 Check that − |−2| = −2

On a graphing calculator, "abs" stands for absolute value. We find |−3| in Fig. 13.

Example 2 Finding Absolute Values of Numbers

Calculate.

1. |2|
2. |−2|
3. − |2|
4. − |−2|

Solution

1. |2| = 2, because 2 is a distance of 2 units from 0 (see Fig. 14).

Figure 14 |2| = 2

2. |−2| = 2, because −2 is a distance of 2 units from 0 (see Fig. 15).
3. − |2| = −(2) |2| = 2
   = −2   *Write without parentheses.*
4. − |−2| = −(2) |−2| = 2
   = −2   *Write without parentheses.*

We can use a graphing calculator to check that − |−2| = −2 (see Fig. 16).

Addition of Two Numbers with the Same Sign

Thinking about credit card balances or the number line can help us see how to add numbers with the same sign.

Example 3 Finding the Sum of Two Numbers with the Same Sign

1. A person has a credit card balance of 0 dollars. If she uses her credit card to make two purchases, one for $2 and one for $5, what is the new balance?
2. Write a sum that is related to the computation in Problem 1.
3. Use a number line to illustrate the sum found in Problem 2.

Solution

1. By making purchases for $2 and $5, the person now owes $7. So, the new balance is −7 dollars.
2. Here is the sum:

Spend $2   Spend $5   Owe $7
$$-2 \; + \; (-5) \; = \; -7$$

3. Using the number line, imagine moving 2 units to the left of 0 and then 5 more units to the left. Figure 17 illustrates that −2 + (−5) = −7.

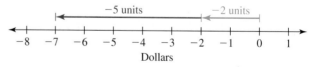

Figure 17 Illustration of −2 + (−5) = −7

In Example 3, we found that −2 + (−5) = −7. To get this result, we added the debts of 2 and 5 to get a total debt of 7. Note that 2 and 5 are the absolute values of −2 and -5. Note also that the result, -7, of the original sum has the same sign as both -2 and -5. These observations suggest the following procedure:

Adding Two Numbers with the Same Sign

To add two numbers with the same sign,

1. Add the absolute values of the numbers.
2. The sum of the original numbers has the same sign as the sign of the original numbers.

Example 4   Finding the Sum of Two Numbers with the Same Sign

Find the sum.

1. $-3 + (-6)$

2. $-\dfrac{1}{5} + \left(-\dfrac{3}{5}\right)$

Solution

1. First we add the absolute values of the numbers -3 and -6: $3 + 6 = 9$. Since both -3 and -6 are negative, their sum is negative. So, $-3 + (-6) = -9$.

2. By adding the absolute values of the fractions, we have $\dfrac{1}{5} + \dfrac{3}{5} = \dfrac{4}{5}$. Since both original fractions are negative, their sum is negative. So,

$$-\frac{1}{5} + \left(-\frac{3}{5}\right) = -\frac{4}{5}$$

∎

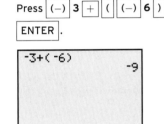

Press $(-)$ **3** $+$ $($ $(-)$ **6** $)$

ENTER .

Figure 18 Calculating $-3 + (-6)$

Toward the start of this course, your instructor will likely have you add real numbers without using a calculator. Once you have mastered that skill, your instructor will likely encourage you to use a calculator for longer computations.

After completing an exercise in this section's homework by hand, you can use a graphing calculator to check your work. For example, we can check our work for $-3 + (-6)$, Problem 1 in Example 4 (see Fig. 18).

Addition of Two Numbers with Different Signs

Thinking about exchanges of money or the number line can also help us see how to add numbers with different signs.

Example 5   Finding the Sum of Two Numbers with Different Signs

1. A brother owes his sister $5. If he then pays her back $2, how much does he still owe her?
2. Write a sum that is related to your work in Problem 1.
3. Use a number line to illustrate the sum you found in Problem 2.

Solution

1. By owing his sister $5 and paying her back $2, the brother now owes his sister $3.
2. Here's the sum:

$$\overset{\text{Owe \$5}}{\overbrace{-5}} \; + \; \overset{\text{Pay back \$2}}{\overbrace{2}} \; = \; \overset{\text{Now owe \$3}}{\overbrace{-3}}$$

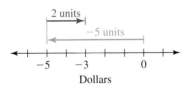

Figure 19 Illustration of $-5 + 2 = -3$

3. Using the number line, imagine moving 5 units to the left of 0 and then 2 units to the right of -5. Figure 19 illustrates that $-5 + 2 = -3$.

∎

In Problem 2 in Example 5, we found that $-5 + 2 = -3$. We can get this result by first finding the difference of 5 and 2:

$$5 - 2 = 3$$

We can think of this operation as lowering a debt of 5 dollars by 2 dollars to get a debt of 3 dollars, so the result is -3. Note that the result, -3, has the same sign as

−5, which has a larger absolute value than 2. These observations suggest the following procedure:

Adding Two Numbers with Different Signs

To add two numbers with different signs,

1. Find the absolute values of the numbers. Then subtract the smaller absolute value from the larger absolute value.

2. The sum of the original numbers has the same sign as the original number with the larger absolute value.

Example 6   Finding the Sum of Two Numbers with Different Signs

Find the sum.

1. $-4 + 7$ **2.** $3 + (-9)$ **3.** $-\dfrac{5}{6} + \dfrac{2}{3}$

Solution

1. First, we find that $7 - 4 = 3$. Since 7 has a larger absolute value than −4, and since 7 is positive, the sum is positive: $-4 + 7 = 3$.
2. First, we find that $9 - 3 = 6$. Since −9 has a larger absolute value than 3, and since −9 is negative, the sum is negative: $3 + (-9) = -6$.
3. First, we write the fractions so that the denominators are the same.

$$-\frac{5}{6} + \frac{2}{3} = -\frac{5}{6} + \frac{2}{3} \cdot \frac{2}{2} \qquad \text{LCD is 6.}$$

$$= -\frac{5}{6} + \frac{4}{6} \qquad \text{Multiply numerators and multiply denominators: } \frac{a}{b} \cdot \frac{c}{d} = \frac{ac}{bd}$$

Next, we subtract the smaller absolute value from the larger absolute value:

$$\frac{5}{6} - \frac{4}{6} = \frac{1}{6}$$

Since $-\dfrac{5}{6}$ has a larger absolute value than the fraction $\dfrac{4}{6}$, and since $-\dfrac{5}{6}$ is negative, the sum is negative:

$$-\frac{5}{6} + \frac{4}{6} = -\frac{1}{6} \qquad \blacksquare$$

We have discussed three ways to add real numbers: thinking in terms of money, the number line, or absolute value. **It is good practice to use one method to find a sum and then use another method (or a calculator) as a check.**

Example 7   Translating from English to Mathematics

Let x be a number. Translate the phrase "the sum of −4 and the number" into a mathematical expression. Then evaluate the expression for $x = -2$.

Solution

The expression is $-4 + x$. We substitute −2 for x in the expression $-4 + x$ and then find the sum:

$$-4 + (-2) = -6 \qquad \blacksquare$$

Applications

Knowing how to add real numbers is a useful skill when you work with quantities that can be negative, such as balances of checking accounts and temperature readings.

Example 8   Applications of Adding Real Numbers

1. A person bounces several checks and is charged service fees such that the balance of the checking account is −90.75 dollars. If the person then deposits 300 dollars, what is the balance?

2. Three hours ago, the temperature was $-11°F$. If the temperature has increased by $5°F$ in the last three hours, what is the current temperature?

Solution

1. The balance is $-90.75 + 300$ dollars. To find this sum, we first find the difference $300 - 90.75 = 209.25$. Since 300 has a larger absolute value than -90.75, and since 300 is positive, the sum is positive: $-90.75 + 300 = 209.25$. So, the balance is \$209.25.
2. The temperature is $-11 + 5$ degrees Fahrenheit. To find this sum, we first find the difference $11 - 5 = 6$. Since -11 has a larger absolute value than 5, and since -11 is negative, the sum is negative: $-11 + 5 = -6$. So, the current temperature is $-6°F$. ∎

Modeling with Expressions

In Example 9, we will use an expression to describe an authentic quantity.

Example 9   Finding and Evaluating an Expression

A person fills up his 15-gallon gasoline tank. Let c be the amount of gasoline (in gallons) consumed after the tank has been filled up.

1. Use a table to help find an expression that describes the amount of gasoline (in gallons) that remains in the tank.
2. Evaluate the expression found in Problem 1 for $c = 7$. What does the result mean in this situation?

Table 13 Gasoline Remaining

Gasoline Consumed (gallons)	Gasoline Remaining (gallons)
1	$15 + (-1)$
2	$15 + (-2)$
3	$15 + (-3)$
4	$15 + (-4)$
c	$15 + (-c)$

Solution

1. First we create Table 13. We show the arithmetic to help us see the pattern. From the last row of the table, we see that the expression $15 + (-c)$ represents the amount of gasoline (in gallons) that remains.
2. We evaluate $15 + (-c)$ for $c = 7$:

$$15 + (-(7)) = 15 + (-7) = 8$$

If 7 gallons of gasoline are consumed, 8 gallons of gasoline will remain. ∎

group exploration

Adding a number and its opposite

1. Evaluate $a + (-a)$ for the given values of a.
   a. $a = 2$ b. $a = 3$ c. $a = 5$
2. Evaluate $a + (-a)$ for the given values of a. [**Hint:** Use $-(-a) = a$.]
   a. $a = -2$ b. $a = -3$ c. $a = -5$
3. What do your results in Problems 1 and 2 suggest about $a + (-a)$?

group exploration

Looking ahead: Subtracting numbers

1. Find the difference $6 - 4$ and the sum $6 + (-4)$, and compare your results.
2. Find the difference $7 - 3$ and the sum $7 + (-3)$, and compare your results.
3. Find the difference $9 - 2$ and the sum $9 + (-2)$, and compare your results.
4. In Problems 1–3, for each difference there is a related sum that gives the same result. Write $a - b$ as a sum.
5. In Problem 4, you wrote $a - b$ as a sum. Use this method to find the given difference.
   a. $8 - 3$ b. $2 - 5$ c. $-4 - 3$

TIPS FOR SUCCESS: Affirmations

Do you have difficulty with math? If so, do you ever tell yourself (or others) that you are not good at it? This is called *negative self-talk*. The more you say this, the more likely your subconscious will believe it—and you *will* do poorly in math.

You can counteract years of negative self-talk by telling yourself with conviction that you are good at math.

It might seem strange to state that something is true that hasn't happened yet, but it works! Such statements are called *affirmations*.

There are four guiding principles for getting the most out of saying affirmations:

1. Say affirmations which imply that the desired event is currently happening. For example, say

"I am good at algebra," not "I will be good at algebra."

2. Say that your desired result is continuing to improve. For example, say

"I am good at algebra and I continue to improve at it."

3. Say affirmations in the positive rather than in the negative. For example, say

"I attend each class," not "I don't cut classes."

4. Say affirmations with conviction.

If you would like to learn more about affirmations, the book *Creative Visualization* (Bantam Books, 1985), by Shakti Gawain, is an excellent resource.

HOMEWORK 2.3

FOR EXTRA HELP ▶

Student Solutions Manual · PH Math/Tutor Center · Math XL, MathXL® · MyMathLab

Compute by hand. Then use a calculator to check your work.

1. $-(-4)$ **2.** $-(-9)$ **3.** $-(-(-7))$
4. $-(-(-2))$ **5.** $|3|$ **6.** $|6|$
7. $|-8|$ **8.** $|-1|$ **9.** $-|4|$
10. $-|5|$ **11.** $-|-7|$ **12.** $-|-9|$

Find the sum by hand. Then use a calculator to check your work.

13. $2 + (-7)$ **14.** $5 + (-3)$ **15.** $-1 + (-4)$
16. $-3 + (-2)$ **17.** $7 + (-5)$ **18.** $6 + (-9)$
19. $-8 + 5$ **20.** $-3 + 4$ **21.** $-7 + (-3)$
22. $-9 + (-5)$ **23.** $4 + (-7)$ **24.** $8 + (-2)$
25. $1 + (-1)$ **26.** $8 + (-8)$ **27.** $-4 + 4$
28. $-7 + 7$ **29.** $12 + (-25)$ **30.** $17 + (-14)$
31. $-39 + 17$ **32.** $-89 + 57$
33. $-246 + (-899)$ **34.** $-347 + (-594)$
35. $25{,}371 + (-25{,}371)$ **36.** $127{,}512 + (-127{,}512)$
37. $-4.1 + (-2.6)$ **38.** $-3.7 + (-9.9)$
39. $-5 + 0.2$ **40.** $-0.3 + 7$
41. $2.6 + (-99.9)$ **42.** $37.05 + (-19.26)$
43. $\dfrac{5}{7} + \left(-\dfrac{3}{7}\right)$ **44.** $\dfrac{2}{5} + \left(-\dfrac{1}{5}\right)$
45. $-\dfrac{5}{8} + \dfrac{3}{8}$ **46.** $-\dfrac{5}{6} + \dfrac{1}{6}$
47. $-\dfrac{1}{4} + \left(-\dfrac{1}{2}\right)$ **48.** $-\dfrac{2}{3} + \left(-\dfrac{5}{6}\right)$

49. $\dfrac{5}{6} + \left(-\dfrac{1}{4}\right)$ **50.** $\dfrac{2}{3} + \left(-\dfrac{3}{4}\right)$

Use a calculator to find the sum. Round the result to two decimal places.

51. $-325.89 + 6547.29$ **52.** $-7498.34 + 6435.28$
53. $-17{,}835.69 + (-79{,}735.45)$
54. $-38{,}487.26 + (-83{,}205.87)$
55. $-\dfrac{34}{983} + \left(-\dfrac{19}{251}\right)$ **56.** $-\dfrac{37}{642} + \left(-\dfrac{25}{983}\right)$

Evaluate the expression for $a = -4$, $b = 3$, and $c = -2$.

57. $a + b$ **58.** $b + a$ **59.** $a + c$ **60.** $b + c$

For Exercises 61–64, let x be a number. Translate the English phrase into a mathematical expression. Then evaluate the expression for $x = -6$.

61. Two more than the number

62. The number increased by 3

63. The sum of -4 and the number

64. The number plus -8

65. A person bounces several checks and is charged service fees such that the balance of the checking account is -75 dollars. If the person then deposits 250 dollars, what is the balance?

66. A person bounces several checks and is charged service fees such that the balance of the checking account is

−112.50 dollars. If the person then deposits 170 dollars, what is the balance?

67. A check register is shown in Table 14. Find the final balance of the checking account.

Table 14 Check Register

Check No.	Date	Description of Transaction	Payment	Deposit	Balance
					−89.00
	7/18	Transfer		300.00	
3021	7/22	State Farm	91.22		
3022	7/22	MCI	44.26		
	7/31	Paycheck		870.00	

68. A check register is shown in Table 15. Find the final balance of the checking account.

Table 15 Check Register

Check No.	Date	Description of Transaction	Payment	Deposit	Balance
					−135.00
	2/31	Paycheck		549.00	
253	3/2	FedEx Kinko's	10.74		
	3/3	ATM	21.50		
254	3/7	Barnes and Noble	17.19		

69. A person has a credit card balance of −5471 dollars. If she sends a check to the credit card company for $2600, what is the new balance?

70. A person has a credit card balance of −2739 dollars. If he sends a check to the credit card company for $530, what is the new balance?

71. A student has a credit card balance of −3496 dollars. If he sends a check to the credit card company for $2500 and then uses his credit card to purchase a camera for $629 and some film for $8, what is the new balance?

72. A student has a credit card balance of −873 dollars. If she sends a check to the credit card company for $500 and then uses the card to buy a tennis racquet for $249 and a tennis outfit for $87, what is the new balance?

73. Three hours ago, it was −5°F. If the temperature has increased by 9°F in the last three hours, what is the current temperature?

74. Four hours ago, it was −12°F. If the temperature has increased by 8°F in the last four hours, what is the current temperature?

75. A person just lost 20 pounds on a diet.
   a. Complete Table 16 to help find an expression that describes the person's current weight (in pounds) if the person's weight before the diet was B pounds. Show the arithmetic to help you see a pattern.

Table 16 Weights before and after the Diet

Weight before Diet (pounds)	Weight after Diet (pounds)
160	
165	
170	
175	
B	

   b. Evaluate the expression you found in part (a) for $B = 169$. What does your result mean in this situation?

76. An electronics store is offering a weekend sale of $15 off the retail price of any of its DVD players.
   a. Complete Table 17 to help find an expression that describes the sale price (in dollars) if the retail price is r dollars. Show the arithmetic to help you see a pattern.

Table 17 Retail and Sale Prices

Retail Price (dollars)	Sale Price (dollars)
125	
150	
175	
200	
r	

   b. Evaluate the expression you found in part (a) for $r = 173$. What does your result mean in this situation?

77. The balance in a person's checking account is −80 dollars.
   a. Complete Table 18 to help find an expression that describes the new balance (in dollars) if the person deposits d dollars. Show the arithmetic to help you see a pattern.

Table 18 Deposits and New Balances

Deposit (dollars)	New Balance (dollars)
50	
100	
150	
200	
d	

   b. Evaluate the expression you found in part (a) for $d = 125$. What does your result mean in this situation?

78. One hour ago, the temperature was −2°F.
   a. Complete Table 19 to help find an expression that describes the current temperature (in degrees Fahrenheit) if the temperature decreased by x degrees Fahrenheit in the past hour. Show the arithmetic to help you see a pattern.

Table 19 Decreases in Temperature and Current Temperatures

Decrease in Temperature (degrees Fahrenheit)	Current Temperature (degrees Fahrenheit)
1	
2	
3	
4	
x	

   b. Evaluate the expression you found in part (a) for $x = 7$. What does your result mean in this situation?

79. If a is negative and b is negative, what can you say about the sign of $a + b$? Use a number line to show this property.

80. If a is positive, b is negative, and b is larger in absolute value than a, what can you say about the sign of $a + b$? Use a number line to show this property.

81. If $a + b = 0$, what can you say about a and b?

82. If $a + b$ is positive, what can you say about a and b?

83. a. Evaluate $-a$ for $a = -3$.
  b. Evaluate $-a$ for $a = -4$.
  c. Evaluate $-a$ for $a = -6$.
  d. A student says that $-a$ represents only negative numbers because $-a$ has a negative sign. Is the student correct? Explain.

84. a. Evaluate $a + b$ for $a = -2$ and $b = 5$.
  b. Evaluate $b + a$ for $a = -2$ and $b = 5$.
  c. Compare your results from parts (a) and (b).
  d. Evaluate $a + b$ and $b + a$ for $a = -4$ and $b = -9$, and then compare the results.
  e. Evaluate $a + b$ and $b + a$ for values of your choosing for a and b, and then compare the results.
  f. Is the statement $a + b = b + a$ true for all numbers a and b? Explain.

2.4 CHANGE IN A QUANTITY AND SUBTRACTING REAL NUMBERS

Objectives

▶ Find the change in a quantity.

▶ Subtract real numbers.

▶ Know the sign of the change for an increasing or decreasing quantity.

In this section, we discuss how to use subtraction to compute how much a quantity has changed. For example, we can compute the increase in a population of wolves or the decrease in the percentage of eligible voters who voted in an election.

Change in a Quantity

If the value of a stock increases from \$5 to \$8, we say that the value *changed* by \$3. Finding the change in a quantity is a very important concept in mathematics and has many applications. A company is extremely focused on the change in its profits. During an operation, a surgeon keeps a close eye on the change in a patient's blood pressure. You probably care deeply about a change in your GPA.

Example 1 Finding the Change in a Quantity

1. If a student's CD collection increases from 2 CDs to 7 CDs, find the change in the number of CDs.

2. Write a difference that is related to the computation in Problem 1.

Solution

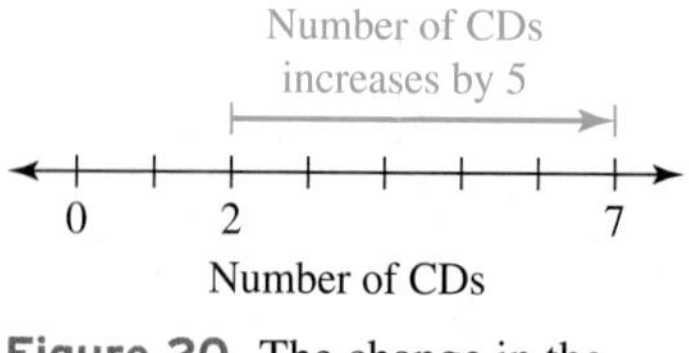

1. If the number of CDs increases from 2 CDs to 7 CDs, then the change in the number of CDs is 5 CDs (see Fig. 20).

2. Here is the difference:

Number of CDs increases by 5

Number of CDs

Figure 20 The change in the number of CDs is 5 CDs

$$
\underbrace{5}_{\substack{\text{Change in} \\ \text{the number of CDs}}} = \underbrace{7}_{\substack{\text{Ending number} \\ \text{of CDs}}} - \underbrace{(2)}_{\substack{\text{Beginning number} \\ \text{of CDs}}}
$$

In Example 1, we found the change in the number of CDs by finding the difference of the ending number of CDs and the beginning number of CDs.

Change in a Quantity
The change in a quantity is the ending amount minus the beginning amount: Change in the quantity = Ending amount − Beginning amount

In Example 2, we find the changes in a quantity from one year to the next.

Example 2 Finding Changes in a Quantity

Table 20 Percentages of ATMs That Are Privately Owned

Year	Percent
1996	7
1997	12
1998	16
1999	21
2000	27
2001	31
2002	34
2003	37

Source: *Dove Consulting*

In 2001, two brothers bought some ATMs to determine people's account information illegally. For a little over a year, they used the information to withdraw $3.5 million from various ATMs in New York, California, and Florida (Source: The New York Times). The percentages of ATMs that are privately owned are shown in Table 20 for various years.

1. Find the change in the percentage of ATMs that were privately owned from 1996 to 1997.
2. Find the change in the percentage of ATMs that were privately owned from one year to the next, beginning in 1996.
3. From which year to the next did the percentage increase the most?

Solution

1. We find the difference of the percent in 1997 (ending) and the percent in 1996 (beginning):

$$\underset{\text{Ending percent}}{12} - \underset{\text{Beginning percent}}{7} = \underset{\text{Change in percent}}{5}$$

So, the percentage from 1996 to 1997 changed by 5 percentage points.

2. The changes in percentages from one year to the next are listed in Table 21. The changes were found by computing differences similar to the one found in Problem 1.

Table 21 Changes in the Percentages from Year to Year

Year	Percent	Percent Change (current minus previous)
1996	7	—
1997	12	$12 - 7 = 5$
1998	16	$16 - 12 = 4$
1999	21	$21 - 16 = 5$
2000	27	$27 - 21 = 6$
2001	31	$31 - 27 = 4$
2002	34	$34 - 31 = 3$
2003	37	$37 - 34 = 3$

3. The percentage changed by 6 percentage points from 1999 to 2000, the greatest change from any year to the next. ∎

Subtraction of Real Numbers

Exploring the change in a quantity can help us see how to subtract real numbers.

Example 3 Finding the Difference of Two Real Numbers

1. A college's enrollment decreases from 7 thousand students to 2 thousand students. What is the change in the enrollment?
2. Write a difference that is related to the computation in Problem 1.

Solution

1. Since the enrollment has decreased from 7 thousand students to 2 thousand students, the change is −5 thousand students. The change is negative because the enrollment is decreasing (see Fig. 21).
2. The change in the enrollment is the difference of the ending enrollment and the beginning enrollment:

$$\underset{\text{Ending enrollment}}{2} - \underset{\text{Beginning enrollment}}{7} = \underset{\text{Change in enrollment}}{-5}$$ ∎

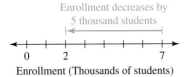

Enrollment decreases by 5 thousand students

0   2   7
Enrollment (Thousands of students)

Figure 21 Enrollment decreases from 7 thousand students to 2 thousand students

In Example 3, we found that

$$2 - 7 = -5$$

Note that $2 + (-7)$ gives the same result:

$$2 + (-7) = -5$$

This means that

$$2 - 7 = 2 + (-7)$$

which suggests that subtracting a number is the same as adding the opposite of that number:

$$\overbrace{2 - 7}^{\text{Subtract 7.}} = \overbrace{2 + (-7)}^{\text{Add the opposite of 7.}}$$

Subtracting a Real Number

$$a - b = a + (-b)$$

In words: To subtract a number, add its opposite.

To subtract real numbers, we first write the difference as a related sum and then find the sum.

Example 4   Finding Differences of Real Numbers

Find the difference.

1. $4 - 6$ $\qquad\qquad\qquad\qquad\qquad$ **2.** $\dfrac{2}{9} - \dfrac{5}{9}$

Solution

1. $\overbrace{4 - 6}^{\text{Subtract 6.}} = \overbrace{4 + (-6)}^{\text{Add the opposite of 6.}} = -2$ $\qquad a - b = a + (-b)$

2. $\dfrac{2}{9} - \dfrac{5}{9} = \dfrac{2}{9} + \left(-\dfrac{5}{9}\right)$ $\qquad$ Add the opposite of $\dfrac{5}{9}$: $a - b = a + (-b)$

$\qquad\qquad = -\dfrac{3}{9}$

$\qquad\qquad = -\dfrac{1}{3}$ $\qquad$ Simplify. $\qquad\blacksquare$

Considering the change in a quantity can also help us see how to subtract a negative number.

Example 5   Subtracting a Negative Number

1. The temperature increases from $-2°F$ to $7°F$. Find the change in temperature.
2. Write a difference that is related to the work in Problem 1.
3. Find the difference obtained in Problem 2 by using the rule $a - b = a + (-b)$.

Solution

1. Since the temperature increased from $-2°F$ to $7°F$, the change in temperature is $9°F$ (see Fig. 22).
2. The change in temperature is the ending temperature minus the beginning temperature:

$$\overbrace{7}^{\text{Ending temperature}} - \overbrace{(-2)}^{\text{Beginning temperature}} = \overbrace{9}^{\text{Change in temperature}}$$

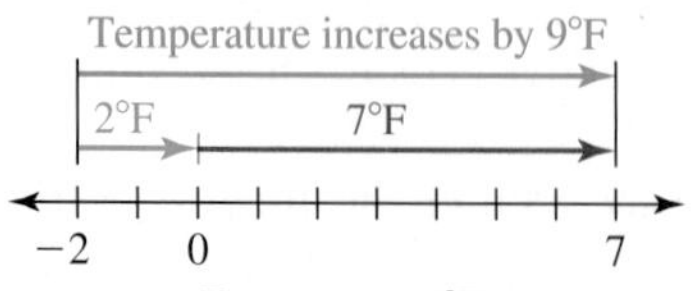

Temperature increases by 9°F

2°F $\qquad$ 7°F

$\begin{array}{ccccc} -2 & 0 & & 7 \end{array}$

Temperature (°F)

Figure 22 Temperature increases by 9°F in going from $-2°F$ to $7°F$

3.

$$
\overbrace{7-(-2)}^{\text{Subtract }-2.} = \overbrace{7+2}^{\substack{\text{Add the opposite}\\ \text{of }-2.\text{ (So, add 2.)}}} \qquad a-b=a+(-b)
$$
$$
= 9 \qquad \text{Add.}
$$

Note that to find the difference $7-(-2)$, we add 2 and 7. This makes sense, because, in going from $-2°\text{F}$ to $0°\text{F}$, the temperature increases by $2°\text{F}$, and in continuing from $0°\text{F}$ to $7°\text{F}$, the temperature increases by another $7°\text{F}$ (see Fig. 22). ■

Press 7 $\boxed{-}$ $\boxed{(}$ $\boxed{(-)}$ $\boxed{2}$ $\boxed{)}$ $\boxed{\text{ENTER}}$.

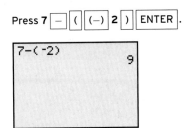

Figure 23 Calculating $7-(-2)$

We can use a calculator to check our work in Example 5 (see Fig. 23). Recall from Section 2.3 that we use the button $\boxed{-}$ for subtraction and the button $\boxed{(-)}$ for negative numbers and for taking opposites.

It is good practice to do homework exercises first by hand and then by using a calculator to check your hand results.

Example 6 Subtracting a Negative Number

Find the difference.

1. $4-(-6)$ **2.** $-9-(-3)$

Solution

1.

$$
\overbrace{4-(-6)}^{\text{Subtract }-6.} = \overbrace{4+6}^{\substack{\text{Add the opposite}\\ \text{of }-6.\text{ (So, add 6.)}}} \qquad a-b=a+(-b)
$$
$$
= 10 \qquad \text{Add.}
$$

2.

$$
\overbrace{-9-(-3)}^{\text{Subtract }-3.} = \overbrace{-9+3}^{\substack{\text{Add the opposite}\\ \text{of }-3.\text{ (So, add 3.)}}} \qquad a-b=a+(-b)
$$
$$
= -6 \qquad \text{Add.} \qquad ■
$$

Example 7 Translating from English to Mathematics

Translate the phrase "the difference of a and b" into a mathematical expression. Then evaluate the expression for $a=3$ and $b=-7$.

Solution

The expression is $a-b$. We substitute 3 for a and -7 for b in the expression $a-b$ and then find the difference:

$$
(3)-(-7) = 3+7 \qquad a-b=a+(-b)
$$
$$
= 10 \qquad \text{Add.} \qquad ■
$$

Elevation

In Example 8, you will work with *elevation*. An object that has a *positive* elevation of 200 ft is 200 ft *above* sea level (see Fig. 24). An object that has a *negative* elevation of -200 ft is 200 feet *below* sea level (see Fig. 24).

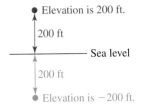

Figure 24 Elevations of 200 ft and -200 ft

Example 8 Finding a Change in Elevation

The Golden Gate Bridge has two towers that support the two main cables of the bridge (see Fig. 25). The top of each tower is at an elevation of 746 ft, and the foot of each tower is at an elevation of -136 ft (136 feet below sea level). Find the height of each tower.

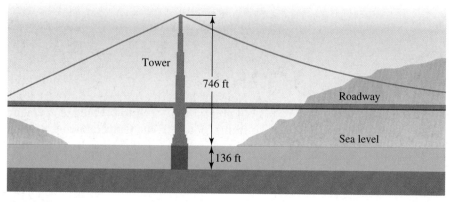

Figure 25 Golden Gate Bridge

Solution

We can find the height of each tower by computing the change in elevation from the bottom of each tower to the top:

Top elevation   Bottom elevation

$$\overbrace{746} - \overbrace{(-136)} = 746 + 136 \qquad a - b = a + (-b)$$
$$= 882 \qquad \text{Add.}$$

So, the height of each tower is 882 ft.

Changes of Increasing and Decreasing Quantities

An increasing quantity has a positive change. For instance, in Example 5 the temperature *increased* from $-2°F$ to $7°F$ and the change in temperature was *positive* ($9°F$).

A decreasing quantity has a negative change. For instance, in Example 3 the college's enrollment *decreased* from 7 thousand students to 2 thousand students and the change in enrollment was *negative* (-5 thousand students).

Changes of Increasing and Decreasing Quantities
• An increasing quantity has a positive change.
• A decreasing quantity has a negative change.

In Example 9, we will consider the meaning of a quantity with a positive or negative change.

Example 9 Finding Changes in Quantities

The numbers of AIDS deaths in the United States are shown in Table 22 for various years.

1. Find the change in the number of AIDS deaths from 1996 to 1997. What does your result mean in terms of the number of AIDS deaths?
2. Find the change in the number of AIDS deaths from 2002 to 2003. What does your result mean in terms of the number of AIDS deaths?
3. Find the change in the number of AIDS deaths from one year to the next, beginning in 1996.
4. Great strides have been made in fighting the AIDS epidemic. However, this progress has brought complacency. Research has shown that some people are now less concerned about getting infected and that risky behaviors, such as unprotected sex and needle sharing, have increased.[†] Do the changes you calculated in Problem 3 support the research findings?

Table 22 Numbers of AIDS Deaths in the United States

Year	Number of AIDS Deaths
1996	37,787
1997	21,923
1998	17,930
1999	16,273
2000	15,245
2001	14,175
2002	16,371
2003	18,017

Source: *U.S. Centers for Disease Control and Prevention*

[†]"Combating Complacency in HIV Prevention," *Body Positive*, Feb. 2001, XIV(2).

Solution

1. Since $21,923 - 37,787 = -15,864$, we conclude that the number of AIDS deaths from 1996 to 1997 decreased by 15,864 deaths.
2. Since $18,017 - 16,371 = 1646$, we conclude that the number of AIDS deaths from 2002 to 2003 increased by 1646 deaths.
3. The changes in number of AIDS deaths from one year to the next are listed in Table 23. The changes were found by computing differences similar to those found in Problems 1 and 2.

Table 23 Changes in AIDS Deaths from Year to Year

Year	Number of AIDS Deaths	Change in the Number of AIDS Deaths (current minus previous)
1996	37,787	—
1997	21,923	$21,923 - 37,787 = -15,864$
1998	17,930	$17,930 - 21,923 = -3993$
1999	16,273	$16,273 - 17,930 = -1657$
2000	15,245	$15,245 - 16,273 = -1028$
2001	14,175	$14,175 - 15,245 = -1070$
2002	16,371	$16,371 - 14,175 = 2196$
2003	18,017	$18,017 - 16,371 = 1646$

4. Yes, the changes do support the research findings. The annual declines generally decreased from 15,864 for the period 1996–1997 to only 1070 for the period 2000–2001, and the number of cases increased for the period 2001–2003. An increase in risky behaviors might explain the reversal. ■

In Example 10, we will find an expression that describes the change in a quantity.

Example 10 Finding an Expression Describing Change

In 2006, a financial planner had 132 clients. Let n be the number of clients in 2007.

1. Use a table to help find an expression that describes the change in the number of clients from 2006 to 2007.
2. Evaluate the expression found in Problem 1 for $n = 115$. What does the result mean in this situation?

Table 24 Changes in Number of Clients

Number of Clients in 2007	Change in the Number of Clients from 2006 to 2007
133	$133 - 132$
134	$134 - 132$
135	$135 - 132$
136	$136 - 132$
n	$n - 132$

Solution

1. First we create Table 24. We show the arithmetic to help us see the pattern. From the last row of the table, we see that the expression $n - 132$ represents the change in the number of clients.
2. We evaluate $n - 132$ for $n = 115$:

$$(115) - 132 = 115 + (-132) = -17$$

So, there was a change of -17 clients from 2006 to 2007. In other words, the financial planner's client base decreased by 17 clients. ■

group exploration

Looking ahead: Finding the product of a positive number and a negative number

We can think of the multiplication of two counting numbers as a repeated addition. For example, we can think of $4(3)$ as adding four 3's:

$$4(3) = 3 + 3 + 3 + 3 = 12$$

We can use this idea to help us find the product of a positive number and a negative number.

1. Write each of the products that follow as a repeated sum. Then find the sum.
   a. $3(-2)$ **b.** $5(-4)$ **c.** $7(-1)$

2. Are your results in Problem 1 positive or negative? What can you say about the product of a positive number and a negative number? If you are not sure, try some more multiplications.

3. Explain why the observation you made in Problem 2 makes sense. [**Hint:** What can you say about a negative number plus a negative number?]

group exploration

Looking ahead: Determining whether a line contains some points

1. a. Plot the points shown in Table 25. Is there one line that contains all the points?
   b. Complete the third column of Table 25. Describe any patterns.

2. a. Plot the points shown in Table 26. Is there one line that contains all the points?
   b. Complete the third column of Table 26. Describe any patterns.

Table 25 Some Ordered Pairs

x	y	Change in y (current value minus previous value)
0	1	—
1	3	
2	5	
3	7	
4	9	

Table 26 Some Ordered Pairs

x	y	Change in y (current value minus previous value)
0	13	—
1	10	
2	7	
3	4	
4	1	

3. Describe what you have learned so far in this exploration.

4. Let p be the profit (in millions) of a company at t years since 2000. Some pairs of values of t and p are shown in Table 27.
   a. Complete the third column of Table 27.

Table 27 Profits of a Company

Years since 2000 t	Profit (millions of dollars) p	Change in Profit (millions of dollars)
0	2	—
1	6	
2	10	
3	14	
4	18	

   b. Without plotting the data points, decide whether this situation can be modeled well by a linear model.
   c. Check your response to part (b) by creating a scattergram of the data.

TIPS FOR SUCCESS: Math Journal

Do you tend to make the same mistakes repeatedly throughout a math course? If so, it might help to keep a journal in which you list errors you have made on assignments, quizzes, and tests. For each error you list, include the correct solution, as well as a description of the concept needed to solve the problem correctly. You can review this journal from time to time to help you avoid making these errors.

HOMEWORK 2.4 FOR EXTRA HELP ▶

Student Solutions Manual   PH Math/Tutor Center   MathXL®   MyMathLab

Find the difference by hand. Then use a calculator to check your work.

1. $6 - 8$ **2.** $3 - 7$

3. $-1 - 5$ **4.** $-3 - 9$

5. $2 - (-7)$ **6.** $5 - (-1)$

7. $-3 - (-2)$ **8.** $-7 - (-3)$

9. $4 - 7$ **10.** $-4 - 7$

11. $4 - (-7)$ **12.** $-4 - (-7)$

13. $-3 - 3$ **14.** $-7 - 7$

15. $-54 - 25$ **16.** $-100 - 257$

17. $381 - (-39)$ **18.** $-1939 - (-352)$

19. $2.5 - 7.9$ **20.** $5.8 - 3.7$

21. $-6.5 - 4.8$ **22.** $-1.7 - 7.4$

23. $3.8 - (-1.9)$ **24.** $3.1 - (-3.1)$

25. $13.6 - (-2.38)$ **26.** $-159.24 - (-7.8)$

27. $-\dfrac{1}{3} - \dfrac{2}{3}$ **28.** $-\dfrac{1}{5} - \dfrac{4}{5}$

29. $-\dfrac{1}{8} - \left(-\dfrac{5}{8}\right)$ **30.** $-\dfrac{4}{9} - \left(-\dfrac{7}{9}\right)$

31. $\dfrac{1}{2} - \left(-\dfrac{1}{4}\right)$ **32.** $\dfrac{5}{12} - \left(-\dfrac{1}{6}\right)$

33. $-\dfrac{1}{6} - \dfrac{3}{8}$ **34.** $-\dfrac{2}{3} - \dfrac{2}{5}$

Perform the indicated operation by hand. Then use a calculator to check your work.

35. $-5 + 7$ **36.** $-3 + 9$ **37.** $-6 - (-4)$

38. $-4 - (-3)$ **39.** $\dfrac{3}{8} - \dfrac{5}{8}$ **40.** $-\dfrac{5}{6} + \dfrac{1}{6}$

41. $-4.9 - (-2.2)$ **42.** $-6.4 + 3.5$ **43.** $-2 + (-5)$

44. $-5 + (-8)$ **45.** $10 - 12$ **46.** $5 - 9$

For Exercises 47–52, use a calculator to perform the indicated operation. Round the result to two decimal places.

47. $-234.913 - 2893.26$ **48.** $-6178.39 - 52.387$

49. $29{,}643.52 - (-83{,}284.39)$

50. $83{,}451.6 - (-408.549)$

51. $-\dfrac{17}{89} - \dfrac{51}{67}$ **52.** $-\dfrac{49}{56} - \dfrac{85}{97}$

53. Three hours ago, the temperature was 7°F. If the temperature has decreased by 19°F in the last three hours, what is the current temperature?

54. Four hours ago, the temperature was −12°F. If the temperature has increased by 18°F in the last four hours, what is the current temperature?

55. Three hours ago, the temperature was −4°F. Now the temperature is 7°F. What is the change in temperature for the past three hours?

56. Four hours ago, the temperature was −2°F. Now the temperature is −13°F. What is the change in temperature for the past four hours?

57. Two hours ago, the temperature was 8°F. The temperature is now −4°F.
  a. What is the change in temperature for the past two hours?
  b. Estimate the change in temperature for the past hour.
  c. Explain why your estimate in part (b) may not be the actual change in temperature for the past hour.

58. Three hours ago, the temperature was −6°F. The temperature is now 9°F.
  a. What is the change in temperature for the past three hours?
  b. Estimate the change in temperature for the past hour.
  c. Explain why your estimate in part (b) may not be the actual change in temperature for the past hour.

59. The lowest elevation in the United States is at Death Valley, California (−282 ft), and the highest elevation is at the top of Mount McKinley, Alaska (20,320 ft). Find the change in elevation from Death Valley to Mount McKinley.

60. The lowest elevation on (dry) land in the world is at the edge of the Dead Sea, along the Israel–Jordan border (−1312 ft), and the highest elevation is at the top of Mount Everest, along the Nepal–Tibet border (29,035 ft). Find the change in the elevation from the Dead Sea to Mount Everest.

61. The U.S. presidential election in 2000 was the closest presidential race in the electoral vote since 1876. Yet, only a little over half of eligible voters chose to cast a vote (see Table 28).

Table 28 Presidential Election Voter Turnout

Year	Percent of Eligible Voters Who Voted
1980	59.2
1984	59.9
1988	57.4
1992	61.9
1996	54.2
2000	54.7
2004	60.7

Source: *U.S. Census Bureau, Current Population Study*

  a. For the years listed in Table 28, find the changes in percent turnout from one presidential election to the next.
  b. What was the greatest increase in percent turnout?
  c. In 1993, in an attempt to increase the number of eligible voters, a "motor voter" law was passed that made voter registration a part of the process of applying for a driver's license. As a result, about 11 million new voters were registered. Compare the change in percent turnout between 1992 and 1996 with other changes you found in part (a). On the basis of the information in Table 28 alone, we cannot know for sure, but does it seem that many of these 11 million people voted? Explain.

62. In the 1930s, the gray wolf was hunted to near extinction across the western United States. In 1995, 14 wolves were reintroduced to Yellowstone National Park. In the following year, 17 more wolves were released into the park. By the end of 1996, there had been 20 births and 11 mortalities, leaving

40 wolves. The wolf population in the Greater Yellowstone Area is shown in Table 29 for various years.

Table 29 Wolf Population

Year	Population
1996	40
1997	86
1998	112
1999	118
2000	177
2001	218
2002	273

Source: *National Park Service, Yellowstone National Park*

a. For the years listed in Table 29, find the changes in population from each year to the next.
b. From what year(s) to the next is the change in population the greatest? What is that change?
c. From what year(s) to the next is the change in population the least? What is that change?
d. From 1997 to 1998, the change in the population was 26 wolves. Does that mean that there were 26 births? Explain.

63. The changes in retail sales (in billions of dollars) of licensed NBA merchandise in the United States and Canada from one year to the next are given in Table 30.

Table 30 Changes in Retail Sales of NBA Merchandise

Years	Changes in Retail Sales (billions of dollars)
1996–1997	0.0
1997–1998	−1.1
1998–1999	−0.1
1999–2000	−0.2
2000–2001	0.0
2001–2002	0.3
2002–2003	0.6

Source: *The Licensing Letter*

a. If there were $2.6 billion in sales in 1996, what were the sales in 2003?
b. During which period(s) were the retail sales increasing?
c. During which period(s) were the retail sales decreasing?

64. The changes in Honda Accord® car sales (in thousands of cars) from one year to the next are shown in Table 31.

Table 31 Changes in Honda Accord Sales

Years	Changes in Sales (thousands of cars)
1995–1996	42
1996–1997	2
1997–1998	16
1998–1999	3
1999–2000	1
2000–2001	10
2001–2002	−16
2002–2003	−1
2003–2004	−12

Source: Automotive News

a. If 340 thousand cars were sold in 1995, what were the sales in 2004?
b. During which period(s) were sales increasing?
c. During which period(s) were sales decreasing?

65. A student scored 87 points on the first exam of the semester.
a. Complete Table 32 to help find an expression that describes the change in score (in points) from the first exam to the second exam if the student scored p points on the second exam. Show the arithmetic to help you see a pattern.

Table 32 Scores on the Second Exam and Changes in Scores

Score on the Second Exam (points)	Change in Score (points)
80	
85	
90	
95	
p	

b. Evaluate the expression you found in part (a) for $p = 81$. What does your result mean in this situation?

66. A year ago, the value of a stock was $35.
a. Complete Table 33 to help find an expression that describes the change in the stock's value (in dollars) if the current value is x dollars. Show the arithmetic to help you see a pattern.

Table 33 Current Values and Changes in Values

Current Value (dollars)	Change in Value (dollars)
30	
35	
40	
45	
x	

b. Evaluate the expression you found in part (a) for $x = 44$. What does your result mean in this situation?

67. Last year the enrollment at a college was 24,500 students.
a. Complete Table 34 to help find an expression that describes the current enrollment if the *change* in enrollment in the past year is c students. Show the arithmetic to help you see a pattern.

Table 34 Changes in Enrollments and Current Enrollments

Change in Enrollment	Current Enrollment
100	
200	
300	
400	
c	

b. Evaluate the expression you found in part (a) for $c = -700$. What does your result mean in this situation?

68. Last year there were 820 deer in a state park.
  a. Complete Table 35 to help find an expression that describes the current deer population if the *change* in population in the past year is c deer. Show the arithmetic to help you see a pattern.

Table 35 Changes in Population and Current Populations

Change in Population	Current Population
10	
20	
30	
40	
c	

  b. Evaluate the expression you found in part (a) for $c = -25$. What does your result mean in this situation?

69. A student tries to find the difference $7 - (-5)$:

$$7 - (-5) = 7 - 5 = 2$$

Describe any errors. Then find the difference correctly.

70. A student tries to find the difference $2 - 6$:

$$2 - 6 = 6 - 2 = 4$$

Describe any errors. Then find the sum correctly.

71. A quantity increases from amount a to amount b.
  a. Find the change in the quantity.
    i. $a = 3, b = 5$
    ii. $a = 1, b = 9$
    iii. $a = 2, b = 7$
  b. By referring to your work in part (a), explain why it makes sense that if a quantity increased, then the change will be positive.

72. A quantity decreases from amount a to amount b.
  a. Find the change in the quantity.
    i. $a = 8, b = 2$
    ii. $a = 9, b = 3$
    iii. $a = 5, b = 1$
  b. By referring to your work in part (a), explain why it makes sense that if a quantity decreased, then the change will be negative.

Evaluate the expression for $a = -5$, $b = 2$, and $c = -7$.

73. $a + b$ **74.** $a + c$
75. $a - b$ **76.** $c - a$
77. $b - c$ **78.** $b - a$

For Exercises 79–84, let x be a number. Translate the English phrase or sentence into a mathematical expression. Then evaluate the expression for $x = -5$.

79. -3 minus the number

80. The number decreased by 4

81. 8 less than the number

82. Subtract 5 from the number.

83. Subtract -2 from the number.

84. The difference of the number and -6

85. **a.** Use a calculator to find each product of two numbers with different signs:
    i. $-2(5)$
    ii. $-4(6)$
    iii. $-7(9)$
  b. What pattern do you notice? If you do not see a pattern, continue finding products of two numbers with different signs until you do.
  c. Without using a calculator, find the product $-3(7)$. Then use a calculator to check your work.
  d. State a general rule for how to multiply two numbers with different signs.

86. **a.** Use a calculator to find each product of two negative numbers.
    i. $-2(-5)$
    ii. $-4(-6)$
    iii. $-7(-9)$
  b. What pattern do you notice? If you do not see a pattern, continue finding products of two negative numbers until you do.
  c. Without using a calculator, find the product $-3(-7)$. Then use a calculator to check your work.
  d. State a general rule for how to multiply two negative numbers.

87. **a.** Evaluate $a - b$ for $a = 8$ and $b = 5$.
  b. Evaluate $b - a$ for $a = 8$ and $b = 5$.
  c. Compare your results from parts (a) and (b).
  d. Evaluate $a - b$ and $b - a$ for $a = -2$ and $b = 4$, and compare your results.
  e. Evaluate $a - b$ and $b - a$ for values of your choosing for a and b, and compare your results.
  f. From your work on parts (a) through (e), what connection do you notice between $a - b$ and $b - a$?

88. If the temperature increases from $3°$F to $5°$F, we find the change in temperature (in degrees Fahrenheit) by performing a subtraction:

$$5 - 3 = 2$$

If the temperature increases from $-3°$F to $5°$F, we find the change in temperature (in degrees Fahrenheit) by eventually performing an addition:

$$5 - (-3) = 5 + 3 = 8$$

Explain why it makes sense that in the first situation we subtract and in the second we eventually add. [**Hint:** How far is -3 from 0 on the number line? How far is 5 from 0 on the number line?]

RATIOS, PERCENTS, AND MULTIPLYING AND 2.5 DIVIDING REAL NUMBERS

Objectives

▹ Find the ratio of two quantities.

▹ Know the meaning of *percent*.

▹ Convert percentages to and from decimal numbers.

▹ Find the percentage of a quantity.

▹ Multiply and divide real numbers.

▹ Know which fractions with negative signs are equal to each other.

In this section, we will use ratios and percents to describe various quantities. Ratios and percents are important tools used in many fields, including business, cooking, chemistry, political science, aeronautics, fire science technology, humanities, journalism, and psychology. We will also discuss how to multiply and divide real numbers.

The Ratio of Two Quantities

Recall from Section 2.1 that the ratio of a to b is the quotient $a \div b$. Usually, we write the ratio of a to b as the fraction $\frac{a}{b}$ or as $a : b$.

We can use a ratio to compare two quantities. For example, if a person has 6 cats and 2 dogs, then the ratio of cats to dogs is

$$\frac{6 \text{ cats}}{2 \text{ dogs}} = \frac{3 \text{ cats}}{1 \text{ dog}}$$

We say that there are "3 cats to 1 dog." This means that there are 3 cats per dog. Or we can say that there are 3 times as many cats as dogs.

The ratio of 3 cats to 1 dog is an example of a unit ratio. A **unit ratio** is a ratio written as $\frac{a}{b}$ with $b = 1$ or as $a : b$ with $b = 1$.

Example 1 Finding a Unit Ratio

In 2003, the average annual charge for tuition and fees was $4698 at public four-year colleges and $1479 at public two-year colleges (Source: *U.S. National Center for Education Statistics*). Find the unit ratio of the average annual charge at public four-year colleges to the average annual charge at public two-year colleges. What does the result mean?

Solution

We divide the average annual charge at public four-year colleges by the average annual charge at public two-year colleges:

$$\begin{array}{r} \text{public four-year colleges} \longrightarrow \\ \text{public two-year colleges} \longrightarrow \end{array} \frac{\$4698}{\$1479} \approx \frac{3.18}{1}$$

So, the average annual charge for tuition and fees at public four-year colleges is about 3.18 times the average annual charge for tuition and fees at public two-year colleges. ■

Example 2 Comparing Ratios

The median sales prices of existing homes and the median incomes in 2001 are shown in Table 36 for four regions of the United States.

1. Find the unit ratio of the median sales price of existing homes to the median income in the Northeast. What does the result mean?

Just think, our equity's growing by $100 per mile!

Welcome to California

Table 36 Median Sales Prices of Existing Homes and Median Incomes

Region	Median Sales Price of Existing Homes (dollars)	Median Income (dollars)
Northeast	146,500	57,000
Midwest	130,200	54,096
South	137,400	46,688
West	194,500	51,966

Sources: *U.S. Census; National Association of Realtors®*

2. For each of the four regions, find the unit ratio of the median sales price of existing homes to the median income. Taking into account the median income of each region, list the regions in order of affordability of existing homes, from greatest to least.
3. A person believes that existing homes in the South are more affordable than in the Northeast, because the median price of existing homes is lower in the South than in the Northeast. What would you tell that person?

Solution

1. We divide the median sales price of existing homes in the Northeast by the median income in the Northeast:

$$\text{Median sales price of existing homes} \longrightarrow \frac{\$146,500}{\$57,000} \approx \frac{2.57}{1} \longleftarrow \text{Median income}$$

So, the median sales price of an existing home in the Northeast is about 2.57 times the median income in that region.

2. We find the unit ratios for each region by dividing the region's median sales price of existing homes by the region's median income (see Table 37).

Table 37 Unit Ratios of Median Sales Prices of Existing Homes to Median Incomes

Region	Median Sales Price of Existing Homes (dollars)	Median Income (dollars)	Unit Ratio of Median Sales Price of Existing Homes to Median Income
Northeast	146,500	57,000	$\frac{146,500}{57,000} \approx \frac{2.57}{1}$
Midwest	130,200	54,096	$\frac{130,200}{54,096} \approx \frac{2.41}{1}$
South	137,400	46,688	$\frac{137,400}{46,688} \approx \frac{2.94}{1}$
West	194,500	51,966	$\frac{194,500}{51,966} \approx \frac{3.74}{1}$

The lower the unit ratio, the more affordable the existing homes are in the region. So, the regions, in order of affordability of existing homes, from greatest to least, are Midwest, Northeast, South, West.

3. Although the median sales price of homes in the South is lower than that in the Northeast, the median income in the Northeast is so much higher than it is in the South that existing homes are actually more affordable in the Northeast. We can tell because the approximate unit ratio of the median sales price of existing homes to median income is smaller in the Northeast (2.57:1) than in the South (2.94:1). See Table 37. ∎

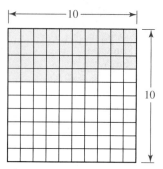

Figure 26 The area of the shaded region is 37% of the area of the large square

Meaning of Percent

Suppose there are 53 women in a class of 100 students. Then the ratio of the number of women to the total number of students is $\frac{53}{100}$. We say that 53% of the students are women.

DEFINITION Percent

Percent means "for each hundred": $a\% = \dfrac{a}{100}$

For example, 37% means 37 for each 100 (the ratio $\frac{37}{100}$, or the unit ratio $\frac{0.37}{1}$). In Fig. 26, the area of the shaded region is 37% of the area of the large square, because 37 of 100 parts of equal area are shaded.

Converting Percentages to and from Decimal Numbers

Since 37% is the ratio $\frac{37}{100}$, 37% is 37 hundredths, or 0.37:

$$37\% = \frac{37}{100} = 0.\underbrace{\overset{\text{tenth's place}}{3}\ \overset{\text{hundredth's place}}{7}}_{37\text{ hundredths}}$$

So, to write 37% as a decimal number, first we remove the percent symbol. Then we divide 37 by 100, which is equivalent to moving the decimal point two places to the left:

$$37\% = 37.0\% = 0.37$$

two places to the left

To write 0.37 as a percentage, first we multiply 0.37 by 100, which is equivalent to moving the decimal point two places to the right. Then we insert a percent symbol:

$$0.37 = 37.0\% = 37\%$$

two places to the right

Converting Percentages to and from Decimal Numbers

- To write a percentage as a decimal number, remove the percent symbol and divide the number by 100 (move the decimal point two places to the left).
- To write a decimal number as a percentage, multiply the number by 100 (move the decimal point two places to the right) and insert a percent symbol.

Example 3 Converting Percentages and Decimal Numbers

Write each percentage as a decimal number, and write each decimal number as a percentage.

1. 86%   **2.** 7%   **3.** 0.125

Solution

1. To write 86% as a decimal number, we remove the percent symbol and move the decimal point two places to the left:

$$86\% = 86.0\% = 0.86$$

two places to the left

2. To write 7% as a decimal number, we remove the percent symbol and move the decimal point two places to the left, using 0 in the tenths place as a placeholder:

$$7\% = 7.0\% = 0.07$$

two places to the left

3. To write 0.125 as a percentage, we move the decimal point two places to the right and insert a percent symbol:

$$0.125 = 12.5\%$$

two places to the right ∎

WARNING   From Problem 2 in Example 3, we see that 7% is *not* equal to 0.7. Rather, 7% is equal to 0.07. Remember to move the decimal point *two* places to the left, using 0 in the tenths place as a placeholder.

Percentage of a Quantity

How do we find the percentage of a quantity? For example, consider 75% of 4. That is the same as $\frac{75}{100}$ of 4. To find a fraction of a number, we *multiply* the fraction by that number:

$$\frac{75}{100} \text{ of } 4 = \frac{75}{100} \cdot 4 = \frac{3}{4} \cdot \frac{4}{1} = \frac{3}{1} = 3$$

So, using decimal notation, we find 75% of 4 by *multiplying* 0.75 by 4:

$$75\% \text{ of } 4 = 0.75(4) = 3$$

To see whether our result makes sense, we first form a large square made up of 4 medium-size squares of equal area (see Fig. 27). To find 75% of the four squares, we divide the large square into 100 small squares of equal area and shade 75 of them. The shaded region contains 3 of the 4 medium-size squares, which checks with our earlier computations.

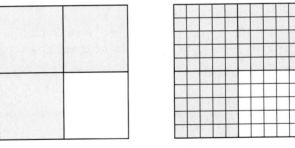

Figure 27 75% of 4 medium-size squares is made up of 75 small squares (in blue), or 3 medium-size squares (in blue)

Finding the Percentage of a Quantity

To find the percentage of a quantity, multiply the decimal form of the percentage and the quantity.

Example 4   Finding the Percentage of a Quantity

1. Find 6% of 3000 students.   **2.** Find 3.5% of $6500.

Solution

1. 0.06(3000) = 180; so, 6% of 3000 students is 180 students.
2. 0.035(6500) = 227.5; so, 3.5% of $6500 is $227.50. ∎

By definition, 100% means 100 for each 100. In other words, 100% of a quantity is *all* of the quantity. For example, 100% of 21 guitars is 21 guitars.

One Hundred Percent of a Quantity
One hundred percent of a quantity is *all* of the quantity.

We will continue to work with ratios and percents as we discuss how to multiply and divide real numbers.

Multiplication of Two Numbers with Different Signs

We can think of multiplication as repeated addition. For example, 3(5) is equal to the sum of three 5's:

$$3(5) = 5 + 5 + 5 = 15$$

Also, 3(5) is equal to the sum of five 3's:

$$3(5) = 3 + 3 + 3 + 3 + 3 = 15$$

We can use the idea of repeated addition to help us find the product of two numbers with different signs.

Example 5 Finding the Product of Two Numbers with Different Signs

Find the product.

  1. 4(−2)   **2.** (−6)(3)

Solution

  1. We write 4(−2) as the sum of four −2's:

$$4(-2) = (-2) + (-2) + (-2) + (-2) = -8$$

This result makes sense in terms of money. If you borrow 2 dollars from a friend four times, you will owe the friend 8 dollars.

  2. We write (−6)(3) as the sum of three −6's:

$$(-6)(3) = (-6) + (-6) + (-6) = -18$$ ∎

In Example 5, we found the products of two numbers with different signs. Note that both results were negative:

$$\underbrace{4(-2)}_{\text{Different signs}} = \underbrace{-8}_{\text{Negative}}$$

$$\underbrace{(-6)(3)}_{\text{Different signs}} = \underbrace{-18}_{\text{Negative}}$$

Multiplying Two Numbers with Different Signs
The product of two numbers that have different signs is negative.

Example 6 Finding the Product of Two Numbers with Different Signs

Find the product.

  1. 7(−4)   **2.** (−0.2)(0.3)

Solution

  1. Since the signs of 7 and −4 are different, their product is negative: $7(-4) = -28$.
  2. Since the signs of −0.2 and 0.3 are different, their product is negative: $(-0.2)(0.3) = -0.06$. ∎

Multiplication of Two Numbers with the Same Sign

We have discussed how to multiply numbers with different signs. What if the signs are the same? To begin this investigation, consider the following pattern:

This factor decreases by 1.
$$3(-5) = -15$$
$$2(-5) = -10$$
$$1(-5) = -5$$
$$0(-5) = 0$$
The product increases by 5.

It turns out that this pattern continues. So, we have

This factor decreases by 1.
$$-1(-5) = 5$$
$$-2(-5) = 10$$
$$-3(-5) = 15$$
The product increases by 5.

Note that for each of the last three computations, the product of the two negative numbers is positive. This is, in fact, always true. Here we find another product of two negative numbers:

$$\overbrace{(-7)(-9)}^{\text{Same signs}} = \overbrace{63}^{\text{Positive}}$$

Multiplying Two Numbers with the Same Sign

The product of two numbers that have the same sign is positive.

Example 7 Finding the Product of Two Numbers with the Same Sign

Find the product.

1. $-5(-6)$

2. $\left(-\dfrac{3}{2}\right)\left(-\dfrac{5}{7}\right)$

Solution

1. Since -5 and -6 have the same sign, their product is positive: $-5(-6) = 30$.

2. Since $-\dfrac{3}{2}$ and $-\dfrac{5}{7}$ have the same sign, their product is positive:

$$\left(-\frac{3}{2}\right)\left(-\frac{5}{7}\right) = \frac{15}{14}$$ ∎

In Fig. 28, we show a multiplication table for some specific numbers. In Fig. 29, we summarize the multiplication sign rules for all nonzero real numbers.

·	4	−4
2	8	−8
−2	−8	8

·	+	−
+	+	−
−	−	+

Figure 28 Multiplication table for 2, −2, 4, and −4

Figure 29 Multiplication table for all nonzero real numbers

Example 8 Application of Multiplying Real Numbers

A person's credit card balance is −2340 dollars. If the person pays off 30% of the balance, what is the new balance?

Solution

If the person pays off 30% of the balance, then $100\% - 30\% = 70\%$ of the balance remains. We find 70% of −2340:

$$0.70(-2340) = -1638$$

The new balance is −1638 dollars. ∎

Division of Real Numbers

We can get an idea of how to divide real numbers by writing multiplications as related divisions. For example, consider this statement:

$$2 \cdot 3 = 6 \text{ implies that } 6 \div 3 = 2.$$

We now write similar statements for $(-2)(-3)$ and $-2 \cdot 3$:

$$(-2)(-3) = 6 \text{ implies that } \overbrace{6 \div (-3)}^{\substack{\text{Different} \\ \text{signs}}} = \overbrace{-2}^{\text{Negative}}.$$

$$-2 \cdot 3 = -6 \text{ implies that } \overbrace{-6 \div 3}^{\substack{\text{Different} \\ \text{signs}}} = \overbrace{-2}^{\text{Negative}}.$$

These statements suggest that the quotient of two numbers with different signs is negative.

Now consider the following statement:

$$2(-3) = -6 \text{ implies that } \overbrace{-6 \div (-3)}^{\text{Same signs}} = \overbrace{2}^{\text{Positive}}.$$

This statement suggests that the quotient of two numbers with the same sign is positive.

All three statements suggest that the sign rules for dividing real numbers are similar to those for multiplying real numbers.

Multiplying or Dividing Real Numbers

The product or quotient of two numbers that have different signs is negative. The product or quotient of two numbers that have the same sign is positive.

Example 9   Finding Quotients of Real Numbers

Find the quotient.

1. $-10 \div 2$ **2.** $-\dfrac{1}{6} \div \left(-\dfrac{3}{5}\right)$

Solution

1. Since -10 and 2 have different signs, the quotient is negative: $-10 \div 2 = -5$. This makes sense in terms of money. If we divide a debt of $10 by 2, the result is a debt of $5.

2. The quotient of two negative numbers is positive. To find the result, we divide the absolute value of the fractions:

$$\frac{1}{6} \div \frac{3}{5} = \frac{1}{6} \cdot \frac{5}{3} \qquad \text{Multiply by reciprocal of } \frac{3}{5}, \text{ which is } \frac{5}{3}:$$
$$\frac{a}{b} \div \frac{c}{d} = \frac{a}{b} \cdot \frac{d}{c}$$
$$= \frac{5}{18} \qquad \text{Multiply numerators and multiply denominators.} \quad \blacksquare$$

We can use a graphing calculator to check our work for Problem 2 in Example 9 (see Fig. 30).

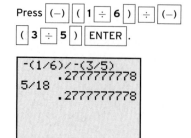

Press $\boxed{(-)}$ $\boxed{(}$ $\boxed{1}$ $\boxed{\div}$ $\boxed{6}$ $\boxed{)}$ $\boxed{\div}$ $\boxed{(-)}$ $\boxed{(}$ $\boxed{3}$ $\boxed{\div}$ $\boxed{5}$ $\boxed{)}$ $\boxed{\text{ENTER}}$.

```
-(1/6)/-(3/5)
           .2777777778
5/18
           .2777777778
```

Figure 30 Verify the work

Example 10   Application of a Ratio of Two Real Numbers

A person has credit card balances of -3950 dollars on a Visa® account and -1225 dollars on a MasterCard® account.

1. Find the unit ratio of the Visa balance to the MasterCard balance.
2. If the person wishes to pay off both accounts gradually in the same amount of time, describe how the result in Problem 1 can help guide the person in making his next payment.

Solution

1. We divide the Visa balance by the MasterCard balance:

$$\frac{-3950}{-1225} \approx \frac{3.22}{1}$$

So, the Visa balance is about 3.22 times the MasterCard balance.

2. For each \$1 the person pays to his MasterCard account, he should pay about \$3.22 to his Visa account. (The ratio will need to be recalculated each month to take into account recent purchases, cash advances, and so on, as well as possible differences in interest rates on the cards.) ∎

Equal Fractions with Negative Signs

We know that $\dfrac{a}{b}$ means that a is divided by b, for $b \neq 0$:

$$\frac{a}{b} = a \div b$$

For example,

$$\frac{-4}{2} = -4 \div 2 = -2$$

Since $\dfrac{4}{-2}$ and $-\dfrac{4}{2}$ are also equal to -2, we can write

$$\frac{-4}{2} = \frac{4}{-2} = -\frac{4}{2}.$$

This suggests the following property:

Equal Fractions with Negative Signs

If $b \neq 0$, then

$$\frac{-a}{b} = \frac{a}{-b} = -\frac{a}{b}$$

If the result of a computation is a negative fraction, we write the result in the form $-\dfrac{a}{b}$ rather than $\dfrac{-a}{b}$ or $\dfrac{a}{-b}$.

Example 11   Simplifying Fractions and Adding Fractions

Perform the indicated operation.

1. $\dfrac{-20}{-5}$
2. $\dfrac{3}{-5} + \dfrac{1}{5}$

Solution

1. $\dfrac{-20}{-5} = -20 \div (-5)$ $\dfrac{a}{b} = a \div b$

   $= 4$ *Quotient of two numbers with same sign is positive.*

2. $\dfrac{3}{-5} + \dfrac{1}{5} = \dfrac{-3}{5} + \dfrac{1}{5}$ $\dfrac{a}{-b} = \dfrac{-a}{b}$

$= \dfrac{-3+1}{5}$ Write numerators as a sum and keep common denominator: $\dfrac{a}{b} + \dfrac{c}{b} = \dfrac{a+c}{b}$.

$= \dfrac{-2}{5}$ Find sum.

$= -\dfrac{2}{5}$ $\dfrac{-a}{b} = -\dfrac{a}{b}$

group exploration

Looking ahead: Order of operations

If an expression has more than one operation, we can use parentheses to indicate which operation to do first.

1. Perform the indicated operations in $(2 + 3) \cdot 4$ by first doing the addition and then doing the multiplication.
2. Perform the indicated operations in $2 + (3 \cdot 4)$ by first doing the multiplication and then doing the addition.
3. Compare your results for Problems 1 and 2. Does it matter in which order we add and multiply?

For each object, determine whether the area is $(2 + 3) \cdot 4$ or $2 + (3 \cdot 4)$. Explain.

4.
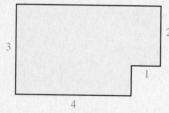

5.

TIPS FOR SUCCESS: Use Your Instructor's Office Hours

Helping students during office hours is part of an instructor's job. Keep in mind that your instructor wants you to succeed and hopes that you take advantage of all opportunities to learn.

It is a good idea to come prepared to office visits. For example, if you are having trouble with a concept, attempt some related exercises and bring your work so that your instructor can see where you are having difficulty. If you miss a class, it is helpful to read the material, borrow class notes, and try completing assigned exercises before visiting your instructor, so that you get the most out of the visit.

HOMEWORK 2.5

FOR EXTRA HELP ▶

 Student Solutions Manual

 PH Math/Tutor Center

Math **XL** MathXL®

 MyMathLab MyMathLab

For Exercises 1–10, write the percentage as a decimal number or write the decimal number as a percentage, as appropriate.

1. 63%   2. 91%   3. 9%   4. 4%
5. 0.08   6. 0.01   7. 7.3%   8. 3.8%
9. 0.052   10. 0.089

11. Find 35% of $8.   12. Find 67% of $4.
13. Find 5% of 2500 students.
14. Find 8% of 4000 students.
15. Find 2.5% of 7000 cars.
16. Find 6.4% of 3500 cars.

Perform the indicated operation by hand. Then use a calculator to check your work.

17. $-2(6)$ **18.** $-5(4)$ **19.** $-3(-6)$

20. $-8(-9)$ **21.** $1(-1)$ **22.** $5(-2)$

23. $-40 \div 5$ **24.** $-63 \div 7$ **25.** $25 \div (-5)$

26. $24 \div (-3)$ **27.** $-56 \div (-7)$ **28.** $-1 \div (-1)$

29. $-15(-37)$ **30.** $-124(-29)$ **31.** $936 \div (-24)$

32. $1008 \div (-21)$ **33.** $-0.2(-0.4)$ **34.** $-0.3(-0.3)$

35. $2.5(-0.39)$ **36.** $3.7(-5.24)$ **37.** $-0.06 \div 0.2$

38. $-0.12 \div 0.3$ **39.** $\dfrac{36}{-4}$ **40.** $\dfrac{9}{-3}$

41. $\dfrac{-32}{-8}$ **42.** $\dfrac{-72}{-8}$

43. $\dfrac{1}{2}\left(-\dfrac{1}{5}\right)$ **44.** $\dfrac{1}{3}\left(-\dfrac{7}{5}\right)$

45. $\left(-\dfrac{4}{9}\right)\left(-\dfrac{3}{20}\right)$ **46.** $\left(-\dfrac{7}{25}\right)\left(-\dfrac{5}{21}\right)$

47. $-\dfrac{3}{4} \div \dfrac{7}{6}$ **48.** $-\dfrac{5}{7} \div \dfrac{15}{8}$

49. $-\dfrac{24}{35} \div \left(-\dfrac{16}{25}\right)$ **50.** $-\dfrac{3}{8} \div \left(-\dfrac{9}{20}\right)$

Perform the indicated operation by hand. Then use a calculator to check your work.

51. $6 + (-9)$ **52.** $-9 + (-4)$ **53.** $-39 \div (-3)$

54. $-49 \div 7$ **55.** $4 - (-2)$ **56.** $-2 - 7$

57. $10(-10)$ **58.** $-5(-9)$ **59.** $-\dfrac{3}{4} + \dfrac{1}{2}$

60. $-\dfrac{8}{3} + \left(-\dfrac{5}{9}\right)$ **61.** $\left(-\dfrac{10}{7}\right)\left(-\dfrac{14}{15}\right)$

62. $\dfrac{9}{2}\left(-\dfrac{4}{21}\right)$ **63.** $\dfrac{3}{4} - \dfrac{5}{3}$ **64.** $-\dfrac{3}{8} - \left(-\dfrac{1}{10}\right)$

65. $-\dfrac{3}{8} \div \dfrac{5}{6}$ **66.** $-\dfrac{22}{9} \div \left(-\dfrac{33}{18}\right)$

Simplify.

67. $\dfrac{-16}{20}$ **68.** $\dfrac{-15}{35}$ **69.** $\dfrac{-18}{-24}$ **70.** $\dfrac{-35}{-21}$

Perform the indicated operation by hand. Then use a calculator to check your work.

71. $\dfrac{3}{-4} + \dfrac{1}{4}$ **72.** $\dfrac{5}{-6} + \dfrac{1}{6}$ **73.** $\dfrac{4}{7} - \left(\dfrac{3}{-7}\right)$

74. $\dfrac{2}{3} - \left(\dfrac{1}{-3}\right)$ **75.** $\dfrac{5}{6} + \dfrac{7}{-8}$ **76.** $\dfrac{1}{4} + \dfrac{5}{-6}$

Use a calculator to perform the indicated operation. Round the result to two decimal places.

77. $-26.87(-381.572)$ **78.** $-489.2(-8.39)$

79. $222.045 \div (-32.76)$ **80.** $64.958 \div (-3.716)$

81. $-\dfrac{11}{18}\left(-\dfrac{15}{19}\right)$ **82.** $-\dfrac{169}{175}\left(-\dfrac{64}{71}\right)$

83. $-\dfrac{59}{13} \div \dfrac{27}{48}$ **84.** $-\dfrac{75}{22} \div \dfrac{13}{48}$

Evaluate the expression for $a = -6$, $b = 4$, and $c = -8$.

85. ab **86.** ac **87.** $\dfrac{a}{b}$ **88.** $\dfrac{b}{a}$

89. $-ac$ **90.** $-bc$ **91.** $-\dfrac{b}{c}$ **92.** $-\dfrac{a}{c}$

Let w be a number. Translate the English phrase into a mathematical expression. Then evaluate the expression for $w = -8$.

93. The quotient of the number and 2

94. The number divided by 4

95. The product of the number and -5

96. -2 times the number

For Exercises 97 and 98, write the ratio as a fraction.

97. the ratio of 6 to 8   **98.** the ratio of 9 to 15

99. The proposed 1776-foot-tall Freedom Tower at the site where New York's World Trade Center once stood would be the world's tallest full-service building (not counting antennas). The John Hancock Tower in Boston is 790 feet tall. Find the unit ratio of the height of the Freedom Tower to the height of the John Hancock Tower. What does your result mean in this situation?

100. About 8.52 million Americans attend one or more baseball games each month during the baseball season, and about 1.43 million Americans attend one or more bowling competitions each month (Source: *Mediamark Research, Inc.*). Find the unit ratio of the number of Americans who attend baseball games to the number of Americans who attend bowling competitions each month. What does your result mean in this situation?

101. There were 313 U.S. billionaires in 2004 and 228 U.S. billionaires in 2002 (Source: Forbes). Find the unit ratio of the number of U.S. billionaires in 2004 to the number of U.S. billionaires in 2002. What does your result mean in this situation?

102. In 2004–2005, the average number of viewers was 6 million viewers per day for the TV show *Good Morning America* and 2.9 million viewers for the competing *This Morning Show/Early Show* (Source: *Nielson Media Research*). Find the unit ratio of the average number of viewers per day of *Good Morning America* to the average number of viewers per day of *This Morning Show/Early Show*. What does your result mean in this situation?

103. A recipe for roasted red-pepper pasta calls for 4 red bell peppers and 5 black olives. Calculate the given unit ratio. What does your result mean in this situation?
a. The unit ratio of the number of red bell peppers to the number of black olives
b. The unit ratio of the number of black olives to the number of red bell peppers

104. A recipe for beef stroganoff calls for 2 cups of sliced mushrooms and 4 cups of cooked noodles. Calculate the given unit ratio. What does your result mean in this situation?
a. The unit ratio of the number of cups of sliced mushrooms to the number of cups of cooked noodles
b. The unit ratio of the number of cups of cooked noodles to the number of cups of sliced mushrooms

105. The *full-time equivalent enrollment* (FTE enrollment) at a college is the number of full-time students it would take for their total credits (units or hours) to equal the total credits in which both part-time and full-time students combined are enrolled in one semester. The number of *full-time equivalent faculty* (FTE faculty) is the number of full-time faculty it would take to teach all the courses that are taught by both part-time and full-time faculty combined. The 2003 FTE enrollments and number of FTE faculty are shown in Table 38 for various colleges.

Table 38 FTE Enrollments and Numbers of FTE Faculty

College	FTE Enrollment	Number of FTE Faculty
Butler University	4168.0	335.1
St. Olaf College	2951.0	248.7
Stonehill College	2351.0	178.0
University of Massachusetts–Amherst	17,016.2	982.4
Texas A&M University	37,682.49	1785.9

Sources: *Butler University, St. Olaf College, Stonehill College, University of Massachusetts, Texas A&M University*

a. Find the unit ratio of FTE enrollment at Texas A&M University to the FTE enrollment at St. Olaf College. What does your result mean in this situation?

b. Find the unit ratio of the number of FTE faculty at the University of Massachusetts–Amherst to the number of FTE faculty at Butler University. What does your result mean in this situation?

c. Find the unit ratio of FTE enrollment to the number of FTE faculty at each of the colleges listed in Table 38.

d. Which college listed in Table 38 has the largest ratio of FTE enrollment to the number of FTE faculty? Which has the smallest?

e. A person believes that the ratio of FTE enrollment to the number of FTE faculty is lower at Stonehill College than at St. Olaf College, because Stonehill College has the lower FTE enrollment. Is that person correct? Explain.

106. The 2004 populations and land areas are shown in Table 39 for various states.

Table 39 Populations and Land Areas

State	Population	Land Area (square miles)
Alaska	655,435	571,951
California	35,893,799	155,959
Michigan	10,112,620	56,804
New Jersey	8,698,879	7417
New York	19,227,088	47,214

Sources: *U.S. Census Bureau; Infoplease*

a. Find the unit ratio of New York's population to New Jersey's population. What does your result mean in this situation?

b. Find the unit ratio of Alaska's land area to California's land area. What does your result mean in this situation?

c. The unit ratio of population to land area is called the *population density*. Find the population density of each state listed in Table 39.

d. Which state listed in Table 39 has the greatest population density? Which has the least?

e. A person believes that Michigan has a greater population density than New Jersey, because Michigan's population is more than New Jersey's population. Is that person correct? Explain.

107. A person has credit card balances of -4360 dollars on a Discover® account and -1825 dollars on a MasterCard® account.
  a. Find the unit ratio of the Discover balance to the MasterCard balance.
  b. If the person wishes to pay off both accounts gradually in the same amount of time, describe how the result in part (a) can help guide the person in making her next payment.

108. A person has credit card balances of -6810 dollars on a Visa® account and -2950 dollars on a Sears® account.
  a. Find the unit ratio of the Visa balance to the Sears balance.
  b. If the person wishes to pay off both accounts gradually in the same amount of time, describe how the result in part (a) can help guide the person in making his next payment.

109. A person's credit card balance is -3720. If the person pays off 15% of the balance, what is the new balance?

110. A person's credit card balance is -1590. If the person pays off 35% of the balance, what is the new balance?

111. A student has zero balance on a credit card. The student uses the credit card to buy 12.3 gallons of gasoline at a cost of $2.40 per gallon. What is the new balance?

112. A person has zero balance on a credit card. The person uses the credit card to buy three CDs at a cost of $13.99 per CD. What is the new balance?

113. a. Find the sum $-2 + (-4)$.
  b. Find the product $-2(-4)$.
  c. Consider the following statements:
    • Two negative numbers make a positive.
    • A negative number times a negative number is equal to a positive number.
    Which statement is clearer? Explain.
  d. Compare the sign rule for adding two negative numbers with the sign rule for multiplying two negative numbers.

114. a. Find the sum $-2 + 3$.
  b. Find the product $-2(3)$.
  c. Consider the following statements:
    • A negative number times a positive number is equal to a negative number.
    • A negative and a positive makes a negative.
    Which statement is clearer? Explain.
  d. Compare the sign rule for adding numbers with different signs with the sign rule for multiplying numbers with different signs.

115. Which of the following fractions are equal? (There may be more than one pair of answers.)

$$\frac{a}{b} \qquad \frac{-a}{b} \qquad \frac{a}{-b} \qquad -\frac{a}{b} \qquad \frac{-a}{-b} \qquad -\frac{-a}{-b}$$

116. a. Is $\dfrac{12}{-4}$ positive or negative? Explain.
  b. If a is positive and b is negative, is $\dfrac{a}{b}$ positive or negative? Explain.

c. A student says that $\dfrac{a}{b}$ is positive because it has no negative signs. Is the student correct? Explain.

117. Discuss in terms of repeated addition why it makes sense that $4(-5)$ is negative.

118. Discuss in terms of repeated addition why it makes sense that $(-4)(3)$ is negative.

119. If ab is negative, what can you say about a or b?

120. If ab is positive, what can you say about a or b?

121. If $ab = 0$, what can you say about a or b?

122. If $\dfrac{a}{b}$ is negative, what can you say about a or b?

123. **a.** Perform the indicated operations in $(8 \div 2) \cdot 4$ by first doing the division and then doing the multiplication.
   b. Perform the indicated operations in $8 \div (2 \cdot 4)$ by first doing the multiplication and then doing the division.
   c. Compare your results for parts (a) and (b). Does it matter in which order we multiply and divide? Explain.

124. **a.** Find each product.
   i. $(-1)(-1)$
   ii. $(-1)(-1)(-1)$
   iii. $(-1)(-1)(-1)(-1)$
   iv. $(-1)(-1)(-1)(-1)(-1)$
   b. Describe any patterns that you notice in your results in part (a). If you don't see a pattern, find some more products of -1's until you do.
   c. Find the product:
   $$\underbrace{(-1)(-1)(-1)\cdots(-1)}_{524 \text{ of the } -1\text{'s}}$$
   d. Find the product:
   $$\underbrace{(-1)(-1)(-1)\cdots(-1)}_{847 \text{ of the } -1\text{'s}}$$

2.6 EXPONENTS AND ORDER OF OPERATIONS

Objectives

▹ Know the meaning of *exponent*.

▹ Use the rules for order of operations to perform computations and evaluate expressions.

In this section, we will discuss an operation called *exponentiation*. We will also discuss the order in which we should perform various operations.

Exponents

The notation x^2 stands for $x \cdot x$. So, $7^2 = 7 \cdot 7 = 49$. The notation x^3 stands for $x \cdot x \cdot x$. So, $2^3 = 2 \cdot 2 \cdot 2 = 8$.

DEFINITION Exponent

For any counting number n,

$$x^n = \underbrace{x \cdot x \cdot x \cdot \ldots \cdot x}_{n \text{ factors of } x}$$

We refer to x^n as the **power,** the **nth power of x,** or **x raised to the nth power.** We call x the **base** and n the **exponent.**

The expression 2^5 is a power. It is the 5th power of 2, or 2 raised to the 5th power. For 2^5, the base is 2 and the exponent is 5. Here, we label the base and the exponent of 2^5 and compute the power:

$$2^5 = \underbrace{2 \cdot 2 \cdot 2 \cdot 2 \cdot 2}_{5 \text{ factors of } 2} = 32$$

where the exponent is 5 and the base is 2.

When we calculate a power, we say that we are performing **exponentiation.**
Notice that the notation b^1 stands for one factor of b, so $b^1 = b$.

Two powers of x have specific names. We refer to x^2 as the **square of x or x squared.** We refer to x^3 as the **cube of x or x cubed.**

For an expression of the form $-a^n$, we calculate a^n before taking the opposite. For example,

$$-3^4 = -(3^4) = -(3 \cdot 3 \cdot 3 \cdot 3) = -81$$

For -3^4, the base is 3. If we want the base to be -3, we enclose -3 in parentheses:

$$(-3)^4 = (-3)(-3)(-3)(-3) = 81$$

We can use a graphing calculator to check both computations (see Fig. 31).

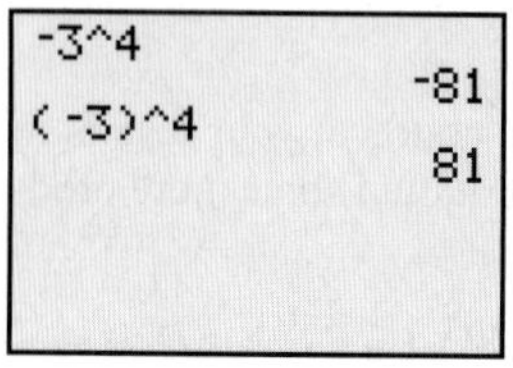

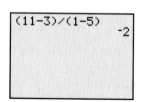

Figure 31 Compute -3^4 and $(-3)^4$

Example 1 Calculating Expressions That Have Exponents

Perform the exponentiation.

1. 5^4 **2.** -3^2 **3.** $(-3)^2$ **4.** $\left(\dfrac{4}{5}\right)^3$

Solution

1. $5^4 = 5 \cdot 5 \cdot 5 \cdot 5 = 625$ The base is 5.
2. $-3^2 = -(3 \cdot 3) = -9$ The base is 3.
3. $(-3)^2 = (-3)(-3) = 9$ The base is -3.
4. $\left(\dfrac{4}{5}\right)^3 = \dfrac{4}{5} \cdot \dfrac{4}{5} \cdot \dfrac{4}{5} = \dfrac{64}{125}$ The base is $\dfrac{4}{5}$. ■

Order of Operations

We can establish the order of operations by using *grouping symbols,* such as parentheses ( ), absolute-value symbols ||, and fraction bars. We do operations that lie within grouping symbols before we perform other operations.

But does it matter in which order we perform operations? From the following calculations, it is clear that it *does* matter:

$$(3 + 2) \cdot 4 = 5 \cdot 4 = 20 \qquad \text{First add; then multiply.}$$
$$3 + (2 \cdot 4) = 3 + 8 = 11 \qquad \text{First multiply; then add.}$$

With a fraction such as $\dfrac{7 + 3}{3 - 1}$, the following use of parentheses is assumed:

$$\frac{7 + 3}{3 - 1} = \frac{(7 + 3)}{(3 - 1)} = \frac{10}{2} = 5$$

So, we compute both the numerator and the denominator before we divide.

Example 2 Performing Operations

Perform the indicated operations.

1. $(7 + 2)(3 - 8)$ **2.** $\dfrac{11 - 3}{1 - 5}$

Solution

1. $(7 + 2)(3 - 8) = (9)(-5) = -45$

2. $\dfrac{11 - 3}{1 - 5} = \dfrac{8}{-4} = -2$

We can use a graphing calculator to check our result (see Fig. 32). ■

Figure 32 Verify the result

For an expression such as $3 + 2 \cdot 4$, where grouping symbols do not specify the order of operations for all operations in the expression, there is an understood order of operations.

Order of Operations

We perform operations in the following order:

1. First, perform operations within parentheses or other grouping symbols, starting with the innermost group.
2. Then perform exponentiations.
3. Next, perform multiplications and divisions, going from left to right.
4. Last, perform additions and subtractions, going from left to right.

So, for the expression $3 + 2 \cdot 4$, we multiply before adding:

$$3 + 2 \cdot 4 = 3 + 8 = 11$$

Example 3 Performing Operations

Perform the indicated operations.

1. $9 - 8 \div 4$ **2.** $10 \div 5 + (6 - 4) \cdot 5$

Solution

1. $9 - 8 \div 4 = 9 - 2$ *Divide before subtracting.*

$\qquad\qquad\quad\; = 7$ *Subtract.*

2. $10 \div 5 + (6 - 4) \cdot 5 = 10 \div 5 + 2 \cdot 5$ *Subtract within parentheses.*

$\qquad\qquad\qquad\qquad = 2 + 2 \cdot 5$ *Divide, because the division is to the left of the multiplication.*

$\qquad\qquad\qquad\qquad = 2 + 10$ *Multiply before adding.*

$\qquad\qquad\qquad\qquad = 12$ *Add.*

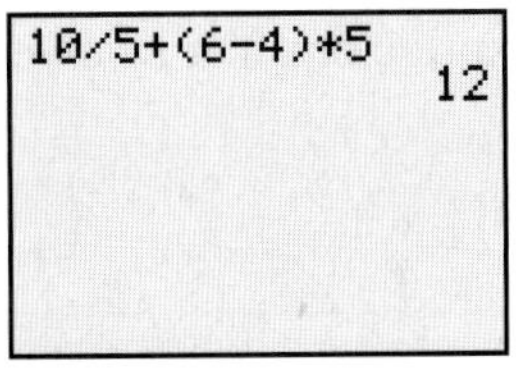

We use a graphing calculator to verify our result (see Fig. 33). ∎

```
10/5+(6-4)*5
            12
```

Figure 33 Verify the result

There is a connection between the order of operations and the strengths of the operations. We explore the strengths of exponentiation, multiplication, and addition by performing these operations on a pair of 10s:

Operation	Computation with 10s
Exponentiation	$10^{10} = 10{,}000{,}000{,}000$
Multiplication	$10 \cdot 10 = 100$
Addition	$10 + 10 = 20$

Exponentiation is much more powerful than multiplication, which in turn is more powerful than addition. Since dividing by a number is the same as multiplying by the reciprocal of the number, division is as powerful as multiplication. Since subtracting a number is the same as adding the opposite of the number, subtraction is as powerful as addition. Here is a summary of the strengths of the operations:

Operation	Strength of Operation
Exponentiation	Most Powerful
Multiplication and Division	Next Most Powerful
Addition and Subtraction	Weakest

Order of Operations and the Strengths of Operations

After we have performed operations in parentheses, the order of operations goes from the most powerful operation, exponentiation, to the next-most-powerful operations, multiplication and division, to the weakest operations, addition and subtraction.

Knowing the relationship between the order of operations and the strengths of the operations will likely help you remember the order of operations.

Example 4   Performing Operations

Perform the indicated operations.

1. $3 + 2^3 + 4(5)$ **2.** $7 + (2-6)^3 - 8 \div (-2)$

Solution

1. $\begin{aligned} 3 + 2^3 + 4(5) &= 3 + 8 + 4(5) \\ &= 3 + 8 + 20 \\ &= 31 \end{aligned}$

Perform exponentiation first: $2^3 = 2 \cdot 2 \cdot 2 = 8$

Multiply before adding.

Add.

2.
$$7 + (2-6)^3 - 8 \div (-2) = 7 + (-4)^3 - 8 \div (-2)$$

Work within parentheses first.

$$= 7 + (-64) - 8 \div (-2)$$

Perform exponentiation: $(-4)^3 = (-4) \cdot (-4) \cdot (-4) = -64$

$$= 7 + (-64) - (-4)$$

Divide.

$$= 7 - 64 + 4$$

Simplify.

$$= -57 + 4$$

Subtract.

$$= -53$$

Add.

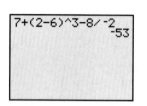

Figure 34 Verify the work

We use a graphing calculator to verify our work (see Fig. 34).   ■

Expressions

Now that we know the order of operations, we can evaluate expressions that involve more than one operation.

Example 5   Evaluating an Expression

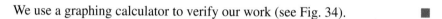

Evaluate $\dfrac{a-b}{c-d}$ for $a = 4$, $b = -2$, $c = -5$, and $d = 4$.

Solution

We begin by substituting 4 for a, -2 for b, -5 for c, and 4 for d:

$$\frac{(4) - (-2)}{(-5) - (4)} = \frac{6}{-9}$$

Subtract.

$$= -\frac{6}{9} \qquad \frac{a}{-b} = -\frac{a}{b}$$

$$= -\frac{2}{3} \qquad \text{Simplify.}$$

■

Example 6   Evaluating Expressions

1. Evaluate $4x^2$ for $x = -3$.
2. Evaluate $b^2 - 4ac$ for $a = 5$, $b = -3$, and $c = -6$.

Solution

1. We begin by substituting -3 for x in the expression $4x^2$:

$$4(-3)^2 = 4(9) \qquad \text{Perform exponentiation first.}$$
$$= 36 \qquad \text{Multiply.}$$

2. We begin by substituting 5 for a, -3 for b, and -6 for c in the expression $b^2 - 4ac$:

$$(-3)^2 - 4(5)(-6) = 9 - 4(5)(-6) \qquad \text{Perform exponentiation first.}$$
$$= 9 - (-120) \qquad \text{Multiply.}$$
$$= 9 + 120 \qquad \text{Simplify.}$$
$$= 129 \qquad \text{Add.}$$

■

Example 7   Using a Table to Find an Expression

The value of a stock was $42 in 2000 and has increased by $3 per year since then.

1. Use a table to help find an expression that stands for the value of the stock (in dollars) at t years since 2000.
2. Evaluate the expression that you found in Problem 1 for $t = 10$. What does your result mean in this situation?

Table 40 Values of a Stock

Years since 2000	Value (dollars)
0	$3 \cdot 0 + 42$
1	$3 \cdot 1 + 42$
2	$3 \cdot 2 + 42$
3	$3 \cdot 3 + 42$
4	$3 \cdot 4 + 42$
t	$3 \cdot t + 42$

Solution

1. We create Table 40. We show the arithmetic to help us see a pattern. From the last row of the table, we see that the value of the stock (in dollars) can be represented by $3t + 42$.
2. We substitute 10 for t in the expression $3t + 42$:

$$3(10) + 42 = 72$$

The value of the stock in $2000 + 10 = 2010$ will be $72. ■

Example 8   Translating from English to Mathematics

Let x be a number. Translate the phrase "6 minus the product of -5 and the number" into a mathematical expression. Then evaluate the expression for $x = -3$.

Solution

First we translate the given phrase into an expression:

$$\underbrace{\text{Six minus}}_{6 \quad -} \quad \underbrace{\text{the product of } -5 \text{ and the number}}_{(-5)x}$$

Then we substitute -3 for x in the expression $6 - (-5)x$ and perform the indicated operations:

$$6 - (-5)(-3) = 6 - 15 \qquad \text{Multiply before subtracting.}$$
$$= -9 \qquad \text{Subtract.}$$

■

group exploration

Looking ahead: Evaluating an expression and plotting points

1. Evaluate the expression $2x + 1$ at each of the given values of x.
   a. $x = -2$ b. $x = -1$ c. $x = 0$
   d. $x = 1$ e. $x = 2$

2. Organize the results that you found in Problem 1 in Table 41. The first row has been done for you.

3. Treat the values of the expression $2x + 1$ as values of the variable y. So, from the first row of Table 41, we see that $y = -3$ when $x = -2$, which can be represented by the ordered pair $(-2, -3)$. List four ordered pairs that are related to the other four rows of Table 41.

4. Plot the ordered pair $(-2, -3)$ and the four ordered pairs that you found in Problem 3 on a coordinate system. Describe any patterns in the position of the points.

5. Repeat the instructions for Problems 1–4, but use the expression $-2x + 3$.

Table 41 Values of x and $2x + 1$

x	$2x + 1$
-2	-3
-1	
0	
1	
2	

> ### TIPS FOR SUCCESS: Study in a Test Environment
>
> Do you ever feel that you understand your homework assignments, yet you perform poorly on quizzes and tests? If so, you may not be studying enough to be ready to solve problems *in a test environment*. For example, although it is a good idea to refer to your lecture notes when you are stumped on a homework exercise, you must continue to solve similar exercises until you can solve them *without* referring to your lecture notes (unless your instructor uses open-notebook tests). The same idea applies to getting help from someone, referring to examples in the text, looking up answers in the back of the text, or any other form of support.
>
> Getting support to help you learn math is a great idea. Just make sure you spend the last part of your study time completing exercises without such support. One way to complete your study time would be to make up a practice quiz or test for you to do in a given amount of time.

HOMEWORK 2.6   FOR EXTRA HELP ▶

Student Solutions Manual   PH Math/Tutor Center   MathXL®   MyMathLab

Perform the exponentiation by hand. Then use a calculator to check your work.

1. 4^3 **2.** 3^4 **3.** 2^5 **4.** 5^3

5. -8^2 **6.** -7^2 **7.** $(-8)^2$ **8.** $(-7)^2$

9. $\left(\dfrac{6}{7}\right)^2$ **10.** $\left(\dfrac{3}{5}\right)^3$

Perform the indicated operations by hand. Then use a calculator to check your work.

11. $3 \cdot (5 - 1)$ **12.** $8 \cdot (2 - 6)$

13. $(2 - 5)(9 - 3)$ **14.** $(2 + 8)(3 - 8)$

15. $4 - (3 - 8) + 1$ **16.** $-6 - (4 - 7) + 5$

17. $\dfrac{1 - 10}{1 + 2}$ **18.** $\dfrac{3 - 9}{2 - 4}$

19. $\dfrac{2 - (-3)}{5 - 7}$ **20.** $\dfrac{4 - 7}{-3 - (-1)}$

21. $\dfrac{4 - (-6)}{-7 - 8}$ **22.** $\dfrac{1 - 9}{2 - (-4)}$

23. $6 + 8 \div 2$ **24.** $2 - 3 \cdot 5$

25. $-5 - 4 \cdot 3$ **26.** $1 + 9 \cdot (-4)$

27. $20 \div (-2) \cdot 5$ **28.** $-16 \div (-4) \cdot 2$

29. $-9 - 4 + 3$ **30.** $3 - 7 + 1$

31. $3(5 - 1) - 4(-2)$ **32.** $-2(2 - 5) + 10 \div 5$

33. $15 \div 3 - (2 - 7)(2)$ **34.** $6(2 + 3) - 5 \cdot 7$

35. $\dfrac{7}{8} - \dfrac{3}{4} \cdot \dfrac{1}{2}$ **36.** $\dfrac{5}{6} + \dfrac{2}{3} \div \dfrac{2}{5}$

37. $2 + 5^2$ **38.** $8 - 3^2$

39. $-3(4)^2$ **40.** $8(-2)^3$

41. $\dfrac{2^4}{4^2}$ **42.** $\dfrac{5^2}{2^5}$

43. $4^3 - 3^4$ **44.** $5^2 + 2^5$

45. $45 \div 3^2$ **46.** $-20 \div 2^2$

47. $(-1)^2 - (-1)^3$ **48.** $4^3 - (-4)^3$

49. $-5(3)^2 + 4$ **50.** $4(-2)^2 - 3$

51. $-4(-1)^2 - 2(-1) + 5$ **52.** $2(-4)^2 + 3(-4) - 7$

53. $\dfrac{9 - 6^2}{12 + 3^2}$ **54.** $\dfrac{10 - (-2)^2}{3^3}$

55. $8 - (9 - 5)^2 - 1$ **56.** $4 + (3 - 6)^2 - 2$

57. $8^2 + 2(4 - 8)^2 \div (-2)$ **58.** $(9 - 7)^2 \cdot (-3) - 2^4$

Use a calculator to perform the indicated operations. Round the result to two decimal places.

59. $13.28 - 35.2(17.9) + 9.43 \div 2.75$

60. $8.53 \div 5.26 + 24.91 - 78.3(45.3)$

61. $5.82 - 3.16^3 \div 4.29$ **62.** $1.98 + 8.22^5 \cdot 5.29$

63. $\dfrac{(25.36)(-3.42) - 17.89}{33.26 + 45.32}$ **64.** $\dfrac{53.25 + 99.83}{31.28 - (6.31)(89.11)}$

Evaluate the expression for $a = -2$, $b = -4$, and $c = 3$.

65. $a + bc$ **66.** $ac - b$

67. $ac - b \div a$ **68.** $c \div a + abc$

69. $a^2 - c^2$ **70.** $c^2 - b^2$

71. $b^2 - 4ac$ **72.** $-cb^2 + a^2$

73. $\dfrac{-b - c^2}{2a}$ **74.** $\dfrac{a^2 - b}{c^2 - b}$

Evaluate $\dfrac{a - b}{c - d}$ for the given values of a, b, c, and d.

75. $a = 3, b = -10, c = -6, d = 1$

76. $a = 5, b = -1, c = -4, d = 7$

77. $a = -3, b = 7, c = 1, d = -3$

78. $a = -2, b = 6, c = 5, d = -1$

79. $a = -8, b = -2, c = -15, d = -5$

80. $a = -3, b = -5, c = -8, d = -3$

Evaluate the expression for $x = -3$.

81. $-3x^2$ **82.** $5x^2$

83. $-x^2 + x$ **84.** $-4x^2 + 4$

85. $2x^2 - 3x + 5$ **86.** $4x^2 + x - 2$

87. Congressional pay in 1975 was $44.6 thousand and has increased by approximately $4 thousand per year since then (Source: *Bureau of Labor Statistics*).

  a. Complete Table 42 to help find an expression that stands for the congressional pay (in thousands of dollars) at t years since 1975. Show the arithmetic to help you see a pattern.

Table 42 Numbers of Years and Congressional Pay	
Years Since 1975	**Congressional Pay (thousands of dollars)**
0	
1	
2	
3	
4	
t	

  b. Evaluate the expression that you found in part (a) for $t = 35$. What does your result mean in this situation?

88. The number of pieces of equipment stolen from California construction sites was 805 in 2001 and has increased by approximately 121 pieces per year since then (Source: *California Attorney General's Office*).

  a. Complete Table 43 to help find an expression that stands for the number of pieces of equipment stolen at t years since 2001. Show the arithmetic to help you see a pattern.

Table 43 Numbers of Years and Pieces of Stolen Equipment	
Years Since 2001	**Number of Pieces of Stolen Equipment**
0	
1	
2	
3	
4	
t	

  b. Evaluate the expression that you found in part (a) for $t = 10$. What does your result mean in this situation?

89. The population of Gary, Indiana, was about 145 thousand in 1980 and has decreased by about 2 thousand per year since then (Source: *U.S. Census Bureau*).

  a. Complete Table 44 to help find an expression that stands for the population of Gary (in thousands) at t years since 1980. Show the arithmetic to help you see a pattern.

Table 44 Populations of Gary	
Years Since 1980	**Population (thousands)**
0	
1	
2	
3	
4	
t	

  b. Evaluate the expression that you found in part (a) for $t = 31$. What does your result mean in this situation?

90. The percentage of companies offering traditional benefit plans was 85% in 1990 and has decreased by approximately 1.2 percentage points per year since then (Source: *Hewitt survey*).

  a. Complete Table 45 to help find an expression that stands for the percentage of companies offering traditional benefit plans at t years since 1990. Show the arithmetic to help you see a pattern.

Table 45 Percentages of Companies Offering Traditional Benefit Plans	
Years Since 1990	**Percent**
0	
1	
2	
3	
4	
t	

  b. Evaluate the expression that you found in part (a) for $t = 20$. What does your result mean in this situation?

For Exercises 91–94, let x be a number. Translate the English phrase or sentence into a mathematical expression. Then evaluate the expression for $x = -4$.

91. 5 more than the product of -6 and the number

92. -3 minus the quotient of 8 and the number

93. Subtract 3 from the quotient of the number and -2.

94. The number plus the product of -5 and the number

95. If a cube has sides of length s feet, then the volume of the cube is s^3 cubic feet. Find the volume of a cubic box with sides of length 2 feet.

96. If the radius of a sphere is r inches, then the volume of the sphere is $\frac{4}{3}\pi r^3$ cubic inches. Find the volume of a sphere with a radius of 3 inches.

97. A student tries to perform the indicated operations in $2(3)^2 + 2(3) + 1$:

$$
\begin{aligned}
2(3)^2 + 2(3) + 1 &= 6^2 + 2(3) + 1 \\
&= 36 + 2(3) + 1 \\
&= 36 + 6 + 1 \\
&= 43
\end{aligned}
$$

Describe any errors. Then perform the operations correctly.

98. A student tries to evaluate $x^2 + 4x + 5$ for $x = -3$:

$$
\begin{aligned}
-3^2 + 4(-3) + 5 &= -9 - 12 + 5 \\
&= -21 + 5 \\
&= -16
\end{aligned}
$$

Describe any errors. Then evaluate the expression correctly.

99. A student thinks that $-4^2 = 16$, because a negative number times a negative number is equal to a positive number. Is the student correct? Explain.

100. A student tries to perform the indicated operations in $16 \div 2 \cdot 4$:

$$16 \div 2 \cdot 4 = 16 \div 8 = 2$$

Describe any errors. Then perform the operations correctly.

101. In Problem 2 of Example 2, we performed the indicated operations in $\dfrac{11 - 3}{1 - 5}$ by the following steps:

$$\frac{11 - 3}{1 - 5} = \frac{8}{-4} = -2$$

a. Perform the indicated operations in $(11 - 3) \div (1 - 5)$.
b. Perform the indicated operations in $11 - 3 \div 1 - 5$.
c. In using a calculator to simplify $\dfrac{11 - 3}{1 - 5}$, a student presses the following buttons:

$$\boxed{11}\ \boxed{-}\ \boxed{3}\ \boxed{\div}\ \boxed{1}\ \boxed{-}\ \boxed{5}$$

The result of this calculation is 3 rather than -2. Describe any errors. Then perform the operations correctly.

102. a. Perform the indicated operations in $(12 \div 3) \cdot 2$.
b. Perform the indicated operations in $12 \div (3 \cdot 2)$.
c. For the expression $12 \div 3 \cdot 2$, does it matter whether we multiply and divide from left to right or multiply and divide from right to left? Explain.
d. Perform the indicated operations in $12 \div 3 \cdot 2$.

103. a. Evaluate $(ab)c$ for $a = 2$, $b = 3$, and $c = 4$.
b. Evaluate $a(bc)$ for $a = 2$, $b = 3$, and $c = 4$.
c. Compare your results for parts (a) and (b).

d. Evaluate $(ab)c$ and $a(bc)$ for $a = 4$, $b = -2$, and $c = 5$, and then compare the results.
e. Evaluate $(ab)c$ and $a(bc)$ for a, b, and c values of your choosing, and then compare the results.
f. Do you think that the statement $(ab)c = a(bc)$ is true for all values of a, b, and c? Explain.
g. What does the statement $(ab)c = a(bc)$ imply about the order of operations?

104. a. Evaluate $(a + b) + c$ for $a = 2$, $b = 3$, and $c = 4$.
b. Evaluate $a + (b + c)$ for $a = 2$, $b = 3$, and $c = 4$.
c. Compare your results for parts (a) and (b).
d. Evaluate $(a + b) + c$ and $a + (b + c)$ for $a = 4$, $b = -2$, and $c = 5$, and then compare the results.
e. Evaluate $(a + b) + c$ and $a + (b + c)$ for a, b, and c values of your choosing, and then compare the results.
f. Do you think that the statement $(a + b) + c = a + (b + c)$ is true for all values of a, b, and c? Explain.
g. What does the statement $(a + b) + c = a + (b + c)$ imply about the order of operations?

105. a. Evaluate x^2 for x equal to -2, -1, 0, 1, and 2.
b. Which of the following describes your results in part (a)?

negative,   nonpositive,   positive,   nonnegative

c. Describe x^2 for any real number x.

106. Describe the order of operations in your own words.

analyze this

GLOBAL WARMING LAB (continued from Chapter 1)

Given the consensus among most scientists that global warming is occurring and that it could have catastrophic results, scientists have been searching for the cause of the warming (see Table 46).

Table 46 Average Surface Temperatures of Earth

Year	Average Temperature (degrees Fahrenheit)	Year	Average Temperature (degrees Fahrenheit)
1900	57.1	1955	57.0
1905	56.7	1960	57.2
1910	56.8	1965	56.9
1915	57.3	1970	57.3
1920	56.9	1975	57.2
1925	56.9	1980	57.7
1930	57.1	1985	57.4
1935	57.1	1990	58.1
1940	57.5	1995	58.0
1945	57.2	2000	57.9
1950	56.9	2003	58.4

Source: *NASA–GISS*

Most scientists believe that global warming is largely the result of carbon emissions (in the form of carbon dioxide) from the burning of fossil fuels such as oil, coal, and natural gas. Carbon emissions in the United States and in the world have increased greatly since 1950 (see Table 47).

Table 47 Carbon Emissions from Burning of Fossil Fuels

Year	Carbon Emissions (billions of metric tons) United States	World
1950	0.7	1.6
1955	0.7	2.0
1960	0.8	2.6
1965	0.9	3.1
1970	1.2	4.1
1975	1.2	4.6
1980	1.3	5.3
1985	1.2	5.4
1990	1.3	6.1
1995	1.4	6.4
2000	1.5	6.6

Source: *U.S. Department of Energy*

Not everyone agrees that carbon dioxide emissions cause global warming, however. For example, Jonathan Adler, professor of environmental law at Case Western Reserve School of Law, argues that two-thirds of the temperature increase occurred in the first half of the 20th century, yet most of the carbon emissions occurred in the second half of the century. He concludes that there is no link between carbon dioxide emissions and global warming. Adler believes that global warming could be due to slight variations in the Sun's output, combined with fluctuations in Earth's orbit.[‡]

Offering another explanation for global warming, Enric Pallé and his colleagues at the Big Bear Solar Observatory and the California Institute of Technology conducted a 2004 study which suggests that the global warming which took place from 1984 to 2000 could be due to reduced cloud cover.[§] Other scientists have raised objections to the study's findings.

Although there have been many dissenters along the way, the vast majority of scientists now believe in the warming–emissions connection. Even in 1997 there was strong enough agreement to motivate the United Nations to negotiate a treaty called the Kyoto Protocol. The treaty's goal is to reduce annual greenhouse gas emissions to about 5% to 7% below 1990 levels by 2012. In November 2004, Russia cast the deciding vote to ratify the protocol, which took effect on February 16, 2005.

To create some flexibility in the treaty's requirements, a country that has exceeded its emissions limit can buy emissions credits from a country that is below its emissions limit. Also, a country can receive emissions credits by financing a project to help lower emissions in another country.

The treaty is legally binding for the 128 countries that have ratified it. Any of these countries that do not meet their goals would have to meet tougher goals in the future, and their ability to receive emissions credits would be suspended.

The United States has declined to ratify the Kyoto Protocol, saying that reducing emissions to the point called for by the treaty would cripple the U.S. economy. If it eventually ratifies the treaty, the United States would have to reduce its 2012 greenhouse gas emissions to 7% below its emissions level in 1990.

Instead of ratifying the treaty, in February 2002 the Bush administration adopted a voluntary program, called the Global Climate Change Initiative (GCCI), that includes tax incentives to motivate companies to reduce their emissions. The Bush administration set a goal of lowering the carbon intensity in 2012 to 18% below its level in 2002. *Carbon intensity* is defined as the ratio of annual carbon emissions to annual economic output.

Critics of the GCCI plan point out that if the U.S. economy is improving, then it is possible for carbon intensity to decrease *even though annual carbon emissions continue to increase.* In fact, even though annual carbon emissions have increased during each of the past three decades, carbon intensity has declined by 18%, 23%, and 16% in those decades! However, the projected decrease for 2002–2012 is only 13%, so modest efforts will have to be implemented to reach the goal of 18%.[¶]

Critics such as the Earth Policy Institute and the Pew Center on Global Climate Change also say that the voluntary plan is unrealistic, because businesses will make modest efforts to reduce emissions when the economy is doing well and will make little or no effort when the economy is doing poorly. These critics argue that the United States would be more likely to respond to the Kyoto Protocol, because it is a legally binding, yet flexible, treaty.

Analyzing the Situation

1. **a.** Find the change in the average global temperature from 1900 to 1950.
   b. Find the change in the average global temperature from 1950 to 2000.
   c. Recall that Adler believes that there is no link between carbon emissions and global warming. What was his argument? Is his reasoning correct? Explain.

2. Let c be U.S. carbon emissions (in billions of metric tons) in the year that is t years since 1950. Draw a careful scattergram of the data.

3. Draw a line that comes close to the points in your scattergram. Use your model to predict U.S. carbon emissions in 2012.

4. Under the Kyoto Protocol, what would be the largest quantity of U.S. carbon emissions allowed in 2012? Find the difference between this quantity and your prediction in Problem 3.

5. The numbers in the list 20, 22, 24, 26, 28 are increasing. Decide whether the ratios in the following list are increasing, decreasing, or neither:

$$\frac{20}{1}, \frac{22}{2}, \frac{24}{4}, \frac{26}{8}, \frac{28}{16}$$

6. Explain why your work in Problem 5 illustrates how it is possible for U.S. carbon intensity to decrease while U.S. annual carbon emissions increase. [**Hint:** Recall that carbon intensity is defined as the *ratio* of annual carbon emissions to annual economic output.]

7. Which seeks to lower U.S. carbon emissions more, the GCCI plan or the Kyoto Protocol? Explain. Taking into account the degree to which American companies would respond to either policy, do you think U.S. carbon emissions would be reduced more by the GCCI plan or by the Kyoto Protocol? Explain.

STOCKS LAB

Imagine that you have $5000 that you plan to invest in five stocks for one week. In this lab, you will explore some possible outcomes of that investment.[*]

[‡]"Global Warming—Hot Problem or Hot Air?" April 1998, *The Freeman*, 48(4), The Foundation for Economic Education, Inc.

[§]"Changes in Earth's Reflectance over the Past Two Decades," May 28, 2004, *Science* 304: 1299–1301.

[¶]"Early Release of the Annual Energy Outlook 2003" (November 2002), Energy Information Administration; and "Projected Greenhouse Gas Emissions," May 2002, *U.S. Climate Action Report 2002*, pp. 70–80, U.S. Department of State, Washington, DC.

[*]Lab suggested by Jim Ryan, State Center Community College District, Clovis Center, Clovis, CA.

Collecting the Data

Use a newspaper or the Internet to select five stocks. For each stock, record the company name and the call letters of the stock. Here are some examples:

> Gateway Computer has the call letters GTW.
> Coca-Cola has the call letters KO.
> EMC corporation has the call letters EMC.

Record the value of one share (the beginning share price) of each of the five stocks. Also, record how you distribute your $5000 investment. For example, you may invest all $5000 in one of the five stocks or $1000 in each of the five stocks, or you may opt for some other distribution of the money. The sum of your investments should equal or be close to $5000 by buying whole amounts of stock. Even if you do not invest money in some of the stocks, still record the information about all five stocks.

After one week, record the new value of one share (the ending share price) of each of the five stocks.

Analyzing the Data

1. Complete Table 48. The *profit* from a stock is the money you collect from the stock, minus the money you invested in the stock.

   What is the total profit from your $5000 investment?

2. For each of the five stocks, create a bar graph displaying the beginning and ending values of one share. You can use the same axes for several bar graphs if the scaling is convenient.

3. Find the change in the share price of each of the five stocks. Which share price had the greatest change? The least? Explain how you can illustrate these changes on your bar graphs.

4. The *percent change* of the value of a stock can be found by dividing the change in value of the stock by the original value of the stock and then converting the decimal result into percent form (by multiplying by 100). For example, suppose that a stock's value increases from $7 to $9. The change in value of this stock is its ending value minus its beginning value: $9 - 7 = 2$ dollars.

Here we find the percent change in value:

$$\text{percent change} = \frac{\text{change in value}}{\text{beginning value}} \cdot 100$$

$$= \frac{9-7}{7} \cdot 100 = \frac{2}{7} \cdot 100 \approx 28.57$$

So, the percent change is about 28.57%.

Now find the percent change in value for each of your five stocks. Which stock had the greatest percent change? The least? Explain how you can at least approximately compare the percent changes by viewing your bar graphs.

5. Among your five stocks, is there a pair of stocks for which one stock has the greater change but the other has the greater percent change? If yes, use these stocks to respond to the questions in parts (a)–(c). If no, then use the following values of fictional stocks A and B:

Stock A	**Stock B**
Increased from $4 to $5	Increased from $20 to $23

   a. Find the profit earned from investing all of the $5000 in the stock with the larger change in value.
   b. Find the profit earned from investing all of the $5000 in the stock with the larger percent change in value.
   c. Which is the better measure of the growth of a stock, change in value or percent change in value? Explain.

6. a. Find the profit earned from each of the following scenarios:
      i. You invest the $5000 in the best-performing stock among the five stocks.
      ii. You invest the $5000 in the worst-performing stock among the five stocks.
      iii. You invest the $5000 by investing $1000 in each of the five stocks.
   b. Describe the benefits and drawbacks to investing your money in a number of stocks rather than in just one stock.

Table 48 Five Stocks' Performances

Call Letters of Stock	Investment in Stock (dollars)	Beginning Share Price (dollars)	Number of Shares	Ending Share Price (dollars)	Money Collected from Stock (dollars)	Profit from Stock (dollars)

Chapter Summary

Key Points
OF CHAPTER 2

Expression (Section 2.1)

An **expression** is a constant, a variable, or a combination of constants, variables, operation symbols, and grouping symbols, such as parentheses.

Evaluate an expression (Section 2.1)

We **evaluate an expression** by substituting a number for each variable in the expression and then calculating the result. If a variable appears more than once in the expression, the same number is substituted for that variable each time.

Division by 0 (Section 2.2)

The fraction $\dfrac{a}{b}$ is undefined if $b = 0$. Division by 0 is undefined.

Simplify a fraction (Section 2.2)

To simplify a fraction,

1. Find the prime factorizations of the numerator and denominator.

2. Find an equal fraction in which the numerator and the denominator do not have common positive factors other than 1 by using the property

$$\frac{ab}{ac} = \frac{a}{a} \cdot \frac{b}{c} = 1 \cdot \frac{b}{c} = \frac{b}{c}$$

where a and c are nonzero.

Simplify results (Section 2.2)

If the result of an exercise is a fraction, simplify it.

Multiplying fractions (Section 2.2)

$\dfrac{a}{b} \cdot \dfrac{c}{d} = \dfrac{ac}{bd}$, where b and d are nonzero.

Dividing fractions (Section 2.2)

$\dfrac{a}{b} \div \dfrac{c}{d} = \dfrac{a}{b} \cdot \dfrac{d}{c}$, where b, c, and d are nonzero.

Adding fractions (Section 2.2)

$\dfrac{a}{b} + \dfrac{c}{b} = \dfrac{a + c}{b}$, where b is nonzero.

Subtracting fractions (Section 2.2)

$\dfrac{a}{b} - \dfrac{c}{b} = \dfrac{a - c}{b}$, where b is nonzero.

How to add or subtract two fractions with different denominators (Section 2.2)

To add (or subtract) two fractions with different denominators, use the fact that $\dfrac{a}{a} = 1$, where a is nonzero, to write an equal sum (or difference) of fractions for which each denominator is the LCD.

The opposite of the opposite of a number (Section 2.3)

$-(-a) = a$

Absolute value (Section 2.3)

The **absolute value** of a number is the distance that the number is from 0 on the number line.

Adding two numbers with the same sign (Section 2.3)

To add two numbers with the same sign,

1. Add the absolute values of the numbers.

2. The sum of the original numbers has the same sign as the sign of the original numbers.

Adding two numbers with different signs (Section 2.3)

To add two numbers with different signs,

1. Find the absolute values of the numbers. Then subtract the smaller absolute value from the larger absolute value.

2. The sum of the original numbers has the same sign as the original number with the larger absolute value.

Change in a quantity (Section 2.4)

The change in a quantity is the ending amount minus the beginning amount:

$$\text{Change in the quantity} = \text{Ending amount} - \text{Beginning amount}$$

Subtracting a number (Section 2.4)

To subtract a number, add its opposite:

$$a - b = a + (-b)$$

Increasing quantity (Section 2.4)

An increasing quantity has a positive change.

Decreasing quantity (Section 2.4)	A decreasing quantity has a negative change.
Unit ratio (Section 2.5)	A **unit ratio** is a ratio written as $\dfrac{a}{b}$ with $b = 1$ or as $a : b$ with $b = 1$.
Percent (Section 2.5)	**Percent** means "for each hundred": $a\% = \dfrac{a}{100}$.
Writing a percentage as a decimal number (Section 2.5)	To write a percentage as a decimal number, remove the percent symbol and divide the number by 100 (move the decimal point two places to the left).
Writing a decimal number as a percentage (Section 2.5)	To write a decimal number as a percentage, multiply the number by 100 (move the decimal point two places to the right) and insert a percent symbol.
Finding the percentage of a quantity (Section 2.5)	To find the percentage of a quantity, multiply the decimal form of the percentage and the quantity.
One hundred percent (Section 2.5)	One hundred percent of a quantity is *all* of the quantity.
The product or quotient of two numbers (Section 2.5)	The product or quotient of two numbers that have different signs is negative. The product or quotient of two numbers that have the same sign is positive.
Equal fractions with negative signs (Section 2.5)	If $b \neq 0$, then $\dfrac{-a}{b} = \dfrac{a}{-b} = -\dfrac{a}{b}$.
Exponent (Section 2.6)	For any counting number n, $$x^n = \underbrace{x \cdot x \cdot x \cdot \ldots \cdot x}_{n \text{ factors of } x}$$ We refer to x^n as the **power**, the **nth power of x**, or **x raised to the nth power.** We call x the **base** and n the **exponent.**
Finding $-a^n$ (Section 2.6)	For an expression of the form $-a^n$, we calculate a^n before taking the opposite.
Order of operations (Section 2.6)	We perform operations in the following order: 1. First, perform operations within parentheses or other grouping symbols, starting with the innermost group. 2. Then, perform exponentiations. 3. Next, perform multiplications and divisions, going from left to right. 4. Last, perform additions and subtractions, going from left to right.
Order of operations and the strengths of operations (Section 2.6)	After we have performed operations in parentheses, the order of operations goes from the most powerful operation, exponentiation, to the next-most-powerful operations, multiplication and division, to the weakest operations, addition and subtraction.

CHAPTER 2 REVIEW EXERCISES

Perform the indicated operations. Then use a calculator to check your work.

1. $8 + (-2)$

2. $(-5) + (-7)$

3. $6 - 9$

4. $8 - (-2)$

5. $8(-2)$

6. $8 \div (-2)$

7. $-24 \div (10 - 2)$

8. $(2 - 6)(5 - 8)$

9. $\dfrac{7 - 2}{2 - 7}$

10. $\dfrac{2 - 8}{3 - (-1)}$

11. $\dfrac{3 - 5(-6)}{-2 - 1}$

12. $3(-5) + 2$

13. $-4 + 2(-6)$

14. $2 - 12 \div 2$

15. $4(-6) \div (-3)$

16. $8 \div (-2) \cdot 5$

17. $2(4 - 7) - (8 - 2)$

18. $-2(3 - 6) + 18 \div (-9)$

19. $-14 \div (-7) - 3(1 - 5)$

20. $-0.3(-0.2)$

21. $4.2 - (-6.7)$

22. $\dfrac{4}{9}\left(-\dfrac{3}{10}\right)$

23. $\left(-\dfrac{8}{15}\right) \div \left(-\dfrac{16}{25}\right)$

24. $\dfrac{5}{9} - \left(-\dfrac{2}{9}\right)$

25. $-\dfrac{5}{6} + \dfrac{7}{8}$

26. $\dfrac{-5}{2} - \dfrac{7}{-3}$

27. $(-8)^2$

28. -8^2

29. 2^4

30. $\left(\dfrac{3}{4}\right)^3$

31. $-6(3)^2$

32. $24 \div 2^3$

33. $(-2)^3 - 4(-2)$

34. $\dfrac{2^3}{3 + 3^2}$

35. $\dfrac{17 - (-3)^2}{5 - 4^2}$

36. $-3(2)^2 - 4(2) + 1$

37. $24 \div (3 - 5)^3$

38. $7^2 - 3(2 - 5)^2 \div (-3)$

Simplify.

39. $\dfrac{-18}{-24}$

40. $\dfrac{-28}{35}$

For Exercises 41 and 42, use a calculator to compute. Round the result to two decimal places.

41. $-5.7 + 2.3^4 \div (-9.4)$ **42.** $\dfrac{3.5(17.4) - 97.6}{54.2 \div 8.4 - 65.3}$

43. A rectangle has a width of $\dfrac{1}{4}$ yard and a length of $\dfrac{5}{6}$ yard. What is the perimeter of the rectangle?

44. A student owes \$4789 to a credit card company. If he sends a check for \$800 and then uses his credit card to purchase a textbook for \$102.99 and a notebook for \$3.50, how much money does he now owe the credit card company?

45. An airplane drops from 32,500 feet to 27,800 feet. Find the change in altitude.

46. Three hours ago the temperature was $4°$F. The temperature is now $-8°$F.
   a. What is the change in temperature for the past three hours?
   b. Estimate the change in temperature for the past hour.
   c. Explain why your estimate in part (b) may not be the actual change in temperature for the past hour.

47. The private contributions (in 2004 dollars) to political conventions are shown in Table 49 for various years.

Table 49 Private Contributions to Political Conventions

Year	Contribution (millions of dollars)	
	Republicans	Democrats
1980	2	1
1984	8	4
1988	3	3
1992	3	9
1996	22	24
2000	22	50
2004	64	50

Source: *Campaign Finance Institute*

   a. Find the change in the private contributions to Democratic conventions from 1996 to 2000.
   b. Find the change in the private contributions to Republican conventions from 1984 to 1988.
   c. Over which four-year period was there the greatest change in private contributions to Democratic conventions? What was that change?
   d. Over which four-year period was there the greatest change in private contributions to Republican conventions? What was that change?

48. There were 63.6 million ring tones for cell phones sold in 2003 and 145.5 million ring tones for phones sold in 2004

(Source: *Jupiter Research*). Find the unit ratio of the number of ring tones sold in 2004 to the number of ring tones sold in 2003. What does your result mean?

For Exercises 49 and 50, write the percentage as a decimal number.

49. 75%   **50.** 2.9%

51. Find 87% of \$43.   **52.** Find 8% of 925 students.

53. A person's credit card balance is -5493 dollars. If the person pays off 20% of the balance, what is the new balance?

Evaluate the expression for $a = 2$, $b = -5$, $c = -4$, and $d = 10$.

54. $ac + c \div a$ **55.** $b^2 - 4ac$

56. $a(b - c)$ **57.** $\dfrac{-b - c^2}{2a}$

58. $2c^2 - 5c + 3$ **59.** $\dfrac{a - b}{c - d}$

For Exercises 60–63, let x be a number. Translate the English phrase into a mathematical expression. Then evaluate the expression for $x = -3$.

60. 5 more than the number

61. The number subtracted from -7

62. 2 minus the product of the number and 4

63. 1 plus the quotient of -24 and the number

64. If T is the total cost (in dollars) for a team to join a softball league and there are n players on the team, then $T \div n$ is the cost (in dollars) per player. Evaluate $T \div n$ for $T = 650$ and $n = 13$. What does your result mean in this situation?

65. A basement is flooded with 400 cubic feet of water. Each hour, 50 cubic feet of water is pumped out of the basement.
   a. Complete Table 50 to help find an expression that stands for the volume (in cubic feet) of water in the basement after water has been pumped out for t hours. Show the arithmetic to help you see a pattern.

Table 50 Volumes of Water

Time (hours)	Volume of Water (cubic feet)
0	
1	
2	
3	
4	
t	

   b. Evaluate the expression that you found in part (a) for $t = 7$. What does your result mean in this situation?

CHAPTER 2 TEST

For Exercises 1–14, perform the indicated operations by hand.

1. $-8 - 5$ **2.** $-7(-9)$

3. $-3 + 9 \div (-3)$ **4.** $(4 - 2)(3 - 7)$

5. $\dfrac{4 - 7}{-1 - 5}$ **6.** $5 - (2 - 10) \div (-4)$

7. $-20 \div 5 - (2 - 9)(-3)$ **8.** $0.4(-0.2)$

9. $-\dfrac{27}{10} \div \dfrac{18}{75}$ **10.** $-\dfrac{3}{10} + \dfrac{5}{8}$

11. 3^4 **12.** -4^2

13. $7 + 2^3 - 3^2$ **14.** $1 - (3 - 7)^2 + 10 \div (-5)$

15. Simplify $\dfrac{84}{-16}$.

16. Two hours ago the temperature was 5°F. If the temperature has decreased by 9°F in the last two hours, what is the current temperature?

17. The chances of being audited by the Internal Revenue Service (IRS) have increased for the first time in years (see Table 51).

Table 51 Tax Audit Rates

Year	Tax Audit Rate (number of audits per 1000 tax returns)
1995	16.7
1997	12.8
1999	9.0
2001	5.8
2003	6.5

Source: *Internal Revenue Service*

  a. Find the change in the tax audit rate from 2001 to 2003.

  b. Find the change in the tax audit rate from 1999 to 2001.

  c. During the late 1990s, the U.S. economy was exceptionally good. During the early 2000s, the economy was poor. On the basis of the information shown in Table 51 alone, under what conditions does the IRS appear to increase the audit rate? Why might the IRS do this?

18. The average ticket price to major league baseball games was $9.14 in 1991 and $19.82 in 2004 (Source: *AP*). Find the unit ratio of the average ticket price in 2004 to the average ticket price in 1991. What does your result mean?

Evaluate the expression for $a = -6$, $b = -2$, $c = 5$, and $d = -1$.

19. $ac - \dfrac{a}{b}$

20. $\dfrac{a - b}{c - d}$

21. $a + b^3 + c^2$

22. $b^2 - 4ac$

Let x be a number. Translate the English phrase into a mathematical expression. Then evaluate the expression for $x = -5$.

23. Twice the number minus the product of 3 and the number

24. 6 subtracted from the quotient of -10 and the number

25. Twenty-five books about obesity were published in 1999. The number of books about obesity that have been published annually has increased by about 7 books per year since 1999 (Source: *Andrew Grabois, R.R. Bowker Co.*).

  a. Complete Table 52 to help find an expression that stands for the number of books about obesity that have been published in the year that is t years since 1999. Show the arithmetic to help you see a pattern.

Table 52 Numbers of Years and Books Published about Obesity

Years since 1999	Number of Books Published
0	
1	
2	
3	
4	
t	

  b. Evaluate the expression that you found in part (a) for $t = 11$. What does your result mean in this situation?

CUMULATIVE REVIEW OF CHAPTERS 1 AND 2

1. A rectangle has a perimeter of 36 inches. Let W be the width, L be the length, and P be the perimeter, all with units in inches.

  a. Sketch three possible rectangles with a perimeter of 36 inches.

  b. Which of the symbols W, L, and P are variables? Explain.

  c. Which of the symbols W, L, and P are constants? Explain.

2. Graph the integers between -2 and 3, inclusive, on a number line.

3. Here are the changes in a stock's value from one month to the next: increase of 1 dollar, decrease of 3 dollars, increase of 4 dollars, and decrease of 2 dollars. Let C be the change of the stock's value (in dollars) from one month to the next. Use points on a number line to describe the changes in value of the stock.

4. What is the x-coordinate of the ordered pair $(-5, 3)$?

5. A person takes a bath. Let V be the volume (in gallons) of water in the bathtub at t minutes after the person pulls out the plug from the drain. Identify the independent variable and the dependent variable.

6. The average price of a standard DVD player has significantly declined since 1997 (see Table 53). Let p be the average price (in dollars) of a DVD player at t years since 1995.

  a. Create a scattergram of the data.

Table 53 Average Price of Standard DVD Player

Year	Average Price (dollars)
1997	625
1998	425
1999	275
2000	205
2001	150
2002	120

Source: *NPD Group*

  b. For which year shown in Table 53 was the average price the most?

  c. For which year shown in Table 53 was the average price the least?

  d. From which year to the next did the average price decrease the most? What was the change in price?

  e. From which year to the next did the average price decrease the least? What was the change in price?

For Exercises 7–10, refer to Fig. 35.

7. Find y when $x = -4$.

8. Find x when $y = 1$.

9. Find the *y*-intercept.   **10.** Find the *x*-intercept.

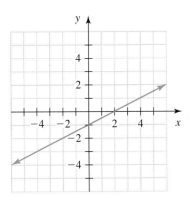

Figure 35 Exercises 7–10

11. A person is laid off from work. Let *B* be the balance (in thousands of dollars) in her checking account at *t* months since she was laid off. Some pairs of values of *t* and *B* are shown in Table 54.

Table 54 Balances in a Checking Account	
t (months)	***B*** (thousands of dollars)
2	16
3	14
5	10
6	8
8	4

a. Create a scattergram of the data. Then draw a linear model.
b. What was the balance 4 months after the person was laid off?
c. When was the balance $6 thousand?
d. What is the *B*-intercept? What does it mean in this situation?
e. What is the *t*-intercept? What does it mean in this situation?

12. Average monthly spending on cable TV per household is shown in Table 55 for various years.

Table 55 Cable TV Costs	
Year	Average Monthly Spending on Cable TV (dollars per household)
1997	26
1998	27
1999	28
2000	30
2001	32
2002	34
2003	35
2004	36

Source: *Veronis Suhler Stevenson Communications Industry Forecast*

Let *c* be the average monthly spending (in dollars) on cable TV per household at *t* years since 1995.
a. Create a scattergram of the data.
b. Draw a line that comes close to the points in your scattergram.

c. What is the *c*-intercept? What does it mean in this situation?
d. Predict when the average monthly spending on cable TV per household will be $43.
e. Predict the average monthly spending on cable TV per household in 2010.

For Exercises 13–18, perform the indicated operations by hand. Then use a calculator to check your work.

13. $\dfrac{3(-8) + 15}{2 - 7(2)}$

14. $-4(3) + 6 - 20 \div (-10)$

15. $\left(-\dfrac{14}{15}\right) \div \left(-\dfrac{35}{27}\right)$

16. $\dfrac{3}{8} - \dfrac{5}{6}$

17. $4 - (7 - 9)^4 + 20 \div (-4)$

18. $\dfrac{5 - 3^2}{4^2 + 2}$

19. Two hours ago the temperature was 5°F. The temperature is now −3°F. What is the change in temperature?

20. A student owes $2692 to a credit card company. If he sends a check to the company for $850 and then uses his credit card to purchase some gasoline for $23, how much money does he now owe?

21. Evaluate $\dfrac{a - b}{c - d}$ for $a = 1$, $b = -4$, $c = -3$, and $d = 7$.

22. Evaluate $b^2 - 4ac$ for $a = 2$, $b = -3$, and $c = -5$.

For Exercises 23 and 24, let x be a number. Translate the English phrase into a mathematical expression. Then evaluate the expression for $x = -4$.

23. The number minus the quotient of −12 and the number

24. 7 more than the product of −2 and the number

25. A stock was worth $42 last year. Let *v* be the value (in dollars) of the stock today. The expression

$$\frac{100(v - 42)}{42}$$

represents the percent growth of the investment. Evaluate the expression for $v = 45$. Round your result to the second decimal place. What does your result mean in this situation?

26. A digital camera company sold 15 thousand cameras in 2000, and sales then increased by 4 thousand cameras each year thereafter.
a. Complete Table 56 to help find an expression that stands for the sales (in thousands of cameras) in the year that is *t* years since 2000. Show the arithmetic to help you see a pattern.

Table 56 Numbers of Years and Sales	
Years since 2000	Sales (thousands of cameras)
0	
1	
2	
3	
4	
t	

b. Evaluate the expression that you found in part (a) for $t = 11$. What does your result mean in this situation?

Chapter **3**

Using the Slope to Graph Linear Equations

We are what we repeatedly do. Excellence then, is not an act, but a habit.

—Aristotle

Table 1 Internet Users in the United States	
Year	**Number of Users (millions)**
1996	39
1997	60
1998	84
1999	105
2000	122
2001	143
2002	166
2003	183
2004	207

Source: *Jupiter MMXI*

Do you use the Internet? The number of Internet users in the United States has increased greatly (see Table 1). In Exercise 27 of Homework 3.5, you will describe how quickly the number of Internet users has increased in the United States over time; this information can be very useful for Internet providers, organizations that offer services on the web, and businesses that advertise on the web.

In Chapter 1, we used a line to describe the relationship between two quantities that are linearly related. In this chapter, we will discuss how to describe such a relationship by a symbolic statement called an *equation*. We will discuss how to use an equation to sketch a line. We will also describe the steepness of a line and how that steepness is related to how quickly one quantity changes in relation to another, such as how quickly the number of Internet users has increased in the United States over time.

3.1 GRAPHING EQUATIONS OF THE FORM $y = mx + b$

Facing it, always facing it, that's the way to get through. Face it.

—Joseph Conrad

Objectives

▹ For an equation in two variables, know the meaning of *solution, satisfy,* and *solution set*.

▹ Know the meaning of the *graph* of an equation.

▹ Graph equations of the form $y = mx + b$.

▹ Know the meaning of b in equations of the form $y = mx + b$.

▹ Know the Rule of Four for equations.

In this section, we begin to work with equations. An **equation** consists of an equality sign "=" with expressions on both sides. Here are some examples of equations:

$$y = 3x - 5, \qquad 2x - 4y = 8, \qquad x = 5$$

In Sections 1.3 and 1.4, we used lines to model data. It turns out that we can describe any line by an equation.

Solutions, Satisfying Equations, and Solution Sets

Consider the equation $y = x + 4$. Let's find y when $x = 3$:

$$y = x + 4 \qquad \text{Original equation}$$
$$y = 3 + 4 \qquad \text{Substitute 3 for } x.$$
$$= 7 \qquad \text{Add.}$$

So, $y = 7$ when $x = 3$. Recall from Section 1.2 that the ordered-pair notation $(3, 7)$ is shorthand for saying that when $x = 3$, $y = 7$.

For the equation $y = x + 4$, we found that $y = 7$ when $x = 3$. This means that the equation $y = x + 4$ becomes a true statement when we substitute 3 for x and 7 for y:

$$y = x + 4 \qquad \text{Original equation}$$
$$7 \stackrel{?}{=} 3 + 4 \qquad \text{Substitute 3 for x and 7 for y.}$$
$$7 \stackrel{?}{=} 7 \qquad \text{Add.}$$
$$\text{true}$$

We say that $(3, 7)$ is a *solution* of the equation $y = x + 4$ and that $(3, 7)$ *satisfies* the equation $y = x + 4$.

A *set* is a container. Much as an egg carton contains eggs, a *solution set* contains solutions.

> **DEFINITION** *Solution, satisfy,* and *solution set* of an equation in two variables
>
> An ordered pair (a, b) is a **solution** of an equation in terms of x and y if the equation becomes a true statement when a is substituted for x and b is substituted for y. We say that (a, b) **satisfies** the equation. The **solution set** of an equation is the set of all solutions of the equation.

Example 1 Identifying Solutions of an Equation

1. Is $(2, 1)$ a solution of $y = 3x - 5$?
2. Is $(4, 9)$ a solution of $y = 3x - 5$?

Solution

1. We substitute 2 for x and 1 for y in the equation $y = 3x - 5$:

$$y = 3x - 5 \qquad \text{Original equation}$$
$$1 \stackrel{?}{=} 3(2) - 5 \qquad \text{Substitute 2 for x and 1 for y.}$$
$$1 \stackrel{?}{=} 6 - 5 \qquad \text{Multiply before subtracting.}$$
$$1 \stackrel{?}{=} 1 \qquad \text{Subtract.}$$
$$\text{true}$$

So, $(2, 1)$ is a solution of $y = 3x - 5$.

2. We substitute 4 for x and 9 for y in the equation $y = 3x - 5$:

$$y = 3x - 5 \qquad \text{Original equation}$$
$$9 \stackrel{?}{=} 3(4) - 5 \qquad \text{Substitute 4 for x and 9 for y.}$$
$$9 \stackrel{?}{=} 12 - 5 \qquad \text{Multiply before subtracting.}$$
$$9 \stackrel{?}{=} 7 \qquad \text{Subtract.}$$
$$\text{false}$$

So, $(4, 9)$ is *not* a solution of $y = 3x - 5$. ∎

Definition of Graph

Next we will learn how to *graph* an equation. As a first step, we plot some solutions of an equation in the next example.

Example 2 Plotting Some Solutions of an Equation

Find five solutions of $y = 2x - 1$ and plot them in the same coordinate system.

Solution

We begin by arbitrarily choosing the values 0, 1, and 2 to substitute for x:

$$
\begin{array}{lll}
y = 2(0) - 1 & y = 2(1) - 1 & y = 2(2) - 1 \\
\quad = 0 - 1 & \quad = 2 - 1 & \quad = 4 - 1 \\
\quad = -1 & \quad = 1 & \quad = 3 \\
\text{Solution: } (0, -1) & \text{Solution: } (1, 1) & \text{Solution: } (2, 3)
\end{array}
$$

The ordered pairs $(-2, -5)$ and $(-1, -3)$ are also solutions. We organize our findings in Table 2. In Fig. 1, we plot the five solutions.

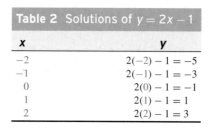

Table 2 Solutions of $y = 2x - 1$	
x	y
-2	$2(-2) - 1 = -5$
-1	$2(-1) - 1 = -3$
0	$2(0) - 1 = -1$
1	$2(1) - 1 = 1$
2	$2(2) - 1 = 3$

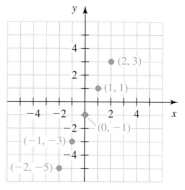

Figure 1 Five solutions of $y = 2x - 1$

In Example 2, we plotted five solutions of $y = 2x - 1$. Note that a line contains these five points (see Fig. 2). It turns out that *every* point on the line represents a solution of the equation $y = 2x - 1$. For example, in Fig. 3 we see that the point $(3, 5)$ lies on the line, and we can show that the ordered pair $(3, 5)$ does satisfy the equation $y = 2x - 1$:

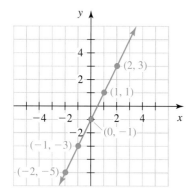

Figure 2 The line contains solutions of $y = 2x - 1$ found in Example 2

$$y = 2x - 1 \qquad \textit{Original equation}$$
$$5 \overset{?}{=} 2(3) - 1 \qquad \textit{Substitute 3 for x and 5 for y.}$$
$$5 \overset{?}{=} 6 - 1 \qquad \textit{Multiply before subtracting.}$$
$$5 \overset{?}{=} 5 \qquad \textit{Add.}$$
$$\text{true}$$

It also turns out that points which do not lie on the line represent ordered pairs that do *not* satisfy the equation. For example, by Fig. 3 we see that the point $(4, 2)$ does not lie on the line, and we can show that the ordered pair $(4, 2)$ does not satisfy the equation $y = 2x - 1$:

$$y = 2x - 1 \qquad \textit{Original equation}$$
$$2 \overset{?}{=} 2(4) - 1 \qquad \textit{Substitute 4 for x and 2 for y.}$$
$$2 \overset{?}{=} 8 - 1 \qquad \textit{Multiply before subtracting.}$$
$$2 \overset{?}{=} 7 \qquad \textit{Subtract.}$$
$$\text{false}$$

We call the line in Fig. 3 the *graph* of the equation $y = 2x - 1$.

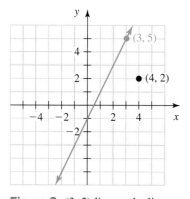

Figure 3 $(3, 5)$ lies on the line, but $(4, 2)$ does not lie on the line

---

DEFINITION Graph

The **graph** of an equation in two variables is the set of points that correspond to all solutions of the equation.

The **graph** of an equation in two variables is a visual description of the solutions of the equation. Every point on the graph represents a solution of the equation. Every point *not* on the graph represents an ordered pair that is *not* a solution.

Graphs of Equations of the Form $y = mx + b$

Directly after Example 2, we found that the graph of the equation $y = 2x - 1$ is a line. The equation $y = 2x - 1$, or $y = 2x + (-1)$, is of the form $y = mx + b$, where $m = 2$ and $b = -1$. It turns out that for any equation of the form $y = mx + b$, where m and b are constants, the graph is a line.

> ### Graph of $y = mx + b$
>
> The graph of an equation of the form $y = mx + b$, where m and b are constants, is a line.

Here are some equations whose graphs are lines:

$$y = 3x + 7, \qquad y = -4x - 5, \qquad y = -4x, \qquad y = x + 3, \qquad y = 2$$

The equation $y = -4x$ is of the form $y = mx + b$ because we can write it as $y = -4x + 0$. The equation $y = 2$ is of the form $y = mx + b$ because we can write it as $y = 0x + 2$.

Example 3   Graphing an Equation

Sketch the graph of $y = -2x + 3$. Also, find the y-intercept.

Solution

Since $y = -2x + 3$ is of the form $y = mx + b$, the graph is a line. Although we can sketch a line from as few as two points, we plot a third point as a check. If the third point is not in line with the other two, then we know that we have computed or plotted at least one of the solutions incorrectly.

To begin, we calculate three solutions of $y = -2x + 3$ in Table 3. Then we plot the three corresponding points and sketch the line through them (see Fig. 4).

We use ZStandard followed by ZSquare to verify our graph (see Fig. 5). See Sections A.3, A.4, and A.6 for graphing calculator instructions.

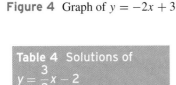

Table 3 Solutions of $y = -2x + 3$

x	y
0	$-2(0) + 3 = 3$
1	$-2(1) + 3 = 1$
2	$-2(2) + 3 = -1$

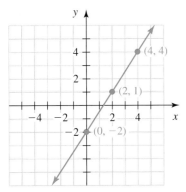

Figure 4 Graph of $y = -2x + 3$

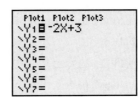

Figure 5 Graph of $y = -2x + 3$

From Table 3 and Fig. 4, we see that the y-intercept is $(0, 3)$.

Example 4   Graphing an Equation

Sketch the graph of $y = \dfrac{3}{2}x - 2$. Also, find the y-intercept.

Solution

In Table 4, we use 0 and multiples of 2 as values of x to avoid fractional values of y. Since $y = \dfrac{3}{2}x - 2$ is of the form $y = mx + b$, the graph is a line. We plot the points that correspond to the solutions we found and sketch the line that contains them in Fig. 6.

We use ZDecimal to verify our graph (see Fig. 7). See Sections A.3, A.4, and A.6 for graphing calculator instructions.

Table 4 Solutions of $y = \dfrac{3}{2}x - 2$

x	y
0	$\dfrac{3}{2}(0) - 2 = -2$
2	$\dfrac{3}{2}(2) - 2 = 1$
4	$\dfrac{3}{2}(4) - 2 = 4$

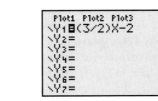

Figure 6 Graph of $y = \dfrac{3}{2}x - 2$

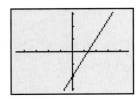

Figure 7 Graph of $y = \dfrac{3}{2}x - 2$

From Table 4 and Fig. 6, we see that the y-intercept is $(0, -2)$.

The Meaning of *b* for an Equation of the Form *y* = *mx* + *b*

For an equation of the form $y = mx + b$, the y-intercept is $(0, b)$. For instance, in Example 3 we found that the line $y = -2x + 3$ has y-intercept $(0, 3)$. In Example 4, we found that the line $y = \frac{3}{2}x - 2$ has y-intercept $(0, -2)$.

Now consider any equation of the form $y = mx + b$. Substituting 0 for x gives

$$y = m(0) + b = 0 + b = b$$

which shows that the y-intercept is $(0, b)$.

---

y-Intercept of the Graph of *y* = *mx* + *b*

The graph of an equation of the form $y = mx + b$ has y-intercept $(0, b)$.

For $y = -5x + 9$, the y-intercept is $(0, 9)$, and for $y = 8x - 4$, the y-intercept is $(0, -4)$.

Rule of Four for Equations

We can describe the solutions of an equation in two variables in four ways. For instance, in Example 4 we described the solutions of the equation $y = \frac{3}{2}x - 2$ by using the equation and a graph (see Fig. 6). We also described some of the solutions by using a table (see Table 4). Finally, we can describe the solutions verbally: For each solution, the y-coordinate is three-halves of the x-coordinate minus 2.

Rule of Four for Solutions of an Equation

We can describe some or all of the solutions of an equation with:

1. an equation,   2. a table,
3. a graph, or   4. words.

These four ways to describe solutions are known as the **Rule of Four.**

Table 5 Solutions of $y = 4x$

x	y
−1	$4(-1) = -4$
0	$4(0) = 0$
1	$4(1) = 4$

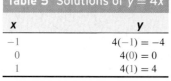

Figure 8 Graph of $y = 4x$

Example 5   Describing Solutions by Using the Rule of Four

1. List some solutions of $y = 4x$ by using a table.
2. Describe the solutions of $y = 4x$ by using a graph.
3. Describe the solutions of $y = 4x$ by using words.

Solution

1. We list three solutions in Table 5.
2. We plot the solutions listed in Table 5 and sketch the line through them (see Fig. 8).
3. For each solution, the y-coordinate is four times the x-coordinate.   ∎

Inputs and Outputs

Recall from Section 1.2 that when we are not describing an authentic situation, we use x as the *independent variable* and y as the *dependent variable*. Recall from Section 1.3 that an *input* is a permitted value of the independent variable that leads to at least one *output*, which is a permitted value of the dependent variable.

In Chapter 1, we used a sketched line to see how an input leads to an output. It will often be easier and more efficient to use an *equation* of a line to perform such a task.

For example, in Fig. 2 we graphed the equation $y = 2x - 1$. From the blue arrows in Fig. 9, we see that the input $x = 3$ leads to the output $y = 5$.

To use the equation, we substitute 3 for x in $y = 2x - 1$:

$$y = 2(3) - 1 = 5$$

This work also shows that the input $x = 3$ leads to the output $y = 5$.

Figure 9 The input $x = 3$ leads to the output $y = 5$

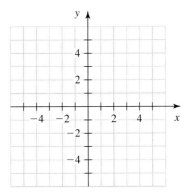

Figure 10 Sketch a graph of $y = x - 3$

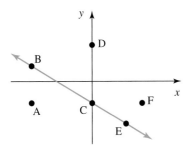

Figure 11 Problem 5

group exploration

Solutions of an equation

Consider the equation $y = x - 3$.

1. Use the coordinate system in Fig. 10 to sketch a graph of $y = x - 3$.
2. Pick three points that lie on the graph of $y = x - 3$. Do the coordinates of these points satisfy the equation $y = x - 3$?
3. Pick three points that do not lie on the graph of $y = x - 3$. Do the coordinates of these points satisfy the equation $y = x - 3$?
4. Which ordered pairs satisfy the equation $y = x - 3$? There are too many to list, but describe them in words. [**Hint:** You should say something about the points that do or do not lie on the line.]
5. The graph of an equation is sketched in Fig. 11. Which of the points A, B, C, D, E, and F represent ordered pairs that satisfy the equation?

TIPS FOR SUCCESS: Use the Graphing Calculator Appendix

Don't forget that Appendix A contains step-by-step graphing calculator instructions for many tools that are helpful for this course. Appendix A includes Section A.23, which describes how to respond to various error messages.

HOMEWORK 3.1   FOR EXTRA HELP ▶

Student Solutions Manual   PH Math/Tutor Center   MathXL®   MyMathLab

---

Which of the given ordered pairs satisfy the given equation?

1. $y = 2x - 4$ $(-3, -10)$, $(1, -3)$, $(2, 0)$
2. $y = 4x - 12$ $(-2, -20)$, $(1, -8)$, $(3, 0)$
3. $y = -3x + 7$ $(-1, 4)$, $(0, 7)$, $(4, -5)$
4. $y = -5x + 8$ $(-2, 3)$, $(0, 8)$, $(3, -7)$

Find the y-intercept. Also, graph the equation by hand. Use a graphing calculator to verify your work.

5. $y = x + 2$
6. $y = x + 4$
7. $y = x - 4$
8. $y = x - 6$
9. $y = 2x$
10. $y = 5x$
11. $y = -3x$
12. $y = -4x$
13. $y = x$
14. $y = -x$
15. $y = \frac{1}{3}x$
16. $y = \frac{1}{2}x$
17. $y = -\frac{5}{3}x$
18. $y = -\frac{3}{2}x$
19. $y = 2x + 1$
20. $y = 3x + 2$
21. $y = 5x - 3$
22. $y = 4x - 1$
23. $y = -3x + 5$
24. $y = -2x + 4$
25. $y = -2x - 3$
26. $y = -4x - 2$
27. $y = \frac{1}{2}x - 3$
28. $y = \frac{1}{3}x - 2$
29. $y = -\frac{2}{3}x + 1$
30. $y = -\frac{4}{3}x + 5$

31. Describe the Rule of Four as applied to the linear equation $y = 2x - 3$:
    a. Describe three solutions of $y = 2x - 3$ by using a table.
    b. Describe the solutions of $y = 2x - 3$ by using a graph.
    c. Describe the solutions of $y = 2x - 3$ by using words.

32. Describe the Rule of Four as applied to the linear equation $y = -4x + 5$:

a. Describe three solutions of $y = -4x + 5$ by using a table.
b. Describe the solutions of $y = -4x + 5$ by using a graph.
c. Describe the solutions of $y = -4x + 5$ by using words.

33. a. For the equation $y = 3x + 1$, find all outputs for the given input. State how many outputs there are for that single input.
    i. the input $x = 2$
    ii. the input $x = 4$
    iii. the input $x = -2$
    b. For $y = 3x + 1$, how many outputs originate from any single input? Explain.
    c. Give an example of an equation of the form $y = mx + b$. Using your equation, find all outputs for the given input. State how many outputs there are for that single input.
    i. the input $x = 3$
    ii. the input $x = 5$
    iii. the input $x = -3$
    d. For your equation, how many outputs originate from any single input? Explain.
    e. For *any* equation of the form $y = mx + b$, how many outputs originate from a single input? Explain.

34. a. Graph $y = 2x - 4$ by hand.
    b. For the equation $y = 2x - 4$, find all outputs for the given input. Explain by using arrows on your graph in part (a). State how many outputs there are for that single input.
    i. the input $x = 3$
    ii. the input $x = 4$
    iii. the input $x = 5$
    c. For $y = 2x - 4$, how many outputs originate from any single input? Explain in terms of drawing arrows.

d. Give an example of an equation of the form $y = mx + b$. Graph your equation by hand.

e. Using your equation, find all outputs for the given input. Explain by using arrows on your graph in part (d). State how many outputs there are for that single input.

   i. the input $x = 1$

   ii. the input $x = 3$

   iii. the input $x = -2$

f. For your equation, how many outputs originate from any single input? Explain in terms of drawing arrows.

g. For *any* equation of the form $y = mx + b$, how many outputs originate from a single input? Explain in terms of drawing arrows.

35. a. Graph the equation by hand. Find all x-intercepts and y-intercepts.

   i. $y = 3x$ **ii.** $y = -2x$ **iii.** $y = \dfrac{2}{5}x$

b. What are the intercepts of the graph of an equation of the form $y = mx$, where $m \neq 0$?

36. Find the intersection point of the lines $y = 4x$ and $y = -5x$. Try to do this without graphing.

37. The graph of an equation is sketched in Fig. 12. Create a table of ordered-pair solutions of this equation. Include at least five ordered pairs.

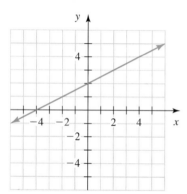

Figure 12 Exercise 37

38. The graph of an equation is sketched in Fig. 13. Create a table of ordered-pair solutions of this equation. Your table should contain at least five ordered pairs.

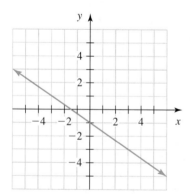

Figure 13 Exercise 38

For Exercises 39–46, refer to the graph sketched in Fig. 14.

39. Find y when $x = -4$.

40. Find y when $x = 0$.

41. Find y when $x = 2$.

42. Find y when $x = -2$.

43. Find x when $y = -1$.

44. Find x when $y = 0$.

45. Find x when $y = 2$.

46. Find x when $y = 3$.

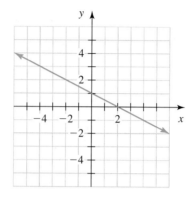

Figure 14
Exercises 39–46

47. The graph of an equation is sketched in Fig. 15. Which of the points A, B, C, D, E, and F represent ordered pairs that satisfy the equation?

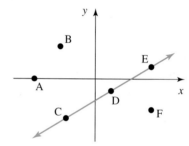

Figure 15 Exercise 47

48. The graphs of equations 1 and 2 are sketched in Fig. 16.

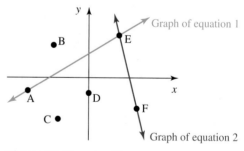

Figure 16 Exercise 48

For each part, decide which one or more of the points A, B, C, D, E, and F represent ordered pairs that

a. satisfy equation 1.

b. satisfy equation 2.

c. satisfy both equations.

d. do not satisfy either equation.

49. Find a solution of $y = x + 2$ that lies in Quadrant II. How many solutions of this equation are in Quadrant II?

50. Find a solution of $y = x - 4$ that lies in Quadrant III. How many solutions of this equation are in Quadrant III?

51. Find an equation of a line that contains the points listed in Table 6. [**Hint:** For each point, what number can be added to the x-coordinate to get the y-coordinate?]

Table 6 Points on a Line (Exercise 51)	
x	**y**
0	3
1	4
2	5
3	6
4	7

52. Find an equation of a line that contains the points listed in Table 7. [**Hint:** For each point, what number can be subtracted from the x-coordinate to get the y-coordinate?]

Table 7 Points on a Line (Exercise 52)	
x	**y**
0	−1
1	0
2	1
3	2
4	3

53. Find an equation of a line that contains the points listed in Table 8.

Table 8 Points on a Line (Exercise 53)	
x	**y**
0	0
1	1
2	2
3	3
4	4

54. Find an equation of a line that contains the points listed in Table 9.

Table 9 Points on a Line (Exercise 54)	
x	**y**
0	0
1	−1
2	−2
3	−3
4	−4

55. The graph of an equation is sketched in Fig. 17.

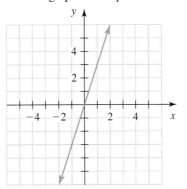

Figure 17 Exercise 55

a. Create a table of ordered-pair solutions of this equation. Your table should contain at least five ordered pairs.

b. Find an equation of the line. [**Hint:** Recognize a pattern from the table you created in part (a).]

56. The graph of an equation is sketched in Fig. 18.

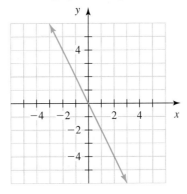

Figure 18 Exercise 56

a. Create a table of ordered-pair solutions of this equation. Your table should contain at least five ordered pairs.

b. Find an equation of the line. [**Hint:** Recognize a pattern from the table you created in part (a).]

57. Graph $x + y = 5$ by hand. [**Hint:** Assume that the graph is a line. Think of pairs of numbers whose sum is 5.]

58. Graph $x - y = 5$ by hand. [**Hint:** Assume that the graph is a line. Think of pairs of numbers whose difference is 5.]

59. Give an example of an equation of the form $y = mx + b$, where m and b are constants. Graph the equation by hand.

60. Describe how to graph an equation of the form $y = mx + b$. (See page 9 for guidelines on writing a good response.)

61. In your own words, describe the Rule of Four for equations. (See page 9 for guidelines on writing a good response.)

3.2 GRAPHING LINEAR MODELS; UNIT ANALYSIS

Objectives

▷ Find an equation of a linear model and make predictions.

▷ Know the Rule of Four for authentic situations.

▷ Perform a unit analysis of a linear model.

▷ Graph equations of the form $y = b$ and $x = a$.

In this section, we will graph equations that describe authentic situations and use such graphs to make predictions. We will find the units of both sides of such equations. We will also graph equations of the form $y = b$ and $x = a$.

Linear Model

In Section 1.3, we defined a linear model as a line that describes two quantities in an authentic situation. An equation of such a line is also called a *linear model*.

Example 1 Using a Linear Model to Make Predictions

A person earns a starting salary of $32 thousand at a company. Each year, he receives a $2 thousand raise. Let s be the person's salary (in thousands of dollars) after he has worked at the company for t years.

1. Use a table to help find an equation for t and s.
2. Substitute 8 for t in the equation. What does the result mean in this situation?
3. Graph the equation.
4. What is the s-intercept? What does it mean in this situation?
5. When will the salary be $42 thousand?

Solution

1. We create Table 10. From the last row of the table, we see that the salary s (in thousands of dollars) can be represented by $2t + 32$. So, $s = 2t + 32$.
2. We substitute 8 for t in the equation $s = 2t + 32$:

$$s = 2(8) + 32 = 48$$

So, the person's salary is $48 thousand after he has worked at the company for 8 years.

3. In Table 11, we substitute values for t in the equation $s = 2t + 32$ to find the corresponding value for s. Then, we plot the points and sketch a line that contains the points (see Fig. 19).

4. The model $s = 2t + 32$ is of the form $s = mt + b$, where $b = 32$. So, the s-intercept is $(0, 32)$. We can also find the s-intercept from Table 11 and Fig. 19. The s-intercept being $(0, 32)$ means that the starting salary is $32 thousand.

5. The red arrows in Fig. 20 show that the output $s = 42$ originates from the input $t = 5$. So, the person's salary will be $42 thousand after he has worked at the company for 5 years.

Table 10 Years at Company and Salaries

Years at Company *t*	Salary (thousands of dollars) *s*
0	$2 \cdot 0 + 32$
1	$2 \cdot 1 + 32$
2	$2 \cdot 2 + 32$
3	$2 \cdot 3 + 32$
4	$2 \cdot 4 + 32$
t	$2 \cdot t + 32$

Table 11 Years Worked and Salaries

t	*s*
0	$2(0) + 32 = 32$
1	$2(1) + 32 = 34$
2	$2(2) + 32 = 36$
3	$2(3) + 32 = 38$
4	$2(4) + 32 = 40$

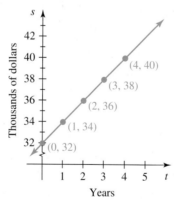

Figure 19 Salary model

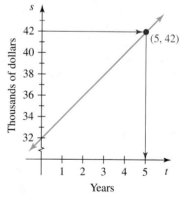

Figure 20 Using the income model to make a prediction

We use a graphing calculator to verify our work (see Fig. 21). See Sections A.4, A.5, and A.7 for graphing calculator instructions.

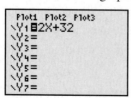
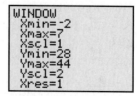
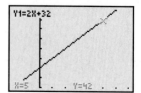

Figure 21 Verify the work

We can add two numbers in either order and get the same result. For example, $2 + 5 = 7$ and $5 + 2 = 7$. We can also multiply two numbers in either order and get the same result. For example, $3 \cdot 8 = 24$ and $8 \cdot 3 = 24$. Depending on how we would have arranged the arithmetic in Table 10, we could have found any one of the following equations:

$$s = 2t + 32 \qquad s = 32 + 2t \qquad s = t(2) + 32 \qquad s = 32 + t(2)$$

All of these equations describe the same relationship between t and s and have the same graph.

Rule of Four for Authentic Situations

Just as we can use equations, tables, graphs, and words to describe solutions of equations, we can use these four ways to describe authentic situations. For instance, in Example 1 we first described an authentic situation by using words:

> A person earns a starting salary of $32 thousand at a company. Each year, he receives a $2 thousand raise.

We also described the situation by using the equation $s = 2t + 32$, a table (see Table 11), and a graph (see Fig. 19).

Rule of Four for Authentic Situations

We can describe an authentic situation with

1. an equation,
2. a table,
3. a graph, or
4. words.

Unit Analysis of a Linear Model

In Example 1, we found the linear model $s = 2t + 32$. We can perform a *unit analysis* of a model by determining the units of the expressions on both sides of the equation:

$$\underbrace{s}_{\text{thousands of dollars}} = \underbrace{2}_{\frac{\text{thousands of dollars}}{\text{year}}} \cdot \underbrace{t}_{\text{years}} + \underbrace{32}_{\text{thousands of dollars}}$$

We can use the fact that $\dfrac{\text{years}}{\text{year}} = 1$ to simplify the units of the expression on the right-hand side of the equation:

$$\frac{\text{thousands of dollars}}{\text{year}} \cdot \text{years} + \frac{\text{thousands}}{\text{of dollars}} = \frac{\text{thousands}}{\text{of dollars}} + \frac{\text{thousands}}{\text{of dollars}}$$

So, the units of the expressions on both sides of the equation are thousands of dollars, which suggests that our equation is correct.

DEFINITION Unit analysis

We perform a **unit analysis** of a model's equation by determining the units of the expressions on both sides of the equation. The units of the expressions on both sides of the equation should be the same.

We can perform a unit analysis of a model's equation to help verify the equation.

Example 2 Using a Linear Model to Make a Prediction

Before starting a diet, a person weighs 160 pounds. While on the diet, she loses 3 pounds per month. Let w be the person's weight (in pounds) after she has been on the diet for t months.

1. Use a table to help find an equation for t and w.
2. Perform a unit analysis of the equation you found in Problem 1.
3. Predict the person's weight after she has been on the diet for 6 months.

Table 12 Numbers of Months and Weights

Time on Diet (months) t	Weight (pounds) w
0	$160 - 3 \cdot 0$
1	$160 - 3 \cdot 1$
2	$160 - 3 \cdot 2$
3	$160 - 3 \cdot 3$
4	$160 - 3 \cdot 4$
t	$160 - 3 \cdot t$

Solution

1. We create Table 12. From the bottom row of Table 12, we see that we can model the situation by using the equation $w = 160 - 3t$.
2. Here is a unit analysis of the equation $w = 160 - 3t$:

$$\underbrace{w}_{\text{pounds}} = \underbrace{160}_{\text{pounds}} - \underbrace{3}_{\dfrac{\text{pounds}}{\text{month}}} \cdot \underbrace{t}_{\text{months}}$$

We can use the fact that $\dfrac{\text{months}}{\text{month}} = 1$ to simplify the units of the expression on the right-hand side of the equation:

$$\text{pounds} - \dfrac{\text{pounds}}{\text{month}} \cdot \text{months} = \text{pounds} - \text{pounds}$$

So, the units on both sides of the equation are pounds, which suggests that our equation is correct.

3. We substitute 6 for t in the equation $w = 160 - 3t$:

$$w = 160 - 3(6) = 142$$

The person will weigh 142 pounds after she has been on the diet for 6 months. ∎

Horizontal Linear Models

We will now model an authentic situation in which one of two quantities does not change.

Example 3   Finding and Graphing a Linear Model

In 1997, the minimum wage was increased to \$5.15. For the period 1997–2006, the minimum wage was constant. At the time of this writing an increase in the minimum wage was under debate for 2007. Let w be the minimum wage (in dollars) at t years since 1997. Find an equation that models the wages for the period 1997–2006. Also, graph the equation.

Solution

For the period 1997–2006, the minimum wage was a constant \$5.15. So, the equation of the model is

$$w = 5.15$$

To graph the equation, we first list some corresponding values of t and w in Table 13. Then we plot the points and sketch a line that contains the points (see Fig. 22).

Table 13 Values of t and w

t (years since 1997)	w (dollars)
0	5.15
1	5.15
2	5.15
3	5.15
4	5.15
5	5.15
6	5.15
7	5.15

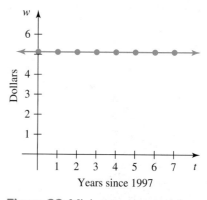

Figure 22 Minimum-wage model

Note that the model is a horizontal line. Recall from Section 1.4 that when a model yields an estimate that is not a good approximation or when it gives a prediction that does not make sense, we say that *model breakdown* has occurred. Model breakdown has occurred in the years before 1997. If the minimum wage does increase in 2007, then model breakdown has also occurred in the years after 2006. ∎

Table 14 Solutions of $y = 3$

x	y
−2	3
−1	3
0	3
1	3
2	3

Horizontal and Vertical Lines

Graphing the horizontal linear model in Example 3 will suggest how to graph another horizontal line in Example 4.

Example 4 Graphing an Equation of the Form $y = b$

Sketch the graph of $y = 3$.

Solution

Note that y must be 3, but x can have any value. Some solutions of $y = 3$ are listed in Table 14. We plot the corresponding points and sketch the line through them (see Fig. 23). The graph of $y = 3$ is a horizontal line.

We can use a graphing calculator to verify our graph (see Fig. 24).

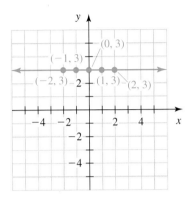

Figure 23 Graph of $y = 3$

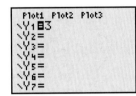 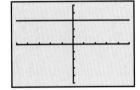

Figure 24 The graph of $y = 3$

In Example 4, we saw that the graph of the equation $y = 3$ is a horizontal line. Any equation that can be written in the form $y = b$, where b is a constant, has a horizontal line as its graph.

Table 15 Solutions of $x = 2$

x	y
2	−2
2	−1
2	0
2	1
2	2

Example 5 Graphing an Equation of the Form $x = a$

Sketch the graph of $x = 2$.

Solution

Note that x must be 2, but y can have any value. Some solutions of $x = 2$ are listed in Table 15. We plot the corresponding points and sketch the line through them (see Fig. 25). The graph of $x = 2$ is a vertical line. ∎

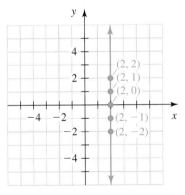

Figure 25 Graph of $x = 2$

In Example 5, we saw that the graph of the equation $x = 2$ is a vertical line. Any equation that can be written in the form $x = a$, where a is a constant, has a vertical line as its graph.

Equations for Horizontal and Vertical Lines

If a and b are constants, then

- The graph of $y = b$ is a horizontal line (see Fig. 26).
- The graph of $x = a$ is a vertical line (see Fig. 27).

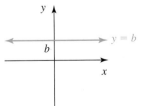 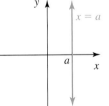

Figure 26 Graph of $y = b$ **Figure 27** Graph of $x = a$

For example, the graphs of the equations $y = 5$ and $y = -2$ are horizontal lines. The graphs of the equations $x = 6$ and $x = -4$ are vertical lines.

> ### Equations Whose Graphs Are Lines
>
> If an equation can be put into the form
>
> $$y = mx + b \quad \text{or} \quad x = a$$
>
> where m, a, and b are constants, then the graph of the equation is a line. We call such an equation a **linear equation in two variables.**

Any equation that can be put into the form $x = a$ has a vertical line as its graph. Any equation that can be put into the form $y = mx + b$ has a nonvertical line as its graph.

group exploration

Looking ahead: Graphical significance of *m* for *y = mx*

1. Use ZDecimal to graph these equations of the form $y = mx$ in order, and describe what you observe:

$$y = x, \qquad y = 2x, \qquad y = 3x, \qquad y = 4x$$

2. Give an example of an equation of the form $y = mx$ whose graph is a line steeper than the lines you sketched in Problem 1.

3. Use ZDecimal to graph these equations in order, and describe what you observe:

$$y = -x, \qquad y = -2x, \qquad y = -3x, \qquad y = -4x$$

4. Describe the graph of $y = mx$ in the following situations:
   a. m is a large positive number
   b. m is a positive number near zero
   c. m is a negative number near zero
   d. m is less than -10
   e. $m = 0$

5. Describe what you have learned in this exploration.

TIPS FOR SUCCESS: Study with a Classmate

It can be helpful to meet with a classmate and discuss what happened in class that day. Not only can you ask questions *of* each other, but you will learn just as much by explaining concepts *to* each other. Explaining a concept to someone else forces you to clarify your own understanding of the concept.

HOMEWORK 3.2

FOR EXTRA HELP ▶

Student Solutions Manual   PH Math/Tutor Center   *Math* XL
MathXL®   **MyMathLab**
MyMathLab

1. A person pays a $3 cover charge to hear a hip-hop band. Let T be the total cost (in dollars) of the cover charge plus d dollars spent on drinks.
   a. Complete Table 16 to help find an equation that describes the relationship between d and T. Show the arithmetic to help you see a pattern.
   b. Perform a unit analysis of the equation you found in part (a).
   c. Graph the equation by hand.
   d. What is the T-intercept of the linear model? What does it mean in this situation?

Table 16 Drink Cost and Total Cost	
Drink Cost (dollars) *d*	Total Cost (dollars) *T*
2	
3	
4	
5	
d	

e. If $10 is spent on drinks, find the total cost of the cover charge and drinks. Explain by using arrows on your graph in part (c).

2. A company offers a $2 mail-in rebate on a shaver. The *retail price* of a shaver is the price paid at the store (not including the $2 rebate). The *net price* is the price of the shaver, taking into account the money saved by the rebate. Let n be the net price (in dollars) of a shaver whose retail price is r dollars.

   a. Complete Table 17 to help find an equation that describes the relationship between r and n. Show the arithmetic to help you see a pattern.

   Table 17 Retail Price and Net Price

Retail Price (in dollars) r	Net Price (in dollars) n
4	
5	
6	
7	
r	

   b. Perform a unit analysis of the equation you found in part (a).

   c. Graph the equation by hand.

   d. If the net price is $6, what is the retail price? Explain by using arrows on your graph in part (c).

3. Chemeketa Community College charged $56 per credit (unit or hour) for tuition in spring semester 2005. Let T be the total cost (in dollars) of tuition for enrolling in c credits of classes.

   a. Complete Table 18 to help find an equation that describes the relationship between c and T. Show the arithmetic to help you see a pattern.

   Table 18 Numbers of Credits and Total Costs

Number of Credits c	Total Cost (in dollars) T
3	
6	
9	
12	
c	

   b. Perform a unit analysis of the equation you found in part (a).

   c. Graph the equation by hand.

   d. What is the total cost of tuition for 15 credits of classes? Explain by using arrows on your graph in part (c).

4. The average gas mileage of a Toyota Camry® is 30 miles per gallon. Let g be the number of gallons used to drive d miles.

   a. Complete Table 19 to help find an equation that describes the relationship between d and g. Show the arithmetic to help you see a pattern.

   b. Perform a unit analysis of the equation you found in part (a).

   c. Graph the equation by hand.

   d. How many gallons are used to travel 75 miles? Explain by using arrows on your graph in part (c).

Table 19 Distances and Gasoline Consumed

Distance (miles) d	Gasoline Consumed (gallons) g
30	
60	
90	
120	
d	

5. A person earns a starting salary of $24 thousand at a company. Each year, she receives a $3 thousand raise. Let s be the salary (in thousands of dollars) after she has worked at the company for t years.

   a. Complete Table 20 to help find an equation for t and s. Show the arithmetic to help you see a pattern.

   Table 20 Years at Company and Salaries

Time at Company (years) t	Salary (thousands of dollars) s
0	
1	
2	
3	
4	
t	

   b. Perform a unit analysis of the equation you found in part (a).

   c. Graph the equation by hand.

   d. What is the s-intercept? What does it mean in this situation?

   e. When will the person's salary be $42 thousand? Explain by using arrows on your graph in part (c).

6. A kitchen appliances company sold 27 thousand ovens in 2000, and sales then increased by 4 thousand ovens each year. Let s be the sales (in thousands of ovens) in the year that is t years since 2000.

   a. Complete Table 21 to help find an equation for t and s. Show the arithmetic to help you see a pattern.

   Table 21 Numbers of Years and Sales

Years since 2000 t	Sales (thousands of ovens) s
0	
1	
2	
3	
4	
t	

   b. Perform a unit analysis of the equation you found in part (a).

   c. Graph the equation by hand.

   d. Find the year when the sales were 51 thousand ovens. Explain by using arrows on your graph in part (c).

7. Just before some bad publicity was released about a company, the company's stock was worth $65. Now that the bad

publicity has been released, the value of the stock has been declining by $5 per week. Let v be the value (in dollars) of the stock t weeks after the publicity was released.

a. Complete Table 22 to help find an equation for t and v. Show the arithmetic to help you see a pattern.

Table 22 Number of Weeks and Stock's Values

Time (weeks) t	Value (dollars) v
0	
1	
2	
3	
4	
t	

b. Perform a unit analysis of the equation you found in part (a).

c. Graph the equation by hand.

d. Find when the stock will be worth $35. Explain by using arrows on your graph in part (c).

8. The percentage of Americans who felt that executing convicted murderers deters others from committing murder was 49% in 1997 and has declined by 2% each year since then (Source: *Harris Interactive*). Let p be the percentage of Americans who feel that executions are a deterrent at t years since 1997.

a. Complete Table 23 to help find an equation for t and p. Show the arithmetic to help you see a pattern.

Table 23 Years and Percents

Years since 1997 t	Percent p
0	
1	
2	
3	
4	
t	

b. Perform a unit analysis of the equation you found in part (a).

c. Graph by hand the equation you found in part (a).

d. What is the p-intercept? What does it mean in this situation?

e. Estimate when 39% of Americans felt that executions were a deterrent. Explain by using arrows on your graph in part (c).

9. To make fudgelike brownies, a person bakes a brownie mix for 5 minutes less than the baking time suggested on the box. Let r be the suggested baking time (in minutes) and a be the actual baking time (in minutes).

a. Find an equation that describes the relationship between r and a. Assume that a is the dependent variable. [**Hint:** If you have trouble finding the equation, create a table of values for r and a.]

b. Perform a unit analysis of the equation you found in part (a).

c. Graph the equation by hand.

d. If the actual baking time is 23 minutes, what is the baking time suggested on the box? Explain by using arrows on your graph in part (c).

10. A person pays $4 for parking at an arts-and-crafts fair. Let T be the total cost (in dollars) of parking plus v dollars spent on a vase.

a. Find an equation that describes the relationship between v and T. [**Hint:** If you have trouble finding the equation, create a table of values for v and T.]

b. Perform a unit analysis of the equation you found in part (a).

c. Graph the equation by hand.

d. If the person spends $25 on the vase, find the total cost of parking and the vase. Explain by using arrows on your graph in part (c).

11. A person drives at a constant speed of 60 miles per hour. Let d be the distance traveled (in miles) after the person has driven for t hours.

a. Find an equation that describes the relationship between t and d. [**Hint:** If you have trouble finding the equation, create a table of values for t and d.]

b. Perform a unit analysis of the equation you found in part (a).

c. Graph the equation by hand.

d. What is the d-intercept? What does it mean in this situation?

e. After how many hours will the person have traveled 150 miles? Explain by using arrows on your graph in part (c).

12. A bicycle shop charges $12 per hour to rent a bicycle. Let C be the total cost (in dollars) of renting a bicycle for t hours.

a. Find an equation that describes the relationship between t and C. [**Hint:** If you have trouble finding the equation, create a table of values for t and C.]

b. Perform a unit analysis of the equation you found in part (a).

c. Graph the equation by hand.

d. If a person paid $18 for renting a bicycle, how long did the person ride the bicycle? Explain by using arrows on your graph in part (c).

13. For spring semester 2006, Cosumnes River College charged $26 per unit (credit or hour) for tuition. All students paid a $1 student representation fee each semester. Students who drove to school paid a $30 parking fee each semester. Let T be the total one-semester cost (in dollars) of tuition and fees for a student who drove to school and took u units of classes.

a. Find an equation for u and T. [**Hint:** If you have trouble finding the equation, try creating a table of values for u and T.]

b. Perform a unit analysis of the equation you found in part (a).

c. What is the total one-semester cost of tuition and fees for a student who drove to school and took 15 units of classes?

14. For spring semester 2005, undergraduates taking 15 hours (units or credits) of courses at Southeastern Louisiana University paid $1516.10 for tuition. Students could rent textbooks for $25 per course. Let T be the total one-semester cost (in dollars) of tuition and renting textbooks for an undergraduate who took n courses for a total of 15 hours.

a. Find an equation for n and T. [**Hint:** If you have trouble finding the equation, try creating a table of values for n and T.]

b. Perform a unit analysis of the equation you found in part (a).

c. What was the total one-semester cost of tuition and renting textbooks for an undergraduate who took five 3-hour courses?

15. A person fills up his car's 11-gallon gasoline tank and then drives at a constant 60 mph. For each hour of driving, the car uses 2 gallons of gasoline. Let g be the amount of gasoline (in gallons) in the tank after the person has driven for t hours.

a. Find an equation for t and g.

b. Perform a unit analysis of the equation you found in part (a).

c. Graph the equation by hand.

d. For how long can the person drive before refueling if he wants to refuel when 1 gallon of gasoline is left in the tank? Explain by using arrows on your graph in part (c).

e. Estimate the amount of gasoline in the tank after 8 hours of driving. Explain by using arrows on your graph in part (c). [**Hint:** Remember, if you think that model breakdown occurs, say so, say where, and explain why.]

16. When a small airplane begins to descend, its altitude is 4 thousand feet. The airplane descends $\frac{1}{2}$ thousand feet each minute. Let h be the altitude (in thousands of feet) of the airplane t minutes after it has begun to descend.

a. Find an equation for t and h.

b. Perform a unit analysis of the equation you found in part (a).

c. Graph the equation by hand.

d. When will the airplane reach an altitude of 1 thousand feet? Explain by using arrows on your graph in part (c).

e. Predict the airplane's altitude 11 minutes after the airplane has begun its descent. Explain by using arrows on your graph in part (c). [**Hint:** Remember, if you think that model breakdown occurs, say so, say where, and explain why.]

17. *Fair-trade coffee* guarantees farmers a minimum fair price for their crops. Although most segments of the U.S. coffee business are stagnant, the sales of fair-trade coffee is widely viewed as the fastest-growing niche. Sales of fair-trade coffee were 4.3 million pounds in 2000 and have increased by about 2.5 million pounds per year since then (Source: *TransFair USA*). Let s be the sales (in millions of pounds) of fair-trade coffee in the year that is t years since 2000. Describe the Rule of Four as applied to this situation:

a. Use a table of values of t and s to describe the situation.

b. Use an equation to describe the situation.

c. Use a graph to describe the situation.

18. U.S. spending on DVDs was $3.3 billion in 2000 and has increased by $3 billion per year since then (Source: *Adams Media Research*). Let s be the spending (in billions of dollars) for the year that is t years since 2000. Describe the Rule of Four as applied to this situation:

a. Use a table of values of t and s to describe the situation.

b. Use an equation to describe the situation.

c. Use a graph to describe the situation.

19. A person has had a constant 45 CDs since 2000. Let n be the number of CDs that the person owns at t years since 2000. Find

an equation that models the number of CDs owned by the person for 2000 and thereafter. Also, graph the equation by hand.

20. A small company has had a constant 35 employees since 2000. Let n be the number of employees at t years since 2000. Find an equation that models the number of employees for 2000 and thereafter. Also, graph the equation by hand.

Graph the equation by hand.

21. $x = 3$ **22.** $x = 6$ **23.** $y = 1$

24. $y = 5$ **25.** $y = -2$ **26.** $y = -4$

27. $x = -1$ **28.** $x = -3$ **29.** $x = 0$

30. $y = 0$

Graph the equation by hand. Use a graphing calculator to verify your work when possible.

31. $y = x - 2$ **32.** $y = x - 4$

33. $y = 2$ **34.** $y = -3$

35. $y = -3x + 1$ **36.** $y = -2x - 2$

37. $y = \frac{3}{5}x$ **38.** $y = \frac{4}{3}x$

39. $y = -\frac{5}{3}x + 1$ **40.** $y = -\frac{3}{2}x + 4$

41. $y = 4x - 3$ **42.** $y = 3x - 5$

43. $x = -4$ **44.** $x = 1$

45. Find an equation of the line sketched in Fig. 28.

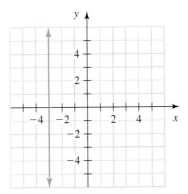

Figure 28 Exercise 45

46. Find an equation of the line sketched in Fig. 29.

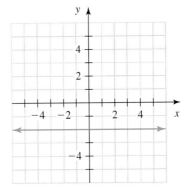

Figure 29 Exercise 46

47. The points $(1, 2)$ and $(5, 2)$ are plotted in Fig. 30.

a. In going from point $(1, 2)$ to point $(5, 2)$, find the change
  i. in the x-coordinate.   **ii.** in the y-coordinate.

b. In going from point $(5, 2)$ to point $(1, 2)$, find the change
  i. in the x-coordinate.   **ii.** in the y-coordinate.

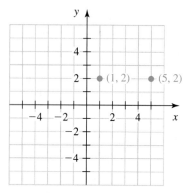

Figure 30   Exercise 47

a. In going from point $(-3, 1)$ to point $(2, 4)$, find the change
   i. in the x-coordinate.   **ii.** in the y-coordinate.
b. In going from point $(2, 4)$ to point $(-3, 1)$, find the change
   i. in the x-coordinate.   **ii.** in the y-coordinate.

50. The points $(1, 4)$ and $(5, -3)$ are plotted in Fig. 33.
  a. In going from point $(1, 4)$ to point $(5, -3)$, find the change
   i. in the x-coordinate.   **ii.** in the y-coordinate.
  b. In going from point $(5, -3)$ to point $(1, 4)$, find the change
   i. in the x-coordinate.   **ii.** in the y-coordinate.

48. The points $(3, -4)$ and $(3, 2)$ are plotted in Fig. 31.
  a. In going from point $(3, -4)$ to point $(3, 2)$, find the change
   i. in the x-coordinate.   **ii.** in the y-coordinate.
  b. In going from point $(3, 2)$ to point $(3, -4)$, find the change
   i. in the x-coordinate.   **ii.** in the y-coordinate.

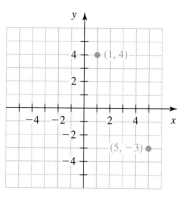

Figure 33   Exercise 50

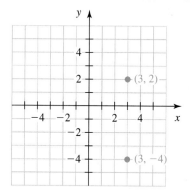

Figure 31   Exercise 48

49. The points $(-3, 1)$ and $(2, 4)$ are plotted in Fig. 32.

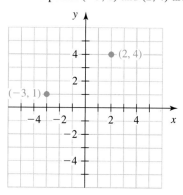

Figure 32   Exercise 49

51. a. Use a graphing calculator to graph the equations $y = x$, $y = 2x$, and $y = 3x$ in the same viewing window.
  b. List the lines $y = x$, $y = 2x$, and $y = 3x$ in order of steepness, from least to greatest.
  c. What pattern does your list in part (b) suggest?
  d. Give an equation of a line that is steeper than the lines $y = x$, $y = 2x$, and $y = 3x$.

52. a. Use a graphing calculator to graph the pair of equations in the same viewing window.
   i. $y = 2x + 1$, $y = 2x + 3$
   ii. $y = 3x - 2$, $y = 3x - 5$
   iii. $y = -2x + 4$, $y = -2x + 3$
  b. Describe what pattern your work in part (a) suggests.
  c. Find an equation whose graph is parallel to the line $y = 4x - 5$.

53. In your own words, describe the Rule of Four for authentic situations.

54. Explain why it makes sense that the graph of an equation of the form $x = a$ is a vertical line and the graph of an equation of the form $y = b$ is a horizontal line.

3.3 SLOPE OF A LINE

Objectives

▷ Use a ratio to compare the steepness of two objects.

▷ Know the meaning of, and how to calculate, the *slope* of a nonvertical line.

▷ Know the sign of the slope of an increasing line and of a decreasing line.

▷ Know that the slope of a horizontal line is zero and that the slope of a vertical line is undefined.

How do we measure the steepness of an object such as a ladder? In this section, we will discuss the *slope* of a line. This key concept has many applications in business, engineering, nursing, surveying, physics, social science, mathematics, and lots of other fields.

Comparing the Steepness of Two Objects

Consider the sketch of two cables (guy wires) running from the ground to a telephone pole in Fig. 34. Which cable is steeper?

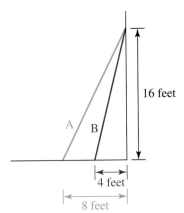

Figure 34 Cable A and cable B running from the ground to a telephone pole

Cable B is steeper than cable A, even though both cables reach a point at the same height on the building. To measure the steepness of each cable, we compare the *vertical* distance from the base of the pole to the top end of the cable with the *horizontal* distance from the bottom end of the cable to the building. In Section 2.5, we used a unit ratio to compare two quantities. Here, we calculate the unit ratio of vertical distance to horizontal distance for cable A:

$$\text{Cable A:} \qquad \frac{\text{vertical distance}}{\text{horizontal distance}} = \frac{16 \text{ feet}}{8 \text{ feet}} = \frac{2}{1}$$

For cable A, the vertical distance is 2 times the horizontal distance.

Now we calculate the unit ratio of vertical distance to horizontal distance for cable B:

$$\text{Cable B:} \qquad \frac{\text{vertical distance}}{\text{horizontal distance}} = \frac{16 \text{ feet}}{4 \text{ feet}} = \frac{4}{1}$$

For cable B, the vertical distance is 4 times the horizontal distance.

These calculations confirm that cable B is steeper than cable A in Fig. 34.

Comparing the Steepness of Two Objects

To compare the steepness of two objects, compute the unit ratio

$$\frac{\text{vertical distance}}{\text{horizontal distance}}$$

for each object. The object with the larger ratio is the steeper object.

Example 1   Comparing the Steepness of Two Objects

A portion of road A climbs steadily for 120 feet over a horizontal distance of 4250 feet. A portion of road B climbs steadily for 95 feet over a horizontal distance of 2875 feet. Which road is steeper? Explain.

Solution

Figure 35 shows sketches of the two roads, but the horizontal distances and vertical distances are not drawn to scale.

Figure 35 Roads A and B

Here, we calculate the unit ratio of the vertical distance to the horizontal distance for each road:

$$\text{Road A:} \qquad \frac{\text{vertical distance}}{\text{horizontal distance}} = \frac{120 \text{ feet}}{4250 \text{ feet}} \approx \frac{0.028}{1}$$

$$\text{Road B:} \qquad \frac{\text{vertical distance}}{\text{horizontal distance}} = \frac{95 \text{ feet}}{2875 \text{ feet}} \approx \frac{0.033}{1}$$

Road B is a little steeper than road A, because road B's ratio of vertical distance to horizontal distance is greater than road A's. ■

The **grade** of a road is the ratio of the vertical distance to the horizontal distance, written as a percentage. Recall from Section 2.5 that to write a decimal number as a percentage, we move the decimal point two places to the right and insert the percent symbol. In Example 1, the grade of road A is about 2.8% and the grade of road B is about 3.3%.

Finding a Line's Slope

To measure the steepness (also called *slope*) of a nonvertical line, we will also use a ratio, but we will work with *changes* in a quantity, which can be negative, rather than with distances, which are always positive.

Consider the line that contains the points $(4, 2)$ and $(6, 5)$ sketched in Fig. 36. To go from point $(4, 2)$ to point $(6, 5)$, we look 2 units to the right and then look 3 units up. So, the horizontal change, called the *run*, is 2 and the vertical change, called the *rise*, is 3. The slope of the line is the ratio of the rise to the run:

$$\text{slope} = \frac{\text{vertical change}}{\text{horizontal change}} = \frac{\text{rise}}{\text{run}} = \frac{3}{2}$$

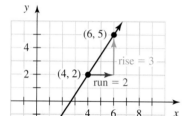

Figure 36 The slope is $\dfrac{3}{2}$

In general, for any line, the **rise** is the vertical change and the **run** is the horizontal change in going from one point on the line to another point on the line. We use the letter m to represent the slope.

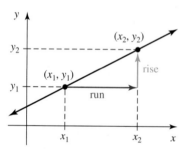

Figure 37 The rise and run between any two points

DEFINITION Slope of a Nonvertical Line

$$m = \text{slope} = \frac{\text{vertical change}}{\text{horizontal change}} = \frac{\text{rise}}{\text{run}}$$

In words: The **slope** of a nonvertical line is equal to the ratio of the rise to the run (in going from one point on the line to another point on the line). See Fig. 37.

Example 2 Finding the Slope of a Line

Find the slope of the line that contains the points $(3, 5)$ and $(6, 1)$.

Solution

We begin by plotting the points $(3, 5)$ and $(6, 1)$ and sketching the line that contains them (see Fig. 38). To go from point $(3, 5)$ to point $(6, 1)$, we must look 3 units to the right and then 4 units down. So, the run is 3 and the rise is -4. The slope of the line is the ratio of the rise to the run:

$$\text{slope} = \frac{\text{vertical change}}{\text{horizontal change}} = \frac{\text{rise}}{\text{run}} = \frac{-4}{3} = -\frac{4}{3} \qquad ■$$

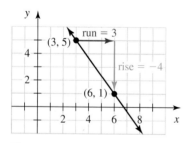

Figure 38 The slope is $\dfrac{-4}{3} = -\dfrac{4}{3}$

In Example 2, we found that, from point $(3, 5)$ to point $(6, 1)$, the run is 3. We can calculate this run by computing the change in the x-coordinates. Recall from Section 2.4 that we can find the change in a quantity by computing the ending amount minus the beginning amount:

$$\text{Change in } x\text{-coordinates} = \frac{x\text{-coordinate}}{\text{of ending point}} - \frac{x\text{-coordinate}}{\text{of beginning point}} = 6 - 3 = 3$$

In Example 2, we also found that, from point $(3, 5)$ to point $(6, 1)$, the rise is -4. We can calculate this rise by computing the change in the y-coordinates:

$$\text{Change in } y\text{-coordinates} = \begin{array}{c} y\text{-coordinate} \\ \text{of ending point} \end{array} - \begin{array}{c} y\text{-coordinate} \\ \text{of beginning point} \end{array} = 1 - 5 = -4$$

How do we calculate the slope of *any* nonvertical line if we are given two points on the line? First, we use the subscript 1 to identify x_1 and y_1 as the coordinates of the first point, (x_1, y_1). Likewise, we identify x_2 and y_2 as the coordinates of the second point, (x_2, y_2). When we look from point (x_1, y_1) to point (x_2, y_2), the run is the difference $x_2 - x_1$ and the rise is the difference $y_2 - y_1$ (see Fig. 39).

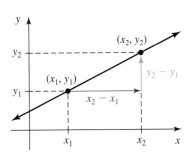

Figure 39 The slope of the line is $m = \dfrac{y_2 - y_1}{x_2 - x_1}$

Calculating Slope

Let (x_1, y_1) and (x_2, y_2) be two distinct points of a nonvertical line. The slope of the line is

$$m = \frac{\text{vertical change}}{\text{horizontal change}} = \frac{y_2 - y_1}{x_2 - x_1}$$

(see Fig. 39).

A *formula* is an equation that contains two or more variables. We will refer to the equation $m = \dfrac{y_2 - y_1}{x_2 - x_1}$ as the **slope formula**.

Example 3   Finding the Slope of a Line

Find the slope of the line that contains the points $(2, 3)$ and $(8, 5)$.

Solution

Using the slope formula with $(x_1, y_1) = (2, 3)$ and $(x_2, y_2) = (8, 5)$, we have

$$m = \frac{y_2 - y_1}{x_2 - x_1} = \frac{5 - 3}{8 - 2} = \frac{2}{6} = \frac{1}{3}$$

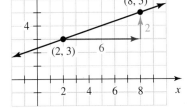

Figure 40 The slope is $\dfrac{2}{6} = \dfrac{1}{3}$

By plotting points, we find that if the run is 6, then the rise is 2 (see Fig. 40). So, the slope is $m = \dfrac{\text{rise}}{\text{run}} = \dfrac{2}{6} = \dfrac{1}{3}$, which is the same as our result from using the formula. ∎

In Example 3, we calculated the slope of a line for $(x_1, y_1) = (2, 3)$ and $(x_2, y_2) = (8, 5)$. Here, we switch the roles of the two points to find the slope when $(x_1, y_1) = (8, 5)$ and $(x_2, y_2) = (2, 3)$:

$$m = \frac{y_2 - y_1}{x_2 - x_1} = \frac{3 - 5}{2 - 8} = \frac{-2}{-6} = \frac{1}{3}$$

The result is the same as that in Example 3. In general, when we use the slope formula with two points on a line, it doesn't matter which point we choose to be first, (x_1, y_1), and which point we choose to be second, (x_2, y_2).

WARNING It is a common error to make incorrect substitutions into the slope formula. Carefully consider why the middle and right-hand formulas are incorrect:

Correct	Incorrect	Incorrect
$m = \dfrac{y_2 - y_1}{x_2 - x_1}$	$m = \dfrac{y_2 - y_1}{x_1 - x_2}$	$m = \dfrac{x_2 - x_1}{y_2 - y_1}$

Example 4 Finding the Slope of a Line

Find the slope of the line that contains the points $(3, 4)$ and $(7, 2)$.

Solution

Using the formula for slope with $(x_1, y_1) = (3, 4)$ and $(x_2, y_2) = (7, 2)$, we have

$$m = \frac{y_2 - y_1}{x_2 - x_1} = \frac{2 - 4}{7 - 3} = \frac{-2}{4} = -\frac{1}{2}$$

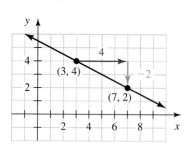

Figure 41 The slope is $\dfrac{-2}{4} = -\dfrac{1}{2}$

By plotting points, we find that when the run is 4, the rise is -2 (see Fig. 41). So, the slope is $\dfrac{-2}{4} = -\dfrac{1}{2}$, which is the same as our result from using the formula. ∎

Increasing and Decreasing Lines

Since the line in Fig. 42 goes upward from left to right, we say that the line (and the graph) is **increasing.** A sign analysis of the rise and run in Fig. 42 shows that the slope of the line is positive.

Since the line in Fig. 43 goes downward from left to right, we say that the line (and the graph) is **decreasing.** A sign analysis of the rise and run in Fig. 43 shows that the slope of the line is negative.

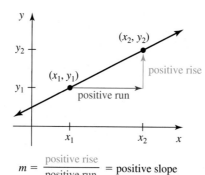

$$m = \frac{\text{positive rise}}{\text{positive run}} = \text{positive slope}$$

Figure 42 Increasing lines have positive slope

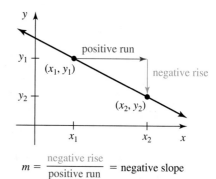

$$m = \frac{\text{negative rise}}{\text{positive run}} = \text{negative slope}$$

Figure 43 Decreasing lines have negative slope

Slopes of Increasing or Decreasing Lines

- An increasing line has positive slope (see Fig. 42).
- A decreasing line has negative slope (see Fig. 43).

When we compute the slope of two points that have negative coordinates, it can help to write first

$$\frac{(\ ) - (\ )}{(\ ) - (\ )}$$

and then insert the coordinates of the two points into the appropriate parentheses.

Example 5   Finding the Slope of a Line

Find the slope of the line that contains the points $(-5, 2)$ and $(3, -4)$.

Solution

$$m = \frac{(-4) - (2)}{(3) - (-5)} = \frac{-4 - 2}{3 + 5} = \frac{-6}{8} = -\frac{6}{8} = -\frac{3}{4}$$

Since the slope is negative, the line is decreasing. ∎

Example 6   Finding the Slope of a Line

Find the approximate slope of the line that contains the points $(-4.9, -3.5)$ and $(-2.3, 5.8)$. Round the result to the second decimal place.

Solution

$$m = \frac{(5.8) - (-3.5)}{(-2.3) - (-4.9)} = \frac{9.3}{2.6} \approx 3.58$$

So, the slope is approximately 3.58. Since the slope is positive, the line is increasing. ∎

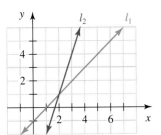

Figure 44 Find the slopes of the two lines

Example 7   Comparing the Slopes of Two Lines

Find the slopes of the two lines sketched in Fig. 44. Which line has the greater slope? Explain why this makes sense in terms of the steepness of a line.

Solution

In Fig. 45, we see that for line l_1, if the run is 1, the rise is 1. We calculate the slope of line l_1:

$$\text{Slope of line } l_1 = \frac{\text{rise}}{\text{run}} = \frac{1}{1} = 1$$

In Fig. 46, we see that for line l_2, if the run is 1, the rise is 3. We calculate the slope of line l_2:

$$\text{Slope of line } l_2 = \frac{\text{rise}}{\text{run}} = \frac{3}{1} = 3$$

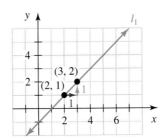

Figure 45 Line with lesser slope

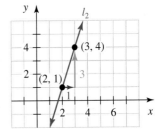

Figure 46 Line with greater slope

Note that the slope of line l_2 is greater than the slope of line l_1, which is what we would expect because line l_2 is steeper than line l_1. ∎

In general, **for two nonparallel increasing lines, the steeper line has the greater slope.**

We show three decreasing lines and three increasing lines and their slopes in Fig. 47.

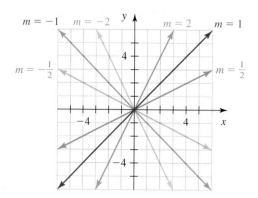

Figure 47 Slopes of some lines

Horizontal and Vertical Lines

So far, we have discussed slopes of lines that are increasing or decreasing. What about the slope of a horizontal line or the slope of a vertical line?

Example 8 Finding the Slope of a Horizontal Line

Find the slope of the line that contains the points (3, 2) and (7, 2).

Solution

We plot the points (3, 2) and (7, 2) and sketch the line that contains the points (see Fig. 48).

The formula for slope gives

$$m = \frac{2-2}{7-3} = \frac{0}{4} = 0$$

So, the slope of this horizontal line is zero. ∎

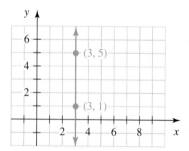

Figure 48 The horizontal line has zero slope

In Example 8, we saw that the horizontal line in Fig. 48 has a slope equal to zero. Note that *any* horizontal line has no rise, so

$$\text{Slope of a horizontal line} = \frac{\text{rise}}{\text{run}} = \frac{0}{\text{run}} = 0$$

The slope of any horizontal line is zero.

Example 9 Finding the Slope of a Vertical Line

Find the slope of the line that contains the points (3, 1) and (3, 5).

Solution

We plot the points (3, 1) and (3, 5) and sketch the line that contains the points (see Fig. 49).

The formula for slope gives

$$m = \frac{5-1}{3-3} = \frac{4}{0}$$

Since division by zero is undefined, the slope of the vertical line is *undefined*. ∎

Figure 49 The vertical line has undefined slope

In Example 9, we saw that the vertical line in Fig. 49 has undefined slope. Note that *any* vertical line has zero run, so

$$\text{Slope of a vertical line} = \frac{\text{rise}}{\text{run}} = \left.\frac{\text{rise}}{0}\right\} \text{undefined}$$

The slope of any vertical line is undefined.

Warning: Undefined slope

Slopes of Horizontal and Vertical Lines

- A horizontal line has a slope equal to zero (see Fig. 50).
- A vertical line has undefined slope (see Fig. 51).

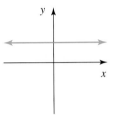

Figure 50 Horizontal lines have a slope equal to zero

Figure 51 Vertical lines have undefined slope

group exploration

Looking ahead: The meaning of *m* in the equation $y = mx + b$

1. **a.** Carefully sketch a graph of the line $y = 2x - 1$.
   b. Using the formula $m = \dfrac{\text{rise}}{\text{run}}$, find the slope of the line you sketched.
   c. What number is multiplied by x in the equation $y = 2x - 1$? How does it compare with the slope you found in part (b)?

2. **a.** Carefully sketch a graph of the line $y = -3x + 5$.
   b. Using the formula $m = \dfrac{\text{rise}}{\text{run}}$, find the slope of the line you sketched.
   c. What number is multiplied by x in the equation $y = -3x + 5$? How does it compare with the slope you found in part (b)?

3. Describe what you have learned in this exploration so far.

4. Without graphing, determine the slope of each line.
   a. $y = 4x - 7$
   b. $y = -2x + 4$
   c. $y = \dfrac{2}{5}x - 3$
   d. $y = x - 2$
   e. $y = 3$

TIPS FOR SUCCESS: Make Changes

If you have not had passing scores on tests and quizzes during the first part of this course, it is time to evaluate what the problem is, what changes you should make, and whether you can commit to making those changes.

Sometimes students must change how they study for a course. For example, Rosie did poorly on exams and quizzes for the first third of the course. It was not clear why she was not passing the course, since she had good attendance, was actively involved in classroom work, and was doing the homework assignments. Suddenly, Rosie started getting A's on every quiz and test. What had happened? Rosie said, "I figured out that, to do well in this course, it was not enough for me to do just the exercises you assigned. So now I do a lot of extra exercises from each section."

1. A portion of road A climbs steadily for 210 feet over a horizontal distance of 3500 feet. A portion of road B climbs steadily for 275 feet over a horizontal distance of 5000 feet. Which road is steeper? Explain.

2. While taking off, airplane A climbs steadily for 4800 feet over a horizontal distance of 9000 feet. Airplane B climbs steadily for 5800 feet over a horizontal distance of 10,500 feet. Which plane is climbing at a greater incline? Explain.

3. Ski run A declines steadily for 80 yards over a horizontal distance of 400 yards. Ski run B declines steadily for 90 yards over a horizontal distance of 600 yards. Which ski run is steeper? Explain.

4. Ski run A declines steadily for 90 yards over a horizontal distance of 500 yards. Ski run B declines steadily for 130 yards over a horizontal distance of 650 yards. Which ski run is steeper? Explain.

Plot the two given points and then sketch the line that contains the points. Find the run and rise in going from the first point listed to the second point listed. Find the slope of the line.

5. (2, 3) and (4, 6)
6. (1, 4) and (6, 5)
7. (3, 6) and (5, 2)
8. (2, 5) and (6, 3)
9. (−4, 1) and (2, 5)
10. (3, −4) and (5, 2)
11. (−4, −2) and (−2, −6)
12. (−3, −2) and (−1, −4)

Use the slope formula to find the slope of the line that passes through the two given points. State whether the line is increasing, decreasing, horizontal, or vertical.

13. (1, 5) and (3, 9)
14. (2, 3) and (5, 12)
15. (3, 10) and (5, 2)
16. (5, 8) and (7, 2)
17. (2, 1) and (8, 4)
18. (3, 2) and (7, 4)
19. (2, 5) and (8, 3)
20. (1, 7) and (9, 5)
21. (−2, 4) and (3, −1)
22. (−3, 4) and (1, −2)
23. (5, −2) and (9, −4)
24. (2, −3) and (8, −6)
25. (−7, −1) and (−2, 9)
26. (−6, −8) and (−4, 2)
27. (−6, −9) and (−2, −3)
28. (−5, −2) and (−1, −3)
29. (6, −1) and (−4, 7)
30. (4, −5) and (−2, 10)
31. (−2, −11) and (7, −5)
32. (−3, −1) and (9, −3)
33. (0, 0) and (4, −2)
34. (−6, −9) and (0, 0)
35. (3, 5) and (7, 5)
36. (−4, −6) and (3, −6)
37. (−3, −1) and (−3, −2)
38. (4, 2) and (4, 7)

Find the approximate slope of the line that contains the two given points. Round your result to the second decimal place. State whether the line is increasing, decreasing, horizontal, or vertical.

39. (−3.2, 5.1) and (−2.8, 1.4)

40. (−1.9, 4.8) and (−3.1, 5.5)
41. (4.9, −2.7) and (6.3, −1.1)
42. (9.7, −6.8) and (4.5, −2.7)
43. (−4.97, −3.25) and (−9.64, −2.27)
44. (−3.22, −8.54) and (−7.29, −6.13)
45. (−2.45, −6.71) and (4.88, −1.53)
46. (−3.99, −2.49) and (1.06, −3.76)

47. Find the slope of the line sketched in Fig. 52.

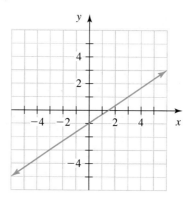

Figure 52 Exercise 47

48. Find the slope of the line sketched in Fig. 53.

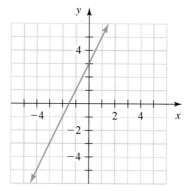

Figure 53 Exercise 48

49. Find the slope of the line sketched in Fig. 54.

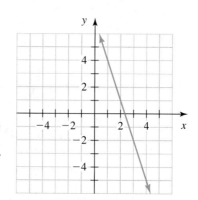

Figure 54 Exercise 49

50. Find the slope of the line sketched in Fig. 55.

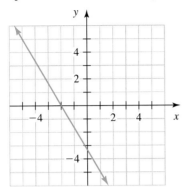

Figure 55 Exercise 50

51. For each line sketched in Fig. 56, determine whether the line's slope is positive, negative, zero, or undefined.

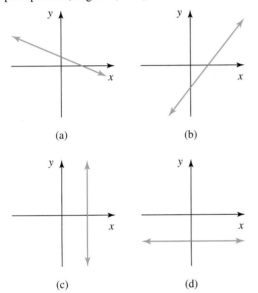

(a)   (b)

(c)   (d)

Figure 56 Exercise 51

52. Find the slope of the line with x-intercept $(-3, 0)$ and y-intercept $(0, 4)$.

Sketch a line that meets the given description. Find the slope of the line.

53. An increasing line that is nearly horizontal

54. A decreasing line that is nearly horizontal

55. A decreasing line that is nearly vertical

56. An increasing line that is nearly vertical

Sketch a line that meets the given description.

57. The slope is a large positive number.

58. The slope is a positive number near zero.

59. The slope is a negative number near zero.

60. The slope is less than -5.

61. A student tries to find the slope of the line that contains the points $(1, 3)$ and $(4, 7)$:

$$\frac{4-1}{7-3} = \frac{3}{4}$$

Describe any errors. Then find the slope correctly.

62. A student tries to find the slope of the line that contains the points $(6, 4)$ and $(3, 9)$:

$$\frac{9-4}{6-3} = \frac{5}{3}$$

Describe any errors. Then find the slope correctly.

63. A student tries to find the slope of the line that contains the points $(-1, -5)$ and $(3, 8)$:

$$\frac{8-5}{3-1} = \frac{3}{2}$$

Describe any errors. Then find the slope correctly.

64. A student tries to find the slope of the line that contains the points $(2, 5)$ and $(6, 7)$:

$$\frac{7-5}{6-2} = \frac{2}{4}$$

Describe any errors. Then find the slope correctly.

65. Sketch a line with a slope of 2 and another line with a slope of 3. Does the steeper line have the greater slope?

66. Sketch a line with a slope of 1 and another line with a slope of $\frac{1}{2}$. Does the steeper line have the greater slope?

67. a. Sketch a line with a slope of -2 and another line with a slope of -3.
b. Does the steeper line have the greater slope?
c. Find the absolute value of the slope of each line. Does the steeper line have the greater absolute value of the slope?
d. Explain why the absolute value of the slope of a line is useful for comparing the steepness of lines.

68. a. Sketch a line with a slope of -2 and another line with a slope of 2.
b. Is the line with a slope of 2 steeper than the line with a slope of -2?
c. Find the absolute value of the slope of each line. Compare the results.
d. Explain why the absolute value of the slope of a line is useful for comparing the steepness of lines.

69. A line contains the points $(2, 1)$ and $(3, 4)$. Find three more points that lie on the line. [**Hint:** Find the slope of the line. Then use the slope and a point on the line to help you find other points on the line.]

70. A line contains the points $(1, 5)$ and $(4, 3)$. Find three more points that lie on the line. [**Hint:** See Exercise 69.]

71. Explain why the slope of a vertical line is undefined.

72. Explain why the slope of a horizontal line is 0.

73. Explain why the slope of an increasing line is positive.

74. Explain why the slope of a decreasing line is negative.

75. Both ladder A and ladder B are leaning against a building. Ladder A reaches a higher point on the building than ladder B does. Can we conclude that ladder A is steeper than ladder B? Explain why this situation suggests that we must take into account both rise and run when we measure the steepness of a line (or a ladder). Draw sketches of rises and runs of lines to illustrate your point.

76. Explain how to find the slope of a line.

USING THE SLOPE TO GRAPH
3.4 LINEAR EQUATIONS

Objectives

▹ Know the meaning of m for an equation of the form $y = mx + b$.

▹ Graph an equation of the form $y = mx + b$ by using the line's slope and y-intercept.

▹ Graph an equation of a linear model by using the model's slope and y-intercept.

▹ Find an equation of a line from its graph.

▹ Know the relationship between slopes of parallel lines.

▹ Know the relationship between slopes of perpendicular lines.

In Sections 3.1 and 3.2, we graphed a linear equation in two variables by first finding solutions of the equation. In this section, we will discuss a more efficient way to graph such equations.

Using the Slope and the y-Intercept to Sketch a Line

We can use the slope and the y-intercept of a line to sketch the line.

Example 1   Sketching a Line

Sketch the line that has slope $m = -\dfrac{2}{5}$ and y-intercept $(0, 3)$.

Solution

We first plot the y-intercept, $(0, 3)$. The slope is $-\dfrac{2}{5} = \dfrac{-2}{5} = \dfrac{\text{rise}}{\text{run}}$. From $(0, 3)$, we count 5 units to the right and 2 units down, where we plot the point $(5, 1)$. See Fig. 57. We then sketch the line that contains these two points (see Fig. 58).

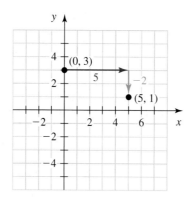

Figure 57 Plot $(0, 3)$. Then count 5 units to the right and 2 units down to plot $(5, 1)$

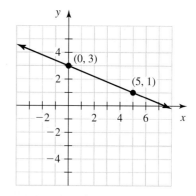

Figure 58 Sketch the line containing $(0, 3)$ and $(5, 1)$

Because the slope is negative, we can verify our work by checking that the line is decreasing. ∎

The Meaning of m for $y = mx + b$

In Example 1, we saw that if we know the slope and the y-intercept of a line, we can sketch that line. Next, we discuss how to determine the slope and y-intercept of a line from the line's equation.

Table 24 Solutions of $y = 2x + 1$

x	y
0	$2(0) + 1 = 1$
1	$2(1) + 1 = 3$
2	$2(2) + 1 = 5$

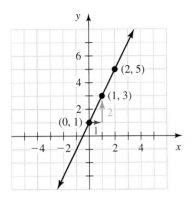

Figure 59 Graph of $y = 2x + 1$

Example 2 Finding the Slope of a Line

Find the slope of the line $y = 2x + 1$.

Solution

We list some solutions in Table 24 and sketch the graph of the equation in Fig. 59.
    If the run is 1, the rise is 2 (see Fig. 59). So, the slope is

$$m = \frac{\text{rise}}{\text{run}} = \frac{2}{1} = 2$$ ∎

In Example 2, we found that the line $y = 2x + 1$ has slope 2. Note that 2 is also the number multiplied by x in the equation $y = 2x + 1$. This observation suggests a general property about a linear equation of the form $y = mx + b$.

Finding the Slope and y-Intercept from a Linear Equation

For a linear equation of the form $y = mx + b$,

- the slope of the line is m and
- the y-intercept of the line is $(0, b)$.

We say that this equation is in **slope–intercept form.**

For example, the equation $y = -4x + 5$ is in slope–intercept form with $m = -4$ and $b = 5$. The graph of this equation is a line with slope -4 and y-intercept $(0, 5)$. The line $y = 8x - 2$ has slope 8 and y-intercept $(0, -2)$.

Graphing Equations of the Form $y = mx + b$

In Example 3, we will graph an equation in slope–intercept form.

Example 3 Graphing an Equation

Sketch the graph of $y = 3x - 4$.

Solution

Note that the y-intercept is $(0, -4)$ and the slope is $3 = \frac{3}{1} = \frac{\text{rise}}{\text{run}}$. To graph the line:

1. Plot the y-intercept $(0, -4)$.
2. From $(0, -4)$, count 1 unit to the right and 3 units up to plot a second point, which we see by inspection is $(1, -1)$. See Fig. 60.
3. Sketch the line that contains these two points (see Fig. 61).

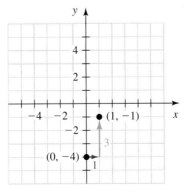

Figure 60 Plot $(0, -4)$. Then count 1 unit to the right and 3 units up to plot $(1, -1)$

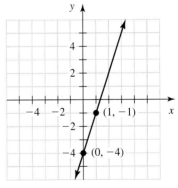

Figure 61 Sketch the line containing $(0, -4)$ and $(1, -1)$

Recall from Section 3.1 that every point on the graph of an equation represents a solution of the equation. We can verify our result by checking that both $(0, -4)$ and $(1, -1)$ are solutions of $y = 3x - 4$:

Check that $(0, -4)$ is a solution

$$y = 3x - 4$$
$$-4 \stackrel{?}{=} 3(0) - 4$$
$$-4 \stackrel{?}{=} 0 - 4$$
$$-4 \stackrel{?}{=} -4$$
true

Check that $(1, -1)$ is a solution

$$y = 3x - 4$$
$$-1 \stackrel{?}{=} 3(1) - 4$$
$$-1 \stackrel{?}{=} 3 - 4$$
$$-1 \stackrel{?}{=} -1$$
true

We check two ordered pairs (rather than just one) because two points determine a line.

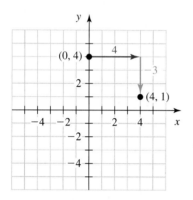

Figure 62 Plot $(0, 4)$. Then count 4 units to the right and 3 units down to plot $(4, 1)$

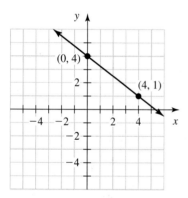

Figure 63 Sketch the line containing $(0, 4)$ and $(4, 1)$

Graphing an Equation in Slope-Intercept Form

To graph an equation of the form $y = mx + b$,

1. Plot the y-intercept $(0, b)$.
2. Use $m = \dfrac{\text{rise}}{\text{run}}$ to plot a second point. For example, if $m = \dfrac{2}{3}$, then count 3 units to the right (from the y-intercept) and 2 units up to plot another point.
3. Sketch the line that passes through the two plotted points.

Example 4   Graphing an Equation

Sketch the graph of $y = -\dfrac{3}{4}x + 4$.

Solution

The y-intercept is $(0, 4)$, and the slope is $-\dfrac{3}{4} = \dfrac{-3}{4} = \dfrac{\text{rise}}{\text{run}}$. To graph the line:

1. Plot the y-intercept $(0, 4)$.
2. From $(0, 4)$, count 4 units to the right and 3 units down to plot a second point, which we see by inspection is $(4, 1)$. See Fig. 62.
3. Sketch the line that contains these two points (see Fig. 63).

We use a graphing calculator to verify our work (see Fig. 64).

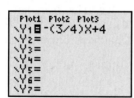 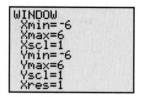 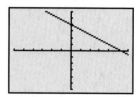

Figure 64 Verify the work

We now know two methods for graphing a linear equation: We can first find solutions of the equation (as discussed in Sections 3.1 and 3.2), or we can first find the slope and y-intercept (as discussed in this section).

Graphing an Equation of a Linear Model

We can use either method to graph an equation of a model.

Example 5   Graphing a Model's Equation

The percentages of people who use tax preparers, rather than calculating their own taxes, are shown in Table 25 for various years.

Table 25 Percentages of People Who Use Tax Preparers

Year	Percent
1980	38.0
1985	45.9
1990	47.9
1995	49.9
2000	57.5
2004	61.0

Source: *National Taxpayers Union*

Little white lies...$25
Twists of truths...$50
Big fat lies...$100

As you can see, our tax preparation rates are quite competitive.

Let p be the percentage of people who use tax preparers at t years since 1980. A reasonable model is

$$p = 0.9t + 39$$

1. Graph the model.
2. Predict when 66% (almost two-thirds) of people will use tax preparers.

Solution

1. The p-intercept is $(0, 39)$, and the slope is $0.9 = \dfrac{0.9}{1} = \dfrac{\text{rise}}{\text{run}}$. It will be easier to graph the model if we multiply the slope by $\dfrac{10}{10} = 1$ so that the rise and run are larger and both are integers:

$$0.9 = \frac{0.9}{1} \cdot \frac{10}{10} = \frac{9}{10} = \frac{\text{rise}}{\text{run}}$$

To graph the model, we first plot the p-intercept $(0, 39)$. From $(0, 39)$, we count 10 units to the right and 9 units up, where we plot the point $(10, 48)$. See Fig. 65. We then sketch the line that contains the two points.

Instead of using the slope to find the point $(10, 48)$, we could have substituted 10 for t in the equation $p = 0.9t + 39$ and solved for p:

$$p = 0.9(10) + 39 = 48$$

So, the point $(10, 48)$ is a point on the linear model.

2. The red arrows in Fig. 65 show that the output $p = 66$ originates from the input $t = 30$. So, 66% of people will use tax preparers in $1980 + 30 = 2010$, according to the model.

To use a graphing calculator to verify our work in Problems 1 and 2, we press WINDOW and set Xmin to be -10 (for 1970), set Xmax to be 40 (for 2020), and use ZoomFit to set the values for Ymin and Ymax automatically (see Fig. 66). Then we use TRACE to check that $(30, 66)$ is a point on the linear model. See Sections A.5, A.6, and A.7 for graphing calculator instructions.

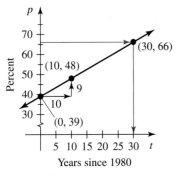

Figure 65 Tax preparer model

 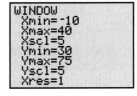 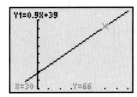

Figure 66 Verify the work

Figure 67 A decreasing line with y-intercept above the origin

Example 6 Interpreting the Signs of *m* and *b*

Graph an equation of the form $y = mx + b$ where m is negative and b is positive.

Solution

We should sketch a decreasing line, because the slope m is negative. We should draw a line whose y-intercept, $(0, b)$, is above the origin, because b is positive. We sketch such a line in Fig. 67.

There is nothing special about the line in Fig. 67—any line that is decreasing and has its y-intercept above the origin will do.

Finding an Equation of a Line

Given the slope and y-intercept of a nonvertical line, we can find an equation of the line.

Example 7 Finding an Equation of a Line from Its Slope and *y*-Intercept

Find an equation of the line that has slope $\frac{2}{3}$ and *y*-intercept $(0, -5)$.

Solution

To find an equation, we substitute $\frac{2}{3}$ for m and -5 for b in the equation $y = mx + b$:

$$y = \frac{2}{3}x + (-5)$$
$$y = \frac{2}{3}x - 5$$

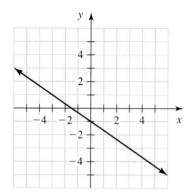

Figure 68 Graph of a line

Given an equation in slope–intercept form, $y = mx + b$, we can find the graph of the equation. We can also go backwards: Given the graph, we can find an equation for it.

Example 8 Finding an Equation of a Line from Its Graph

Find an equation of the line sketched in Fig. 68.

Solution

From Fig. 69, we see that the *y*-intercept of the line is $(0, -1)$. We also see that if the run is 3, then the rise is -2. So, the slope is $\frac{-2}{3} = -\frac{2}{3}$. By substituting $-\frac{2}{3}$ for m and -1 for b in the equation $y = mx + b$, we have $y = -\frac{2}{3}x - 1$.

We can verify our equation by checking that both $(0, -1)$ and $(3, -3)$ satisfy the equation. Or we can use a graphing calculator to verify our work (see Fig. 70).

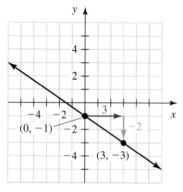

Figure 69 Finding the slope of the line

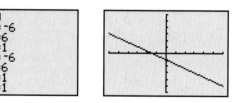

Figure 70 Verify the work

Finding an Equation of a Line from a Graph

To find an equation of a line from a graph,

1. Determine the slope m and the *y*-intercept $(0, b)$ from the graph.
2. Substitute your values for m and b into the equation $y = mx + b$.

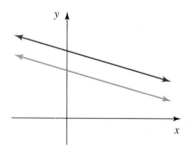

Figure 71 Two parallel lines

Parallel Lines

Two lines are called **parallel** if they do not intersect (see Fig. 71). How do the slopes of two parallel lines compare?

Example 9 Find the Slopes of Two Parallel Lines

Find the slopes of the parallel lines l_1 and l_2 sketched in Fig. 72.

Solution

For both lines, if the run is 2, the rise is 1 (see Fig. 73). So, the slope of both parallel lines is

$$m = \frac{\text{rise}}{\text{run}} = \frac{1}{2}$$

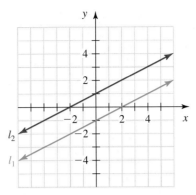

Figure 72 Two parallel lines

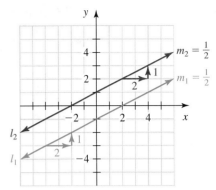

Figure 73 Calculate the slopes of the parallel lines

It makes sense that parallel nonvertical lines have equal slope, since parallel lines have the same steepness.

Slopes of Parallel Lines

If lines l_1 and l_2 are parallel nonvertical lines, then the slopes of the lines are equal:
$$m_1 = m_2$$
Also, if two distinct lines have equal slope, then the lines are parallel.

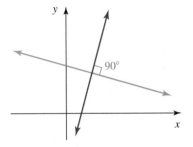

Figure 74 Two perpendicular lines

Perpendicular Lines

Two lines are called **perpendicular** if they intersect at a 90° angle (see Fig. 74). How do the slopes of two perpendicular lines compare?

Example 10   Find the Slopes of Two Perpendicular Lines

Find the slopes of the perpendicular lines l_1 and l_2 sketched in Fig. 75.

Solution

From Fig. 76 we see that the slope of line l_1 is
$$m_1 = \frac{2}{3}$$
and the slope of line l_2 is
$$m_2 = \frac{-3}{2} = -\frac{3}{2}$$

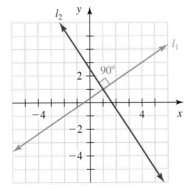

Figure 75 Two perpendicular lines

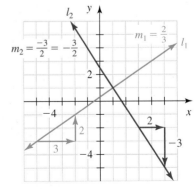

Figure 76 Calculate the slopes of the perpendicular lines

In Example 10, the slope $-\frac{3}{2}$ is the opposite of the reciprocal of the slope $\frac{2}{3}$. A similar relationship applies to any pair of perpendicular nonvertical lines.

Slopes of Perpendicular Lines

If lines l_1 and l_2 are perpendicular nonvertical lines, then the slope of one line is the opposite of the reciprocal of the slope of the other line:

$$m_2 = -\frac{1}{m_1}$$

Also, if the slope of one line is the opposite of the reciprocal of another line's slope, then the lines are perpendicular.

Example 11   Identifying Parallel Lines and Perpendicular Lines

Determine whether the pair of lines is parallel, perpendicular, or neither.

1. $y = 3x - 5$ and $y = 3x + 2$
2. $y = \dfrac{8}{5}x - 3$ and $y = -\dfrac{5}{8}x + 6$

Solution

1. The slope of each line is 3. Since the slopes of the lines are equal, the lines are parallel. We use ZStandard followed by ZSquare to verify this in Fig. 77.

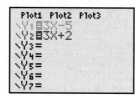 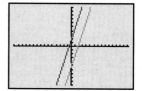

Figure 77 The lines appear to be parallel

2. The slopes of the lines are $\dfrac{8}{5}$ and $-\dfrac{5}{8}$. Since the slope $-\dfrac{5}{8}$ is the opposite of the reciprocal of the slope $\dfrac{8}{5}$, the lines are perpendicular. We use ZStandard followed by ZSquare to verify this in Fig. 78.

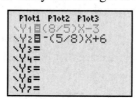 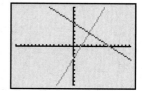

Figure 78 The lines appear to be perpendicular ∎

group exploration

Finding an equation of a line from a graph

In this exploration, you will need to work with one other student.

1. First, both you and your partner should create a linear equation of the form $y = mx + b$, but do not show each other your equations. For m, use a rational number between -3 and 3. For b, use an integer between -3 and 3.

2. Use ZDecimal to graph the equation.

3. Exchange graphing calculators.

4. By inspecting the graph, determine the equation that your partner created. You can use TRACE to find coordinates of a point on your line. Or you can use a grid by pressing ⬛2nd⬛ [FORMAT]. Press ⬛▽⬛ twice, then press ⬛▷⬛, and then press ⬛ENTER⬛. The "GridOn" choice should now be highlighted. Next, press ⬛GRAPH⬛. A grid should now be visible.

5. Press ⬛Y=⬛ to check the equation you found.

TIPS FOR SUCCESS: Visualize

Virtually nothing on Earth can stop a person with a positive attitude who has his goal clearly in sight.

—*Denis Waitley*

To help prepare themselves mentally and physically for competition, many exceptional athletes visualize themselves performing well at their event many times throughout their training period. For example, a runner training for the 100-meter dash might imagine getting set in the starting blocks, taking off right after the gun goes off, being in front of the other runners, and so on, right up until the moment of breaking the tape at the finish line.

In an experiment, three groups of basketball players were used to test the effectiveness of visualization. The first group warmed up by shooting baskets before a game. The second group visualized shooting baskets, but did not shoot any baskets during the pregame warm-up. The third group did not warm up or visualize before the game. The visualization group not only outperformed the group that did not warm up, but also did better than the group that warmed up by shooting baskets!

You can do visualizations, too. Visualize doing all the things you feel you need to do to succeed in the course. If you do this regularly, you will have better follow-through with what you intend to do. You will also feel more confident about succeeding.

HOMEWORK 3.4 — FOR EXTRA HELP ▶

Student Solutions Manual · PH Math/Tutor Center · Math XL · MathXL® · MyMathLab · MyMathLab

Sketch the line that has the given slope and contains the given point.

1. $m = \frac{2}{3}$, $(0, 1)$ **2.** $m = \frac{3}{5}$, $(0, 2)$

3. $m = -\frac{5}{2}$, $(0, 4)$ **4.** $m = -\frac{3}{4}$, $(0, -2)$

5. $m = -\frac{3}{2}$, $(0, 0)$ **6.** $m = -\frac{1}{2}$, $(0, 0)$

7. $m = 2$, $(0, 1)$ **8.** $m = 4$, $(0, -3)$

9. $m = -3$, $(0, -2)$ **10.** $m = -5$, $(0, 4)$

11. $m = -1$, $(0, 3)$ **12.** $m = 1$, $(0 - 2)$

13. $m = 0$, $(4, -5)$ **14.** $m = 0$, $(6, 3)$

15. m is undefined, $(2, -1)$ **16.** m is undefined, $(-1, -3)$

Determine the slope and the y-intercept. Then use the slope and the y-intercept to graph the equation by hand. Verify your work with a graphing calculator.

17. $y = \frac{2}{3}x - 1$ **18.** $y = \frac{1}{5}x + 2$

19. $y = -\frac{1}{3}x + 4$ **20.** $y = -\frac{3}{2}x - 1$

21. $y = \frac{4}{3}x + 2$ **22.** $y = \frac{5}{2}x - 3$

23. $y = -\frac{4}{5}x - 1$ **24.** $y = -\frac{1}{4}x + 2$

25. $y = \frac{1}{2}x$ **26.** $y = \frac{1}{4}x$

27. $y = -\frac{5}{3}x$ **28.** $y = -\frac{4}{5}x$

Determine the slope and the y-intercept. Then use the slope and the y-intercept to graph the equation by hand. Verify your work by checking that two points on your line satisfy the given equation.

29. $y = 4x - 2$ **30.** $y = 2x - 4$

31. $y = -2x + 4$ **32.** $y = -3x + 5$

33. $y = -4x - 1$ **34.** $y = -2x - 2$

35. $y = x + 1$ **36.** $y = x - 4$

37. $y = -x + 3$ **38.** $y = -x + 2$

39. $y = -3x$ **40.** $y = 4x$

41. $y = x$ **42.** $y = -x$

43. $y = -3$ **44.** $y = -2$

45. $y = 0$ **46.** $y = 1$

47. The percentages of degrees in medicine that are earned by women are shown in Table 26 for various years.

Table 26 Percentage of Degrees in Medicine Earned by Women	
Year	**Percent**
1985	30.4
1990	34.2
1995	38.8
2000	42.7
2003	45.3

Source: *U.S. National Center for Education Statistics*

Let p be the percentage of degrees in medicine that are earned by women at t years since 1985. A reasonable model of this situation is

$$p = 0.8t + 30.3$$

a. Graph the model by hand.
b. Predict when half of the degrees in medicine will be earned by women. Explain by using arrows on your graph in part (a).

48. The average monthly employee contribution required to cover a family in an employer-sponsored health plan has increased greatly since 2000 (see Table 27).

Table 27 Costs to Cover a Family in an Employer-Sponsored Health Plan

Year	Average Monthly Employee Contribution (dollars)
2000	135
2001	150
2002	175
2003	202
2004	222

Source: *Kaiser Family Foundation*

Let a be the average monthly employee contribution (in dollars) required to cover a family in an employer-sponsored health plan at t years since 2000. A reasonable model of the situation is

$$a = 23t + 132$$

a. Graph the model by hand.
b. Use your graph to predict when the average monthly employee contribution will be $350. Explain by using arrows on your graph in part (a).

49. J. D. Power's Initial Quality Study surveys about 60,000 car buyers 90 days after they purchase a new car. Participants grade their car or truck on 135 attributes, such as trim quality, wind noise, and fuel economy. The average number of problems incurred per 100 vehicles is shown in Table 28 for various years.

Table 28 Average Number of Vehicle Problems per 100 Vehicles

Year	Average Number of Vehicle Problems per 100 Vehicles
1998	176
1999	166
2000	155
2001	148
2002	133
2003	133
2004	119
2005	118

Source: *J. D. Power & Associates*

Let n be the average number of problems incurred per 100 vehicles at t years since 1995. A reasonable model of the situation is

$$n = -8.6t + 199.4$$

a. Graph the model by hand.
b. In 2005, the brand Lexus® had the fewest problems, 81 per 100 cars. Predict when the average number of problems for all vehicles (of any brand) will reach that level. Explain by using arrows on your graph in part (a).

50. The percentage of travelers visiting another country who visit the United States has declined since 1992 (see Table 29).

Table 29 Percentages of Travelers Visiting Another Country Who Visit the United States

Year	Percent
1992	9.4
1994	8.0
1996	7.8
1998	7.4
2000	7.4
2002	6.0
2004	5.8

Source: *Travel History Association of America*

Let p be the percentage of travelers visiting another country who visit the United States at t years since 1990. A reasonable model of this situation is

$$p = -0.3t + 9.6$$

a. Graph the model by hand.
b. Predict when 3.8% of travelers visiting another country will visit the United States. Explain by using arrows on your graph in part (a).

51. Graphs of four linear equations are shown in Fig. 79. State whether m and b are positive, negative, zero, or undefined for the $y = mx + b$ form of each equation.

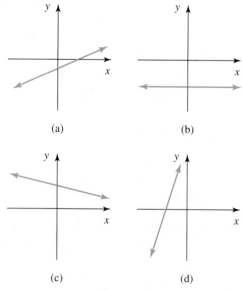

Figure 79 Exercise 51

52. Graphs of four linear equations are shown in Fig. 80. State the signs of the constants m and b for the $y = mx + b$ form of each equation.

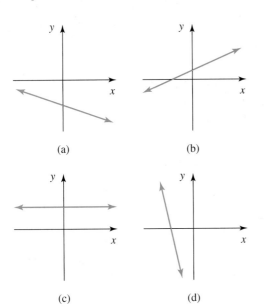

(a)   (b)

(c)   (d)

Figure 80 Exercise 52

Graph by hand an equation of the form $y = mx + b$ that meets the given criteria for m and b. Also, find an equation of each graph.

53. m is positive and b is positive

54. m is positive and b is negative

55. m is negative and b is negative

56. m is negative and $b = 0$

57. $m = 0$ and b is negative

58. $m = 0$ and $b = 0$

Find an equation of a line that has the given slope and contains the given point.

59. $m = 3$, $(0, -4)$ **60.** $m = -2$, $(0, 5)$

61. $m = -\dfrac{6}{5}$, $(0, 3)$ **62.** $m = \dfrac{3}{4}$, $(0, -2)$

63. $m = -\dfrac{2}{7}$, $(0, 0)$ **64.** $m = 0$, $(0, -1)$

65. Find an equation of the line sketched in Fig. 81.

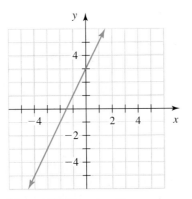

Figure 81 Exercise 65

66. Find an equation of the line sketched in Fig. 82.

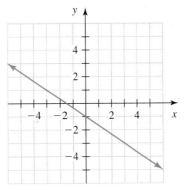

Figure 82 Exercise 66

Determine whether the pair of lines is parallel, perpendicular, or neither. Explain.

67. $y = \dfrac{2}{5}x + 1$ and $y = -\dfrac{5}{2}x - 3$

68. $y = \dfrac{3}{7}x + 5$ and $y = -\dfrac{3}{7}x - 6$

69. $y = 3x - 1$ and $y = -3x + 2$

70. $y = 2x - 1$ and $y = 2x + 6$

71. $y = -4x + 2$ and $y = -4x + 3$

72. $y = -7x + 5$ and $y = \dfrac{1}{7}x - 8$

73. $y = \dfrac{2}{3}x - 1$ and $y = \dfrac{3}{2}x + 3$

74. $y = -\dfrac{4}{11}x + 2$ and $y = \dfrac{11}{4}x + 5$

75. $y = 2$ and $y = -4$ **76.** $x = -2$ and $y = 1$

77. $x = 0$ and $y = 0$ **78.** $x = -2$ and $x = 4$

79. Are the lines sketched in Fig. 83 parallel? Explain.

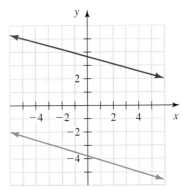

Figure 83 Exercise 79

80. Are the lines sketched in Fig. 84 perpendicular? Explain.

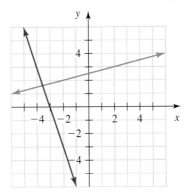

Figure 84 Exercise 80

81. A student says that the slope of the line $y = 2x + 1$ is $2x$. Is the student correct? Explain.

82. A student says that the y-intercept of the line $y = 3x - 2$ is $(0, 2)$. Is the student correct? Explain.

83. Use the slope and y-intercept of the line $y = 2x - 1$ to graph the equation $y = 2x - 1$ by hand. Then choose two points that lie on the graph, and show that both of the corresponding ordered pairs are solutions of the equation $y = 2x - 1$.

84. Use the slope and y-intercept of the line $y = -2x + 3$ to graph the equation $y = -2x + 3$ by hand. Then choose two points that lie on the graph, and show that both of the corresponding ordered pairs are solutions of the equation $y = -2x + 3$.

85. Describe the Rule of Four as applied to the equation $y = \frac{1}{2}x + 2$:

   a. Describe the solutions of $y = \frac{1}{2}x + 2$ by using a graph.

   b. Describe three solutions of $y = \frac{1}{2}x + 2$ by using a table.

   c. Describe the solutions of $y = \frac{1}{2}x + 2$ by using words.

86. Describe the Rule of Four as applied to the equation $y = \frac{1}{3}x - 1$:

   a. Describe the solutions of $y = \frac{1}{3}x - 1$ by using a graph.

   b. Describe three solutions of $y = \frac{1}{3}x - 1$ by using a table.

   c. Describe the solutions of $y = \frac{1}{3}x - 1$ by using words.

87. A student tries to graph the equation $y = -3x + 1$ (see Fig. 85).

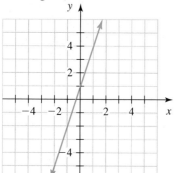

Figure 85 Exercise 87

Choose two points that lie on the line, and show that at least one of these points is not a solution of $y = -3x + 1$. Explain why your work shows that Fig. 85 is incorrect. Then sketch the correct graph.

88. A student tries to graph the equation $y = \frac{3}{2}x - 1$ (see Fig. 86).

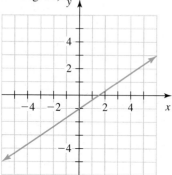

Figure 86 Exercise 88

Choose two points that lie on the line, and show that at least one of these points is not a solution of $y = \frac{3}{2}x - 1$. Explain why your work shows that Fig. 86 is incorrect. Then sketch the correct graph.

For Exercises 89–94, refer to Fig. 87.

89. Find y when $x = -3$.

90. Find x when $y = -3$.

91. Find x when $y = 0$.

92. Find y when $x = 0$.

93. Find the slope of the line.

94. Find an equation of the line.

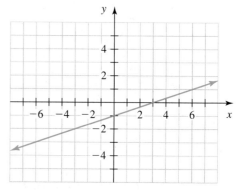

Figure 87 Exercises 89–94

95. A line passes through the point $(-1, 1)$ and has a slope of 2.
   a. Sketch the line.
   b. Find an equation of the line. [**Hint:** Use part (a) to find b in $y = mx + b$.]

96. A line passes through the point $(1, 2)$ and has a slope of -3.
   a. Sketch the line.
   b. Find an equation of the line. [**Hint:** Use part (a) to find b in $y = mx + b$.]

97. a. Find the slope of each line: $y = 3$, $y = -5$, and $y = 0$.
   b. Find the slope of the graph of any linear equation of the form $y = k$, where k is a constant.

98. a. Find the slope of each line: $x = 2$, $x = -4$, and $x = 0$.
   b. Find the slope of the graph of any equation of the form $x = k$, where k is a constant.

99. Graphs of the equations $y = mx + b$ and $y = kx + c$ (where m, b, k, and c are constants) are sketched in Fig. 88.

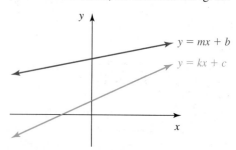

Figure 88 Exercise 99

   a. Which is greater, m or k? Explain.
   b. Which is greater, b or c? Explain.

100. Describe how to use the method discussed in this section to graph an equation of the form $y = mx + b$.

3.5 RATE OF CHANGE

Man's mind stretched to a new idea never goes back to its original dimensions.
—Oliver Wendell Holmes

Objectives

▷ Calculate the rate of change of a quantity.
▷ Understand why slope is a rate of change.
▷ Use the rate of change to help find a linear model.
▷ Know the slope addition property.

In this section, we describe how quickly a quantity changes in relation to another quantity. For example, we can describe how quickly the tuition of a college increases in relation to time or how quickly the altitude of an airplane declines in relation to time.

Calculating Rate of Change

Suppose that the temperature increased *steadily* by 6°F in the past 3 hours. We can compute how much the temperature changed *per hour* by finding the unit ratio of the change in temperature (6°F) to the change in time (3 hours):

$$\frac{6°F}{3\text{ hours}} = \frac{2°F}{1\text{ hour}}$$

So, the temperature increased by 2°F per hour. This is an example of a *rate of change*. We say that the rate of change of temperature with respect to time is 2°F per hour. The rate of change is a *constant* because the temperature increased *steadily*.

Here are some other examples of rates of change:

- A student's CD collection increases by 4 CDs per month.
- The revenue of a company decreases by $3 million per year.
- A college charges $70 per unit (hour or credit).

Example 1   Finding Rates of Change

1. An airplane's altitude increases steadily by 10,000 feet over a 5-minute period. Find the rate of change of altitude with respect to time.
2. The temperature increased steadily from 60°F at 6 A.M. to 72°F at 10 A.M. Find the rate of change of temperature with respect to time.

Solution

1. We find the unit ratio of the change in altitude (10,000 feet) to the change in time (5 minutes):

$$\frac{\text{change in altitude}}{\text{change in time}} = \frac{10,000\text{ feet}}{5\text{ minutes}} \qquad \textit{Find ratio.}$$

$$= \frac{2000\text{ feet}}{1\text{ minute}} \qquad \textit{Find unit ratio.}$$

So, the airplane climbed at a rate of 2000 feet per minute.

2. We find the unit ratio of the change in temperature to the change in time:

$$\frac{\text{change in temperature}}{\text{change in time}} = \frac{72°F - 60°F}{10:00 - 6:00} \qquad \textit{Change in a quantity is ending amount minus beginning amount.}$$

$$= \frac{12°F}{4\text{ hours}} \qquad \textit{Subtract.}$$

$$= \frac{3°F}{1\text{ hour}} \qquad \textit{Find unit ratio.}$$

So, the temperature increased 3°F per hour. ∎

Suppose that the temperature increases by 3°F in one hour, but by 5°F in the next hour. We can find the *average rate of change* of temperature with respect to time by finding the unit ratio of the *total* change in temperature (8°F) to the *total* change in time (2 hours):

$$\frac{8°F}{2 \text{ hours}} = \frac{4°F}{1 \text{ hour}}$$

So, the average rate of change is 4°F per hour.

Formula for Rate of Change and Average Rate of Change

Suppose that a quantity y changes steadily from y_1 to y_2 as a quantity x changes steadily from x_1 to x_2. Then the **rate of change** of y with respect to x is the ratio of the change in y to the change in x:

$$\frac{\text{change in } y}{\text{change in } x} = \frac{y_2 - y_1}{x_2 - x_1}$$

If either quantity does not change steadily, then the preceding formula is the **average rate of change** of y with respect to x.

Example 2 Finding Rates of Change

1. The number of pedestrian fatalities in the United States declined approximately steadily from 5321 fatalities in 1997 to 4763 fatalities in 2000 (Source: *U.S. National Highway Traffic Safety Administration*). Find the average rate of change of the number of pedestrian fatalities per year between 1997 and 2000.
2. The total cost of 12 karate classes and an enrollment fee is $158. The total cost of 20 karate classes and the same enrollment fee is $230. The charge per class is the same, regardless of the number of classes for which you pay. Find the rate of change of the total cost with respect to the number of classes.

Solution

1.

$$\frac{\text{change in number of fatalities}}{\text{change in time}}$$

$$= \frac{4763 \text{ fatalities} - 5321 \text{ fatalities}}{\text{year } 2000 - \text{year } 1997} \qquad \text{Change in a quantity is ending}$$
$$\text{amount minus beginning amount.}$$

$$= \frac{-558 \text{ fatalities}}{3 \text{ years}} \qquad \text{Subtract.}$$

$$= \frac{-186 \text{ fatalities}}{1 \text{ year}} \qquad \text{Find unit ratio.}$$

The average rate of change of the number of pedestrian fatalities was -186 fatalities per year. The number of fatalities had an average yearly decline of 186 fatalities.

2. To be consistent in finding the signs of the changes, we assume that the number of classes increases from 12 to 20 and that the total cost increases from $158 to $230:

$$\frac{\text{change in total cost}}{\text{change in number of classes}} = \frac{230 \text{ dollars} - 158 \text{ dollars}}{20 \text{ classes} - 12 \text{ classes}} \qquad \begin{array}{l}\text{Change in a quantity is} \\ \text{ending amount minus} \\ \text{beginning amount.}\end{array}$$

$$= \frac{72 \text{ dollars}}{8 \text{ classes}} \qquad \text{Subtract.}$$

$$= \frac{9 \text{ dollars}}{1 \text{ class}} \qquad \text{Find unit ratio.}$$

The rate of change of the total cost with respect to the number of classes is $9 per class. So, the cost of each class is $9. ■

From our work in Example 2, we can see a connection between the sign of a rate of change and whether a quantity is increasing or decreasing. In Problem 2, the rate of change was *positive* because the total cost *increases* (as the number of classes increases). In Problem 1, the average rate of change was *negative* because the number of pedestrian fatalities was *decreasing* (as time increases).

> **Increasing and Decreasing Quantities**
>
> Suppose that a quantity p depends on a quantity t. Then
>
> - If p increases steadily as t increases steadily, then the rate of change of p with respect to t is positive.
> - If p decreases steadily as t increases steadily, then the rate of change of p with respect to t is negative.

Slope Is a Rate of Change

You may have noticed that the expression

$$\frac{y_2 - y_1}{x_2 - x_1}$$

we have been using to calculate rate of change is the same expression we use to calculate the slope of a line. In other words, the slope of a linear model is a rate of change. We will explore this important concept in Example 3.

Example 3   Comparing Slope with a Rate of Change

Table 30 Times and Distances

Time (hours) t	Distance (miles) d
0	0
1	50
2	100
3	150
4	200
5	250

Suppose that a student travels at a constant rate on a road trip. Let d be the distance (in miles) that the student can drive in t hours. Some values of t and d are shown in Table 30.

1. Create a scattergram. Then draw a linear model.
2. Find the slope of the linear model.
3. Find the rate of change of the distance per hour in each given period. Compare each result with the slope of the linear model.
   a. From $t = 3$ to $t = 4$ b. From $t = 0$ to $t = 5$

Solution

1. We draw a scattergram and then draw a line that contains the data points (see Fig. 89).

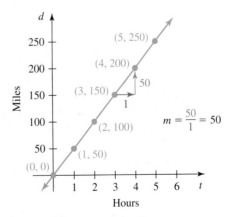

Figure 89 Car model and scattergram

2. Recall from Section 3.3 that the formula for slope is $m = \dfrac{y_2 - y_1}{x_2 - x_1}$. So, with the variables t and d, we have

$$\frac{d_2 - d_1}{t_2 - t_1}$$

We arbitrarily use the points $(3, 150)$ and $(4, 200)$ to calculate the slope of the linear model:

$$\frac{200 - 150}{4 - 3} = \frac{50}{1} = 50$$

So, the slope is 50. This checks with the calculation shown in Fig. 89.

3. a. First we calculate the rate of change of distance with respect to time from $t = 3$ to $t = 4$:

$$\frac{\text{change in distance}}{\text{change in time}} = \frac{200 \text{ miles} - 150 \text{ miles}}{4 \text{ hours} - 3 \text{ hours}}$$

Change in a quantity is ending amount minus beginning amount.

$$= \frac{50 \text{ miles}}{1 \text{ hour}}$$

Subtract.

$$= 50 \text{ miles per hour}$$

Divide.

The rate of change (50 miles per hour) is equal to the slope (50).

b. Now we calculate the rate of change of distance with respect to time from $t = 0$ to $t = 5$:

$$\frac{\text{change in distance}}{\text{change in time}} = \frac{250 \text{ miles} - 0 \text{ miles}}{5 \text{ hours} - 0 \text{ hours}}$$

Change in a quantity is ending amount minus beginning amount.

$$= \frac{250 \text{ miles}}{5 \text{ hours}}$$

Subtract.

$$= \frac{50 \text{ miles}}{1 \text{ hour}}$$

Find unit ratio.

$$= 50 \text{ miles per hour}$$

Divide.

The rate of change (50 miles per hour) is equal to the slope (50). ■

In Example 3, we found that the time t and the distance d are linearly related. We also found that the slope of the linear model is equal to the rate of change of distance with respect to time.

Slope Is a Rate of Change

If there is a linear relationship between the quantities t and p, and if p depends on t, then the slope of the linear model is equal to the rate of change of p with respect to t.

In Problem 3 of Example 3, we calculated the same rate of change (50 miles per hour) for two different periods. In fact, the rate of change is 50 miles per hour for *any* period within the first five hours. This makes sense because the rate of change is equal to the slope of the line (50), which is a constant.

Constant Rate of Change

Suppose that a quantity p depends on a quantity t. Then

- If there is a linear relationship between t and p, then the rate of change of p with respect to t is constant.
- If the rate of change of p with respect to t is constant, then there is a linear relationship between t and p.

Finding an Equation of a Linear Model

We can use what we have learned about rate of change to help us find an equation of a linear model.

Example 4 Finding a Model

In 2000, a college's enrollment is 20 thousand students. Each year, the enrollment increases by 2 thousand students. Let E be the enrollment (in thousands of students) at t years since 2000.

1. Is there a linear relationship between t and E? Explain.
2. Find the E-intercept of a linear model. What does it mean in this situation?
3. Find the slope of the linear model. What does it mean in this situation?
4. Find an equation of the linear model.

Solution

1. Since the rate of change of enrollment per year is a *constant* 2 thousand students per year, the variables t and E are linearly related.
2. We list some values of t and E in Table 31. We plot the corresponding points and sketch the line that contains the points in Fig. 90.

Table 31 College Enrollments	
Years since 2000 t	Enrollment (thousands of students) E
0	20
1	22
2	24
3	26
4	28
5	30

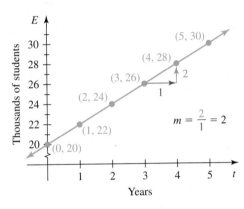

Figure 90 Enrollment scattergram and model

From the table and the graph, we see that the E-intercept is $(0, 20)$. This means that the enrollment was 20 thousand students in the year 2000.

3. The rate of change of enrollment per year is 2 thousand students per year. So, the slope of the linear model is 2. This checks with the calculation shown in Fig. 90.
4. An equation of the line can be written in slope–intercept form, $y = mx + b$. Using t and E, we have $E = mt + b$. Since the slope is 2 and the E-intercept is $(0, 20)$, we have $E = 2t + 20$.

To check our work with a graphing calculator, we begin by entering our model (see Fig. 91). Then we check that the entries in the graphing calculator table in Fig. 92 equal the entries in Table 31. We also check that the graph of our equation contains the points of the scattergram of the data (see Fig. 93).

Figure 91 Enter the model

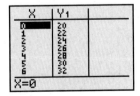

Figure 92 Use a table to verify the model

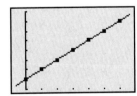

Figure 93 Use a graph to verify the model

For graphing calculator instructions on creating tables, see Section A.13. For instructions on creating scattergrams, see Sections A.8 and A.10. ■

In Example 4, we found the model $E = 2t + 20$. Here is the connection between parts of the equation and the situation:

$$E = \underbrace{2}_{\substack{\text{rate of} \\ \text{change of} \\ \text{enrollment} \\ \text{per year}}} \cdot \; t \; + \underbrace{20}_{\substack{\text{enrollment} \\ \text{at } t = 0}}$$

Example 5 Finding a Model

The temperature at which water boils (the *boiling point*) depends on elevation: the higher you are, the lower the boiling point is. At sea level (elevation 0), water boils at 212°F. The boiling point declines by 5.9°F for each thousand-meter increase in elevation (Source: *Thermodynamics, an Engineering Approach by Cengal & Boles*). Let B be the boiling point (in degrees Fahrenheit) at an elevation of E thousand meters.

1. Is there a linear relationship between E and B? If so, find the slope.
2. Find an equation of the model.
3. Perform a unit analysis of the equation.
4. Mount McKinley, the highest mountain in the United States, reaches 6.194 thousand meters. What is the boiling point of water at the peak?

Solution

1. Since the boiling point depends on the elevation, we will consider the rate of change of the boiling point with respect to elevation. Because this rate of change is a *constant* −5.9°F per thousand meters, the variables E and B are linearly related and the slope is −5.9.
2. We want an equation of the form $B = mE + b$. Since $B = 212$ when $E = 0$, the B-intercept is (0, 212). Recall that $m = -5.9$, so an equation is $B = -5.9E + 212$.
3. Here is a unit analysis of the equation $B = -5.9E + 212$:

$$\underbrace{B}_{\text{degrees Fahrenheit}} = \underbrace{-5.9}_{\substack{\text{degrees Fahrenheit} \\ \text{thousand meters}}} \cdot \underbrace{E}_{\text{thousand meters}} + \underbrace{212}_{\text{degrees Fahrenheit}}$$

We can use the fact that $\dfrac{\text{thousand meters}}{\text{thousand meters}} = 1$ to simplify the units of the expression on the right-hand side of the equation:

$$\frac{\text{degrees Fahrenheit}}{\text{thousand meters}} \cdot \text{thousand meters} + \frac{\text{degrees}}{\text{Fahrenheit}} = \frac{\text{degrees}}{\text{Fahrenheit}} + \frac{\text{degrees}}{\text{Fahrenheit}}$$

So, the units of the expressions on both sides of the equation are degrees Fahrenheit, which suggests that our equation is correct.

4. To find the boiling point, we substitute the input 6.194 for E in the equation $B = -5.9E + 212$:

$$B = -5.9(6.194) + 212 \qquad \text{Substitute 6.194 for E.}$$
$$\approx 175.46 \qquad \text{Perform indicated operations.}$$

So, the boiling point is about 175°F at the peak of Mount McKinley. ■

Example 6 Analyzing a Model

The numbers of residential swimming pools sold in the United States are shown in Table 32 for various years.

Let s be the swimming-pool sales (in thousands of pools) at t years since 1990. A model of the situation is

$$s = 15.9t + 207.7$$

1. Use a graphing calculator to draw a scattergram and the model in the same viewing window. Check whether the line comes close to the data points.
2. What is the slope? What does it mean in this situation?
3. Find the rates of change in sales per year from one year to the next. Compare the rates of change with the result in Problem 2.
4. What is the s-intercept? What does it mean in this situation?
5. Predict the pool sales in 2011.

Table 32 Swimming-Pool Sales

Year	Sales (thousands of pools)
1996	300
1997	319
1998	337
1999	352
2000	369
2001	378

Source: *P. K. Data*

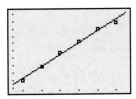

Figure 94 Verify the model

Solution

1. We draw the scattergram and the model in the same viewing window (see Fig. 94). For graphing calculator instructions on drawing scattergrams and models, see Sections A.8 and A.10.

    It appears that the line comes close to the data points, so the model is a reasonable one.
2. The slope is 15.9, because $s = 15.9t + 207.7$ is of the form $y = mx + b$ and $m = 15.9$. According to the model, sales are increasing by 15.9 thousand pools per year.
3. The rates of change in sales per year are shown in Table 33. All of the rates of change are fairly close to the rate of change of 15.9 thousand pools per year, except for the relatively small rate of change of 9 thousand pools per year from 2000 to 2001.

Table 33 Rates of Change in Swimming-Pool Sales

Year	Sales (thousands)	Rate of Change in Sales (from previous year to current year) (thousands of pools per year)
1996	300	–
1997	319	$(319 - 300) \div (1997 - 1996) = 19$
1998	337	$(337 - 319) \div (1998 - 1997) = 18$
1999	352	$(352 - 337) \div (1999 - 1998) = 15$
2000	369	$(369 - 352) \div (2000 - 1999) = 17$
2001	378	$(378 - 369) \div (2001 - 2000) = 9$

4. The s-intercept is $(0, 207.7)$, because $s = 15.9t + 207.7$ is of the form $y = mx + b$ and $b = 207.7$. According to the model, 207.7 thousand pools were sold in 1990.
5. The year 2011 is represented by $t = 2011 - 1990 = 21$. We substitute the input 21 for t in the equation $s = 15.9t + 207.7$:

$$s = 15.9(21) + 207.7$$
$$= 541.6$$

According to the model, about 542 thousand pools will be sold in 2011. ■

In Example 6, we found that the slope of the swimming-pool model is 15.9, which means that, according to the model, swimming-pool sales are increasing by 15.9 thousand pools per year. In reality, the pool sales never increased by 15.9 thousand pools in *any* of the years between 1996 and 2001, inclusive. However, 15.9 thousand pools per year *is* a reasonable estimate of the *average* yearly increase.

> ### Slope Is an Average Rate of Change
>
> If two quantities t and p are approximately linearly related and p depends on t, then the slope of a reasonable linear model is approximately equal to the average rate of change of p with respect to t.

WARNING   A common error in describing the meaning of the slope of a model is vagueness. For example, a description such as

> The slope means that it is increasing.

neither specifies the quantity that is increasing nor the rate of increase. The following statement includes the missing information:

> The slope of 15.9 means that sales are increasing by 15.9 thousand pools per year.

Slope Addition Property

So far, we have discussed rate of change for linear models. Now we will explore rate of change for linear equations that are not used as models.

Table 34 Some Solutions of $y = 2x + 1$

x	y
0	1
1	3
2	5
3	7
4	9

Table 35 Some Solutions of $y = -3x + 8$

x	y
0	8
1	5
2	2
3	−1
4	−4

Example 7   Interpreting Slope as a Rate of Change

What is the slope of the line $y = 2x + 1$? Interpret the slope as a rate of change.

Solution

The slope is 2. So, the rate of change of y with respect to x is 2. This means that if the value of x increases by 1, the value of y increases by 2 (see Table 34).   ■

Example 8   Interpreting Slope as a Rate of Change

What is the slope of the line $y = -3x + 8$? Interpret the slope as a rate of change.

Solution

The slope is −3. So, the rate of change of y with respect to x is −3. This means that if the value of x increases by 1, the value of y decreases by 3 (see Table 35).   ■

Our observations made in Examples 7 and 8 suggest yet another way to think about slope: the **slope addition property.**

Slope Addition Property

For a linear equation of the form $y = mx + b$, if the value of x increases by 1, then the value of y changes by the slope m. In other words, if an input increases by 1, the output changes by the slope.

For the line $y = 4x - 9$, the slope is 4. If the value of x increases by 1, then the value of y changes by 4. For the line $y = -5x - 2$, the slope is −5. If the value of x increases by 1, then the value of y changes by −5.

Example 9   Identifying Possible Linear Equations

Some solutions of four equations are listed in Table 36. Which of the equations could be linear?

Table 36 Some Solutions of Four Equations

Equation 1		Equation 2		Equation 3		Equation 4	
x	y	x	y	x	y	x	y
1	23	4	12	0	3	50	8
2	20	5	17	1	6	51	8
3	17	6	22	2	12	52	8
4	14	7	27	3	24	53	8
5	11	8	32	4	48	54	8

Solution

1. Equation 1 could be linear: Each time the value of x increases by 1, the value of y changes by −3.
2. Equation 2 could be linear: Each time the value of x increases by 1, the value of y changes by 5.
3. Equation 3 is not linear: Each time the value of x increases by 1, the value of y does not change by the same value.
4. Equation 4 could be $y = 8$, which is a linear equation: Each time the value of x increases by 1, the value of y changes by 0.   ■

Table 37 Points on a Line

x	y
0	15
1	9
2	3
3	−3
4	−9

Example 10   Finding an Equation of a Line

A line contains the points listed in Table 37. Find an equation of the line.

Solution

For $y = mx + b$, the graph has y-intercept $(0, b)$. From Table 37, we see that the y-intercept is $(0, 15)$, so $b = 15$. As the value of x increases by 1, the value of y changes by -6. By the slope addition property, we know that $m = -6$. Therefore, an equation of the line is $y = -6x + 15$.

We use a graphing calculator table to verify our work in Fig. 95.

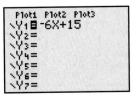 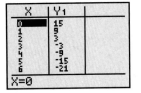

Figure 95 Verify the equation

group exploration

Averaging rates of change

Table 38 Numbers of New Wipe Products

Year	Number of New Wipe Products
1998	33
1999	71
2000	86
2001	110
2002	130

Source: *Marketing Intelligence Service*

The number of new wipe products has almost quadrupled in the past few years (see Table 38). There are now specialty wipes for cleaning everything from pet paws to car interiors.

1. Find the average rate of change of the number of new wipe products per year from 1998 to 2002. [**Hint:** Use the 1998 and 2002 data in Table 38.]

2. Find the rate of change in the number of new wipe products per year from each year to the next, beginning in 1998. [**Hint:** You should find four results.]

3. Find the average of the four rates of change that you found in Problem 2. [**Hint:** Divide the sum of the four rates of change by 4.]

4. Compare your result in Problem 3 with your result in Problem 1.

5. The following expression is an example of a *telescoping sum*:

$$(71 - 33) + (86 - 71) + (110 - 86) + (130 - 110)$$

Explain why the above sum is equal to $130 - 33$.

6. Use your result in Problem 5 to help explain why the following statement is true:

$$\frac{\dfrac{71 - 33}{1999 - 1998} + \dfrac{86 - 71}{2000 - 1999} + \dfrac{110 - 86}{2001 - 2000} + \dfrac{130 - 110}{2002 - 2001}}{4} = \frac{130 - 33}{2002 - 1998}$$

Also, explain how this statement is related to your comparison in Problem 4.

7. Describe what you have learned from doing this exploration.

group exploration

Looking ahead: Laws of operations

1. **a.** Evaluate $a(b + c)$ for $a = 2$, $b = 3$, and $c = 5$.
   b. Evaluate $ab + ac$ for $a = 2$, $b = 3$, and $c = 5$.
   c. Compare your results for Problems 1 and 2.
   d. Evaluate both $a(b + c)$ and $ab + ac$ for $a = 4$, $b = 2$, and $c = 6$, and then compare the results.
   e. Evaluate both $a(b + c)$ and $ab + ac$ for values of your choosing for a, b, and c, and then compare the results.
   f. Make an educated guess as to whether $a(b + c) = ab + ac$ is true for all numbers a, b, and c.

2. Evaluate both $a(bc)$ and $(ab)(ac)$ for values of your choosing for a, b, and c. Is the statement $a(bc) = (ab)(ac)$ true for all numbers a, b, and c?

3. Evaluate both $a(bc)$ and $(ab)c$ for values of your choosing for a, b, and c. Make an educated guess as to whether the statement $a(bc) = (ab)c$ is true for all numbers a, b, and c.

4. Evaluate both $a + b$ and $b + a$ for values of your choosing for a and b. Make an educated guess as to whether the statement $a + b = b + a$ is true for all numbers a and b.

5. What are the main points of this exploration?

TIPS FOR SUCCESS: Get the Most Out of Working Exercises

If you work an exercise by referring to a similar example in your notebook or in the text, it is a good idea to try the exercise again without referring to your source of help. If you need to refer to your source of help to solve the exercise a second time, consider trying the exercise a third time without help. When you finally complete the exercise without help, reflect on which concepts you used to work the exercise, where you had difficulty, and what the key idea was that opened the door of understanding for you.

A similar strategy can be used in getting help from a student, an instructor, or a tutor.

If this sounds like a lot of work, it is! But it is well worth it. Although it is important to complete each assignment, it is also important to learn as much as possible while progressing through it.

HOMEWORK 3.5

FOR EXTRA HELP ▶

Student Solutions Manual

PH Math/Tutor Center

Math XL
MathXL®

MyMathLab
MyMathLab

1. A person's annual salary increases by $12,400 over an 8-year period. Find the average rate of change of the salary per year.

2. A person's CD collection increases by 185 CDs over a 5-year period. Find the average rate of change of the number of CDs per year.

3. An airplane's altitude declines steadily by 24,750 feet over a 15-minute period. Find the rate of change of the airplane's altitude per minute.

4. The temperature decreases steadily by 12°F over a 4-hour period. Find the rate of change of temperature per hour.

5. The number of employers offering health insurance to gay families increased approximately linearly from 96 employers in 1999 to 228 employers in 2004 (Source: *Human Rights Campaign*). Find the average rate of change of the number of employers offering health insurance to gay families per year.

6. Sales of handheld computers in the United States increased approximately linearly from 6.0 million units in 2000 to 8.2 million units in 2003 (Source: *Gartner*). Find the average rate of change of sales per year.

7. The percentage of all e-mail that is spam increased approximately linearly from 13% in 1999 to 30% in 2004 (Source: *IDC*). Find the average rate of change of this percentage per year.

8. The percentage of Americans who say that they or their spouses have saved money for retirement increased approximately linearly from 57% in 1994 to 68% in 2004 (Source: *Employee Benefit Research Institute*). Find the average rate of change of this percentage per year.

9. In Alaska, the Steller's sea lion population decreased approximately linearly from about 53 thousand in 1989 to about 25 thousand in 2000 (Source: *National Marine Mammal Laboratory*). Find the average rate of change of the Steller's sea lion population per year.

10. Department store sales decreased approximately linearly from $234 billion in 2000 to $214 billion in 2004 (Source: *Redbook Research*). Find the average rate of change of sales per year.

11. Iraq's oil exports under the oil-for-food program decreased approximately linearly from 800 million barrels in 1999 to 455 million barrels in 2002 (Source: The New York Times). Find the average rate of change of Iraq's oil exports per year under the oil-for-food program.

12. The percentage of families living below the poverty level decreased approximately linearly from 13.6% in 1993 to 9.6% in 2000 (Source: *U.S. Census Bureau*). Find the average rate of change of this percentage per year.

13. In order for a family living in New York to qualify for the health insurance program Family Health Plus, the family's annual income must be less than a maximum level. In 2006, the maximum income level of a four-person family was $29,925 and the maximum income level of a seven-person family was $44,775 (Source: *New York State Department of Health*). Find the average rate of change of maximum income level with respect to family size.

14. A person stacks some cups of uniform shape and size (one placed inside the next). The height of 3 stacked cups is 17.5 centimeters and of 5 stacked cups is 23.0 centimeters.

Find the average rate of change of the height of the stacked cups with respect to the number of cups.

15. A company's revenue for 2005 is $3 million. Each year its revenue increases by $4 million. Let r be the revenue (in millions of dollars) for the year that is t years since 2005.
  a. Is there a linear relationship between t and r? Explain. If the relationship is linear, find the slope and describe what it means in this situation.
  b. Describe the Rule of Four as applied to this situation:
    i. Use an equation to describe the situation.
    ii. Use a table of values of t and r to describe the situation.
    iii. Use a graph to describe the situation.

16. As of February 1, a garage band knows 5 songs. Each week, the band members learn 2 more songs. Let n be the number of songs that the band knows at t weeks since February 1.
  a. Is there a linear relationship between t and n? Explain. If the relationship is linear, find the slope and describe what it means in this situation.
  b. Describe the Rule of Four as applied to this situation:
    i. Use an equation to describe the situation.
    ii. Use a table of values of t and n to describe the situation.
    iii. Use a graph to describe the situation.

17. In 2004, the number of drive-in movie sites in the United States was 402. The number has been decreasing by about 21.9 sites per year since then (Source: *National Association of Theatre Owners*). Let n be the number of drive-in movie sites at t years since 2004.
  a. Is there an approximate linear relationship between t and n? Explain. If the relationship is approximately linear, find the slope and describe what it means in this situation.
  b. What is the n-intercept of the model? What does it mean in this situation?
  c. Find an equation of the model.
  d. Predict the number of sites in 2011.
  e. Predict the number of sites in 2024.

18. The average fare paid for business air travel was $259 in 2000 and has decreased by about $8 per year since then (Source: *American Express Business Travel Monitor*). Let F be the average fare (in dollars) paid for business air travel in the year that is t years since 2000.
  a. Is there an approximate linear relationship between t and F? Explain. If the relationship is approximately linear, find the slope and describe what it means in this situation.
  b. What is the F-intercept of the model? What does it mean in this situation?
  c. Find an equation of the model.
  d. Perform a unit analysis of the equation you found in part (c).
  e. Predict the average fare in 2010.

19. A student's savings account has a balance of $4700 on September 1. Each month, the balance declines by $650. Let B be the balance (in dollars) at t months since September 1.
  a. Find the slope of the linear model that describes this situation. What does it mean in this situation?
  b. What is the B-intercept of the model? What does it mean in this situation?
  c. Find an equation of the model.
  d. Perform a unit analysis of the equation you found in part (c).
  e. Find the balance on March 1 (6 months after September 1).

20. A person owns a propane-gas barbecue grill with a tank that holds 5 gallons of propane. The person always sets the temperature to $350°F$, which uses 0.125 gallon of propane per hour. Let g be the number of gallons of propane that remain in the tank after t hours of cooking since the tank was last filled.
  a. Find the slope of the linear model that describes this situation. What does it mean in this situation?
  b. What is the g-intercept of the model? What does it mean in this situation?
  c. Find an equation of the model.
  d. The person fills the propane tank and then uses the grill for 3 hours. How much propane remains in the tank?
  e. Estimate the amount of propane that will remain in the tank after 50 hours of cooking since the tank was last filled.

21. For the spring semester 2005, part-time students at Centenary College paid $385 per credit (unit or hour) for tuition and paid a mandatory part-time student fee of $10 per semester. Let T be the total one-semester cost (in dollars) of tuition plus part-time student fee for c credits of classes.
  a. Find the slope of the linear model that describes this situation. What does it mean in this situation?
  b. Find an equation of the model.
  c. Perform a unit analysis of the equation you found in part (b).
  d. What was the total one-semester cost of tuition plus part-time student fee for 9 credits of classes?

22. For the spring semester 2006, California residents paid an enrollment fee of $26 per unit (credit or hour) at Santa Barbara City College. Students were also required to pay a $13 health fee each semester. Let c be the total cost (in dollars) of tuition plus health fee for u units of classes.
  a. Find the slope of the linear model that describes this situation. What does it mean in this situation?
  b. Find an equation of the model.
  c. Perform a unit analysis of the equation you found in part (b).
  d. What was the total one-semester cost of tuition plus fee for 15 units of classes?

23. A person drives her Toyota Prius® on a road trip. At the start of the trip, she fills up the 11.9-gallon tank with gasoline. During the trip, the car uses about 0.02 gallon of gasoline per mile. Let G be the number of gallons of gasoline remaining in the tank after driving d miles.
  a. What is the slope of the linear model that describes this situation? What does it mean in this situation?
  b. What is the G-intercept of the model? What does it mean in this situation?
  c. Find an equation of the model.
  d. Perform a unit analysis of the equation you found in part (c).
  e. If the person drives 525 miles before she refuels the car, how much gasoline is required to fill up the tank?

24. Although the United States and Great Britain use the Fahrenheit ($°F$) temperature scale, most countries use the Celsius ($°C$) scale. The temperature reading $0°C$ is equivalent to the Fahrenheit reading $32°F$. An increase of $1°C$ is equivalent to an increase of $1.8°F$. Let F be the Fahrenheit reading that is equivalent to a Celsius reading of C degrees. Assume that F is the dependent variable.
  a. Find the slope of a linear model that describes this situation. What does it mean in this situation?

b. What is the F-intercept of the model? What does it mean in this situation?

c. Find an equation of the model.

d. Perform a unit analysis of the equation you found in part (c).

e. If the temperature is 30°C, what is the Fahrenheit reading?

25. Let n be the number of U.S. oil refineries at t years since 2000. A reasonable model of the number of U.S. oil refineries is $n = -3.65t + 157.31$ (Source: *Energy Information Administration*).

a. What is the slope? What does it mean in this situation?

b. What is the n-intercept? What does it mean in this situation?

c. Predict the number of refineries in 2011.

26. Let s be the sales of frozen-food bowls (in millions) in the year that is t years since 2000. A reasonable model of frozen-food bowl sales is $s = -34t + 224$ (Source: *Information Resources, Inc.*).

a. What is the slope? What does it mean in this situation?

b. What is the s-intercept? What does it mean in this situation?

c. Estimate the sales in 2006.

27. The number of Internet users in the United States is shown in Table 39 for various years.

Table 39 Internet Users in the United States

Year	Number of Users (millions)
1996	39
1997	60
1998	84
1999	105
2000	122
2001	143
2002	166
2003	183
2004	207

Source: *Jupiter MMXI*

Let n be the number of Internet users (in millions) in the United States at t years since 1995. A model of the situation is $n = 20.7t + 19.6$.

a. Use a graphing calculator to draw a scattergram and the model in the same viewing window. Does the line come close to the data points?

b. What is the slope? What does it mean in this situation?

c. Find the rates of change of the number of users per year from each year to the next. Compare the rates of change with your result in part (b).

d. What is the n-intercept? What does it mean in this situation?

e. Predict the number of users in 2010. If the U.S. Census Bureau's prediction that the U.S. population will be about 309 million in 2010 is correct, has model breakdown occurred? Explain.

28. Traffic congestion on highways has continued to increase over time. The average annual person-hours (the number of hours each person loses due to traffic delays per year) are listed in Table 40 for various years. Let L be the average annual person-hours of lost time to traffic delays for the year that is t years since 1990. A model of the situation is $L = 1.15t + 20$.

Table 40 Average Annual Person-Hours of Lost Time

Year	Average Annual Lost Time (hours)
1992	22
1994	25
1996	27
1998	30
2000	32
2001	32

Source: *Federal Highway Administration*

a. Use a graphing calculator to draw a scattergram and the model in the same viewing window. Does the line come close to the data points?

b. What is the slope? What does it mean in this situation?

c. Find the rates of change of the average person-hours of lost time per year from each year to the next. Compare the rates of change with your result in part (b).

d. Predict the average person-hours of lost time for 2011.

e. What is the L-intercept? What does it mean in this situation?

29. In an attempt to raise the low graduation rates of college football and basketball players, Division I institutions require that prospective athletes must meet new grade point average (GPA) standards (based on a maximum of 4.0) to play as freshmen, and enrolled athletes must meet these new standards to keep playing (see Table 41).

Table 41 Core GPAs Needed to Qualify to Play, for Given SAT Scores

SAT score	Core GPA
620	3.0
700	2.8
780	2.6
860	2.4
940	2.2
1010	2.0

Source: *NCAA*

Let G be the qualifying core GPA for an SAT score of s points. A model of the situation is $G = -0.00254s + 4.58$.

a. Use a graphing calculator to draw a scattergram and the model in the same viewing window. Does the line come close to the data points?

b. If an athlete's SAT score is 400 points, the lowest possible score, estimate the student's qualifying core GPA. The actual qualifying core GPA is 3.55. Compute the error in the estimate.

c. What is the slope? What does it mean in this situation?

d. What is the G-intercept? What does it mean in this situation?

30. The 2005 average selling prices for a Subaru Outback® of various ages are shown in Table 42. Let p be the average price of a Subaru Outback (in dollars) of age a years. A model of the situation is $p = -3284.68a + 30,700.57$.

a. Use a graphing calculator to draw a scattergram and the model in the same viewing window. Does the line come close to the data points?

b. What is the slope? What does it mean in this situation?

Table 42 Average Selling Prices of Subaru Outbacks	
Age (years)	Average Price (dollars)
1	27,080
2	25,240
3	21,112
4	16,950
5	13,400
6	9992
7	9159

Source: *AutoTrader*®

c. Estimate the average price of an 8-year-old Subaru Outback.

31. A person is on a car trip. Let d be the distance (in miles) traveled in t hours of driving. The line in Fig. 96 describes the relationship between t and d.

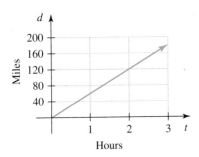

Figure 96 Car model

a. Is the car traveling at a constant speed? Explain.
b. What is the speed of the car?

32. A person is on an airplane. Let d be the distance (in miles) traveled in t hours. The line in Fig. 97 describes the relationship between t and d.

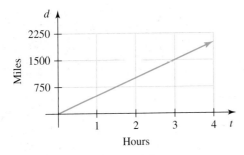

Figure 97 Airplane model

a. Is the airplane traveling at a constant speed? Explain.
b. What is the speed of the airplane?

33. A person is on a car trip. Let V be the volume (in gallons) of gasoline that remains in the gasoline tank after t hours of driving. The line in Fig. 98 describes the relationship between t and V.
a. Is the rate of change of gasoline remaining in the tank per hour constant? Explain.
b. What is the rate of change of gasoline remaining in the tank per hour?

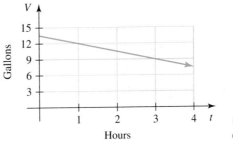

Figure 98
Gasoline model

34. Let F be the temperature (in degrees Fahrenheit) at t hours after noon. The line in Fig. 99 describes the relationship between t and F.

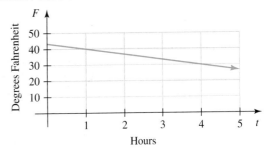

Figure 99 Temperature model

a. Is the rate of change of temperature per hour constant? Explain.
b. What is the rate of change of temperature per hour?

35. A math tutor charges $30 per hour. Let c be the total charge (in dollars) for t hours of tutoring. Are t and c linearly related? If so, find the slope and describe what it means in this situation.

36. Gasoline sells for $3 per gallon at a certain gas station. Let T be the total charge (in dollars) for g gallons of gas. Are g and T linearly related? If so, find the slope and describe what it means in this situation.

37. Some solutions of four equations are listed in Table 43. Which of the equations could be linear?

Table 43 Solutions of Four Equations							
Equation 1		Equation 2		Equation 3		Equation 4	
x	y	x	y	x	y	x	y
0	2	0	5	2	5	5	20
1	5	1	6	3	3	6	19
2	8	2	8	4	1	7	17
3	11	3	9	5	−1	8	14
4	14	4	11	6	−3	9	10

38. Some solutions of four equations are listed in Table 44. Which of the equations could be linear?

Table 44 Solutions of Four Equations							
Equation 1		Equation 2		Equation 3		Equation 4	
x	y	x	y	x	y	x	y
0	2	0	35	4	−7	23	25
1	4	1	31	5	−3	24	23
2	8	2	27	6	1	25	20
3	16	3	23	7	5	26	18
4	32	4	19	8	9	27	15

39. Some solutions of four linear equations are listed in Table 45. Complete the table.

Table 45 Solutions of Four Linear Equations

Equation 1		Equation 2		Equation 3		Equation 4	
x	y	x	y	x	y	x	y
0	3	0	99	21	16	43	
1	8	1	92	22		44	
2		2		23	12	45	23
3		3		24		46	
4		4		25		47	29

40. Some solutions of four linear equations are listed in Table 46. Complete the table.

Table 46 Solutions of Four Linear Equations

Equation 1		Equation 2		Equation 3		Equation 4	
x	y	x	y	x	y	x	y
0	100	0	2	13	32	75	4
1	94	1	9	14		76	
2		2		15		77	
3		3		16	26	78	
4		4		17		79	16

41. Four sets of points are described in Table 47. For each set, find an equation of a line that contains the points.

Table 47 Solutions of Four Equations

Set 1		Set 2		Set 3		Set 4	
x	y	x	y	x	y	x	y
0	5	0	20	0	21	0	9
1	7	1	17	1	29	1	4
2	9	2	14	2	37	2	−1
3	11	3	11	3	45	3	−6
4	13	4	8	4	53	4	−11

42. Four sets of points are described in Table 48. For each set, find an equation of a line that contains the points.

Table 48 Solutions of Four Equations

Set 1		Set 2		Set 3		Set 4	
x	y	x	y	x	y	x	y
0	22	0	3	0	15	0	−12
1	19	1	7	1	6	1	−10
2	16	2	11	2	−3	2	−8
3	13	3	15	3	−12	3	−6
4	10	4	19	4	−21	4	−4

43. For the equation $y = 7x + 2$, describe the change in the value of y as the value of x is increased by 1.

44. For the equation $y = -9x - 5$, describe the change in the value of y as the value of x is increased by 1.

45. For the equation $y = -6x + 40$, use a table of solutions to show that if the value of x is increased by 1, then the value of y changes by the slope.

46. For the equation $y = 8x + 1$, use a table of solutions to show that if the value of x is increased by 1, then the value of y changes by the slope.

47. The Rule of Four states that we can describe solutions of a linear equation by using an equation, a graph, a table, or words. Describe the slope of the line $y = 3x - 4$ in each of these four ways:
a. Explain how you can determine the slope of the line $y = 3x - 4$ from the equation.
b. Describe the slope of the line $y = 3x - 4$ in terms of "run" and "rise."
c. Describe the slope of the line $y = 3x - 4$ in terms of the slope addition property.
d. Describe the meaning of the slope of the line $y = 3x - 4$ in your own words.

48. In your own words, describe the slope addition property.

49. Explain why slope is a rate of change.

analyze this

GLOBAL WARMING LAB (continued from Chapter 2)

Scientists estimate that the average global temperature has increased by about 1°F in the past century (see Table 49). Recall from the Global Warming Lab in Chapter 1 that many scientists believe that an increase as small as 3.6°F could be dangerous. Because the planet's average temperature increased by 1°F in the past century, it might seem that it would take about another two-and-a-half centuries for Earth to reach a dangerous temperature level.

However, the Intergovernmental Panel on Climate Change (IPCC), composed of hundreds of scientists around the world, predicted that Earth's average temperature will rise 2.5°F to 10.4°F degrees Fahrenheit in the coming

Table 49 Average Surface Temperatures of Earth

Year	Average Temperature (degrees Fahrenheit)	Year	Average Temperature (degrees Fahrenheit)
1900	57.1	1955	57.0
1905	56.7	1960	57.2
1910	56.8	1965	56.9
1915	57.3	1970	57.3
1920	56.9	1975	57.2
1925	56.9	1980	57.7
1930	57.1	1985	57.4
1935	57.1	1990	58.1
1940	57.5	1995	58.0
1945	57.2	2000	57.9
1950	56.9	2003	58.4

Source: *NASA–GISS*

century.* Scientists from the United States and Europe have predicted that there is a 9-out-of-10 chance that the global average temperature will increase 3°F to 9°F, with a range of 4°F to 7°F most likely.†

One result of the last century of global warming is that glaciers are receding and ocean levels are rising. For example, NASA scientist Bill Krabill estimates that Greenland's ice cap, the world's second largest, may be losing ice at a rate of 50 cubic kilometers per year.‡ The ice cap contained 2.85 million cubic kilometers of ice in 2000. Climatologist Jonathan Gregory of the University of Reading, in the United Kingdom, believes that by 2050 the ice cap may start an irreversible runaway melting. A total meltdown could take 1000 years.§

Jonathan Overpeck, director of the Institute for the Study of Planet Earth at the University of Arizona, believes that Greenland's ice cap could melt completely in as little as 150 years and that a partial melting of Greenland's ice cap by 2100 could raise ocean levels by more than 1 meter.¶ Coastal engineers estimate that a 1-meter rise would translate into a loss of about 100 meters of land.‖ In some areas of the world, land loss could be much greater.** For example, most of Florida is less than 1 meter above sea level.

Most scientists predict that global warming will also cause the extinction of plants and animals, bring severe water shortages, create extreme weather events, and increase the number of heat-related illnesses and deaths.

Analyzing the Situation

1. Let A be the average global temperature (in degrees Fahrenheit) at t years since 1900. Refer to the scattergram that you drew in Problem 3 of the Global Warming Lab of Chapter 1. (If you did not draw this scattergram earlier, create a *careful* one now, using the data in Table 49.) Does it appear that Earth's average temperature increased much from 1900 to 1965? Explain.

2. **a.** Estimate the average global temperature from 1900 to 1965. [**Hint:** Divide the sum of the temperatures by the number of temperature readings.]
   b. Estimate the average global temperature from 1990 to 2000. [**Hint:** Divide the sum of the temperatures by the number of temperature readings.]
   c. Use your results in parts (a) and (b) to estimate the change in Earth's average temperature in the past century. Compare your result with the scientists' estimate of 1°F.

3. In Problem 5 of the Global Warming Lab of Chapter 1, you showed that the planet's average temperature increased approximately linearly from 1965 to 2000. (If you didn't do this earlier, do so now.) Find the rate of change of the average global temperature from 1965 to 2000.

4. Use the rate of change found in Problem 3 to predict the change in average temperature for the coming century. Does your result fall within the 2.5–10.4°F IPCC range? Does it fall within either the 3–9°F or 4–7°F range predicted by the team of scientists from the United States and Europe? If yes, which one(s)?

5. Explain why, by using terminology such as "9 out of 10" and "most likely," scientists have been able to predict narrower ranges of increase in global average temperatures in the coming century. Explain how this terminology, coupled with narrower ranges, will help policymakers better understand the risks involved in various courses of action or inaction.

6. Let I be the amount of ice (in cubic kilometers) in Greenland's ice cap at t years since 2000. Assuming that Greenland's ice cap continues to melt at the current rate, find an equation that models the situation.

7. **a.** Use your model to predict the amount of ice in Greenland's ice cap in 1000 years. If a total meltdown occurs in 1000 years, as predicted by Gregory, what must happen to the rate of melting?
   b. If a total meltdown occurs in 150 years, as predicted by Overpeck, what must happen to the rate of melting?

WORKOUT LAB

In this lab, you will explore your walking or running speed.

Check with your instructor about whether you should collect your own data or use the data listed in Table 50.

Table 50	Times for Walking 440-Yard Laps	
Lap Number	Distance (yards)	Time (seconds)
0	0	0
1	440	217
2	880	436
3	1320	656
4	1760	878
5	2200	1095
6	2640	1308

Source: *J. Lehmann*

Materials

You will need the following items:

- A timing device
- A pencil or pen
- A small pad of paper

Preparation

Locate a running track on which you can walk or run. For most tracks, one lap is 440 yards; so, four laps are

---

*IPCC, 2001: Summary for Policymakers.

†T. M. L. Wigley and S. C. B. Raper, "Interpretation of high projections for global-mean warming," *Science* 293 (5529):451–454, July 20, 2001.

‡W. Krabill et al., "Greenland ice sheet: High-elevation balance and peripheral thinning," *Science* 289: 428–430, 2000.

§J. M. Gregory et al., "Climatology: Threatened loss of the Greenland ice sheet," *Nature* (April 8, 2004):426:616.

¶American Geophysical Union meeting, October 2002.

‖National Oceanic and Atmospheric Administration.

**R. J. Nicholls and F. M. J. Hoozemans, "Vulnerability to sea-level rise with reference to the Mediterranean region," *Medcoast 95*, Vol. II, October 1995.

1760 yards, or 1 mile. You may select some other type of route, provided that you know the distance of one lap and you can easily complete six laps. Or map out a route in your neighborhood and estimate the distance by measuring it with the odometer in a car.

Recording of Data
Start your timing device and begin walking or running. Complete six laps of your course. Each time you complete a lap, record the *total* elapsed time. It will be easier to have a friend record the times for you. You may go slowly or quickly, but try to move at a constant speed throughout this experiment.

Analyzing the Data
1. Describe your route and the distance of one lap.
2. Use a table to describe the six total elapsed times for the six laps.
3. Let d be the distance (in yards) after you have walked or run for t seconds. Throughout this lab, treat d as the dependent variable and t as the independent variable. Show the six pairs of values of t and d in a table.
4. Create a scattergram for the variables t and d.
5. For each of the six laps, calculate your average speed. Did you move at a steady rate, slow down, speed up, or engage in a combination of these?
6. Explain how your six average speeds are related to the position of the points in your scattergram.
7. Find the average of the six average speeds you found in Problem 5. Compare this result with the average speed for the entire workout.
8. What is your average speed for the entire workout, in units of miles per hour?
9. Use your result from Problem 8 to predict how long it would take you to walk or run 2 miles.
10. Use your result from Problem 8 to predict how long it would take you to walk or run a marathon, which is 26.2 miles long. Has model breakdown occurred? Explain.

BALLOON LAB

In this lab, you will explore how long it takes for air to be released from a balloon when it is inflated with various amounts of air.

Check with your instructor about whether you should collect your own data or use the data listed in Table 51.

Materials
You will need one helper and the following items:

- A balloon
- A timing device
- A bucket, sink, or bathtub
- Water to fill the bucket, sink, or bathtub
- A transparent one-cup or larger measuring cup (a larger cup is more convenient)

Table 51 Average Release Times for a Balloon Inflated with a Single Breath Having a Volume of 20 Ounces

Number of Breaths	Volume (ounces)	Average Release Time (seconds)
0	0	0
5	100	1.9
10	200	3.5
15	300	6.1
20	400	6.0
25	500	7.7
30	600	10.6

Source: *J. Lehmann*

Preparation
Inflate and deflate the balloon fully several times to stretch it out. Each time you inflate the balloon, practice blowing into it with uniform-size breaths. There is no need to breathe deeply into the balloon. Medium-size breaths will work well for this lab.

Recording of Data
Perform the following tasks:

1. Count how many medium-size breaths it takes to fill the balloon.
2. For each trial of this experiment, you will time how long it takes for the balloon to release all the air inside it. Run three trials for each of six different volumes. Decide for which volumes you will have trials. For example, if it takes 30 breaths to fill the balloon, you could run three trials for each of the volumes of 5, 10, 15, 20, 25, and 30 breaths. To run a trial, first fill the balloon to the desired volume. Then begin timing as you release the balloon. Stop timing when the balloon is completely deflated. Record the time and the corresponding volume (in number of breaths).
3. To find the volume of a medium-size breath, fill a bucket, sink, or bathtub with water. Have one person fully submerge the measuring cup, so that there is no air in it, and then turn the cup upside down while it is still under water. The cup should not rest on the bottom of the container. Next, have a second person blow once into the balloon and then carefully release the air into the submerged cup. The air from the balloon will displace the water in the cup. The volume of air in the balloon will likely be more than what can fit in the cup, so the second person will have to stop releasing air from the balloon, and then the first person can empty out the air by resubmerging the cup. The second person should continue releasing air from the balloon into the cup until the balloon is empty. Depending on the size of the breath and the cup, the first person may have to resubmerge the cup several times. Then compute the volume of a medium-size breath by measuring the amount of air in the cup and taking into account how many times you filled the cup with air.

Analyzing the Data

1. For each of the volumes of 5, 10, 15, 20, 25, and 30 breaths, compute the average of the three deflation times. Then record the numbers of breaths and the average deflation times in columns 1 and 3 of a table similar to Table 51. Use the volume of one breath to help you find the entries for the second column of the table.

2. Let T be the time (in seconds) it takes for the balloon to deflate completely when the initial volume is n ounces. Create a scattergram of the data.

3. Draw a line that comes close to the data.

4. What is the T-intercept of the linear model? What does it mean in this situation? If model breakdown occurs, draw a better model.

5. Find the slope of the linear model. What does it mean in this situation?

6. Find an equation of the model.

7. Use the model to estimate the deflation time for a volume other than any of the volumes you used for the trials.

8. As you inflated the balloon, did it take about the same amount of effort to breathe in each time, or did it get progressively easier or harder? Thinking about how that effort is related to deflation times, does it suggest that the data points should lie close to a line or to a curve that bends? Explain. Do the (actual) data points support your theory? Explain.

Chapter Summary

Key Points
OF CHAPTER 3

Solution, satisfy, and solution set of an equation (Section 3.1)	An ordered pair (a, b) is a **solution** of an equation in terms of x and y if the equation becomes a true statement when a is substituted for x and b is substituted for y. We say that (a, b) **satisfies** the equation. The **solution set** of an equation is the set of all solutions of the equation.
Graph (Section 3.1)	The **graph** of an equation in two variables is the set of points that correspond to all solutions of the equation.
Graph of $y = mx + b$ (Section 3.1)	The graph of an equation of the form $y = mx + b$, where m and b are constants, is a line.
Rule of Four (Sections 3.1 and 3.2)	We can describe some or all of the solutions of an equation with an equation, a table, a graph, or words. An authentic situation can also be described in these four ways, which collectively are known as the **Rule of Four.**
Unit analysis (Section 3.2)	We perform a **unit analysis** of a model's equation by determining the units of the expressions on both sides of the equation. The units of the expressions on both sides of the equation should be the same.
Equations of horizontal and vertical lines (Section 3.2)	If a and b are constants, then • The graph of $y = b$ is a horizontal line. • The graph of $x = a$ is a vertical line.
Equations whose graphs are lines (Section 3.2)	If an equation can be put into the form $y = mx + b$ or $x = a$, where m, a, and b are constants, then the graph of the equation is a line. We call such an equation a **linear equation in two variables.**
Comparing the steepness of two objects (Section 3.3)	To compare the steepness of two objects, compute the unit ratio $$\frac{\text{vertical distance}}{\text{horizontal distance}}$$ for each object. The object with the larger ratio is the steeper object.
Slope of a nonvertical line (Section 3.3)	Let (x_1, y_1) and (x_2, y_2) be two distinct points of a nonvertical line. Then the **slope** of the line is $$m = \frac{\text{vertical change}}{\text{horizontal change}} = \frac{\text{rise}}{\text{run}} = \frac{y_2 - y_1}{x_2 - x_1}$$
Slopes of increasing and decreasing lines (Section 3.3)	An increasing line has positive slope. A decreasing line has negative slope.

Slopes of horizontal and vertical lines (Section 3.3)

A horizontal line has a slope equal to zero.

A vertical line has undefined slope.

Slope and y-intercept of a linear equation of the form $y = mx + b$; slope-intercept form (Section 3.4)

For a linear equation of the form $y = mx + b$,
- the slope of the line is m and
- the y-intercept of the line is $(0, b)$.

We say that this equation is in **slope–intercept form.**

Using slope to graph an equation of the form $y = mx + b$ (Section 3.4)

To graph an equation of the form $y = mx + b$,
1. Plot the y-intercept $(0, b)$.
2. Use $m = \dfrac{\text{rise}}{\text{run}}$ to plot a second point.
3. Sketch the line that passes through the two plotted points.

Find an equation of a line from a graph (Section 3.4)

To find an equation of a line from a graph,
1. Determine the slope m and the y-intercept $(0, b)$ from the graph.
2. Substitute your values for m and b into the equation $y = mx + b$.

Slopes of parallel lines (Section 3.4)

If lines l_1 and l_2 are parallel nonvertical lines, then the slopes of the lines are equal:

$$m_1 = m_2$$

Also, if two distinct lines have equal slope, then the lines are parallel.

Slopes of perpendicular lines (Section 3.4)

If lines l_1 and l_2 are perpendicular nonvertical lines, then the slope of one line is the opposite of the reciprocal of the slope of the other line:

$$m_2 = -\frac{1}{m_1}$$

Also, if the slope of one line is the opposite of the reciprocal of another line's slope, then the lines are perpendicular.

Rate of change and average rate of change (Section 3.5)

Suppose that a quantity y changes steadily from y_1 to y_2 as a quantity x changes steadily from x_1 to x_2. Then the **rate of change** of y with respect to x is the ratio of the change in y to the change in x:

$$\frac{\text{change in } y}{\text{change in } x} = \frac{y_2 - y_1}{x_2 - x_1}$$

If either quantity does not change steadily, then the preceding formula is the **average rate of change** of y with respect to x.

Increasing and decreasing quantities (Section 3.5)

Suppose that a quantity p depends on a quantity t. Then
- If p increases steadily as t increases steadily, then the rate of change of p with respect to t is positive.
- If p decreases steadily as t increases steadily, then the rate of change of p with respect to t is negative.

Slope is a rate of change (Section 3.5)

If there is a linear relationship between the quantities t and p, and if p depends on t, then the slope of the linear model is equal to the rate of change of p with respect to t.

Constant rate of change (Section 3.5)

Suppose that a quantity p depends on a quantity t. Then
- If there is a linear relationship between t and p, then the rate of change of p with respect to t is constant.
- If the rate of change of p with respect to t is constant, then there is a linear relationship between t and p.

Slope is an average rate of change (Section 3.5)

If two quantities t and p are approximately linearly related and p depends on t, then the slope of a reasonable linear model is approximately equal to the average rate of change of p with respect to t.

Slope addition property (Section 3.5)

For a linear equation of the form $y = mx + b$, if the value of x increases by 1, then the value of y changes by the slope m. In other words, if an input increases by 1, the output changes by the slope.

CHAPTER 3 REVIEW EXERCISES

1. Which of the ordered pairs $(-3, 9)$, $(1, 2)$, and $(4, -5)$ satisfy the equation $y = -2x + 3$?

For Exercises 2–7, refer to the graph sketched in Fig. 100.

2. Find y when $x = 2$.   **3.** Find y when $x = -2$.

4. Find y when $x = 0$.   **5.** Find x when $y = -3$.

6. Find x when $y = -4$.   **7.** Find x when $y = 0$.

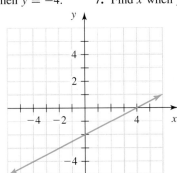

Figure 100 Exercises 2–7

8. While taking off, airplane A climbs steadily for 6500 feet over a horizontal distance of 12,700 feet. Airplane B climbs steadily for 7400 feet over a horizontal distance of 15,600 feet. Which plane is climbing at a greater incline? Explain.

Find the slope of the line passing through the two given points. State whether the line is increasing, decreasing, horizontal, or vertical.

9. $(-3, 1)$ and $(2, 11)$ **10.** $(-2, -4)$ and $(1, -7)$

11. $(4, -3)$ and $(8, -1)$ **12.** $(-6, 0)$ and $(0, -3)$

13. $(-5, 5)$ and $(2, -2)$ **14.** $(-10, -3)$ and $(-4, -5)$

15. $(-5, 2)$ and $(3, -7)$ **16.** $(-4, -1)$ and $(2, -5)$

17. $(-4, 7)$ and $(-4, -3)$ **18.** $(-5, 2)$ and $(-1, 2)$

Find the approximate slope of the line that contains the two given points. Round your result to the second decimal place. State whether the line is increasing, decreasing, horizontal, or vertical.

19. $(5.4, 7.9)$ and $(8.3, -2.6)$

20. $(-8.74, -2.38)$ and $(-1.16, 4.77)$

21. Sketch a line whose slope is a negative number near zero.

Sketch the line that has the given slope and contains the given point.

22. $m = 3$, $(0, -4)$ **23.** $m = \dfrac{4}{3}$, $(0, 1)$

24. $m = 0$, $(2, -3)$

Determine the slope and the y-intercept. Use them to graph the equation by hand. Use a graphing calculator to verify your work.

25. $y = \dfrac{3}{4}x - 1$ **26.** $y = -\dfrac{1}{2}x + 3$

27. $y = -\dfrac{2}{5}x - 1$ **28.** $y = \dfrac{2}{3}x$

29. $y = -4x$ **30.** $y = 2x - 4$

31. $y = -3x + 1$ **32.** $y = x + 2$ **33.** $y = -5$

Graph the equation by hand.

34. $x = -3$ **35.** $y = 2$

36. Describe the Rule of Four as applied to the linear equation $y = -2x + 1$:
   a. Describe three solutions of $y = -2x + 1$ by using a table.
   b. Describe the solutions of $y = -2x + 1$ by using a graph.
   c. Describe the solutions of $y = -2x + 1$ by using words.

37. When a certain person loses his job, the balance in his checking account is $19 thousand. While the person is unemployed, the balance declines by $3 thousand per month.
   a. Let B be the balance (in thousands of dollars) in the checking account after t months of unemployment. Find an equation for t and B.
   b. Graph by hand the equation you found in part (a).
   c. What is the B-intercept? What does it mean in this situation?
   d. How long has the person been unemployed if the balance is $4 thousand? Explain by using arrows on your graph in part (b).

38. Find an equation of the line sketched in Fig. 101.

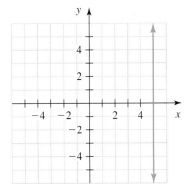

Figure 101 Exercise 38

39. Find an equation of the line sketched in Fig. 102.

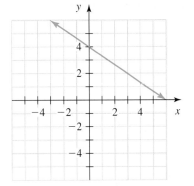

Figure 102 Exercise 39

Determine whether the pair of lines is parallel, perpendicular, or neither. Explain.

40. $y = 3x - 2$ and $y = \dfrac{1}{3}x + 6$

41. $y = \dfrac{4}{7}x + 1$ and $y = -\dfrac{7}{4}x - 5$

42. $x = -2$ and $y = 5$ **43.** $x = -4$ and $x = 1$

44. The temperature declines by 6°F over a 4-hour period. Find the rate of change of temperature per hour.

45. U.S. sales of male impotency drugs increased approximately linearly from $0.9 billion in 2000 to $1.2 billion in 2004

(Source: *IMS Health*). Find the rate of change of the sales per year.

46. In 2005 in New York City, taxis charged a flat fee of $2.50 plus $2 per mile (40 cents per $\frac{1}{5}$ mile). Let c be the total charge (in dollars) for a trip of d miles.
  a. Find an equation for d and c.
  b. Perform a unit analysis of the equation you found in part (a).
  c. What was the cost of a 17-mile trip in 2005?

47. A person weighs 195 pounds when he begins a weight-loss program. After beginning the program, he loses 4 pounds per month. Let w be his weight (in pounds) t months after he has begun the program.
  a. Find the slope of the linear model of the situation. What does it mean in this situation?
  b. Find an equation of the model.
  c. The person reaches his goal after he has dieted for 6 months. What was his goal?

48. The average monthly day-care cost in 2001 was $516. The cost has increased by $17.75 per year (Source: *Runzheimer International*). Costs are based on a three-year-old attending a day-care center eight hours a day, five days a week. Let c be the average monthly cost (in dollars) at t years since 2001.
  a. Find the slope of the linear model of the situation. What does it mean in this situation?
  b. Find an equation of the model.
  c. Predict the average monthly cost in 2011.

49. The suggested retail price for a TI-84 Plus® graphing calculator is $130. Let C be the total cost (in dollars) of purchasing n of these calculators at the retail price. Are n and C linearly related? If so, what is the slope? What does the slope mean in this situation?

50. Some solutions of four equations are listed in Table 52. Which of the equations could be linear?

Table 52 Solutions of Four Equations

Equation 1		Equation 2		Equation 3		Equation 4	
x	y	x	y	x	y	x	y
0	17	0	2	3	4	1	−8
1	14	1	6	4	4	2	−3
2	11	2	9	5	4	3	2
3	8	3	11	6	4	4	7
4	5	4	12	7	4	5	12

51. Some solutions of four linear equations are listed in Table 53. Complete the table.

Table 53 Solutions of Four Linear Equations

Equation 1		Equation 2		Equation 3		Equation 4	
x	y	x	y	x	y	x	y
0	50	0	12	61	25	26	−4
1	41	1	16	62		27	
2		2		63		28	
3		3		64	19	29	
4		4		65		30	8

52. For the equation $y = -6x + 39$, describe the change in the value of y as the value of x is increased by 1.

CHAPTER 3 TEST

For Exercises 1–4, refer to the graph sketched in Fig. 103.

1. Find y when $x = -3$.   **2.** Find x when $y = -1$.

3. Find the y-intercept.   **4.** Estimate the x-intercept.

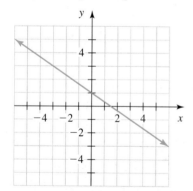

Figure 103
Exercises 1–4

5. Ski run A declines steadily for 115 yards over a horizontal distance of 580 yards. Ski run B declines steadily for 150 yards over a horizontal distance of 675 yards. Which ski run is steeper? Explain.

Find the slope of the line passing through the two given points. State whether the line is increasing, decreasing, horizontal, or vertical.

6. $(3, -8)$ and $(5, -2)$ **7.** $(-4, -1)$ and $(2, -4)$

8. $(-5, 4)$ and $(1, 4)$ **9.** $(-2, -7)$ and $(-2, 3)$

10. Find the approximate slope of the line that contains the points $(-5.99, -3.27)$ and $(2.83, 8.12)$. Round your result to the second decimal place. State whether the line is increasing, decreasing, horizontal, or vertical.

11. Sketch the line that has a slope of $\frac{2}{5}$ and contains the point $(0, -3)$.

Determine the slope and the y-intercept. Use them to graph the equation by hand.

12. $y = -\frac{3}{2}x + 2$ **13.** $y = \frac{5}{6}x$ **14.** $y = 3x - 4$

15. $y = 2$ **16.** $y = -2x + 3$

17. Find an equation of the line sketched in Fig. 104.

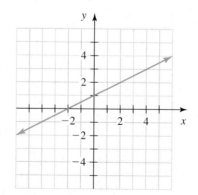

Figure 104 Exercise 17

18. A used car is worth $17 thousand. Each year, the car's value declines by $2 thousand. Let v be the car's value (in thousands of dollars) after t years.
  a. Find an equation for t and v.
  b. Graph by hand the equation you found in part (a).
  c. What is the v-intercept? What does it mean in this situation?
  d. When will the value of the car be $5 thousand dollars? Explain by using arrows on your graph in part (b).

19. Graphs of four equations are shown in Fig. 105. State whether m and b are positive, negative, zero, or undefined for the $y = mx + b$ form of each equation.

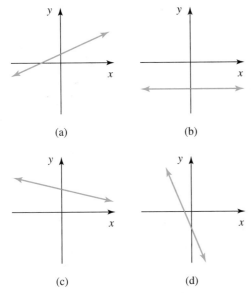

      (a)   (b)

      (c)   (d)

Figure 105 Exercise 19

For Exercises 20 and 21, determine whether the pair of lines is parallel, perpendicular, or neither. Explain.

20. $y = \dfrac{2}{5}x + 3$ and $y = \dfrac{5}{2}x - 7$

21. $y = -3x + 8$ and $y = -3x - 1$

22. The balance in a savings account declines by $3150 over a six-month period. Find the average rate of change of the balance per month.

23. The number of flu vaccine doses given in the United States increased approximately linearly from 77.9 million doses in 2000 to 100 million doses in 2004 (Source: *Centers for Disease Control and Prevention*). Find the average rate of change of the number of doses per year.

24. The total U.S. sales of vegetarian foods and dairy alternatives were $1.6 billion in 2003. The total sales have increased by about $0.17 billion per year since then (Source: *Mintel International*). Let s be the total sales (in billions of dollars) at t years since 2003.
  a. Find the slope of the model of the situation. What does it mean in this situation?
  b. Find an equation that models the situation.
  c. Predict the total sales in 2010.

25. The cooking times of a turkey in a 325°F oven are shown in Table 54 for various weights.

Table 54 Cooking Times of a Turkey in a 325°F Oven			
Weight Group (pounds)	Weight Used to Represent Weight Group (pounds)	Cooking-Time Group (hours)	Time Used to Represent Cooking-Time Group (hours)
6–8	7	3.0–3.5	3.25
8–12	10	3.5–4.5	4
12–16	14	4.5–5.5	5
16–20	18	5.5–6.5	6
20–24	22	6.5–7.0	6.75

Source: *About, Inc.*

Let T be the cooking time (in hours) of a turkey that weighs w pounds. A model of the situation is $T = 0.24w + 1.64$.
  a. Use a graphing calculator to draw a scattergram and the model in the same viewing window. Does the line come close to the data points?
  b. What is the slope? What does it mean in this situation?
  c. What is the T-intercept? What does it mean in this situation?
  d. Estimate the cooking time of a 19-pound turkey.

26. A person goes out for a run. Let d be the distance (in miles) traveled in t minutes of running. The graph in Fig. 106 describes the relationship between t and d.

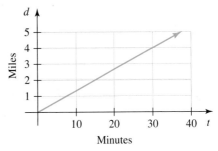

Figure 106 Runner model

  a. Is the person running at a constant speed? Explain.
  b. How fast is the person running? Use units of miles per hour.

27. Four sets of points are described in Table 55. For each set, find an equation of a line that contains the points.

Table 55 Solutions of Four Equations							
Set 1		Set 2		Set 3		Set 4	
x	y	x	y	x	y	x	y
0	25	0	2	0	12	0	47
1	22	1	6	1	7	1	53
2	19	2	10	2	2	2	59
3	16	3	14	3	-3	3	65
4	13	4	18	4	-8	4	71

28. For the equation $y = 3x - 8$, describe the change in the value of y as the value of x is increased by 1.

4

Simplifying Expressions and Solving Equations

That which we persist in doing becomes easier—not that the nature of the task has changed, but our ability to do it has increased.

—*Ralph Waldo Emerson*

Table 1 College Freshmen Whose Average Grade in High School Was an A

Year	Percent
1990	29.4
1995	36.1
1998	39.8
2001	44.1
2004	47.5

Source: *Higher Education Research Institute*

What was your average grade in high school? The percentage of college freshmen whose average grade in high school was an A has increased greatly since 1990 (see Table 1). Do you think that the increase is due to students learning more or to teachers lowering their standards? In Exercise 51 of Homework 4.4, you will estimate when half of all college freshmen earned an average grade of A in high school.

In this chapter, we will discuss how to write expressions more simply. We will also discuss how to find solutions of equations in *one* variable. Once we have learned these skills, we will be able to make predictions efficiently about the independent variable (or the dependent variable) in authentic situations, such as the grade data just described.

4.1 COMMUTATIVE, ASSOCIATIVE, AND DISTRIBUTIVE LAWS

Objectives

▷ Know the commutative, associative, and distributive laws.
▷ Know the meaning of *equivalent expressions*.
▷ Know how to *simplify expressions*.
▷ Subtract two expressions.
▷ Know how to show that a statement is false.
▷ Compare the distributive and associative laws for multiplication.

In this section, we will discuss three laws of operations that will help us write expressions more simply.

Commutative Laws

Consider the equations $2 + 6 = 6 + 2$ and $2 \cdot 6 = 6 \cdot 2$. These true statements suggest the *commutative laws*.

Commutative Laws for Addition and Multiplication

Commutative law for addition: $a + b = b + a$
Commutative law for multiplication: $ab = ba$

In words: We can add two numbers in either order and get the same result, and we can multiply two numbers in either order and get the same result.

Example 1 Using the Commutative Laws

Use a commutative law to write the expression in another form.

1. $3 + x$ **2.** $x(6)$

Solution

1. $3 + x = x + 3$ Commutative law for addition: $a + b = b + a$

2. $x(6) = 6x$ Commutative law for multiplication: $ab = ba$ ■

A **term** is a constant, a variable, or a product of a constant and one or more variables raised to powers. Here are some terms:

$$4x \qquad -7 \qquad y \qquad -5xy \qquad \frac{3x}{y}$$

The expression $7xy + 4x - 5y + 3$ has four terms: $7xy, 4x, 5y,$ and 3. Note that, in the expression, the terms are separated by addition and subtraction symbols.

Variable terms are terms that contain variables. **Constant terms** are terms that do not contain variables. For example, $-5x$ is a variable term, and 8 is a constant term. We usually write a sum of both variable and constant terms with the variable terms to the left of the constant terms. So, for $5 - 8x$, we write

$$5 - 8x = 5 + (-8x) \qquad a - b = a + (-b)$$
$$= -8x + 5 \qquad \text{Commutative law for addition: } a + b = b + a$$

Associative Laws

As we discussed in Section 2.6, we work from left to right when we perform additions and subtractions. However, if a sum is of the form $a + b + c$, we can get the same result by performing the addition on the left first or the one on the right first. For example,

$$(4 + 2) + 3 = 6 + 3 = 9$$
$$4 + (2 + 3) = 4 + 5 = 9$$

A similar law is true for an expression of the form abc:

$$(4 \cdot 2) \cdot 3 = 8 \cdot 3 = 24$$
$$4 \cdot (2 \cdot 3) = 4 \cdot 6 = 24$$

These examples suggest the *associative laws*.

Associative Laws for Addition and Multiplication

Associative law for addition: $a + (b + c) = (a + b) + c$
Associative law for multiplication: $a(bc) = (ab)c$

In words: For an expression $a + b + c$, we get the same result by performing the addition on the left first or the one on the right first. For an expression abc, we get the same result by performing the multiplication on the left first or the one on the right first.

It is important to know the difference between the associative laws and the commutative laws. The associative laws change the order of *operations,* whereas the commutative laws change the order of *terms* or even *expressions.*

For example, consider the expression $(3 + x) + 1$. By the associative law of addition, we can change the order of the additions:

$$(3 + x) + 1 = 3 + (x + 1)$$

By the commutative law of addition, we can change the order of the terms 3 and x:

$$(3 + x) + 1 = (x + 3) + 1$$

We can also use the commutative law to change the order of the expressions $x + 3$ and 1:

$$(x + 3) + 1 = 1 + (x + 3)$$

Example 2   Using the Associative Laws

Use an associative law to write the expression in another form.

1. $(x + 2) + 5$ **2.** $4(3x)$

Solution

1. $(x + 2) + 5 = x + (2 + 5)$ Associative law for addition: $(a + b) + c = a + (b + c)$

$\qquad\qquad\quad = x + 7$ Add.

2. $4(3x) = (4 \cdot 3)x$ Associative law for multiplication: $a(bc) = (ab)c$

$\qquad\quad = 12x$ Multiply.   ∎

We can use a combination of the commutative law and the associative law, both for addition, to write the terms of $a + b + c$ in different orders:

$$a + c + b \quad b + a + c \quad b + c + a \quad c + a + b \quad c + b + a$$

We **rearrange the terms** of an expression by writing the terms in a different order.

Likewise, we **rearrange the factors** of an expression by writing the factors in a different order. Here, we rearrange the factors of abc to get the following products:

$$acb \quad bac \quad bca \quad cab \quad cba$$

Example 3   Rearranging Terms

Rearrange the terms of $9 - 2x - 5 + 3$ so that the numbers can be added.

Solution

$9 - 2x - 5 + 3 = 9 + (-2x) + (-5) + 3$ $a - b = a + (-b)$

$\qquad\qquad\qquad = -2x + 9 + (-5) + 3$ Rearrange terms.

$\qquad\qquad\qquad = -2x + 7$ Add constant terms.   ∎

It is helpful to practice problems such as the one in Example 3 until you can do each problem in one step.

Distributive Law

For the expression $2(3 + 4)$, we perform the addition before the multiplication. However, compare the result with that for computing $2 \cdot 3 + 2 \cdot 4$:

$$2(3 + 4) = 2(7) = 14$$
$$2 \cdot 3 + 2 \cdot 4 = 6 + 8 = 14$$

Since both results are equal to 14, we can write

$$2(3 + 4) = 2 \cdot 3 + 2 \cdot 4$$

The blue curves indicate what numbers we multiply by 2. We say that we *distribute* the 2 to both the 3 and the 4.

Finally, we compare $2(7 - 3)$ with $2 \cdot 7 - 2 \cdot 3$:

$$2(7 - 3) = 2(4) = 8$$
$$2 \cdot 7 - 2 \cdot 3 = 14 - 6 = 8$$

Since both results are equal to 8, we can write

$$2\overparen{(7 - 3)} = 2 \cdot 7 - 2 \cdot 3$$

Here we have distributed the 2 to both the 7 and the 3. These examples suggest the **distributive law.**

Distributive Law
$$a(b + c) = ab + ac$$
In words: To find $a(b + c)$, distribute a to both b and c.

Example 4 Using the Distributive Law

Find the product.

1. $3(x + 5)$ **2.** $5(2x + 4y)$

Solution

1. $3\overparen{(x + 5)} = 3x + 3 \cdot 5$ Distributive law: $a(b + c) = ab + ac$

            $= 3x + 15$ Multiply.

2. $5\overparen{(2x + 4y)} = 5 \cdot 2x + 5 \cdot 4y$ Distributive law: $a(b + c) = ab + ac$

               $= 10x + 20y$ Multiply. ∎

WARNING   In Problem 1 of Example 4, we found that $3(x + 5) = 3x + 15$. In applying the distributive law to an expression such as $3(x + 5)$, remember to distribute the 3 to *every* term in the parentheses. For example, the expression $3x + 5$ is an incorrect result.

Since subtracting a number is the same as adding the opposite of the number, it is also true that

$$a(b - c) = ab - ac$$

We can use the distributive law and the commutative law for multiplication to show that we can "distribute from the right" as well as from the left (see Exercises 83 and 84):

$$(a + b)c = ac + bc$$
$$(a - b)c = ac - bc$$

Example 5 Using the Distributive Law

Find the product.

1. $4(3x - 2y)$ **2.** $(x - 6)(3)$ **3.** $-2(6 + t)$ **4.** $-5(w - 3)$

Solution

1. $4\overparen{(3x - 2y)} = 4 \cdot 3x - 4 \cdot 2y$ Distributive law: $a(b - c) = ab - ac$

               $= 12x - 8y$ Multiply.

2. $\overparen{(x - 6)}(3) = x \cdot 3 - 6 \cdot 3$ Distributive law: $(a - b)c = ac - bc$

             $= 3x - 18$ Commutative law for multiplication: $ab = ba$; multiply.

3. $-2\overparen{(6 + t)} = -2(6) + (-2t)$ Distributive law: $a(b + c) = ab + ac$

             $= -12 + (-2t)$ Multiply.

             $= -2t + (-12)$ Commutative law for addition: $a + b = b + a$

             $= -2t - 12$ $a + (-b) = a - b$

4. $-5\overparen{(w - 3)} = -5w - (-5)(3)$ Distributive law: $a(b - c) = ab - ac$

             $= -5w + 15$ Multiply. ∎

It is extremely helpful to practice problems such as those in Example 5 until you can do them in one step.

We can also use the distributive law when there are more than two terms inside the parentheses.

Example 6 Distributive Law

Find the product $3(2t - 5w + 4)$.

Solution

$$3(2t - 5w + 4) = 3 \cdot 2t - 3 \cdot 5w + 3 \cdot 4 \qquad \text{Distributive law}$$
$$= 6t - 15w + 12 \qquad \text{Multiply.} \qquad \blacksquare$$

Equivalent Expressions

In Problem 1 of Example 4, we found that $3(x + 5) = 3x + 15$. In Example 7, we will explore the meaning of this statement.

Example 7 Evaluating Expressions

Evaluate both of the expressions $3(x + 5)$ and $3x + 15$ for the given values of x.
1. $x = 2$
2. $x = 4$
3. $x = 0$, $x = 1$, $x = 2$, $x = 3$, $x = 4$, $x = 5$, and $x = 6$

Solution

1. First we evaluate $3(x + 5)$ for $x = 2$:

$$3(2 + 5) = 3(7) = 21$$

Then we evaluate $3x + 15$ for $x = 2$:

$$3(2) + 15 = 6 + 15 = 21$$

Both results are equal to 21.
2. First we evaluate $3(x + 5)$ for $x = 4$:

$$3(4 + 5) = 3(9) = 27$$

Then we evaluate $3x + 15$ for $x = 4$:

$$3(4) + 15 = 12 + 15 = 27$$

Both results are equal to 27.
3. We use a graphing calculator to create a table for the equations $y = 3(x + 5)$ and $y = 3x + 15$ (see Fig. 1). See Section A.14 for graphing calculator instructions.

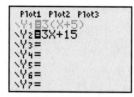 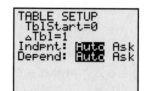

Figure 1 Verify the work

For each value of x in the table, the values of y for the two equations are equal. So for each of these values of x, the two results of evaluating the two expressions are equal. $\blacksquare$

In Example 7, we found that for each value that we used to evaluate the expressions $3(x + 5)$ and $3x + 15$, the two results were equal.

DEFINITION Equivalent expressions

Two or more expressions are **equivalent expressions** if, when each variable is evaluated for *any* real number (for which all the expressions are defined), the expressions all give equal results.

Simplifying an Expression

We **simplify** an expression by using the laws of operations to remove any parentheses and to rearrange terms so that we can add any constant terms. We will discuss other ways to simplify expressions throughout this text. The result of simplifying an expression is a **simplified expression,** which is equivalent to the original expression.

Simplifying an expression often makes it easier to evaluate the expression and to graph an equation that contains the expression. We will see other benefits of simplifying expressions throughout the text.

Example 8 Simplifying Expressions

Simplify.

1. $4(2t - 3) - 7$ **2.** $5 - 6(x - 2)$

Solution

1. $\overparen{4(2t - 3)} - 7 = 8t - 12 - 7$ Distributive law: $a(b - c) = ab - ac$
$\qquad\qquad\qquad = 8t - 19$ Subtract constant terms.

2. $5 - \overparen{6(x - 2)} = 5 - 6x + 12$ Distributive law: $a(b - c) = ab - ac$
$\qquad\qquad\qquad = -6x + 5 + 12$ Rearrange terms.
$\qquad\qquad\qquad = -6x + 17$ Add constant terms.

We use a graphing calculator table to verify our work (see Fig. 2). For each value of x in the table, the two results of evaluating the two expressions are equal; this is strong evidence that the expressions are equivalent.

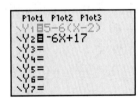

Figure 2 Verify the work

After we simplify an expression, it is wise to use a graphing calculator table to verify that the result and the original expression are indeed equivalent.

Example 9 Translating from English to Mathematics

Let x be a number. Translate the phrase "1, plus 3 times the difference of twice the number and 4" to an expression. Then simplify the expression.

Solution

The expression is

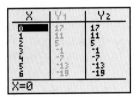

Next, we simplify the expression:

$$1 + 3 \cdot (2x - 4) = 1 + 6x - 12 \qquad \text{Distributive law: } a(b - c) = ab - ac$$
$$= 6x + 1 - 12 \qquad \text{Rearrange terms.}$$
$$= 6x - 11 \qquad \text{Subtract constant terms.} \qquad \blacksquare$$

The products $-1 \cdot 2 = -2$, $-1 \cdot 5 = -5$, and $-1 \cdot 6 = -6$ suggest the following property:

Multiplying a Number by −1

$$-1a = -a$$

In words: -1 times a number is equal to the opposite of the number.

To remove parentheses in $-(x + 3)$, we use the fact that $-a = -1a$ to write

$$-(x + 3) = -1(x + 3) \qquad -a = -1a$$
$$= -1x + (-1)3 \qquad \text{Distributive law: } a(b + c) = ab + ac$$
$$= -x - 3 \qquad -1a = -a$$

Example 10   Simplifying an Expression

Simplify $-(x - 4y + 7)$.

Solution

$$-(x - 4y + 7) = -1(x - 4y + 7) \qquad -a = -1a$$
$$= -1x - (-1)4y + (-1)7 \qquad \text{Distributive law: } a(b - c) = ab - ac$$
$$= -x + 4y - 7 \qquad -1a = -a \qquad \blacksquare$$

Subtracting Two Expressions

Note that $a - 1b = a - b$. We can use this fact, written $a - b = a - 1b$, to help us subtract two expressions.

Example 11   Simplifying an Expression

Simplify $7 - (x + 9)$.

Solution

$$7 - (x + 9) = 7 - 1(x + 9) \qquad a - b = a - 1b$$
$$= 7 - 1x + (-1)9 \qquad \text{Distributive law: } a(b + c) = ab + ac$$
$$= 7 - x - 9 \qquad -1a = -a$$
$$= -x - 2 \qquad \text{Rearrange terms; subtract constant terms.}$$

We use a graphing calculator table to verify our work (see Fig. 3).

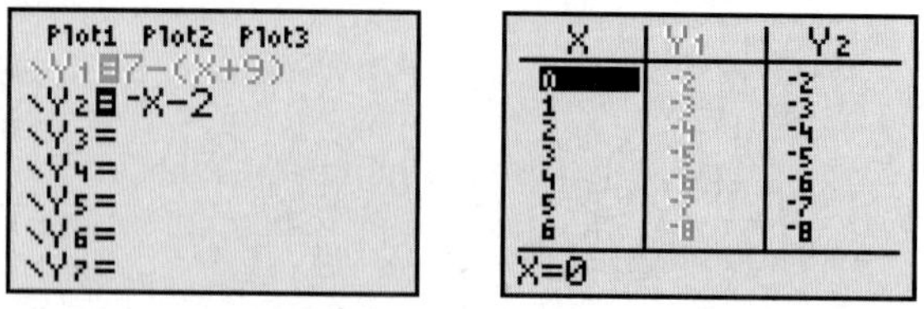

Figure 3 Verify the work $\blacksquare$

False Statements

We have discussed the commutative laws for addition and multiplication. Are there similar laws for subtraction and division? Example 12 will show that there are no such laws. **To show that an equation is false, we need find only *one* set of numbers that, when substituted for any variables, give a result of the form $a = b$, where a and b are different real numbers.**

Example 12 Showing That a Statement Is False

Show that the statement is false.

1. $a - b = b - a$
2. $a \div b = b \div a$

Solution

1. We substitute the arbitrary number 5 for a and the arbitrary number 2 for b in the statement $a - b = b - a$:

$$a - b = b - a$$
$$5 - 2 \stackrel{?}{=} 2 - 5 \qquad \text{Substitute 5 for } a \text{ and 2 for } b.$$
$$3 \stackrel{?}{=} -3 \qquad \text{Subtract.}$$
$$\text{false}$$

So, the statement $a - b = b - a$ is false.

2. We substitute the arbitrary number 6 for a and the arbitrary number 3 for b in the statement $a \div b = b \div a$:

$$a \div b = b \div a$$
$$6 \div 3 \stackrel{?}{=} 3 \div 6 \qquad \text{Substitute 6 for } a \text{ and 3 for } b.$$
$$2 \stackrel{?}{=} \frac{3}{6} \qquad \text{Divide; } a \div b = \frac{a}{b}.$$
$$2 \stackrel{?}{=} \frac{1}{2} \qquad \text{Simplify.}$$
$$\text{false}$$

So, the statement $a \div b = b \div a$ is false. ■

In Example 12, we showed that there is no commutative law for either subtraction or division. In Exercises 79 and 80, you will show that there is no associative law for either subtraction or division as well.

Comparing the Distributive Law with the Associative Law for Multiplication

It is helpful to compare the distributive law with the associative law for multiplication to avoid confusing the two laws. For the expression $a(b + c)$, we distribute a to both of the terms b and c:

$$a(b + c) = ab + ac \qquad \text{Distributive law}$$

For the expression $a(bc)$, we do *not* distribute the a to both of the factors b and c. Rather, we distribute a to just one of the factors:

$$a(bc) = (ab)c \qquad \text{Associative law for multiplication}$$

group exploration

Equivalent expressions

1. Evaluate both $3(x + 2)$ and $x + 6$ for $x = 0$.

2. Your results in Problem 1 should be equal. If not, check your work. If the results are equal, does that mean that $3(x + 2)$ and $x + 6$ are equivalent? Explain.

3. Evaluate both $3(x + 2)$ and $x + 6$ for $x = 5$. Compare your results and explain what your comparison tells you.

4. Explain why it is good practice to evaluate expressions several times, each time for a different value of x, to be more confident that the expressions are equivalent.

5. Simplify $3(x + 2)$. Then evaluate both $3(x + 2)$ and your simplified result for several values of x to check that your work is correct.

TIPS FOR SUCCESS: Desire and Faith

To accomplish anything worthwhile, including succeeding in this course, requires substantial effort *and* faith that you will succeed. The following quote describes what it takes to achieve a goal:

> *The secret of making something work in your lives is, first of all, the deep desire to make it work. Then the faith and belief that it can work. Then to hold that clear definite vision in your consciousness and see it working out step by step without one thought of doubt or disbelief.*
>
> —Eileen Caddy, *Footprints on the Path* (1991)

Your deep desire to succeed in this course might be to earn a degree so that you can earn more money. Or perhaps you will be motivated to learn algebra for the love of learning or to experience setting a goal and reaching it. Your faith and belief can come from knowing that you, your instructor, and your college will do everything possible to ensure your success. To hold your vision of success "without one thought of doubt or disbelief" is a tall order, but the more you look for ways to succeed rather than feel discouraged, the better are your chances of success.

HOMEWORK 4.1   FOR EXTRA HELP ▶

Student Solutions Manual   PH Math/Tutor Center   MathXL®   *MyMathLab* MyMathLab

Simplify. Use a graphing calculator table to verify your work when possible.

1. $2(5x)$ **2.** $3(6x)$ **3.** $-4(-9x)$

4. $-5(-7x)$ **5.** $\frac{1}{2}(-8x)$ **6.** $\frac{2}{3}(-12x)$

7. $7\left(\dfrac{x}{4}\right)$ **8.** $6\left(\dfrac{x}{5}\right)$ **9.** $3(x+9)$

10. $6(x+7)$ **11.** $(x-5)2$ **12.** $(x-8)4$

13. $-2(t+5)$ **14.** $-4(w+3)$ **15.** $-5(6-2x)$

16. $-8(5-3x)$ **17.** $(4x+7)(-6)$ **18.** $(8x+2)(-3)$

19. $2(3x-5y)$ **20.** $4(2x-8y)$

21. $-5(4x+3y-8)$ **22.** $-6(2x-4y+5)$

23. $3+2(x+1)$ **24.** $1+5(x+3)$

25. $-0.3(x+0.2)$ **26.** $-0.5(x+0.4)$

27. $4(a+3)+7$ **28.** $2(t+4)+1$

29. $-3(4x-2)+3$ **30.** $-5(2x-3)+7$

31. $4-3(3a-5)$ **32.** $-2-6(2p-3)$

33. $-3.7+4.2(2.5x-8.3)$ **34.** $6.4+3.9(6.1x-4.4)$

35. $-(t+2)$ **36.** $-(a+5)$

37. $-(8x-9y)$ **38.** $-(4x-3y)$

39. $-(5x+8y-1)$ **40.** $-(3x-7y+4)$

41. $-(3x+2)+5$ **42.** $-(4x+1)+3$

43. $8-(x+3)$ **44.** $4-(x+1)$

45. $-2-(2t-5)$ **46.** $-3-(4a-2)$

47. $\dfrac{1}{2}(4x+8)$ **48.** $\dfrac{2}{3}(6x+9)$

Simplify the expression. Then evaluate both the expression and your simplified result for $x=2$ to check your work. Finally,

evaluate both expressions for a value of x of your choosing (other than 2) to be even more confident that your work is correct.

49. $2(4x)$ **50.** $3(5x)$ **51.** $5(x-7)$

52. $4(x-3)$ **53.** $5-3(x+4)$ **54.** $1-6(x+3)$

Let x be a number. Translate the English phrase to a mathematical expression. Then simplify the expression.

55. 3 times the sum of the number and 2

56. -5 times the difference of the number and 7

57. -4 times the difference of twice the number and 5

58. 2 times the sum of twice the number and 6

59. 5, minus 2 times the difference of the number and 4

60. 6, minus 3 times the sum of the number and 1

61. -2, plus 7 times the sum of twice the number and 1

62. 3, minus 2 times the difference of twice the number and 5

63. a. Simplify $2(x-3)+4$.
   b. Evaluate $2(x-3)+4$ for $x=5$.
   c. Evaluate your result in part (a) for $x=5$.
   d. Compare your results in parts (b) and (c). If the results are equal, what does this tell you? If the results are not equal, what does that tell you?

64. a. Simplify $7-4(x+2)$.
   b. Evaluate $7-4(x+2)$ for $x=3$.
   c. Evaluate your result in part (a) for $x=3$.
   d. Compare your results in parts (b) and (c). If the results are equal, what does this tell you? If the results are not equal, what does that tell you?

65. A student tries to simplify $3(x+4)$:

$$3(x+4)=3x+4$$

Evaluate both $3(x+4)$ and $3x+4$ for $x=2$ and compare your results. What does this comparison tell you?

66. A student tries to simplify $-2(x-5)$:

$$-2(x-5) = -2x - 10$$

Evaluate both $-2(x-5)$ and $-2x-10$ for $x = 3$ and compare your results. What does this comparison tell you?

67. It is a common error to write $a(bc) = (ab)(ac)$.
  a. Evaluate both $a(bc)$ and $(ab)(ac)$ for $a = 2$, $b = 3$, and $c = 4$.
  b. Explain why your results show that the statement $a(bc) = (ab)(ac)$ is false.
  c. Write a true statement that involves $a(bc)$.

68. It is a common error to write $a(b+c) = ab + c$.
  a. Evaluate both $a(b+c)$ and $ab + c$ for $a = 2$, $b = 3$, and $c = 4$.
  b. Explain why your results show that the statement $a(b+c) = ab + c$ is false.
  c. Write a true statement that involves $a(b+c)$.

Simplify the right-hand side of the equation. Then graph the equation by hand. Finally, use graphing calculator graphs to verify your work.

69. $y = 2(x-3)$ **70.** $y = -4(x+2)$

71. $y = -3(x+1)$ **72.** $y = -2(x-2)$

73. $y = -2(3x)$ **74.** $y = 4(2x)$

75. Describe the meaning of *equivalent expressions*. (See page 9 for guidelines on writing a good response.)

76. Give an example of two equivalent expressions.

77. Give three examples of expressions that are equivalent to the expression $2x - 6$.

78. Give an example of two expressions that are not equivalent, but that happen to give the same result when they are evaluated for $x = 0$.

79. Show that the statement $a - (b - c) = (a - b) - c$ is false by substituting 7 for a, 5 for b, and 1 for c. Explain why your work shows that the statement is false and that there is no associative law for subtraction.

80. Show that the statement $a \div (b \div c) = (a \div b) \div c$ is false by substituting 8 for a, 4 for b, and 2 for c. Explain why your

work shows that the statement is false and why there is no associative law for division.

81. Explain what law was used in each step to rearrange terms of $a + b + c$ as follows:

$$
\begin{aligned}
a + b + c &= (a + b) + c && \text{Add from left to right.}\\
&= (b + a) + c\\
&= b + (a + c)\\
&= b + (c + a)\\
&= (b + c) + a\\
&= b + c + a && \text{Add from left to right.}
\end{aligned}
$$

82. Explain what law was used in each step to rearrange factors of abc as follows:

$$
\begin{aligned}
abc &= (ab)c && \text{Multiply from left to right.}\\
&= a(bc)\\
&= a(cb)\\
&= (ac)b\\
&= (ca)b\\
&= cab && \text{Multiply from left to right.}
\end{aligned}
$$

83. Explain what law was used in each step to show that $(a + b)c = ac + bc$:

$$
\begin{aligned}
(a + b)c &= c(a + b)\\
&= ca + cb\\
&= ac + bc
\end{aligned}
$$

84. Use the distributive law $a(b - c) = ab - ac$ and other laws to show that $(a - b)c = ac - bc$. [**Hint:** See Exercise 83.]

85. In Section 2.5, we learned that the product of two numbers with the same sign is positive. Explain what property or law was used in each step to show why $(-2)(-3) = 6$:

$$
\begin{aligned}
(-2)(-3) &= (-1 \cdot 2)(-3)\\
&= -1(2 \cdot (-3))\\
&= -1(-6)\\
&= -(-6)\\
&= 6
\end{aligned}
$$

4.2 SIMPLIFYING EXPRESSIONS

Objectives

▹ Learn to combine like terms.

▹ Simplify expressions.

▹ Locate errors in simplifying expressions by evaluating expressions.

In Section 4.1, we discussed several ways to simplify an expression. In this section, we discuss yet another technique used to simplify an expression: *combining like terms*.

Combining Like Terms

The **coefficient** of a variable term is the constant factor of the term. For example, the coefficient of the variable term $-3x$ is -3. Since $x = 1 \cdot x$, the coefficient of the variable term x is 1. Since $-x = -1x$, the coefficient of the variable term $-x$ is -1.

Like terms are either constant terms or variable terms that contain the same variable(s) raised to exactly the same power(s). For example, 5 and 9 are like terms; so are $3x$ and $8x$. Also, $2y^3$, $5y^3$, and $-6y^3$ are like terms.

If terms are not like terms, we say that they are **unlike terms.** For example, $3x$ and $8y$ are unlike terms, because $3x$ contains an x but $8y$ does not and because $8y$ contains a y but $3x$ does not. The terms $4x^2$ and $7x^3$ are unlike terms, because the exponents of x are different.

For a sum of like terms, such as $2x + 4x$, we can use the distributive law to write the sum as one term:

$$2x + 4x = (2 + 4)x \quad \text{Distributive law: } ac + bc = (a + b)c$$
$$= 6x \quad \text{Add.}$$

When we write a sum or difference of like terms as one term, we say that we have **combined like terms.**

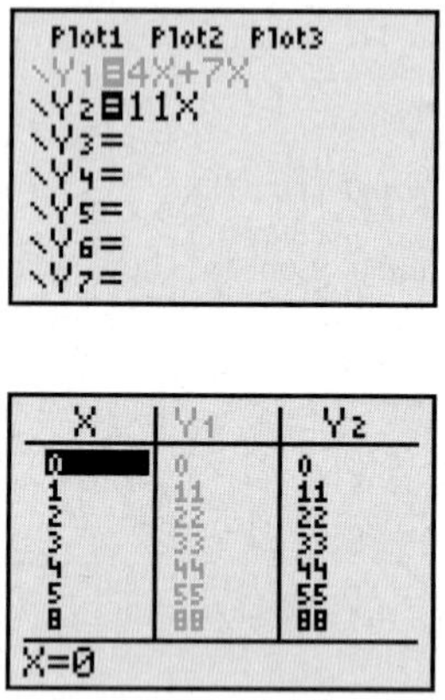

Example 1 Combining Two Like Terms

Combine like terms.

1. $4x + 7x$ **2.** $8x - 5x$ **3.** $6a + a$

Solution

1. $4x + 7x = (4 + 7)x \quad$ Distributive law: $ac + bc = (a + b)c$
$\qquad\qquad = 11x \quad$ Add.

We use a graphing calculator table to verify our work (see Fig. 4).

2. $8x - 5x = (8 - 5)x \quad$ Distributive law: $ac - bc = (a - b)c$
$\qquad\qquad = 3x \quad$ Subtract.

3. $6a + a = 6a + 1a \quad a = 1a$
$\qquad\quad = (6 + 1)a \quad$ Distributive law: $ac + bc = (a + b)c$
$\qquad\quad = 7a \quad$ Add. ■

Figure 4 Verify the work

In Problem 1 of Example 1, we found that $4x + 7x = 11x$. We can find the sum $4x + 7x$ in one step by adding the coefficients of $4x$ and $7x$ (that is, 4 and 7), to get the coefficient of the result, 11:

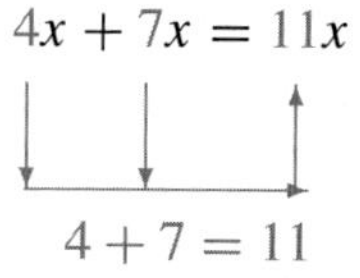

$$4x + 7x = 11x$$

$$4 + 7 = 11$$

In Problem 2 of Example 1, we found that $8x - 5x = 3x$. Note that if we write $8x - 5x = 8x + (-5x)$, then we can add the coefficients of $8x$ and $-5x$ (that is, 8 and -5), to get the coefficient of the result, 3:

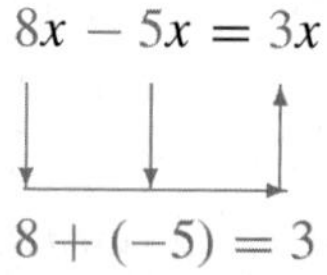

$$8x - 5x = 3x$$

$$8 + (-5) = 3$$

Combining Like Terms

To combine like terms, add the coefficients of the terms and keep the same variable factors.

Simplifying an Expression

We simplify an expression by removing parentheses and combining like terms.

Example 2   Simplifying Expressions

Simplify.

1. $2x + 5y - 6x + 2 + 3y + 7$ 2. $-2(x + 5y) - 3(2x - 4y)$
3. $4(x - 2) - (x + 3)$

Solution

1. $2x + 5y - 6x + 2 + 3y + 7 = 2x - 6x + 5y + 3y + 2 + 7$ Rearrange terms.
$$= -4x + 8y + 9$$ Combine like terms.

2. $-2(x + 5y) - 3(2x - 4y) = -2x - 10y - 6x + 12y$ Distributive law
$$= -2x - 6x - 10y + 12y$$ Rearrange terms.
$$= -8x + 2y$$ Combine like terms.

3. We write $4(x - 2) - (x + 3)$ as $4(x - 2) - 1(x + 3)$ so that we can distribute the -1:

$$4(x - 2) - (x + 3) = 4(x - 2) - 1(x + 3)$$ $a - b = a - 1b$
$$= 4x - 8 - 1x - 3$$ Distributive law
$$= 4x - 1x - 8 - 3$$ Rearrange terms.
$$= 3x - 11$$ Combine like terms.

We use a graphing calculator table to verify the work (see Fig. 5). ∎

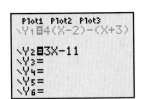

Figure 5 Verify the work

Example 3   Translating from English to Mathematics

Let x be a number. Translate the phrase "3 times the number, minus 4, plus twice the number" to an expression. Then simplify the expression.

Solution

The expression is

3 times the number,	minus 4,	plus	twice the number
$3x$	-4	$+$	$2x$

Next, we simplify the expression:

$$3x - 4 + 2x = 3x + 2x - 4$$ Rearrange terms.
$$= 5x - 4$$ Combine like terms. ∎

In Example 3, we translated an English phrase into a mathematical expression. In Example 4, we will translate a mathematical expression into an English phrase.

Example 4   Translating from Mathematics to English

Let x be a number. Translate the expression $x - 5 \cdot (x - 2)$ into an English phrase. Then simplify the expression.

Solution

Here is the translation:

x $-$	$5 \cdot$	$(x$ $-$ $2)$
the number, minus	5 times	the difference of the number and 2

One of many possible correct translations is "the number, minus 5 times the difference of the number and 2." Next, we simplify the expression:

$$x - 5 \cdot (x - 2) = x - 5x + 10$$ Distributive law
$$= -4x + 10$$ Combine like terms. ∎

Locate Errors in Simplifying Expressions

So far, we have used a graphing calculator table to verify our work in simplifying an expression. If we determine that we have made an error, we still have to find it to correct it. In Example 5, we will discuss how to pinpoint the step(s) in which any errors have been made.

Example 5   Locating an Error

Consider the following incorrect work:

$$5x + 6 + 3(x + 4) = 5x + 6 + 3x + 4 \quad \text{Incorrect}$$
$$= 5x + 3x + 6 + 4$$
$$= 8x + 10$$

1. Use a graphing calculator table to show that the work is incorrect.
2. Pinpoint where the error was made by evaluating the expression in each step for $x = 2$.

Solution

1. From Fig. 6, we can tell that the expressions $5x + 6 + 3(x + 4)$ and $8x + 10$ are *not* equivalent. So, an error has been made.

Figure 6   The work is incorrect

2. We evaluate each expression for $x = 2$:

Expression	Evaluate the Expression for $x = 2$
$5x + 6 + 3(x + 4)$	$5(2) + 6 + 3(2 + 4) = 34$
$= 5x + 6 + 3x + 4$	$5(2) + 6 + 3(2) + 4 = 26$
$= 5x + 3x + 6 + 4$	$5(2) + 3(2) + 6 + 4 = 26$
$= 8x + 10$	$8(2) + 10 = 26$

Since the result of evaluating $5x + 6 + 3(x + 4)$ for $x = 2$ is different from the result of evaluating $5x + 6 + 3x + 4$ for $x = 2$, we conclude that an error was made in writing $5x + 6 + 3(x + 4) = 5x + 6 + 3x + 4$. (Note that the right-hand side of the equation *should* be $5x + 6 + 3x + 12$.) ∎

Locating an Error in Simplifying an Expression

After simplifying an expression, use a graphing calculator table to check whether the result is equivalent to the original expression. If the expressions are not equivalent, an error was made. To pinpoint the error, evaluate all of the expressions by using the same number(s) and find the pair of consecutive expressions for which the results are different.

---

group exploration

Laws of operations

For the work shown, carefully describe any errors.

1. $3(x + 5) + 4x = 3x + 5 + 4x$
$$= 3x + 4x + 5$$
$$= 7x + 5$$

2. $2(xy) = (2x)(2y)$
$= 2(2)xy$
$= 4xy$

3. $5 + x - 3 = 5 + 3 - x$
$= 8 - x$
$= x - 8$

TIPS FOR SUCCESS: Read Your Response

After writing a response to a question, read your response to make sure that it says what you intended it to say. Reading your response aloud, if you are in a place where you feel comfortable doing so, can help.

HOMEWORK 4.2   FOR EXTRA HELP ▶

Student Solutions Manual   PH Math/Tutor Center   MathXL®   MyMathLab

Simplify. Use a graphing calculator table to verify your work when possible.

1. $2x + 5x$
2. $3x + 6x$

3. $9x - 4x$
4. $6x - 5x$

5. $-8w - 5w$
6. $-4p + 7p$

7. $-t + 5t$
8. $a + 3a$

9. $6.6x - 7.1x$
10. $-4.5x - 2.9x$

11. $\frac{2}{3}x + \frac{5}{3}x$
12. $\frac{9}{5}x - \frac{2}{5}x$

13. $2 + 4x - 5 - 7x$
14. $-8x - 1 + 3x - 4$

15. $-3p + 2 + p - 9$
16. $7 - w - 10 - 2w$

17. $3y + 5x - 2y - 2x + 1$
18. $5 - 3x + 7y + 6x$

19. $-4.6x + 3.9y + 2.1 - 5.3x - 2.8y$

20. $4.7 - 3.5y + 8.8x - 6.2y + 1.9x$

21. $-3(a - 5) + 2a$
22. $5(t - 8) - 6t$

23. $5.2(8.3x + 4.9) - 2.4$
24. $-3.8(2.7x - 5.5) - 8.4$

25. $4(3a - 2b) - 5a$
26. $-2(8m + 4n) - 3n$

27. $8 - 2(x + 3) + x$
28. $3 - 4(x + 1) - x$

29. $6x - (4x - 3y) - 5y$
30. $8x - (3x + 7y) - 2y$

31. $2t - 3(5t + 2) + 1$
32. $3a - 2(3a + 4) + 5$

33. $6 - 2(x + 3y) + 2y$
34. $4 - 5(x + 2y) + 7y$

35. $-3(x - 2) - 5(x + 4)$
36. $-2(x - 3) - 7(x + 2)$

37. $6(2x - 3y) - 4(9x + 5y)$

38. $2(7x - 5y) - 5(3x + y)$

39. $-(x - 1) - (1 - x)$

40. $-(6x - 7) - (7 - 6x)$

41. $2x - 5y - 3(2x - 4y + 7)$

42. $4x + 3y - 2(5x + 2y + 8)$

43. $5(2x - 4y) - (3x - 7y + 2)$

44. $3(4x - 3y) - (9x + 2y - 5)$

45. $\frac{2}{7}(a + 1) - \frac{4}{7}(a - 1)$
46. $\frac{3}{5}(t - 1) + \frac{1}{5}(t + 1)$

47. $5x - \frac{1}{2}(4x + 6)$
48. $7x - \frac{1}{3}(6x + 9)$

Let x be a number. Translate the English phrase into a mathematical expression. Then simplify the expression.

49. The number plus the product of 5 and the number

50. The number minus the product of the number and 3

51. 4 times the difference of the number and 2

52. -6 times the sum of the number and 4

53. The number, plus 3 times the difference of the number and 7

54. The number, minus 5 times the sum of the number and 2

55. Twice the number, minus 4 times the sum of the number and 6

56. Twice the number, plus 9 times the difference of the number and 4

For Exercises 57–64, let x be a number. Translate the expression into an English phrase. Then simplify the expression.

57. $2x + 6x$
58. $3x - 8x$

59. $7(x - 5)$
60. $-2(x + 4)$

61. $x + 5(x + 1)$
62. $x - 8(x - 3)$

63. $2x - 3(x - 9)$
64. $2x + 6(x - 4)$

65. Find the sum of $3x - 7$ and $5x + 2$.

66. Find the sum of $6x - 3$ and $8x - 9$.

67. Find the difference of $4x + 8$ and $7x - 1$.

68. Find the difference of $5x - 9$ and $2x + 6$.

For Exercises 69–72, simplify. Then evaluate both the expression and your result for $x = 4$ to check your work. Finally, evaluate both expressions for an x-value of your choosing (other than 4) to be even more confident that your work is correct.

69. $-2x + 5 - 3 + 7x$
70. $4x - 8 - 2 + x$

71. $4(x + 2) - (x - 3)$
72. $-(x - 7) + 4(x + 2)$

73. a. Simplify $3(x + 4) + 5x$.
b. Evaluate $3(x + 4) + 5x$ for $x = 2$.
c. Evaluate your result in part (a) for $x = 2$.
d. Compare your results in parts (b) and (c). If the results are equal, what does this tell you? If the results are not equal, what does that tell you?

74. a. Simplify $4(x - 2) - 3(x + 1)$.
  b. Evaluate $4(x - 2) - 3(x + 1)$ for $x = 3$.
  c. Evaluate your result in part (a) for $x = 3$.
  d. Compare your results in parts (b) and (c). If the results are equal, what does this tell you? If the results are not equal, what does that tell you?

75. Which of the following expressions are equivalent?

$$-2x - 3 \qquad -2(x - 3) \qquad 2(3 - x) \qquad -2x - 6$$
$$-3(x - 2) + x \qquad -2x + 6$$

76. Which of the following expressions are equivalent?

$$-(2x - 3) \qquad 3 - 2x \qquad -(3 - 2x) \qquad -2x + 3$$
$$-(3x - 2) + x + 1 \qquad -(x - 1) - (x - 2)$$

77. A student incorrectly simplifies $4(x - 3) + 5x - 1$:

Expression	Evaluate the Expression for $x = 2$
$4(x - 3) + 5x - 1$	
$= 4x - 3 + 5x - 1$	
$= 4x + 5x - 3 - 1$	
$= 9x - 4$	

Evaluate each of the four expressions for $x = 2$ to pinpoint where the error was made. Then simplify $4(x - 3) + 5x - 1$ correctly.

78. A student incorrectly simplifies $2(x + 1) + x - 4$:

Expression	Evaluate the Expression for $x = 3$
$2(x + 1) + x - 4$	
$= 2x + 2 + x - 4$	
$= 2x + x + 2 - 4$	
$= 2x - 2$	

Evaluate each of the four expressions for $x = 3$ to pinpoint where the error was made. Then simplify $2(x + 1) + x - 4$ correctly.

79. Give three examples of expressions equivalent to the expression $2(x - 5) + 3(x + 1)$.

80. Give three examples of expressions equivalent to x.

Simplify the right-hand side of the equation. Then graph the equation by hand. Finally, use graphing calculator graphs to verify your work.

81. $y = 3x - 5x$ **82.** $y = x + 2x$

83. $y = 9x - 4 - 7x$ **84.** $y = 4x + 3 - 6x$

85. $y = 4(2x - 1) - 5x$ **86.** $y = -2(3x - 2) + 5x$

87. Describe how to simplify an expression.

4.3 SOLVING LINEAR EQUATIONS IN ONE VARIABLE

The successful man will profit from his mistakes and try again in a different way.

—Dale Carnegie

Objectives

▸ Know the meaning of *linear equation in one variable*.

▸ For an equation in one variable, know the meaning of *satisfy, solution, solution set,* and *solve*.

▸ Know three roles of variables.

▸ Know the meaning of *equivalent equations*.

▸ Know the addition property of equality and the multiplication property of equality.

▸ Solve a linear equation in one variable.

▸ Use graphing or a table to solve a linear equation in one variable.

▸ Think of solving an equation in terms of "undoing" operations.

So far, we have discussed how to graph linear equations in *two* variables, such as

$$y = x + 2 \qquad y = -5x \qquad y = 4x - 7 \qquad y = 2x - 6$$

In this section and Section 4.4, we will work with *linear equations in one variable,* such as:

$$0 = x + 2 \qquad 9 = -5x \qquad 4x - 7 = 3 \qquad x + 1 = 2x - 6$$

We will see in this section and in Section 4.4 that we can put each of these four equations in the form $mx + b = 0$, where m and b are constants and $m \neq 0$.

> **DEFINITION** Linear equation in one variable
>
> A **linear equation in one variable** is an equation that can be put into the form
>
> $$mx + b = 0$$
>
> where m and b are constants and $m \neq 0$.

Working with linear equations in one variable will enable us to make efficient estimates and predictions about authentic situations.

Meaning of Satisfy, Solution, Solution Set, and Solve

Consider the linear equation

$$x + 1 = 6$$

This equation becomes a false statement if we substitute 2 for x:

$$
\begin{aligned}
x + 1 &= 6 && \text{Original equation} \\
(2) + 1 &\overset{?}{=} 6 && \text{Substitute 2 for } x. \\
3 &\overset{?}{=} 6 && \text{Add.} \\
&\text{false}
\end{aligned}
$$

However, the equation $x + 1 = 6$ becomes a true statement if we substitute 5 for x:

$$
\begin{aligned}
x + 1 &= 6 && \text{Original equation} \\
(5) + 1 &\overset{?}{=} 6 && \text{Substitute 5 for } x. \\
6 &\overset{?}{=} 6 && \text{Add.} \\
&\text{true}
\end{aligned}
$$

We say that 5 *satisfies* the equation $x + 1 = 6$ and that 5 is a *solution* of the equation. In fact, 5 is the only solution of $x + 1 = 6$, because 5 is the only number that, when increased by 1, is equal to 6. We call the set containing only this number the *solution set* of the equation.

DEFINITION   *Solution, satisfy, solution set,* and *solve* for an equation in one variable

A number is a **solution** of an equation in one variable if the equation becomes a true statement when the number is substituted for the variable. We say that the number **satisfies** the equation. The set of all solutions of the equation is called the **solution set** of the equation. We **solve** the equation by finding its solution set.

Example 1   Identifying Solutions of an Equation

1. Is 3 a solution of the equation $5(x - 1) = 10 + 2x$?
2. Is 5 a solution of the equation $5(x - 1) = 10 + 2x$?

Solution

1. We begin by substituting 3 for x in $5(x - 1) = 10 + 2x$:

$$
\begin{aligned}
5(x - 1) &= 10 + 2x && \text{Original equation} \\
5(3 - 1) &\overset{?}{=} 10 + 2(3) && \text{Substitute 3 for } x. \\
5(2) &\overset{?}{=} 10 + 6 && \text{Simplify.} \\
10 &\overset{?}{=} 16 && \text{Simplify.} \\
&\text{false}
\end{aligned}
$$

So, 3 is not a solution of the equation $5(x - 1) = 10 + 2x$.

2. We begin by substituting 5 for x in $5(x - 1) = 10 + 2x$:

$$
\begin{aligned}
5(x - 1) &= 10 + 2x && \text{Original equation} \\
5(5 - 1) &\overset{?}{=} 10 + 2(5) && \text{Substitute 5 for } x. \\
5(4) &\overset{?}{=} 10 + 10 && \text{Simplify.} \\
20 &\overset{?}{=} 20 && \text{Simplify.} \\
&\text{true}
\end{aligned}
$$

So, 5 is a solution of the equation $5(x - 1) = 10 + 2x$.

Roles of a Variable

In Section 2.1, we discussed two roles of a variable. In one role, a variable represents a quantity that can vary. For example, if we let s represent the speed of an airplane, then the variable s can vary between 0 and 2200 miles per hour. In another role, a variable can serve as a placeholder in an expression. For example, the variable x is a placeholder for a number in the expression $2x + 5$.

A variable can be used in yet another way: In an equation, a variable is used to represent any number that is a solution of the equation. For instance, in Example 1, the variable x is used to represent any number that satisfies the equation $5(x-1) = 10 + 2x$. In Problem 2 of Example 1, we found that 5 is such a number.

Roles of a Variable

Here are three roles of a variable:

1. A variable represents a quantity that can vary.
2. In an expression, a variable is a placeholder for a number.
3. In an equation, a variable represents any number that is a solution of the equation.

Here, we emphasize the third role of a variable.

Equivalent Equations

Consider the equation $x = 2$. We add 5 to both sides of the equation:

$$x = 2 \qquad \text{Original equation}$$
$$x + 5 = 2 + 5 \qquad \text{Add 5 to both sides.}$$
$$x + 5 = 7 \qquad \text{Add.}$$

Note that 2 satisfies all three equations:

Equation	Does 2 satisfy the equation?	
$x = 2$	$(2) \overset{?}{=} 2$	true
$x + 5 = 2 + 5$	$(2) + 5 \overset{?}{=} 2 + 5$	true
$x + 5 = 7$	$(2) + 5 \overset{?}{=} 7$	true

In fact, 2 is the *only* number that satisfies any of these equations. So, the equations $x = 2$, $x + 5 = 2 + 5$, and $x + 5 = 7$ have the same solution set. We say that the three equations are *equivalent*.

DEFINITION Equivalent equations

Equivalent equations are equations that have the same solution set.

Addition Property of Equality

The fact that the equations $x = 2$ and $x + 5 = 2 + 5$ have the same solution set suggests that adding a number to both sides of an equation does not change an equation's solution set. This property is called the **addition property of equality.**

Addition Property of Equality

If A and B are expressions and c is a number, then the equations $A = B$ and $A + c = B + c$ are equivalent.

To solve an equation in one variable, x, we can sometimes use the addition property of equality to get x alone on one side of the equation. Then we can identify solutions of the equation. For example, for the equation $x = 3$, we can see that the solution is 3.

Example 2   Solving an Equation by Adding a Number to Both Sides

Solve $x - 2 = 3$.

Solution

To get x alone on the left side, we add 2 to *both* sides:

$$x - 2 = 3 \qquad \text{Original equation}$$
$$x - 2 + 2 = 3 + 2 \qquad \text{Addition property of equality: Add 2 to both sides.}$$
$$x + 0 = 5 \qquad \text{Simplify: } -a + a = 0$$
$$x = 5 \qquad \text{Simplify: } a + 0 = a$$

Next, we check that 5 satisfies the original equation, $x - 2 = 3$:

$$x - 2 = 3 \qquad \text{Original equation}$$
$$(5) - 2 \overset{?}{=} 3 \qquad \text{Substitute 5 for } x.$$
$$3 \overset{?}{=} 3 \qquad \text{Subtract.}$$
$$\text{true}$$

So, the solution is 5. ∎

After solving an equation, check that all of your results satisfy the equation.

In Example 2, we worked with the equations $x - 2 = 3$, $x - 2 + 2 = 3 + 2$, $x + 0 = 5$, and $x = 5$. Although we could find the solution set with any of these equivalent equations, it is easiest to determine the solution set from the equation $x = 5$, which has the variable alone on the left side. **Our strategy in solving linear equations in one variable will be to use properties to get the variable alone on one side of the equation.**

Assume that A and B are expressions and that c is a number. Since subtracting a number is the same as adding the opposite of the number, the addition property of equality implies that the equations $A = B$ and $A - c = B - c$ are equivalent.

Example 3   Solving an Equation by Subtracting a Number from Both Sides

Solve $x + 4 = 6$.

Solution

To get x alone on the left side, we subtract 4 from *both* sides:

$$x + 4 = 6 \qquad \text{Original equation}$$
$$x + 4 - 4 = 6 - 4 \qquad \text{Subtract 4 from both sides.}$$
$$x = 2 \qquad a + 0 = a$$

Next, we check that 2 satisfies the original equation, $x + 4 = 6$:

$$x + 4 = 6 \qquad \text{Original equation}$$
$$(2) + 4 \overset{?}{=} 6 \qquad \text{Substitute 2 for } x.$$
$$6 \overset{?}{=} 6 \qquad \text{Add.}$$
$$\text{true}$$

The solution is 2. ∎

Multiplication Property of Equality

Consider the equation $x = 3$. We multiply both sides of $x = 3$ by 7:

$$x = 3 \qquad \text{Original equation}$$
$$7 \cdot x = 7 \cdot 3 \qquad \text{Multiply both sides by 7.}$$
$$7x = 21 \qquad \text{Multiply.}$$

Note that 3 satisfies all three equations:

Equation	**Does 3 satisfy the equation?**	
$x = 3$	$(3) \overset{?}{=} 3$	true
$7 \cdot x = 7 \cdot 3$	$7 \cdot (3) \overset{?}{=} 7 \cdot 3$	true
$7x = 21$	$7(3) \overset{?}{=} 21$	true

In fact, 3 is the *only* number that satisfies any of these equations. So the equations $x = 3$, $7 \cdot x = 7 \cdot 3$, and $7x = 21$ are equivalent. The fact that the equations $x = 3$ and $7 \cdot x = 7 \cdot 3$ are equivalent suggests the **multiplication property of equality.**

Multiplication Property of Equality

If A and B are expressions and c is a nonzero number, then the equations $A = B$ and $Ac = Bc$ are equivalent.*

The multiplication property of equality is often helpful in solving equations that contain fractions. It is useful to know that a fraction times its reciprocal is equal to 1. For example,

$$\frac{2}{7} \cdot \frac{7}{2} = \frac{14}{14} = 1$$

Example 4 Solving an Equation by Multiplying Both Sides by a Number

Solve $\frac{4}{5}x = 8$.

Solution

To get x alone on the left side, we multiply *both* sides by the reciprocal of $\frac{4}{5}$, which is $\frac{5}{4}$:

$$\frac{4}{5}x = 8 \qquad \text{Original equation}$$

$$\frac{5}{4} \cdot \frac{4}{5}x = \frac{5}{4} \cdot 8 \qquad \text{Multiplication property of equality: Multiply both sides by } \frac{5}{4}.$$

$$1x = 10 \qquad \text{Simplify.}$$

$$x = 10 \qquad 1a = a$$

The solution is 10. We use a graphing calculator table to check that, when 10 is substituted for x in the expression $\frac{4}{5}x$, the result is 8 (see Fig. 7). For graphing calculator instructions on using "Ask" in a table, see Section A.15.

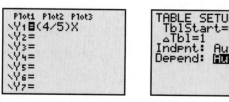

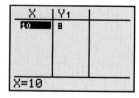

Figure 7 Verify the work ∎

Assume that A and B are expressions and that c is a nonzero number. Since dividing by a number is the same as multiplying by its reciprocal, the multiplication property of equality implies that the equations $A = B$ and $\frac{A}{c} = \frac{B}{c}$ are equivalent.

---

*If $c = 0$, the equations may not be equivalent. For example, $x = 5$ and $0 \cdot x = 0 \cdot 5$ (or $0 = 0$) are not equivalent. The only solution of $x = 5$ is 5, but the material in Section 4.5 will help us see that the solution set of $0 = 0$ is the set of all real numbers.

Example 5   Solving an Equation by Dividing Both Sides by a Number

Solve $-12 = -3t$.

Solution

The variable term $-3t$ is on the right-hand side this time. To get t alone on this side, we divide *both* sides by -3:

$$-12 = -3t \qquad \text{Original equation}$$

$$\frac{-12}{-3} = \frac{-3t}{-3} \qquad \text{Divide both sides by } -3.$$

$$4 = t \qquad \text{Simplify.}$$

The solution is 4. We can check that 4 satisfies the original equation (try it).   ∎

The last step of Example 5 is $4 = t$. Note that the equation $t = 4$ is equivalent to $4 = t$ and that we can see from either equation that the solution is 4.

Example 6   Solving an Equation by Multiplying Both Sides by -1

Solve $-w = 5$.

Solution

To get w alone on the left side, we multiply both sides by -1:

$$-w = 5 \qquad \text{Original equation}$$

$$-1(-w) = -1(5) \qquad \text{Multiply both sides by } -1.$$

$$w = -5 \qquad \text{Multiply.}$$

Next, we check that -5 satisfies the original equation, $-w = 5$:

$$-w = 5 \qquad \text{Original equation}$$

$$-(-5) \stackrel{?}{=} 5 \qquad \text{Substitute } -5 \text{ for } w.$$

$$5 \stackrel{?}{=} 5 \qquad -(-a) = a$$

$$\text{true}$$

The solution is -5.   ∎

Example 7   Solving an Equation by Multiplying Both Sides by a Number

Solve $\dfrac{2x}{3} = \dfrac{5}{6}$.

Solution

Since $\dfrac{2}{3}x = \dfrac{2}{3} \cdot \dfrac{x}{1} = \dfrac{2x}{3}$, we first write $\dfrac{2x}{3}$ as $\dfrac{2}{3}x$. Then we multiply both sides by the reciprocal of $\dfrac{2}{3}$, which is $\dfrac{3}{2}$, to get x alone on the left side:

$$\frac{2x}{3} = \frac{5}{6} \qquad \text{Original equation}$$

$$\frac{2}{3}x = \frac{5}{6} \qquad \text{Write } \frac{2x}{3} \text{ as } \frac{2}{3}x.$$

$$\frac{3}{2} \cdot \frac{2}{3}x = \frac{3}{2} \cdot \frac{5}{6} \qquad \text{Multiply both sides by } \frac{3}{2}.$$

$$1x = \frac{3 \cdot 5}{2 \cdot 2 \cdot 3} \qquad \text{Simplify; } \frac{a}{b} \cdot \frac{c}{d} = \frac{ac}{bd}$$

$$x = \frac{5}{4} \qquad \text{Simplify: } \frac{3}{3} = 1$$

Figure 8 Verify the work.

The solution is $\frac{5}{4}$. We use a graphing calculator table to check that, when $\frac{5}{4} = 1.25$ is substituted for x in the expression $\frac{2x}{3}$, the result is $\frac{5}{6} \approx 0.83333$ (see Fig. 8). ∎

Using Graphing to Solve an Equation in One Variable

In Example 2, we showed that the solution of the equation $x - 2 = 3$ is 5. How can we use graphing to solve this equation? Here are three steps:

Step 1. We set y equal to the left side, $x - 2$, to form the equation $y = x - 2$, and we set y equal to the right side, 3, to form the equation $y = 3$. Then we graph the two equations $y = x - 2$ and $y = 3$ in the same coordinate system (see Fig. 9).

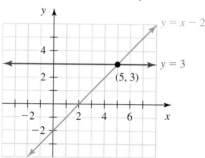

Figure 9 The intersection point is $(5, 3)$

Step 2. We find the intersection point of the two graphs, which is the point $(5, 3)$.

Step 3. The solution of the original equation $x - 2 = 3$ is the x-coordinate of the intersection point $(5, 3)$. So, the solution is 5.

In general, the solutions are all of the x-coordinates of any intersection points.

Using Graphing to Solve an Equation in One Variable

To use graphing to solve an equation $A = B$ in one variable, x, where A and B are expressions,

1. Graph the equations $y = A$ and $y = B$ in the same coordinate system. (For example, if the original equation is $5x - 9 = 3x + 7$, then we would graph the equations $y = 5x - 9$ and $y = 3x + 7$.)
2. Find all intersection points.
3. The x-coordinates of those intersection points are the solutions of the equation $A = B$.

Example 8   Solving an Equation in One Variable by Graphing

The graphs of $y = \frac{3}{2}x + 1$, $y = 4$, and $y = -5$ are shown in Fig. 10. Use these graphs to solve the given equation.

1. $\frac{3}{2}x + 1 = 4$ 2. $\frac{3}{2}x + 1 = -5$

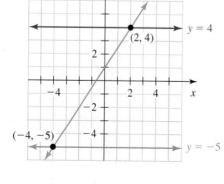

Figure 10 Solving $\frac{3}{2}x + 1 = 4$ and $\frac{3}{2}x + 1 = -5$

Solution

1. The graphs of $y = \frac{3}{2}x + 1$ and $y = 4$ intersect only at the point $(2, 4)$. The intersection point, $(2, 4)$, has x-coordinate 2. So, 2 is the solution of $\frac{3}{2}x + 1 = 4$. We can verify our work by checking that 2 satisfies the equation (try it).

2. The graphs of $y = \frac{3}{2}x + 1$ and $y = -5$ intersect only at the point $(-4, -5)$. The intersection point, $(-4, -5)$, has x-coordinate -4. So, -4 is the solution of $\frac{3}{2}x + 1 = -5$. We can verify our work by checking that -4 satisfies the equation (try it). ■

Example 9   Solving an Equation in One Variable by Graphing

Use "intersect" on a graphing calculator to solve the equation $-2x + 4 = x - 5$.

Solution

We use a graphing calculator to graph the equations $y = -2x + 4$ and $y = x - 5$ in the same coordinate system and then use "intersect" to find the intersection point, which turns out to be $(3, -2)$. See Fig. 11. See Section A.17 for graphing calculator instructions.

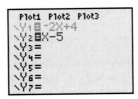 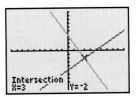

Figure 11 Using "intersect" to solve an equation in one variable

The intersection point, $(3, -2)$, has x-coordinate 3. So, the solution of the original equation is 3. We can verify our work by checking that 3 satisfies the equation $-2x + 4 = x - 5$ (try it). ■

Using a Table to Solve an Equation

We can use a table of solutions of an equation in two variables to help us solve an equation in one variable.

Example 10   Solving an Equation in One Variable by Using a Table

Table 2 Solutions of $y = 3x - 5$

x	y
-1	-8
0	-5
1	-2
2	1
3	4

Some solutions of $y = 3x - 5$ are shown in Table 2. Use the table to solve the equation $1 = 3x - 5$.

Solution

If we substitute 1 for y in the equation $y = 3x - 5$, the result is the equation $1 = 3x - 5$, which is what we are trying to solve. From Table 2, we see that the output $y = 1$ originates from the input $x = 2$. The ordered pair $(2, 1)$ satisfies the equation $y = 3x - 5$:

$$1 = 3(2) - 5$$

This means that 2 is a solution of the equation $1 = 3x - 5$. ■

We have solved an equation by getting the variable alone on one side of the equation, by using graphing, and by using a table. All three methods give the same results.

Undoing Operations

Consider the following four equations and the first step to get x alone on the left side of the equation:

Equation	First Step in Solving Equation	How to Get x Alone on the Left Side
$x + 5 = 2$	Subtract 5 from both sides.	Undo addition by subtracting.
$x - 3 = 6$	Add 3 to both sides.	Undo subtraction by adding.
$\dfrac{x}{2} = 3$	Multiply both sides by 2.	Undo division by multiplying.
$4x = 8$	Divide both sides by 4.	Undo multiplication by dividing.

For the left side of each equation, we get x alone by "undoing" the operation that involves x and the number. For example, we "undo" the addition of 5 to x by subtracting 5 from that sum:

$$(x + 5) - 5 = x$$

This logic also works for equations such as $\dfrac{4}{5}x = 8$. We undo the multiplication of x by $\dfrac{4}{5}$ by dividing that product by $\dfrac{4}{5}$. But dividing by $\dfrac{4}{5}$ is the same as multiplying by the reciprocal of $\dfrac{4}{5}$, which is $\dfrac{5}{4}$. In fact, we did just that in Example 4.

Remember that **whether adding, subtracting, multiplying, or dividing, what is done to one side of the equation must also be done to the other side of the equation.**

group exploration
Locating an error in solving an equation

A student tries to solve the equation $x - 3 = 5$:

$$x - 3 = 5$$
$$x - 3 - 3 = 5 - 3$$
$$x - 0 = 2$$
$$x = 2$$

Substitute 2 for x in each of the four equations to determine which equations have 2 as a solution. Explain how your work shows that the student made an error and how your work helps you pinpoint the step in which the error was made.

group exploration
Looking ahead: Solving linear equations

1. Solve $x + 3 = 11$.
2. Solve $2x = 8$.
3. Solve $2x + 3 = 11$. [**Hint:** First, subtract 3 from both sides.]
4. Solve $6x - 4x + 3 = 11$. [**Hint:** First, combine like terms on the left side of the equation.]
5. Solve the equation $2x + 4 + 3x = 14$.

> **TIPS FOR SUCCESS: Review Material**
>
> At various times throughout this course, you can improve your understanding of algebra by reviewing material that you have learned so far. Your review should include solving problems, redoing explorations, and reexamining concepts and techniques from previous sections.

HOMEWORK 4.3

FOR EXTRA HELP ▶

Student Solutions Manual   PH Math/Tutor Center   MathXL®   MyMathLab

Determine whether 2 is a solution of the equation.

1. $3x + 1 = 7$ **2.** $-2x + 5 = 4$

3. $5(2x - 1) = 0$ **4.** $-3(x - 2) = 0$

5. $12 - x = 2(4x - 3)$ **6.** $5 - x = 3(2x - 1)$

Solve. Verify that your result satisfies the equation.

7. $x - 3 = 2$ **8.** $x - 4 = 9$ **9.** $x + 6 = -8$

10. $x + 1 = -5$ **11.** $t - 9 = 15$ **12.** $a - 5 = 12$

13. $x + 11 = -17$ **14.** $x + 18 = -13$ **15.** $-5 = x - 2$

16. $-3 = x - 4$ **17.** $x - 3 = 0$ **18.** $x - 5 = 0$

19. $6r = 18$ **20.** $4w = 24$ **21.** $-3x = 12$

22. $-5x = 20$ **23.** $15 = 3x$ **24.** $24 = 2x$

25. $6x = 8$ **26.** $4x = 6$ **27.** $-10x = -12$

28. $-14x = -6$ **29.** $-2x = 0$ **30.** $-5x = 0$

31. $\dfrac{1}{3}t = 5$ **32.** $\dfrac{1}{2}w = 4$ **33.** $-\dfrac{2}{7}x = -3$

34. $-\dfrac{4}{9}x = -5$ **35.** $-9 = \dfrac{3x}{4}$ **36.** $-8 = \dfrac{2x}{5}$

37. $\dfrac{2}{5}p = -\dfrac{4}{3}$ **38.** $\dfrac{4}{7}b = -\dfrac{5}{21}$ **39.** $-\dfrac{3x}{8} = -\dfrac{9}{4}$

40. $-\dfrac{5x}{4} = -\dfrac{15}{8}$ **41.** $-x = 3$ **42.** $-x = 2$

43. $-\dfrac{1}{2} = -x$ **44.** $-\dfrac{3}{4} = -x$ **45.** $x + 4.3 = -6.8$

46. $x + 7.5 = -2.8$ **47.** $25.17 = x - 16.59$

48. $5.27 = x - 28.85$ **49.** $-3.7r = -8.51$

50. $-2.9w = 13.34$

For Exercises 51–54, use the graph of $y = -\dfrac{1}{2}x + 1$, shown in Fig. 12, to solve the given equation.

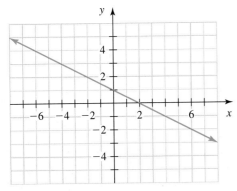

Figure 12 Exercises 51–54

51. $-\dfrac{1}{2}x + 1 = 3$ **52.** $-\dfrac{1}{2}x + 1 = 2$

53. $-\dfrac{1}{2}x + 1 = -1$ **54.** $-\dfrac{1}{2}x + 1 = -2$

Use "intersect" on a graphing calculator to solve the equation.

55. $x + 2 = 7$ **56.** $x - 3 = 4$

57. $2x - 3 = 5$ **58.** $-3x + 5 = -7$

59. $-4(x - 1) = -8$ **60.** $2(x + 5) = 6$

61. $\dfrac{2}{3}t - \dfrac{3}{2} = -\dfrac{7}{2}$ **62.** $\dfrac{5}{2}w - \dfrac{5}{3} = \dfrac{10}{3}$

For Exercises 63–66, solve the given equation by referring to the solutions of $y = 5x - 3$ shown in Table 3.

63. $5x - 3 = 12$ **64.** $5x - 3 = 7$

65. $5x - 3 = -13$ **66.** $5x - 3 = -8$

Table 3 Exercises 63-66	
x	**y**
−3	−18
−2	−13
−1	−8
0	−3
1	2
2	7
3	12

*Solve. [**Hint:** Combine like terms on the left side.]*

67. $2x + 5x = 14$ **68.** $3x + x = 20$

69. $4x - 5x = -2$ **70.** $2x - 8x = -4$

71. A student tries to solve the equation $2x = 10$:

$$2x = 10$$
$$2x - 2 = 10 - 2$$
$$x = 8$$

Describe any errors. Then solve the equation correctly.

72. A student tries to solve the equation $x + 6 = 9$:

$$x + 6 = 9$$
$$x + 6 - 6 = 9$$
$$x + 0 = 9$$
$$x = 9$$

Describe any errors. Then solve the equation correctly.

73. A student solves the equation $4x = 12$:

$$4x = 12$$
$$\frac{4x}{4} = \frac{12}{4}$$
$$x = 3$$

Show that 3 satisfies each of the three equations.

74. A student solves the equation $x - 4 = 7$:

$$x - 4 = 7$$
$$x - 4 + 4 = 7 + 4$$
$$x + 0 = 11$$
$$x = 11$$

Show that 11 satisfies each of the four equations.

75. A student tries to solve the equation $x + 2 = 7$:

$$x + 2 = 7$$
$$x + 2 - 7 = 7 - 7$$
$$x - 5 = 0$$
$$x - 5 + 5 = 0 + 5$$
$$x = 5$$

Is the work correct? If yes, show a way to solve the equation in fewer steps. If no, describe the student's error(s) and then solve the equation correctly.

76. A student tries to solve the equation $\frac{3}{7}x = 2$:

$$\frac{3}{7}x = 2$$
$$7 \cdot \frac{3}{7}x = 7 \cdot 2$$
$$\frac{7}{1} \cdot \frac{3}{7}x = 14$$
$$3x = 14$$
$$\frac{3x}{3} = \frac{14}{3}$$
$$x = \frac{14}{3}$$

Is the work correct? If yes, show a way to solve the equation in fewer steps. If no, describe the student's error(s) and then solve the equation correctly.

77. Give an example of an equation for which it is helpful to add 7 to both sides. Then solve the equation.

78. Give an example of an equation for which it is helpful to subtract 7 from both sides. Then solve the equation.

79. Give an example of an equation for which it is helpful to multiply both sides by 7. Then solve the equation.

80. Give an example of an equation for which it is helpful to divide both sides by 7. Then solve the equation.

81. Are the equations $\frac{x}{3} = 2$ and $x - 1 = 4$ equivalent?

82. Are the equations $2x = 10$ and $x + 2 = 7$ equivalent?

83. Give three examples of equations that have 4 as their solution.

84. Give three examples of equations that have -2 as their solution.

85. Give an example of three equations that are equivalent. Explain.

86. Give an example of two equations that are not equivalent. Explain.

87. Are the equations $\frac{x-5}{2} = \frac{3}{x+1}$ and $\frac{x-5}{2} + 6 = \frac{3}{x+1} + 6$ equivalent? Explain.

88. Are the equations $x(x + 1) = 6$ and $2x(x + 1) = 12$ equivalent? Explain.

89. a. Solve $x + 2 = 7$.
   b. Solve $x + 5 = 9$.
   c. Solve $x + b = k$, where b and k are constants. [**Hint:** Refer to parts (a) and (b). The solution will be in terms of b and k.]

90. a. Solve $2x = 7$.
   b. Solve $5x = 9$.
   c. Solve $mx = p$, where m and p are constants and m is nonzero. [**Hint:** Refer to parts (a) and (b). The solution will be in terms of m and p.]

91. Describe in your own words what the addition property of equality means.

92. Describe in your own words what the multiplication property of equality means.

4.4 SOLVING MORE LINEAR EQUATIONS IN ONE VARIABLE

Objectives

▷ Solve a linear equation in one variable.

▷ Use a model to make predictions.

▷ Translate English sentences to and from mathematical equations.

▷ Use graphing or a table to solve a linear equation in one variable.

▷ Know the meaning of *conditional equation, inconsistent equation,* and *identity* and how to solve these types of equations.

▷ Locate errors in solving linear equations.

In this section, we will solve more complicated linear equations in one variable than those in Section 4.3. We will solve such equations to help us make predictions about authentic situations.

Solving Linear Equations in One Variable

In Section 4.3, we used either the addition property of equality or the multiplication property of equality to solve a linear equation. In this section, we will use a combination of these properties to solve linear equations.

Example 1 Using the Addition and Multiplication Properties of Equality

Solve $2x - 3 = 5$.

Solution

We begin by adding 3 to both sides to get $2x$ alone on the left side:

$$2x - 3 = 5 \qquad \text{Original equation}$$
$$2x - 3 + 3 = 5 + 3 \qquad \text{Add 3 to both sides.}$$
$$2x = 8 \qquad \text{Combine like terms.}$$
$$\frac{2x}{2} = \frac{8}{2} \qquad \text{Divide both sides by 2.}$$
$$x = 4 \qquad \text{Simplify.}$$

Next, we check that 4 satisfies the equation $2x - 3 = 5$:

$$2x - 3 = 5 \qquad \text{Original equation}$$
$$2(4) - 3 \stackrel{?}{=} 5 \qquad \text{Substitute 4 for } x.$$
$$8 - 3 \stackrel{?}{=} 5 \qquad \text{Multiply.}$$
$$5 \stackrel{?}{=} 5 \qquad \text{Subtract.}$$
$$\text{true}$$

So, the solution is 4. ∎

It is wise to check that a result (or the results) of solving an equation does indeed satisfy the equation.

In Example 1, we used the addition property of equality so that a variable term was alone on one side of the equation and a constant term was on the other side. Then we used the multiplication property of equality to get the variable alone on one side of the equation.

We will do the same thing in Example 2, but first we will simplify one side of the equation.

Example 2 Combining Like Terms to Help Solve an Equation

Solve $4x - 7x + 2 = 17$.

Solution

First, we combine like terms on the left side:

$$4x - 7x + 2 = 17 \qquad \text{Original equation}$$
$$-3x + 2 = 17 \qquad \text{Combine like terms.}$$
$$-3x + 2 - 2 = 17 - 2 \qquad \text{Subtract 2 from both sides to get } -3x \text{ alone on left side.}$$
$$-3x = 15 \qquad \text{Combine like terms.}$$
$$\frac{-3x}{-3} = \frac{15}{-3} \qquad \text{Divide both sides by } -3 \text{ to get } x \text{ alone on left side.}$$
$$x = -5 \qquad \text{Simplify.}$$

The solution is -5. We use a graphing calculator table to check that if -5 is substituted for x in the expression $4x - 7x + 2$, the result is 17 (see Fig. 13). ∎

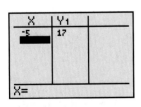

Figure 13 Verify the work

In Example 3, we will solve an equation that contains both variable terms and constant terms on each side of the equation. To solve such an equation, we first use the addition property of equality to write the variable terms on one side of the equation and constant terms on the other side.

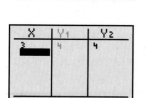

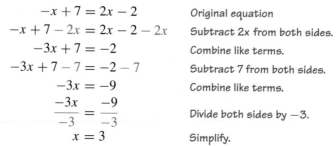

Figure 14 Verify the work

Example 3   Solving an Equation with Variable Terms on Both Sides

Solve $-x + 7 = 2x - 2$.

Solution

First, we subtract $2x$ from both sides to get all of the variable terms on the left side:[‡]

$-x + 7 = 2x - 2$	Original equation
$-x + 7 - 2x = 2x - 2 - 2x$	Subtract 2x from both sides.
$-3x + 7 = -2$	Combine like terms.
$-3x + 7 - 7 = -2 - 7$	Subtract 7 from both sides.
$-3x = -9$	Combine like terms.
$\dfrac{-3x}{-3} = \dfrac{-9}{-3}$	Divide both sides by -3.
$x = 3$	Simplify.

The solution is 3. We use a graphing calculator table to check that if 3 is substituted for x in the expressions $-x + 7$ and $2x - 2$, the two results are equal (see Fig. 14).   ■

When possible, we simplify each side of an equation before using properties such as the addition property of equality or the multiplication property of equality.

Example 4   Using the Distributive Law to Help Solve an Equation

Solve $-2(a - 3) + 4 = 4(a - 2)$.

Solution

First, we use the distributive law on each side:

$-2(a - 3) + 4 = 4(a - 2)$	Original equation
$-2a + 6 + 4 = 4a - 8$	Distributive law
$-2a + 10 = 4a - 8$	Combine like terms.
$-2a + 10 - 4a = 4a - 8 - 4a$	Subtract 4a from both sides.
$-6a + 10 = -8$	Combine like terms.
$-6a + 10 - 10 = -8 - 10$	Subtract 10 from both sides.
$-6a = -18$	Combine like terms.
$\dfrac{-6a}{-6} = \dfrac{-18}{-6}$	Divide both sides by -6.
$a = 3$	Simplify; divide.

Next, we check that 3 satisfies $-2(a - 3) + 4 = 4(a - 2)$:

$-2(a - 3) + 4 = 4(a - 2)$	Original equation
$-2(3 - 3) + 4 \stackrel{?}{=} 4(3 - 2)$	Substitute 3 for a.
$-2(0) + 4 \stackrel{?}{=} 4(1)$	Simplify.
$4 \stackrel{?}{=} 4$	Simplify.
true	

So, 3 is the solution.   ■

A key step in solving an equation that contains fractions is to multiply both sides of the equation by the LCD so that there are no fractions on either side of the equation. An equation that is "cleared of fractions" will be easier to solve than the original equation.

[‡]The addition property of equality implies that we can add a number to both sides of an equation without changing the equation's solution set. We can also add (or subtract) a variable term to both sides of an equation without changing the equation's solution set.

Example 5 Solving an Equation That Contains Fractions

Solve $\dfrac{2}{3}x + \dfrac{1}{6} = \dfrac{3}{4}$.

Solution

To find the LCD of the three fractions in the equation, we list the multiples of 3, the multiples of 6, and the multiples of 4:

$$\text{Multiples of 3:}\quad 3, 6, 9, 12, 15, 18, 21, \ldots$$
$$\text{Multiples of 6:}\quad 6, 12, 18, 24, 30, 36, 42, \ldots$$
$$\text{Multiples of 4:}\quad 4, 8, 12, 16, 20, 24, 28, \ldots$$

The LCD is 12. Next, we multiply both sides of $\dfrac{2}{3}x + \dfrac{1}{6} = \dfrac{3}{4}$ by 12 to clear the equation of fractions:

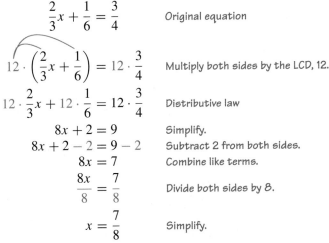

$\dfrac{2}{3}x + \dfrac{1}{6} = \dfrac{3}{4}$	Original equation
$12 \cdot \left(\dfrac{2}{3}x + \dfrac{1}{6}\right) = 12 \cdot \dfrac{3}{4}$	Multiply both sides by the LCD, 12.
$12 \cdot \dfrac{2}{3}x + 12 \cdot \dfrac{1}{6} = 12 \cdot \dfrac{3}{4}$	Distributive law
$8x + 2 = 9$	Simplify.
$8x + 2 - 2 = 9 - 2$	Subtract 2 from both sides.
$8x = 7$	Combine like terms.
$\dfrac{8x}{8} = \dfrac{7}{8}$	Divide both sides by 8.
$x = \dfrac{7}{8}$	Simplify.

The solution is $\dfrac{7}{8}$. We check our work by using ZDecimal to graph the equations $y = \dfrac{2}{3}x + \dfrac{1}{6}$ and $y = \dfrac{3}{4}$ in the same coordinate system and by using "intersect" to find the intersection point, $(0.875, 0.75)$. See Fig. 15. So, the solution of the original equation is $\dfrac{7}{8} = 0.875$, which checks. ∎

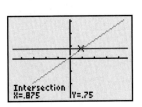

Figure 15 Verify the work

Example 6 Solving an Equation That Contains Fractions

Solve $\dfrac{3x - 1}{2} = \dfrac{4x + 2}{3}$.

Solution

We multiply both sides of $\dfrac{3x - 1}{2} = \dfrac{4x + 2}{3}$ by the LCD, 6, to clear the equation of fractions:

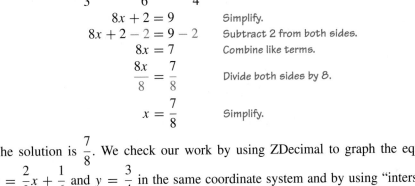

$\dfrac{3x - 1}{2} = \dfrac{4x + 2}{3}$	Original equation
$6 \cdot \dfrac{3x - 1}{2} = 6 \cdot \dfrac{4x + 2}{3}$	Multiply both sides by the LCD, 6.
$3(3x - 1) = 2(4x + 2)$	Simplify.
$9x - 3 = 8x + 4$	Distributive law
$9x - 3 - 8x = 8x + 4 - 8x$	Subtract 8x from both sides.
$x - 3 = 4$	Combine like terms.
$x - 3 + 3 = 4 + 3$	Add 3 to both sides.
$x = 7$	Combine like terms.

We can use "intersect" on a graphing calculator to check our work. ∎

To solve some true-to-life problems, we will need to solve equations that contain decimal numbers. When solving such problems, we round results to two decimal places.

Example 7 Solving an Equation That Contains Decimal Numbers

Solve $2.71t = -3.4(5.9t - 4.8)$. Round any solutions to two decimal places.

Solution

$$
\begin{array}{ll}
2.71t = -3.4(5.9t - 4.8) & \text{Original equation} \\
2.71t = -20.06t + 16.32 & \text{Distributive law} \\
2.71t + 20.06t = -20.06t + 16.32 + 20.06t & \text{Add 20.06t to both sides.} \\
22.77t = 16.32 & \text{Combine like terms.} \\
\dfrac{22.77t}{22.77} = \dfrac{16.32}{22.77} & \text{Divide both sides by 22.77.} \\
t \approx 0.72 & \text{Round right-hand side} \\
 & \text{to two decimal places.}
\end{array}
$$

The solution is approximately 0.72. We can check that 0.72 approximately satisfies the original equation (try it). ∎

Using a Model to Make Predictions

Now that we know how to solve linear equations in one variable, we can make predictions about the dependent variable of a linear model.

Example 8 Making Predictions

The percentage of workers who telecommute at least once a week was 26% in 2000 and increases by about 3.6 percentage points per year (Source: *Society for Human Resource Management*). Let p be the percentage of workers who telecommute at least once a week at t years since 2000.

1. Find a model of the situation.
2. Predict what percentage of workers will telecommute at least once a week in 2011.
3. Predict when half of all workers will telecommute at least once a week.

Solution

1. Since the rate of change of the percentage per year is increasing by about a *constant* 3.6 percentage points per year, the variables t and p are approximately linearly related. We want an equation of the form $p = mt + b$. Because the slope is 3.6 and the p-intercept is $(0, 26)$, a reasonable model is

$$p = 3.6t + 26$$

2. We substitute the input 11 for t in the equation $p = 3.6t + 26$ and solve for p:

$$p = 3.6(11) + 26 = 65.6$$

About 66% of workers will telecommute at least once a week in 2011, according to the model.

3. Half of all the workers is 50%. So, we substitute the output 50 for p in the equation $p = 3.6t + 26$ and solve for t:

$$
\begin{array}{ll}
50 = 3.6t + 26 & \text{Substitute 50 for p.} \\
50 - 26 = 3.6t + 26 - 26 & \text{Subtract 26 from both sides.} \\
24 = 3.6t & \text{Combine like terms.} \\
\dfrac{24}{3.6} = \dfrac{3.6t}{3.6} & \text{Divide both sides by 3.6.} \\
6.67 \approx t & \text{Round left side to two decimal places.}
\end{array}
$$

In $2000 + 6.67 \approx 2007$, half of all workers will telecommute at least once a week, according to the model. We verify our work in Problems 2 and 3 by using a graphing calculator table (see Fig. 16). ∎

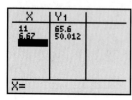

Figure 16 Verify the work

Translating Sentences to and from Equations

Let x represent a number. Here are some sentences that have the same meaning as $x = 4$:

- The number is 4.
- The number is equal to 4.
- The number is the same as 4.

Example 9 Translating an English Sentence to an Equation

Five, minus 2 times the sum of a number and 3 is 11. What is the number?

Solution

Let x be the number. Next, we translate the given information into an equation:

$$\underbrace{5}_{\text{Five, minus}} - \underbrace{2}_{\text{2 times}} \cdot \underbrace{(x + 3)}_{\substack{\text{the sum of a} \\ \text{number and 3}}} = \underbrace{11}_{\text{is 11.}}$$

Then, we solve the equation.

$$
\begin{aligned}
5 - 2(x + 3) &= 11 & &\text{Original equation} \\
5 - 2x - 6 &= 11 & &\text{Distributive law} \\
-2x - 1 &= 11 & &\text{Combine like terms.} \\
-2x - 1 + 1 &= 11 + 1 & &\text{Add 1 to both sides.} \\
-2x &= 12 & &\text{Combine like terms.} \\
\frac{-2x}{-2} &= \frac{12}{-2} & &\text{Divide both sides by } -2. \\
x &= -6 & &\text{Simplify.}
\end{aligned}
$$

So, the number is -6. ∎

In Example 9, we translated an English sentence into a mathematical equation, solved the equation, and then translated the result into an English sentence.

Example 10 Translating an Equation to an English Sentence

Let x be a number. Translate the equation $7x - 1 = 4x + 5$ into an English sentence.

Solution

$$\underbrace{7x}_{\substack{\text{Seven times} \\ \text{the number,}}} \underbrace{- 1}_{\text{minus 1,}} \underbrace{=}_{\substack{\text{is equal} \\ \text{to}}} \underbrace{4x}_{\substack{\text{4 times} \\ \text{the number,}}} \underbrace{+ 5}_{\text{plus 5.}}$$

One of many correct translations is "Seven times the number, minus 1, is equal to 4 times the number, plus 5." The solution of the equation is 2 (try it). ∎

Using Graphing to Solve an Equation in One Variable

In Section 4.3, we used graphing to solve some linear equations in one variable. We can use graphing to solve more complicated linear equations in one variable.

Example 11 Solving an Equation in One Variable by Graphing

The graphs of $y = -\frac{1}{2}x - 1$ and $y = -\frac{5}{4}x + 2$ are shown in Fig. 17. Use these graphs to solve the equation $-\frac{1}{2}x - 1 = -\frac{5}{4}x + 2$.

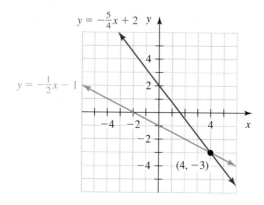

Figure 17 The intersection point is $(4, -3)$

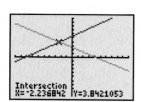

Figure 18 Using "intersect" to solve an equation in one variable

Solution

The two lines intersect only at $(4, -3)$, whose x-coordinate is 4. So, 4 is the solution of the equation $-\dfrac{1}{2}x - 1 = -\dfrac{5}{4}x + 2$. ∎

Example 12   Solving an Equation in One Variable by Graphing

Use "intersect" on a graphing calculator to solve the equation $-\dfrac{3}{5}x + \dfrac{5}{2} = \dfrac{2}{3}x + \dfrac{16}{3}$, with the solution rounded to the second decimal place.

Solution

We use a graphing calculator to graph the equations $y = -\dfrac{3}{5}x + \dfrac{5}{2}$ and $y = \dfrac{2}{3}x + \dfrac{16}{3}$ on the same coordinate system and then use "intersect" to find the approximate intersection point, $(-2.24, 3.84)$. See Fig. 18. See Section A.17 for graphing calculator instructions.

The approximate intersection point $(-2.24, 3.84)$ has x-coordinate -2.24. So, the approximate solution of the original equation is -2.24. ∎

Using Tables to Solve an Equation in One Variable

Recall from Section 4.3 that we can also solve equations by using tables.

Example 13   Solving an Equation in One Variable by Using Tables

The solutions of $y = 2x - 5$ and $y = -4x + 13$ are shown in Tables 4 and 5, respectively. Use the tables to solve the equation $2x - 5 = -4x + 13$.

Table 4 Solutions of $y = 2x - 5$	
x	**y**
0	−5
1	−3
2	−1
3	1
4	3

Table 5 Solutions of $y = -4x + 13$	
x	**y**
0	13
1	9
2	5
3	1
4	−3

Solution

From Tables 4 and 5, we see that for both of the equations $y = 2x - 5$ and $y = -4x + 13$, the input 3 leads to the output 1:

$$2(3) - 5 = 1 \quad \text{and} \quad -4(3) + 13 = 1$$

It follows that

$$2(3) - 5 = -4(3) + 13$$

which means that 3 is a solution of the equation $2x - 5 = -4x + 13$. ∎

Three Types of Equations

Each of the linear equations we solved in Section 4.3 and in this section so far has exactly one solution. In fact, **any linear equation in one variable has exactly one solution.** (You will show this in the first Exploration.)

A linear equation in one variable is an example of a conditional equation. A **conditional equation** is sometimes true and sometimes false, depending on which values are substituted for the variable(s). There are two other types of equations: inconsistent equations and identities.

If an equation does not have any solutions, we call the equation **inconsistent** and say that the solution set is the **empty set.** For example, the equation $x = x + 1$ is inconsistent, because no number is 1 more than itself. Here, we subtract x from both sides of the equation:

$$x = x + 1 \qquad \text{Original equation}$$
$$x - x = x + 1 - x \qquad \text{Subtract x from both sides.}$$
$$0 = 1 \qquad \text{Combine like terms.}$$
$$\text{false}$$

When we apply the usual steps to solve an inconsistent equation, the result is always a false statement.

An equation that is true for all permissible values of the variable(s) it contains is called an **identity.** For example, $x + 5 = x + 2 + 3$ is an identity, because it is true for all real numbers. Here, we add the numbers 2 and 3 and then subtract x from both sides:

$$x + 5 = x + 2 + 3 \qquad \text{Original equation}$$
$$x + 5 = x + 5 \qquad \text{Add.}$$
$$x + 5 - x = x + 5 - x \qquad \text{Subtract x from both sides.}$$
$$5 = 5 \qquad \text{Combine like terms.}$$
$$\text{true}$$

When we apply the usual steps to solve an identity, the result is always a true statement of the form $a = a$, where a is a constant.

Example 14 Solving Three Types of Equations

Solve the equation. State whether the equation is a conditional equation, an inconsistent equation, or an identity.

1. $2x + 1 = 9$
2. $3x + 2 = 3x + 7$
3. $2(x + 4) + 1 = 2x + 9$

Solution

1.
$$2x + 1 = 9 \qquad \text{Original equation}$$
$$2x + 1 - 1 = 9 - 1 \qquad \text{Subtract 1 from both sides.}$$
$$2x = 8 \qquad \text{Combine like terms.}$$
$$\frac{2x}{2} = \frac{8}{2} \qquad \text{Divide both sides by 2.}$$
$$x = 4 \qquad \text{Simplify.}$$

The number 4 is the only solution, so the equation is conditional.

2.
$$3x + 2 = 3x + 7 \qquad \text{Original equation}$$
$$3x + 2 - 3x = 3x + 7 - 3x \qquad \text{Subtract 3x from both sides.}$$
$$2 = 7 \qquad \text{Combine like terms.}$$
$$\text{false}$$

Since $2 = 7$ is a false statement, we conclude that the original equation is inconsistent and its solution set is the empty set.

3.

$$2(x + 4) + 1 = 2x + 9 \qquad \text{Original equation}$$
$$2x + 8 + 1 = 2x + 9 \qquad \text{Distributive law}$$
$$2x + 9 = 2x + 9 \qquad \text{Combine like terms.}$$
$$2x + 9 - 2x = 2x + 9 - 2x \qquad \text{Subtract 2x from both sides.}$$
$$9 = 9 \qquad \text{Combine like terms.}$$
$$\text{true}$$

Since $9 = 9$ is a true statement (of the form $a = a$), we conclude that the original equation is an identity and its solution set is the set of all real numbers. ∎

Locating Errors in Solving Linear Equations in One Variable

In Example 15, we will check whether a result satisfies equations. Doing this will help us locate an error.

Example 15 Locating an Error

Consider the following incorrect work in an attempt to solve the equation $2x + 1 = 5$:

$$2x + 1 = 5$$
$$2x + 1 - 1 = 5 - 1$$
$$2x = 4$$
$$\frac{2x}{2} = 4 \qquad \text{Incorrect}$$
$$x = 4$$

1. Show that the work is incorrect by substituting the result 4 for x in the original equation.
2. Pinpoint the error by substituting 4 for x in all five equations.

Solution

1. We substitute 4 for x in the equation $2x + 1 = 5$:

$$2x + 1 = 5 \qquad \text{Original equation}$$
$$2(4) + 1 \overset{?}{=} 5 \qquad \text{Substitute 4 for x.}$$
$$9 \overset{?}{=} 5 \qquad \text{Simplify.}$$
$$\text{false}$$

Since 4 does not satisfy the original equation $2x + 1 = 5$, we know that an error was made in solving that equation.

2. We substitute 4 for x in all five equations and note which pair of consecutive equations gives a false statement followed by a true statement.

Equation	Does 4 satisfy the equation?	
$2x + 1 = 5$	$2(4) + 1 \overset{?}{=} 5$	false
$2x + 1 - 1 = 5 - 1$	$2(4) + 1 - 1 \overset{?}{=} 5 - 1$	false
$2x = 4$	$2(4) \overset{?}{=} 4$	false
$\dfrac{2x}{2} = 4$	$\dfrac{2(4)}{2} \overset{?}{=} 4$	true
$x = 4$	$(4) \overset{?}{=} 4$	true

Since substituting 4 for x in $2x = 4$ gives a false statement, but substituting 4 for x in the next equation, $\dfrac{2x}{2} = 4$, gives a true statement, we know that an error was made when only the left side of the equation $2x = 4$ was divided by 2:

$$2x = 4 \qquad \text{Third equation of the work}$$
$$\frac{2x}{2} = 4 \qquad \text{Incorrect}$$

Instead, *both* sides of the equation $2x = 4$ should be divided by 2:

$$2x = 4 \qquad \text{Third equation of the work}$$
$$\frac{2x}{2} = \frac{4}{2} \qquad \text{Divide both sides by 2.}$$
$$x = 2 \qquad \text{Simplify.}$$

So, the solution is 2, not 4. We can check that 2 satisfies the original equation $2x + 1 = 5$ (try it). ■

Locating an Error in Solving a Linear Equation

If the result of an attempt to solve a linear equation in one variable does not satisfy the equation, an error was made. To pinpoint the error, substitute the result into all of the equations of the work and find which pair of consecutive equations gives a false statement followed by a true statement.

---

group exploration

Any linear equation in one variable has exactly one solution.

1. Solve $7x + 5 = 0$.
2. Solve $4x + 3 = 0$.
3. Solve $5x + 2 = 0$.
4. Solve $mx + b = 0$, where m and b are constants and $m \neq 0$. [**Hint:** Perform steps similar to those you followed for Problems 1–3.]
5. Describe a linear equation in one variable.
6. Does using the addition property of equality or the multiplication property of equality change an equation's solution set? Explain.
7. Explain why your responses to Problems 4–6 show that any linear equation in one variable has exactly one solution.
8. Use your result from Problem 4 to solve $8x + 3 = 0$ in one step.

group exploration

Looking ahead: Comparing expressions and equations

1. Simplify the expression $2(x + 3)$.
2. Solve the equation $2(x + 3) = 0$.
3. Compare your results from Problems 1 and 2. Also, explain how to substitute values for x to check your work in the problems.
4. Simplify the expression $3x + 4x$.
5. Solve the equation $3x + 4x = 15 + 2x$.
6. Compare your results from Problems 4 and 5. Also, explain how to substitute values for x to check your work in the problems.
7. In general, compare the results of simplifying an expression with the results of solving an equation.

> ### TIPS FOR SUCCESS: Choose a Good Time and Place to Study
>
> To improve your effectiveness at studying, consider taking stock of when and where you are best able to study. Tracy, a student who lives in a sometimes distracting household, completes her assignments at the campus library just after she attends her classes. Gerome, a morning person, gets up early so that he can study before classes. Being consistent in the time and location for studying can help, too. Studies have shown that, after repeating a daily activity for about 21 days, the activity becomes a habit. Even if it takes some willpower to shuffle your schedule so that you can study at your prime time and location, things will start to feel comfortable and familiar within three weeks.

HOMEWORK 4.4

FOR EXTRA HELP ▶

Student Solutions Manual   PH Math/Tutor Center   MathXL®   MyMathLab

Solve. Use a graphing calculator to verify your result.

1. $3x - 2 = 13$ **2.** $5x - 1 = 9$ **3.** $-4x + 6 = 26$

4. $-2x + 7 = 23$ **5.** $-5 = 6x + 3$ **6.** $-7 = 4x - 1$

7. $8 - x = -4$ **8.** $2 - x = -9$ **9.** $2x + 6 - 7x = -4$

10. $4x + 3 - 9x = -22$ **11.** $5x + 4 = 3x + 16$

12. $7x + 5 = 4x + 17$ **13.** $-3r - 1 = 2r + 24$

14. $-6w - 3 = 4w + 17$ **15.** $9 - x - 5 = 2x - x$

16. $8 - 2x - 2 = 3x + x$ **17.** $2(x + 3) = 5x - 3$

18. $-3(x - 4) = 2x + 2$ **19.** $1 - 3(5b - 2) = 4 - (7b + 3)$

20. $3 - 4(3p + 2) = 7 - (9p - 1)$

21. $4x = 3(2x - 1) + 5$

22. $2x = 5(2x + 9) + 3$

23. $3(4x - 5) - (2x + 3) = 2(x - 4)$

24. $-2(5x + 3) - (4x - 1) = 5(x + 2)$

25. $\dfrac{x}{2} - \dfrac{3}{4} = \dfrac{1}{2}$ **26.** $\dfrac{x}{9} - \dfrac{1}{3} = \dfrac{2}{9}$

27. $\dfrac{5x}{6} + \dfrac{2}{3} = 2$ **28.** $\dfrac{3x}{8} - \dfrac{1}{2} = 1$

29. $\dfrac{5}{6}k = \dfrac{3}{4}k + \dfrac{1}{2}$ **30.** $\dfrac{3}{8}t = \dfrac{5}{6}t - \dfrac{1}{4}$

31. $\dfrac{7}{12}x - \dfrac{5}{3} = \dfrac{7}{4} + \dfrac{5}{6}x$ **32.** $\dfrac{5}{4} + \dfrac{9}{2}x = \dfrac{3}{8}x - \dfrac{1}{4}$

33. $\dfrac{4}{3}x - 2 = 3x + \dfrac{5}{2}$ **34.** $\dfrac{2}{5}x - 4 = 2x - \dfrac{3}{4}$

35. $\dfrac{3(x - 4)}{5} = -2x$ **36.** $\dfrac{5(x - 2)}{3} = -4x$

37. $\dfrac{4x + 3}{5} = \dfrac{2x - 1}{3}$ **38.** $\dfrac{3x + 2}{2} = \dfrac{6x - 3}{5}$

39. $\dfrac{4m - 5}{2} - \dfrac{3m + 1}{3} = \dfrac{5}{6}$

40. $\dfrac{2p + 4}{3} - \dfrac{5p - 7}{6} = \dfrac{11}{12}$

Solve. Round the result to the second decimal place. Use a graphing calculator to verify your work.

41. $0.3x + 0.2 = 0.7$

42. $0.6x - 0.1 = 0.4$

43. $5.27x - 6.35 = 2.71x + 9.89$

44. $8.25x - 17.56 + 4.38x = 25.86$

45. $0.4x - 1.6(2.5 - x) = 3.1(x - 5.4) - 11.3$

46. $3.2x + 0.5(7.3 - x) = 4.7 - 6.4(x - 2.1)$

47. In 2000, there was an average of 483 high school students per guidance counselor. The average number of students per counselor is decreasing by about 5.6 students per counselor each year (Source: *National Center for Education Statistics*). Let a be the average number of students per guidance counselor at t years since 2000.

　a. Find an equation of a linear model to describe the situation.

　b. Predict the average number of students per counselor in 2011.

　c. The recommended guideline for the ratio of students per counselor is 250 students per counselor. Predict when that ratio will be reached.

48. The average salary for a hockey player in the National Hockey League was $1.36 million in 2000 and has increased by about $0.13 million per year since then (Source: *National Hockey League Players Association*). Let s be the average salary (in millions of dollars) at t years since 2000.

　a. Find an equation of a linear model to describe the situation.

　b. Predict the average salary in 2010.

　c. Predict when the average salary will be $3 million.

49. Martha Stewart went to prison for selling nearly 4000 shares of ImClone Systems stock because she had inside information that the company was about to announce bad news that would result in the stock's value plummeting. Surprisingly, the value of a share of Martha Stewart Living Omnimedia stock increased by about $3.36 per month after she was sentenced. The value of a share of her stock was $8.16 just before she was sentenced (Source: *Bloomberg Financial Markets*). Let v be the value (in dollars) of a share of the stock at t months since Stewart was sentenced.

　a. Find an equation of a model to describe the situation.

　b. Estimate the value of a share of the stock 3 months after the sentencing, which is about when Stewart reported to prison.

　c. Estimate when the value of a share of the stock reached $28.32.

50. Due to the cost of health insurance coverage for its employees, General Motors (GM) charges more for its cars than Japanese competitors do. As a result, GM's share of new-vehicle sales in the United States is decreasing by about 0.73 percentage point per year. In 2004, GM's share was 25.1% (Source: *GM*). Let s be GM's share of new-vehicle sales in the United States at t years since 2004.
   a. Find an equation of a model to describe the situation.
   b. Predict when GM's share of new-vehicle sales will be 15%.
   c. Predict GM's share of new-vehicle sales in 2011.

51. The percentages of college freshmen whose average grade in high school was an A are shown in Table 6 for various years.

Table 6 College Freshmen Whose Average Grade in High School Was an A

Year	Percent
1990	29.4
1995	36.1
1998	39.8
2001	44.1
2004	47.5

Source: *Higher Education Research Institute*

Let p be the percentage of college freshmen whose average grade in high school was an A at t years since 1990. A model of the situation is $p = 1.30t + 29.49$.
   a. Use a graphing calculator to draw a scattergram and the model in the same viewing window. Does the line come close to the data points?
   b. What is the slope? What does it mean in this situation?
   c. Estimate when half of all college freshmen earned an average grade of A in high school.
   d. Predict the percentage of college freshmen in 2010 who will have earned an average grade of A in high school.
   e. Give at least two possible explanations of why the percentage of college freshmen who have earned an average grade of A is increasing.

52. The average selling prices of a home sold in San Bruno, California, are shown in Table 7 for various square footages.

Table 7 Average Selling Prices of Homes

Number of Square Feet	Number Used to Represent Square Feet	Average Selling Price (thousands of dollars)
500–1000	750	632
1001–1500	1250	733
1501–2000	1750	814
2001–2500	2250	894
2501–3000	2750	1025

Source: *Green Banker*

Let p be the average selling price (in thousands of dollars) of a home measuring s square feet. A model of the situation is $p = 0.19s + 488.15$.
   a. Use a graphing calculator to draw a scattergram and the model in the same viewing window. Does the line come close to the data points?
   b. What is the slope? What does it mean in this situation?
   c. Estimate the square footage for which the average selling price is $950 million.

For Exercises 53–60, find the number.

53. Three more than the product of 5 and a number is 18.

54. Four more than the product of 3 and a number is 11.

55. Three times the difference of a number and 2 is -18.

56. Six times the sum of a number and 2 is -12.

57. Three subtracted from a number is equal to twice the number plus 1.

58. Four, plus 7 times a number, is equal to 5 times the number, minus 3.

59. One, minus 4 times the sum of a number and 5 is 9.

60. Three, plus 5 times the difference of a number and 2 is 18.

Let x be a number. Translate the given equation into an English sentence. Then solve the equation.

61. $2x - 3 = 7$ **62.** $3x + 8 = 2$

63. $6x - 3 = 8x - 4$ **64.** $2x + 9 = 7x + 12$

65. $2(x - 4) = 10$ **66.** $3(x + 6) = 12$

67. $4 - 7(x + 1) = 2$ **68.** $9 + 5(x - 3) = 4$

For Exercises 69–72, solve the given equation by referring to the solutions of $y = -3x + 7$ and $y = 5x + 15$ shown in Tables 8 and 9, respectively.

69. $-3x + 7 = 5x + 15$ **70.** $-3x + 7 = 4$

71. $5x + 15 = 5$ **72.** $5x + 15 = 25$

Table 8 Solutions of $y = -3x + 7$

x	y
−2	13
−1	10
0	7
1	4
2	1

Table 9 Solutions of $y = 5x + 15$

x	y
−2	5
−1	10
0	15
1	20
2	25

Use "intersect" on a graphing calculator to solve the equation. Round the solution to the second decimal place.

73. $-4x + 8 = 2x - 9$ **74.** $-2x - 7 = x + 1$

75. $2.5x - 6.4 = -1.7x + 8.1$

76. $-1.5x - 9.3 = 3.1x + 2.1$

77. $\dfrac{1}{3}x - \dfrac{7}{3} = -\dfrac{3}{5}x - \dfrac{15}{2}$

78. $-\dfrac{3}{2}x + \dfrac{5}{4} = \dfrac{2}{3}x - \dfrac{13}{5}$

For Exercises 79–84, solve the given equation by referring to the graphs of $y = -\dfrac{3}{2}x + 2$ and $y = \dfrac{1}{2}x - 2$ shown in Fig. 19.

79. $-\dfrac{3}{2}x + 2 = \dfrac{1}{2}x - 2$ **80.** $-\dfrac{3}{2}x + 2 = -4$

81. $-\dfrac{3}{2}x + 2 = 5$ **82.** $\dfrac{1}{2}x - 2 = -3$

83. $\dfrac{1}{2}x - 2 = 0$ **84.** $-\dfrac{3}{2}x + 2 = 2$

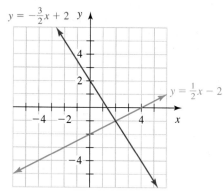

Figure 19 Exercises 79–84

For Exercises 85–90, solve the given equation by referring to the graphs of $y = -\dfrac{1}{3}x + \dfrac{5}{3}$ and $y = \dfrac{3}{2}x + \dfrac{7}{2}$ shown in Fig. 20.

85. $-\dfrac{1}{3}x + \dfrac{5}{3} = \dfrac{3}{2}x + \dfrac{7}{2}$ **86.** $-\dfrac{1}{3}x + \dfrac{5}{3} = 3$

87. $-\dfrac{1}{3}x + \dfrac{5}{3} = 1$ **88.** $\dfrac{3}{2}x + \dfrac{7}{2} = -4$

89. $\dfrac{3}{2}x + \dfrac{7}{2} = -1$ **90.** $-\dfrac{1}{3}x + \dfrac{5}{3} = 0$

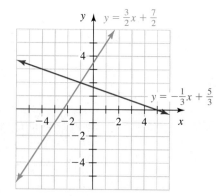

Figure 20 Exercises 85–90

Solve the equation. State whether the equation is a conditional equation, an inconsistent equation, or an identity.

91. $3x + 4x = 7x$

92. $9x - 3x = 6x$

93. $4x - 5 - 2x = 2x - 1$

94. $3x + 8x - 7 = 15x - 3 - 4x$

95. $3k + 10 - 5k = 4k - 2$

96. $5 - r + 8r = 3r + 1$

97. $2(x + 3) - 2 = 2x + 4$

98. $5(x - 2) + 1 = 5x - 9$

99. $5(2x - 3) - 4x = 3(2x - 1) + 6$

100. $4(3x + 2) - 6x = 2(3x - 4) + 1$

101. A student incorrectly solves $3(x - 3) = 12$:

$$3(x - 3) = 12$$
$$3x - 9 = 12$$
$$3x - 9 - 9 = 12 - 9$$
$$3x = 3$$
$$\frac{3x}{3} = \frac{3}{3}$$
$$x = 1$$

Substitute 1 for x in each of the six equations, and explain how the results help you pinpoint the error. Describe the error. Then solve the equation correctly.

102. A student incorrectly solves $-2x + 1 = 3$:

$$-2x + 1 = 3$$
$$-2x + 1 - 1 = 3 - 1$$
$$-2x = 2$$
$$\frac{-2x}{-2} = 2$$
$$x = 2$$

Substitute 2 for x in each of the five equations, and explain how the results help you pinpoint the error. Describe the error. Then solve the equation correctly.

103. A student tries to solve $2(x - 5) = x - 3$:

$$2(x - 5) = x - 3$$
$$2x - 10 = x - 3$$
$$2x - 10 + 10 = x - 3 + 10$$
$$2x = x + 7$$
$$\frac{2x}{2} = \frac{x + 7}{2}$$
$$x = \frac{x + 7}{2}$$

Describe any errors. Then solve the equation correctly.

104. A student tries to solve the equation $\dfrac{1}{2}x + 3 = \dfrac{5}{2}$:

$$\frac{1}{2}x + 3 = \frac{5}{2}$$
$$2 \cdot \frac{1}{2}x + 3 = 2 \cdot \frac{5}{2}$$
$$x + 3 = 5$$
$$x + 3 - 3 = 5 - 3$$
$$x = 2$$

Describe any errors. Then solve the equation correctly.

105. Two students try to solve the equation $5x + 3 = 3x + 7$:

Student A	Student B
$5x + 3 = 3x + 7$	$5x + 3 = 3x + 7$
$5x + 3 - 3x = 3x + 7 - 3x$	$5x + 3 - 3 = 3x + 7 - 3$
$2x + 3 = 7$	$5x = 3x + 4$
$2x + 3 - 3 = 7 - 3$	$5x - 3x = 3x + 4 - 3x$
$2x = 4$	$2x = 4$
$\dfrac{2x}{2} = \dfrac{4}{2}$	$\dfrac{2x}{2} = \dfrac{4}{2}$
$x = 2$	$x = 2$

Compare the methods of solving the equation. Is one method better than the other? Explain.

106. Two students try to solve the equation $2x + 4 = 0$:

Student 1	Student 2
$2x + 4 = 0$	$2x + 4 = 0$
$2x + 4 + 6 = 0 + 6$	$2x + 4 - 4 = 0 - 4$
$2x + 10 = 6$	$2x = -4$
$2x + 10 - 10 = 6 - 10$	$\dfrac{2x}{2} = \dfrac{-4}{2}$
$2x = -4$	$x = -2$
$2x \cdot 3 = -4 \cdot 3$	
$6x = -12$	
$\dfrac{6x}{6} = \dfrac{-12}{6}$	
$x = -2$	

Did either student, both students, or neither student solve the equation correctly? Explain.

107. Solve $5(x - 2) = 2x - 1$. Show that your result satisfies each of the equations in your work.

108. Solve $3(x + 1) + 2 = x + 7$. Show that your result satisfies each of the equations in your work.

109. Consider the following equations:

$$x = 3$$
$$x + 2 = 3 + 2$$
$$x + 2 = 5$$
$$3(x + 2) = 3 \cdot 5$$
$$3(x + 2) = 15$$
$$3(x + 2) + 1 = 15 + 1$$
$$3(x + 2) + 1 = 16$$

What is the solution of the equation $3(x + 2) + 1 = 16$? What is the solution of the equation $3(x + 2) = 15$? Explain how you can respond to these questions without doing any work besides the work already included in this problem.

4.5 COMPARING EXPRESSIONS AND EQUATIONS

Objectives

▹ Compare the meanings of *simplifying expressions* and *solving equations*.

▹ Translate English phrases and sentences to mathematical expressions and equations.

In this section, we compare the meanings of mathematical expressions and equations.

Comparing Expressions and Equations

First, we define a *linear expression in one variable*.

DEFINITION Linear expression in one variable

A **linear expression in one variable** is an expression that can be put into the form

$$mx + b$$

where m and b are constants and $m \neq 0$.

Next, we recall the definition of a linear equation from Section 4.3: A *linear equation in one variable* is an equation that can be put into the form $mx + b = 0$, where m and b are constants and $m \neq 0$.

For example, $4x + 7$ is a linear expression in one variable and $4x + 7 = 0$ is a linear equation in one variable. Also, the expression $2(x - 5) + 3x$ is linear, because it can be put into the form $mx + b$, where $m \neq 0$. (Try it.) The equation $2(x - 5) + 3x = 8x - 4$ is linear, because it can be put into the form $mx + b = 0$, where $m \neq 0$. (Try it.)

Example 1   Simplifying an Expression and Solving an Equation

1. Simplify $5(x + 2) - 2x$.
2. Solve $5(x + 2) - 2x = 2x + 14$.
3. Use a graphing calculator table to check the work in Problem 1.
4. Use a graphing calculator table to check the work in Problem 2.

Solution

1.
$$5(x + 2) - 2x = 5x + 10 - 2x \qquad \text{Distributive law}$$
$$= 3x + 10 \qquad \text{Combine like terms.}$$

The simplified expression is $3x + 10$.

2.
$$5(x + 2) - 2x = 2x + 14 \qquad \text{Original equation}$$
$$5x + 10 - 2x = 2x + 14 \qquad \text{Distributive law}$$
$$3x + 10 = 2x + 14 \qquad \text{Combine like terms.}$$
$$3x + 10 - 2x = 2x + 14 - 2x \qquad \text{Subtract 2x from both sides.}$$
$$x + 10 = 14 \qquad \text{Combine like terms.}$$
$$x + 10 - 10 = 14 - 10 \qquad \text{Subtract 10 from both sides.}$$
$$x = 4 \qquad \text{Simplify.}$$

The solution is 4.

3. We use a graphing calculator to find a table for $y = 5(x + 2) - 2x$ and $y = 3x + 10$, using the original expression and the simplified result, respectively (see Fig. 21). For each value of x in the table, the values of y for both expressions are equal. So, we are confident that the two expressions are equivalent.

Figure 21 Verify the work

4. We use a graphing calculator table to check that if 4 is substituted for x in the expressions $5(x + 2) - 2x$ and $2x + 14$ (both sides of the original equation), the two results are equal (see Fig. 22).

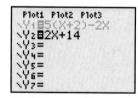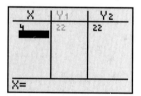

Figure 22 Verify the work

Note that if any other number besides 4 is substituted for x in the expressions $5(x + 2) - 2x$ and $2x + 14$, the two results will *not* be equal. That is because 4 is the *only* solution of the equation $5(x + 2) - 2x = 2x + 14$.   ■

From our work in Example 1, we can make two important observations: First, the result of simplifying the *expression* $5(x + 2) - 2x$ is $3x + 10$, an *expression;* second, the result of solving the *equation* $5(x + 2) - 2x = 2x + 14$ is 4, a *number*.

Results of Simplifying an Expression and Solving an Equation

The result of simplifying an expression is an expression. The result of solving a linear equation in one variable is a number.

Example 2   Identifying an Error in Solving an Equation

A student tries to solve the equation $2(x - 3) = x$:

$$2(x - 3) = x$$
$$2x - 6 = x$$
$$2x - 6 + 6 = x + 6$$
$$2x = x + 6$$
$$x = \frac{x + 6}{2}$$

What would you tell the student?

Solution

The student found an equation that is equivalent to the original equation, but the student did not find the *solution* of the original equation. The solution of a linear equation is a number, not an expression that contains a variable. So, the solution cannot be $\frac{x + 6}{2}$. Notice that x still appears on both sides of the equation. By getting x alone on one side of the equation, we can show that the solution is 6. (Try it.)   ■

In Example 1, we simplified $5(x + 2) - 2x$ to $3x + 10$. So, if we use the same number to evaluate both expressions, the two results will be equal. This implies that *any* number satisfies the equation $5(x + 2) - 2x = 3x + 10$. For example, in Fig. 21 we see that the numbers 0, 1, 2, 3, 4, 5, and 6 are solutions of this equation. Since any number satisfies the equation, the equation is an identity.

We also found that 4 is the solution of the linear equation $5(x + 2) - 2x = 2x + 14$. This means that 4 is the *only* number which satisfies that equation. In general, a linear equation in one variable has exactly one solution.

Comparing Simplifying an Expression with Solving an Equation

- If an expression A in one variable is simplified to an expression B, then every real number (for which both expressions are defined) is a solution of the equation $A = B$. In other words, the equation $A = B$ is an identity.
- Exactly one number is a solution of a linear equation in one variable.

Example 3   Comparing Expressions and Equations

1. The simplified form of $4(2a - 5) - 2(3a + 4)$ is $2a - 28$. What does this mean?
2. The solution of $4(2a - 5) - 2(3a + 4) = 2$ is 15. What does this mean?

Solution

1. This means that the linear expressions $4(2a - 5) - 2(3a + 4)$ and $2a - 28$ are equivalent. It also means that every real number is a solution of the equation $4(2a - 5) - 2(3a + 4) = 2a - 28$. So, the equation is an identity.
2. This means that 15 is the only number that satisfies the linear equation $4(2a - 5) - 2(3a + 4) = 2$. (Try it.)   ■

Example 4   Comparing Expressions and Equations

1. Simplify $\frac{2}{5}x + \frac{1}{3}x$.
2. Solve $\frac{2}{5}x + \frac{1}{3}x = \frac{4}{5}$.
3. Compare the first steps of your work in Problems 1 and 2.

Solution

1. $\dfrac{2}{5}x + \dfrac{1}{3}x = \dfrac{3}{3} \cdot \dfrac{2}{5}x + \dfrac{5}{5} \cdot \dfrac{1}{3}x$ The LCD of $\dfrac{2}{5}$ and $\dfrac{1}{3}$ is 15.

$\qquad\qquad = \dfrac{6}{15}x + \dfrac{5}{15}x$ Multiply numerators and multiply denominators: $\dfrac{a}{b} \cdot \dfrac{c}{d} = \dfrac{ac}{bd}$

$\qquad\qquad = \dfrac{11}{15}x$ Combine like terms.

2. $\qquad\quad \dfrac{2}{5}x + \dfrac{1}{3}x = \dfrac{4}{5}$ Original equation

$15 \cdot \left(\dfrac{2}{5}x + \dfrac{1}{3}x \right) = 15 \cdot \dfrac{4}{5}$ Multiply both sides by 15.

$15 \cdot \dfrac{2}{5}x + 15 \cdot \dfrac{1}{3}x = 15 \cdot \dfrac{4}{5}$ Distributive law

$\qquad\qquad 6x + 5x = 12$ Simplify.

$\qquad\qquad\quad 11x = 12$ Combine like terms.

$\qquad\qquad\quad \dfrac{11x}{11} = \dfrac{12}{11}$ Divide both sides by 11.

$\qquad\qquad\qquad x = \dfrac{12}{11}$ Simplify.

3. To start simplifying the expression in Problem 1, we multiplied the term $\dfrac{2}{5}x$ by $\dfrac{3}{3} = 1$ and multiplied the term $\dfrac{1}{3}x$ by $\dfrac{5}{5} = 1$. To start solving the equation in Problem 2, we multiplied both sides of the equation by 15, which is *not* equal to 1. ∎

Problem 3 in Example 4 suggests the following comparison of multiplying an expression by a number and multiplying both sides of an equation by a number:

Multiplying Expressions and Both Sides of Equations by Numbers

In simplifying an expression, the only number that we can multiply the expression or part of it by is 1. In solving an equation, we can multiply both sides of the equation by *any* number except 0.

By the property $a \cdot 1 = a$, we know that multiplying an expression by 1 gives an equivalent expression. By the multiplication property of equality, we know that multiplying both sides of an equation by *any* nonzero number gives an equivalent equation.

Translating from English to Mathematics

In Section 2.1, we discussed how to translate an English phrase or sentence into a mathematical expression, and in Section 4.4 we discussed how to translate an English sentence into an equation. We use an expression to represent a quantity, and we use an equation to state that one quantity is equal to another.

Example 5   Translating from English to Mathematics

Let x be a number. Translate the following into an expression or an equation, as appropriate, and then simplify the expression or solve the equation:

1. Five, minus 3 times the sum of the number and 4
2. The product of 6 and the number is equal to the number subtracted from 2.

Solution

1. The phrase describes a single quantity. So, we translate the phrase to an expression:

$$\underbrace{5}_{\text{Five, minus}} \underbrace{-}_{} \underbrace{3}_{\text{3 times}} \cdot \underbrace{(x + 4)}_{\substack{\text{the sum of the} \\ \text{number and 4.}}}$$

Next, we simplify the expression:

$$5 - 3 \cdot (x + 4) = 5 - 3x - 12 \qquad \text{Distributive law}$$
$$= -3x - 7 \qquad \text{Combine like terms.}$$

2. The sentence states that one quantity is equal to another. So, we translate the sentence to an equation:

$$\underbrace{6x}_{\substack{\text{The product of 6} \\ \text{and the number}}} \underbrace{=}_{\text{is equal to}} \underbrace{2 - x}_{\substack{\text{the number subtracted} \\ \text{from 2.}}}$$

Next, we solve the equation:

$$6x = 2 - x \qquad \text{Original equation}$$
$$6x + x = 2 - x + x \qquad \text{Add x to both sides.}$$
$$7x = 2 \qquad \text{Combine like terms.}$$
$$\frac{7x}{7} = \frac{2}{7} \qquad \text{Divide both sides by 7.}$$
$$x = \frac{2}{7} \qquad \text{Simplify.}$$

So, the solution is $\frac{2}{7}$. ■

group exploration

Simplifying versus solving

1. Two students try to solve the equation $2 = \frac{x}{2} + \frac{x}{3}$:

Student A

$$2 = \frac{x}{2} + \frac{x}{3}$$
$$6 \cdot 2 = 6\left(\frac{x}{2} + \frac{x}{3}\right)$$
$$12 = 6 \cdot \frac{x}{2} + 6 \cdot \frac{x}{3}$$
$$12 = 3x + 2x$$
$$12 = 5x$$
$$\frac{12}{5} = \frac{5x}{5}$$
$$\frac{12}{5} = x$$

Student B

$$2 = \frac{x}{2} + \frac{x}{3}$$
$$= \frac{x}{2} + \frac{x}{3} - 2$$
$$= \frac{x}{2} \cdot \frac{3}{3} + \frac{x}{3} \cdot \frac{2}{2} - 2$$
$$= \frac{3x}{6} + \frac{2x}{6} - 2$$
$$= \frac{3x + 2x}{6} - 2$$
$$= \frac{5x}{6} - 2$$

Did either student, both students, or neither student solve the equation correctly? Explain.

2. Three students try to simplify the expression $\dfrac{x}{2} + \dfrac{x}{3}$:

Student C

$$\frac{x}{2} + \frac{x}{3} = 6\left(\frac{x}{2} + \frac{x}{3}\right)$$
$$= 6 \cdot \frac{x}{2} + 6 \cdot \frac{x}{3}$$
$$= 3x + 2x$$
$$= 5x$$

Student D

$$\frac{x}{2} + \frac{x}{3} = \frac{3}{3} \cdot \frac{x}{2} + \frac{2}{2} \cdot \frac{x}{3}$$
$$= \frac{3x}{6} + \frac{2x}{6}$$
$$= \frac{3x + 2x}{6}$$
$$= \frac{5x}{6}$$

Student E

$$\frac{x}{2} + \frac{x}{3} = 0$$
$$6\left(\frac{x}{2} + \frac{x}{3}\right) = 6 \cdot 0$$
$$6 \cdot \frac{x}{2} + 6 \cdot \frac{x}{3} = 0$$
$$3x + 2x = 0$$
$$5x = 0$$
$$\frac{5x}{5} = \frac{0}{5}$$
$$x = 0$$

Which students, if any, simplified the expression correctly? Explain.

TIPS FOR SUCCESS: Ask Questions

When you have a question during class time, do you ask it? Many students are reluctant to ask questions for fear of slowing the class down or feeling embarrassed. If you tend to shy away from asking questions, keep in mind that the main idea of school is for you to learn through open communication with your instructor and other students.

If you are confused about some concept, it's highly likely that other students in your class are confused, too. If you ask your question, everyone else who is confused will be grateful that you did so. Most instructors want students to ask questions. It helps an instructor know when students understand the material and when they are having trouble.

HOMEWORK 4.5   FOR EXTRA HELP ▶   Student Solutions Manual   PH Math/Tutor Center   MathXL®   MyMathLab

Is the following a linear expression or a linear equation?

1. $3x + 7x = 8$

2. $2x + 4x - 7$

3. $3x + 7x$

4. $2x + 4x - 7 = 5$

5. $4 - 2(x - 9)$

6. $3(x - 1) = 7$

7. $4 - 2(x - 9) = 5$

8. $3(x - 1)$

Simplify the expression or solve the equation, as appropriate. Use a graphing calculator to verify your work.

9. $3x + 4x = 14$

10. $2x - 7x = 15$

11. $3x + 4x$

12. $2x - 7x$

13. $b - 5(b - 1)$

14. $-6k + 2(k + 4)$

15. $b - 5(b - 1) = 0$

16. $-6k + 2(k + 4) = 0$

17. $3(3x - 5) + 2(5x + 4) = 0$

18. $-4(2x + 1) - 2(3x - 6) = 0$

19. $3(3x - 5) + 2(5x + 4)$

20. $-4(2x + 1) - 2(3x - 6)$

21. $3(x - 2) - (7x + 2) = 4(3x + 1)$

22. $5(2x + 3) - (3x - 8) - 2(4x - 5)$

23. $3(x - 2) - (7x + 2) - 4(3x + 1)$

24. $5(2x + 3) - (3x - 8) = 2(4x - 5)$

25. $7.2p - 4.5 - 1.3p$

26. $8.3t + 9.2 - 7.7t$

27. $7.2k - 4.5 - 1.3k = 20.5 - 6.6k$

28. $8.3c + 9.2 - 7.7c = 3.5 - 1.2c$

29. $-3.5(x - 8) - 2.6(x - 2.8) = 13.93$

30. $-4.8(x + 3) + 6.5(x - 1.2)$

31. $-3.5(x - 8) - 2.6(x - 2.8)$

32. $-4.8(x + 3) + 6.5(x - 1.2) = -25.09$

33. $-\dfrac{6w}{8}$

34. $-\dfrac{15t}{10}$

35. $-\dfrac{6w}{8} = \dfrac{3}{2}$

36. $-\dfrac{15t}{10} = -\dfrac{5}{4}$

37. $\dfrac{5x}{6} + \dfrac{1}{2} - \dfrac{3x}{4}$

38. $\dfrac{2x}{3} - \dfrac{5x}{2} - 1 = 0$

39. $\dfrac{5x}{6} + \dfrac{1}{2} - \dfrac{3x}{4} = 0$

40. $\dfrac{2x}{3} - \dfrac{5x}{2} - 1$

41. $\dfrac{7}{2}x - \dfrac{5}{6} = \dfrac{1}{3} + \dfrac{3}{4}x$

42. $\dfrac{3}{5}x + 2x - \dfrac{5}{2} - \dfrac{7}{10}x$

43. $\dfrac{7}{2}x - \dfrac{5}{6} - \dfrac{1}{3} + \dfrac{3}{4}x$

44. $\dfrac{3}{5}x + 2x = \dfrac{5}{2} - \dfrac{7}{10}x$

For Exercises 45 and 46, give an example of each. Then simplify or solve, as appropriate.

45. linear expression

46. linear equation in one variable

47. Two students try to solve the equation $\dfrac{3}{4}x - \dfrac{5}{6} = \dfrac{1}{3}$:

Student A

$\dfrac{3}{4}x - \dfrac{5}{6} = \dfrac{1}{3}$

$\dfrac{3}{4}x = \dfrac{1}{3} + \dfrac{5}{6}$

$\dfrac{3}{3} \cdot \dfrac{3}{4}x = \dfrac{4}{4} \cdot \dfrac{1}{3} + \dfrac{2}{2} \cdot \dfrac{5}{6}$

$\dfrac{9}{12}x = \dfrac{4}{12} + \dfrac{10}{12}$

$\dfrac{9}{12}x = \dfrac{14}{12}$

$\dfrac{3}{4}x = \dfrac{7}{6}$

$\dfrac{4}{3} \cdot \dfrac{3}{4}x = \dfrac{4}{3} \cdot \dfrac{7}{6}$

$1x = \dfrac{2 \cdot 2 \cdot 7}{3 \cdot 2 \cdot 3}$

$x = \dfrac{14}{9}$

Student B

$\dfrac{3}{4}x - \dfrac{5}{6} = \dfrac{1}{3}$

$12\left(\dfrac{3}{4}x - \dfrac{5}{6}\right) = 12 \cdot \dfrac{1}{3}$

$12 \cdot \dfrac{3}{4}x - 12 \cdot \dfrac{5}{6} = 4$

$9x - 10 = 4$

$9x = 14$

$x = \dfrac{14}{9}$

Compare the methods of solving the equation. Is one method better than the other? Explain.

48. A student believes that 5 is the solution (and the only solution) of the equation $-2(x + 4) = -2x - 8$, because 5 satisfies

that equation:

$$-2(5 + 4) \overset{?}{=} -2(5) - 8$$
$$-18 \overset{?}{=} -18 \qquad \text{true}$$

Is the student correct? Explain.

49. A student believes that $x + 6$ is the solution of an equation. Is the student correct? Explain.

50. A student tries to simplify an expression. The student writes "The solution of the expression is 5." Is the student correct? Explain.

51. A student tries to simplify the expression $\dfrac{1}{4}x + \dfrac{1}{3}x$:

$$\dfrac{1}{4}x + \dfrac{1}{3}x = 12\left(\dfrac{1}{4}x + \dfrac{1}{3}x\right)$$
$$= 12 \cdot \dfrac{1}{4}x + 12 \cdot \dfrac{1}{3}x$$
$$= 3x + 4x$$
$$= 7x$$

Describe any errors. Then simplify the expression correctly.

52. A student tries to simplify the expression $3x + 5x - 16$:

$$3x + 5x - 16 = 0$$
$$8x - 16 = 0$$
$$8x = 16$$
$$x = 2$$

Describe any errors. Then simplify the expression correctly.

53. The simplified form of $7 + 2(x + 3)$ is $2x + 13$. What does this mean?

54. The solution of $7 + 2(x + 3) = 19$ is 3. What does this mean?

55. A student tries to simplify $2(5 + x)$:

$$2(5 + x) = 10 + x$$

To check the work, the student evaluates $2(5 + x)$ and $10 + x$, using 0 for x:

Evaluate $2(5 + x)$ for $x = 0$: $2(5 + 0) = 10$

Evaluate $10 + x$ for $x = 0$: $10 + 0 = 10$

Since the results of evaluating the expressions are equal, the student decides that the work is correct. What would you tell the student about checking the work?

56. A student evaluates both of the expressions $3x + 1$ and $21 - 2x$, using 4 for x:

Evaluate $3x + 1$ for $x = 4$: $3(4) + 1 = 13$

Evaluate $21 - 2x$ for $x = 4$: $21 - 2(4) = 13$

Since the results of evaluating the expressions are equal, the student concludes that the expressions are equivalent. Is the student correct? Explain.

57. Give three examples of an equation whose solution is 5.

58. Give three examples of an expression equivalent to the expression 5.

For Exercises 59–70, let x be a number. Translate each of the following into an expression or an equation, as appropriate, and then simplify the expression or solve the equation:

59. The sum of 3 and twice the number is -10.

60. The difference of 3 times the number and 5 is -12.

61. Four, minus 6 times the difference of the number and 2

62. Six, plus 3 times the sum of the number and 8

63. The product of -9 and the number is equal to the difference of the number and 5.

64. The sum of the number and 4 is equal to the number subtracted from 1.

65. The quotient of the number and 2 is equal to 3 times the difference of the number and 5.

66. The quotient of the number and 3 is equal to 2 plus the quotient of the number and 6.

67. The number plus the product of the number and 6

68. The number minus the product of the number and -4

69. The number plus the quotient of the number and 2

70. Four minus the quotient of the number and 5

71. Explain the meaning of simplifying an expression. (See page 9 for guidelines on writing a good response.)

72. Explain the meaning of solving an equation. (See page 9 for guidelines on writing a good response.)

4.6 FORMULAS

Objectives

▶ Know area and perimeter formulas of a rectangle and a total-value formula.

▶ Use formulas to solve various types of problems.

▶ Translate an English sentence into a formula.

▶ Solve a formula for a variable.

Recall from Section 3.3 that a **formula** is an equation that contains two or more variables. We can use formulas to find quantities such as the area and perimeter of rectangular objects, the total value of a group of objects, and the average test score of a group of tests.

Area Formula of a Rectangle

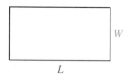

Figure 23 The length L and width W of a rectangle

Recall from Section 2.1 that the area of a rectangle is equal to the length times the width. We can write $A = LW$, where A is the area, L is the length, and W is the width (see Fig. 23). The equation $A = LW$ is a formula.

Example 1 Using the Area Formula of a Rectangle

An architect is designing a school building so that each classroom contains 35 student desks. Fire codes require that there be 18 square feet per student desk. If there is enough room for classrooms of length 28 feet, what is the smallest width permitted by the fire codes?

Figure 24 The length is 28 feet, and the width is W feet

Solution

To find the area of the floor of one room, we multiply the area needed per student desk by the number of student desks: $18(35) = 630$ square feet. Next, we must find the width of a rectangle that has an area of 630 square feet and a length of 28 feet (see Fig. 24). To do so, we substitute 630 for A and 28 for L in the formula $A = LW$, where the units of L and W are feet and the units of A are square feet. Then we solve for W:

$$A = LW \qquad \text{Area formula of a rectangle}$$
$$630 = (28)W \qquad \text{Substitute 630 for } A \text{ and 28 for } L.$$
$$\frac{630}{28} = \frac{28W}{28} \qquad \text{Divide both sides by 28.}$$
$$22.5 = W \qquad \text{Simplify.}$$

The smallest width permitted is 22.5 feet.

To check, we find the product of the length and width: $28(22.5) = 630$, which is equal to the area. ∎

In Example 1, we substituted values for A and L in the formula $A = LW$ and then solved for the variable W. In general, **to find a single value of a variable in a formula, we often substitute numbers for all of the other variables and then solve for that one variable.**

Perimeter Formula of a Rectangle

The **perimeter** of a polygon, which is a geometrical object such as a triangle, rectangle, or trapezoid, is the total distance around the object. For example, consider a rectangle with length L and width W (see Fig. 25).

Figure 25 The length L and width W of a rectangle

The perimeter P of a rectangle is equal to the sum of the lengths of the four sides:

$$P = L + W + L + W$$

Combining like terms on the right-hand side of the formula gives

$$P = 2L + 2W$$

Area and Perimeter of a Rectangle

For a rectangle with length L, width W, area A, and perimeter P,

- $A = LW$
- $P = 2L + 2W$

Example 2   Using the Perimeter Formula of a Rectangle

A landscaper has budgeted enough money for 100 feet of fencing to enclose a rectangular garden. If the length of the garden is to run the full length of the property, which is 37.5 feet, what will be the width of the garden (see Fig. 26)?

37.5 feet

Figure 26 The length is 37.5 feet, and the width is W feet

Solution

Since the 100-foot-long fencing is on the border of the garden, the perimeter of the garden is 100 feet. So, we substitute 100 for P and 37.5 for L in the formula $P = 2L + 2W$, where the units of P, L, and W are feet. Then we solve for W:

$$P = 2L + 2W \qquad \text{Perimeter formula of a rectangle}$$
$$100 = 2(37.5) + 2W \qquad \text{Substitute 100 for } P \text{ and 37.5 for } L.$$
$$100 = 75 + 2W \qquad \text{Simplify.}$$
$$100 - 75 = 75 + 2W - 75 \qquad \text{Subtract 75 from both sides.}$$
$$25 = 2W \qquad \text{Combine like terms.}$$
$$\frac{25}{2} = \frac{2W}{2} \qquad \text{Divide both sides by 2.}$$
$$12.5 = W \qquad \text{Write } \frac{25}{2} \text{ as a decimal number.}$$

The width of the garden will be 12.5 feet.

To check, we find twice the length of 37.5 feet plus twice the width of 12.5 feet: $2(37.5) + 2(12.5) = 100$ feet, which is equal to the perimeter. ■

When describing an authentic quantity, we use a decimal form rather than a fractional form. For example, in Example 2 we say that the width is 12.5 feet rather than $\frac{25}{2}$ feet.

Total-Value Formula

Three quarters are worth $25 \cdot 3 = 75$ cents. Note that we found the total value of the quarters by multiplying the value of one quarter (25 cents) by the number of quarters (3).

> **Using the Total-Value Formula**
>
> If n objects each have value v, then their total value T is given by
>
> $$T = vn$$
>
> In words: The total value is equal to the value of one object times the number of objects.

Example 3 Total-Value Formula

In 2005, Chicago Cubs baseball tickets for upper-deck reserved seating at Wrigley Field cost $14 each.

1. Find a formula of the total cost T (in dollars) of n of these tickets.
2. Perform a unit analysis of the formula found in Problem 1.
3. Substitute 84 for T in the formula from Problem 1 and solve for n. What does the result mean in this situation?

Solution

1. We substitute 14 for v in the formula $T = vn$:

$$T = 14n$$

2. Here is a unit analysis of the formula $T = 14n$:

$$\underbrace{T}_{\text{dollars}} = \underbrace{14}_{\frac{\text{dollars}}{\text{ticket}}} \quad \underbrace{n}_{\text{number of tickets}}$$

The units of the expressions on both sides of the equation are dollars, which checks.

3.
$$
\begin{aligned}
T &= 14n && \text{Total-value formula} \\
84 &= 14n && \text{Substitute 84 for } T. \\
6 &= n && \text{Divide both sides by 14.}
\end{aligned}
$$

The total cost of 6 upper-deck reserved tickets at Wrigley Field is $84. ∎

Translating from English to a Mathematics Formula

In the next example, we will translate some information into a formula and then use the formula to find a quantity.

Example 4 Translating from English to Mathematics

A Celsius temperature reading is equal to $\dfrac{5}{9}$ times the difference of the Fahrenheit temperature reading and 32.

1. Write a formula of the Celsius temperature in terms of the Fahrenheit temperature.
2. Convert 50°F to the equivalent Celsius temperature.
3. Use a graphing calculator to convert 50°F, 59°F, 68°F, 77°F, and 86°F to the equivalent Celsius temperatures.

Solution

1. We let C be the Celsius reading and F be the equivalent Fahrenheit reading. Next, we translate the given information into a formula:

$$
\underbrace{\text{A Celsius reading}}_{C} \;\; \underbrace{\text{is equal to}}_{=} \;\; \underbrace{\dfrac{5}{9}}_{\frac{5}{9}} \;\; \underbrace{\text{times}}_{\cdot} \;\; \underbrace{\text{the difference of the Fahrenheit reading and 32.}}_{(F \;-\; 32)}
$$

So, the formula is $C = \dfrac{5}{9}(F - 32)$.

2. We substitute 50 for F in the formula $C = \dfrac{5}{9}(F - 32)$ and solve for C:

$$C = \frac{5}{9}(50 - 32) = 10$$

So, $10°C$ is equivalent to $50°F$.

3. We use a graphing calculator table to substitute the inputs 50, 59, 68, 77, and 86 for F in the formula $C = \dfrac{5}{9}(F - 32)$. See Fig. 27. See Section A.13 for graphing calculator instructions.

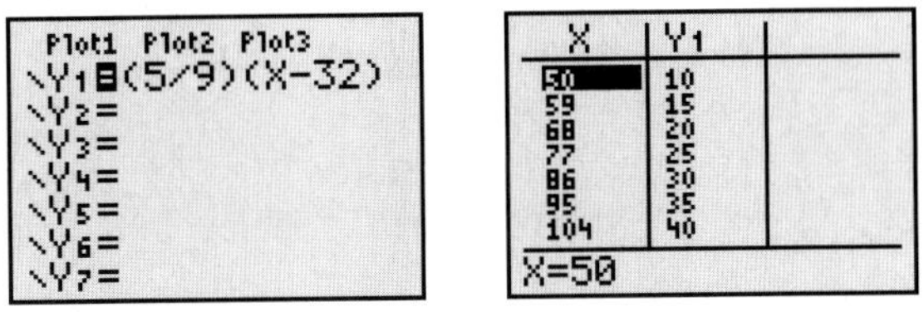

Figure 27 Finding Celsius temperatures

By viewing the second column of the graphing calculator table, we see that the values of C are 10, 15, 20, 25, and 30. So, $50°F$, $59°F$, $68°F$, $77°F$, and $86°F$ are equivalent to $10°C$, $15°C$, $20°C$, $25°C$, and $30°C$, respectively. ■

Solving a Formula

In Problem 3 of Example 4, we converted five Fahrenheit temperature readings to equivalent Celsius temperature readings. How can we convert several Celsius temperature readings to equivalent Fahrenheit temperature readings? We must first solve $C = \dfrac{5}{9}(F - 32)$ for F and then use a graphing calculator to do the conversions. We will do just that in Example 8, but first we work with some simpler formulas in Examples 5, 6, and 7.

Example 5 Solving a Formula for One of Its Variables

Solve $A = LW$ for W.

Solution

We can solve the formula $A = LW$ for W in much the same way that we solved the equation $630 = (28)W$ in Example 1:

$$
\begin{array}{ll|ll}
630 = (28)W & & A = LW & \text{Area formula} \\[2mm]
\dfrac{630}{28} = \dfrac{28W}{28} & & \dfrac{A}{L} = \dfrac{LW}{L} & \text{Divide both sides by } L. \\[3mm]
22.5 = W & & \dfrac{A}{L} = W & \text{Simplify.}
\end{array}
$$

The result is $W = \dfrac{A}{L}$. We are done, because W is alone on one side of the formula and does not appear on the other side. ■

Example 6   Solving a Formula for One of Its Variables

Solve $P = 2L + 2W$ for W.

Solution

We can solve the formula $P = 2L + 2W$ for W in much the same way that we solved the equation $100 = 75 + 2W$ in Example 2:

$100 = 75 + 2W$	$P = 2L + 2W$	Perimeter formula
$100 - 75 = 75 + 2W - 75$	$P - 2L = 2L + 2W - 2L$	Subtract $2L$ from both sides.
$25 = 2W$	$P - 2L = 2W$	Combine like terms.
$\dfrac{25}{2} = W$	$\dfrac{P - 2L}{2} = W$	Divide both sides by 2.

The result is $W = \dfrac{P - 2L}{2}$. ∎

The equation $2x - 5y = 10$ is a formula, because it is an equation that contains two (or more) variables. In Example 7, we will solve this equation for y.

Example 7   Writing a Linear Equation in Slope-Intercept Form

Write the equation $2x - 5y = 10$ in slope–intercept form ($y = mx + b$).

Solution

We solve $2x - 5y = 10$ for y:

$2x - 5y = 10$	Original equation
$2x - 5y - 2x = 10 - 2x$	Subtract $2x$ from both sides to get $-5y$ alone on left side.
$-5y = -2x + 10$	Combine like terms; rearrange terms.
$\dfrac{-5y}{-5} = \dfrac{-2x}{-5} + \dfrac{10}{-5}$	Divide both sides by -5 to get y alone on left side.
$y = \dfrac{2}{5}x - 2$	Simplify. ∎

In Section 5.1, we will perform work similar to our work in Example 7 to help us graph linear equations.

Example 8   Solving a Formula for One of Its Variables

1. Solve the Fahrenheit–Celsius model $C = \dfrac{5}{9}(F - 32)$ for F.
2. Convert $10°C$ to the equivalent Fahrenheit temperature.
3. Use a graphing calculator to convert $10°C$, $15°C$, $20°C$, $25°C$, and $30°C$ to the equivalent Fahrenheit temperatures.

Solution

1.

$C = \dfrac{5}{9}(F - 32)$	Celsius formula
$\dfrac{9}{5} \cdot C = \dfrac{9}{5} \cdot \dfrac{5}{9}(F - 32)$	Multiply both sides by $\dfrac{9}{5}$.
$\dfrac{9}{5}C = F - 32$	Simplify.
$\dfrac{9}{5}C + 32 = F - 32 + 32$	Add 32 to both sides.
$\dfrac{9}{5}C + 32 = F$	Combine like terms.

The result is $F = \dfrac{9}{5}C + 32$.

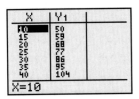

Figure 28 Finding Fahrenheit temperatures

2. We substitute 10 for C in the formula $F = \frac{9}{5}C + 32$ and solve for F:

$$F = \frac{9}{5}(10) + 32 = 50$$

So, $10°C$ is equivalent to $50°F$.

3. We use a graphing calculator table to substitute 10, 15, 20, 25, and 30 for C in the formula $F = \frac{9}{5}C + 32$ (see Fig. 28).

By viewing the second column of the graphing calculator table, we see that the values of F are 50, 59, 68, 77, and 86. So, $10°C$, $15°C$, $20°C$, $25°C$, and $30°C$ are equivalent to $50°F$, $59°F$, $68°F$, $77°F$, and $86°F$, respectively. ∎

In Example 4, we used the formula $C = \frac{5}{9}(F - 32)$ to convert five Fahrenheit temperature readings to five equivalent Celsius temperature readings. In Example 8, we used the formula $F = \frac{9}{5}C + 32$ to convert the same five Celsius temperature readings to the five equivalent Fahrenheit temperature readings. We can make two observations:

1. Both formulas describe the same relationship between F and C. In general, **solving for a variable in a formula will not change the relationship between the variables in the formula.**

2. To find Celsius readings, it is convenient to use the formula $C = \frac{5}{9}(F - 32)$, because C is alone on one side of the equation (and does not appear on the other side). Similarly, to find Fahrenheit readings, it is convenient to use the formula $F = \frac{9}{5}C + 32$. In general, **to find several values of a variable in a formula, we usually solve the formula for that variable before we make any substitutions.**

Example 9 Solving a Linear Model for One of Its Variables

The average ATM fee was $2.55 in 2000. The average fee has increased by approximately $0.06 per year since then (Source: *Bankrate.com*). Let F be the average ATM fee (in dollars) at t years since 2000.

1. Find an equation of a linear model to describe the situation.
2. Solve the equation that we found in Problem 1 for t.
3. Predict when the average ATM fee was $2.90.
4. Use a graphing calculator table to predict in which years the average ATM fee will be $3.00, $3.10, $3.20, $3.30, and $3.40.

Solution

1. Since the average ATM fee increases by about $0.06 each year, we can model the situation by using a linear model with slope 0.06. Since the average ATM fee was $2.55 in 2000, the F-intercept is $(0, 2.55)$. So, a reasonable model is

$$F = 0.06t + 2.55$$

2.
$$F = 0.06t + 2.55 \qquad \text{Original equation}$$
$$F - 2.55 = 0.06t + 2.55 - 2.55 \qquad \text{Subtract 2.55 from both sides.}$$
$$F - 2.55 = 0.06t \qquad \text{Combine like terms.}$$
$$\frac{F - 2.55}{0.06} = \frac{0.06t}{0.06} \qquad \text{Divide both sides by 0.06.}$$
$$\frac{F - 2.55}{0.06} = t \qquad \text{Simplify: } \frac{0.06}{0.06} = 1$$

The result is the formula $t = \dfrac{F - 2.55}{0.06}$.

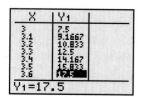

Figure 29 Finding the years

3. We substitute 2.90 for F in $t = \dfrac{F - 2.55}{0.06}$:

$$t = \frac{2.90 - 2.55}{0.06} \approx 5.83$$

The average ATM fee was \$2.90 in $2000 + 6 = 2006$.

4. We use a graphing calculator table to substitute 3.00, 3.10, 3.20, 3.30, and 3.40 for F in the equation $t = \dfrac{F - 2.55}{0.06}$ (see Fig. 29).

By viewing the second column of the graphing calculator table, we see that the approximate values of t (rounded to the one's place) are 8, 9, 11, 13, and 14. The average ATM fee will be \$3.00, \$3.10, \$3.20, \$3.30, and \$3.40 in the years 2008, 2009, 2011, 2013, and 2014, respectively. ■

group exploration

Looking ahead: Graphing linear equations

1. a. Solve the equation $3x + 4y = 8$ for y.
  b. Use the result you found in part (a) to help you graph $3x + 4y = 8$ by hand.
2. a. Solve the equation $5x - 3y - 9 = 0$ for y.
  b. Use the result you found in part (a) to help you graph $5x - 3y - 9 = 0$ by hand.

TIPS FOR SUCCESS: Relax during a Test

If you get flustered during a test, close your eyes, take a couple of deep breaths, and think about something pleasant or nothing at all for a moment. This short break from the test might give you some perspective and help you relax.

HOMEWORK 4.6 FOR EXTRA HELP ▶

Student Solutions Manual   PH Math/Tutor Center   MathXL®   MyMathLab

Find a formula of the perimeter P of the polygon.

1. See Fig. 30.

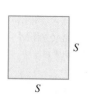

Figure 30 Exercise 1

2. See Fig. 31.

Figure 31 Exercise 2

3. See Fig. 32.

4. See Fig. 33.

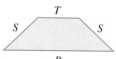

Figure 32 Exercise 3   **Figure 33** Exercise 4

5. See Fig. 34.

Figure 34 Exercise 5

6. See Fig. 35.

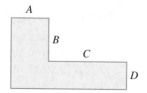

Figure 35 Exercise 6

For Exercises 7–16, substitute the given values for the variables and then solve the equation for the remaining variable. Round approximate results to the second decimal place.

7. $P = VI$; $P = 20$, $I = 4$ (power in an electrical circuit)
8. $PV = nRT$; $P = 0.97$, $V = 2.85$, $R = 0.082$, $T = 295$ (pressure, volume, and temperature of a gas)
9. $A = \dfrac{1}{2}BH$; $A = 6$, $H = 3$ (area of a triangle)

10. $U = -\dfrac{GmM}{r}$; $U = -50$, $G = 9.8$, $M = 7$, $r = 2$ (gravitational potential energy)

11. $v = gt + v_0$; $v = 80$, $g = 32.2$, $v_0 = 20$ (speed of a projectile)

12. $A = P + Prt$; $A = 850$, $P = 500$, $r = 0.1$ (simple interest)

13. $S = 2WL + 2WH + 2LH$; $S = 52$, $W = 2$, $H = 4$ (surface area of a rectangular box)

14. $S = 2WL + 2WH + 2LH$; $S = 76$, $L = 2$, $H = 4$ (surface area of a rectangular box)

15. $A = \dfrac{a+b+c}{3}$; $A = 5$, $a = 2$, $c = 6$ (average length of a side of a triangle)

16. $\dfrac{x}{a} + \dfrac{y}{a} = 1$; $a = 2$, $y = 6$ (equation of a line)

17. A rectangular carpet has an area of 116 square feet and a width of 8 feet. Find the length of the carpet.

18. A rectangular room has an area of 207 square feet and a length of 18 feet. Find the width of the room.

19. A rectangle has a perimeter of 52 inches and a width of 10 inches. Find the length of the rectangle.

20. A rectangle has a perimeter of 88 inches and a length of 25 inches. Find the width of the rectangle.

21. A portable volleyball court includes a 177-foot cord to use as the rectangular boundary of the official-size court. The official width of a volleyball court is 29.5 feet. What is the official length of the court?

22. A photographer plans to use 55 inches of wood to make a rectangular frame for a photo. If the width of the frame is to be 12.5 inches, what will its length be?

23. A rectangle has a width of 3 inches and a length of x inches.
    a. Find a formula of the area A (in square inches) of the rectangle.
    b. Find a formula of the perimeter P (in inches) of the rectangle.
    c. Perform a unit analysis of the formula you found in part (b).

24. A rectangle has a width of x feet and a length of 7 feet.
    a. Find a formula of the area A (in square feet) of the rectangle.
    b. Find a formula of the perimeter P (in feet) of the rectangle.
    c. Perform a unit analysis of the formula you found in part (b).

25. A landscaper is going to dig a rectangular garden and enclose the garden with 40 feet of fencing. He can plant one flower per square foot of ground.
    a. Give three examples of rectangular gardens that could be enclosed with 40 feet of fencing. State the length and width of each garden.
    b. Find the area of each of the three gardens you described in part (a).
    c. Which of the gardens you described in part (a) would hold the greatest number of flowers? Explain.

26. A landscaper is going to dig a small rectangular garden that has an area of 36 square feet.
    a. Give three examples of rectangular gardens that have an area of 36 square feet. State the length and width of each garden.
    b. Find the perimeter of each of the three gardens you described in part (a).
    c. The landscaper plans to enclose the garden with fencing. Which of the gardens you described in part (a) would require the least amount of fencing? How much fencing is that?

27. a. What is the value (in cents) of 3 dimes?
    b. What is the value (in cents) of 4 dimes?
    c. Find a formula of the total value T (in cents) of d dimes.
    d. Perform a unit analysis of the formula you found in part (c).

28. a. What is the value (in dollars) of 3 nickels?
    b. What is the value (in dollars) of 4 nickels?
    c. Find a formula of the total value T (in dollars) of n nickels.
    d. Perform a unit analysis of the formula you found in part (c).

29. Find a formula of the total sales T (in dollars) of selling n music CDs at $12.95 per CD.

30. Find a formula of the total sales T (in dollars) of selling n gallons of gasoline at $3.15 per gallon.

31. Morrissey, Sonic Youth, and Le Tigre were 3 of 16 bands that played in the Lollapalooza concert in 2004. The average general-admission ticket price was $32.50.
    a. Assuming that all tickets at the concert cost $32.50 each, find a formula of the total sales T (in dollars) if x people bought tickets.
    b. Substitute 601,965 for T in the formula you found in part (a), and solve for x. What does your result mean in this situation?

32. AOL® charged $23.90 per month for unlimited hours of service in 2006.
    a. Find a formula of the total charge T (in dollars) for m months of service.
    b. Substitute $286.80 for T in the formula you found in part (a), and solve for m. What does your result mean in this situation?

33. Baseball tickets at Yankee Stadium in the Bronx ranged from $10 to $95 in 2005.
    a. Find a formula of the total cost C (in dollars) of k tickets that sold for $10 per ticket.
    b. Find a formula of the total cost E (in dollars) of n tickets that sold for $95 per ticket.
    c. Find a formula of the total cost T (in dollars) of k tickets that sold for $10 per ticket and n tickets that sold for $95 per ticket.
    d. Use the formula you found in part (c) to find the value of n when $k = 8000$ and $T = 270,000$. What does your result mean in this situation?

34. Football tickets at Boston College ranged from $25 to $45 in 2005.
    a. Find a formula of the total cost C (in dollars) of k tickets that sold for $25 per ticket.
    b. Find a formula of the total cost E (in dollars) of n tickets that sold for $45 per ticket.
    c. Find a formula of the total cost T (in dollars) of k tickets that sold for $25 per ticket and n tickets that sold for $45 per ticket.
    d. Use the formula you found in part (c) to find the value of n when $k = 11,200$ and $T = 538,750$. What does your result mean in this situation?

35. One cubic foot is shown in Fig. 36. The *volume,* in cubic feet, of an object is the number of cubic feet that it takes to fill that object. Let L be the length, W be the width, and H be the height, all in feet, of a rectangular box (see Fig. 37). The volume V (in cubic feet) of the rectangular box is equal to the length times the width times the height of the box.

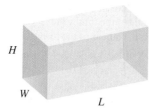

Figure 36 One cubic foot

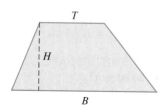

Figure 37 The length L, width W, and height H of a rectangular box.

a. Write a formula of the volume of the rectangular box.
b. Find the height of the rectangular box if the volume is 48 cubic feet, the length is 3 feet, and the width is 2 feet.

36. Let B be the length of the base, T be the length of the top, and H be the height of a trapezoid (see Fig. 38). The area of a trapezoid is equal to $\frac{1}{2}$ of the height times the sum of the lengths of the base and top.

Figure 38 The base B, top T, and height H of a trapezoid

a. Write a formula of the area A of a trapezoid.
b. Find the base of a trapezoid with an area of 20 square inches, a height of 4 inches, and a top length of 3 inches.

37. The interest I (in dollars) earned from a simple-interest bank account is equal to the money P (in dollars) invested, times the decimal annual interest rate r, times the number of years t of the investment.
a. Write a formula of I.
b. Find the interest earned from investing $5000 in a 4% simple-interest account for 3 years. [**Hint:** If the interest rate is 4%, then the decimal interest rate is 0.04.]
c. The balance B (in dollars) of a bank account is equal to the original money invested plus the interest. Find a formula of B in terms of P, r, and t. [**Hint:** Build on your formula from part (a).]
d. Find the balance after 4 years in a 5% simple-interest account in which $2000 was originally invested.

38. The force F (in newtons) needed to lift an object is equal to the product of the object's mass m (in kilograms) and the constant a (in meters per second squared), which is the object's acceleration due to gravity.

a. Write a formula of the force needed to lift an object.
b. Describe the unit newton (N) in terms of kilograms (kg), meters (m), and seconds (s). [**Hint:** Perform a unit analysis of the right-hand side of the formula you found in part (a).]
c. If a force of 25 newtons is required to lift an object and the constant a is 9.8 meters per second squared, find the mass of the object.

39. The average test score A of five test scores t_1, t_2, t_3, t_4, and t_5 can be computed by using the formula

$$A = \frac{t_1 + t_2 + t_3 + t_4 + t_5}{5}$$

If a student has test scores of 74, 81, 79, and 84, what score does she need on the fifth test so that her five-test average is 80?

40. The average test score A of four test scores t_1, t_2, t_3, and t_4 can be computed by using the formula

$$A = \frac{t_1 + t_2 + t_3 + t_4}{4}$$

If a student has test scores of 87, 92, and 86, what score does he need on the fourth test so that his four-test average is 90?

Solve the equation for the specified variable.

41. $A = LW$, for W **42.** $P = VI$, for I

43. $PV = nRT$, for T **44.** $PV = nRT$, for n

45. $U = -\dfrac{GmM}{r}$, for M **46.** $U = -\dfrac{GmM}{r}$, for m

47. $A = \dfrac{1}{2}BH$, for B **48.** $A = \dfrac{1}{2}BH$, for H

49. $v = gt + v_0$, for t **50.** $y = mx + b$, for x

51. $A = P + Prt$, for r **52.** $A = P + Prt$, for t

53. $A = \dfrac{a+b+c}{3}$, for b **54.** $A = \dfrac{a+b+c}{3}$, for c

55. $y - k = m(x - h)$, for x **56.** $y - k = m(x - h)$, for h

57. $\dfrac{x}{a} + \dfrac{y}{a} = 1$, for y **58.** $\dfrac{x}{a} + \dfrac{y}{a} = 1$, for x

Write the linear equation in slope–intercept form.

59. $3x + 4y = 16$ **60.** $2x + 3y = 24$

61. $2x + 4y - 8 = 0$ **62.** $7x + 3y - 21 = 0$

63. $5x - 2y = 6$ **64.** $4x - 5y = 30$

65. $-3x - 7y = 5$ **66.** $-5x - 9y = 2$

67. Due to recent natural disasters such as severe hurricanes, federal flood insurance coverage has increased greatly (see Table 10).

Table 10 Federal Flood Insurance Coverage	
Year	**Federal Flood Insurance Coverage (billions of dollars)**
1996	401
1998	498
2000	568
2002	654
2004	765

Source: *Federal Emergency Management Agency*

Let c be federal flood insurance coverage (in billions of dollars) at t years since 1990. A model of the situation is $c = 44.2t + 135.2$.

a. Use a graphing calculator to draw a scattergram and the model in the same viewing window. Does the line come close to the data points?

b. Solve the equation $c = 44.2t + 135.2$ for t.

c. Use the equation that you found in part (b) to estimate when federal flood insurance coverage was $800 billion.

d. Use a graphing calculator table to predict in which years federal flood insurance coverage will be 850, 900, 950, 1000, and 1050, all in billions of dollars.

68. Computer and video game sales (in billions of dollars) in the United States are shown in Table 11 for various years.

Table 11 Computer and Video Game Sales

Year	Computer and Video Game Sales (billions of dollars)
2000	6.0
2001	6.4
2002	6.9
2003	7.0
2004	7.3

Source: NPD Group

Let s be computer and video game sales (in billions of dollars) in the year that is t years since 2000. A model of the situation is $s = 0.32t + 6.08$.

a. Use a graphing calculator to draw a scattergram and the model in the same viewing window. Does the line come close to the data points?

b. Solve the equation $s = 0.32t + 6.08$ for t.

c. Use the equation you found in part (b) to estimate in which year computer and video game sales were $8 billion.

d. Use a graphing calculator table to predict in which years computer and video game sales will be 9, 9.5, 10, 10.5, and 11, all in billions of dollars.

69. Kia® sold 155 thousand automobiles in 2000, and sales are increasing by about 28.25 thousand automobiles per year (Source: Automotive News). Let s be sales (in thousands of automobiles) in the year that is t years since 2000.

a. Find an equation of a linear model to describe the situation.

b. Solve the equation you found in part (a) for t.

c. Kia has a goal of selling 500 thousand automobiles in 2010. According to the model, will Kia reach that goal? If not, in which year does the model predict that will happen?

d. Use a graphing calculator table to predict in which years Kia's automobile sales will be 400, 450, 500, 550, and 600, all in thousands of automobiles.

70. In 2000, 12% of travel bookings were made online. The percentage of travel bookings made online is increasing by 6.13 percentage points per year (Source: PhoCus Wright). Let p be the percentage of travel bookings made online at t years since 2000.

a. Find an equation of a linear model to describe the situation.

b. Solve the equation you found in part (a) for t.

c. Use the equation you found in part (b) to estimate in which year 43% of travel bookings were made online.

d. Use a graphing calculator table to help you predict in which years the percentage of travel bookings made online will be 60%, 70%, 80%, 90%, and 100%.

71. The number of immigrant visas issued to orphans entering the United States was 17.7 thousand in 2000 and is increasing by 1.3 thousand per year (Source: U.S. State Department). Let n be the number of immigrant visas (in thousands) issued to orphans entering the United States in the year that is t years since 2000.

a. Find an equation of a linear model to describe the situation.

b. Use the equation you found in part (a) to predict when 32 thousand immigrant visas will be issued to orphans entering the United States.

c. Solve the equation you found in part (a) for t.

d. Use the equation you found in part (c) to predict when 32 thousand immigrant visas will be issued to orphans entering the United States.

e. Compare your results from parts (b) and (d). Which equation was easier to use? Explain.

f. Which equation is easier to use to predict the number of visas issued to orphans entering the United States in 2010? Explain. Then find that number of visas.

72. Some 155 million pounds of fireworks were used in the United States in 2000, and that amount is increasing by about 8.55 million pounds per year (Source: American Pyrotechnics Association). Let F be the amount of fireworks used (in millions of pounds) in the year that is t years since 2000.

a. Find an equation of a linear model to describe the situation.

b. Use the equation you found in part (a) to predict when 250 million pounds of fireworks will be used.

c. Solve the equation you found in part (a) for t.

d. Use the equation you found in part (c) to predict when 250 million pounds of fireworks will be used.

e. Compare your results from parts (b) and (d). Which equation was easier to use? Explain.

f. Which equation is easier to use to predict the amount of fireworks used in 2010? Explain. Then find that amount.

73. A person travels d miles at a constant speed s (in miles per hour) for t hours.

a. Complete Table 12 to help find a formula that describes the relationship among s, t, and d. Show the arithmetic to help you see a pattern.

Table 12 Speed, Time, and Distance

Speed (miles per hour) s	Time (hours) t	Distance (miles) d
50	4	
70	3	
65	2	
55	5	
s	t	

b. Perform a unit analysis of the formula you found in part (a).

c. Solve the formula you found in part (a) for t.

d. Use the formula you found in part (c) to find the value of t when $d = 315$ and $s = 70$. What does your result mean in this situation?

e. A student plans to drive from Albuquerque Technical Vocational Institute in Albuquerque, New Mexico, to Denver, Colorado. The speed limit is 75 mph in New Mexico and 65 mph in Colorado. The trip involves 229.9 miles of travel in New Mexico, followed by 219.2 miles in Colorado. If the student drives at the posted speed limits, estimate the driving time.

74. a. Solve the equation $mx + b = 0$, where $m \neq 0$, for x.
   b. Explain why a linear equation in one variable has exactly one solution.

75. An object travels d miles at a constant speed s (in miles per hour) for t hours. A student believes that the formula of s is

$$s = dt \quad \text{Incorrect}$$

Perform a unit analysis to show that the formula is incorrect.

76. A student believes that a formula of the total value T (in dollars) of n objects, each worth v dollars, is

$$n = Tv \quad \text{Incorrect}$$

Perform a unit analysis to show that the formula is incorrect.

77. a. How does doubling the width and doubling the length of a rectangle affect the perimeter of the rectangle?
   b. How does doubling the width and doubling the length of a rectangle affect the area of the rectangle?

78. a. How does increasing the length of a rectangle by 1 inch affect the perimeter (in inches) of the rectangle?
   b. How does increasing the length of a rectangle by 1 inch affect the area (in square inches) of the rectangle?

79. In what cases would you first solve a formula for a variable and then make substitutions for any other variables, and in what cases would you first make substitutions for all but one variable and then solve for the remaining variable?

80. Give an example of an equation in one variable and an example of a formula. How are these equations different?

Chapter Summary

Key Points
OF CHAPTER 4

Commutative law for addition (Section 4.1)	$a + b = b + a$
Commutative law for multiplication (Section 4.1)	$ab = ba$
Term (Section 4.1)	A **term** is a constant, a variable, or a product of a constant and one or more variables raised to powers.
Associative law for addition (Section 4.1)	$a + (b + c) = (a + b) + c$
Associative law for multiplication (Section 4.1)	$a(bc) = (ab)c$
Distributive law (Section 4.1)	$a(b + c) = ab + ac$
Equivalent expressions (Section 4.1)	Two or more expressions are **equivalent expressions** if, when each variable is evaluated for *any* real number (for which all the expressions are defined), the expressions all give equal results.
Multiplying a number by -1 (Section 4.1)	$-1a = -a$
Showing that an equation is false (Section 4.1)	To show that an equation is false, we need find only *one* set of numbers that, when substituted for any variables, give a result of the form $a = b$, where a and b are different real numbers.
Like terms (Section 4.2)	**Like terms** are either constant terms or variable terms that contain the same variable(s) raised to exactly the same power(s).
Combining like terms (Section 4.2)	To combine like terms, add the coefficients of the terms and keep the same variable factors.
Simplifying an expression (Section 4.2)	We simplify an expression by removing parentheses and combining like terms.
Locating an error in simplifying an expression (Section 4.2)	After simplifying an expression, use a graphing calculator table to check whether the result is equivalent to the original expression. If the expressions are not equivalent, an error was made. To pinpoint the error, evaluate all of the expressions by using the same number(s) and find the pair of consecutive expressions for which the results are different.
Linear equation in one variable (Section 4.3)	A **linear equation in one variable** is an equation that can be put into the form $mx + b = 0$, where m and b are constants and $m \neq 0$.
Solution, satisfy, solution set, and solve for an equation in one variable (Section 4.3)	A number is a **solution** of an equation in one variable if the equation becomes a true statement when the number is substituted for the variable. We say that the number **satisfies** the equation. The set of all solutions of the equation is called the **solution set** of the equation. We **solve** the equation by finding its solution set.

Roles of a variable (Section 4.3)

Here are three roles of a variable:

1. A variable represents a quantity that can vary.

2. In an expression, a variable is a placeholder for a number.

3. In an equation, a variable represents any number that is a solution of the equation.

Equivalent equations (Section 4.3)

Equivalent equations are equations that have the same solution set.

Addition property of equality (Section 4.3)

If A and B are expressions and c is a number, then the equations $A = B$ and $A + c = B + c$ are equivalent.

Checking results after solving an equation (Section 4.3)

After solving an equation, check that all of your results satisfy the equation.

Multiplication property of equality (Section 4.3)

If A and B are expressions and c is a nonzero number, then the equations $A = B$ and $Ac = Bc$ are equivalent.

Using graphing to solve an equation in one variable (Section 4.3)

To use graphing to solve an equation $A = B$ in one variable, x, where A and B are expressions,

1. Graph the equations $y = A$ and $y = B$ in the same coordinate system.

2. Find all intersection points.

3. The x-coordinates of those intersection points are the solutions of the equation $A = B$.

Getting the variable alone on one side of the equation (Section 4.3)

Our strategy in solving linear equations in one variable will be to use properties to get the variable alone on one side of the equation. For example, to solve the equation $x + 3 = 7$, $x - 3 = 7$, $3x = 7$, or $\frac{x}{3} = 7$, we get x alone by "undoing" the operation that involves x and 3.

Solving an equation that contains fractions (Section 4.4)

A key step in solving an equation that contains fractions is to multiply both sides of the equation by the LCD so that there are no fractions on either side of the equation.

Number of solutions (Section 4.4)

Any linear equation in one variable has exactly one solution.

Conditional equation, inconsistent equation, and identity (Section 4.4)

There are three types of equations:

1. A **conditional equation** is sometimes true and sometimes false, depending on which values are substituted for the variable(s).

2. If an equation does not have any solutions, we call the equation **inconsistent** and say that the solution is the **empty set.** When we apply the usual steps to solve an inconsistent equation, the result is always a false statement.

3. An equation that is true for all permissible values of the variable(s) it contains is called an **identity.** When we apply the usual steps to solve an identity, the result is always a true statement of the form $a = a$, where a is a constant.

Locating an error in solving a linear equation (Section 4.4)

If the result of an attempt to solve a linear equation in one variable does not satisfy the equation, an error was made. To pinpoint the error, substitute the result into all of the equations in the work and find which pair of consecutive equations gives a false statement followed by a true statement.

Linear expression in one variable (Section 4.5)

A **linear expression in one variable** is an expression that can be put into the form $mx + b$, where m and b are constants and $m \neq 0$.

Results of simplifying an expression and solving an equation (Section 4.5)

The result of simplifying an expression is an expression. The result of solving a linear equation in one variable is a number.

Connection between simplifying an expression and an identity (Section 4.5)

If an expression A in one variable is simplified to an expression B, then every real number (for which both expressions are defined) is a solution of the equation $A = B$. In other words, the equation $A = B$ is an identity.

Multiplying expressions and both sides of equations by numbers (Section 4.5)

In simplifying an expression, the only number that we can multiply the expression or part of it by is 1. In solving an equation, we can multiply both sides of the equation by *any* number except 0.

Formula (Section 4.6)

A **formula** is an equation that contains two or more variables.

Finding a single value of a variable in a formula (Section 4.6)

If we want to find a single value of a variable in a formula, we often substitute numbers for all of the other variables and then solve for that one variable.

Perimeter (Section 4.6)

The **perimeter** of a polygon, such as a triangle, rectangle, or trapezoid, is the total distance around the object.

Area and perimeter of a rectangle (Section 4.6)	For a rectangle with length L, width W, area A, and perimeter P, • $A = LW$ **Area formula** • $P = 2L + 2W$ **Perimeter formula**
Total-value formula (Section 4.6)	If n objects each have value v, then their total value T is given by $T = vn$.
Finding several values of a variable in a formula (Section 4.6)	If we want to find several values of a variable in a formula, we usually solve the formula for that variable before making any substitutions.

CHAPTER 4 REVIEW EXERCISES

Simplify. Use a graphing calculator table to verify your work when possible.

1. $-5(4x)$

2. $-3(8x + 4)$

3. $\frac{4}{5}(15y - 35)$

4. $-(3x - 6y - 8)$

5. $\frac{2}{9}x + \frac{5}{9}x$

6. $5a + 2 - 13b - a + 4b - 9$

7. $-5y - 3(4x + y) - 6x$

8. $-2.6(3.1x + 4.5) - 8.5$

9. $-(2m - 4) - (3m + 8)$

10. $4(3a - 7b) - 3(5a + 4b)$

For Exercises 11 and 12, let x be a number. Translate the English phrase to a mathematical expression. Then simplify the expression.

11. -4 times the difference of the number and 7

12. -7, plus 2 times the sum of the number and 8

13. Use values for a, b, and c to show that $a(b + c) = ab + c$ is false. Then write a true statement that involves $a(b + c)$.

14. Give three examples of expressions that are equivalent to $3x - 9$.

15. Which of the following expressions are equivalent?

$$-5x - 20 \qquad -5(x - 4) \qquad 5(4 - x) \qquad -5x - 4$$
$$-2(x - 10) - 3x \qquad -5x + 20$$

For Exercises 16 and 17, graph the equation by hand.

16. $y = 2x + 3 - 4x$

17. $y = -3(x - 2)$

18. Determine whether 3 is a solution of the linear equation $2 - 5x = -3(4x - 7)$.

Solve. Use a graphing calculator to verify your result.

19. $a + 5 = 12$

20. $-4x = 20$

21. $-p = -3$

22. $-\frac{7}{3}a = 14$

23. $4.5x - 17.2 = -5.05$

24. $5x - 9x + 3 = 17$

25. $8m - 3 - m = 2 - 4m$

26. $8x = -7(2x - 3) + x$

27. $6(4x - 1) - 3(2x + 5) = 2(5x - 3)$

28. $\frac{w}{8} - \frac{3}{4} = \frac{5}{6}$

29. $\frac{3p - 4}{2} = \frac{5p + 2}{4} + \frac{7}{6}$

30. A student tries to solve the equation $x - 5 = 2$:

$$x - 5 = 2$$
$$x - 5 + 5 = 2$$
$$x + 0 = 2$$
$$x = 2$$

Describe any errors. Then solve the equation correctly.

31. Give three examples of equations that have the solution -6.

32. Solve $-2.5(3.8x - 1.9) = 83.7$. Round the result to two decimal places.

33. The number of prisoners in the United States was 1.9 million inmates in 2000 and increases by about 0.06 million inmates per year (Source: *U.S. Bureau of Justice Statistics*). Let n be the number of prisoners (in millions) at t years since 2000.
   a. Find an equation of a model.
   b. Predict how many prisoners there will be in 2010.
   c. Predict when there will be 2.6 million prisoners.

For Exercises 34 and 35, find the number.

34. Four times the difference of a number and 6 is 15.

35. Two, minus 3 times the sum of a number and 8, is 95.

36. Use "intersect" on a graphing calculator to solve the equation $\frac{1}{2}x + \frac{5}{3} = -\frac{2}{3}x - \frac{1}{4}$. Round the solution to the second decimal place.

For Exercises 37–40, solve the given equation by referring to the solutions of $y = -2x + 17$ and $y = 5x - 4$ shown in Tables 13 and 14.

37. $-2x + 17 = 5x - 4$

38. $-2x + 17 = 15$

39. $5x - 4 = 6$

40. $5x - 4 = -4$

Table 13 Solutions of $y = -2x + 17$		Table 14 Solutions of $y = 5x - 4$	
x	**y**	**x**	**y**
0	17	0	-4
1	15	1	1
2	13	2	6
3	11	3	11
4	9	4	16

Solve the equation. State whether the equation is a conditional equation, an inconsistent equation, or an identity.

41. $7x - 4 + 3x = 2 + 10x - 6$

42. $6(2x - 3) - (5x + 2) = -2(4x - 1)$

43. $2(x - 5) + 3 = 2x - 4$

44. A student incorrectly solves $4(x - 5) = 28$:

$$4(x - 5) = 28$$
$$4x - 20 = 28$$
$$4x - 20 - 20 = 28 - 20$$
$$4x = 8$$
$$\frac{4x}{4} = \frac{8}{4}$$
$$x = 2$$

Substitute 2 for x in each of the six equations, and explain how the results help you pinpoint the error. Describe the error. Then solve the equation correctly.

Is the following a linear expression or a linear equation?

45. $8 - 3(x + 5)$

46. $8 - 3(x + 5) = 4x$

Simplify the expression or solve the equation, as appropriate.

47. $6t - 8t$

48. $0.1 + 0.5a - 0.3a = 0.7$

49. $6t - 8t = 10$

50. $0.1 + 0.5a - 0.3a$

51. $9(2p - 5) - 3(7p + 3)$

52. $\frac{5}{6}r - \frac{3}{4} = \frac{1}{6} + \frac{7}{2}r$

53. $9(2p - 5) - 3(7p + 3) = 0$

54. $\frac{5}{6}r - \frac{3}{4} - \frac{1}{6} + \frac{7}{2}r$

55. A student is trying to simplify an expression. The student writes "The solution of the expression is 2." Is the student correct? Explain.

56. A student tries to simplify the expression $\frac{2}{3}x + \frac{7}{5}$:

$$\frac{2}{3}x + \frac{7}{5} = 15\left(\frac{2}{3}x + \frac{7}{5}\right)$$
$$= 15 \cdot \frac{2}{3}x + 15 \cdot \frac{7}{5}$$
$$= 10x + 21$$

Describe any errors. Then simplify the expression correctly.

For Exercises 57 and 58, let x be a number. Translate the English phrase or sentence into an expression or an equation. Then simplify the expression or solve the equation, as appropriate.

57. Four times the difference of 6 and the number is 17.

58. The number minus the quotient of the number and 2

59. Find a formula of the perimeter P of the polygon shown in Fig. 39.

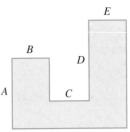

Figure 39 Exercise 59

60. a. Find a formula of the total cost T (in dollars) of n tickets that sell for \$15 per ticket and w tickets that sell for \$25 per ticket.

b. Use the formula you found in part (a) to find the value of w when $n = 370$ and $T = 11{,}050$. What does your result mean in this situation?

Solve the equation for the specified variable.

61. $C = 2\pi r$, for r

62. $P = a + b + c$, for c

63. $3x - 6y = 18$, for y

64. $A = \frac{1}{2}H(B + T)$, for T

65. The number of visits to U.S. libraries was 1.17 billion in 2000 and increases by 0.04 billion visits per year (Source: *American Library Association*). Let v be the number of visits (in billions) to U.S. libraries in the year that is t years since 2000.

a. Find an equation of a linear model to describe the situation.

b. Solve the equation you found in part (a) for t.

c. Use the equation you found in part (a) to predict in which year there will be 1.6 billion visits.

d. Use the equation you found in part (b) to predict in which year there will be 1.6 billion visits.

e. Compare your results from parts (c) and (d). Which equation was easier to use? Explain.

f. Which equation is easier to use to predict the number of U.S. library visits in 2010? Explain. Then find that number of visits.

CHAPTER 4 TEST

Simplify.

1. $-\frac{2}{3}(6x - 9)$

2. $9.36 - 2.4(1.7x + 3.5)$

3. $-5(2w - 7) - 3(4w - 6)$

4. $-(3a + 7b) - (8a - 4b + 2)$

5. A student incorrectly simplifies $3(x - 2) - (5x + 4)$:

Expression	Evaluate the Expression at $x = 3$
$3(x - 2) - (5x + 4)$	
$= 3x - 6 - 5x + 4$	
$= 3x - 5x - 6 + 4$	
$= -2x - 2$	

Evaluate each of the four expressions for $x = 3$ to pinpoint where the error was made. Then simplify $3(x - 2) - (5x + 4)$ correctly.

6. Graph $y = -2(x + 1)$ by hand.

Solve.

7. $6x - 3 = 19$

8. $\frac{3}{5}x = 6$

9. $9a - 5 = 8a + 2$

10. $8 - 2(3t - 1) = 7t$

11. $3(2x - 5) - 2(7x + 9) = 49$

12. $\frac{7}{8}x + \frac{3}{10} = \frac{1}{4}x - \frac{1}{2}$

13. Solve $8.21x = 3.9(4.4x - 2.7)$. Round your result to two decimal places.

14. Four, minus 2 times the sum of a number and 7 is 54. Find the number.

Simplify the expression or solve the equation, as appropriate.

15. $9(3x + 2) - (4x - 6)$

16. $9(3x + 2) - (4x - 6) = x$

17. A student believes that $x - 3$ is the solution of an equation. Is the student correct? Explain.

18. Give three examples of an expression equivalent to the expression 4.

For Exercises 19 and 20, let x be a number. Translate the following into an expression or an equation, as appropriate. Then simplify the expression or solve the equation.

19. Five times the difference of the number and 2 is 29.

20. Two, plus 4 times the sum of 3 and the number

For Exercises 21–24, solve the given equation by referring to the graphs of $y = \frac{3}{2}x - 4$ and $y = \frac{1}{2}x - 2$ shown in Fig. 40.

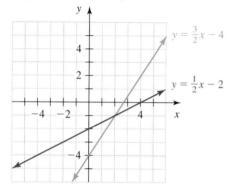

Figure 40 Exercises 21–24

21. $\frac{3}{2}x - 4 = \frac{1}{2}x - 2$

22. $\frac{3}{2}x - 4 = 2$

23. $\frac{1}{2}x - 2 = -3$

24. $\frac{1}{2}x - 2 = 0$

25. The number of U.S. patent applications was 303 thousand in 2000 and increases by about 30 thousand per year (Source: *U.S. Patent and Trademark Office*). Let n be the number of patent applications (in thousands) during the year that is t years since 2000.
   a. Find an equation of a model.
   b. Predict the number of patent applications in 2010.
   c. Predict when there will be 650 thousand patent applications.

26. A person plans to enclose a rectangular garden with fencing. If the width of the garden is 8 feet, and if 52 feet of fencing is required to enclose the garden, what is the length of the garden?

27. Solve the equation $A = \dfrac{a + b}{2}$ for the variable a.

CUMULATIVE REVIEW OF CHAPTERS 1-4

1. Choose a variable name for the number of pages in a book. Give two numbers that the variable represents and two numbers that it does not represent.

2. Graph the numbers -3, $-\frac{5}{2}$, 1, and $\frac{3}{2}$ on one number line.

3. Let n be the number (in millions) of unique visitors to the website YouTube in the year that is t years since 2000. What does the ordered pair $(6, 72.1)$ represent?

For Exercises 4–7, perform the indicated operations by hand. Then use a calculator to check your work.

4. $4 + 3(-2)$

5. $-8 \div 4 - 2(7 - 10)$

6. $\frac{15}{8} \cdot \left(-\frac{4}{25}\right)$

7. $\left(-\frac{3}{10}\right) + \left(-\frac{7}{8}\right)$

8. Simplify $\frac{27}{-45}$.

9. A rectangle has a width of $\frac{3}{4}$ foot and a length of $\frac{5}{6}$ foot. What is the perimeter of the rectangle?

10. A student scores 92 points on one test and 85 points on the next test. What is the change in the scores?

11. Evaluate the expression $a(b - c)$ for $a = -3$, $b = 5$, and $c = -4$.

For Exercises 12 and 13, find the slope of the line passing through the two given points. State whether the line is increasing, decreasing, horizontal, or vertical.

12. $(-5, -2)$ and $(-1, -4)$ **13.** $(-4, -5)$ and $(-4, 3)$

14. Road A climbs steadily for 150 feet over a horizontal distance of 5000 feet. Road B climbs steadily for 95 feet over a horizontal distance of 3500 feet. Which road is steeper? Explain.

For Exercises 15–18, graph the equation by hand.

15. $y = -\frac{2}{3}x + 4$

16. $y = 2x - 3$

17. $x = -5$

18. $y = 3$

19. Find an equation of the line sketched in Fig. 41.

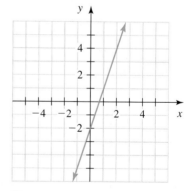

Figure 41 Exercise 19

20. A person's savings account balance decreased from $7500 to $2840 in 5 months. Find the average rate of change of the balance per month.

21. The number of reports filed with the U.S. Food and Drug Administration (FDA) about patients' adverse reactions to drugs increased from 264 thousand reports in 2000 to 422 thousand reports in 2004 (Source: *FDA*). Find the rate of change of the number of reports per year.

22. The number of convertibles sold in the United States was 242 thousand convertibles in 2000 and increases by 16.7 thousand cars per year (Source: *R. L. Polk*). Let *s* be the sales (in thousands) of convertibles in the year that is *t* years since 2000.

   a. Find the slope of the linear model that describes the situation. What does it mean in this situation?

   b. Find an equation of the model.

   c. Predict the sales of convertibles in 2011.

   d. Predict when sales will reach 450 thousand convertibles.

23. Some solutions of four equations are listed in Table 15. Which of the equations could be linear?

Table 15 Solutions of Four Equations

Equation 1		Equation 2		Equation 3		Equation 4	
x	y	x	y	x	y	x	y
0	11	0	56	3	35	1	1
1	13	1	53	4	44	2	1
2	16	2	50	5	53	3	1
3	20	3	47	6	62	4	1
4	25	4	44	7	71	5	1

24. The numbers of live births from fertility treatments are shown in Table 16 for various years.

Table 16 Numbers of Live Births from Fertility Treatments

Year	Number of Live Births from Fertility Treatments (thousands)
1996	20.8
1997	25.4
1998	29.6
1999	30.8
2000	35.6
2001	41.0
2002	45.8

Source: *U.S. Centers for Disease Control and Prevention*

Let *n* be the number of live births (in thousands) from fertility treatments in the year that is *t* years since 1995. A model of the situation is $n = 4.01t + 16.69$.

   a. Use a graphing calculator to draw a scattergram and the model in the same viewing window. Does the line come close to the data points?

   b. What is the slope? What does it mean in this situation?

   c. What is the *n*-intercept? What does it mean in this situation?

   d. Predict the number of live births from fertility treatments in 2010.

   e. Predict when there will be 85 thousand live births from fertility treatments.

Simplify the expression or solve the equation, as appropriate. Use a graphing calculator to verify your work.

25. $3r + 4 = 7r - 8$

26. $2(3x - 2) = 4(3x + 5) - 3x$

27. $4a - 5b + 6 - 2b - 7a$

28. $7 - 2(3p - 5) + 5(4p - 2)$

29. $\dfrac{2}{3}r - \dfrac{5}{6} = \dfrac{1}{2}$

30. $-(2a + 5) - (4a - 1)$

31. Solve $25.93 - 7.6(2.1x + 8.7) = 53.26$. Round the result to the second decimal place.

For Exercises 32 and 33, let x be a number. Translate the following into an expression or an equation, as appropriate. Then simplify the expression or solve the equation.

32. The number, plus 9 times the quotient of the number and 3

33. Two times the difference of 7 and twice the number is 87.

Solve the equation for the specified variable.

34. $A = 2\pi rh$, for *h*

35. $4x - 6y = 12$, for *y*

Chapter 5

Linear Equations in Two Variables

Whether you think that you can, or that you can't, you are usually right.

—*Henry Ford*

Table 1 Percentages of Children Living with Parents in "Very Happy" Marriages

Year	Percent
1976	51
1981	47
1986	43
1991	41
1996	37
2002	37

Source: *National Opinion Research Center, University of Chicago*

Did many of your friends grow up with parents who were very happily married? The percentage of children living with parents in marriages described as "very happy" has declined since 1976 (see Table 1). In Exercise 7 of Homework 5.4, you will predict when the percentage of children living with parents in "very happy" marriages will be 31% (less than one-third).

In Chapter 4, we discussed how to simplify expressions and solve equations. In this chapter, we will use these skills to help us increase our effectiveness in graphing equations, finding equations, and modeling authentic situations such as the data on "very happy" marriages.

5.1 GRAPHING LINEAR EQUATIONS

Objectives

▹ Graph a linear equation by solving for y.

▹ Graph a linear equation by finding the intercepts of its graph.

▹ Find coordinates of solutions of a linear equation.

▹ Know the difference between equations in one variable and equations in two variables.

In this section, we will discuss how to graph linear equations by three methods: solving for y, finding intercepts, and finding three points.

Graphing Equations by Solving for y

In Section 3.4, we discussed how to use the slope to graph an equation of the form $y = mx + b$. In Example 1, we will discuss how to graph equations that are not in this form, but that can be put into it.

Example 1   Graphing an Equation by Solving for y

Sketch the graph of $2x + 3y = 9$.

Solution

We will put the equation in $y = mx + b$ form so that we can use the y-intercept and the slope to help us graph the equation. To begin, we get y alone on one side of the equation:

$$2x + 3y = 9$$ Original equation

$$2x + 3y - 2x = 9 - 2x$$ Subtract $2x$ from both sides to get $3y$ alone on left-hand side.

$$3y = -2x + 9$$ Combine like terms; rearrange right-hand side.

$$\frac{3y}{3} = \frac{-2x}{3} + \frac{9}{3}$$ Divide both sides by 3 to get y alone on left-hand side.

$$y = -\frac{2}{3}x + 3$$ Simplify.

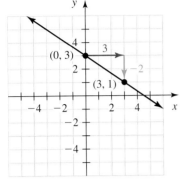

Since $2x + 3y = 9$ can be put into the form $y = mx + b$ (as $y = -\frac{2}{3}x + 3$), we know that the graph of $2x + 3y = 9$ is a line. The y-intercept is $(0, 3)$ and the slope is $-\frac{2}{3}$. The graph is shown in Fig. 1.

We can verify our result by checking that both $(0, 3)$ and $(3, 1)$ are solutions of $2x + 3y = 9$. ■

Figure 1 Graph of $2x + 3y = 9$

Before we can use the y-intercept and the slope to graph a linear equation, we must solve for y to put the equation into the form $y = mx + b$.

WARNING It is a common error to think that the slope of an equation such as $2x + 3y = 9$ is 2, because 2 is the coefficient of x. However, we must solve for y (and get $y = -\frac{2}{3}x + 3$) to determine that the slope is $-\frac{2}{3}$ (see Example 1).

Finding Intercepts

Another way to graph a linear equation is to use the intercepts of its graph. In Fig. 2, we show the x-intercept and y-intercept of a line.

Recall from Section 1.3 that for an x-intercept, the y-coordinate is 0 (see Fig. 2). For a y-intercept, the x-coordinate is 0 (see Fig. 2).

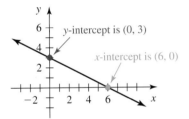

Figure 2 Intercepts of a line

Intercepts of the Graph of an Equation

For an equation containing the variables x and y,

• To find the x-coordinate of each x-intercept, substitute 0 for y and solve for x.
• To find the y-coordinate of each y-intercept, substitute 0 for x and solve for y.

Example 2   Graphing an Equation by Using Intercepts

Consider the equation $5y - 2x = 10$.

1. Find the x-intercept.
2. Find the y-intercept.
3. Sketch the graph of the equation.

Solution

1. To find the x-intercept, we substitute 0 for y and solve for x:

$$5y - 2x = 10$$ Original equation

$$5(0) - 2x = 10$$ Substitute 0 for y.

$$-2x = 10$$ Simplify.

$$\frac{-2x}{-2} = \frac{10}{-2}$$ Divide both sides by -2.

$$x = -5$$ Simplify.

The x-intercept is $(-5, 0)$.

2. To find the y-intercept, we substitute 0 for x and solve for y:

$$5y - 2x = 10 \qquad \text{Original equation}$$
$$5y - 2(0) = 10 \qquad \text{Substitute 0 for x.}$$
$$5y = 10 \qquad \text{Simplify.}$$
$$\frac{5y}{5} = \frac{10}{5} \qquad \text{Divide both sides by 5.}$$
$$y = 2 \qquad \text{Simplify.}$$

The y-intercept is $(0, 2)$.

3. The equation $5y - 2x = 10$ can be put into the form $y = mx + b$ (as $y = \frac{2}{5}x + 2$; try it). So, the graph of the equation $5y - 2x = 10$ is a line.

Before sketching the line, we find another solution to check our work. Here, we substitute 3 for x and solve for y:

$$5y - 2x = 10 \qquad \text{Original equation}$$
$$5y - 2(3) = 10 \qquad \text{Substitute 3 for x.}$$
$$5y - 6 = 10 \qquad \text{Multiply.}$$
$$5y = 16 \qquad \text{Add 6 to both sides.}$$
$$y = \frac{16}{5} \qquad \text{Divide both sides by 5.}$$

So, $\left(3, \dfrac{16}{5}\right)$ is a solution.

Finally, in Table 2 we list the three solutions of $5y - 2x = 10$ that we found, and in Fig. 3 we sketch the graph.

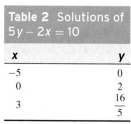

Table 2 Solutions of $5y - 2x = 10$

x	y
−5	0
0	2
3	$\frac{16}{5}$

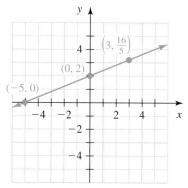

Figure 3 Graph of $5y - 2x = 10$

We can verify our work by checking that the ordered pairs $(-5, 0)$ and $(0, 2)$ satisfy the equation $5y - 2x = 10$. ■

In Example 2, we used the intercepts of the line $5y - 2x = 10$ and another point on the line to graph the equation $5y - 2x = 10$.

Using Intercepts to Graph an Equation

To graph a linear equation whose graph has exactly two intercepts,

1. Find the intercepts.

2. Plot the intercepts and a third point on the line, and graph the line that contains the three points.

When working with models, we will sometimes want to find intercepts that have decimal coordinates.

Example 3   Finding Approximate Intercepts

Find the indicated intercept of $y = -2.67x + 8.95$. Round the coordinates to the second decimal place.

1. x-intercept

2. y-intercept

Solution

1. To find the x-intercept, we substitute 0 for y and solve for x:

$$y = -2.67x + 8.95 \qquad \text{Original equation}$$
$$0 = -2.67x + 8.95 \qquad \text{Substitute 0 for y.}$$
$$0 + 2.67x = -2.67x + 8.95 + 2.67x \qquad \text{Add 2.67x to both sides.}$$
$$2.67x = 8.95 \qquad \text{Combine like terms.}$$
$$\frac{2.67x}{2.67} = \frac{8.95}{2.67} \qquad \text{Divide both sides by 2.67.}$$
$$x \approx 3.35 \qquad \begin{array}{l}\text{Round right-hand side to second}\\ \text{decimal place.}\end{array}$$

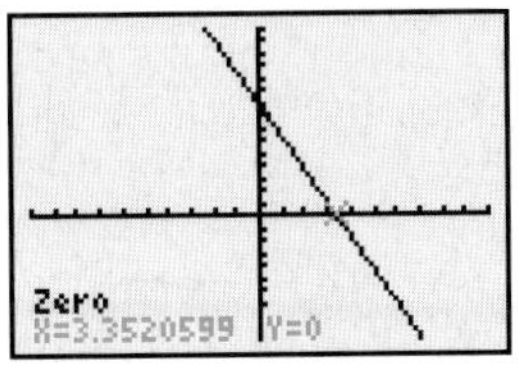

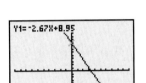

Figure 4 Verify the work

The approximate x-intercept is $(3.35, 0)$. We use "zero" on a graphing calculator to verify our work (see Fig. 4). See Section A.19 for graphing calculator instructions.

2. For an equation of the form $y = mx + b$, as is $y = -2.67x + 8.95$, the graph has y-intercept $(0, b)$. So, the y-intercept is $(0, 8.95)$. We use TRACE on a graphing calculator to verify our work (see Fig. 5). See Section A.5 for graphing calculator instructions. ∎

Figure 5 Verify the work

Finding Coordinates of Solutions

In Section 3.1, we graphed equations of the form $y = mx + b$ by plotting points. After learning how to solve equations in one variable in Chapter 4, we can now graph a linear equation in any form by plotting points.

Example 4 Graphing an Equation by Plotting Points

Complete the following steps to graph the equation $3x + 2y = 8$:

1. Find y when $x = 4$.
2. Find x when $y = 1$.
3. Sketch the graph of $3x + 2y = 8$.

Solution

1. We substitute 4 for x in the equation $3x + 2y = 8$ and solve for y:

$$3x + 2y = 8 \qquad \text{Original equation}$$
$$3(4) + 2y = 8 \qquad \text{Substitute 4 for x.}$$
$$12 + 2y = 8 \qquad \text{Multiply.}$$
$$12 + 2y - 12 = 8 - 12 \qquad \text{Subtract 12 from both sides.}$$
$$2y = -4 \qquad \text{Combine like terms.}$$
$$\frac{2y}{2} = \frac{-4}{2} \qquad \text{Divide both sides by 2.}$$
$$y = -2 \qquad \text{Simplify.}$$

So, $y = -2$ when $x = 4$. The ordered pair $(4, -2)$ is a solution.

2. We substitute 1 for y in the equation $3x + 2y = 8$ and solve for x:

$$3x + 2y = 8 \qquad \text{Original equation}$$
$$3x + 2(1) = 8 \qquad \text{Substitute 1 for y.}$$
$$3x + 2 = 8 \qquad \text{Multiply.}$$
$$3x + 2 - 2 = 8 - 2 \qquad \text{Subtract 2 from both sides.}$$
$$3x = 6 \qquad \text{Combine like terms.}$$
$$\frac{3x}{3} = \frac{6}{3} \qquad \text{Divide both sides by 3.}$$
$$x = 2 \qquad \text{Simplify.}$$

So, $x = 2$ when $y = 1$. The ordered pair $(2, 1)$ is a solution.

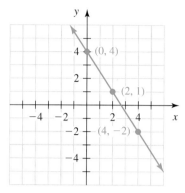

Figure 6 Graph of $3x + 2y = 8$

3. First, we find that the y-intercept is $(0, 4)$. (Try it.) Then we plot the points $(4, -2)$, $(2, 1)$, and $(0, 4)$ and sketch the line that contains them (see Fig. 6). ■

Using Three Solutions to Graph a Linear Equation

To graph any linear equation,

1. Find three solutions of the equation.
2. Plot the three solutions and graph the line that contains them.

Examples 1, 2, and 4 show three ways to graph an equation in which y is not alone on one side of the equation. For an equation in which y is alone on one side, the y-intercept and the slope are usually best to use to graph the equation.

Comparing Equations in One and Two Variables

In Example 5, we will compare solutions of an equation in one variable with solutions of an equation in two variables.

Example 5   Comparing Equations in One and Two Variables

1. Describe the solution(s) of the equation $y = x + 1$, an equation in two variables.
2. Describe the solution(s) of the equation $3 = x + 1$, an equation in one variable.

Solution

1. The equation $y = x + 1$ is a linear equation in *two* variables, x and y. We can use the graph of the equation (a line) to describe the solutions (see Fig. 7). Since there is an infinite number of points on the line, there is an infinite number of solutions of the equation $y = x + 1$.
2. The equation $3 = x + 1$ is a linear equation in *one* variable, x. So, exactly one number is a solution:

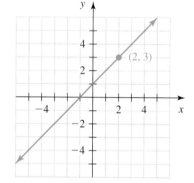

Figure 7 Graph of $y = x + 1$

$$3 = x + 1 \qquad \text{Original equation}$$
$$3 - 1 = x + 1 - 1 \qquad \text{Subtract 1 from both sides.}$$
$$2 = x \qquad \text{Combine like terms.}$$

The solution is the number 2. ■

There is a connection between the equations $y = x + 1$ and $3 = x + 1$. If we substitute 3 for y in the equation $y = x + 1$, the result is the equation $3 = x + 1$, whose solution is $x = 2$. So, $(2, 3)$ is one of an infinite number of solutions of $y = x + 1$ (see Fig. 7).

The graph of any linear equation in two variables is a line, which contains an infinite number of points. So, a linear equation in two variables has an infinite number of (ordered-pair) solutions. Recall from Section 4.4 that a linear equation in one variable has exactly one (real-number) solution.

Comparing Linear Equations in One and Two Variables

A linear equation in two variables has an infinite number of (ordered-pair) solutions.
A linear equation in one variable has exactly one (real-number) solution.

When modeling, we use an equation in two variables to describe the relationship between two quantities. After substituting a value for one of the variables, we solve the equation in one variable to make an estimate or prediction.

group exploration

Comparing two graphing techniques

In this exploration, you will compare two methods of graphing an equation.

1. Consider the equation $4x + 5y = 20$.
   a. Graph the equation by finding the slope and the y-intercept.
   b. Graph the equation by finding both intercepts.
   c. Compare your graphs from parts (a) and (b). Decide which method you prefer for graphing the equation $4x + 5y = 20$.

2. Consider the equation $y = \frac{2}{5}x - 1$.
   a. Graph the equation by finding the slope and the y-intercept.
   b. Graph the equation by finding both intercepts.
   c. Compare your graphs from parts (a) and (b). Decide which method you prefer for graphing the equation $y = \frac{2}{5}x - 1$.

3. In general, which types of linear equations do you prefer to graph by using the slope and the y-intercept? Which types do you prefer to graph by finding the intercepts? Explain.

TIPS FOR SUCCESS: Keep a Positive Attitude

If I were asked to give what I consider the single most useful bit of advice for all humanity it would be this: Expect trouble as an inevitable part of life and when it comes, hold your head high, look it squarely in the eye and say, "I will be bigger than you. You cannot defeat me."

—Ann Landers

Sometimes when students have difficulty in doing mathematics, they take this as evidence that they cannot learn it. However, getting stumped when trying to solve a math problem is something everyone has experienced—even your math instructor! Students who are successful at mathematics realize that this is part of the process of learning the subject. When they have trouble learning something, they keep a positive attitude and continue working hard, knowing that they will eventually learn the material.

HOMEWORK 5.1

FOR EXTRA HELP ▶

 Student Solutions Manual
 PH Math/Tutor Center
 *Math* XL MathXL®
MyMathLab MyMathLab

Determine the slope and the y-intercept. Use the slope and the y-intercept to graph the equation by hand. Use a graphing calculator to verify your work.

1. $y = 2x - 3$
2. $y = 4x - 3$
3. $y = -3x + 5$
4. $y = -4x + 1$
5. $y = -\frac{3}{5}x - 2$
6. $y = -\frac{2}{3}x - 1$
7. $y = \frac{1}{2}x + 2$
8. $y = \frac{1}{3}x + 4$
9. $y = -3x$
10. $y = 2x$
11. $y = -4$
12. $y = 0$
13. $y + x = 3$
14. $y + x = 1$
15. $y + 2x = 4$
16. $y + 3x = -2$
17. $y - 2x = -1$
18. $y - 3x = 2$
19. $3y = 2x$
20. $2y = -5x$
21. $2y = 5x - 6$
22. $3y = -4x + 6$
23. $5y = 4x - 15$
24. $2y = 5x - 6$
25. $3x - 4y = 8$
26. $4x + 3y = 9$
27. $6x - 15y = 30$
28. $6x - 8y = 16$
29. $x + 4y = 4$
30. $-x + 5y = -20$
31. $-4 = x + 2y$
32. $9 = x + 3y$
33. $4x + y + 2 = 0$
34. $2x - y - 3 = 0$
35. $6x - 4y + 8 = 0$
36. $15x + 12y - 36 = 0$
37. $0 = 5x + 3y$
38. $0 = 4x - 6y$
39. $y - 3 = 0$
40. $y + 5 = 0$

Find the x-intercept and y-intercept. Then graph the equation by hand.

41. $x - 3y = 6$ **42.** $2x - y = 6$

43. $15 = 3x + 5y$ **44.** $20 = 4x - 5y$

45. $2x - 3y + 12 = 0$ **46.** $2x - 4y - 8 = 0$

47. $y = -3x + 6$ **48.** $y = x + 2$

49. $\dfrac{1}{2}x + \dfrac{1}{3}y = 2$ **50.** $\dfrac{1}{4}x - \dfrac{1}{2}y = -1$

51. $\dfrac{x}{3} + \dfrac{y}{5} = 1$ **52.** $\dfrac{x}{4} + \dfrac{y}{3} = 1$

Find the approximate x-intercept and the approximate y-intercept. Round the coordinates to the second decimal place.

53. $6.2x + 2.8y = 7.5$ **54.** $8.1x + 3.9y = 14.2$

55. $6.62x - 3.91y = -13.55$ **56.** $-7.29x + 4.72y = 26.36$

57. $y = -4.5x + 9.32$ **58.** $y = 3.5x - 4.8$

59. $y = -2.49x - 37.21$ **60.** $y = -8.79x - 92.58$

Graph the equation by hand. Use any method.

61. $2x - y = 5$ **62.** $3x + y = 1$

63. $3y = 4x - 3$ **64.** $5y = 3x + 10$

65. $4y - 3x = 0$ **66.** $2y + 5x = 0$

67. $2x - 3y - 12 = 0$ **68.** $3x + 4y + 24 = 0$

69. Consider the equation $6x + 5y = -13$:
  a. Find y when $x = -3$.
  b. Find x when $y = -5$.
  c. Use your results from parts (a) and (b) to help you graph the equation $6x + 5y = -13$ by hand.

70. Consider the equation $y = -2x + 4$:
  a. Find y when $x = 3$.
  b. Find x when $y = 6$.
  c. Use your results from parts (a) and (b) to help you graph the equation $y = -2x + 4$ by hand.

71. A student thinks that the graph of $3y + 2x = 6$ has slope 2 because the coefficient of the $2x$ term is 2. Is the student correct? If yes, explain why. If no, find the slope.

72. A student says that the y-intercept of the line $2y = 3x + 4$ is $(0, 4)$ because the constant term is 4. Is the student correct? Explain.

73. Describe the Rule of Four as applied to the equation $3x - 5y = 10$:
  a. Describe the solutions of the equation by using a graph.
  b. Describe three solutions of the equation by using a table.
  c. Describe the solutions of the equation by using words.

74. Describe the Rule of Four as applied to the equation $3x + 4y = 8$:
  a. Describe the solutions of the equation by using a graph.
  b. Describe three solutions of the equation by using a table.
  c. Describe the solutions of the equation by using words.

75. a. Find the intercepts of $\dfrac{x}{5} + \dfrac{y}{7} = 1$.

  b. Find the intercepts of $\dfrac{x}{4} + \dfrac{y}{6} = 1$.

  c. Find the intercepts of $\dfrac{x}{a} + \dfrac{y}{b} = 1$, where a and b are nonzero constants. [**Hint:** Do the same steps as in parts (a) and (b).]

d. Find an equation of a line whose x-intercept is $(2, 0)$ and whose y-intercept is $(0, 5)$. [**Hint:** See your work in part (c).]

76. a. Find the slope of the line $3x + 5y = 7$.
  b. Find the slope of the line $2x + 7y = 3$.
  c. Find the slope of the line $ax + by = c$, where a, b, and c are constants and b is nonzero. [**Hint:** Do the same steps as in parts (a) and (b).]

77. Explain why the y-coordinate of the x-intercept is zero. Explain why the x-coordinate of the y-intercept is zero.

78. a. Solve the equation $2y - 6 = 4x$ for y.
  b. Find two solutions of the equation $y = 2x + 3$.
  c. Show that the two ordered pairs you found in part (b) satisfy both of the equations $2y - 6 = 4x$ and $y = 2x + 3$.
  d. Explain why it makes sense that the graphs of $2y - 6 = 4x$ and $y = 2x + 3$ are the same.

79. If the graph of a linear equation has a defined slope, describe how to find the slope and the y-intercept, and use this information to graph the equation. (See page 9 for guidelines on writing a good response.)

80. If the graph of a linear equation has exactly two intercepts, describe how to find the intercepts, and use the intercepts to graph the equation. (See page 9 for guidelines on writing a good response.)

Related Review

For Exercises 81–88, refer to the graph sketched in Fig. 8.

81. Find y when $x = 6$.   **82.** Find y when $x = -2$.

83. Find x when $y = -1$.   **84.** Find x when $y = -5$.

85. Find the x-intercept.   **86.** Find the y-intercept.

87. Find the slope of the line.

88. Find the equation of the line.

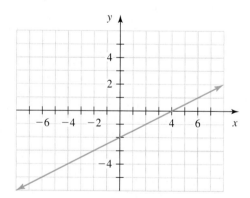

Figure 8 Exercises 81–88

Expressions, Equations, and Graphs

Perform the indicated instruction. Then use words such as linear, one variable, *and* two variables *to describe the expression or equation. For instance, to describe $2x = 8$, you could say "$2x = 8$ is a linear equation in one variable."*

89. Simplify $5 - 3(2x - 7)$.

90. Graph $y = -2x + 6$ by hand.

91. Solve $5 - 3(2x - 7) = 8$.

92. Solve $10 = -2x + 4$.

5.2 FINDING LINEAR EQUATIONS

Objectives

▷ Find an equation of a line by using the slope and a point.
▷ Find an equation of a line by using two points.
▷ Know the point–slope form of a linear equation.

In this section, we will discuss two methods of finding an equation of a line. We will first use the slope–intercept form. Then we will use another form of an equation of a line called the *point–slope form*.

Method 1: Using the Slope-Intercept Form

In Example 1, we will use the concept that a point which lies on a line satisfies an equation of the line.

Example 1   Using the Slope and a Point to Find a Linear Equation

Find an equation of the line that has slope $m = 2$ and contains the point $(4, 3)$.

Solution

An equation of a nonvertical line can be put into the form $y = mx + b$. Since $m = 2$, we have

$$y = 2x + b$$

To find b, recall from Section 3.1 that any point on the graph of an equation represents a solution of that equation. In particular, the point $(4, 3)$ should satisfy the equation $y = 2x + b$:

$y = 2x + b$	Slope is $m = 2$.
$3 = 2(4) + b$	Substitute 4 for x and 3 for y.
$3 = 8 + b$	Multiply.
$3 - 8 = 8 + b - 8$	Subtract 8 from both sides.
$-5 = b$	Combine like terms.

Now we substitute -5 for b in $y = 2x + b$:

$$y = 2x - 5$$

To verify our work, we check that the coefficient of the variable term $2x$ is 2 (the given slope) and we see whether $(4, 3)$ satisfies the equation $y = 2x - 5$:

$y = 2x - 5$	The equation we found
$3 \stackrel{?}{=} 2(4) - 5$	Substitute 4 for x and 3 for y.
$3 \stackrel{?}{=} 8 - 5$	Multiply.
$3 \stackrel{?}{=} 3$	Subtract.
true	

■

Finding an Equation of a Line by Using the Slope, a Point, and the Slope-Intercept Form

To find an equation of a line by using the slope and a point,

1. Substitute the given value of the slope m into the equation $y = mx + b$.
2. Substitute the coordinates of the given point into your equation from step 1 and solve for b.
3. Substitute the value of b from step 2 into your equation from step 1.
4. Check that the graph of your equation contains the given point.

Example 2   Using the Slope and a Point to Find a Linear Equation

Find an equation of the line that has slope $m = -\frac{2}{5}$ and contains the point $(-4, 1)$.

Solution

Since the slope is $-\frac{2}{5}$, the equation has the form $y = -\frac{2}{5}x + b$. To find b, we substitute the coordinates of the point $(-4, 1)$ into the equation $y = -\frac{2}{5}x + b$ and solve for b:

$$y = -\frac{2}{5}x + b \qquad \text{Slope is } -\frac{2}{5}.$$

$$1 = -\frac{2}{5}(-4) + b \qquad \text{Substitute } -4 \text{ for } x \text{ and } 1 \text{ for } y.$$

$$1 = \frac{8}{5} + b \qquad \text{Simplify.}$$

$$5 = 8 + 5b \qquad \text{Multiply both sides by LCD, 5, to clear equation of fractions.}$$

$$5 - 8 = 8 + 5b - 8 \qquad \text{Subtract 8 from both sides.}$$

$$-3 = 5b \qquad \text{Combine like terms.}$$

$$-\frac{3}{5} = b \qquad \text{Divide both sides by 5.}$$

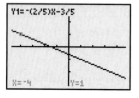

Figure 9 Check that the line contains $(-4, 1)$

So, the equation is $y = -\frac{2}{5}x - \frac{3}{5}$. In Fig. 9, we use ZDecimal followed by TRACE on a graphing calculator to check that the line $y = -\frac{2}{5}x - \frac{3}{5}$ contains the point $(-4, 1)$. ∎

In Examples 1 and 2, we found an equation of a line by using the slope of the line and a point. We can also find an equation of a line by using two points.

Example 3   Using Two Points to Find a Linear Equation

Find an equation of the line that passes through $(-1, 4)$ and $(2, -5)$.

Solution

First, we find the slope of the line:

$$m = \frac{y_2 - y_1}{x_2 - x_1} = \frac{-5 - 4}{2 - (-1)} = \frac{-5 - 4}{2 + 1} = \frac{-9}{3} = -3$$

So, we have $y = -3x + b$. Next, we will find the value of b. Since the line contains the point $(-1, 4)$, we substitute -1 for x and 4 for y:

$$y = -3x + b \qquad \text{Slope is } m = -3.$$

$$4 = -3(-1) + b \qquad \text{Substitute } -1 \text{ for } x \text{ and } 4 \text{ for } y.$$

$$4 = 3 + b \qquad \text{Multiply.}$$

$$4 - 3 = 3 + b - 3 \qquad \text{Subtract 3 from both sides.}$$

$$1 = b \qquad \text{Combine like terms.}$$

The equation is $y = -3x + 1$. To verify our equation, we check that both $(-1, 4)$ and $(2, -5)$ satisfy the equation $y = -3x + 1$:

Check that $(-1, 4)$ is a solution

$$y = -3x + 1$$
$$4 \overset{?}{=} -3(-1) + 1$$
$$4 \overset{?}{=} 3 + 1$$
$$4 \overset{?}{=} 4$$
$$\text{true}$$

Check that $(2, -5)$ is a solution

$$y = -3x + 1$$
$$-5 \overset{?}{=} -3(2) + 1$$
$$-5 \overset{?}{=} -6 + 1$$
$$-5 \overset{?}{=} -5$$
$$\text{true}$$
∎

In Example 3, we substituted the coordinates of the given point $(-1, 4)$ into the equation $y = -3x + b$ to help us find the constant b. If we had used the other given point, $(2, -5)$, we would have found the same value of b. (Try it.)

Finding an Equation of a Line by Using Two Points and the Slope-Intercept Form

To find an equation of the line that passes through two given points whose x-coordinates are different,

1. Use the formula $m = \dfrac{y_2 - y_1}{x_2 - x_1}$, or use a graph to determine $\dfrac{\text{rise}}{\text{run}}$, to find the slope of the line containing the two points.

2. Substitute the m value you found in step 1 into the equation $y = mx + b$.

3. Substitute the coordinates of one of the given points into the equation from step 2 and solve for b.

4. Substitute the b value you found in step 3 into your equation from step 2.

5. Check that the graph of your equation contains the two given points.

Example 4   Using Two Points to Find a Linear Equation

Find an equation of the line that passes through the points $(-9, -2)$ and $(-3, 7)$.

Solution

We begin by finding the slope of the line:

$$m = \frac{y_2 - y_1}{x_2 - x_1} = \frac{7 - (-2)}{-3 - (-9)} = \frac{9}{6} = \frac{3}{2}$$

So, we have $y = \dfrac{3}{2}x + b$. To find b, we substitute the coordinates of $(-3, 7)$ into the equation $y = \dfrac{3}{2}x + b$ and solve for b:

$$
\begin{aligned}
y &= \frac{3}{2}x + b && \text{Slope is } \tfrac{3}{2}. \\
7 &= \frac{3}{2}(-3) + b && \text{Substitute } -3 \text{ for } x \text{ and } 7 \text{ for } y. \\
7 &= -\frac{9}{2} + b && \text{Simplify.} \\
14 &= -9 + 2b && \text{Multiply both sides by LCD, 2, to clear equation of fractions.} \\
14 + 9 &= -9 + 2b + 9 && \text{Add 9 to both sides.} \\
23 &= 2b && \text{Combine like terms.} \\
\frac{23}{2} &= b && \text{Divide both sides by 2.}
\end{aligned}
$$

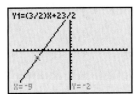

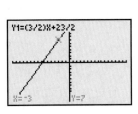

Figure 10 Check that the line contains both $(-9, -2)$ and $(-3, 7)$

The equation is $y = \dfrac{3}{2}x + \dfrac{23}{2}$. We use ZStandard followed by ZSquare to check that the line $y = \dfrac{3}{2}x + \dfrac{23}{2}$ contains the points $(-9, -2)$ and $(-3, 7)$. See Fig. 10. For graphing calculator instructions, see Sections A.3–A.6. ∎

In Example 5, we will work with decimal numbers to prepare us to find equations of models in Section 5.3.

Example 5   Finding an Approximate Equation of a Line

Find an approximate equation of the line that contains the points $(-3.1, 5.7)$ and $(1.6, -4.8)$.

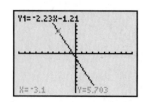

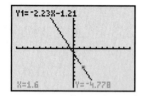

Figure 11 Check that the line comes very close to $(-3.1, 5.7)$ and $(1.6, -4.8)$

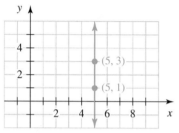

Figure 12 The line that contains $(5, 1)$ and $(5, 3)$

Solution

First, we find the slope of the line:

$$m = \frac{y_2 - y_1}{x_2 - x_1} = \frac{-4.8 - 5.7}{1.6 - (-3.1)} = \frac{-10.5}{4.7} \approx -2.23$$

So, the equation has the form $y = -2.23x + b$. To find b, we substitute the coordinates of the point $(-3.1, 5.7)$ into the equation $y = -2.23x + b$ and solve for b:

$$y = -2.23x + b \qquad \text{Slope is approximately } -2.23.$$
$$5.7 = -2.23(-3.1) + b \qquad \text{Substitute } -3.1 \text{ for } x \text{ and } 5.7 \text{ for } y.$$
$$5.7 = 6.913 + b \qquad \text{Multiply.}$$
$$5.7 - 6.913 = 6.913 + b - 6.913 \qquad \text{Subtract } 6.913 \text{ from both sides.}$$
$$-1.21 \approx b \qquad \text{Combine like terms.}$$

The approximate equation is $y = -2.23x - 1.21$.

We use a graphing calculator to check that the line $y = -2.23x - 1.21$ comes very close to the points $(-3.1, 5.7)$ and $(1.6, -4.8)$. See Fig. 11. ∎

Example 6 Finding an Equation of a Vertical Line

Find an equation of the line that contains the points $(5, 1)$ and $(5, 3)$.

Solution

Since the x-coordinates of the given points are equal (both 5), the line that contains the points is vertical (see Fig. 12). An equation of the line is $x = 5$. ∎

In Example 7, we will find an equation of a line that contains a given point and is perpendicular to a given line. Recall from Section 3.4 that if two nonvertical lines are perpendicular, then the slope of one line is the opposite reciprocal of the slope of the other line.

Example 7 Finding an Equation of a Line Perpendicular to a Given Line

Find an equation of the line l that contains the point $(4, 2)$ and is perpendicular to the line $x + 3y = -6$.

Solution

First, we write $x + 3y = -6$ in slope–intercept form:

$$x + 3y = -6 \qquad \text{Line perpendicular to line } l$$
$$x + 3y - x = -6 - x \qquad \text{Subtract } x \text{ from both sides.}$$
$$3y = -x - 6 \qquad \text{Combine like terms.}$$
$$\frac{3y}{3} = \frac{-x}{3} - \frac{6}{3} \qquad \text{Divide both sides by 3.}$$
$$y = -\frac{1}{3}x - 2 \qquad \text{Simplify.}$$

For the line $y = -\frac{1}{3}x - 2$, the slope is $-\frac{1}{3}$. The slope of the line l is the opposite reciprocal of $-\frac{1}{3}$, or 3. An equation of line l is $y = 3x + b$. To find b, we substitute the coordinates of the given point $(4, 2)$ into $y = 3x + b$ and solve for b:

$$2 = 3(4) + b \qquad \text{Substitute 4 for } x \text{ and 2 for } y.$$
$$2 = 12 + b \qquad \text{Multiply.}$$
$$2 - 12 = 12 + b - 12 \qquad \text{Subtract 12 from both sides.}$$
$$-10 = b \qquad \text{Combine like terms.}$$

An equation of line l is $y = 3x - 10$. We use ZStandard followed by ZSquare to verify our work (see Fig. 13).

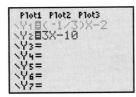
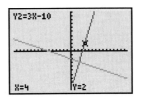

Figure 13 Check that the line contains $(4, 2)$ and is perpendicular to $x + 3y = -6$

Method 2: Using the Point-Slope Form

We can also find an equation of a line by yet another method. Suppose that a nonvertical line has slope m and contains the point (x_1, y_1). Then if (x, y) represents a different point on the line, the slope of the line is

$$\frac{y - y_1}{x - x_1} = m$$

Multiplying both sides of the equation by $x - x_1$ gives

$$\frac{y - y_1}{x - x_1} \cdot (x - x_1) = m \cdot (x - x_1)$$
$$y - y_1 = m(x - x_1)$$

We say that this linear equation is in **point–slope form.**[*]

Point-Slope Form of an Equation of a Line

If a nonvertical line has slope m and contains the point (x_1, y_1), then an equation of the line is

$$y - y_1 = m(x - x_1)$$

Example 8 Using the Point-Slope Form to Find an Equation of a Line

Use the point–slope form to find an equation of the line that has slope $m = -3$ and contains the point $(-4, 2)$. Then write the equation in slope–intercept form.

Solution

We begin by substituting $x_1 = -4$, $y_1 = 2$, and $m = -3$ into the point-slope form $y - y_1 = m(x - x_1)$:

$y - y_1 = m(x - x_1)$	Point–slope form
$y - 2 = -3(x - (-4))$	Substitute $x_1 = -4$, $y_1 = 2$, and $m = -3$.
$y - 2 = -3(x + 4)$	Simplify.
$y - 2 = -3x - 12$	Distributive law
$y - 2 + 2 = -3x - 12 + 2$	Add 2 to both sides.
$y = -3x - 10$	Combine like terms.

So, the equation is $y = -3x - 10$.

---

[*]Although we assumed that (x, y) is different from (x_1, y_1), note that (x_1, y_1) is a solution of the equation $y - y_1 = m(x - x_1)$: $y_1 - y_1 = m(x_1 - x_1)$, or $0 = 0$, a true statement.

Example 9   Using the Point-Slope Form to Find an Equation of a Line

Use the point–slope form to find an equation of the line that contains the points $(2, -6)$ and $(5, -4)$. Then write the equation in slope–intercept form.

Solution

We begin by finding the slope of the line:

$$m = \frac{-4 - (-6)}{5 - 2} = \frac{2}{3}$$

Then we substitute $x_1 = 2$, $y_1 = -6$, and $m = \frac{2}{3}$ into the equation $y - y_1 = m(x - x_1)$:

$$y - y_1 = m(x - x_1) \qquad \text{Point–slope form}$$

$$y - (-6) = \frac{2}{3}(x - 2) \qquad \text{Substitute } x_1 = 2,\ y_1 = -6,\ \text{and } m = \frac{2}{3}.$$

$$y + 6 = \frac{2}{3}x - \frac{4}{3} \qquad \text{Simplify; distributive law}$$

$$y + 6 - 6 = \frac{2}{3}x - \frac{4}{3} - 6 \qquad \text{Subtract 6 from both sides.}$$

$$y = \frac{2}{3}x - \frac{22}{3} \qquad \text{Combine like terms; } -\frac{4}{3} - 6 = -\frac{4}{3} - \frac{18}{3} = -\frac{22}{3}.$$

So, the equation is $y = \frac{2}{3}x - \frac{22}{3}$. ■

group exploration

Finding equations of lines

The objective of this game is to earn 19 credits. You earn credits by finding equations of lines that pass through one or more of the following points:

$(-3, 2)$, $(-3, 0)$, $(-2, -7)$, $(-2, -1)$, $(-1, 4)$, $(0, 2)$, $(1, -1)$, $(2, -2)$, $(3, 1)$, $(3, 3)$

If a line passes through exactly one point, then you earn one credit. If a line passes through exactly two points, then you earn three credits. If a line passes through exactly three points, then you earn five credits. You may use five equations. You may use points more than once. [**Hints:** First, plot the points. After finding your equations, use your graphing calculator to check that they are correct.]

TIPS FOR SUCCESS: Show What You Know

Even if you don't know how to do one step of a problem, you can still show the instructor that you understand the other steps of the problem. Depending on how your instructor grades tests, you may earn partial credit even though you pick an incorrect number to be the result for a particular step, as long as you then show what you would do with that number in the remaining steps of the solution. Check with your instructor first.

For example, suppose you want to find an equation of the line that passes through the points $(1, 5)$ and $(2, 8)$, but you have forgotten how to find the slope. You could still write,

I've drawn a blank on finding slope. However, assuming that the slope is 2, then

$$y = 2x + b$$
$$5 = 2(1) + b$$
$$5 = 2 + b$$
$$5 - 2 = 2 + b - 2$$
$$3 = b$$

Therefore, $y = 2x + 3$.

You could point out that you know your result is incorrect, because the graph of $y = 2x + 3$ does *not* pass through the point $(2, 8)$. Also, seeing your result (with the graph) may jog your memory about finding the slope and allow you to go back and do the problem correctly.

HOMEWORK 5.2

FOR EXTRA HELP ▶

Student Solutions Manual

PH Math/Tutor Center

MathXL®

MyMathLab
MyMathLab

Find an equation of the line that has the given slope and contains the given point. If possible, write your equation in slope–intercept form. Check that the ordered pair that represents the given point satisfies your equation.

1. $m = 2, (3, 5)$ **2.** $m = 3, (2, 4)$

3. $m = -3, (1, -2)$ **4.** $m = -5, (3, -8)$

5. $m = 2, (-4, -6)$ **6.** $m = 4, (-5, -1)$

7. $m = -6, (-2, -3)$ **8.** $m = -1, (-7, -4)$

9. $m = \dfrac{2}{5}, (3, 1)$ **10.** $m = \dfrac{1}{2}, (5, 3)$

11. $m = -\dfrac{3}{4}, (-2, -5)$ **12.** $m = -\dfrac{5}{3}, (-4, -2)$

13. $m = 0, (5, 3)$ **14.** $m = 0, (-1, -3)$

15. m is undefined, $(-2, 4)$ **16.** m is undefined, $(3, -2)$

Find an approximate equation of the line that has the given slope and contains the given point. Write your equation in slope–intercept form. Round the constant term to two decimal places. Use a graphing calculator to verify your equation.

17. $m = 2.1, (3.7, -5.9)$ **18.** $m = 5.8, (2.6, -4.3)$

19. $m = -5.6, (-4.5, 2.8)$ **20.** $m = -1.3, (-6.6, 3.8)$

21. $m = -6.59, (-2.48, -1.61)$

22. $m = -2.07, (-4.73, -9.60)$

Find an equation of the line that passes through the two given points. If possible, write your equation in slope–intercept form. Check that the graph of your equation contains the given points.

23. $(3, 2)$ and $(5, 6)$ **24.** $(1, 4)$ and $(2, 1)$

25. $(1, 6)$ and $(3, 2)$ **26.** $(4, 7)$ and $(6, 15)$

27. $(-1, -7)$ and $(2, 8)$ **28.** $(-2, -10)$ and $(3, 5)$

29. $(-5, -7)$ and $(-2, -4)$ **30.** $(-3, -8)$ and $(-1, -2)$

31. $(0, 0)$ and $(1, 1)$ **32.** $(0, 0)$ and $(1, -1)$

33. $(0, 9)$ and $(2, 1)$ **34.** $(-3, 1)$ and $(0, -5)$

35. $(3, 2)$ and $(5, 2)$ **36.** $(-5, -3)$ and $(-1, -3)$

37. $(-4, -1)$ and $(-4, 3)$ **38.** $(7, 1)$ and $(7, 6)$

39. $(4, 3)$ and $(8, 5)$ **40.** $(2, 3)$ and $(6, 1)$

41. $(-4, -3)$ and $(5, 3)$ **42.** $(-6, -2)$ and $(4, 6)$

43. $(-3, 2)$ and $(3, 1)$ **44.** $(2, -2)$ and $(6, -5)$

45. $(-2, 1)$ and $(5, -1)$ **46.** $(-6, 5)$ and $(-2, 2)$

47. $(-4, -2)$ and $(6, 4)$ **48.** $(-1, -2)$ and $(5, 6)$

49. $(-4, -8)$ and $(-2, -5)$ **50.** $(-6, -9)$ and $(-2, -4)$

Find an approximate equation of the line that passes through the two given points. Write your equation in slope–intercept form. Round the slope and the constant term to two decimal places. Use a graphing calculator to verify your equation.

51. $(2.3, 5.8)$ and $(4.5, 3.1)$

52. $(1.9, 5.6)$ and $(2.4, 8.3)$

53. $(-4.5, 2.2)$ and $(1.2, -7.5)$

54. $(-8.1, -5.3)$ and $(-3.3, -2.7)$

55. $(2.46, -1.84)$ and $(5.87, -5.29)$

56. $(3.57, -9.41)$ and $(5.98, -2.84)$

57. $(-4.57, -8.29)$ and $(7.17, -2.69)$

58. $(-8.99, 4.82)$ and $(-5.85, -3.92)$

Find an equation of the line that contains the given point and is parallel to the given line. Use a graphing calculator to verify your equation.

59. $(2, 3), y = 2x - 7$ **60.** $(5, 1), y = -3x + 4$

61. $(-3, 5), 4x + y = 1$ **62.** $(4, -2), -2x + y = 3$

Find an equation of the line that contains the given point and is perpendicular to the given line. Use ZStandard followed by ZSquare with a graphing calculator to verify your equation.

63. $(3, 4), y = \dfrac{1}{2}x + 5$

64. $(1, 2), y = \dfrac{1}{4}x - 1$

65. $(-2, -5), x + 4y = 12$

66. $(-3, -1), x - 3y = 9$

67. Find an equation of the line sketched in Fig. 14. [**Hint:** Choose two points whose coordinates appear to be integers.]

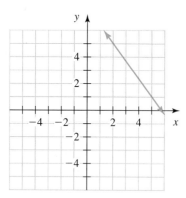

Figure 14 Exercise 67

68. Find an equation of the line sketched in Fig. 15. [**Hint:** Choose two points whose coordinates appear to be integers.]

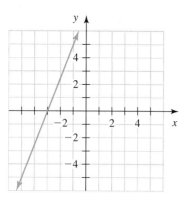

Figure 15 Exercise 68

69. Describe the Rule of Four as it applies to the line that contains the points $(-2, -6)$ and $(4, 3)$:
  a. Find an equation of the line.
  b. Sketch the line on a coordinate system.
  c. Use a table to list five ordered pairs that correspond to points on the line.

70. Describe the Rule of Four as it applies to the line that contains the points $(-3, 4)$ and $(6, -2)$:
  a. Find an equation of the line.
  b. Sketch the line on a coordinate system.
  c. Use a table to list five ordered pairs that correspond to points on the line.

71. Decide whether it is possible for a line to have the indicated number of x-intercepts. If it is possible, find an equation of such a line. If it is not possible, explain why not.
  a. no x-intercepts
  b. exactly one x-intercept
  c. exactly two x-intercepts
  d. an infinite number of x-intercepts

72. Decide whether it is possible for a line to have the indicated number of y-intercepts. If it is possible, find an equation of such a line. If it is not possible, explain why not.
  a. no y-intercepts
  b. exactly one y-intercept
  c. exactly two y-intercepts
  d. an infinite number of y-intercepts

73. Let E be the enrollment (in thousands of students) at a college t years after the college opens. Some pairs of values of t and E are listed in Table 3.

Table 3 Enrollments	
Age of College (years) t	**Enrollment (thousands of students)** E
2	9
4	13
7	19
9	23
12	29

Find an equation that describes the relationship between t and E.

74. Let s be a person's savings (in thousands of dollars) at t years since 1995. Some pairs of values of t and s are listed in Table 4.

Table 4 A Person's Savings	
Years t	**Savings (thousands of dollars)** s
1	5
4	14
5	17
7	23
10	32

Find an equation that describes the relationship between t and s.

75. a. Use each of the following forms to find an equation of the line that contains the points $(2, 1)$ and $(4, 7)$. Write each result in slope–intercept form.
  i. slope–intercept form
  ii. point–slope form
  b. Compare your results from parts (ai) and (aii).

76. a. Use each of the following forms to find an equation of the line that contains the points $(-4, 3)$ and $(2, -5)$. Write each result in slope–intercept form.
  i. slope–intercept form
  ii. point–slope form
  b. Compare your results from parts (ai) and (aii).

77. a. Find an equation of a line with slope -2. [**Hint:** There are *many* correct answers.]
  b. Find an equation of a line with y-intercept $(0, 4)$.
  c. Find an equation of a line that contains the point $(3, 5)$.
  d. Determine whether there is a line that has slope -2 and y-intercept $(0, 4)$ and contains the point $(3, 5)$. Explain.

78. Describe how to find an equation of a line that contains two given points. Also, explain how you can check that the graph of your equation contains the two points.

Related Review

79. Four sets of points are described in Table 5. For each set, find an equation of a line that contains the points.

Table 5 Four Sets of Points

Set 1		Set 2		Set 3		Set 4	
x	*y*	*x*	*y*	*x*	*y*	*x*	*y*
0	25	0	12	0	77	0	3
1	23	1	16	1	72	1	3
2	21	2	20	2	67	2	3
3	19	3	24	3	62	3	3
4	17	4	28	4	57	4	3

80. Sketch a vertical line on a coordinate system. Find an equation of the line. What is the slope of the line?

81. Sketch a decreasing line that is nearly horizontal on a coordinate system. Find an equation of the line.

82. Find an equation of a line that has no solutions in Quadrant I.

83. a. Graph $y = 2x - 3$ by hand.

b. Choose two points that lie on the graph. Then use the two points to find an equation of the line that contains them. Compare your equation with the equation $y = 2x - 3$.

84. A line contains the points $(2, 4)$ and $(4, 1)$. Find three more points that lie on the line.

Expressions, Equations, and Graphs

Perform the indicated instruction. Then use words such as linear, one variable, *and* two variables *to describe the expression or equation. For instance, to describe* $2x + 7x$, *you could say* "$2x + 7x$ *is a linear expression in one variable.*"

85. Graph $3x + 2y = 6$ by hand.

86. Simplify $\dfrac{5x}{8} + \dfrac{3}{4} - \dfrac{7x}{2}$.

87. Evaluate $3x + 2y$ for $x = 4$ and $y = -5$.

88. Solve $\dfrac{5x}{8} + \dfrac{3}{4} = \dfrac{7x}{2}$.

5.3 FINDING EQUATIONS OF LINEAR MODELS

There is no failure except in no longer trying.

—Elbert Hubbard

Objectives

▷ Find an equation of a linear model by using data described in words.

▷ Find an equation of a linear model by using data displayed in a table.

In Section 5.2, we used two given points to find an equation of a line. In this section, we use this skill to find an equation of a linear model.

Finding a Model by Using Data Described in Words

In Example 1, we will use data described in words to find an equation of a linear model.

Example 1 Finding an Equation of a Linear Model

The number of U.S. reserve soldiers has been decreasing approximately linearly, with 53 thousand in 1992 and 27 thousand in 2003 (Source: *U.S. Army Recruiting Command*). Let n be the number of reserve soldiers (in thousands) at t years since 1990.

1. Find an equation of a linear model.
2. What is the slope? What does it mean in this situation?
3. What is the n-intercept? What does it mean in this situation?

Solution

Table 6 Known Values of t and n

Years since 1990 t	Number of Reserve Soldiers (thousands) n
2	53
13	27

1. Known values of t and n are shown in Table 6. A linear equation can be put into the form $y = mx + b$, where y depends on x. Since the variables t and n are approximately linearly related and n depends on t, we will find an equation of the form $n = mt + b$. We can use the data points $(2, 53)$ and $(13, 27)$ to find the values of m and b.

First, we use $(2, 53)$ and $(13, 27)$ to find the slope:

$$m = \frac{27 - 53}{13 - 2} \approx -2.36$$

So, we can substitute -2.36 for m in the equation $n = mt + b$:

$$n = -2.36t + b$$

To find the constant b, we substitute the coordinates of the point $(2, 53)$ into the equation $n = -2.36t + b$ and then solve for b:

$$53 = -2.36(2) + b \qquad \text{Substitute 2 for } t \text{ and 53 for } n.$$
$$53 = -4.72 + b \qquad \text{Multiply.}$$
$$53 + 4.72 = -4.72 + b + 4.72 \qquad \text{Add 4.72 to both sides.}$$
$$57.72 = b \qquad \text{Combine like terms.}$$

Now we can substitute 57.72 for b in the equation $n = -2.36t + b$:

$$n = -2.36t + 57.72$$

We verify our equation by using a graphing calculator to check that our line approximately contains the points $(2, 53)$ and $(13, 27)$. See Fig. 16.

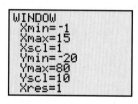 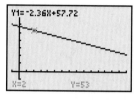 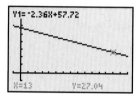

Figure 16 Checking that the model approximately contains both $(2, 53)$ and $(13, 27)$

2. The slope is -2.36. According to the model, the number of reserve soldiers decreases by 2.36 thousand soldiers per year.
3. Since the model $n = -2.36t + 57.72$ is in slope–intercept form, we see that the n-intercept is $(0, 57.72)$. So, the model estimates that there were 57.72 thousand reserve soldiers in 1990. ∎

Table 7 Average SAT Math Scores

Year	Average Math Score (points)
1982	493
1985	500
1990	501
1995	506
2000	514
2002	516
2004	518

Source: *The College Board*

Finding a Linear Model by Using Data Displayed in a Table

We begin by finding an equation of a model by using data shown in a table.

Example 2   Finding an Equation of a Linear Model

The average SAT math scores of college-bound high school seniors are shown in Table 7 for various years. Let s be the average SAT math score (in points) at t years since 1980. Find an equation of a line that comes close to the points in the scattergram of the data.

Solution

We begin by viewing the positions of the points in the scattergram (see Fig. 17).

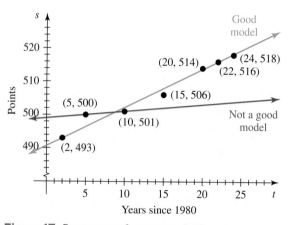

Figure 17 Scattergram for average SAT math scores

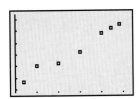

Figure 18 Graphing calculator scattergram

To save time and improve the accuracy of plotting points, we can use a graphing calculator to view the scattergram (see Fig. 18). For graphing calculator instructions on drawing scattergrams, see Sections A.8, A.9, and A.12.

Our task is to find an equation of a line that comes close to the data points. It is not necessary to use two *data* points to find an equation, although it will often be convenient and satisfactory to do so.

The red line that contains the points (5, 500) and (10, 501) does *not* come close to the other data points (see Fig. 17). However, the blue line that passes through the points (2, 493) and (22, 516) appears to come close to the rest of the points. We will find the equation of this line.

To obtain an equation of the form $s = mt + b$, we use the points (2, 493) and (22, 516) to find the values of m and b. First, we calculate the slope of this line (see Fig. 19):

$$m = \frac{516 - 493}{22 - 2} = \frac{23}{20} = 1.15$$

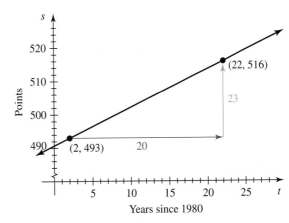

Figure 19 Points (2, 493) and (22, 516) of the SAT scattergram

So, 1.15 can be substituted for m in the equation $s = mt + b$:

$$s = 1.15t + b$$

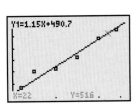

The constant b can be found by substituting the coordinates of the point (2, 493) into the equation $s = 1.15t + b$ and then solving for b:

$493 = 1.15(2) + b$	Substitute 2 for t and 493 for s.
$493 = 2.3 + b$	Multiply.
$493 - 2.3 = 2.3 + b - 2.3$	Subtract 2.3 from both sides.
$490.7 = b$	Combine like terms.

Now 490.7 can be substituted for b in the equation $s = 1.15t + b$:

$$s = 1.15t + 490.7$$

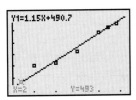

Figure 20 Checking that the model contains both (2, 493) and (22, 516)

We can check the correctness of our equation by using a graphing calculator to verify that our line contains the points (2, 493) and (22, 516). See Fig. 20. For graphing calculator instructions, see Sections A.10 and A.11.

We also see that our model fits the data well, which is our objective. ■

We benefit in many ways by viewing a scattergram of data. First, we can determine whether the data are approximately linearly related. Second, if the data are approximately linearly related, viewing a scattergram helps us choose two good points with which to find an equation of a linear model. Third, by graphing the model with the scattergram, we can assess whether the model fits the data reasonably well.

Example 2 outlines four steps to take to find an equation of a linear model.

Finding an Equation of a Linear Model

To find an equation of a linear model, given some data,

1. Create a scattergram of the data.
2. Determine whether there is a line that comes close to the data points. If so, choose two points (not necessarily data points) that you can use to find the equation of a linear model.
3. Find an equation of the line.
4. Use a graphing calculator to verify that the graph of your equation comes close to the points of the scattergram.

What should you do if you discover that a model does not fit a data set well? A good first step is to check for any graphing or calculation errors. If your work appears to be correct, then one option is to try using different points to derive your equation. Another option is to increase or decrease the slope m and/or the constant term b until the fit is good. You can practice this "trial-and-error" process by completing the exploration in this section.

Example 3   Finding an Equation of a Linear Model

According to a poll performed by the National Consumers League, people are more concerned about privacy issues than about health care, education, crime, or taxes. The percentages of Americans of various age groups who consider their home address to be personal information are listed in Table 8.

Table 8 Americans Who Consider Their Home Address to Be Personal Information

Age Group	Age Used to Represent Age Group	Percent
18–24	21	84
25–34	29.5	80
35–44	39.5	74
45–54	49.5	74
55–64	59.5	67
over 64	75	60

Source: *American Demographics*

Let p be the percentage of Americans at age a years who consider their home address to be personal information. Find an equation that models the data well.

Solution

We see from the scattergram in Fig. 21 that a line containing the data points (21, 84) and (59.5, 67) comes close to the rest of the data points.

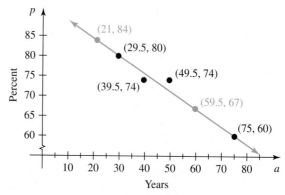

Figure 21 Personal information scattergram and linear model

For an equation of the form $p = ma + b$, we first use the points $(21, 84)$ and $(59.5, 67)$ to find m:

$$m = \frac{67 - 84}{59.5 - 21} = -0.44$$

So, the equation has the form

$$p = -0.44a + b$$

To find b, we use the point $(21, 84)$ and substitute 21 for a and 84 for p:

$84 = -0.44(21) + b$	Substitute 21 for a and 84 for p.
$84 = -9.24 + b$	Multiply.
$84 + 9.24 = -9.24 + b + 9.24$	Add 9.24 to both sides.
$93.24 = b$	Combine like terms.

So, the equation is $p = -0.44a + 93.24$.

We can use a graphing calculator to verify that the linear model contains the points $(21, 84)$ and $(59.5, 67)$ and comes close to the other data points (see Fig. 22). ■

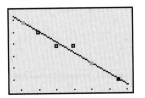

Figure 22 Verifying the information model

group exploration

Adjusting the fit of a model

The winning times for the men's Olympic 100-meter freestyle swimming event are shown in Table 9 for various years.

Table 9 Winning Times for the Men's Olympic 100-Meter Freestyle

Year	Swimmer	Country	Winning Time (seconds)
1980	Jörg Woithe	E. Germany	50.40
1984	Rowdy Gaines	USA	49.80
1988	Matt Biondi	USA	48.63
1992	Aleksandr Popov	Unified Team	49.02
1996	Aleksandr Popov	Russia	48.74
2000	Pieter van den Hoogenband	Netherlands	48.30
2004	Pieter van den Hoogenband	Netherlands	48.17

Source: The New York Times 2006 Almanac

Let w be the winning time (in seconds) at t years since 1980.

1. Use a graphing calculator to draw a scattergram of the data. Do the variables t and w appear to be approximately linearly related?

2. The linear model $w = -0.0325t + 48.95$ can be found by using the data points $(20, 48.30)$ and $(24, 48.17)$. Draw the line and the scattergram in the same viewing window. Check that the line contains these two points.

3. The model $w = -0.0325t + 48.95$ does not fit the seven data points very well. Adjust the equation by increasing or decreasing the slope -0.0325 and/or the constant term 48.95 so that your new model will fit the data better. Keep adjusting the model until it fits the data points reasonably well.

4. Use your improved model to predict the winning time in the 2012 Olympics.

TIPS FOR SUCCESS: Verify Your Work

Remember to use your graphing calculator to verify your work. In this section, for example, you can use your graphing calculator to check your equations. Checking your work increases your chances of catching errors and thus will likely improve your performance on homework assignments, quizzes, and tests.

HOMEWORK 5.3   FOR EXTRA HELP ▶

Student Solutions Manual   PH Math/Tutor Center   MathXL   MyMathLab

1. The number of African-American federal and state legislators has increased approximately linearly from 179 positions in 1970 to 633 positions in 2001 (Source: *Joint Center for Political and Economic Studies*). Let *n* be the number of African-American legislators at *t* years since 1970. Find an equation of a linear model to describe the data.

2. The total online spending in the United States has increased approximately linearly from \$51 billion in 2001 to \$145 billion in 2004 (Source: *Forrester Research*). Let *s* be the online spending (in billions of dollars) at *t* years since 2000. Find an equation of a linear model to describe the data.

3. The number of inmates younger than 18 held in state prisons has decreased approximately linearly from 3147 inmates in 2001 to 2266 inmates in 2005 (Source: *U.S. Department of Justice*). Let *n* be the number of inmates younger than 18 held in state prisons at *t* years since 2000. Find an equation of a linear model to describe the data.

4. The average number of hours an American spends reading a daily newspaper has decreased approximately linearly from 177 hours in 2001 to 169 hours in 2004 (Source: *Census Bureau*). Let *n* be the average number of hours an American spends reading a daily newspaper in the year that is *t* years since 2000. Find an equation of a linear model to describe the data.

5. The percentage of sexual harassment charges filed by men has increased approximately linearly from 9.1% in 1992 to 14.7% in 2003 (Source: *Equal Employment Opportunity Commission*). Let *p* be the percentage of sexual harassment charges filed by men at *t* years since 1990.
   a. Find an equation of a linear model to describe the data.
   b. What is the slope? What does it mean in this situation?
   c. What is the *p*-intercept? What does it mean in this situation?

6. The number of publishers in the United States has increased approximately linearly from 40,964 publishers in 1993 to 73,200 publishers in 2003 (Source: *Andrew Grabois, R. R. Bowker*). Let *n* be the number of publishers at *t* years since 1990.
   a. Find an equation of a linear model to describe the data.
   b. What is the slope? What does it mean in this situation?
   c. What is the *n*-intercept? What does it mean in this situation?

7. The number of U.S. flag desecrations (intentionally defacing or dishonoring the flag) has declined approximately linearly from 14 in 2001 to 3 in 2004 (Source: *Citizens Flag Alliance*). Let *n* be the number of U.S. flag desecrations in the year that is *t* years since 2000.
   a. Find an equation of a linear model to describe the data.
   b. What is the slope? What does it mean in this situation?
   c. What is the *n*-intercept? What does it mean in this situation?

8. The market share of American automakers has decreased approximately linearly from 66% in 2000 to 59% in 2005 (Source: *Ward's AutoInfoBank*). Let *p* be American automakers' market share at *t* years since 2000.
   a. Find an equation of a linear model to describe the data.

b. What is the slope? What does it mean in this situation?
c. What is the *p*-intercept? What does it mean in this situation?

For Exercises 9 and 10, consider the scattergram of data and the graph of the model $y = mt + b$ in the indicated figure. Sketch the graph of a linear model that describes the data better, and then explain how you would adjust the values of m and b of the original model so that it would describe the data better.

9. See Fig. 23.

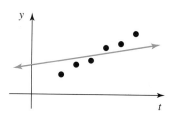

Figure 23 Exercise 9

10. See Fig. 24.

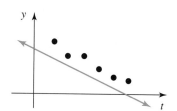

Figure 24 Exercise 10

11. Find an equation of a line that comes close to the points listed in Table 10. Then use a graphing calculator to check that your line comes close to the points.

Table 10 Find an Equation	
x	**y**
4	8
5	10
6	13
7	15
8	18

12. Find an equation of a line that comes close to the points listed in Table 11. Then use a graphing calculator to check that your line comes close to the points.

Table 11 Find an Equation	
x	**y**
1	6
5	13
7	14
10	17
12	22

13. Find an equation of a line that comes close to the points listed in Table 12. Then use a graphing calculator to check that your line comes close to the points.

Table 12 Find an Equation

x	y
3	18
7	14
9	9
12	7
16	4

14. Find an equation of a line that comes close to the points listed in Table 13. Then use a graphing calculator to check that your line comes close to the points.

Table 13 Find an Equation

x	y
2	14
5	10
6	8
8	4
11	1

15. Three students are to find a linear model of the data in Table 14. Student A uses the points (4, 15.2) and (5, 11.5), student B uses the points (7, 9.2) and (8, 5.7), and student C uses the points (6, 10.5) and (9, 4.5). Which student seems to have made the best choice of points? Explain.

Table 14 Three Students Model Data

x	y
4	15.2
5	11.5
6	10.5
7	9.2
8	5.7
9	4.5
10	2.5

16. Three students are to find a linear model of the data in Table 15. Student A uses the points (1, 11.9) and (2, 7.1), student B uses the points (1, 11.9) and (7, 17.0), and student C uses the points (2, 7.1) and (7, 17.0). Which student seems to have made the best choice of points? Explain.

Table 15 Three Students Model Data

x	y
1	11.9
2	7.1
3	8.9
4	10.8
5	13.1
6	15.2
7	17.0

17. Airlines are scheduling flights so that more seats are filled to offset sharp increases in fuel costs. The percentages of all seats that are filled in May for the six largest airlines are shown in Table 16 for various years.

Table 16 Percentage of Seats Filled in May for the Six Largest Airlines

Year	Percent
2001	71.0
2002	72.9
2003	75.2
2004	76.3
2005	79.2

Source: *Back Aviation Solutions*

Let p be the percentage of airline seats that are filled in May of the year that is t years since 2000 for the six largest airlines.

a. Use a graphing calculator to draw a scattergram of the data.

b. Find an equation of a linear model to describe the data.

c. Draw your line and the scattergram in the same viewing window. Verify that the line passes through the two points that you chose in finding the equation in part (b) and that it comes close to all of the data points.

18. The average per-person annual consumption of sports drinks in the United States is shown in Table 17 for various years.

Table 17 U.S. per-Person Annual Consumption of Sports Drinks

Year	Annual Consumption (gallons per person)
1998	1.9
1999	2.1
2000	2.2
2001	2.3
2002	2.5

Source: *Beverage Digest*

Let C be the average per-person annual consumption of sports drinks (in gallons per person) at t years since 1990.

a. Use a graphing calculator to draw a scattergram of the data.

b. Find an equation of a linear model to describe the data.

c. Draw your line and the scattergram in the same viewing window. Verify that the line passes through the two points that you chose in finding the equation in part (b) and that it comes close to all of the data points.

19. Table 18 shows how old a 20- to 50-pound dog is in relation to human years.

Table 18 Dog Years Compared with Human Years

Dog Years	Human Years
1	15
2	24
3	29
5	38
7	47
9	56
11	65
13	74
15	83

Source: *Fred Metzger, State College, PA*

Let H be the number of human years that is equivalent to d dog years.
a. Use a graphing calculator to draw a scattergram of the data.
b. Find an equation of a linear model to describe the data.
c. Draw your line and the scattergram in the same viewing window. Verify that the line passes through the two points that you chose in finding the equation in part (b) and that it comes close to all of the data points.

20. The enrollments in an elementary algebra course at the College of San Mateo are shown in Table 19 for various Tuesdays leading up to the first day of class on Tuesday, January 17.

Table 19 Enrollments in an Elementary Algebra Course

Date	Number of Weeks Since November 22	Enrollment
November 22	0	8
November 29	1	9
December 6	2	12
December 13	3	13
December 20	4	16
December 27	5	18
January 3	6	19

Source: *J. Lehmann*

Let E be the enrollment in the elementary algebra course at t weeks since November 22.
a. Use a graphing calculator to draw a scattergram of the data.
b. Find an equation of a linear model to describe the data.
c. Draw your line and the scattergram in the same viewing window. Verify that the line passes through the two points that you chose in finding the equation in part (b) and that it comes close to all of the data points.

21. The percentages of American households that own a computer are shown in Table 20 for various income groups.

Table 20 Percentage of Households That Own a Computer

Income Group (thousands of dollars)	Income Used to Represent Income Group (thousands of dollars)	Percentage That Own a Computer
5–10	7.5	15
10–15	12.5	22
15–20	17.5	28
20–25	22.5	31
25–35	30.0	45
35–50	42.5	59
50–75	62.5	73

Source: *U.S. Dept. of Commerce*

Let p be the percentage of households with income d (in thousands of dollars) that own a computer.
a. Use a graphing calculator to draw a scattergram of the data.
b. Find an equation of a linear model to describe the data.
c. Draw your line and the scattergram in the same viewing window. Verify that the line passes through the two points that you chose in finding the equation in part (b) and that it comes close to all of the data points.

22. The cost of bringing fiber-optic cable to households has declined greatly (see Table 21).

Table 21 Cost of Bringing Fiber-Optic Cable to the Home

Year	Cost (thousands of dollars)
2000	4.50
2001	3.60
2002	2.80
2003	2.10
2004	1.65

Source: *Render, Vanderslice, & Associates*

Let c be the cost (in thousands of dollars) to bring fiber-optic cable to a household at t years since 2000.
a. Use a graphing calculator to draw a scattergram of the data.
b. Find an equation of a linear model to describe the data.
c. Draw your line and the scattergram in the same viewing window. Verify that the line passes through the two points that you chose in finding the equation in part (b) and that it comes close to all of the data points.

23. The percentages of Americans who have earned a college degree are shown in Table 22 for various years.

Table 22 Percentages of Americans Who Have Earned a College Degree

Year	Percent
1960	7.7
1970	10.7
1980	16.2
1990	21.3
2000	25.6
2003	27.2

Source: *U.S. Census Bureau*

Let p be the percentage of Americans who have earned a college degree at t years since 1960.
a. Use a graphing calculator to draw a scattergram of the data.
b. Find an equation of a linear model to describe the data.
c. Draw your line and the scattergram in the same viewing window. Verify that the line passes through the two points that you chose in finding the equation in part (b) and that it comes close to all of the data points.
d. What is the slope? What does it mean in this situation?
e. What is the p-intercept? What does it mean in this situation?

24. The amounts of farmland in the United States are shown in Table 23 for various years.

Table 23 Farmland in the United States

Year	Amount of Farmland (millions of acres)
1993	968
1995	961
1997	956
1999	948
2001	941
2004	937

Source: *Agricultural Statistics Board*

Let F be the amount of farmland (in millions of acres) in the United States at t years since 1990.
a. Use a graphing calculator to draw a scattergram of the data.

b. Find an equation of a linear model to describe the data.

c. Draw your line and the scattergram in the same viewing window. Verify that the line passes through the two points that you chose in finding the equation in part (b) and that it comes close to all of the data points.

d. What is the slope? What does it mean in this situation?

e. What is the F-intercept? What does it mean in this situation?

25. World land speed records are shown in Table 24 for various years.

Table 24 Land Speed Records

Year	Record (miles per hour)
1904	91
1914	124
1927	175
1935	301
1947	394
1960	407
1970	622
1983	633
1997	763

Source: *Fédération International de l'Automobile*

Let r be the land speed record (in miles per hour) at t years since 1900.

a. Use a graphing calculator to draw a scattergram of the data.

b. Find an equation of a linear model to describe the data.

c. Draw your line and the scattergram in the same viewing window. Verify that the line passes through the two points that you chose in finding the equation in part (b) and that it comes close to all of the data points.

d. What is the slope? What does it mean in this situation?

e. What is the r-intercept? What does it mean in this situation?

26. Sales of prescription sleep aids in the United States have increased greatly since 2000 (see Table 25).

Table 25 Sales of Prescription Sleep Aids

Year	Sales of Prescription Sleep Aids (millions of dollars)
2000	27
2001	30
2002	33
2003	35
2004	38

Source: *TNS Media Intelligence*

Let s be sales of prescription sleep aids (in millions of dollars) in the year that is t years since 2000.

a. Use a graphing calculator to draw a scattergram of the data.

b. Find an equation of a linear model to describe the data.

c. Draw your line and the scattergram in the same viewing window. Verify that the line passes through the two points that you chose in finding the equation in part (b) and that it comes close to all of the data points.

d. What is the slope? What does it mean in this situation?

e. What is the s-intercept? What does it mean in this situation?

27. As a nurse's patient load increases, so does the chance that his or her patients will die. Table 26 lists the percent increase in patient mortality for various patient-to-nurse ratios compared with a patient-to-nurse ratio of 4 (that is, 4 to 1).

Table 26 Nurse Staffing and Patient Mortality Rates

Patient-to-Nurse Ratio	Percent Increase in Patient Mortality
4	0
5	7
6	14
7	23
8	31

Source: *University of Pennsylvania*

Let p be the percent increase in patient mortality if the patient-to-nurse ratio is r to 1 rather than 4 to 1.

a. Use a graphing calculator to draw a scattergram of the data.

b. Find an equation of a linear model to describe the data.

c. Draw your line and the scattergram in the same viewing window. Verify that the line passes through the two points that you chose in finding the equation in part (b) and that it comes close to all of the data points.

28. The numbers of firearm-related deaths in the United States are shown in Table 27 for various years.

Table 27 Firearm-Related U.S. Deaths

Year	Firearm-Related Deaths (thousands of deaths)
1993	40
1995	36
1997	32
1999	29
2000	28

Source: *Centers for Disease Control and Prevention*

Let n be the number (in thousands) of firearm-related deaths for the year that is t years since 1990.

a. Use a graphing calculator to draw a scattergram of the data.

b. Find an equation of a linear model to describe the data.

c. Draw your line and the scattergram in the same viewing window. Verify that the line passes through the two points that you chose in finding the equation in part (b) and that it comes close to all of the data points.

29. Explain how to find an equation of a linear model for a given situation. Also, explain how you can verify that the line models the situation reasonably well. (See page 9 for guidelines on writing a good response.)

30. Describe the benefits of viewing a scattergram of data. (See page 9 for guidelines on writing a good response.)

Related Review

31. In 2005, a stock is worth $10. Each year thereafter, the value of the stock increases by $2. Let V be the value (in dollars) of the stock at t years since 2005. Describe the Rule of Four as it applies to this situation:

a. Use an equation to describe the situation. Then perform a unit analysis of the equation.

b. Use a graph to describe the situation.

c. Use a table of values of t and V to describe the situation.

32. A person is on a car trip. At the start of the trip, there are 12 gallons of gasoline in the gasoline tank. The car consumes 3 gallons of gasoline per hour of driving. Let V be the volume of gasoline (in gallons) in the gasoline tank after t hours of driving. Describe the Rule of Four as it applies to this situation:

a. Use an equation to describe the situation. Then perform a unit analysis of the equation.

b. Use a graph to describe the situation.

c. Use a table of values of t and V to describe the situation.

33. The average attendance at college bowl games was 50,575 in 2003 and is decreasing by about 298 spectators per year (Source: *NCAA*). Let n be the number of spectators at t years since 2003.

a. Find an equation of a linear model to describe the situation.

b. What is the slope? What does it mean in this situation?

c. Perform a unit analysis of the equation you found in part (a).

34. In 2002, 13% of the U.S. workforce had blue-collar jobs. The percentage of the workforce having blue-collar jobs is decreasing by about 0.43 percentage point per year (Source: *Economic Report of the President*). Let p be the percentage of the workforce that has blue-collar jobs at t years since 2002.

a. Find an equation of a linear model to describe the data.

b. What is the slope? What does it mean in this situation?

c. Perform a unit analysis of the equation that you found in part (a).

Expressions, Equations, and Graphs

Perform the indicated instruction. Then use words such as linear, one variable, and two variables to describe the expression or equation. For instance, to describe $y = 4x$, you could say "$y = 4x$ is a linear equation in two variables."

35. Solve $3 = -\dfrac{2}{5}x + 4$.

36. Simplify $4x - 3(2x - 5) + 1$.

37. Graph $y = -\dfrac{2}{5}x + 4$ by hand.

38. Solve $4x - 3(2x - 5) + 1 = 0$.

5.4 USING LINEAR EQUATIONS TO MAKE ESTIMATES AND PREDICTIONS

Objectives

▷ Using equations of linear models to make estimates and predictions.

▷ Know a four-step modeling process.

In Section 5.3, we discussed how to find an equation of a linear model. In this section, we will use such an equation to make estimates and predictions.

Example 1 Making Estimates and Predictions

In Example 3 of Section 5.3, we found the equation $p = -0.44a + 93.24$, where p is the percentage of Americans at age a years who consider their home address to be personal information (see Table 28).

I'm so glad you can come to our party! I'll fax over the security paperwork for you to sign before I reveal our address.

Table 28 Americans Who Consider Their Home Address to be Personal Information

Age Group	Age Used to Represent Age Group	Percent
18–24	21	84
25–34	29.5	80
35–44	39.5	74
45–54	49.5	74
55–64	59.5	67
over 64	75	60

Source: *American Demographics*

1. Estimate the percentage of 19-year-old Americans who consider their home address to be personal information.

2. Estimate the age at which 70% of Americans consider their home address to be personal information.

3. Find the p-intercept. What does it mean in this situation?

4. Find the a-intercept. What does it mean in this situation?

5. What is the slope? What does it mean in this situation?

Solution

1. We substitute 19 for a in the equation $p = -0.44a + 93.24$:

$$p = -0.44(19) + 93.24 = 84.88$$

So, $p = 84.88$ when $a = 19$. According to the model, 84.9% of 19-year-old Americans consider their home address to be personal information.

2. Here, we substitute 70 for p in the equation $p = -0.44a + 93.24$ and solve for a:

$70 = -0.44a + 93.24$	Substitute 70 for p.
$70 - 93.24 = -0.44a + 93.24 - 93.24$	Subtract 93.24 from both sides.
$-23.24 = -0.44a$	Combine like terms.
$\dfrac{-23.24}{-0.44} = \dfrac{-0.44a}{-0.44}$	Divide both sides by -0.44.
$52.82 \approx a$	Divide; simplify.

According to the model, 70% of 53-year-old Americans consider their home address to be personal information.

We can verify our work in Problems 1 and 2 by using a graphing calculator graph (see Fig. 25). The values of a in these two Problems are 19 and 52.82. To set up the window, we select a number (say, -10) that is less than 19 to be Xmin. We select a number (say, 80) that is more than 52.82 to be Xmax. Then we use ZoomFit to adjust Ymin and Ymax.

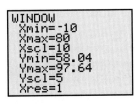 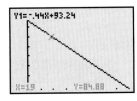 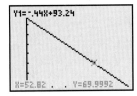

Figure 25 Verify the work

3. Since the model $p = -0.44a + 93.24$ is in slope–intercept form, we see that the p-intercept is $(0, 93.24)$. So, the model estimates that 93.24% of newborn Americans consider their home address to be personal information. Model breakdown has occurred.

4. To find the a-intercept, we substitute 0 for p and solve for a:

$0 = -0.44a + 93.24$	Substitute 0 for p.
$0 + 0.44a = -0.44a + 93.24 + 0.44a$	Add $0.44a$ to both sides.
$0.44a = 93.24$	Combine like terms.
$\dfrac{0.44a}{0.44} = \dfrac{93.24}{0.44}$	Divide by 0.44.
$a \approx 211.91$	Simplify; divide.

The a-intercept is $(211.91, 0)$. So, the model estimates that no 212-year-old Americans consider their home address to be personal information. This estimate is correct, but only because no one lives to be 212 years old!

We can verify our work in Problems 1–4 by using a graphing calculator table (see Figs. 26 and 27). The expression "-4E-4" stands for the number -0.0004, which is close to 0. For graphing calculator instructions, see Section A.15.

5. The slope is -0.44. According to the model, the percentage of Americans who consider their home address to be personal information decreases by 0.44 percentage point per year of age. ∎

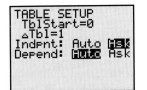

Figure 26 Putting the table in "Ask" mode

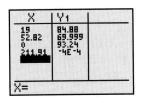

Figure 27 Verify the work

We now describe in general how to use an equation of a model to make predictions (or estimates).

> ### Using an Equation of a Linear Model to Make Predictions
>
> - When making a prediction about the dependent variable of a linear model, substitute a chosen value for the independent variable in the model and then solve for the dependent variable.
> - When making a prediction about the independent variable of a linear model, substitute a chosen value for the dependent variable in the model and then solve for the independent variable.

We next describe in general how to use an equation of a model to find the intercepts of the model.

> ### Using an Equation of a Linear Model to Find Intercepts
>
> Suppose that an equation of the form $p = mt + b$, where $m \neq 0$, is used to model a situation. Then
>
> - The p-intercept is $(0, b)$.
> - To find the t-coordinate of the t-intercept, substitute 0 for p in the model's equation and then solve for t.

In Section 5.3 and this section, we have discussed how to find linear models and how to use these models to make estimates and predictions. Here is a summary of this process.

> ### Four-Step Modeling Process
>
> To find a linear model and then make estimates and predictions,
>
> 1. Create a scattergram of the data to determine whether there is a nonvertical line that comes close to the points. If so, choose two points (not necessarily data points) that you can use to find an equation of a linear model.
> 2. Find an equation of your model.
> 3. Verify your equation by checking that the graph of your model contains the two chosen points and comes close to all of the data points.
> 4. Use the equation of your model to make estimates, make predictions, and draw conclusions.

In most application problems in this text so far, variable names and their definitions have been provided. In Example 2, a key step will be to create variable names and define the variables.

Example 2 Making a Prediction

The percentage of American mothers who want additional time for themselves has increased approximately linearly from 21% in 1975 to 56% in 2004 (Source: *Harris Interactive*). Predict when 65% of American mothers will want additional time for themselves.

Solution

Let p be the percentage of mothers who want additional time for themselves at t years since 1970. Known values of t and p are shown in Table 29. Since the variables t and p are approximately linearly related, we want an equation of the form $p = mt + b$. We can use the data points $(5, 21)$ and $(34, 56)$ to find the values of m and b.

Table 29 Known Values of t and p

Years since 1970 t	Percent p
5	21
34	56

First, we find the slope of the model:

$$m = \frac{56 - 21}{34 - 5} \approx 1.21$$

So, we can substitute 1.21 for m in the equation $p = mt + b$:

$$p = 1.21t + b$$

We can find the constant b by substituting the coordinates of the point $(5, 21)$ into the equation $p = 1.21t + b$ and then solving for b:

$21 = 1.21(5) + b$	Substitute 5 for t and 21 for p.
$21 = 6.05 + b$	Multiply.
$21 - 6.05 = 6.05 + b - 6.05$	Subtract 6.05 from both sides.
$14.95 = b$	Combine like terms.

Now we can substitute 14.95 for b in the equation $p = 1.21t + b$:

$$p = 1.21t + 14.95$$

Finally, we can predict when 65% of mothers will want additional time for themselves by substituting 65 for p in the equation $p = 1.21t + 14.95$ and solving for t:

$65 = 1.21t + 14.95$	Substitute 65 for p.
$65 - 14.95 = 1.21t + 14.95 - 14.95$	Subtract 14.95 from both sides.
$50.05 = 1.21t$	Combine like terms.
$\dfrac{50.05}{1.21} = \dfrac{1.21t}{1.21}$	Divide both sides by 1.21.
$41.36 \approx t$	Divide; simplify.

The model predicts that 65% of mothers will want additional time for themselves in $1970 + 41 = 2011$. We can verify our work by using a graphing calculator table (see Fig. 28). ∎

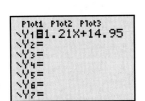

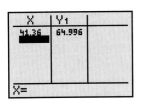

Figure 28 Verify the work

In Example 2, we defined t to be the number of years since *1970*. If we had defined t to be the number of years since *1900* (or any other year), we still would have obtained the same prediction for 2011, although the equation of our model would have been different. So, **if an exercise does not state a year from which to begin counting, you may choose any year for that purpose.**

group exploration

Looking ahead: Properties of inequalities

The symbol "$<$" means *is less than*. For example, the statement $2 < 5$ means that 2 is less than 5, which is true. The statement $-5 < -3$ means that -5 is less than -3, which is also true.

The symbol "$>$" means *is greater than*. For example, the statement $6 > 1$ means that 6 is greater than 1, which is true.

Statements of the form $a < b$ or $a > b$ are examples of *inequalities*.

1. Decide whether each inequality is true.
   a. $4 < 9$ **b.** $3 > 8$ **c.** $-1 < -7$ **d.** $-2 > -6$
2. Decide whether performing the given operation on both sides of the true inequality $4 < 6$ will give a true statement.
   a. Add 2 to both sides.
   b. Add -2 to both sides.
   c. Multiply both sides by 2.
   d. Multiply both sides by -2.

3. For any true inequality (such as $3 < 7$), can you add the given type of number to both sides of the inequality and then still have a true inequality?
  a. positive number
  b. negative number
  c. zero

4. For any true inequality, can you multiply both sides of the inequality by the given type of number and then still have a true inequality?
  a. positive number
  b. negative number
  c. zero

5. Recall from Section 2.4 that subtracting a number is the same as adding the opposite of that number. What does this tell you about subtracting a number from both sides of an inequality?

6. Recall from Section 2.2 that dividing by a nonzero number is the same as multiplying by the reciprocal of that number. What does this tell you about dividing both sides of an inequality by a nonzero number?

TIPS FOR SUCCESS: Stick with It

If you are having difficulty doing an exercise, don't panic! Reread the exercise and reflect on what you have already sorted out about the problem—what you know and where you need to go. Your solution to the problem may be just around the corner.

HOMEWORK 5.4   FOR EXTRA HELP ▶

Student Solutions Manual   PH Math/Tutor Center   MathXL®   MyMathLab

---

1. In Exercise 17 of Homework 5.3, you found an equation close to $p = 1.98t + 68.98$, where p is the percentage of all airline seats that are filled in May of the year that is t years since 2000 for the six largest airlines (see Table 30).

Table 30 Percentage of All Seats Filled in May for the Six Largest Airlines

Year	Percent
2001	71.0
2002	72.9
2003	75.2
2004	76.3
2005	79.2

Source: *Back Aviation Solutions*

a. Predict the percentage of seats that will be filled in May of 2011 for the six largest airlines.
b. Estimate in which year all seats will be filled in May for the six largest airlines.
c. What is the p-intercept? What does it mean in this situation?
d. What is the slope? What does it mean in this situation?

2. In Exercise 18 of Homework 5.3, you found an equation close to $C = 0.14t + 0.8$, where C is the average per-person annual U.S. consumption of sports drinks (in gallons per person) at t years since 1990 (see Table 31).

Table 31 U.S. per Person Annual Consumption of Sports Drinks

Year	Annual Consumption (gallons per person)
1998	1.9
1999	2.1
2000	2.2
2001	2.3
2002	2.5

Source: *Beverage Digest*

a. Predict the average annual per-person sports drink consumption in 2010.
b. Predict when the average annual per-person sports drink consumption will reach 4 gallons per person.
c. What is the slope? What does it mean in this situation?
d. Find the C-intercept. What does it mean in this situation?
e. Find the t-intercept. What does it mean in this situation?

3. In Exercise 21 of Homework 5.3, you found an equation close to $p = 1.08d + 8.81$, where p is the percentage of American households with income d (in thousands of dollars) that own a computer (see Table 32).

a. Estimate the income at which half of households own a computer.

Table 32 Percentage of Households That Own a Computer

Income Group (thousands of dollars)	Income Used to Represent Income Group (thousands of dollars)	Percentage That Own a Computer
5–10	7.5	15
10–15	12.5	22
15–20	17.5	28
20–25	22.5	31
25–35	30.0	45
35–50	42.5	59
50–75	62.5	73

Source: *U.S. Dept. of Commerce*

b. Find the slope. What does it mean in this situation?

c. Estimate the percentage of households with incomes between $0 and $5 thousand that own a computer. The actual percentage is 22%. What is the error in your estimate? Explain why it is surprising that the percentage is 22%. [**Hint:** See Table 32.]

4. In Exercise 22 of Homework 5.3, you found an equation close to $c = -0.72t + 4.37$, where c is the cost (in thousands of dollars) of bringing fiber-optic cable to a household at t years since 2000 (see Table 33).

Table 33 Cost of Bringing Fiber-Optic Cable to the Home

Year	Cost (thousands of dollars)
2000	4.50
2001	3.60
2002	2.80
2003	2.10
2004	1.65

Source: *Render, Vanderslice, & Associates*

a. Estimate when it will cost $0.75 thousand to bring cable to a household.

b. What is the slope? What does it mean in this situation?

c. Find the c-intercept. What does it mean in this situation?

d. Find the t-intercept. What does it mean in this situation?

5. In Exercise 19 of Homework 5.3, you found an equation close to $H = 4.5d + 15.3$, where H is the number of "human years" that is equivalent to d "dog years" for dogs that weigh between 20 pounds and 50 pounds, inclusive (see Table 34).

Table 34 Dog Years Compared with Human Years

Dog Years	Human Years
1	15
2	24
3	29
5	38
7	47
9	56
11	65
13	74
15	83

Source: *Fred Metzger, State College, PA*

a. The life expectancy of an American is about 78 years. Use the model to estimate the life expectancy of dogs.

b. The oldest dog ever was Bluey, an Australian cattle dog, who died at age 29 years. How long did Bluey live in human years?

c. The oldest American ever was Sarah Knauss, who died at age 119 years. How long did Knauss live in dog years?

d. What is the slope? What does it mean in this situation?

e. Although the information in Table 34 is much more accurate, it has long been believed that each dog year is equivalent to 7 human years. Use this old rule of thumb to find another model that describes the relationship between d and H.

f. What is the slope of the graph of your equation in part (e)? What does it mean in this situation? Compare this slope with the slope you found in part (d).

6. In Exercise 20 of Homework 5.3, you found an equation close to $E = 1.96t + 7.68$, where E is the enrollment in an elementary algebra course at the College of San Mateo at t weeks since November 22 (see Table 35).

Table 35 Enrollments in an Elementary Algebra Course

Date	Number of Weeks Since November 22	Enrollment
November 22	0	8
November 29	1	9
December 6	2	12
December 13	3	13
December 20	4	16
December 27	5	18
January 3	6	19

Source: *J. Lehmann*

a. Find the E-intercept. What does it mean in this situation?

b. What is the slope? What does it mean in this situation?

c. Estimate when the enrollment reached 35 students, which is the maximum number of students allowed. The last day to add a class was Monday, January 31, which was 10 weeks after November 22. Has model breakdown occurred? Explain.

d. Estimate the enrollment on the first day of class, January 17, which was 8 weeks after November 22. The actual enrollment on the first day was 32 students. Explain why it is not surprising that your estimation is an underestimate.

7. The percentage of children living with parents in marriages described as "very happy" has declined since 1976 (see Table 36).

Table 36 Percentages of Children Living with Parents in "Very Happy" Marriages

Year	Percent
1976	51
1981	47
1986	43
1991	41
1996	37
2002	37

Source: *National Opinion Research Center, University of Chicago*

Let p be the percentage of children living with parents in a "very happy" marriage at t years since 1970.
a. Use a graphing calculator to draw a scattergram of the data.
b. Find an equation of a linear model to describe the data.
c. Use your model to predict when the percentage of children living with parents in "very happy" marriages will be 31%.
d. Find the p-intercept. What does it mean in this situation?
e. Find the t-intercept. What does it mean in this situation?

8. The out-of-district tuition rates at Austin Community College are shown in Table 37 for various years.

Table 37 Out-of-District Tuition Rates at Austin Community College

End of Academic Year	Tuition Rate (dollars per credit hour)
1997	51
1999	63
2001	81
2003	84
2005	97

Source: *Texas School Performance Review*

Let T be the out-of-district tuition rate (in dollars per credit hour) at t years since 1990.
a. Use a graphing calculator to draw a scattergram of the data.
b. Find an equation of a linear model to describe the data.
c. Estimate the out-of-district tuition rate for 2004.
d. Predict when the out-of-district tuition rate will reach $150 per credit hour.
e. What is the slope? What does it mean in this situation?

9. During the summer of 2003, a math professor was on a weight-loss program (see Table 38).

Table 38 Weights of the Math Professor

Number of Weeks on Weight-Loss Program	Weight (pounds)
0	159.0
1	157.0
2	157.0
3	155.0
4	154.5
5	152.0

Source: *J. Lehmann*

Let w be the weight (in pounds) of the math professor at t weeks since he started the weight-loss program.
a. Use a graphing calculator to draw a scattergram of the data.
b. Find an equation of a linear model to describe the data.
c. What is the slope? What does it mean in this situation?
d. Estimate when the math professor reached his goal of 145 pounds.
e. Find the t-intercept. What does it mean in this situation?

10. The percentages of Americans who always wear seat belts are shown in Table 39 for various years.

Table 39 Percentages of Americans Who Always Wear Seat Belts

Year	Percent
1992	70
1994	71
1996	75
1998	77
2000	79
2002	81
2004	83
2005	86

Source: *Harris Interactive*

Let p be the percentage of Americans who always wear seat belts in the year that is t years since 1990.
a. Use a graphing calculator to draw a scattergram of the data.
b. Find an equation of a linear model to describe the data.
c. What is the p-intercept? What does it mean in this situation?
d. Predict the percentage of Americans in 2011 who will always wear seat belts.
e. Use your model to predict when every American will always wear seat belts.

11. The percentages of high school students who dropped out of school are shown in Table 40 for various years.

Table 40 Percentages of High School Students Who Dropped Out of School

Year	Percent
1980	12.0
1985	10.6
1990	10.1
1995	9.9
2000	9.1
2003	8.4

Source: *U.S. Census Bureau*

Let p be the percentage of high school students who have dropped out of school at t years since 1980.
a. Use a graphing calculator to draw a scattergram of the data.
b. Find an equation of a linear model to describe the data.
c. Predict when 7.5% of high school students will drop out of school.
d. What is the p-intercept? What does it mean in this situation?
e. What is the t-intercept? What does it mean in this situation?

12. If you could stop time and live forever in good health at a particular age, what age would you choose? The average ideal ages chosen by various age groups are shown in Table 41.

Table 41 Ideal Ages

Age Group (years)	Age Used to Represent Age Group (years)	Average Ideal Age (years)
18–24	21	27
25–29	27	31
30–39	34.5	37
40–49	44.5	40
50–64	57	44
over 64	75	59

Source: *Harris Poll*

Let I be the average ideal age (in years) chosen by people whose actual age is a years.

a. Use a graphing calculator to draw a scattergram of the data.

b. Find an equation of a linear model to describe the data.

c. Use your model to estimate the average ideal age chosen by 18-year-olds.

d. What is the slope? What does it mean in this situation?

e. What is the age of people whose ideal age is equal to their actual age?

13. In Exercise 27 of Homework 5.3, you found an equation close to $p = 7.8r - 31.8$, where p is the percent increase in patient mortality if the patient-to-nurse ratio were r to 1 rather than 4 to 1 (see Table 42).

Table 42 Nurse Staffing and Patient Mortality Rates

Patient-to-Nurse Ratio	Percent Increase
4	0
5	7
6	14
7	23
8	31

Source: *University of Pennsylvania*

a. What is the slope? What does it mean in this situation?

b. In January 2004, California passed the first law in the United States that restricts the patient-to-nurse ratio. When the law was first being discussed, some people in the hospital industry felt that the ratio should be 10-to-1. Estimate the percent increase in patient mortality if the ratio were 10-to-1 rather than 4-to-1.

c. Beginning in 2005, the law requires a 5-to-1 ratio (the required ratio will be lower for intensive care and children's wards). Use the model to estimate the percent increase in patient mortality if the ratio is 5 to 1 rather than 4 to 1. Is your result an underestimate or an overestimate? Explain.

d. How does lowering the patient-to-nurse ratio affect the number of patients who can be admitted, the total labor costs for nurses, and the mortality rates of patients?

14. Lake Tahoe in California is a huge lake—1600 feet deep with a perimeter of 72 miles. Long cherished for its beauty, this lake has gotten murkier since 1970 (see Table 43).

Table 43 Average Depth of Visibility in Lake Tahoe

Year	Average Depth of Visibility (feet)
1970	98
1975	88
1980	85
1985	78
1990	80
1995	72
2000	68

Source: *U.C. Davis Tahoe Research Group*

Let A be average depth of visibility (in feet) at t years since 1970.

a. Use a graphing calculator to draw a scattergram of the data.

b. Find an equation of a linear model to describe the data.

c. Use your model to estimate by how much visibility is changing each year.

d. According to *USA Today* (Aug. 1, 2003), "During the past 20 years, water clarity has declined about a foot a year." Is the situation a bit better or a bit worse than what the article claims? Explain.

e. Charles Goldman was the first scientist to warn about Lake Tahoe's decline in clarity. According to Goldman, if the lake's clarity continues to decline at the same rate up until 2013, it may reach a point of no return. Predict the average depth of visibility in that year.

f. Estimate the visibility in Lake Tahoe in 2002. The actual visibility was 78 feet. Since 1997, federal funds have financed environmental improvement programs for the Lake Tahoe area. Assuming that these programs have improved the lake's visibility, by how many feet have they improved the visibility compared with what it would have been without the programs?

15. The number of cable TV networks has increased approximately linearly from 106 networks in 1994 to 339 networks in 2003 (Source: *FCC Annual Report on the Status of Video Competition*). Predict when there will be 520 cable networks.

16. The average American household income (in 2000 dollars) has increased approximately linearly from $41,910 in 1980 to $57,045 in 2000 (Source: *U.S. Census Bureau, Money in the United States: 2000*). Predict when the average American household income will be $66,000 (in 2000 dollars).

17. The percentage of mothers who smoke cigarettes during pregnancy has declined approximately linearly from 13.9% in 1995 to 12.0% in 2001 (Source: *Centers for Disease Control and Prevention*). Predict the percentage in 2010.

18. The purchasing power of the minimum wage has declined approximately linearly since 1997. The minimum wage was $5.15 in 1997 and $4.31 (in 1997 dollars) in 2005 (Source: *Bureau of Labor Statistics*). Estimate the minimum wage (in 1997 dollars) in 2006.

19. The salary of a U.S. senator has increased approximately linearly from $141,300 in 2000 to $162,100 in 2005 (Source: *U.S. Senate*). Predict when the salary will be $200,000.

20. The percentage of Americans who are Protestant has decreased approximately linearly from 67% in 1960 to 50% in 2004 (Source: *Gallup Organization*). Predict when the percentage will be 47%.

21. The price for a 30-second ad during the Academy Awards has increased approximately linearly from $0.6 million in 1994 to $1.6 million in 2005 (Source: *Nielsen Media Research*). Predict when the price will be $2.2 million.

22. Describe in your own words the four-step modeling process.

Related Review

23. The number of Americans who live alone was 27.2 million in 2000 and has increased by about 0.55 million each year since then (Source: *U.S. Census Bureau*). Let n be the number (in millions) of Americans who live alone at t years since 2000.

a. Find an equation of a linear model to describe the data.

b. What is the n-intercept? What does it mean in this situation?

c. Perform a unit analysis of the equation you found in part (a).

d. The combined population of Texas, the second most populous state, and New York, the third most populous state, is 41.4 million. Predict when 41.4 million people will live alone.

24. The median age of buyers of Harley-Davidson motorcycles was 45.6 years in 2000 and has increased by about 0.84 year each year since then (Source: *Harley-Davidson Motor Co.*). Let a be the median age (in years) of buyers at t years since 2000.
   a. Find an equation of a linear model to describe the data.
   b. Perform a unit analysis of the equation you found in part (a).
   c. Predict when the median age of buyers will be 50 years.
   d. Estimate the median age of buyers in 2010.

25. Annual sales of echinacea were $152 million in 2004 and have decreased by about $20.67 million per year since then (Source: *Nutrition Business Journal*). Let s be the annual echinacea sales (in millions of dollars) in the year that is t years since 2004.
   a. Find an equation of a linear model to describe the data.
   b. What is the s-intercept? What does it mean in this situation?
   c. Perform a unit analysis of the equation you found in part (a).

d. Estimate when annual echinacea sales were $111 million.

26. The number of killer whales in Puget Sound was 78 in 2001 and has decreased by an average of 3.63 whales per year since then (Source: *Center for Whale Research*). Let p be the population of killer whales in Puget Sound at t years since 2001.
   a. Find an equation of a linear model to describe the data.
   b. Perform a unit analysis of the equation you found in part (a).
   c. Predict when the killer whale population in Puget Sound will be 50 whales.
   d. Predict the killer whale population in Puget Sound in 2011.

Expressions, Equations, and Graphs

Perform the indicated instruction. Then use words such as linear, one variable, *and* two variables *to describe the expression or equation.*

27. Solve $-4(3x - 5) = 3(2x + 1)$.
28. Graph $4x - 3y + 3 = 0$ by hand.
29. Simplify $-4(3x - 5) - 3(2x + 1)$.
30. Evaluate $4x - 3y + 3$ for $x = -2$ and $y = -3$.

5.5 SOLVING LINEAR INEQUALITIES IN ONE VARIABLE

Objectives

▸ Know the meaning of *inequality symbols* and *inequality*.

▸ Know how to graph an inequality.

▸ Know the properties of inequalities.

▸ Know the meaning of *satisfy, solution,* and *solution set* for a *linear inequality in one variable*.

▸ Solve a linear inequality in one variable, and graph the solution set.

▸ Use linear inequalities to make estimates and predictions about authentic situations.

In Section 5.4, we predicted when a quantity would reach a certain amount. In this section, we will discuss how to use a model to predict when a quantity will be more than (or less than) a certain amount. For instance, in Example 11 we will predict the years when guitar sales will be more than 2.5 million guitars.

Inequality Symbols

We will use the **inequality symbols** $<$, $\leq$, $>$, and $\geq$ to compare the sizes of two quantities. Here are the meanings of these symbols and some examples of *inequalities:*

Symbol	Meaning	Examples of Inequalities
$<$	is less than	$2 < 5,\ 0 < 5,\ -6 < -1$
$\leq$	is less than or equal to	$4 \leq 7,\ 2 \leq 2,\ -3 \leq 0$
$>$	is greater than	$9 > 2,\ -4 > -6,\ 2 > 0$
$\geq$	is greater than or equal to	$8 \geq 3,\ 5 \geq 5,\ -2 \geq -8$

An **inequality** contains one of the symbols $<$, $\leq$, $>$, and $\geq$ with expressions on both sides. Here are some more examples of inequalities:

$$2x + 5 < 3x - 9 \qquad x \leq 5 \qquad 7 > 2 \qquad 4x - 1 \geq 6$$

Example 1 Inequalities

Decide whether the inequality statement is true or false.

1. $3 \le 6$ **2.** $-5 > -2$

3. $8 \ge 8$ **4.** $9 < 9$

Solution

1. Since 3 is less than 6, the statement $3 \le 6$ is true.
2. Since -5 lies to the left of -2 on the number line, -5 is less than -2. So, -5 is *not* greater than -2, and the statement $-5 > -2$ is false.
3. Since 8 is equal to itself, the statement $8 \ge 8$ is true.
4. Since 9 is not less than itself, the statement $9 < 9$ is false. ■

Graphing Inequalities

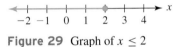

Figure 29 Graph of $x \le 2$

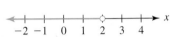

Figure 30 Graph of $x < 2$

Consider the inequality $x \le 2$. This inequality says that the values of x are less than or equal to 2. We can represent these values graphically on a number line by shading the part of the number line that lies to the left of 2 (see Fig. 29). We draw a *filled-in* circle at 2 to indicate that 2 is a value of x, too.

To graph the inequality $x < 2$, we shade the part of the number line that lies to the left of 2, but draw an *open* circle at 2 to indicate that 2 is *not* a value of x (see Fig. 30).

We use **interval notation** to describe a set of numbers. For example, we describe the numbers greater than 3 by $(3, \infty)$. We describe the numbers greater than or equal to 3 by $[3, \infty)$. We describe the set of real numbers by $(-\infty, \infty)$. More examples of inequalities and interval notation are shown in Fig. 31.

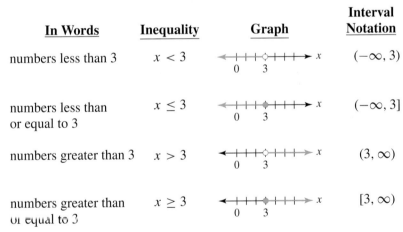

In Words	Inequality	Graph	Interval Notation
numbers less than 3	$x < 3$		$(-\infty, 3)$
numbers less than or equal to 3	$x \le 3$		$(-\infty, 3]$
numbers greater than 3	$x > 3$		$(3, \infty)$
numbers greater than or equal to 3	$x \ge 3$		$[3, \infty)$

Figure 31 Words, inequalities, graphs, and interval notation

Example 2 Graphing an Inequality

Write the inequality $x > -2$ in interval notation, and graph the values of x.

Solution

The inequality $x > -2$ means that the values of x are greater than -2. We describe these numbers in interval notation by $(-2, \infty)$. To graph the values of x, we shade the part of the number line that lies to the right of -2 and draw an open circle at -2 (see Fig. 32).

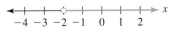

Figure 32 Graph of $x > -2$ ■

Addition Property of Inequalities

What happens if we add 3 to both sides of the inequality $4 < 6$?

$$4 < 6 \qquad \text{Original inequality}$$
$$4 + 3 \overset{?}{<} 6 + 3 \qquad \text{Add 3 to both sides.}$$
$$7 \overset{?}{<} 9 \qquad \text{Simplify.}$$
$$\text{true}$$

What happens if we add -3 to both sides of the inequality $4 < 6$?

$$4 < 6 \qquad \text{Original inequality}$$
$$4 + (-3) \overset{?}{<} 6 + (-3) \qquad \text{Add } -3 \text{ to both sides.}$$
$$1 \overset{?}{<} 3 \qquad \text{Simplify.}$$
$$\text{true}$$

These examples suggest the following property:

Addition Property of Inequalities
If $a < b$, then $a + c < b + c$.
Similar properties hold for $\leq$, $>$, and $\geq$.

Similar rules hold for subtraction, since subtracting a number is the same as adding the opposite of the number.

We can use a number line to illustrate that if $a < b$, then $a + c < b + c$. From Figs. 33 and 34, we see that if a lies to the left of b ($a < b$), then $a + c$ lies to the left of $b + c$ ($a + c < b + c$). In Fig. 33, c is negative; in Fig. 34, c is positive.

Figure 33 Adding c where c is negative

Figure 34 Adding c where c is positive

Multiplication Property of Inequalities

What if we multiply both sides of the inequality $4 < 6$ by 3?

$$4 < 6 \qquad \text{Original inequality}$$
$$4(3) \overset{?}{<} 6(3) \qquad \text{Multiply both sides by 3.}$$
$$12 \overset{?}{<} 18 \qquad \text{Simplify.}$$
$$\text{true}$$

Finally, what happens if we multiply both sides of $4 < 6$ by -3?

$$4 < 6 \qquad \text{Original inequality}$$
$$4(-3) \overset{?}{<} 6(-3) \qquad \text{Multiply both sides by } -3.$$
$$-12 \overset{?}{<} -18 \qquad \text{Simplify.}$$
$$\text{false}$$

The result is the false statement $-12 < -18$. We can get a *true* statement if we *reverse the inequality symbol* when we multiply both sides of $4 < 6$ by -3:

$$4 < 6 \qquad \text{Original inequality}$$
$$4(-3) \overset{?}{>} 6(-3) \qquad \text{Reverse inequality symbol.}$$
$$-12 \overset{?}{>} -18 \qquad \text{Simplify.}$$
$$\text{true}$$

So, when we multiply both sides of an inequality by a *negative* number, we *reverse* the inequality symbol.

Multiplication Property of Inequalities

- For a *positive* number c,   if $a < b$, then $ac < bc$.
- For a *negative* number c,   if $a < b$, then $ac > bc$.

Similar properties hold for $\leq$, $>$, and $\geq$. In words: When we multiply both sides of an inequality by a positive number, we keep the inequality symbol. When we multiply both sides by a negative number, we reverse the inequality symbol.

Similar rules apply for division, since dividing by a nonzero number is the same as multiplying by its reciprocal. Therefore, **when we multiply or divide both sides of an inequality by a negative number, we reverse the inequality symbol.**

Consider multiplying both sides of $a < b$ by -1:

$$a < b \qquad \text{Original inequality}$$
$$-1a > -1b \qquad \text{Reverse inequality symbol.}$$
$$-a > -b \qquad -1a = -a$$

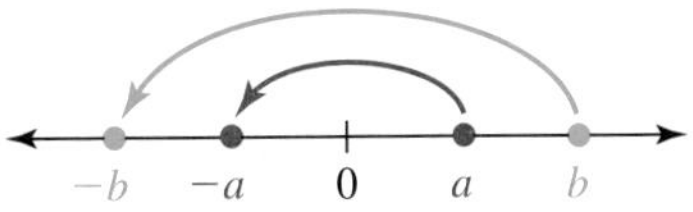

So, if $a < b$, then $-a > -b$. We can use a number line to illustrate this fact. To plot the point for $-a$, we move the point for a to the other side of the origin so that the points for $-a$ and a are the same distance from the origin (see Fig. 35).

From Fig. 35 we see that if the point for a lies to the *left* of the point for b ($a < b$), then the point for $-a$ lies to the *right* of the point for $-b$ ($-a > -b$).

Figure 35 The points for a, b, $-a$, and $-b$

Solving Linear Inequalities in One Variable

Here are some examples of *linear inequalities in one variable:*

$$3x + 5 < 8, \qquad 2x \leq 5, \qquad x - 5 > 4 - 2x, \qquad 5(x - 3) \geq 1$$

DEFINITION   Linear inequality in one variable

A **linear inequality in one variable** is an inequality that can be put into one of the forms

$$mx + b < 0, \qquad mx + b \leq 0, \qquad mx + b > 0, \qquad mx + b \geq 0$$

where m and b are constants and $m \neq 0$.

We say that a number **satisfies** an inequality in one variable if the inequality becomes a true statement after we have substituted the number for the variable.

Example 3   Identifying Solutions of an Inequality

1. Does the number 4 satisfy the inequality $2x - 3 < 7$?
2. Does the number 6 satisfy the inequality $2x - 3 < 7$?

Solution

1. We substitute 4 for x in the inequality $2x - 3 < 7$:

$$2(4) - 3 \overset{?}{<} 7 \qquad \text{Substitute 4 for } x.$$
$$8 - 3 \overset{?}{<} 7 \qquad \text{Multiply.}$$
$$5 \overset{?}{<} 7 \qquad \text{Subtract.}$$
$$\text{true}$$

So, 4 satisfies the inequality $2x - 3 < 7$.

2. We substitute 6 for x in the inequality $2x - 3 < 7$:

$$2(6) - 3 \overset{?}{<} 7 \qquad \text{Substitute 6 for } x.$$
$$12 - 3 \overset{?}{<} 7 \qquad \text{Multiply.}$$
$$9 \overset{?}{<} 7 \qquad \text{Subtract.}$$
$$\text{false}$$

So, 6 does not satisfy the inequality $2x - 3 < 7$. ∎

> **DEFINITION** *Solution, solution set,* and *solve* for an inequality in one variable
>
> We say that a number is a **solution** of an inequality in one variable if it satisfies the inequality. The **solution set** of an inequality is the set of all solutions of the inequality. We **solve** an inequality by finding its solution set.

To solve a linear inequality in one variable, we apply properties of inequalities to get the variable alone on one side of the inequality.

Example 4 Solving a Linear Inequality

Solve $2x - 3 < 7$. Describe the solution set as an inequality, in interval notation, and in a graph.

Solution

We get x alone on one side of the inequality:

$$2x - 3 < 7 \qquad \text{Original inequality}$$
$$2x - 3 + 3 < 7 + 3 \qquad \text{Add 3 to both sides to get } 2x \text{ alone on left side.}$$
$$2x < 10 \qquad \text{Combine like terms.}$$
$$\frac{2x}{2} < \frac{10}{2} \qquad \text{Divide both sides by 2 to get } x \text{ alone on left side.}$$
$$x < 5 \qquad \text{Simplify.}$$

Figure 36 Graph of $x < 5$

The solution set is the set of all numbers less than 5, which we describe in interval notation as $(-\infty, 5)$. We graph the solution set on a number line in Fig. 36. ∎

In Example 3, we found that 4 is a solution of the inequality $2x - 3 < 7$ but 6 is not. This checks with our work in Example 4, because 4 is on the graph in Fig. 36 and 6 is not.

Example 5 Solving a Linear Inequality

Solve the inequality $-3x \geq -12$. Describe the solution set as an inequality, in interval notation, and in a graph.

Solution

We divide both sides of the inequality by -3, a negative number:

$$-3x \geq -12 \qquad \text{Original inequality}$$

$$\frac{-3x}{-3} \leq \frac{-12}{-3} \qquad \text{Divide both sides by } -3; \text{ reverse inequality symbol.}$$

$$x \leq 4 \qquad \text{Simplify.}$$

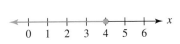

Figure 37 Graph of $x \leq 4$

Since we divided both sides of the inequality by a negative number, we reversed the inequality symbol. The solution set is $(-\infty, 4]$. We graph the solution set in Fig. 37.

WARNING   It is a common error to forget to reverse an inequality symbol when multiplying or dividing both sides of an inequality by a negative number. For instance, in Example 5, it is important that we reversed the inequality symbol $\geq$ when we divided both sides of the inequality $-3x \geq -12$ by -3.

Example 6 Solving a Linear Inequality

Solve the inequality $3x \geq -12$. Describe the solution set as an inequality, in interval notation, and in a graph.

Solution

We divide both sides of the inequality by 3, a positive number:

$$3x \geq -12 \qquad \text{Original inequality}$$

$$\frac{3x}{3} \geq \frac{-12}{3} \qquad \text{Divide both sides by 3.}$$

$$x \geq -4 \qquad \text{Simplify.}$$

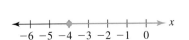

Figure 38 Graph of $x \geq -4$

Since we divided both sides of the inequality by a positive number, we did *not* reverse the inequality symbol. We graph the solution set, $[-4, \infty)$, in Fig. 38.

Example 7 Solving a Linear Inequality

Solve $2x - 5 > 6x + 3$. Describe the solution set as an inequality, in interval notation, and in a graph.

Solution

$$2x - 5 > 6x + 3 \qquad \text{Original inequality}$$

$$2x - 5 - 6x > 6x + 3 - 6x \qquad \text{Subtract } 6x \text{ from both sides.}$$

$$-4x - 5 > 3 \qquad \text{Combine like terms.}$$

$$-4x - 5 + 5 > 3 + 5 \qquad \text{Add 5 to both sides.}$$

$$-4x > 8 \qquad \text{Combine like terms.}$$

$$\frac{-4x}{-4} < \frac{8}{-4} \qquad \text{Divide both sides by } -4; \text{ reverse inequality symbol.}$$

$$x < -2 \qquad \text{Simplify.}$$

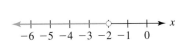

Figure 39 Graph of $x < -2$

We graph the solution set, $(-\infty, -2)$, in Fig. 39.

To verify our result, we check that for some values of x less than -2, the value of $2x - 5$ is greater than the value of $6x + 3$ (see Fig. 40). We do this by setting up the table so that x begins at -2 and increases by 1. Then, we scroll up 3 rows so that we can view values of x that are less than -2 and values of x that are greater than -2.

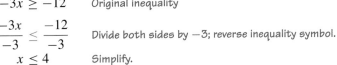

Figure 40 Verify the work

Example 8 Solving a Linear Inequality

Solve $3(x - 2) < 5x - 3$. Describe the solution set as an inequality, in interval notation, and in a graph.

Solution

$3(x - 2) < 5x - 3$	Original inequality
$3x - 6 < 5x - 3$	Distributive law
$3x - 6 - 5x < 5x - 3 - 5x$	Subtract 5x from both sides.
$-2x - 6 < -3$	Combine like terms.
$-2x - 6 + 6 < -3 + 6$	Add 6 to both sides.
$-2x < 3$	Combine like terms.
$\dfrac{-2x}{-2} > \dfrac{3}{-2}$	Divide both sides by -2; reverse inequality symbol.
$x > -\dfrac{3}{2}$	Simplify.

We graph the solution set, $\left(-\dfrac{3}{2}, \infty\right)$, in Fig. 41.

Figure 41 Graph of $x > -\dfrac{3}{2}$

To verify our result, we check that for some values of x greater than $-\dfrac{3}{2}$, the value of $3(x - 2)$ is less than the value of $5x - 3$ (see Fig. 42). We do this by setting up the table so that x begins at -1.5 and increases by 1. Then, we scroll up 3 rows so that we can view values of x that are less than -1.5 and values of x that are greater than -1.5.

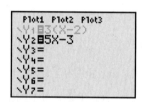

Figure 42 Verify the work

Example 9 Solving a Linear Inequality That Contains Decimals

Solve $-2.9x + 4.1 \le -6.05$. Describe the solution set as an inequality, in interval notation, and in a graph.

Solution

$-2.9x + 4.1 \le -6.05$	Original inequality
$-2.9x + 4.1 - 4.1 \le -6.05 - 4.1$	Subtract 4.1 from both sides.
$-2.9x \le -10.15$	Combine like terms.
$\dfrac{-2.9x}{-2.9} \ge \dfrac{-10.15}{-2.9}$	Divide both sides by -2.9; reverse inequality symbol.
$x \ge 3.5$	Simplify.

We graph the solution set, $[3.5, \infty)$, in Fig. 43.

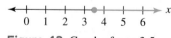

Figure 43 Graph of $x \ge 3.5$

We can use graphing calculator tables to check our work.

Example 10 Solving a Linear Inequality That Contains Fractions

Solve $\dfrac{2}{3}w + \dfrac{1}{2} \le \dfrac{5}{6}$. Describe the solution set as an inequality, in interval notation, and in a graph.

Solution

We begin by multiplying both sides by the LCD, 6, to clear the fractions:

$$\frac{2}{3}w + \frac{1}{2} \leq \frac{5}{6}$$ Original inequality

$$6 \cdot \left(\frac{2}{3}w + \frac{1}{2}\right) \leq 6 \cdot \frac{5}{6}$$ Multiply both sides by 6.

$$6 \cdot \frac{2}{3}w + 6 \cdot \frac{1}{2} \leq 6 \cdot \frac{5}{6}$$ Distributive law

$$4w + 3 \leq 5$$ Simplify.

$$4w + 3 - 3 \leq 5 - 3$$ Subtract 3 from both sides.

$$4w \leq 2$$ Simplify.

$$\frac{4w}{4} \leq \frac{2}{4}$$ Divide both sides by 4.

$$w \leq \frac{1}{2}$$ Simplify.

We graph the solution set, $\left(-\infty, \frac{1}{2}\right]$, in Fig. 44.

$$\overset{\longleftarrow}{\underset{-3 \quad -2 \quad -1 \quad 0 \quad 1 \quad 2 \quad 3}{\rule{0pt}{0pt}}} \; w$$

Figure 44 Graph of $w \leq \frac{1}{2}$

We can use graphing calculator tables to check our work. ■

Using Linear Inequalities to Make Predictions

When working with a linear model, we can make certain types of estimates and predictions by solving a linear inequality that is related to the linear model.

Example 11 Making a Prediction

Sales of guitars in the United States are shown in Table 44 for various years. Let s be the sales (in millions of guitars) in the year that is t years since 1990. A reasonable model is

$$s = 0.093t + 0.54$$

Predict the years when sales will be more than 2.5 million guitars.

Solution

To predict when sales will be more than 2.5 million guitars, we find the values of t such that the expression $0.093t + 0.54$ is more than 2.5:

$$0.093t + 0.54 > 2.5$$ Original inequality

$$0.093t + 0.54 - 0.54 > 2.5 - 0.54$$ Subtract 0.54 from both sides.

$$0.093t > 1.96$$ Combine like terms.

$$\frac{0.093t}{0.093} > \frac{1.96}{0.093}$$ Divide both sides by 0.093.

$$t > 21.08$$ Simplify; divide.

So, sales will be more than 2.5 million guitars after 2011 ($t > 21$). ■

Table 44 Guitar Sales	
Year	**Sales (millions of guitars)**
1992	0.6
1993	0.9
1994	1.0
1995	1.1
1996	1.1
1997	1.1
1998	1.2
1999	1.3
2000	1.6

Source: *Music Trades*

Stage-fright control. It plays the entire solo in "Free bird" in a pinch.

What's the button for?

group exploration

Looking ahead: Using a system of equations to model a situation

Table 45 Annual Fuel Consumption

	Annual Fuel Consumption (billions of gallons)	
Year	Cars	Light Trucks
1992	65.6	40.9
1994	68.1	44.1
1996	69.2	47.4
1998	71.7	50.5
2000	73.1	52.9
2002	75.5	55.2
2003	74.6	56.3

Source: *U.S. Federal Highway Administration*

Fuel consumption by cars and light trucks (vans, pickups, and SUVs) is described in Table 45 for various years. Annual fuel consumption (in billions of gallons) C for cars and L for light trucks is modeled by the system

$$C = 0.87t + 64.27$$
$$L = 1.40t + 38.62$$

where t is the number of years since 1990.

1. Find the C-intercept of the car model and the L-intercept of the light-truck model. What do these intercepts mean in this situation?

2. Find the slopes of both models. What do these slopes mean in this situation?

3. Your responses to Problems 1 and 2 should suggest an event that will happen in the future. Describe that event.

4. By using Zoom Out and "intersect" on a graphing calculator, estimate the coordinates of the point where the graphs of the models intersect. (See Sections A.6 and A.17 for graphing calculator instructions.) In terms of fuel consumption, what does it mean that the graphs intersect at this point? State the year and fuel consumption for this event.

TIPS FOR SUCCESS: Scan Test Problems

When you take a test, scan the test problems quickly, pick the ones with which you feel most comfortable, and complete those problems first. By doing so, you will have warmed up and gained confidence, and you may do better on the rest of the test. Also, you will probably have a better idea of how to allot your time on the remaining problems.

HOMEWORK 5.5

FOR EXTRA HELP ▶

Student Solutions Manual PH Math/Tutor Center *Math XL* MathXL® *MyMathLab* MyMathLab

---

Decide whether the given inequality is true or false.

1. $-3 > -5$ **2.** $-6 \leq -2$ **3.** $4 \geq 4$ **4.** $-5 < -5$

Sketch the graph of the given inequality.

5. $x < 4$ **6.** $x < -5$ **7.** $x \geq -1$ **8.** $x \geq 1$

9. $x \leq -2$ **10.** $x \leq 4$ **11.** $x > 6$ **12.** $x > -3$

13. Use words, inequalities, graphs, and interval notation to complete Fig. 45.

In Words	Inequality	Graph	Interval Notation

numbers less than or equal to -2			
	$x > -5$		$(-\infty, 1)$

Figure 45 Exercise 13

14. Use words, inequalities, graphs, and interval notation to complete Fig. 46.

In Words	Inequality	Graph	Interval Notation
	$x \leq -2$		
numbers greater than 1			
			$[-4, \infty)$

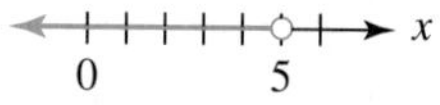

Figure 46 Exercise 14

Which of the given numbers satisfy the given inequality?

15. $3x + 5 \geq 14$; 2, 3, 6

16. $-4x - 7 > 1$; $-3, -2, 0$

17. $2x < x + 2$; $-4, 2, 3$

18. $x - 9 \leq 4x$; $-3, 2, 5$

Solve the inequality. Describe the solution set as an inequality, in interval notation, and in a graph. Then, use a graphing calculator table to verify your result.

19. $x + 2 > 3$ **20.** $x + 5 \leq 9$ **21.** $x - 1 < -4$

22. $x - 3 \geq -1$ **23.** $2x \leq 6$ **24.** $3x > 9$

25. $4x \geq -8$ **26.** $2x < -10$ **27.** $-3t \geq 6$

28. $-2w \leq 2$ **29.** $-5x < 20$ **30.** $-7x > 21$

31. $-2x > 1$ **32.** $-4x < -2$ **33.** $5x \leq 0$

34. $-3x > 0$ **35.** $-x < 2$ **36.** $-x \leq -1$

37. $\frac{1}{2}a < 3$ **38.** $\frac{1}{3}r > -2$ **39.** $-\frac{2}{3}x \geq 2$

40. $-\frac{5}{2}x \leq 10$ **41.** $3x - 1 \geq 2$ **42.** $4x + 7 < 15$

43. $5 - 3x < -7$ **44.** $8 - 2x \leq 6$

45. $-4x - 3 \geq 5$ **46.** $-9x + 7 > -11$

47. $3c - 6 \leq 5c$ **48.** $7w + 4 > 3w$

49. $5x \geq x - 12$ **50.** $4x < 6 - 2x$

51. $-3.8x + 1.9 > -7.6$ **52.** $-2.4x + 5.8 \leq 8.92$

53. $3b + 2 > 7b - 6$ **54.** $5k - 1 \leq 2k + 8$

55. $4 - 3x < 9 - 2x$ **56.** $8 - x \geq 2 - 3x$

57. $2(x + 3) \leq 8$ **58.** $5(x - 2) \geq 15$

59. $-(a - 3) > 4$ **60.** $-(t + 5) < -2$

61. $3(2x - 1) \leq 2(2x + 1)$ **62.** $4(3x - 5) > 5(2x - 3)$

63. $4(2x - 3) + 1 \geq 3(4x - 5) - x$

64. $-2(5x + 3) - 2x < -3(2x - 4) + 2$

65. $4.3(1.5 - x) \geq 13.76$ **66.** $3.1(2.7 - x) > -1.55$

67. $\frac{1}{2}y + \frac{2}{3} \geq \frac{3}{2}$ **68.** $\frac{3}{4}t - \frac{1}{2} \leq \frac{1}{4}$

69. $\frac{5}{3} - \frac{1}{6}x < \frac{1}{2}$ **70.** $\frac{1}{4} - \frac{2}{3}x > \frac{7}{12}$

71. $\frac{2(3 - x)}{3} > -4x$ **72.** $\frac{3(4 - x)}{2} \geq -6x$

73. $\frac{3r - 5}{4} \leq \frac{2r - 3}{3}$ **74.** $\frac{4k + 6}{7} > \frac{3k + 4}{5}$

75. Table 46 shows the percentages of teachers who say they are "very satisfied" with teaching for various years.

Table 46 Teachers Who Are "Very Satisfied" with Teaching

Year	Percent
1984	40
1989	44
1995	54
2001	52
2003	57

Source: *Harris Interactive*

Let p be the percentage of teachers who are "very satisfied" with teaching at t years since 1980. A reasonable model is $p = 0.83t + 37.43$.
   a. What is the slope? What does it mean in this situation?
   b. In which years will more than 60% of teachers be "very satisfied" with teaching?

76. The percentages of obstetricians and gynecologists who are women are shown in Table 47 for various years.

Table 47 Percentages of Obstetricians and Gynecologists Who Are Women

Year	Percent
1980	12
1985	19
1990	23
1995	30
2000	34
2002	38

Source: *American Medical Association*

Let p be the percentage of obstetricians and gynecologists who are women at t years since 1980. A reasonable model is $p = 1.13t + 12.41$.
   a. What is the slope? What does it mean in this situation?
   b. In which years will more than half of physicians specializing in obstetrics and gynecology be women?

77. The numbers of married-couple households (in millions) and the percentages of households that are married-couple households are shown in Table 48 for various years.

Table 48 Married-Couple Households

Year	Number of Married-Couple Households (millions)	Percent of Households That Are Married-Couple Households
1963	40.9	74.0
1973	46.3	67.8
1983	49.9	59.5
1993	53.1	55.1
2003	57.3	51.5

Source: *U.S. Census Bureau*

Let p be the percentage of households that are married-couple households at t years since 1960.
   a. Find an equation of a model to describe t and p.
   b. In which years were more than 53% of households married-couple households?
   c. Although the number of married-couple households is *increasing*, the percentage of households that are married-couple households is *decreasing*. Explain how this is possible.

78. As sales of digital cameras continue to grow, the percentages of U.S. households that own traditional cameras continue to decline (see Table 49).

Table 49 Households That Own Traditional Cameras

Year	Percent
1996	93
1998	88
2000	80
2002	73
2003	70

Source: *Photo Marketing Association*

Let p be the percentage of households that own a traditional camera at t years since 1990.
   a. Find an equation of a model to describe the data.

b. The Photo Marketing Association (PMA) predicted that 67% of households would own a traditional camera in 2004. Is it possible that PMA used a linear model to make this estimate? Explain.

c. Predict the years when less than half of households will own a traditional camera.

79. The teenage birthrate in the United States has declined since 1994 (see Table 50).

Table 50 American Teenage Birthrate

Year	Birthrate (number of births per 1000 women ages 15–19)
1994	58.2
1996	53.5
1998	50.3
2000	47.7
2002	43.0
2003	41.7

Source: *U.S. National Center for Health Statistics*

Let r be the American teenage birthrate (number of births per 1000 women ages 15–19) at t years since 1990.

a. Find an equation of a model to describe the data.

b. Predict the teenage birthrate in 2010. Then predict the *number* of births to women ages 15–19 in 2010. Use the U.S. Census Bureau's prediction that there will be 10,398,000 women ages 15–19 in that year.

c. The American teenage birthrate is 2 to 10 times larger than that in other Western countries. For example, the birthrate in France is 10 births per 1000 women ages 15–19. Predict in which years the American birthrate will be less than 10 births per 1000 women ages 15–19.

80. A dollar store is a no-frills retailer that sells many goods for one dollar. In 2002, there were 13,000 dollar stores in the United States—three times as many as in 1993. The percentages of households in various income groups that shop at dollar stores are shown in Table 51.

Table 51 Households That Shop at Dollar Stores

Household Income Group (thousands of dollars)	Income Used to Represent Income Group (thousands of dollars)	Percent of Households That Shop at Dollar Stores
0–19.999	10	74
20–29.999	25	71
30–39.999	35	67
40–49.999	45	64
50–69.999	60	58
70+	100	45

Source: *ACNielsen Homescan Panel*

Let p be the percentage of households with an income of d thousand dollars that shop at dollar stores.

a. Find an equation of a model to describe the data.

b. What percentage of households with an income of $27 thousand shop at dollar stores?

c. At what incomes do more than half of households shop at dollar stores?

81. A student tries to solve $-3x < 15$:

$$-3x < 15$$
$$\frac{-3x}{-3} < \frac{15}{-3}$$
$$x < -5$$

Describe any errors. Then solve the inequality correctly.

82. A student tries to solve $4x < -24$:

$$4x < -24$$
$$\frac{4x}{4} > \frac{-24}{4}$$
$$x > -6$$

Describe any errors. Then solve the inequality correctly.

83. a. List three numbers that satisfy $3x - 7 < 5$.
   b. List three numbers that do not satisfy $3x - 7 < 5$.

84. a. List three numbers that satisfy $2(x + 3) > 17$.
   b. List three numbers that do not satisfy $2(x + 3) > 17$.

85. Give an example of a linear inequality of the form $mx + b \leq c$, where m, b, and c are constants. Then solve the inequality. Describe the solution set as an inequality, in interval notation, and in a graph.

86. Describe how to solve a linear inequality in one variable. Describe when you need to reverse an inequality symbol. Explain why it is necessary to reverse the symbol in this case. Finally, explain what you have accomplished by solving an inequality.

Related Review

Solve the equation or inequality. If the statement is an inequality, then describe the solution set as an inequality, in interval notation, and in a graph.

87. $-2x + 6 = 3x - 14$

88. $-3(2x - 5) = 21$

89. $-2x + 6 > 3x - 14$

90. $-3(2x - 5) \leq 21$

Let x be a number. Translate the English sentence into an inequality. Then solve the inequality. Describe the solution set as an inequality, in interval notation, and in a graph.

91. The sum of the number and 5 is greater than 2.

92. The difference of 7 and the number is less than 3.

93. Twice the number is less than or equal to 5 times the number, minus 6.

94. Three times the number is greater than or equal to the difference of the number and 8.

Expressions, Equations, and Graphs

Give an example of the following. Then solve, simplify, or graph, as appropriate.

95. linear equation in one variable

96. linear expression in one variable with four terms

97. linear inequality in one variable

98. linear equation in two variables

analyze this

GLOBAL WARMING LAB (continued from Chapter 3)

Recall from the Global Warming Lab in Chapter 1 that many scientists believe that an increase in average global temperature as small as 3.6°F could be a dangerous climate change (see Table 52). Recall from the Global Warming Lab in Chapter 2 that most scientists believe that global warming is largely the result of carbon emissions from the burning of fossil fuels (see Table 53).

Table 52 Average Surface Temperatures of Earth

Year	Average Temperature (degrees Fahrenheit)	Year	Average Temperature (degrees Fahrenheit)
1900	57.1	1955	57.0
1905	56.7	1960	57.2
1910	56.8	1965	56.9
1915	57.3	1970	57.3
1920	56.9	1975	57.2
1925	56.9	1980	57.7
1930	57.1	1985	57.4
1935	57.1	1990	58.1
1940	57.5	1995	58.0
1945	57.2	2000	57.9
1950	56.9	2003	58.4

Source: *NASA–GISS*

Table 53 Carbon Emissions from Burning of Fossil Fuels

Year	Carbon Emissions (billions of metric tons) United States	World
1950	0.6	1.6
1955	0.7	2.0
1960	0.8	2.6
1965	1.0	3.1
1970	1.2	4.1
1975	1.2	4.6
1980	1.3	5.3
1985	1.2	5.4
1990	1.4	6.1
1995	1.4	6.4
2000	1.5	6.5

Source: *U.S. Department of Energy*

Finally, recall from the Global Warming Lab in Chapter 2 that the goal of the Kyoto Protocol is to cut developed countries' carbon emissions to about 5% to 7% below 1990 levels by 2012. This will be a crucial first step, but the Intergovernmental Panel on Climate Change (IPCC) is calling for carbon emissions in 2050 to be 60% less than carbon emissions in 1990.

Many countries are condemning the United States because the Bush administration has refused to ratify the Kyoto Protocol. Although the U.S. population in 2000 was only 5% of the world population, 23% of world carbon emissions that year was produced by the United States (see Table 54).

Table 54 United States and World Populations

Year	Population (billions) United States	World
1960	0.18	3.04
1970	0.21	3.71
1980	0.23	4.45
1990	0.25	5.28
2000	0.28	6.07

Source: *U.S. Census Bureau*

Critics of the Kyoto Protocol say that a fairer pact would be for all countries to commit to the same level of carbon emissions *per person*. Using the IPCC's recommendation for 2050 carbon emissions, coupled with the United Nations' prediction of 8.92 billion people in 2050, carbon emissions should be about 0.3 metric ton per person that year.[†]

In 2000, annual carbon emissions were about 1.1 metric tons of carbon per person. Average annual carbon emissions for developing countries is 0.4 metric ton per person, which is close to IPCC's recommendation. However, average annual carbon emissions for developed countries is 2.7 metric tons per person, far above IPCC's recommendation (see Table 55). GNP, the gross national product, is a measure of a country's economic strength.

Table 55 GNP Ranks and per-Person Carbon Emissions

Country	1998 GNP Rank	1998 GNP per-Person Rank	2000 per-Person Carbon Emissions (metric tons)
Sweden	20	14	1.4
Switzerland	18	3	1.5
France	4	17	1.7
Italy	6	25	2.0
Austria	21	12	2.1
Denmark	23	6	2.3
Japan	2	7	2.6
Germany	3	13	2.6
United Kingdom	5	22	2.6
Belgium	19	15	2.7
Norway	25	4	3.0
Australia	14	24	4.9
United States	1	10	5.4
Netherlands	12	18	12.6

Source: *The World Bank Group*

[†]United Nations Population Division.

With annual carbon emissions of 5.4 metric tons per person, the United States would have to reduce emissions by 94% to meet the standard of 0.3 metric ton per person. This means that Americans could emit only 6% of the carbon that they currently emit. Imagine driving your car, heating and cooling your home, using your home appliances (including your refrigerator), using your computer, using your lights, and watching your television only 6% (about one-twentieth) of the time that you currently do.

So far, the voluntary program has not slowed the increase of carbon emissions. In fact, the U.S. Energy Information Administration predicts that U.S. annual carbon emissions will reach 6.0 metric tons per person in 2010.[‡]

The Bush administration says it will not take stronger action because the economy would suffer. Some experts, however, such as the engineer Alan Pears, codirector of the environmental consultancy Sustainable Solutions, believe that it is possible for emissions to be significantly reduced without harming a country's economy. Switzerland, Denmark, Japan, and Norway, for instance, have a better GNP per-person ranking than the United States has, as well as significantly lower per-person carbon emissions. In fact, with the exception of the Netherlands, all of the countries listed in Table 55 have strong economies and significantly lower carbon emissions than the United States has.

Many states have taken the matter into their own hands by adopting policies to reduce carbon emissions. And by using alternative sources of energy, many countries have slowed or reversed the growth of carbon emissions in recent years.[*]

In addition to national, state, and even corporate actions, individuals can help lower carbon emissions by purchasing hybrid automobiles, major appliances with the Energy Star logo, solar thermal systems to help provide hot water, and compact fluorescent light bulbs. Individuals can also car pool or use public transportation.

Analyzing the Situation

1. **a.** Let A be the average global temperature (in degrees Fahrenheit) at t years since 1900. Use a graphing calculator to draw a scattergram of the data in Table 52 and then find an equation of a model *for the years 1965 to 2000*. Then, verify that your model fits the data well for those years.
   b. In Problem 2a of the Global Warming Lab in Chapter 3, you found the average global temperature from 1900 to 1965. If you didn't do this, do so now. [**Hint:** Divide the sum of the temperatures by the number of temperature readings.] Use your model to predict when the planet's average temperature will have increased by 3.6°F—a potentially dangerous climate change.

2. Use Tables 53 and 54 to verify the claims that although the U.S. population in 2000 was only 5% of world population, 23% of annual world carbon emissions was produced by the United States in that year.

3. Use the United Nations' prediction that the world population will be 8.92 billion in 2050 to verify the claim that per-person carbon emissions that year should be about 0.3 metric ton per person for the IPCC recommendation of a 60% reduction by then.

4. Let p be the U.S population (in billions) at t years since 1950. Create a scattergram of the data and then find a model of the situation. Finally, verify that your model fits the data well.

5. Let c be U.S. carbon emissions (in billions of metric tons) in the year that is t years since 1950. Use a graphing calculator to draw a scattergram of the data and then find a model of the situation. Finally, verify that your model fits the data well.

6. **a.** Use your U.S. population model of Problem 4 to predict U.S. population in 2010.
   b. Use your U.S. carbon emissions model of Problem 5 to predict U.S. carbon emissions in 2010.
   c. Use your results from parts (a) and (b) to predict U.S. *per-person* carbon emissions in 2010. Compare your result with the U.S. Energy Information Administration's prediction of 6.0 metric tons per person.
   d. Is the United States heading in the right direction to meet IPCC's recommendation of 0.3 metric ton of carbon per person per year? Explain.

ROPE LAB

In this lab, you will explore the relationship between the number of knots tied in a rope and the rope's length.

Check with your instructor whether you should collect your own data or use the data listed in Table 56.

Table 56 Lengths of a Rope with Diameter about 7 Millimeters

Number of Knots	Length of Rope (centimeters)
0	60.0
1	53.2
2	45.8
3	38.3
4	30.6

Source: *J. Lehmann*

Materials

1. A 60-centimeter-long piece of rope with diameter about 7 millimeters

2. A meterstick or other measuring device with units of millimeters

[‡]U.S. Energy Information Administration, *International Energy Outlook, 2002.*

[*]Pamela Person, "Reducing Greenhouse Gas Emissions," Maine Center for Economic Policy, *Choices*, VII(9), Oct. 11, 2001.

Recording of Data

Pull the rope taut and measure its length (in centimeters). Then, tie a knot close to one end of the rope and measure the length of the rope again. Next, tie another knot next to the first one and measure the length of the rope. Continue tying knots, working your way along the rope, and measuring the rope's length after you have tied each knot. Tie a total of four knots.

Analyzing the Situation

1. Display your data in a table or use the data in Table 56.
2. Let L be the length (in centimeters) of the rope with n knots. Use a graphing calculator to draw a scattergram of the data. Copy the scattergram by hand.
3. Find an equation of a linear model to describe the data.
4. Use a graphing calculator to draw a graph of your equation and the scattergram in the same viewing window. Copy the graph of the equation and the scattergram by hand.
5. Is your line increasing or decreasing? What does that mean in this situation?
6. Find the L-intercept of your model. What does it mean in this situation?
7. Find the slope of your model. What does it mean in this situation?
8. Use your model to estimate the length of the rope with five knots.
9. Check whether your result in Problem 8 is an underestimate or an overestimate by tying a fifth knot in the rope and then measuring the rope's length.
10. Continue tying knots in the rope. When does model breakdown first occur? Explain.
11. Find the n-intercept of your model. What does it mean in this situation?

SHADOW LAB

In this lab, you will compare the relationship between an object's height and the length of its shadow.

Check with your instructor whether you should collect your own data or use the data listed in Table 57.

Table 57 Heights of Objects and the Lengths of Their Shadows

Object	Height (inches)	Length of Shadow (inches)
Nothing	0	0
Wine bottle	15.5	10.3
Toy putter	20.8	13.0
Box	36.5	23.6
Mop	47.8	31.0
Person's shoulder	55.0	36.5
Person	63.0	41.5

Source: *J. Lehmann*

Materials

1. Six objects of various heights up to 7 feet
2. A building, pole, tree, or other tall object with height greater than 15 feet
3. A tape measure or other measuring device

Recording of Data

Run the experiment when the objects (including the tall object) have noticeable and measurable shadows. For each object, measure its height and the length of its shadow. Record the beginning and ending time of the experiment. Also, record the length of the shadow of the tall object. It is important that you record all the data quickly.

Analyzing the Situation

1. Display your data in a table similar to Table 57. Those data were collected from 2:10 P.M. to 2:20 P.M., and the tall object is a tree whose shadow has a length of 49.5 feet.
2. Let L be the height (in inches) of an object and s be the length (in inches) of the object's shadow. Use a graphing calculator to draw a scattergram of the data. Copy the scattergram by hand.
3. Find an equation of a linear model to describe the data.
4. Use a graphing calculator to draw a graph of your equation and the scattergram in the same viewing window. Copy the graph of the equation and the scattergram by hand.
5. Is your line increasing or decreasing? What does that mean in this situation?
6. Find the s-intercept of your model. What does it mean in this situation?
7. Find the slope of your model. What does it mean in this situation?
8. Use the length of the shadow of the tall object to estimate the object's height.
9. Explain why it was important that you run the experiment quickly.
10. Suppose you had run the experiment half an hour later. How would that have affected the slope of your model? Explain. (If the Sun would have set by then, describe the impact on the slope if the experiment had been performed half an hour earlier.) Would this change in time have resulted in a different estimate of the height of the tall object? Explain.

LINEAR EQUATION LAB: TOPIC OF YOUR CHOICE

Your objective in this lab is to use a linear model to describe some authentic situation. Find some data on two quantities which describe a situation that has not been discussed in this text. Almanacs, newspapers, magazines, scientific journals, and the Internet are good resources. Or you can conduct an experiment. Choose something that interests you!

Analyzing the Situation

1. What two quantities did you explore? Define variables for the quantities. Include units in your definitions.

2. Which variable is the dependent variable? Which variable is the independent variable? Explain.

3. Describe how you found your data. If you conducted an experiment, provide a careful description with specific details of how you ran your experiment. If you didn't conduct an experiment, state the source of your data.

4. Include a table of your data.

5. Use a graphing calculator to draw a scattergram of your data. (If your data are not approximately linear, find some data that are.)

6. Find an equation of a linear model to describe the data.

7. What is the slope of your linear model? What does it mean in this situation?

8. Does it make sense that your variables are approximately linearly related in terms of the situation you chose to model? Explain.

9. Choose a value for your independent variable. On the basis of that chosen value, use your model to find a value for your dependent variable. Describe what your result means in the situation you are modeling.

10. Choose a value for your dependent variable. On the basis of that chosen value, use your model to find a value for your independent variable. Describe what your result means in the situation you are modeling.

11. Find the intercepts of your linear model. What do they mean in the situation you are modeling? Has model breakdown occurred at the intercepts?

12. Comment on your lab experience.
    a. For example, you might address whether the lab was enjoyable, insightful, and so on.
    b. Were you surprised by any of your findings? If so, which ones?
    c. How would you improve your process for this lab if you were to do it again?
    d. How would you improve your process if you had more time and money?

Chapter Summary

Key Points
OF CHAPTER 5

Solving for y to graph (Section 5.1)	Before we can use the y-intercept and the slope to graph a linear equation, we must solve for y to put the equation into the form $y = mx + b$.
Intercepts of the graph of an equation (Section 5.1)	For an equation containing the variables x and y, <ul><li>To find the x-coordinate of each x-intercept, substitute 0 for y and solve for x.</li><li>To find the y-coordinate of each y-intercept, substitute 0 for x and solve for y.</li></ul>
Using intercepts to graph an equation (Section 5.1)	To graph a linear equation whose graph has exactly two intercepts, 1. Find the intercepts. 2. Plot the intercepts and a third point on the line, and graph the line that contains the three points.
Comparing linear equations in one and two variables (Section 5.1)	A linear equation in two variables has an infinite number of (ordered-pair) solutions. A linear equation in one variable has exactly one (real-number) solution.
Finding an equation of a line by using the slope and a point (Section 5.2)	To find an equation of a line by using the slope and a point, 1. Substitute the given value of the slope m into the equation $y = mx + b$. 2. Substitute the coordinates of the given point into your equation from step 1 and solve for b. 3. Substitute the value of b from step 2 into your equation from step 1. 4. Check that the graph of your equation contains the given point.
Finding an equation of a line by using two points (Section 5.2)	To find an equation of the line that passes through two given points whose x-coordinates are different, 1. Use the formula $m = \dfrac{y_2 - y_1}{x_2 - x_1}$, or use a graph to determine $\dfrac{\text{rise}}{\text{run}}$, to find the slope of the line containing the two points.

2. Substitute the m value you found in step 1 into the equation $y = mx + b$.

3. Substitute the coordinates of one of the given points into the equation from step 2 and solve for b.

4. Substitute the b value you found in step 3 into your equation from step 2.

5. Check that the graph of your equation contains the two given points.

Point–slope form (Section 5.2)

If a nonvertical line has slope m and contains the point (x_1, y_1), then an equation of the line is $y - y_1 = m(x - x_1)$.

Finding an equation of a linear model (Section 5.3)

To find an equation of a linear model, given some data,

1. Create a scattergram of the data.

2. Determine whether there is a line that comes close to the data points. If so, choose two points (not necessarily data points) that you can use to find the equation of a linear model.

3. Find an equation of the line.

4. Use a graphing calculator to verify that the graph of your equation comes close to the points of the scattergram.

Using an equation of a linear model to make predictions (Section 5.4)

When making a prediction about the dependent variable of a linear model, substitute a chosen value for the independent variable in the model and then solve for the dependent variable.

When making a prediction about the independent variable of a linear model, substitute a chosen value for the dependent variable in the model and then solve for the independent variable.

Using an equation of a linear model to find intercepts (Section 5.4)

Suppose that an equation of the form $p = mt + b$, where $m \neq 0$, is used to model a situation. Then

- The p-intercept is $(0, b)$.
- To find the t-coordinate of the t-intercept, substitute 0 for p in the model's equation and then solve for t.

Four-step modeling process (Section 5.4)

To find a linear model and then make estimates and predictions,

1. Create a scattergram of the data to determine whether there is a nonvertical line that comes close to the points. If so, choose two points (not necessarily data points) that you can use to find an equation of a linear model.

2. Find an equation of your model.

3. Verify your equation by checking that the graph of your model contains the two chosen points and comes close to all of the data points.

4. Use the equation of your model to make estimates, make predictions, and draw conclusions.

Addition property of inequalities (Section 5.5)

If $a < b$, then $a + c < b + c$. Similar properties hold for $\leq$, $>$, and $\geq$.

Multiplication property of inequalities (Section 5.5)

- For a *positive* number c,   if $a < b$, then $ac < bc$.
- For a *negative* number c,   if $a < b$, then $ac > bc$.

Similar properties hold for $\leq$, $>$, and $\geq$.

Linear inequality in one variable (Section 5.5)

A **linear inequality in one variable** is an inequality that can be put into one of the forms

$$mx + b < 0, \qquad mx + b \leq 0, \qquad mx + b > 0, \qquad mx + b \geq 0$$

where m and b are constants and $m \neq 0$.

Solution, solution set, and solve for an inequality in one variable (Section 5.5)

We say that a number is a **solution** of an inequality in one variable if it satisfies the inequality. The **solution set** of an inequality is the set of all solutions of the inequality. We **solve** an inequality by finding its solution set.

CHAPTER 5 REVIEW EXERCISES

Determine the slope and the y-intercept. Use the slope and the y-intercept to graph the equation by hand. Use a graphing calculator to verify your work.

1. $y = 4x - 5$

2. $y = -3x + 4$

3. $y = \dfrac{1}{2}x + 1$

4. $y = -\dfrac{2}{3}x - 2$

5. $y = \dfrac{5}{3}x$

6. $y = -5$

7. $2x - y = 5$

8. $3x - 2y = -6$

9. $4x + 5y = 10$

10. $x + 3y = 6$

11. $2x + 5y - 20 = 0$

12. $y - 4 = 0$

Find the x-intercept and y-intercept. Then graph the equation by hand.

13. $4x - 5y = 20$

14. $3x + 2y = 6$

15. $3x + 4y + 12 = 0$

16. $y = 2x - 4$

17. $y = -x + 3$

18. $\dfrac{1}{3}x - \dfrac{1}{2}y = 1$

Find the approximate x-intercept and approximate y-intercept. Round the coordinates to the second decimal place.

19. $9.2x - 3.8y = 87.2$

20. $y = 2.56x + 97.25$

21. a. Find the intercepts of the graph of $y = 3x + 7$.
b. Find the intercepts of the graph of $y = 2x + 9$.
c. Find the intercepts of the graph of $y = mx + b$, where b is a constant and m is a nonzero constant.

22. Consider $2x - 4y = 8$.
a. Graph the equation by finding the slope and the y-intercept.
b. Graph the equation by finding the intercepts.
c. Compare your graphs from parts (a) and (b). Decide which method you prefer for graphing the equation $2x - 4y = 8$.

Find an equation of the line that has the given slope and contains the given point. If possible, write your equation in slope–intercept form. Check that the graph of your equation contains the given point.

23. $m = -4$, $(2, -1)$

24. $m = 2$, $(-9, -3)$

25. $m = -3$, $(-4, 5)$

26. $m = -\dfrac{2}{5}$, $(-3, 6)$

27. $m = \dfrac{3}{7}$, $(2, 9)$

28. $m = -\dfrac{2}{3}$, $(-6, -4)$

29. m is undefined, $(2, 5)$

30. $m = 0$, $(-1, -4)$

Find an approximate equation of the line that has the given slope and contains the given point. Write your equation in slope–intercept form. Round the constant term to the second decimal place. Use a graphing calculator to verify your equation.

31. $m = -5.29$, $(-4.93, 8.82)$

32. $m = 1.45$, $(-2.79, -7.13)$

Find an equation of the line that passes through the two given points. If possible, write your equation in slope–intercept form. Check that the graph of your equation contains the given points.

33. $(2, 1)$ and $(5, 7)$

34. $(-2, 9)$ and $(1, 3)$

35. $(-2, -7)$ and $(1, 2)$

36. $(2, -5)$ and $(4, 5)$

37. $(-5, 8)$ and $(-1, 2)$

38. $(-8, -5)$ and $(4, -2)$

39. $(-3, 9)$ and $(6, -6)$

40. $(-4, -10)$ and $(-2, -7)$

41. $(5, -3)$ and $(5, 2)$

42. $(-4, -3)$ and $(-1, -3)$

Find an approximate equation of the line that passes through the two given points. Write your equation in slope–intercept form. Round the slope and the constant term to two decimal places. Use a graphing calculator to verify your equation.

43. $(3.5, 9.2)$ and $(8.7, 4.8)$

44. $(-5.22, 2.49)$ and $(1.83, -3.99)$

45. Find an equation of the line sketched in Fig. 47.

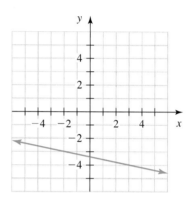

Figure 47 Exercise 45

46. Find an equation of a line that comes close to the points listed in Table 58. Then use a graphing calculator to check that your line comes close to those points.

Table 58 Find an Equation	
x	**y**
1	28
4	23
6	16
9	13
10	8

47. The percentage of restaurant patrons who order dessert at dinner has declined since 1990 (see Table 59).

Table 59 Percentages of Restaurant Patrons Who Order Dessert	
Year	**Percent**
1989	15
1992	14
1995	13
1997	13
2000	12
2003	11

Source: *NPD Group*

Let p be the percentage of restaurant patrons who have ordered dessert at t years since 1980.

a. Use a graphing calculator to draw a scattergram of the data.
b. Find an equation of a linear model to describe the data.
c. Use your model to predict when 9% of restaurant patrons will order dessert.
d. Find the t-intercept. What does it mean in this situation?
e. Find the p-intercept. What does it mean in this situation?

48. Sales of fitness equipment have more than doubled over the past decade (see Table 60).

Table 60 Fitness Equipment Sales	
Year	Sales (billions of units)
1992	2.1
1994	2.8
1996	3.2
1998	3.2
2000	3.6
2002	4.3

Source: *National Sporting Goods Association*

Let s be sales of fitness equipment (in billions of units) in the year that is t years since 1990.
a. Use a graphing calculator to draw a scattergram of the data.
b. Find an equation of a linear model to describe the data.
c. What is the slope? What does it mean in this situation?
d. Predict the sales in 2009.
e. Predict when the sales will be 6 billion units.

49. The number of living Americans who have been diagnosed with cancer is increasing by 0.3 million Americans per year. In 2000, there were 9.6 million living Americans who had been diagnosed with cancer at some time in their lives. Let n be the number of living Americans (in millions) at t years since 2000 who will have been diagnosed with cancer at some time during their lives.
a. Find an equation of a linear model to describe the data.
b. What is the slope? What does it mean in this situation?
c. Predict the number of living Americans in 2010 who will have been diagnosed with cancer at some time during their lives.
d. Predict when 14 million living Americans will have been diagnosed with cancer at some time during their lives.

50. Grades have always been college admissions counselors' top criteria, but the percentage of counselors who consider admission tests as being "considerably important" has increased approximately linearly from 39% in 1992 to 57% in 2002 (Source: *National Association for College Admission Counseling*). Predict when the percentage will be 75%.

Solve the inequality. Describe the solution set as an inequality, in interval notation, and in a graph. Then, use a graphing calculator table to verify your result.

51. $x + 7 > 10$

52. $x - 3 \geq -4$

53. $3w \leq -15$

54. $-\dfrac{4}{3}p < -8$

55. $-4x < 8$

56. $5x - 3 > 3x - 9$

57. $-3(2a + 5) + 5a \geq 2(a - 3)$

58. $\dfrac{2b - 4}{3} \leq \dfrac{3b - 4}{4}$

59. The violent-crime rate in the United States declined during the 1990s, when the economy was performing well. Many experts were surprised when the violent-crime rate continued to decline during the early 2000s, when the economy did poorly (see Table 61).

Table 61 Numbers of Violent Crimes	
Year	Number of Violent Crimes per 1000 People Age 12 or Older
1994	50
1996	42
1998	36
2000	28
2002	23
2003	22

Source: *Bureau of Justice*

Let r be the violent-crime rate (number of violent crimes per 1000 people age 12 or older) at t years since 1990.
a. Find an equation of a linear model to describe the data.
b. What is the slope? What does it mean in this situation?
c. Predict the years when the violent-crime rate will be less than 11 violent crimes per 1000 people age 12 or older.

CHAPTER 5 TEST

Determine the slope and the y-intercept. Use the slope and the y-intercept to graph the equation by hand.

1. $y = -3x - 1$ **2.** $2x - 5y = 10$ **3.** $y - 5 = 0$

4. Find the x-intercept and y-intercept of the graph of the equation $6x - 3y = 18$. Then graph the equation by hand.

5. Find the approximate x-intercept and approximate y-intercept of the graph of $5.93x - 4.81y = 43.79$. Round the coordinates to the second decimal place.

6. Find the x-intercept and y-intercept of the graph of the equation $\dfrac{x}{2} + \dfrac{y}{7} = 1$.

7. A student thinks that the line $5x - 4y = 20$ has slope 5 because the coefficient of the $5x$ term is 5. Is the student correct? If yes, explain why. If no, then find the slope.

Find an equation of the line that has the given slope and contains the given point. If possible, write your equation in slope–intercept form.

8. $m = 7, (-2, -4)$

9. $m = -\dfrac{2}{3}, (6, -1)$

10. m is undefined, $(2, -4)$

11. Find an equation of the line that passes through the points $(-4, 6)$ and $(2, 3)$.

12. Find an approximate equation of the line that passes through the points $(-3.4, 2.9)$ and $(1.8, -7.1)$. Write your equation in slope–intercept form. Round the slope and the constant term to two decimal places.

13. Find an equation of the line sketched in Fig. 48.

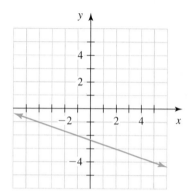

Figure 48 Exercise 13

14. Consider the scattergram of data and the graph of the model $y = mt + b$ in Fig. 49. Sketch a linear model that describes the data better, and then explain how you would adjust the values of m and b of the original model so that it would describe the data better.

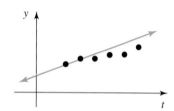

Figure 49 Exercise 14

15. Find an equation of a line that comes close to the points listed in Table 62. Then use a graphing calculator to check that your line comes close to those points.

Table 62 Find an Equation	
x	**y**
1	7
3	11
7	16
12	20
14	23

16. The percentages of employers who offer a pension are shown in Table 63 for various years.

Table 63 Percentages of Employers Who Offer a Pension	
Year	**Percent**
1985	91
1990	85
1995	80
2000	73
2003	68

Source: *Hewitt Associates*

Let p be the percentage of employers who offer a pension at t years since 1980.

a. Use a graphing calculator to draw a scattergram of the data.

b. Find an equation of a linear model to describe the data.

c. What is the slope? What does it mean in this situation?

d. What is the p-intercept? What does it mean in this situation?

e. What is the t-intercept? What does it mean in this situation?

f. When did all companies offer a pension, according to the model?

g. Predict in which years less than half of all employers will offer pensions.

Solve the inequality. Describe the solution set as an inequality, in interval notation, and in a graph.

17. $3(2x + 1) \le 4(x + 2) - 1$

18. $\dfrac{5}{6} - \dfrac{2}{3}x > -\dfrac{1}{2}$

Systems of Linear Equations

This course has really changed my viewpoint on math. I admit that I used to be one of those people who said, "Algebra? What do I possibly need that for in real life?" I will also admit that I thought as soon as I was done with my college math classes, I would never use it again. But this course taught me that math is everywhere. Most of the time it works behind the scenes, but once you know what to look for, and how to apply it, math, and algebra, can tell us a lot about the world and how it works.

—Brendan M., student

Did you know that although men's world record times are better than women's world record times for running and swimming events, women's record times are declining at a greater rate than men's record times? Table 1 shows the world record times for the 200-meter run. In Exercise 39 of Homework 6.1, you will predict when the women's record time for the 200-meter run will be equal to the men's record time.

Table 1 200-Meter Run World Record Times			
Women		**Men**	
Year	Record Time (seconds)	Year	Record Time (seconds)
1973	22.38	1951	20.6
1974	22.21	1963	20.3
1978	22.06	1967	20.14
1984	21.71	1979	19.72
1988	21.34	1996	19.32

Source: *WR Progression*

In Chapters 3–5, we took a close look at linear equations in two variables. In this chapter, we will work with *two* such equations together. This work will enable us to predict when two quantities will be equal, such as the women's record time and the men's record time in the 200-meter run.

6.1 USING GRAPHS AND TABLES TO SOLVE SYSTEMS

Objectives

▷ Know the meaning of *solution* and *solution set* of a *system of linear equations in two variables*.

▷ Use a graphical approach to solve systems of linear equations.

▷ Use graphing to make estimates and predictions about situations that can be modeled by two linear models.

▷ Know the three types of linear systems of two equations.

▷ Find the solution of a system of linear equations from tables of solutions of such equations.

In this section, we will work with graphs and tables that describe solutions of two linear equations in two variables. We will also work with situations that can be modeled by using two linear models.

Systems of Two Linear Equations

A **system of linear equations in two variables,** or a **linear system,** for short, consists of two or more linear equations in two variables. Here is an example of a system of two linear equations in two variables:

$$y = 3x - 1$$
$$y = -2x + 4$$

We will work with such systems throughout this chapter.

Recall from Section 3.1 that every point on the graph of an equation represents a solution of the equation and that every point *not* on the graph represents an ordered pair that is *not* a solution. Knowing the meaning of a graph will help us greatly in this section.

Example 1   Finding Ordered Pairs That Satisfy Both of Two Given Equations

Find all ordered pairs that satisfy both of the equations in the system

$$y = 3x - 1$$
$$y = -2x + 4$$

Solution

To begin, we graph each equation on the same coordinate system (see Fig. 1).

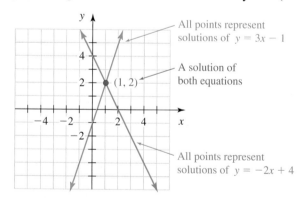

Figure 1 The intersection point is (1, 2)

For an ordered pair to be a solution of *both* equations, it must represent a point that lies on *both* lines. The intersection point (1, 2) is the only point that lies on both lines. So, the ordered pair (1, 2) is the only ordered pair that satisfies both equations.

We can verify that $(1, 2)$ satisfies both equations:

$$y = 3x - 1 \qquad\qquad y = -2x + 4$$
$$2 \stackrel{?}{=} 3(1) - 1 \qquad\qquad 2 \stackrel{?}{=} -2(1) + 4$$
$$2 \stackrel{?}{=} 2 \qquad\qquad\qquad 2 \stackrel{?}{=} 2$$
$$\text{true} \qquad\qquad\qquad\quad \text{true} \qquad\qquad \blacksquare$$

Solution Set of a System

In Example 1, we worked with the system

$$y = 3x - 1$$
$$y = -2x + 4$$

We found that the only point whose coordinates satisfy both equations is the intersection point $(1, 2)$. We call the set containing only $(1, 2)$ the *solution set of the system*.

> **DEFINITION**   *Solution, solution set*, and *solve* for a system
>
> We say that an ordered pair (a, b) is a **solution** of a system of two equations in two variables if it satisfies both equations. The **solution set** of a system is the set of all solutions of the system. We **solve** a system by finding its solution set.

In general, **the solution set of a system of two linear equations can be found by locating any intersection point(s) of the graphs of the two equations.**

Example 2   Solving a System of Two Linear Equations by Graphing

Solve the system

$$y = 2x + 1$$
$$y = \frac{1}{2}x - 2$$

Solution

The graphs of the equations are sketched in Fig. 2.

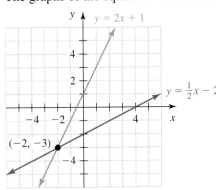

Figure 2 The intersection point is $(-2, -3)$

The intersection point is $(-2, -3)$. So, the solution is the ordered pair $(-2, -3)$.

We can check that $(-2, -3)$ satisfies both equations:

$$y = 2x + 1 \qquad\qquad\qquad y = \frac{1}{2}x - 2$$
$$-3 \stackrel{?}{=} 2(-2) + 1 \qquad -3 \stackrel{?}{=} \frac{1}{2}(-2) - 2$$
$$-3 \stackrel{?}{=} -3 \qquad\qquad -3 \stackrel{?}{=} -1 - 2$$
$$\text{true} \qquad\qquad\qquad -3 \stackrel{?}{=} -3$$
$$\text{true}$$

We can also check our work by using "intersect" on a graphing calculator (see Fig. 3). See Section A.17 for graphing calculator instructions. $\blacksquare$

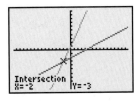

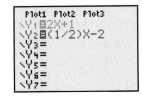

Figure 3 Verify that the intersection point is $(-2, -3)$

WARNING After solving a system of two linear equations, you commit a common error if you check that a result satisfies only one of the two equations. It is important to check that your result satisfies *both* equations.

In Example 2, we solved a system in which both equations are in $y = mx + b$ form. If any equations in a system are not in $y = mx + b$ form, we usually begin by writing them in $y = mx + b$ form.

Example 3 Solving a System of Two Linear Equations by Graphing

Solve the system

$$3x + 2y = -4 \qquad \text{Equation (1)}$$

$$y = \frac{1}{2}x + 2 \qquad \text{Equation (2)}$$

Solution

First, we write equation (1) in slope–intercept form by solving the equation for y:

$$3x + 2y = -4 \qquad \text{Equation (1)}$$

$$2y = -3x - 4 \qquad \text{Subtract 3x from both sides.}$$

$$\frac{2y}{2} = \frac{-3x}{2} - \frac{4}{2} \qquad \text{Divide both sides by 2.}$$

$$y = -\frac{3}{2}x - 2 \qquad \text{Simplify.}$$

Next, we sketch a graph of the equations $y = -\frac{3}{2}x - 2$ and $y = \frac{1}{2}x + 2$ (see Fig. 4).

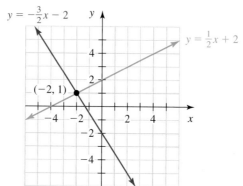

Figure 4 The solution is $(-2, 1)$

The intersection point is $(-2, 1)$. So, the solution is the ordered pair $(-2, 1)$.

Although we could check our work by using "intersect" on a graphing calculator with the equations $y = -\frac{3}{2}x - 2$ and $y = \frac{1}{2}x + 2$, this check would not verify that we solved $3x + 2y = -4$ for y correctly. To check all of our work, we check that $(-2, 1)$ satisfies both of the *original* equations:

$$3x + 2y = -4 \qquad\qquad y = \frac{1}{2}x + 2$$

$$3(-2) + 2(1) \overset{?}{=} -4 \qquad\qquad 1 \overset{?}{=} \frac{1}{2}(-2) + 2$$

$$-6 + 2 \overset{?}{=} -4 \qquad\qquad 1 \overset{?}{=} -1 + 2$$

$$-4 \overset{?}{=} -4 \qquad\qquad 1 \overset{?}{=} 1$$

$$\text{true} \qquad\qquad\qquad \text{true} \qquad\blacksquare$$

Using a Graphing Calculator to Solve a System

So far, we have solved systems with simple equations that can be accurately graphed by hand in a reasonable amount of time. In Example 4, we will use a graphing calculator to help us solve a system with equations that are more difficult to graph.

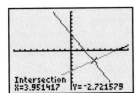

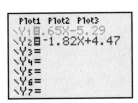

Figure 5 The approximate solution is $(3.95, -2.72)$

Example 4   Using a Graphing Calculator to Solve a System

Use a graphing calculator to find any approximate solutions of the system

$$y = 0.65x - 5.29$$
$$y = -1.82x + 4.47$$

Solution

We use "intersect" to find the approximate solution $(3.95, -2.72)$. See Fig. 5.
  We check that $(3.95, -2.72)$ approximately satisfies both of the original equations:

$$y = 0.65x - 5.29 \qquad\qquad y = -1.82x + 4.47$$
$$-2.72 = 0.65(3.95) - 5.29 \qquad -2.72 = -1.82(3.95) + 4.47$$
$$-2.72 \approx -2.7225 \qquad\qquad -2.72 \approx -2.719$$ ∎

Using Two Models to Make a Prediction

We can use graphing to make estimates and predictions about some authentic situations that can be modeled by two linear equations.

Example 5   Making a Prediction

To control costs during the past three decades, colleges have steadily increased the number of instructors that work part time (see Table 2).

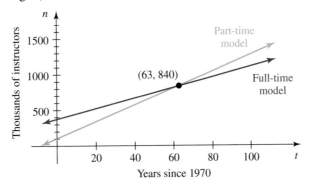

Table 2 Numbers of Part-Time and Full-Time Instructors		
	Number of Instructors (thousands)	
Year	**Part Time**	**Full Time**
1970	104	369
1975	188	440
1980	236	450
1985	256	459
1991	291	536
1995	381	551
1999	437	591
2003	543	632

Source: *U.S. National Center for Education Statistics*

Let n be the number of college instructors (in thousands) at t years since 1970. Reasonable models for part-time instructors and full-time instructors are

$$n = 11.75t + 101.77 \qquad \text{Part-time instructors}$$
$$n = 7.38t + 376.15 \qquad \text{Full-time instructors}$$

Use graphing to predict when the number of part-time instructors will equal the number of full-time instructors. What is that number of instructors?

Solution

We begin by sketching graphs of the two models on the same coordinate system (see Fig. 6).

Figure 6 Part-time and full-time models

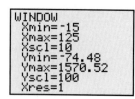

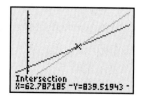

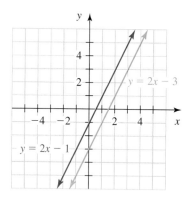

Figure 7 Verify the work

Note that the intersection point is about (63, 840). So, the models predict that there will be the same number of part-time instructors as full-time instructors (840 thousand of each) in $1970 + 63 = 2033$. We are not very confident about this prediction, however, because it is so far into the future.

We verify our work by using "intersect" on a graphing calculator (see Fig. 7). See Sections A.7 and A.17 for graphing calculator instructions. ∎

Intersection Point of the Graphs of Two Models

If the independent variable of two models represents time, then an intersection point of the graphs of the two models indicates when the quantities represented by the dependent variables were or will be equal.

Three Types of Linear Systems of Two Equations

Each of the systems in Examples 1–5 has one solution. Some systems do not have exactly one solution, however.

Example 6   A System Whose Solution Is the Empty Set

Solve the system

$$y = 2x - 1$$
$$y = 2x - 3$$

Solution

Since the distinct lines have equal slopes, the lines are parallel (see Fig. 8).

Parallel lines do not intersect, so there is no ordered pair that satisfies both equations. The solution set is the empty set. ∎

A linear system whose solution is the empty set is called an **inconsistent system.**

Example 7   A System That Has an Infinite Number of Solutions

Solve the system

$$y = 3x - 5 \qquad \text{Equation (1)}$$
$$6x - 2y = 10 \qquad \text{Equation (2)}$$

Solution

We write equation (2) in slope–intercept form:

$$6x - 2y = 10 \qquad \text{Equation (2)}$$
$$-2y = -6x + 10 \qquad \text{Subtract 6x from both sides.}$$
$$y = 3x - 5 \qquad \text{Divide both sides by } -2.$$

So, the graph of $6x - 2y = 10$ and the graph of $y = 3x - 5$ are the same line. The solution set of the system is the set of the infinite number of ordered pairs represented by points on the line $y = 3x - 5$ and the (same) line $6x - 2y = 10$. ∎

A linear system that has an infinite number of solutions is called a **dependent system.**

In Examples 1, 6, and 7, we have seen three types of linear systems. We now describe them.

Figure 8 The solution set is the empty set

Types of Linear Systems

There are three types of linear systems of two equations:

1. **One-solution system:** The lines intersect in one point. The solution of the system is the ordered pair that corresponds to that point. See Fig. 9.

2. **Inconsistent system:** The lines are parallel. The solution set of the system is the empty set. See Fig. 10.

3. **Dependent system:** The lines are identical. The solution set of the system is the set of the infinite number of solutions represented by points on the same line. See Fig. 11.

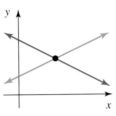

Figure 9 Graphs of the equations in a system with one solution

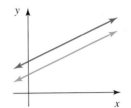

Figure 10 Graphs of the equations in an inconsistent system

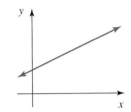

Figure 11 Graph of the equations in a dependent system

Finding the Solution of a System from Tables of Solutions

By the Rule of Four, we know that we can use tables to describe some solutions of a linear equation in two variables. In Example 8, we will use such tables to help us find the solution of a system.

Example 8   Using Tables to Solve a System

Some solutions of two linear equations are shown in Tables 3 and 4. Find the solution of the system of the two equations.

Table 3 Solutions of One Equation	
x	**y**
0	3
1	7
2	11
3	15
4	19

Table 4 Solutions of the Other Equation	
x	**y**
0	21
1	19
2	17
3	15
4	13

Solution

Since the ordered pair (3, 15) is listed in both tables, we know that it is a solution of both equations. So, it is a solution of the system of equations.

The graphs of the two equations are two lines with slopes 4 and −2. Since the lines have different slopes, there is only one intersection point. So, the ordered pair (3, 15) is the *only* solution of the system. ∎

In Example 8, we used tables of solutions of two linear equations to help us find the solution of the linear system. **If an ordered pair is listed in both tables of solutions of two linear equations, then that ordered pair is a solution of that system.**

group exploration

Three types of systems

1. Solve the system

$$y = 2x + 1$$
$$y = -3x + 6$$

2. **a.** Graph by hand the equations in the system

$$y = 2x + 1$$
$$y = 2x + 3$$

   b. Are there any intersection points of the two lines? Explain.
   c. Are there any solutions of the system? If yes, find them. If no, explain why not.

3. **a.** Graph by hand the equations in the system

$$y = 2x + 1$$
$$y = 2x + 1$$

   b. Are there any intersection points of the two lines? If yes, describe them. If no, explain why not.
   c. Are there any solutions of the system? If yes, describe them. If no, explain why not.

4. Problems 1–3 suggest that there are three types of systems of two linear equations. Describe these three types.

5. Solve the system

$$y = -2x + 3$$
$$4x + 2y = 10$$

6. Solve the system

$$y = 3x - 1$$
$$9x - 3y = 3$$

TIPS FOR SUCCESS: Take a Break

Have you ever had trouble solving a problem but returned to the problem hours later and found it easy to solve? By taking a break, you can return to the exercise with a different perspective and renewed energy. You've also given your unconscious mind a chance to reflect on the problem while you take your break. You can strategically take advantage of this phenomenon by allocating time to complete your homework assignment at two different points in your day.

HOMEWORK 6.1

FOR EXTRA HELP ▶

 Student Solutions Manual   PH Math/Tutor Center   Math XL MathXL®   MyMathLab MyMathLab

*Determine which of the given ordered pairs is a solution of the given system. [**Hint:** There is no need to graph.]*

1. $(2, 3)$, $(1, -1)$, $(-4, 6)$
   $y = 4x - 5$
   $y = -2x + 1$

2. $(-3, 5)$, $(-7, 3)$, $(3, -7)$
   $y = -x - 4$
   $y = -3x + 2$

3. $(-1, 8)$, $(3, -2)$, $(7, 1)$
   $5x + 2y = 11$
   $3x - 4y = 17$

4. $(3, -1)$, $(-4, -2)$, $(-2, -5)$
   $4x - 5y = 17$
   $3x + 2y = -16$

Find the solution of the system by graphing the equations by hand. Use "intersect" on a graphing calculator to verify your work.

5. $y = 3x - 5$
   $y = -2x + 5$

6. $y = 4x - 2$
   $y = -2x + 4$

7. $y = \dfrac{1}{2}x + 2$
   $y = -\dfrac{3}{2}x - 2$

8. $y = -\dfrac{2}{3}x + 4$
   $y = \dfrac{5}{3}x - 3$

Find the solution of the system by graphing the equations by hand. Verify any solution by checking that it satisfies both equations.

9. $y = -3x$
$y = 4x$

10. $y = x$
$y = -x$

11. $3x + y = 2$
$2x - y = 8$

12. $4x - y = -9$
$2x + y = -3$

13. $4y = -3x$
$x - 2y = 10$

14. $3y = 2x$
$5x - 3y = -9$

15. $4x + 3y = 6$
$2x - 3y = 12$

16. $x + 2y = -4$
$-3x + 2y = 4$

17. $6y + 15x = 12$
$2y - x = -8$

18. $6y + 10x = -18$
$3y - x = 9$

19. $x = 3$
$y = -2$

20. $x = 0$
$y = 0$

21. $x = \frac{1}{3}y$
$y = -2x + 5$

22. $y = x - 2$
$x = \frac{1}{2}y$

Use "intersect" on a graphing calculator to solve the system, with the coordinates of the solution rounded to the second decimal place. Check that your result approximately satisfies both equations.

23. $y = 2.18x - 5.34$
$y = -3.53x + 1.29$

24. $y = 4.95x + 7.51$
$y = -0.84x - 5.38$

25. $y = \frac{5}{4}x + 2$
$y = -\frac{1}{4}x - 5$

26. $y = -\frac{7}{3}x + 3$
$y = -\frac{2}{3}x - 2$

27. $y = \frac{3}{4}x - 8$
$5x + 3y = 6$

28. $y = -\frac{2}{3}x - 5$
$7y - 3x = 7$

29. $-2x + 5y = 15$
$6x + 14y = -14$

30. $12x + 9y = 18$
$-x - 4y = 20$

Find the solution of the system by graphing the equations by hand. If the system is inconsistent or dependent, say so.

31. $y = -2x + 4$
$6x + 3y = 12$

32. $y = \frac{3}{5}x + 1$
$3x - 5y = -5$

33. $y = -2x - 1$
$3x - y = 6$

34. $y = 3x + 2$
$3x - 2y = 2$

35. $y = 3x - 4$
$6x - 2y = 6$

36. $y = \frac{2}{3}x - 2$
$2x - 3y = 9$

37. The numbers of women and men who earned a bachelor's degree are listed in Table 5 for various years. Let n be the number of people (in thousands) who earned a bachelor's degree in the year that is t years since 1980. Reasonable models for women and men are

$$n = 13.28t + 440.09 \quad \text{Women}$$
$$n = 3.42t + 468.14 \quad \text{Men}$$

Use "intersect" on a graphing calculator with the window shown in Fig. 12 to estimate when the number of women who earned a bachelor's degree was equal to the number of men who earned a bachelor's degree. What is that number of people?

Table 5 Women and Men Who Earned a Bachelor's Degree

| | Number of People Who Earned a Bachelor's Degree (thousands) | |
Year	Women	Men
1980	456	474
1985	497	483
1990	560	492
1995	634	526
2000	708	530
2002	742	550

Source: *U.S. National Center for Education Statistics*

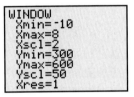

Figure 12 Window for bachelor's-degree models

38. The numbers of morning daily newspapers and evening daily newspapers are shown in Table 6 for various years.

Table 6 Numbers of Morning Dailies and Evening Dailies

| | Number of Daily Newspapers | |
Year	Morning	Evening
1980	387	1388
1985	482	1220
1990	559	1084
1995	656	891
2000	766	727
2004	813	653

Source: *Editor & Publisher Co.*

Let n be the number of daily newspapers at t years since 1980. Reasonable models for morning newspapers and evening newspapers are

$$n = 18.13t + 386.86 \quad \text{Morning newspapers}$$
$$n = -31.52t + 1382.53 \quad \text{Evening newspapers}$$

Use "intersect" on a graphing calculator with the window shown in Fig. 13 to estimate when the number of morning daily newspapers was equal to the number of evening daily newspapers. What is that number of newspapers?

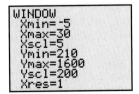

Figure 13 Window for newspaper models

39. World record times for the 200-meter run are listed in Table 7 for various years.

Table 7 200-Meter-Run World Record Times			
Women		**Men**	
Year	**Record Time (seconds)**	**Year**	**Record Time (seconds)**
1973	22.38	1951	20.6
1974	22.21	1963	20.3
1978	22.06	1967	20.14
1984	21.71	1979	19.72
1988	21.34	1996	19.32

Source: *WR Progression*

a. Let r be the women's record time (in seconds) at t years since 1900. Find an equation of a model to describe the data.

b. Let r be the men's record time (in seconds) at t years since 1900. Find an equation of a model to describe the data.

c. Use "intersect" on a graphing calculator with the window shown in Fig. 14 to predict when the women's record time will equal the men's record time. What is that record time?

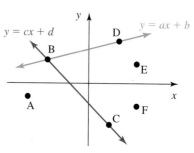

Figure 14 Window for 200-meter-run models

40. The winning times for the Olympic 500-meter race in speed skating have generally been decreasing over the past three decades (see Table 8).

Table 8 Olympic 500-Meter Speed-Skating Winning Times		
	Winning Time (seconds)	
Year	**Women**	**Men**
1972	43.33	39.44
1976	42.76	39.17
1980	41.78	38.03
1984	41.02	38.19
1988	39.10	36.45
1992	40.33	37.14
1994	39.25	36.33
1998	38.21	35.59
2002	37.375	34.615

Source: The Universal Almanac

a. Let w be the women's winning time (in seconds) at t years since 1970. Find an equation of a model to describe the data.

b. Let w be the men's winning time (in seconds) at t years since 1970. Find an equation of a model to describe the data.

c. Use "intersect" on a graphing calculator with the window shown in Fig. 15 to predict when the women's winning time will be equal to the men's winning time. What is that winning time?

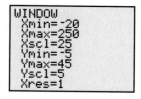

Figure 15 Window for speed-skating models

41. The graphs of $y = ax + b$ and $y = cx + d$ are sketched in Fig. 16.

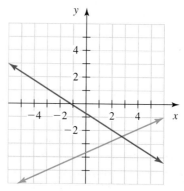

Figure 16 Exercise 41

Which of the points A, B, C, D, E, and F represent an ordered pair that

a. satisfies the equation $y = ax + b$?

b. satisfies the equation $y = cx + d$?

c. satisfies both equations?

d. does not satisfy either equation?

42. Consider the system

$$y = 4x - 5$$
$$y = -3x + 2$$

Find an ordered pair that

a. satisfies $y = 4x - 5$, but does not satisfy $y = -3x + 2$.

b. satisfies $y = -3x + 2$, but does not satisfy $y = 4x - 5$.

c. satisfies both equations.

d. does not satisfy either equation.

43. Figure 17 shows the graphs of two linear equations. Estimate, to the first decimal place, the coordinates of the solution of the system.

Figure 17 Exercise 43

44. Figure 18 shows the graphs of two linear equations. Estimate, to the first decimal place, the coordinates of the solution of the system.

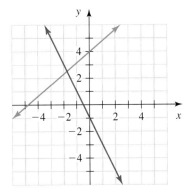

Figure 18 Exercise 44

45. Figure 19 shows the graphs of two linear equations. Estimate to the nearest integer, the coordinates of the solution of the system. Explain. [**Hint:** Use the slope of each line.]

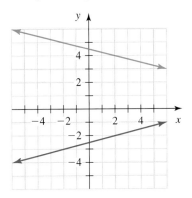

Figure 19 Exercise 45

46. Figure 20 shows the graphs of two linear equations. Estimate, to the nearest integer, the coordinates of the solution of the system. Explain. [**Hint:** Use the slope of each line.]

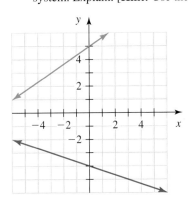

Figure 20 Exercise 46

47. Some solutions of two linear equations are shown in Tables 9 and 10. Find the solution of the system of the two equations.

Table 9 Solutions of One Equation		Table 10 Solutions of the Other Equation	
x	y	x	y
0	3	0	21
1	5	1	17
2	7	2	13
3	9	3	9
4	11	4	5

48. Some solutions of two linear equations are shown in Tables 11 and 12. Find the solution of the system of the two equations.

Table 11 Solutions of One Equation		Table 12 Solutions of the Other Equation	
x	y	x	y
0	5	0	21
1	8	1	16
2	11	2	11
3	14	3	6
4	17	4	1

49. Create a system of two linear equations whose solution is $(1, 3)$. Verify your system graphically.

50. Create a system of two linear equations whose solution is $(-4, 2)$. Verify your system graphically.

51. A student tries to solve the system

$$y = -x + 5$$
$$y = 2x - 4$$

After graphing the equations, the student believes that the solution is $(4, 1)$. The student then checks whether $(4, 1)$ satisfies $y = -x + 5$:

$$y = -x + 5$$
$$1 \overset{?}{=} -(4) + 5$$
$$1 \overset{?}{=} 1$$
$$\text{true}$$

The student concludes that $(4, 1)$ is the solution of the system. Is the student correct? Explain.

52. Explain why the solution(s) of a system of two linear equations can be found by locating intersection point(s) of the graphs of the two equations. (See page 9 for guidelines on writing a good response.)

Related Review

53. a. Determine whether the lines $y = 3x - 1$ and $y = 3x + 2$ are parallel.
 b. Do parallel lines that are not the same line have an intersection point?
 c. Solve the system

$$y = 3x - 1$$
$$y = 3x + 2$$

54. a. Determine whether the lines $3x - 7y = 14$ and $y = \frac{3}{7}x - 2$ are parallel.
 b. Do parallel lines that are not the same line have an intersection point? Explain.
 c. Solve the system

$$3x - 7y = 14$$
$$y = \frac{3}{7}x - 2$$

55. Create a system of two linear equations for which the slopes of both lines are negative and the solution of the system corresponds to a point in Quadrant IV.

56. Create a system of two linear equations for which the slopes of both lines are positive and the solution of the system corresponds to a point in Quadrant II.

Expressions, Equations, and Graphs

Perform the indicated instruction. Then use words such as linear, one variable, and two variables to describe the expression, equation, or system.

57. Solve the following:

$$y = 2x + 5$$
$$y = -3x - 5$$

58. Simplify $(2x + 5) - (-3x - 5)$.

59. Solve $2x + 5 = -3x - 5$.

60. Graph $y = -3x - 5$ by hand.

6.2 USING SUBSTITUTION TO SOLVE SYSTEMS

Objectives

▸ Use substitution to solve a system of two linear equations.

▸ Isolate a variable in an equation to help solve a system by substitution.

▸ Solve inconsistent and dependent systems by substitution.

In Section 6.1, we used graphs to solve systems. In this section, we use *equations* to solve systems.

The Substitution Method for Solving Systems

In Example 1, we discuss how to solve a system by using a technique called *substitution*.

Example 1   Making a Substitution for *y*

Solve the system

$$y = x + 3 \qquad \text{Equation (1)}$$
$$2x + 3y = 14 \qquad \text{Equation (2)}$$

Solution

From equation (1), we know that for any solution of the system, the value of y is equal to the value of $x + 3$. So, we substitute $x + 3$ for y in equation (2):

$$2x + 3y = 14 \qquad \text{Equation (2)}$$
$$2x + 3(x + 3) = 14 \qquad \text{Substitute } x + 3 \text{ for } y.$$

Note that, by making this substitution, we now have an equation in *one* variable. Next, we solve that equation for x:

$$2x + 3(x + 3) = 14$$
$$2x + 3x + 9 = 14 \qquad \text{Distributive law}$$
$$5x + 9 = 14 \qquad \text{Combine like terms.}$$
$$5x = 5 \qquad \text{Subtract 9 from both sides.}$$
$$x = 1 \qquad \text{Divide both sides by 5.}$$

Thus the x-coordinate of the solution is 1. To find the y-coordinate, we substitute 1 for x in either of the original equations and solve for y:

$$y = x + 3 \qquad \text{Equation (1)}$$
$$y = 1 + 3 \qquad \text{Substitute 1 for } x.$$
$$y = 4 \qquad \text{Add.}$$

So, the solution is $(1, 4)$. We can check that $(1, 4)$ satisfies both of the system's equations:

$$y = x + 3 \qquad\qquad 2x + 3y = 14$$
$$4 \stackrel{?}{=} 1 + 3 \qquad\qquad 2(1) + 3(4) \stackrel{?}{=} 14$$
$$4 \stackrel{?}{=} 4 \qquad\qquad 2 + 12 \stackrel{?}{=} 14$$
$$\text{true} \qquad\qquad\qquad \text{true}$$

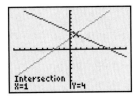

Figure 21 Verify that the solution is $(1, 4)$

Or we can verify that $(1, 4)$ is the solution by graphing equations (1) and (2) and checking that $(1, 4)$ is the intersection point of the two lines (see Fig. 21). To do so on a graphing calculator, we must first solve equation (2) for y:

$$2x + 3y = 14 \qquad \text{Equation (2)}$$
$$3y = -2x + 14 \qquad \text{Subtract 2x from both sides.}$$
$$\frac{3y}{3} = \frac{-2x}{3} + \frac{14}{3} \qquad \text{Divide both sides by 3.}$$
$$y = -\frac{2}{3}x + \frac{14}{3} \qquad \text{Simplify.} \qquad ■$$

Using Substitution to Solve a Linear System

To use **substitution** to solve a system of two linear equations,

1. Isolate a variable to one side of either equation.
2. Substitute the expression for the variable found in step 1 into the other equation.
3. Solve the equation in one variable found in step 2.
4. Substitute the solution found in step 3 into one of the original equations, and solve for the other variable.

A system of equations can be solved by substitution as well as by graphing. The methods give the same result.

In Example 1, we solved a system by making a substitution for y. In Example 2, we will discuss how to solve a system by making a substitution for x.

Example 2   Making a Substitution for x

Solve the system

$$x = 2y - 8 \qquad \text{Equation (1)}$$
$$3x + 4y = 6 \qquad \text{Equation (2)}$$

Solution

Since x is alone on one side of the equation $x = 2y - 8$, we begin by substituting $2y - 8$ for x in equation (2):

$$3x + 4y = 6 \qquad \text{Equation (2)}$$
$$3(2y - 8) + 4y = 6 \qquad \text{Substitute 2y − 8 for x.}$$
$$6y - 24 + 4y = 6 \qquad \text{Distributive law}$$
$$10y - 24 = 6 \qquad \text{Combine like terms.}$$
$$10y = 30 \qquad \text{Add 24 to both sides.}$$
$$y = 3 \qquad \text{Divide both sides by 10.}$$

The y-coordinate of the solution is 3. To find the x-coordinate, we substitute 3 for y in equation (1) and solve for x:

$$x = 2y - 8 \qquad \text{Equation (1)}$$
$$x = 2(3) - 8 \qquad \text{Substitute 3 for y.}$$
$$x = -2 \qquad \text{Simplify.}$$

The solution is $(-2, 3)$. We could then check that $(-2, 3)$ satisfies *both* of the original equations. ■

In Example 1, we made a substitution for y because y was alone on one side of one of the given equations. In Example 2, we made a substitution for x because x was alone on one side of one of the given equations.

In Example 3, we will solve a system of equations that have decimal coefficients. This will help prepare us for Section 6.4, where we will use systems to model situations.

Example 3 Solving a System of Equations

Solve the system

$$y = 1.72x - 4.38 \qquad \text{Equation (1)}$$
$$y = -0.53x + 6.94 \qquad \text{Equation (2)}$$

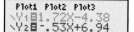

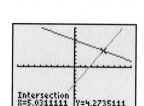

Figure 22 Verify that the approximate solution is (5.03, 4.27)

Solution

We start by substituting $1.72x - 4.38$ for y in equation (2):

$$y = -0.53x + 6.94 \qquad \text{Equation (2)}$$
$$1.72x - 4.38 = -0.53x + 6.94 \qquad \text{Substitute } 1.72x - 4.38 \text{ for } y.$$
$$2.25x - 4.38 = 6.94 \qquad \text{Add } 0.53x \text{ to both sides.}$$
$$2.25x = 11.32 \qquad \text{Add } 4.38 \text{ to both sides.}$$
$$x \approx 5.03 \qquad \text{Divide both sides by } 2.25.$$

Next, we substitute 5.03 for x in equation (1):

$$y = 1.72x - 4.38 \qquad \text{Equation (1)}$$
$$y = 1.72(5.03) - 4.38 \qquad \text{Substitute 5.03 for } x.$$
$$y \approx 4.27 \qquad \text{Simplify.}$$

So, the approximate solution is (5.03, 4.27). We use "intersect" on a graphing calculator to check our work (see Fig. 22). ■

Isolating a Variable

If no variable is alone on one side of either equation in a system of two linear equations, then we must first solve for a variable in one of the equations before we can make a substitution.

Example 4 Isolating a Variable and Then Using Substitution

Solve the system

$$3x + y = -7 \qquad \text{Equation (1)}$$
$$2x - 5y = 1 \qquad \text{Equation (2)}$$

Solution

We begin by solving for one of the variables in one of the equations. We can avoid fractions by choosing to solve equation (1) for y:

$$3x + y = -7 \qquad \text{Equation (1)}$$
$$y = -3x - 7 \qquad \text{Subtract } 3x \text{ from both sides.}$$

Next, we substitute $-3x - 7$ for y in equation (2) and solve for x:

$$2x - 5y = 1 \qquad \text{Equation (2)}$$
$$2x - 5(-3x - 7) = 1 \qquad \text{Substitute } -3x - 7 \text{ for } y.$$
$$2x + 15x + 35 = 1 \qquad \text{Distributive law}$$
$$17x + 35 = 1 \qquad \text{Combine like terms.}$$
$$17x = -34 \qquad \text{Subtract 35 from both sides.}$$
$$x = -2 \qquad \text{Divide both sides by 17.}$$

Finally, we substitute -2 for x in the equation $y = -3x - 7$ and solve for y:

$$y = -3(-2) - 7 \qquad \text{Substitute } -2 \text{ for } x.$$
$$y = 6 - 7 \qquad \text{Multiply.}$$
$$y = -1 \qquad \text{Subtract.}$$

The solution is $(-2, -1)$. We could verify our work by checking that $(-2, -1)$ satisfies both of the original equations. ■

Solving Inconsistent and Dependent Systems by Substitution

Each system in Examples 1–4 has one solution. What happens when we solve an inconsistent system (empty solution set) or a dependent system (infinitely many solutions) by substitution?

Example 5   Applying Substitution to an Inconsistent System

Consider the linear system

$$y = 3x + 2 \qquad \text{Equation (1)}$$
$$y = 3x + 4 \qquad \text{Equation (2)}$$

The graphs of the equations are parallel lines (why?), so the system is inconsistent and the solution set is the empty set. What happens when we solve this system by substitution?

Solution

We substitute $3x + 2$ for y in equation (2) and solve for x:

$$y = 3x + 4 \qquad \text{Equation (2)}$$
$$3x + 2 = 3x + 4 \qquad \text{Substitute } 3x + 2 \text{ for } y.$$
$$2 = 4 \qquad \text{Subtract } 3x \text{ from both sides.}$$
$$\text{false}$$

We get the *false* statement $2 = 4$.   ∎

Inconsistent System of Equations

If the result of applying substitution to a system of equations is a false statement, then the system is inconsistent; that is, the solution set is the empty set.

Example 6   Applying Substitution to a Dependent System

In Example 7 of Section 6.1, we found that the system

$$y = 3x - 5 \qquad \text{Equation (1)}$$
$$6x - 2y = 10 \qquad \text{Equation (2)}$$

is dependent and that the solution set is the infinite set of solutions of the equation $y = 3x - 5$. What happens when we solve this system by substitution?

Solution

We substitute $3x - 5$ for y in equation (2) and solve for x:

$$6x - 2y = 10 \qquad \text{Equation (2)}$$
$$6x - 2(3x - 5) = 10 \qquad \text{Substitute } 3x - 5 \text{ for } y.$$
$$6x - 6x + 10 = 10 \qquad \text{Distributive law}$$
$$10 = 10 \qquad \text{Combine like terms.}$$
$$\text{true}$$

We get the *true* statement $10 = 10$.   ∎

Dependent System of Two Linear Equations

If the result of applying substitution to a linear system of two equations is a true statement that can be put into the form $a = a$, then the system is dependent; that is, the solution is the set of ordered pairs represented by every point on the (same) line.

---

group exploration

Comparing techniques for solving systems

1. Consider the system

$$y = -2x + 3$$
$$y = 4x + 3$$

a. Solve the system by graphing the equations by hand.
b. Now solve the system by substitution.
c. Compare your results from parts (a) and (b).
d. Decide which method you prefer for this system. Explain.

2. Consider the system

$$3x - 2y = -4$$
$$y = -2x + 9$$

a. Solve the system by graphing the equations by hand.
b. Now solve the system by substitution.
c. Compare your results from parts (a) and (b).
d. Decide which method you prefer for this system. Explain.

3. In general, are there any systems that you prefer to solve by graphing by hand? If yes, describe them and give an example. If no, explain why not.

4. In general, are there any systems that you prefer to solve by substitution? If yes, describe them and give an example. If no, explain why not.

TIPS FOR SUCCESS: Write a Summary

Consider writing, after each class meeting, a summary of what you have learned. Your summaries will increase your understanding, as well as your memory, of concepts and procedures and will also serve as a good reference for quizzes and exams.

HOMEWORK 6.2

FOR EXTRA HELP ▶

Student Solutions Manual   PH Math/Tutor Center   MathXP MathXL®   MyMathLab MyMathLab

Solve the system by substitution. Verify your solution by checking that it satisfies both equations of the system.

1. $y = 2x$
      $3x + y = 10$

2. $y = 4x$
      $2x + y = 18$

3. $x - 4y = -3$
      $x = 2y - 1$

4. $x + 2y = 4$
      $x = 3y - 1$

5. $2x + 3y = 5$
      $y = x + 5$

6. $4x - 3y = 13$
      $y = x - 4$

7. $-5x - 2y = 17$
      $x = 4y + 1$

8. $3x + 2y = -9$
      $x = 1 - 2y$

9. $2x - 5y - 3 = 0$
      $y = 2x - 7$

10. $4x + 3y + 2 = 0$
      $y = 1 - 3x$

11. $x = 2y + 6$
   $-4x + 5y + 12 = 0$

12. $x = -2y + 5$
   $7x + 2y + 13 = 0$

Solve the system by substitution. Verify your solution by using "intersect" on a graphing calculator.

13. $y = -2x - 1$
      $y = 3x + 9$

14. $y = 4x + 2$
      $y = -3x - 5$

15. $y = 2x$
      $y = 3x$

16. $y = x$
      $y = -x$

17. $y = 2(x - 4)$
      $y = -3(x + 1)$

18. $y = 5(x - 1)$
      $y = 2(x + 2)$

Use substitution to solve the system, with coordinates of solutions rounded to the second decimal place. Verify your work by using "intersect" on a graphing calculator.

19. $y = 2.57x + 7.09$
      $y = -3.61x - 5.72$

20. $y = -1.45x - 6.18$
      $y = 2.63x - 2.73$

21. $y = -3.17x + 8.92$
      $y = 1.65x - 7.24$

22. $y = -0.51x - 2.64$
      $y = -2.79x + 5.94$

23. $y = -1.82x + 3.95$
      $y = 1.57x + 4.68$

24. $y = -2.49x - 6.17$
      $y = 4.81x + 3.45$

Solve the system by substitution. Verify your solution by checking that it satisfies both equations of the system.

25. $2x + y = -9$
      $5x - 3y = 5$

26. $-3x + y = -14$
      $2x - 7y = 22$

27. $4x - 7y = 15$
      $x - 3y = 5$

28. $2x + 3y = -1$
      $x + 3y = -5$

29. $4x + 3y = 5$
$x - 2y = -7$

30. $6x - 5y = -8$
$x + 3y = 14$

31. $2x - y = 1$
$5x - 3y = 5$

32. $4x - 3y = -7$
$3x - y = -9$

33. $3x + 2y = -3$
$2x = y + 5$

34. $4x - 5y = -2$
$3x = y - 7$

Solve the system by substitution. If the system is inconsistent or dependent, say so.

35. $x = 4 - 3y$
$2x + 6y = 8$

36. $x = 2y - 1$
$4x - 8y = -4$

37. $5x - 2y = 18$
$y = -3x + 2$

38. $-3x - 2y = -10$
$y = 4 - 2x$

39. $y = -5x + 3$
$15x + 3y = 6$

40. $y = 2x - 5$
$-4x + 2y = -6$

41. $-4x + 12y = 4$
$x = 3y - 1$

42. $3x + 12y = 6$
$x = -4y + 2$

43. $y = 3x + 2$
$12x - 4y = 9$

44. $y = 4x - 1$
$-8x + 2y = -5$

45. Some solutions of two linear equations are shown in Tables 13 and 14. Find the solution of the system of two equations. [**Hint:** You could begin by finding two equations.]

Table 13 Solutions of One Equation	
x	**y**
0	5
1	8
2	11
3	14
4	17

Table 14 Solutions of the Other Equation	
x	**y**
0	90
1	88
2	86
3	84
4	82

46. Some solutions of two linear equations are shown in Tables 15 and 16. Find the solution of the system of two equations. [**Hint:** You could begin by finding two equations.]

Table 15 Solutions of One Equation	
x	**y**
-2	73
-1	67
0	61
1	55
2	49

Table 16 Solutions of the Other Equation	
x	**y**
-2	-17
-1	-13
0	-9
1	-5
2	-1

47. Find the coordinates of the points A, B, C, and D as shown in Fig. 23. The equations of lines l_1 and l_2 are provided, but no

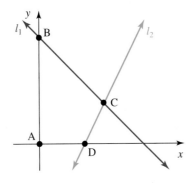

$l_1: y = -x + 8$
$l_2: y = 2x - 7$

Figure 23 Exercise 47

attempt has been made to sketch the lines accurately except for showing the intersection points. Use TRACE and "intersect" on a graphing calculator to verify your work.

48. Find the coordinates of the points A, B, C, and D as shown in Fig. 24. The equations of lines l_1 and l_2 are provided, but no attempt has been made to sketch the lines accurately except for showing the intersection points. Use TRACE and "intersect" on a graphing calculator to verify your work.

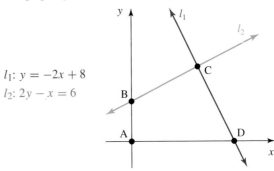

$l_1: y = -2x + 8$
$l_2: 2y - x = 6$

Figure 24 Exercise 48

49. Consider the system

$$x + y = 3$$
$$3x + 2y = 7$$

a. Isolate x to one side of one of the equations, and then solve the system by substitution.

b. Isolate y to one side of one of the equations, and then solve the system by substitution.

c. Compare your results in parts (a) and (b).

50. Consider the system

$$2x + y = 8$$
$$x + 3y = 9$$

a. Isolate x to one side of one of the equations, and then solve the system by substitution.

b. Isolate y to one side of one of the equations, and then solve the system by substitution.

c. Compare your results in parts (a) and (b).

51. Describe how to solve a system by substitution.

52. Is it more reliable to use substitution or graphing by hand to find exact solutions of a system? Explain.

Related Review

53. a. Use "intersect" on a graphing calculator to solve the system

$$y = 2x - 3$$
$$y = -3x + 7$$

b. Solve the same system by substitution.

c. Solve the equation $2x - 3 = -3x + 7$.

d. Compare your result in part (c) with the x-coordinate of the result in part (b).

e. Use "intersect" to help you solve $x - 1 = -2x + 8$. Then solve this equation without using graphing. Compare your results.

f. Summarize the main concepts addressed in this exercise. Explain how you can use "intersect" to verify work.

54. a. Find an equation of the line that has slope 3 and contains the point $(3, 4)$.

   b. Find an equation of the line that has slope -2 and contains the point $(3, 4)$.

   c. Solve the system of the two linear equations you found in parts (a) and (b).

   d. Explain how your work in part (c) is related to your work in parts (a) and (b).

Expressions, Equations, and Graphs

Perform the indicated instruction. Then use words such as lin-
ear, one variable, and two variables to describe the expression,
equation, or system.

55. Graph $y = -3x + 5$ by hand.

56. Solve $7 = -3x + 5$.

57. Solve the following:

$$9x + 2y = 1$$
$$y = -3x + 5$$

58. Simplify $9x + 2y - 3x + 5$.

6.3 USING ELIMINATION TO SOLVE SYSTEMS

Judge a man by his questions rather than his answers.

—*Voltaire*

Objectives

▹ Use elimination to solve a system of two linear equations.

▹ Compare three ways to solve a system.

So far, we have solved systems by graphing and by substitution. In this section, we will use a third technique, called *elimination* or the *addition method.*

The Elimination Method for Solving Systems

To solve systems by elimination, we will need to use the following property:

Adding Left Sides and Right Sides of Two Equations
If $a = b$ and $c = d$, then $$a + c = b + d$$ In words, the sum of the left sides of two equations is equal to the sum of the right sides.

For example, if we add the left sides and add the right sides of the equations $3 = 3$ and $5 = 5$, we obtain the true statement $3 + 5 = 3 + 5$.

Example 1   Solving a System by Elimination

Solve the system

$$3x + 2y = 9 \qquad \text{Equation (1)}$$
$$-3x + 5y = 12 \qquad \text{Equation (2)}$$

Solution

We begin by adding the left sides and adding the right sides of the two equations:

$$3x + 2y = 9 \qquad \text{Equation (1)}$$
$$\underline{-3x + 5y = 12} \qquad \text{Equation (2)}$$
$$0 + 7y = 21 \qquad \text{Add left sides and add right sides; combine like terms.}$$

By "eliminating" the variable x, we now have an equation in *one* variable. Next, we solve that equation for y:

$$0 + 7y = 21$$
$$7y = 21 \qquad 0 + a = a$$
$$y = 3 \qquad \text{Divide both sides by 7.}$$

Then, we substitute 3 for y in either of the original equations and solve for x:

$$3x + 2y = 9 \qquad \text{Equation (1)}$$
$$3x + 2(3) = 9 \qquad \text{Substitute 3 for } y.$$
$$3x + 6 = 9 \qquad \text{Multiply.}$$
$$3x = 3 \qquad \text{Subtract 6 from both sides.}$$
$$x = 1 \qquad \text{Divide both sides by 3.}$$

The solution is $(1, 3)$. We check that $(1, 3)$ satisfies *both* equations (1) and (2):

$$3x + 2y = 9 \qquad\qquad -3x + 5y = 12$$
$$3(1) + 2(3) \overset{?}{=} 9 \qquad\qquad -3(1) + 5(3) \overset{?}{=} 12$$
$$3 + 6 \overset{?}{=} 9 \qquad\qquad -3 + 15 \overset{?}{=} 12$$
$$\text{true} \qquad\qquad\qquad \text{true}$$

∎

Example 2   Solving a System by Elimination

Solve the system

$$-5x + 3y = -9 \qquad \text{Equation (1)}$$
$$3x + 6y = 21 \qquad \text{Equation (2)}$$

Solution

If we add the left sides and add the right sides of the equations as given to us, neither variable would be eliminated. Therefore, we first multiply both sides of equation (1) by -2, yielding the system

$$10x - 6y = 18 \qquad \text{Multiply both sides of equation (1) by } -2.$$
$$3x + 6y = 21 \qquad \text{Equation (2)}$$

Now that the coefficients of the y terms are equal in absolute value and opposite in sign, we add the left sides and add the right sides of the equations and solve for x:

$$\begin{aligned} 10x - 6y &= 18 \\ 3x + 6y &= 21 \\ \hline 13x + 0 &= 39 \qquad \text{Add left sides and add right sides; combine like terms.} \\ 13x &= 39 \qquad a + 0 = a \\ x &= 3 \qquad \text{Divide both sides by 13.} \end{aligned}$$

We substitute 3 for x in equation (1) and solve for y:

$$-5x + 3y = -9 \qquad \text{Equation (1)}$$
$$-5(3) + 3y = -9 \qquad \text{Substitute 3 for } x.$$
$$-15 + 3y = -9 \qquad \text{Simplify.}$$
$$3y - 6 \qquad \text{Add 15 to both sides.}$$
$$y = 2 \qquad \text{Divide both sides by 3.}$$

The solution is $(3, 2)$. We could check that $(3, 2)$ satisfies both of the original equations.

∎

Using Elimination to Solve a Linear System

To use **elimination** to solve a system of two linear equations,

1. Use the multiplication property of equality (Section 4.3) to get the coefficients of one variable to be equal in absolute value and opposite in sign.
2. Add the left sides and add the right sides of the equations to eliminate one of the variables.
3. Solve the equation in one variable found in step 2.
4. Substitute the solution found in step 3 into one of the original equations, and solve for the other variable.

Example 3 Solving a System by Elimination

Solve the system

$$5x - 4y = 13 \qquad \text{Equation (1)}$$
$$9x + 3y = 3 \qquad \text{Equation (2)}$$

Solution

To eliminate the y terms, we multiply both sides of equation (1) by 3 and multiply both sides of equation (2) by 4, yielding the system

$$15x - 12y = 39 \qquad \text{Multiply both sides of equation (1) by 3.}$$
$$36x + 12y = 12 \qquad \text{Multiply both sides of equation (2) by 4.}$$

The coefficients of the y terms are now equal in absolute value and opposite in sign. Next, we add the left sides and add the right sides of the equations and solve for x:

$$
\begin{array}{l}
15x - 12y = 39 \\
36x + 12y = 12 \\
\hline
\end{array}
$$
$$51x + 0 = 51 \qquad \text{Add left sides and add right sides; combine like terms.}$$
$$51x = 51 \qquad a + 0 = a$$
$$x = 1 \qquad \text{Divide both sides by 51.}$$

Substituting 1 for x in equation (1) gives

$$5x - 4y = 13 \qquad \text{Equation (1)}$$
$$5(1) - 4y = 13 \qquad \text{Substitute 1 for } x.$$
$$5 - 4y = 13 \qquad \text{Simplify.}$$
$$-4y = 8 \qquad \text{Subtract 5 from both sides.}$$
$$y = -2 \qquad \text{Divide both sides by } -4.$$

The solution is $(1, -2)$. ∎

In Example 3, we eliminated y by getting the coefficients of the y terms to be equal in absolute value and opposite in sign. Note that this process is similar to finding a least common multiple.

Solving Inconsistent Systems and Dependent Systems by Elimination

When we apply elimination to an inconsistent system or a dependent system, the results are similar to the results obtained from applying substitution to such systems.

Solving Inconsistent Systems and Dependent Systems by Elimination

If the result of applying elimination to a linear system of two equations is

- a false statement, then the system is inconsistent; that is, the solution set is the empty set.
- a true statement that can be put into the form $a = a$, then the system is dependent; that is, the solution set is the set of ordered pairs represented by every point on the (same) line.

Example 4 Using Elimination to Solve a Dependent System

Solve

$$2x - 7y = 5 \qquad \text{Equation (1)}$$
$$6x - 21y = 15 \qquad \text{Equation (2)}$$

Solution

To eliminate the x terms (and the y terms), we multiply both sides of equation (1) by -3, yielding the system

$$-6x + 21y = -15 \qquad \text{Multiply both sides of equation (1) by } -3.$$
$$6x - 21y = 15 \qquad \text{Equation (2)}$$

Now that the coefficients of the x terms (and those of the y terms) are equal in absolute value and opposite in sign, we add the left sides and add the right sides of the equations:

$$-6x + 21y = -15$$
$$\underline{6x - 21y = 15}$$
$$0 + 0 = 0 \qquad \text{Add left sides and add right sides; combine like terms.}$$
$$0 = 0 \qquad \text{Simplify.}$$

Since $0 = 0$ is a true statement of the form $a = a$, we conclude that the system is dependent and that the solution set of the system is the set of ordered pairs represented by the points on the line $2x - 7y = 5$ and the (same) line $6x - 21y = 15$. ∎

Comparing Three Ways to Solve a System

In the next example, we solve the same system that we solved by substitution in Example 3 of Section 6.2, but now we solve it by elimination.

Example 5   Solving a System by Elimination

Solve the system

$$y = 1.72x - 4.38 \qquad \text{Equation (1)}$$
$$y = -0.53x + 6.94 \qquad \text{Equation (2)}$$

Solution

First we multiply both sides of equation (1) by -1, yielding the system

$$-y = -1.72x + 4.38 \qquad \text{Multiply equation (1) by } -1.$$
$$y = -0.53x + 6.94 \qquad \text{Equation (2)}$$

Now that the coefficients of y are equal in absolute value and opposite in sign, we add the left sides and add the right sides of the equations and solve for x:

$$-y = -1.72x + 4.38$$
$$\underline{y = -0.53x + 6.94}$$
$$0 = -2.25x + 11.32 \qquad \text{Add left sides and add right sides; combine like terms.}$$
$$2.25x = 11.32 \qquad \text{Add 2.25x to both sides.}$$
$$x \approx 5.03 \qquad \text{Divide both sides by 2.25.}$$

Then we substitute 5.03 for x in equation (1):

$$y = 1.72x - 4.38 \qquad \text{Equation (1)}$$
$$y = 1.72(5.03) - 4.38 \qquad \text{Substitute 5.03 for x.}$$
$$y \approx 4.27 \qquad \text{Simplify.}$$

So, the approximate solution is (5.03, 4.27), which is the same result that we obtained in Example 3 of Section 6.2. ∎

In both Sections 6.2 and 6.3, we solved the system

$$y = 1.72x - 4.38$$
$$y = -0.53x + 6.94$$

by graphing, substitution, and elimination. We obtained the same approximate solution by all three methods. In fact, **any linear system of two equations can be solved by graphing, substitution, or elimination. All three methods give the same result.**

How do we determine which method to use? When we solve a system of two equations, if a variable is alone or is easy to isolate to one side of either equation, substitution is often convenient. Otherwise, elimination is usually easiest. With most systems, we avoid solving by graphing or by using tables, except as a check, because these methods often lead only to approximate solutions.

group exploration

Comparing techniques of solving systems

Consider the system

$$x + 2y = 4$$
$$y = 3x - 5$$

1. Use substitution to solve the system.
2. Use elimination to solve the system.
3. Solve the system by graphing the equations by hand.
4. Compare your results from Problems 1, 2, and 3.
5. Give an example of a system that is easiest to solve by substitution. Also, give an example of a system that is easiest to solve by elimination. Finally, give an example of a system that is easiest to solve by graphing by hand. Explain. Solve your three systems.

TIPS FOR SUCCESS: Cross-Checks

If you finish a quiz or an exam early, it pays to verify your answers with cross-checks. For example, suppose you determine by elimination that the solution of the system

$$5x + 2y = 9$$
$$3x + 4y = 11$$

is $(1, 2)$. There are several ways to verify your answer. You could check that $(1, 2)$ satisfies both equations. You could graph each equation and check that the intersection point is $(1, 2)$. Or you could solve the system by substitution.

HOMEWORK 6.3

FOR EXTRA HELP ▶

Student Solutions Manual   PH Math/Tutor Center   MathXL®   MyMathLab

Solve the system by elimination. Verify your solution by checking that it satisfies both equations in the system.

1. $2x + 3y = 7$
$-2x + 5y = 1$

2. $-4x + 5y = 11$
$4x + 3y = 13$

3. $5x - 2y = 2$
$-3x + 2y = 2$

4. $3x + 5y = 19$
$2x - 5y = -4$

5. $x + 2y = -4$
$3x - 4y = 18$

6. $x - 4y = 5$
$5x - 2y = -11$

7. $2x - 3y = 8$
$5x + 6y = -7$

8. $3x + 4y = -5$
$-5x + 2y = 17$

9. $6x - 5y = 4$
$2x + 3y = -8$

10. $4x - 7y = 25$
$8x + 3y = -1$

11. $5x + 7y = -16$
$2x - 5y = 17$

12. $6x + 5y = -14$
$-4x - 7y = 2$

13. $-8x + 3y = 1$
$3x - 4y = 14$

14. $5x - 2y = 8$
$2x - 3y = 1$

15. $y = 3x - 6$
$y = -4x + 1$

16. $y = 2x + 7$
$y = -x - 8$

17. $3x + 6y - 18 = 0$
$17 = 7x + 9y$

18. $3 = -2x + 9y$
$5x - 5y - 10 = 0$

19. $\dfrac{2}{9}x + \dfrac{1}{3}y = 4$
$\dfrac{1}{2}x - \dfrac{2}{5}y = -\dfrac{5}{2}$

20. $\dfrac{1}{2}x - \dfrac{5}{4}y = \dfrac{9}{2}$
$\dfrac{3}{8}x + \dfrac{1}{2}y = \dfrac{1}{2}$

Solve the system by elimination. If the system is inconsistent or dependent, say so.

21. $4x - 7y = 3$
$8x - 14y = 6$

22. $4x + 6y = 10$
$-6x - 9y = -15$

23. $8x - 6y = 4$
$12x - 9y = 5$

24. $3x + 2y = 5$
$-12x - 8y = -17$

25. $3x - 2y = -14$
$6x + 5y = -19$

26. $8x - 3y = -2$
$5x + 12y = -29$

27. $6x - 15y = 7$
$-4x + 10y = -5$

28. $10x - 12y = 5$
$-15x + 18y = -8$

29. $3x - 9y = 12$
$-4x + 12y = -16$

30. $9x - 6y = 15$
$12x - 8y = 20$

Use elimination to solve the system, with coordinates of solutions rounded to the second decimal place. Verify your work by using "intersect" on a graphing calculator.

31. $y = 4.29x - 8.91$
$y = -1.26x + 9.75$

32. $y = 1.28x + 2.05$
$y = 3.94x - 8.83$

33. $y = -2.15x + 8.38$
$y = 1.67x + 2.57$

34. $y = 3.28x + 1.43$
$y = 0.56x + 6.72$

35. $y = -2.62x + 7.24$
$y = 1.89x - 6.44$

36. $y = 1.64x + 5.07$
$y = -2.57x - 3.39$

Solve the system by either elimination or substitution. Verify your work by using "intersect" on your graphing calculator or by checking that your result satisfies both equations of the system.

37. $4x - y = -12$
$3x + 5y = 14$

38. $6x + y = -13$
$3x - 2y = -19$

39. $-2x + 7y = -3$
$x = 3y + 2$

40. $4x - 5y = 23$
$y = 1 - 2x$

41. $2x + 7y = 13$
$3x - 4y = -24$

42. $5x - 2y = 7$
$4x - 3y = 7$

43. $2x - 7y = -1$
$-x - 3y = 7$

44. $3x - 8y = -4$
$x - 5y = -6$

45. $y = -2x - 3$
$y = 3x + 7$

46. $y = -4x - 9$
$y = 2x + 3$

47. $8x + 5y = 7$
$7y = -6x + 15$

48. $-3x + 2y = 19$
$5y = -4x - 10$

49. $3(2x - 5) + 4y = 11$
$5x - 2(3y + 1) = 1$

50. $4(2x - 7) + 3y = 7$
$7x - 5(2y + 3) = 3$

Solve the system of equations three times, once by each of the three methods: graphing by hand, substitution, and elimination. Decide which method you prefer for this system. Explain.

51. $3x + 2y = 8$
$2x - y = 3$

52. $y = -2x + 7$
$5x - 2y = 4$

53. Consider the system

$$5x + 3y = 11$$
$$2x - 4y = -6$$

   a. Solve the system by eliminating the x terms.
   b. Solve the system by eliminating the y terms.
   c. Compare your results in parts (a) and (b).

54. Consider the system

$$4x - 7y = 15$$
$$5x + 3y = 7$$

   a. Solve the system by eliminating the x terms.
   b. Solve the system by eliminating the y terms.
   c. Compare your results in parts (a) and (b).

55. Some solutions of two linear equations are shown in Tables 17 and 18. Find the solution of the system of the two equations. [**Hint:** You could begin by finding two equations.]

Table 17 Solutions of One Equation		Table 18 Solutions of the Other Equation	
x	**y**	**x**	**y**
−2	99	−2	−26
−1	96	−1	−24
0	93	0	−22
1	90	1	−20
2	87	2	−18

56. Some solutions of two linear equations are shown in Tables 19 and 20. Find the solution of the system of the two equations. [**Hint:** You could begin by finding two equations.]

Table 19 Solutions of One Equation		Table 20 Solutions of the Other Equation	
x	**y**	**x**	**y**
−2	−84	−2	91
−1	−80	−1	88
0	−76	0	85
1	−72	1	82
2	−68	2	79

57. Find the coordinates of the points A, B, C, and D as shown in Fig. 25. The equations of lines l_1 and l_2 are provided, but no attempt has been made to sketch the lines accurately except for showing the intersection points. Use TRACE and "intersect" on a graphing calculator to verify your work.

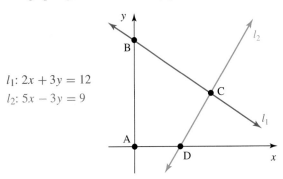

$l_1: 2x + 3y = 12$
$l_2: 5x - 3y = 9$

Figure 25 Exercise 57

58. Find the coordinates of the points A, B, C, and D as shown in Fig. 26. The equations of lines l_1 and l_2 are provided, but no attempt has been made to sketch the lines accurately except for showing the intersection points. Use TRACE and "intersect" on a graphing calculator to verify your work.

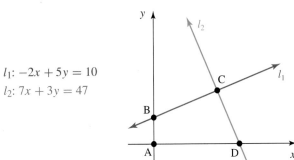

$l_1: -2x + 5y = 10$
$l_2: 7x + 3y = 47$

Figure 26 Exercise 58

59. Describe how to solve a system by elimination. (See page 9 for guidelines on writing a good response.)

60. Describe how to determine whether to solve a system by graphing, substitution, or elimination. (See page 9 for guidelines on writing a good response.)

Related Review

61. In this exercise, you will solve a system of equations to help find an equation of the line that contains the points $(2, 5)$ and $(4, 9)$.
  a. Substitute 2 for x and 5 for y in the equation $y = mx + b$.
  b. Substitute 4 for x and 9 for y in the equation $y = mx + b$.
  c. Consider the system of equations formed by your results from parts (a) and (b). Solve this system to find values of m and b.
  d. Substitute the values you found for m and b in the equation $y = mx + b$.
  e. Verify your equation by using a graphing calculator.

62. In this exercise, you will solve a system of equations to help find an equation of the line that contains the points $(-1, 5)$ and $(2, -4)$.
  a. Substitute -1 for x and 5 for y in the equation $y = mx + b$.
  b. Substitute 2 for x and -4 for y in the equation $y = mx + b$.
  c. Consider the system of equations formed by your results from parts (a) and (b). Solve this system to find values of m and b.
  d. Substitute the values you found for m and b in the equation $y = mx + b$.
  e. Verify your equation by using a graphing calculator.

Expressions, Equations, and Graphs

Perform the indicated instruction. Then use words such as linear, one variable, *and* two variables *to describe the expression, equation, or system.*

63. Simplify $2(5x + 4) - 6(3x + 2)$.

64. Graph $5x + 4y = 8$ by hand.

65. Solve $2(5x + 4) - 6(3x + 2) = 0$.

66. Solve the following:

$$5x + 4y = 8$$
$$3x + 2y = 2$$

6.4 USING SYSTEMS TO MODEL DATA

Objectives

▷ Use substitution and elimination to make predictions about situations described by a table of data.

▷ Use substitution and elimination to make predictions about situations described by rates of change.

In Section 6.1, we used graphing to help us find the intersection point of the graphs of two models. In this section, we discuss how to use substitution and elimination to find such an intersection point.

Using a Table of Data to Find a System for Modeling

In Example 1, we will work with a table of data and the two linear models we used in Section 6.1.

Example 1 Using a System to Make a Prediction

Let n be the number of college instructors (in thousands) at t years since 1970 (see Table 21). In Example 5 of Section 6.1, we used the following reasonable models for part-time instructors and full-time instructors:

$$n = 11.75t + 101.77 \quad \text{(1) Part-time instructors}$$
$$n = 7.38t + 376.15 \quad \text{(2) Full-time instructors}$$

Table 21 Numbers of Part-Time and Full-Time Instructors

Year	Number of Instructors (thousands)	
	Part Time	Full Time
1970	104	369
1975	188	440
1980	236	450
1985	256	459
1991	291	536
1995	381	551
1999	437	591
2003	543	632

Source: *U.S. National Center for Education Statistics*

Use a nongraphical approach to predict when the number of part-time instructors will equal the number of full-time instructors. Also, find that number of instructors.

Solution

From Example 5 of Section 6.1, we saw that the intersection point of the graphs of the two models indicates the year the numbers of part-time instructors and full-time instructors will be the same. Instead of graphing, we can find this intersection point by substitution (or elimination). To solve by substitution, we substitute $11.75t + 101.77$ for n in equation (2) and solve for t:

$$n = 7.38t + 376.15 \qquad \text{Equation (2)}$$
$$11.75t + 101.77 = 7.38t + 376.15 \qquad \text{Substitute } 11.75t + 101.77 \text{ for } n.$$
$$4.37t + 101.77 = 376.15 \qquad \text{Subtract } 7.38t \text{ from both sides.}$$
$$4.37t = 274.38 \qquad \text{Subtract } 101.77 \text{ from both sides.}$$
$$t \approx 62.79 \qquad \text{Divide both sides by } 4.37.$$

Next, we substitute 62.79 for t in equation (1):

$$n = 11.75(62.79) + 101.77 \approx 839.55$$

So, the approximate solution of the system is (62.79, 839.55). According to the models, there will be the same number of part-time instructors as full-time instructors (840 thousand of each) in 2033. This is the same conclusion we came to in Example 5 of Section 6.1. ∎

We can find an intersection point of two linear models by graphing, substitution, or elimination. For the homework exercises, you will use substitution or elimination to find such intersection points and then use "intersect" on a graphing calculator to verify your work.

Example 2 Using a System to Make a Prediction

World record times for the 800-meter run are listed in Table 22.

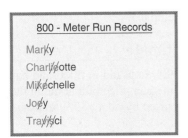

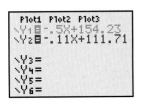

Table 22 800-Meter-Run Record Times

Women		Men	
Year	Record Time (seconds)	Year	Record Time (seconds)
1952	128.5	1938	108.4
1960	124.3	1939	106.6
1967	121.0	1955	105.7
1976	116.0	1968	104.3
1983	113.28	1976	103.5
		1997	101.11

Source: *Matti Koskimies/Peter Larrson*

Let r be the 800-meter record time (in seconds) at t years since 1900. Reasonable models for the women's record time and the men's record time are

$$r = -0.50t + 154.23 \qquad \text{Women}$$
$$r = -0.11t + 111.71 \qquad \text{Men}$$

1. Check that the models fit the data well.
2. Predict when the women's record time and the men's record time will be equal.

Solution

1. To keep track of the two types of data points, we use square "Marks" for the women's record scattergram and use plus-sign "Marks" for the men's record scattergram (see Fig. 27). See Section A.16 for graphing calculator instructions. The models appear to fit the data quite well.

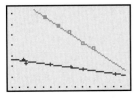

Figure 27 Check that the models fit the data well

2. We solve the system

$$r = -0.50t + 154.23 \qquad \text{Equation (1)}$$
$$r = -0.11t + 111.71 \qquad \text{Equation (2)}$$

by substitution. We do so by substituting $-0.50t + 154.23$ for r in equation (2):

$$r = -0.11t + 111.71 \qquad \text{Equation (2)}$$
$$-0.50t + 154.23 = -0.11t + 111.71 \qquad \text{Substitute } -0.50t + 154.23 \text{ for } r.$$
$$-0.39t + 154.23 = 111.71 \qquad \text{Add } 0.11t \text{ to both sides.}$$
$$-0.39t = -42.52 \qquad \text{Subtract } 154.23 \text{ from both sides.}$$
$$t \approx 109.03 \qquad \text{Divide both sides by } -0.39.$$

Next, we substitute 109.03 for t in equation (1) and solve for r:

$$r = -0.50(109.03) + 154.23 \approx 99.72$$

So, according to the models, the winning times for both women and men will be 99.72 seconds in 2009. We can verify our result by using "intersect" on a graphing calculator (see Fig. 28).

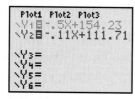 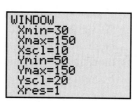 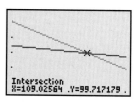

Figure 28 Verify that the intersection point is about (109.03, 99.72)

Using Rate of Change to Find a System for Modeling

Recall from Section 3.5 that if the rate of change of the dependent variable with respect to the independent variable is constant, then there is a linear relationship between the variables. Also, for the linear model that describes that relationship, the slope is equal to the constant rate of change. We use these ideas in Example 3.

Example 3 Using a System to Make an Estimate

The median price of a home in the Northeast was $139.4 thousand in 2000 and has since increased by about $3.3 thousand each year. The median price of a home in the South was $123.6 thousand in 2000 and has since increased by about $5.9 thousand each year (Source: *National Association of Realtors®*).

1. Let p be the median price (in thousands of dollars) of a home in the Northeast at t years since 2000. Find an equation of a model to describe the data.
2. Let p be the median price (in thousands of dollars) of a home in the South at t years since 2000. Find an equation of a model to describe the data.
3. When were median home prices for the two regions equal? What is that price?

Solution

1. Since the median price increases by about $3.3 thousand each year, we can model the situation by a linear equation. The slope (rate of change) is 3.3 thousand dollars per year. Since the price is $139.4 thousand in 2000, the p-intercept is (0, 139.4). So, a reasonable model is

$$p = 3.3t + 139.4$$

2. By the same reasoning, the model is $p = 5.9t + 123.6$.
3. To find when the median home prices for the two regions were equal, we solve the system

$$p = 3.3t + 139.4 \qquad \text{(1) Northeast}$$
$$p = 5.9t + 123.6 \qquad \text{(2) South}$$

To solve the system by substitution, we substitute $3.3t + 139.4$ for p in equation (2):

$$3.3t + 139.4 = 5.9t + 123.6$$

Then we solve for t:

$$-2.6t + 139.4 = 123.6 \qquad \text{Subtract 5.9}t \text{ from both sides.}$$
$$-2.6t = -15.8 \qquad \text{Subtract 139.4 from both sides.}$$
$$t \approx 6.08 \qquad \text{Divide both sides by } -2.6.$$

Next, we substitute 6.08 for t in equation (1):

$$p = 3.3(6.08) + 139.4 \approx 159.46$$

So, the two regions had the same median home price of \$159.5 thousand in 2006, according to the models. We verify our work by using "intersect" on a graphing calculator (see Fig. 29).

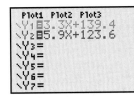 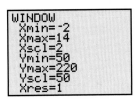 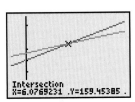

Figure 29 Verify that the intersection point is about (6.08, 159.46)

group exploration

Using a difference to make a prediction

1. Suppose that A and B are real numbers.
   a. If $A - B = 2$, which is larger, A or B? How much larger?
   b. If $A - B = -3$, which is larger, A or B? How much larger?
   c. If $A - B = 0$, what is true about A and B?

2. In Example 3, we worked with the models

$$p = 3.3t + 139.4 \qquad \text{Northeast}$$
$$p = 5.9t + 123.6 \qquad \text{South}$$

   Find the difference of the expressions $3.3t + 139.4$ and $5.9t + 123.6$. [**Hint:** Use parentheses and then simplify.]

3. Evaluate the result in Problem 2 for $t = 3$. What does your result mean in this situation?

4. Evaluate the result in Problem 2 for $t = 10$. What does your result mean in this situation?

5. For what values of t is the result in Problem 2 equal to 0? What does your result mean in this situation? [**Hint:** Solve an equation.]

6. Compare your result in Problem 5 with the result in Example 3.

TIPS FOR SUCCESS: Reread a Problem

After you think you have solved a problem, reread it to make sure you have answered its question(s). Also, reread the problem with the solution in mind. If what you read makes sense, then you've provided another check of your result(s).

1. Let n be the number of people (in thousands) who earned a bachelor's degree in the year that is t years since 1980 (see Table 23). In Exercise 37 of Homework 6.1, you worked with the following reasonable models for the numbers of women and men:

$$n = 13.28t + 440.09 \quad \text{Women}$$
$$n = 3.42t + 468.14 \quad \text{Men}$$

Table 23 Women and Men Who Earned a Bachelor's Degree

	Number of People Who Earned a Bachelor's Degree (thousands)	
Year	Women	Men
1980	456	474
1985	497	483
1990	560	492
1995	634	526
2000	708	530
2002	742	550

Source: *U.S. National Center for Education Statistics*

Use substitution or elimination to estimate when the number of women who earned a bachelor's degree was equal to the number of men who earned a bachelor's degree. What is that number of people?

2. Let n be the number of daily newspapers at t years since 1980 (see Table 24). In Exercise 38 of Homework 6.1, you worked with the following reasonable models for the numbers of morning daily newspapers and evening daily newspapers:

$$n = 18.13t + 386.86 \quad \text{Morning newspapers}$$
$$n = -31.52t + 1382.53 \quad \text{Evening newspapers}$$

Table 24 Numbers of Morning Dailies and Evening Dailies

	Number of Daily Newspapers	
Year	Morning	Evening
1980	387	1388
1985	482	1220
1990	559	1084
1995	656	891
2000	766	727
2004	813	653

Source: *Editor & Publisher Co.*

Use substitution or elimination to estimate when the number of morning daily newspapers was equal to the number of evening daily newspapers. What is that number of newspapers?

3. Let r be the record time (in seconds) for the 200-meter run at t years since 1900 (see Table 25). In Exercise 39 of Homework 6.1, you worked with the following reasonable models for the women's world record times and the men's world record times:

$$r = -0.064t + 27.00 \quad \text{Women}$$
$$r = -0.029t + 22.10 \quad \text{Men}$$

Table 25 200-Meter-Run World Record Times

Women		Men	
Year	Record Time (seconds)	Year	Record Time (seconds)
1973	22.38	1951	20.6
1974	22.21	1963	20.3
1978	22.06	1967	20.14
1984	21.71	1979	19.72
1988	21.34	1996	19.32

Source: *WR Progression*

Use substitution or elimination to predict when the women's record time will equal the men's record time. What is that record time?

4. Let w be the winning time (in seconds) for the Olympic 500-meter race in speed skating at t years since 1970 (see Table 26). In Exercise 40 of Homework 6.1, you worked with the following reasonable models for the women's winning times and the men's winning times:

$$w = -0.19t + 43.73 \quad \text{Women}$$
$$w = -0.16t + 39.90 \quad \text{Men}$$

Table 26 Olympic 500-Meter Speed-Skating Winning Times

	Winning Time (seconds)	
Year	Women	Men
1972	43.33	39.44
1976	42.76	39.17
1980	41.78	38.03
1984	41.02	38.19
1988	39.10	36.45
1992	40.33	37.14
1994	39.25	36.33
1998	38.21	35.59
2002	37.375	34.615

Source: *The Universal Almanac*

Use substitution or elimination to predict when the women's winning time will be equal to the men's winning time. What is that winning time?

5. The percentages of U.S. households with an Internet connection and of U.S. households with a computer are shown in Table 27 for various years.

Table 27 Households with an Internet Connection; Households with a Computer

	Percent of U.S. Households	
Year	Internet Connection	Computer
1997	18.6	36.6
1998	26.2	42.1
2000	41.5	51.0
2001	50.3	56.2
2003	54.6	61.8

Source: *U.S. Census Bureau*

a. Let p be the percentage of U.S. households with an Internet connection at t years since 1990. Find an equation of a model to describe the data.

b. Let p be the percentage of U.S. households with a computer at t years since 1990. Find an equation of a model to describe the data.

c. Use substitution or elimination to estimate when the percentage of households with a computer will equal the percentage of households with an Internet connection. What is that percentage of households?

6. The numbers of Home Depot® stores and Lowe's® stores are shown in Table 28 for various years.

Table 28 Numbers of Home Depot Stores and Lowe's Stores

| Year | Number of Stores | |
	Home Depot	Lowe's
1997	624	446
1998	761	484
2000	1134	650
2002	1532	867
2004	1890	952

Source: *SEC filings, Home Depot, Lowe's*

a. Let n be the number of Home Depot stores at t years since 1990. Find an equation of a model to describe the data.

b. Let n be the number of Lowe's stores at t years since 1990. Find an equation of a model to describe the data.

c. Use substitution or elimination to estimate when the number of Home Depot stores was equal to the number of Lowe's stores. What is that number of stores?

7. Ad-supported cable television has gained ground steadily on broadcast television (see Table 29).

Table 29 Broadcast and Ad-Supported Cable Television

| | Prime-Time Household Viewing Shares (percent) | |
Year	Broadcast	Ad-Supported Cable
1988	78.8	11.6
1990	74.8	16.4
1992	72.1	19.8
1994	70.6	21.3
1996	64.9	26.7
1998	58.2	32.5
2000	54.0	35.2
2003	46.5	42.4

Source: *Cabletelevision Advertising Bureau (analysis of Nielsen data)*

a. Let s be the prime-time household viewing share for all broadcast TV stations at t years since 1980. Find an equation of a model to describe the data.

b. Let s be the prime-time household viewing share for all ad-supported cable TV stations at t years since 1980. Find an equation of a model to describe the data.

c. Use substitution or elimination to estimate when the prime-time household viewing shares for all broadcast stations and all ad-supported cable stations were equal. What is that viewing share?

d. In addition to broadcast and ad-supported cable stations, there are PBS, pay cable, and other types of cable stations. Predict the combined prime-time household viewing share for PBS, pay cable, and other types of cable stations in 2011.

8. Energy consumption by North America and combined energy consumption by Asia and Oceania are shown in Table 30 for various years. (88.3 quadrillion Btu is 88,300,000,000,000,000 British thermal units.)

Table 30 Energy Consumption

| | Energy Consumption (quadrillion Btu) | |
Year	North America	Asia and Oceania
1985	91.2	59.3
1990	100.8	74.1
1995	108.8	95.6
2000	118.3	108.9
2003	119.1	120.1

Source: *U.S. Energy Information Administration*

a. Let c be North America's energy consumption (in quadrillion Btu) at t years since 1980. Find an equation of a model to describe the data.

b. Let c be the energy consumption (in quadrillion Btu) in Asia and Oceania at t years since 1980. Find an equation of a model to describe the data.

c. Use substitution or elimination to predict when North America's energy consumption was equal to energy consumption in Asia and Oceania. What is that energy consumption?

9. The numbers of women and men in the House of Representatives are shown in Table 31 for various years.

Table 31 Numbers of Women and Men in the House of Representatives

| | | Number of Representatives | |
Congress	Year	Women	Men
98th	1983	21	413
99th	1985	22	412
100th	1987	23	412
101st	1989	25	408
102nd	1991	28	407
103rd	1993	47	388
104th	1995	47	388
105th	1997	52	383
106th	1999	56	379
107th	2001	59	376
108th	2003	59	376
109th	2005	69	366

Source: *U.S. Census Bureau*

Let n be the number of representatives at t years since 1980. Models for the numbers of female and male representatives are

$$n = 2.34t + 9.54 \quad \text{Women}$$
$$n = -2.29t + 424.40 \quad \text{Men}$$

a. Use substitution or elimination to predict when the same number of women and men will be in the House of Representatives.

b. Find the greatest change in women from one Congress to the next. When did this change occur? Is the change about the same as the other changes in the number of women from one Congress to the next?

c. Here are models that fit the data well from the 103rd Congress to the 109th Congress:

$$n = 1.73t + 22.66 \quad \text{Women}$$
$$n = -1.73t + 412.34 \quad \text{Men}$$

Using these models, predict when the same number of women and men will be in the House of Representatives.

d. Explain why it makes sense that your predicted year in part (c) is later than your predicted year in part (a). Refer to the scattergrams of the data and/or the slopes of the models.

10. The numbers of women and men who live alone are shown in Table 32 for various years.

Table 32 Numbers of Women and Men Who Live Alone

Year	Number Living Alone (millions)	
	Women	**Men**
1980	11.3	7.0
1985	12.7	7.9
1990	14.0	9.0
1995	14.6	10.1
2000	15.6	11.2
2004	17.0	12.6

Source: *U.S. Census Bureau*

a. Let n be the number (in millions) of women who live alone at t years since 1980. Find an equation of a model to describe the data.

b. Let n be the number (in millions) of men who live alone at t years since 1980. Find an equation of a model to describe the data.

c. Compare the slopes of your two models. What do they mean in this situation?

d. Will the number of men who live alone ever equal the number of women who live alone, according to your models? If so, explain why this event will not happen until far into the future. If not, explain why not.

11. The number of students who earned a bachelor's degree in communications was 56.9 thousand in 2000 and has since increased by 1.6 thousand degrees each year. The number of students who earned a bachelor's degree in computer science was 36.2 thousand students in 2000 and has since increased by 2.4 thousand degrees each year (Source: *U.S. National Center for Education Statistics*).

a. Let d be the number (in thousands) of bachelor's degrees earned in communications at t years since 2000. Find an equation of a model to describe the data.

b. Let d be the number (in thousands) of bachelor's degrees earned in computer science at t years since 2000. Find an equation of a model to describe the data.

c. Use substitution or elimination to predict when the same number of bachelor's degrees will be earned in communications and computer science. What is that number of degrees?

12. In 1998, Mexico exported $94 billion worth of products to the United States, and the exports have since increased by

$9.8 billion each year. In 1998, China exported $72.2 billion worth of products to the United States, and the exports have since increased by $13.2 billion each year (Source: *U.S. Census Bureau*).

a. Let E be the value of exports (in billions of dollars) from Mexico at t years since 1998. Find an equation of a model to describe the data.

b. Let E be the value of exports (in billions of dollars) from China at t years since 1998. Find an equation of a model to describe the data.

c. Use substitution or elimination to estimate when the value of exports from Mexico and China were equal. What is that value?

13. Digital-camera sales were $4.6 million in 2000 and have since increased by $2.5 million each year. Traditional-camera sales were $20 million in 2000 and have since decreased by $2.6 million each year (Source: *Photo Marketing Association International*).

a. Let s be digital-camera sales (in millions of dollars) at t years since 2000. Find an equation of a model to describe the data.

b. Let s be traditional-camera sales (in millions of dollars) at t years since 2000. Find an equation of a model to describe the data.

c. Use substitution or elimination to estimate when sales of digital cameras were equal to sales of traditional cameras. What are those sales?

14. In 2002, a one-year-old Saturn® L100 sedan was worth about $8000, with a depreciation of about $916 per year. A one-year-old Nissan® Altima sedan was worth about $12,041, with a depreciation of about $1436 per year (Source: *Edmund's Automobile Buyer's Guide*).

a. Let V be the value (in dollars) of the Saturn L100 at t years since 2002. Find an equation of a model.

b. Let V be the value (in dollars) of the Nissan Altima at t years since 2002. Find an equation of a model.

c. Use substitution or elimination to predict when the two cars will have the same value. What is that value?

15. In 2002, a one-year-old Acura® CL coupe was worth about $18,249, with a depreciation of about $1903 per year. A one-year-old Subaru® Legacy sedan was worth about $14,564, with a depreciation of about $1225 per year (Source: *Edmund's Automobile Buyer's Guide*).

a. Let V be the value (in dollars) of the Acura CL at t years since 2002. Find an equation of a model to describe the data.

b. Let V be the value (in dollars) of the Subaru Legacy at t years since 2002. Find an equation of a model to describe the data.

c. Use substitution or elimination to predict when the two cars will have the same value. What is that value?

16. Company A earned $4.8 million in profits in 2005, and its profits have been increasing by $1.1 million each year. Company B earned $17.5 million in profits in 2005, and its profits have been decreasing by $1.5 million each year.

a. Let p be company A's profit (in millions of dollars) at t years since 2005. Find an equation of a model to describe the data.

b. Let p be company B's profit (in millions of dollars) at t years since 2005. Find an equation of a model to describe the data.

c. Use substitution or elimination to predict when both companies' profits will be equal. What is that profit?

17. Company A sold $29.5 million of software products in 2005, and its sales have since declined by $1.2 million each year. Company B sold $12.1 million of software products in 2005, and its sales have since increased by $1.7 million each year.

  a. Let s be company A's sales (in millions of dollars) at t years since 2005. Find an equation of a model to describe the data.

  b. Let s be company B's sales (in millions of dollars) at t years since 2005. Find an equation of a model to describe the data.

  c. Use substitution or elimination to predict when both companies' sales will be equal. What are those sales?

18. The enrollment at college A was 12,532 students in 2005 and has been increasing by 570 students each year. The enrollment at college B was 15,130 students in 2005 and has been increasing by 135 students each year.

  a. Let E be the enrollment (in number of students) at college A at t years since 2005. Find an equation of a model to describe the data.

  b. Let E be the enrollment (in number of students) at college B at t years since 2005. Find an equation of a model to describe the data.

  c. Use substitution or elimination to predict when the two colleges' enrollments will be equal. What is that enrollment?

19. The tuition at college A was $5425 in 2005 and has been increasing by $850 each year. The tuition at college B was $6557 in 2005 and has been increasing by $570 each year.

  a. Let c be the tuition (in dollars) at college A at t years since 2005. Find an equation of a model to describe the data.

  b. Let c be the tuition (in dollars) at college B at t years since 2005. Find an equation of a model to describe the data.

  c. Use substitution or elimination to predict when both colleges will have the same tuition. What is that tuition?

20. Describe how you can find a system of linear equations to model a situation. Also, explain how you can use the system to make an estimate or prediction about the situation.

Related Review

21. The average daily calorie consumption by men has increased approximately linearly from 2450 calories per day in 1980 to 2618 calories per day in 2000. The average for women has increased approximately linearly from 1521 calories per day in 1980 to 1877 calories per day in 2000 (Source: *Centers for Disease Control and Prevention*). Predict when the average will be the same for both genders. What is that average?

22. Clogged arteries can be treated with angioplasty, in which a balloon is inserted in an artery, or stents, in which a wire cage is implanted to keep the artery open. The number of angioplasty procedures increased approximately linearly from 0.86 million procedures in 2000 to 1.06 million procedures in 2003. The number of stent procedures increased approximately linearly from 0.67 million procedures in 2000 to 0.95 million procedures in 2003 (Source: *Merrill Lynch*). Predict in which year the same number of both procedures will be performed.

Expressions, Equations, and Graphs

Perform the indicated instruction. Then use words such as linear, one variable, *and* two variables *to describe the expression, equation, or system.*

23. Solve the following:

$$y = -2x + 6$$
$$y = 3x + 1$$

24. Evaluate $-2x + 6$ for $x = 1$.

25. Solve $-2x + 6 = 3x + 1$.

26. Simplify $(-2x + 6) + (3x + 1)$.

6.5 PERIMETER, VALUE, INTEREST, AND MIXTURE PROBLEMS

Objectives

▶ Know a five-step problem-solving method.

▶ Solve perimeter, value, interest, and mixture problems.

In this section, we will use a five-step method to solve problems that involve the perimeter of a rectangle, the (dollar) value of an object, the interest from an investment, and the percentage of a substance in a mixture.

Five-Step Problem-Solving Method

To solve some problems in which we want to find two quantities, it is useful to perform the following five steps:

- *Step 1: Define each variable.* For each quantity that we are trying to find, we usually define a variable to represent that unknown quantity.

- *Step 2: Write a system of two equations.* We find a system of two equations by using the variables from step 1. We can usually write each equation either by translating the information stated in the problem into mathematics or by making a substitution into a formula.

- *Step 3: Solve the system.* We solve the system of equations from step 2.
- *Step 4: Describe each result.* We use a complete sentence to describe the quantities we found.
- *Step 5: Check.* We reread the problem and check that the quantities we found agree with the given information.

Perimeter Problems

Recall from Section 4.6 that the formula of the perimeter P of a rectangle with length L and width W is $P = 2L + 2W$.

Example 1 Solving a Perimeter Problem

Throughout history, rectangles whose length is about 1.62 times their width, called *golden rectangles,* have been viewed as the most pleasing to the eye (see Fig. 30). Many famous structures, including the Parthenon in Athens and the Great Pyramid of Giza in Cairo, incorporate golden rectangles into their design. For Leonardo Da Vinci's *Mona Lisa,* the edges of the painting form a golden rectangle.

Suppose that an artist wants a piece of canvas in the shape of a golden rectangle with a perimeter of 9 feet. Find the dimensions of the canvas.

Figure 30 A golden rectangle drawn to scale

Solution

Step 1: Define each variable. Let W be the width (in feet) and L be the length (in feet). See Fig. 30.

Step 2: Write a system of two equations. Since the length must be 1.62 times the width, our first equation is

$$L = 1.62W$$

Because the perimeter is 9 feet, we find our second equation by substituting 9 for P in the perimeter formula $P = 2L + 2W$:

$$9 = 2L + 2W$$

The system is

$$L = 1.62W \qquad \text{Equation (1)}$$
$$2L + 2W = 9 \qquad \text{Equation (2)}$$

Step 3: Solve the system. We can solve the system by substitution. We substitute $1.62W$ for L in equation (2) and then solve for W:

$$
\begin{aligned}
2L + 2W &= 9 && \text{Equation (2)} \\
2(1.62W) + 2W &= 9 && \text{Substitute 1.62W for L.} \\
3.24W + 2W &= 9 && \text{Simplify.} \\
5.24W &= 9 && \text{Combine like terms.} \\
\frac{5.24W}{5.24} &= \frac{9}{5.24} && \text{Divide both sides by 5.24.} \\
W &\approx 1.72 && \text{Round } \frac{9}{5.24} \text{ to second decimal place.}
\end{aligned}
$$

To find the approximate length, we substitute $\dfrac{9}{5.24}$ for W in equation (1):

$$L = 1.62\left(\frac{9}{5.24}\right) \approx 2.78$$

Step 4: Describe each result. The approximate width is 1.72 feet, and the approximate length is 2.78 feet.

Step 5: Check. We add the lengths of the four sides: $1.72 + 2.78 + 1.72 + 2.78 = 9$, which checks for a perimeter of 9 feet. We can also check that 2.78 is about 1.62 times 1.72 (try it). ■

Value Problems

Recall from Section 4.6 that if n objects each have value v, then their total value T is given by

$$T = vn$$

When some objects are sold, we refer to the total money collected as the **revenue** from selling the objects.

Example 2   Solving Value Problems

A vendor charges $5 per hamburger and $3 per hot dog.

1. What is the vendor's total revenue from selling 40 hamburgers and 85 hot dogs?
2. If the vendor sells a total of 135 hamburgers and hot dogs for a total revenue of $495, how many hamburgers and hot dogs did he sell?

Solution

1. The revenue from hamburgers is equal to the price per hamburger times the number of hamburgers sold: $5 \cdot 40 = 200$ dollars. The revenue from hot dogs is equal to the price per hot dog times the number of hot dogs sold: $3 \cdot 85 = 255$ dollars. We add the revenue of the hamburgers and that of the hot dogs to find the total revenue:

$$\underbrace{5}_{\substack{\text{dollars} \\ \text{hamburger}}} \cdot \underbrace{40}_{\text{hamburgers}} + \underbrace{3}_{\substack{\text{dollars} \\ \text{hot dog}}} \cdot \underbrace{85}_{\text{hot dogs}} = \underbrace{455}_{\substack{\text{total revenue} \\ \text{in dollars}}}$$

So, the total revenue from hamburgers and hot dogs is $455. The units of the expressions on both sides of the equation are dollars, which suggests that our work is correct.

2. **Step 1: Define each variable.**   Let x be the number of hamburgers sold and y be the number of hot dogs sold.

   Step 2: Write a system of two equations.   Our work in Problem 1 suggests that the formula of the total revenue T (in dollars) is

$$T = \underbrace{5}_{\substack{\text{dollars} \\ \text{hamburger}}} \cdot \underbrace{x}_{\text{hamburgers}} + \underbrace{3}_{\substack{\text{dollars} \\ \text{hot dog}}} \cdot \underbrace{y}_{\text{hot dogs}}$$

   To obtain our first equation, we substitute 495 for T:

$$495 = 5x + 3y$$

   Since the vendor sells a total of 135 hamburgers and hot dogs, our second equation is

$$x + y = 135$$

   The system is

$$5x + 3y = 495 \qquad \text{Equation (1)}$$
$$x + y = 135 \qquad \text{Equation (2)}$$

   Step 3: Solve the system.   We can use elimination to solve the system. To eliminate the y terms, we multiply both sides of equation (2) by -3, yielding the system

$$5x + 3y = 495 \qquad \text{Equation (1)}$$
$$-3x - 3y = -405 \qquad \text{Multiply both sides of equation (2) by } -3.$$

Then we add the left sides and add the right sides of the equations and solve for x:

$$5x + 3y = 495$$
$$\underline{-3x - 3y = -405}$$
$$2x + 0 = 90 \qquad \text{Add left sides and right sides; combine like terms.}$$
$$2x = 90 \qquad a + 0 = a$$
$$x = 45 \qquad \text{Divide both sides by 2.}$$

Next, we substitute 45 for x in equation (2) and solve for y:

$$x + y = 135 \qquad \text{Equation (2)}$$
$$45 + y = 135 \qquad \text{Substitute 45 for } x.$$
$$y = 90 \qquad \text{Subtract 45 from both sides.}$$

Step 4: Describe each result.   The vendor sold 45 hamburgers and 90 hot dogs.

Step 5: Check.   First, we find the sum $45 + 90 = 135$, which is equal to the total number of hamburgers and hot dogs sold. Next, we find the total revenue from selling 45 hamburgers and 90 hot dogs: $5 \cdot 45 + 3 \cdot 90 = 495$, which checks.   ■

Notice that the arithmetic $5 \cdot 40 + 3 \cdot 85 = 455$ in Problem 1 of Example 2 suggests the equation $5x + 3y = 495$ that we formed in Problem 2. **If you have difficulty forming an equation, try making up numbers for unknown quantities and performing arithmetic to compute a related quantity. Your computation may suggest how to form the desired equation.**

Example 3   Solving a Value Problem

An auditorium has 500 balcony seats and 2100 main-level seats. If tickets for balcony seats cost $18 less than tickets for main-level seats, what should the prices be for each type of ticket so that the total revenue from a sellout performance will be $128,800?

Solution

Step 1: Define each variable.   Let b be the price (in dollars) for balcony seats and m be the price (in dollars) for main-level seats.

Step 2: Write a system of two equations.   Since tickets for balcony seats will cost $18 less than tickets for main-level seats, our first equation is

$$\underset{b}{\underbrace{\text{balcony ticket price}}} \quad \underset{=}{\underbrace{\text{is}}} \quad \underset{m - 18}{\underbrace{\text{\$18 less than main-level ticket price}}}$$

Since the total revenue is $128,800, our second equation is

$$\underset{b}{\underbrace{\frac{\text{dollars}}{\text{balcony ticket}}}} \cdot \underset{500}{\underbrace{\begin{array}{c}\text{balcony} \\ \text{tickets}\end{array}}} + \underset{m}{\underbrace{\frac{\text{dollars}}{\text{main-level ticket}}}} \cdot \underset{2100}{\underbrace{\begin{array}{c}\text{main-level} \\ \text{tickets}\end{array}}} = \underset{128,800}{\underbrace{\begin{array}{c}\text{total revenue} \\ \text{in dollars}\end{array}}}$$

The units of the expressions on both sides of the equation are dollars, which suggests that our work is correct. The system is

$$b = m - 18 \qquad \text{Equation (1)}$$
$$500b + 2100m = 128,800 \qquad \text{Equation (2)}$$

Step 3: Solve the system.   We can use substitution to solve the system. We substitute $m - 18$ for b in equation (2) and solve for m:

$$500b + 2100m = 128{,}800 \qquad \text{Equation (2)}$$
$$500(m - 18) + 2100m = 128{,}800 \qquad \text{Substitute } m - 18 \text{ for } b.$$
$$500m - 9000 + 2100m = 128{,}800 \qquad \text{Distributive law}$$
$$2600m - 9000 = 128{,}800 \qquad \text{Combine like terms.}$$
$$2600m = 137{,}800 \qquad \text{Add 9000 to both sides.}$$
$$m = 53 \qquad \text{Divide both sides by 2600.}$$

Then we substitute 53 for m in equation (1) and solve for b:

$$b = 53 - 18 = 35$$

Step 4: Describe each result.   Tickets for balcony seats should be priced at $35 per ticket, and tickets for main-level seats should be priced at $53 per ticket.

Step 5: Check.   First we find the difference in the ticket prices: $53 - 35 = 18$ dollars, which checks. Then we compute the total revenue from selling 500 of the $35 tickets and 2100 of the $53 tickets: $35 \cdot 500 + 53 \cdot 2100 = 128{,}800$ dollars, which checks. ■

Interest Problems

Money deposited in an account, such as a savings account, certificate of deposit (CD), or mutual fund, is called **principal.** A person invests money in hopes of later getting back the principal plus additional money, called **interest,** which is a percentage of the principal (see Fig. 31). The **annual simple-interest rate** is the percentage of the principal that equals the interest earned per year. So, if we invest $100 and earn $5 per year, then the annual simple-interest rate is 5%.

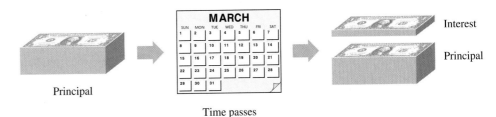

Figure 31 Invest the principal; time passes; get back the principal plus interest

Example 4   Finding the Interest from an Investment

How much interest will a person earn by investing $2500 in an account at 6% annual interest for one year?

Solution

We find 6% of 2500:

$$0.06(2500) = 150$$

The person will earn $150 in interest. ■

Some people invest in a variety of accounts, including lower-risk accounts and some higher-risk accounts. These investors might earn large amounts of interest from the higher-risk accounts, but will have a safety net of principal and interest from the lower-risk accounts.

Example 5   Solving Interest Problems

1. How much interest will a person earn from investing $6000 in a Presidential Bank Internet CD account at 2.5% annual interest and $1000 in a John Hancock Health Sciences A mutual fund account at 12% annual interest for one year?

2. A person plans to invest a total of $7000. She will invest in the two accounts described in Problem 1. How much money should she invest in each account to earn a total interest of $400 in one year?

Solution

1. We add the interest from both accounts to compute the total interest:

$$\underbrace{0.025(6000)}_{\substack{\text{interest from}\\ \text{2.5\% account}}} + \underbrace{0.12(1000)}_{\substack{\text{interest from}\\ \text{12\% account}}} = \underbrace{270}_{\text{total interest}}$$

2. *Step 1: Define each variable.* Let x be the money (in dollars) invested at 2.5% annual interest, and let y be the money (in dollars) invested at 12% annual interest.

Step 2: Write a system of two equations. Our work in Problem 1 suggests the first equation:

$$\underbrace{0.025x}_{\substack{\text{interest from}\\ \text{2.5\% account}}} + \underbrace{0.12y}_{\substack{\text{interest from}\\ \text{12\% account}}} = \underbrace{400}_{\text{total interest}}$$

Since the total investment is $7000, our second equation is

$$x + y = 7000$$

The system is

$$0.025x + 0.12y = 400 \qquad \text{Equation (1)}$$
$$x + y = 7000 \qquad \text{Equation (2)}$$

Step 3: Solve the system. We can use elimination to solve the system. To eliminate the x terms, we multiply both sides of equation (2) by -0.025, yielding the system

$$0.025x + 0.12y = 400 \qquad \text{Equation (1)}$$
$$-0.025x - 0.025y = -175 \qquad \text{Multiply both sides of equation (2) by } -0.025.$$

Then we add the left sides and add the right sides of the equations and solve for y:

$$
\begin{array}{l}
0.025x + 0.12y = 400 \\
\underline{-0.025x - 0.025y = -175} \\
\qquad 0 + 0.095y = 225 \qquad \text{Add left sides and right sides; combine like terms.} \\
\qquad\quad 0.095y = 225 \qquad 0 + a = a \\
\qquad\qquad\quad y = 2368.42 \qquad \text{Divide both sides by 0.095.}
\end{array}
$$

Next, we substitute 2368.42 for y in equation (2) and solve for x:

$$
\begin{array}{l}
x + y = 7000 \qquad \text{Equation (2)} \\
x + 2368.42 = 7000 \qquad \text{Substitute 2368.42 for } y. \\
\qquad\quad x = 4631.58 \qquad \text{Subtract 2368.42 from both sides.}
\end{array}
$$

Step 4: Describe each result. The person should invest $4631.58 at 2.5% annual interest and $2368.42 at 12% annual interest.

Step 5: Check. First, we find the sum $4631.58 + 2368.42 = 7000$, which checks with the total money invested. Next, we calculate the total interest earned from investing $4631.58 at 2.5% and $2368.42 at 12%:

$$0.025(4631.58) + 0.12(2368.42) \approx 400$$

which checks. ∎

Mixture Problems

Many fields, including chemistry, cooking, and medicine, involve mixing different substances. Suppose that 2 ounces of lime juice is mixed with 6 ounces of water to

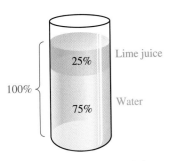

Figure 32 A 25% lime-juice solution

make 8 ounces of unsweetened limeade. Note that $\frac{2}{8} = 0.25 = 25\%$ of the limeade is lime juice. We call the limeade a *25% lime-juice solution.*

Then, the remaining $\frac{6}{8} = 0.75 = 75\%$ of the limeade is water. The percentage of the solution that is lime juice plus the percentage of the solution that is water is equal to 100% (all) of the solution: 25% + 75% = 100%. See Fig. 32.

Therefore, in a 20% lime-juice solution, 20% of the solution is lime juice and 100% − 20% = 80% of the solution is water. In general, **for an $x\%$ solution of two substances that are mixed, $x\%$ of the solution is one substance and $(100 - x)\%$ is the other substance.**

Example 6 Solving Mixture Problems

1. How much lime juice is in 6 ounces of a 15% lime-juice solution?
2. If 5 gallons of a 40% antifreeze solution is mixed with 3 gallons of a 75% antifreeze solution, how much pure antifreeze is in the mixture?

Solution

1. We find 15% of 6:

$$0.15(6) = 0.9$$

   There is 0.9 ounce of lime juice in the 6 ounces of lime-juice solution.
2. The total amount of pure antifreeze doesn't change, regardless of how it is distributed in the two solutions. To find the amount of pure antifreeze in the mixture, we add the pure amounts of antifreeze in the 40% solution and the 75% solution:

number of gallons of pure antifreeze in 40% solution		number of gallons of pure antifreeze in 75% solution		number of gallons of pure antifreeze in mixture
$0.40(5)$	$+$	$0.75(3)$	$=$	4.25

   There are 4.25 gallons of pure antifreeze in the mixture. ∎

Our work in Problem 2 of Example 6 suggests how to find an equation for step 2 of our work in Example 7.

Example 7 Solving a Mixture Problem

A chemist needs 8 liters of a 20% alcohol solution, but she has only a 15% alcohol solution and a 35% alcohol solution. How many liters of the 15% alcohol solution should she mix with the 35% alcohol solution to make the desired 8 liters of 20% alcohol solution?

Solution

Step 1: Define each variable. Let x be the number of liters of 15% alcohol solution, and let y be the number of liters of 35% alcohol solution.

Step 2: Write a system of two equations. Since she wants 8 liters of the total mixture, our first equation is

$$x + y = 8$$

We find our second equation from the fact that the sum of the amounts of pure alcohol in the 15% alcohol solution and the 35% alcohol solution is equal to the amount of pure alcohol in the mixture:

pure alcohol in 15% solution		pure alcohol in 35% solution		pure alcohol in mixture
$0.15x$	$+$	$0.35y$	$=$	$0.20(8)$

The system is

$$x + y = 8 \qquad \text{Equation (1)}$$
$$0.15x + 0.35y = 1.6 \qquad \text{Equation (2)}$$

Step 3: Solve the system.   We can solve the system by substitution. First, we solve equation (1) for y:

$$x + y = 8 \qquad \text{Equation (1)}$$
$$y = 8 - x \qquad \text{Subtract x from both sides.}$$

Then, we substitute $8 - x$ for y in equation (2) and solve for x:

$$0.15x + 0.35y = 1.6 \qquad \text{Equation (2)}$$
$$0.15x + 0.35(8 - x) = 1.6 \qquad \text{Substitute 8 − x for y.}$$
$$0.15x + 2.8 - 0.35x = 1.6 \qquad \text{Distributive law}$$
$$-0.20x + 2.8 = 1.6 \qquad \text{Combine like terms.}$$
$$-0.20x = -1.2 \qquad \text{Subtract 2.8 from both sides.}$$
$$x = 6 \qquad \text{Divide both sides by −0.20.}$$

Next, we substitute 6 for x in the equation $y = 8 - x$ and solve for y:

$$y = 8 - 6 = 2$$

Step 4: Describe each result.   Six liters of the 15% alcohol solution and 2 liters of the 35% alcohol solution are required.

Step 5: Check.   First, we compute the total amount (in liters) of pure alcohol in 6 liters of the 15% alcohol solution and 2 liters of the 35% alcohol solution:

$$0.15(6) + 0.35(2) = 1.6$$

Next, we compute the amount (in liters) of pure alcohol in 8 liters of the 20% alcohol solution:

$$0.20(8) = 1.6$$

Since the two results are equal, they check. Also, $6 + 2 = 8$, which checks with the chemist wanting 8 liters of the 20% solution.   ■

Example 8   Solving a Mixture Problem

A chemist needs 6 ounces of a 5% acid solution, but has only a 40% acid solution. How much of the 40% solution and water should he mix to form the desired 6 ounces of the 5% solution?

Solution

Step 1: Define each variable.   Let x be the number of ounces of the 40% solution, and let y be the number of ounces of water.

Step 2: Write a system of two equations.   Since he wants 6 ounces of the total mixture, our first equation is

$$x + y = 6$$

There is no acid in pure water, so we find our second equation from the fact that the amount of pure acid in the 40% acid solution is equal to the amount of pure acid in the mixture:

$$\underbrace{0.40x}_{\substack{\text{amount of pure} \\ \text{acid in 40% solution}}} = \underbrace{0.05(6)}_{\substack{\text{amount of pure} \\ \text{acid in mixture}}}$$

The system is

$$x + y = 6 \qquad \text{Equation (1)}$$
$$0.40x = 0.3 \qquad \text{Equation (2)}$$

Step 3: Solve the system. We begin by solving equation (2) for x:

$$0.40x = 0.3 \qquad \text{Equation (2)}$$
$$x = 0.75 \qquad \text{Divide both sides by 0.40.}$$

Next, we substitute 0.75 for x in equation (1):

$$x + y = 6 \qquad \text{Equation (1)}$$
$$0.75 + y = 6 \qquad \text{Substitute 0.75 for } x.$$
$$y = 5.25 \qquad \text{Subtract 0.75 from both sides.}$$

Step 4: Describe each result. The chemist needs to mix 0.75 ounce of the 40% acid solution with 5.25 ounces of water.

Step 5: Check. First, we find $0.75 + 5.25 = 6$, which checks with the chemist wanting 6 ounces of the 5% solution. Next, we compute the amount of pure acid in 0.75 ounce of the 40% solution: $0.40(0.75) = 0.3$ ounce. Finally, we compute the amount of pure acid in the 5% solution: $0.05(6) = 0.3$ ounce. The computed amounts of pure acid in the 40% solution and the 5% solution are equal, so they check. ∎

group exploration

Looking ahead: Graphing an inequality in two variables

1. Graph $y = x + 1$ carefully by hand.
2. Find 4 points that lie above the line, 2 points that lie on the line, and 4 points that lie below the line. Choose your 10 points so that there are at least 2 points in each of the four quadrants. List the ordered pairs for these points so that it is clear which points are above, below, or on the line.
3. Consider the inequality $y > x + 1$. We say that the ordered pair $(2, 5)$ *satisfies* the inequality, because the inequality becomes a true statement when we substitute 2 for x and 5 for y:

$$y > x + 1$$
$$5 \overset{?}{>} 2 + 1$$
$$5 \overset{?}{>} 3$$
$$\text{true}$$

   Which of the ordered pairs found in Problem 2 satisfy the given inequality or equation?
   a. $y > x + 1$ **b.** $y < x + 1$ **c.** $y = x + 1$

4. Describe all of the ordered pairs (not just those in Problem 2) that satisfy the given inequality or equation. Include in your description any of the three categories (above, below, or on the line) that are relevant.
   a. $y > x + 1$ **b.** $y < x + 1$ **c.** $y = x + 1$

5. Draw a dashed line where the graph of $y = -\dfrac{1}{2}x - 1$ would be in a coordinate system. Then use shading to indicate points whose ordered pairs satisfy the inequality $y < -\dfrac{1}{2}x - 1$.

HOMEWORK 6.5 FOR EXTRA HELP ▶

Student Solutions Manual

PH Math/Tutor Center

Math XL
MathXL®

MyMathLab
MyMathLab

1. The length of a golden rectangle is equal to about 1.62 times the width. If an architect wants to design the base of a building to be a golden rectangle, what are the dimensions of the base if the perimeter is to be 600 feet?

2. The length of a golden rectangle is equal to about 1.62 times the width. An architect wants to design a room so that the floor is a golden rectangle with a perimeter of 50 feet. What are the dimensions of the floor?

3. A landscaper plans to dig a rectangular garden whose length is to be 5 feet more than the width. If the landscaper has 42 feet of fencing to enclose the garden, what should be the dimensions of the garden?

4. An official football field is a rectangle whose length (including end zones) is 2.25 times the width. If a little-league football field is to be constructed with a perimeter of 200 yards, what should be the width and length of the field?

5. An official tennis court is a rectangle whose length is 6 feet more than twice the width. The perimeter of the court is 228 feet. Find the dimensions of the court.

6. An official table tennis (ping pong) table is a rectangle whose length is 1 foot less than twice the width. The perimeter of the table is 28 feet. Find the dimensions of the table.

7. The length of a rectangle is 3 inches less than twice the width. If the perimeter is 108 inches, find the dimensions of the rectangle.

8. The length of a rectangle is 5 inches more than twice the width. If the perimeter is 112 inches, find the dimensions of the rectangle.

9. The length of a rectangle is 1 yard more than 3 times the width. If the perimeter is 146 yards, find the length and width of the rectangle.

10. The length of a rectangle is 2 yards less than 3 times the width. If the perimeter is 148 yards, find the length and width of the rectangle.

11. A 2000-seat theater has tickets for sale at $15 and $22. How many tickets should be sold at each price for a sellout performance to generate a total revenue of $33,500?

12. A 10,000-seat amphitheater has tickets for sale at $45 and $65. How many tickets should be sold at each price for a sellout performance to generate a total revenue of $560,000?

13. The CD *Amnesiac,* by Radiohead, has a list price of $17.98, and the CD *Hail to the Thief,* by Radiohead, has a list price of $18.98. If a store's total sales of both CDs is 253 CDs and the total revenue of both CDs is $4762.94, how many of each CD were sold?

14. Worldwide sales of *Significant Other,* by Limp Bizkit, have reached 6 million CDs and cassettes. The list price of the CD is $18.97, and the list price of the cassette is $12.98. If a store's total sales of both formats is 400 CDs and cassettes, and the total revenue from both formats is $7378.35, how many CDs and how many cassettes are sold?

15. *Harry Potter and the Half-Blood Prince* (Book 6 of the series), by J. K. Rowling, sells as a hardcover book with a list price of $29.99 and as an audio CD with a list price of $75.00. If a bookstore sold 1500 of these books and CDs at the list prices for a total revenue of $53,987, how many books and how many CDs were sold?

16. *A Million Little Pieces,* by James Frey, sells as a hardcover book with a list price of $22.95 and as a paperback book with a list price of $14.95. If a bookstore sold 680 of these books at the list prices for a total revenue of $11,206, how many of each type of book were sold?

17. An auditorium has 300 balcony seats and 1400 main-level seats. If tickets for balcony seats are priced at $12 less than the price of main-level seats, what should the prices be for each type of ticket so that the total revenue from a sellout performance will be $40,600?

18. An auditorium has 450 balcony seats and 1700 main-level seats. If tickets for balcony seats are priced at $15 less than the price of main-level seats, what should the prices be for each type of ticket so that the total revenue from a sellout performance will be $100,750?

19. An amphitheater has 8000 general seats and 4000 reserved seats. If tickets for general seats are priced at $25 less than the price of reserved seats, what should the prices be for each type of ticket so that the total revenue from a sellout performance will be $544,000?

20. An amphitheater has 9500 general seats and 3200 reserved seats. If tickets for general seats are priced at $17 less than the price of reserved seats, what should the prices be for each type of ticket so that the total revenue from a sellout performance will be $435,400?

21. A person invests the given amount of money in an account at 8% annual interest. Find the interest in one year.
   a. 2500 dollars   b. 3500 dollars   c. *d* dollars

22. A person invests the given amount of money in an account at 5% annual interest. Find the interest in one year.
   a. 4000 dollars   b. 5000 dollars   c. *d* dollars

23. Find the total interest earned in one year from the given investments and annual interest rates.
   a. A person invests $2000 in a 3% account and $7000 in a 6% account.
   b. A person invests $4000 in a 3% account and $5000 in a 6% account.
   c. A person invests *x* dollars in a 3% account and *y* dollars in a 6% account.

24. Find the total interest earned in one year from the given investments and annual interest rates.
   a. A person invests $3000 in a 4% account and $6000 in a 12% account.
   b. A person invests $4000 in a 4% account and $8000 in a 12% account.
   c. A person invests *x* dollars in a 4% account and 2*x* dollars in a 12% account.

25. A person plans to invest a total of $20,000 in a First Funds TN Tax-Free I account at 6% annual interest and a W & R International Growth C account at 11% annual interest. How

much should she invest in each account so that the total interest in one year will be $1500?

26. A person plans to invest a total of $3500 in a cdbank.com account at 3% annual interest and a Turner Small Cap Value account at 16% annual interest. How much should he invest in each account so that the total interest in one year will be $200?

27. A person plans to invest a total of $8500 in a Middlesex Savings Bank CD account at 3.6% annual interest and a First Funds Growth & Income I account at 15% annual interest. How much should he invest in each account so that the total interest in one year will be $990?

28. A person plans to invest a total of $6500 in a Connecticut Bank CD account at 3% annual interest and an Artisan International account at 14% annual interest. How much should she invest in each account so that the total interest in one year will be $400?

29. A person plans to invest three times as much in a Limited Term NY Municipal X account at 5% interest as in a Calvert Income A account at 10% annual interest. How much will the person have to invest in each account to earn a total of $625 in one year?

30. A person plans to invest twice as much money in a BankUSA CD account at 3.5% annual interest as in a Fidelity Select Health Care account at 17% annual interest. How much will the person have to invest in each account to earn a total of $700 in one year?

31. A person plans to invest an equal amount of money in a USAA Tax Exempt Short-Term account at 5% annual interest and a Putnam Global Equity B account at 9% annual interest. How much should the person invest in each account so that the total interest in one year will be $700?

32. A person plans to invest twice as much money in a LaSalle Bank CD account at 2.5% annual interest as in a Franklin Strategic Mortgage account at 8% annual interest. How much should the person invest in each account so that the total interest for one year will be $650?

33. A salad dressing is a 65% oil solution. Determine the amount of oil in the given amount of oil solution.
   a. 2 ounces of oil solution
   b. 3 ounces of oil solution
   c. x ounces of oil solution

34. Some milk is a 2% butterfat solution. Determine the amount of butterfat in the given amount of 2% milk.
   a. 4 cups of milk
   b. 8 cups of milk
   c. M cups of milk

35. a. If 4 ounces of a 35% alcohol solution is mixed with 3 ounces of a 10% alcohol solution, how much pure alcohol is in the mixture?
   b. How many ounces of a 35% acid solution and a 10% acid solution need to be mixed to make 15 ounces of a 20% solution?

36. a. If 2 liters of a 24% alcohol solution is mixed with 5 liters of an 8% alcohol solution, how much pure alcohol is in the mixture?
   b. How many liters of a 24% acid solution and an 8% acid solution need to be mixed to make 8 liters of a 20% solution?

37. A chemist wants to mix a 20% acid solution and a 30% acid solution to make a 22% acid solution. How many quarts of each solution must be mixed to make 5 quarts of the 22% solution?

38. A chemist wants to mix a 30% acid solution and a 50% acid solution to make a 42% acid solution. How many quarts of each solution must be mixed to make 10 quarts of the 42% solution?

39. How many gallons of a 10% antifreeze solution and a 25% antifreeze solution need to be mixed to make 3 gallons of a 20% antifreeze solution?

40. How many liters of a 25% antifreeze solution and a 40% antifreeze solution need to be mixed to make 6 liters of a 30% antifreeze solution?

41. A chemist wants to mix a 15% acid solution and a 35% acid solution to make a 30% acid solution. How many quarts of each solution must be mixed to make 4 quarts of the 30% solution?

42. A chemist wants to mix a 5% acid solution and a 20% acid solution to make a 10% acid solution. How many quarts of each solution must be mixed to make 6 quarts of the 10% solution?

43. How many gallons of a 12% acid solution and a 24% acid solution need to be mixed to make 8 gallons of a 21% acid solution?

44. How many quarts of a 10% acid solution and a 45% acid solution need to be mixed to make 5 quarts of a 31% acid solution?

45. A chemist needs 5 ounces of a 12% alcohol solution, but has only a 20% alcohol solution. How much 20% solution and water should she mix to make the desired 5 ounces of 12% solution?

46. A chemist needs 6 liters of a 25% alcohol solution, but has only a 30% alcohol solution. How much 30% solution and water should he mix to make the desired 6 liters of 25% solution?

47. For a sellout performance at a 300-seat theater,
   a. Find the total revenue if all tickets sell for $10.
   b. Find the total revenue if all tickets sell for $15.
   c. If some, none, or all tickets sell for $10 and the remaining tickets, if any, sell for $15, is it possible for the total revenue to be the amount that follows? If yes, how many of each ticket must be sold? If no, explain.
      i. $2500
      ii. $4875
      iii. $3250

48. A person plans to invest a total of $3000, part of it in an account at 5% annual interest and the rest in an account at 10% annual interest.
   a. Find the interest earned in one year from the following investments:
      i. The person invests all $3000 in the 5% account.
      ii. The person invests all $3000 in the 10% account.
   b. Is it possible for her to earn the given amount of interest in one year? If yes, how much should she invest in each account? If no, explain why not.
      i. $120
      ii. $330
      iii. $180

In many examples, we wrote a system of two equations that helped us find quantities for an authentic situation. For the following exercises, you will work backward. For example, given the system

$$xy = 48$$
$$y = x - 2$$

we can relate the following to the system:

A person wants to build a rectangular garden with an area of 48 square feet in which the width is 2 feet less than the length. Find the dimensions of the garden.

Describe an authentic situation for the given equation. Find the unknown quantities for your situation.

49. $y = x + 3$
     $2x + 2y = 50$

50. $x + y = 500$
     $15x + 25y = 8300$

51. $y = 2x$
     $0.03x + 0.07y = 340$

52. $x + y = 8$
     $0.35x + 0.55y = 0.40(8)$

Related Review

Solve the equation for the specified variable.

53. $P = 2(L + W)$, for L

54. $P = 2(L + W)$, for W

55. $A = P(1 + rt)$, for r

56. $A = P(1 + rt)$, for P

Expressions, Equations, and Graphs

Perform the indicated instruction. Then use words such as linear, one variable, *and* two variables *to describe the expression, equation, or system.*

57. Simplify $-4(7x - 3)$.

58. Solve the following:

$$3x + 2y = 6$$
$$y = 2x - 11$$

59. Solve $-4(7x - 3) = 5x + 2$.

60. Graph $3x + 2y = 6$ by hand.

6.6 LINEAR INEQUALITIES IN TWO VARIABLES; SYSTEMS OF LINEAR INEQUALITIES IN TWO VARIABLES

Objectives

▷ Graph a linear inequality in two variables.

▷ Solve a system of linear inequalities in two variables.

▷ Use a system of linear inequalities in two variables to make estimates.

Recall from Section 3.2 that a linear equation in two variables is an equation that can be put into the form $y = mx + b$ or $x = a$, where m, a, and b are constants. A **linear inequality in two variables** is an inequality of the form

$$y < mx + b \quad \text{or} \quad x < a$$

(or with $<$ replaced with $\leq$, $>$, or $\geq$), where m, a, and b are constants. Some examples are $y < 3x - 5$, $y \leq -2x + 4$, $5x + 2y > 10$, and $x \geq 2$. In the first part of this section, we will work with one linear inequality in two variables. In the second part, we will work with two or more such inequalities.

Linear Inequality in Two Variables

We use the terms *solution, satisfy,* and *solution set* for an inequality in two variables in much the same way that we have used them for equations in one or two variables and inequalities in one variable.

> **DEFINITION** *Satisfy, solution, solution set, and solve* for an inequality in two variables
>
> If an inequality in the two variables x and y becomes a true statement when a is substituted for x and b is substituted for y, we say that the ordered pair (a, b) **satisfies** the inequality and call (a, b) a **solution** of the inequality. The **solution set** of an inequality is the set of all solutions of the inequality. We **solve** the inequality by finding its solution set.

We describe the solution set of an inequality by graphing all of the solutions.

Example 1 Sketching the Graph of an Inequality

Graph $y > x - 1$.

Solution

We begin by sketching the graph of $y = x - 1$ (see Fig. 33).

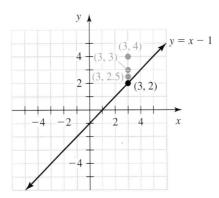

Figure 33 Some solutions of $y > x - 1$
(in blue)

To investigate how to solve $y > x - 1$, we choose a value of x (say, 3) and find several solutions with an x-coordinate of 3. For $y = x - 1$, if $x = 3$, then $y = (3) - 1 = 2$. So, the point $(3, 2)$ is on the line $y = x - 1$.

For $y > x - 1$, if $x = 3$, then

$$y > (3) - 1$$
$$y > 2$$

So, if $x = 3$, some possible values of y are $y = 2.5$, $y = 3$, and $y = 4$. Note that the points $(3, 2.5)$, $(3, 3)$, and $(3, 4)$ lie *above* the point $(3, 2)$, which is on the line $y = x - 1$ (see Fig. 33).

We could choose other values of x besides 3 and go through a similar argument. These investigations would suggest that the solutions of $y > x - 1$ lie *above* the graph of $y = x - 1$. This is, in fact, true. In Fig. 34, we shade the region that contains all of the points that represent solutions of $y > x - 1$.

We make the line $y = x - 1$ dashed in Fig. 34 to indicate that its points are *not* solutions of $y > x - 1$. For example, the point $(3, 2)$ on the line $y = x - 1$ does *not* satisfy the inequality $y > x - 1$:

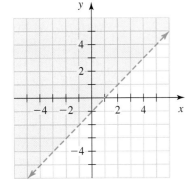

Figure 34 Graph of $y > x - 1$

$y > x - 1$	Original inequality
$2 \overset{?}{>} (3) - 1$	Substitute 3 for x and 2 for y.
$2 \overset{?}{>} 2$	Subtract.
false	

We can draw a graph of $y > x - 1$ by using a graphing calculator (see Fig. 35), but we have to imagine that the border $y = x - 1$ is drawn with a dashed line. To shade above a line, press $\boxed{Y=}$ and then press $\boxed{\triangleleft}$ twice. Next press $\boxed{\text{ENTER}}$ as many times as necessary for the triangle shown in Fig. 35 to appear.

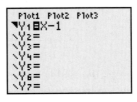 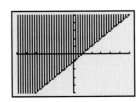

Figure 35 Graph of $y > x - 1$ (where we imagine that the border line is dashed)

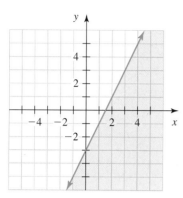

Figure 36 Graph of $y \leq 2x - 3$

Example 2 Sketching the Graph of an Inequality

Sketch a graph of $y \leq 2x - 3$.

Solution

The graph of $y \leq 2x - 3$ is the line $y = 2x - 3$, as well as the region below that line (see Fig. 36). We use a *solid* line along the border $y = 2x - 3$ to indicate that the points on the line $y = 2x - 3$ are solutions of $y \leq 2x - 3$. ∎

Graph of an Inequality in Two Variables

- The graph of an inequality of the form $y > mx + b$ is the region above the line $y = mx + b$. The graph of an inequality of the form $y < mx + b$ is the region below the line $y = mx + b$. For either inequality, we use a dashed line to show that $y = mx + b$ is not part of the graph.
- The graph of an inequality of the form $y \geq mx + b$ is the line $y = mx + b$, as well as the region above that line. The graph of an inequality of the form $y \leq mx + b$ is the line $y = mx + b$, as well as the region below that line.

To sketch a graph of an inequality in two variables, if the variable y is not alone on one side of the inequality, we begin by isolating it. Recall from Section 5.5 that when we multiply or divide both sides of an inequality by a negative number, we must reverse the inequality symbol.

Example 3 Sketching the Graph of an Inequality

Sketch a graph of $-2x - 5y > 10$.

Solution

First, we get y alone on one side of the inequality:

$$-2x - 5y > 10 \qquad \text{Original inequality}$$
$$-2x - 5y + 2x > 10 + 2x \qquad \text{Add 2x to both sides.}$$
$$-5y > 2x + 10 \qquad \text{Combine like terms.}$$
$$\frac{-5y}{-5} < \frac{2x}{-5} + \frac{10}{-5} \qquad \text{Divide both sides by } -5; \text{ reverse inequality symbol.}$$
$$y < -\frac{2}{5}x - 2 \qquad \text{Simplify.}$$

The graph of $y < -\dfrac{2}{5}x - 2$ is the region below the line $y = -\dfrac{2}{5}x - 2$ (see Fig. 37).

To verify our work, we choose a point on our graph—say, $(-4, -2)$—and check that it satisfies the original inequality:

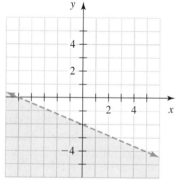

Figure 37 Graph of $-2x - 5y > 10$

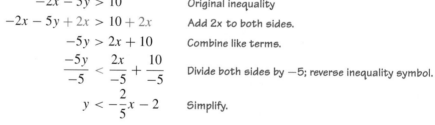

$$-2x - 5y > 10 \qquad \text{Original inequality}$$
$$-2(-4) - 5(-2) \stackrel{?}{>} 10 \qquad \text{Substitute } -4 \text{ for } x \text{ and } -2 \text{ for } y.$$
$$18 \stackrel{?}{>} 10 \qquad \text{Simplify.}$$
$$\text{true}$$

To check further, we could choose several other points on the graph and check that each point satisfies the inequality.

We could also choose several points that are *not* on the graph and check that each point does *not* satisfy the inequality. For example, note that $(0, 0)$ is not on the graph and that this point does not satisfy the original inequality $-2x - 5y > 10$:

$$-2x - 5y > 10 \qquad \textit{Original inequality}$$

$$-2(0) - 5(0) \overset{?}{>} 10 \qquad \textit{Substitute 0 for x and 0 for y.}$$

$$0 \overset{?}{>} 10 \qquad \textit{Simplify.}$$

$$\text{false} \qquad \blacksquare$$

WARNING   It is a common error to think that the graph of an inequality such as $-2x - 5y > 10$ is the region *above* the line $-2x - 5y = 10$, because the symbol ">" means "is greater than." However, we must first isolate y on the left side of a linear inequality before we can determine whether the graph includes the region that is above or below a line. In fact, in Example 3 we wrote the inequality $-2x - 5y > 10$ as $y < -\frac{2}{5}x - 2$ and concluded that the graph is the region *below* the line $y = -\frac{2}{5}x - 2$.

Example 4   Sketching the Graph of an Inequality

Sketch the graph of the inequality.

1. $y \geq -2$ **2.** $x < 3$

Solution

1. The graph of $y \geq -2$ is the horizontal line $y = -2$, as well as the region above that line (see Fig. 38).
2. Ordered pairs with x-coordinates *less* than 3 are represented by points that lie to the *left* of the vertical line $x = 3$. So, the graph of $x < 3$ is the region to the left of $x = 3$ (see Fig. 39).

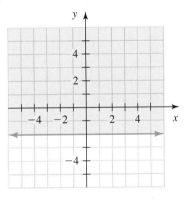
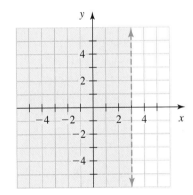

Figure 38 Graph of $y \geq -2$ **Figure 39** Graph of $x < 3$ $\blacksquare$

System of Inequalities

A **system of inequalities in two variables** consists of two or more linear inequalities in two variables. Here is an example:

$$y > x - 1$$
$$y \leq -\frac{2}{3}x - 2$$

DEFINITION   *Solution, solution set,* and *solve* for a system of inequalities in two variables

We say that an ordered pair (a, b) is a **solution** of a system of inequalities in two variables if it satisfies all of the inequalities of the system. The **solution set** of a system is the set of all solutions of the system. We **solve** a system by finding its solution set.

Recall from Section 6.1 that we can find the solution set of a system of two linear *equations* in two variables by locating any intersection point(s) of the graphs of the two equations. Similarly, **the solution set of a system of inequalities in two variables can be found by locating the intersection of the graphs of all of the inequalities.** This makes sense, because a solution is an ordered pair that satisfies all of the inequalities, which means that the point which represents the ordered pair lies on the graphs of all of the inequalities.

In this text, we use a graph to describe the solution set of a system of inequalities.

Example 5 Graphing the Solution Set of a System of Inequalities

Solve the system

$$y > x - 1$$
$$y \leq -\frac{2}{3}x - 2$$

Solution

First, we sketch the graph of $y > x - 1$ (see Fig. 40, blue region) and the graph of $y \leq -\frac{2}{3}x - 2$ (see Fig. 40, red region). The graph of the solution set of the system is the intersection of the graphs of the inequalities, which is shown in Fig. 41.

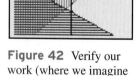

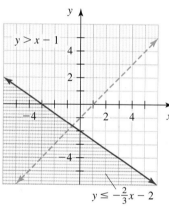

Figure 40 Graphs of $y > x - 1$ (in blue) and $y \leq -\frac{2}{3}x - 2$ (in red)

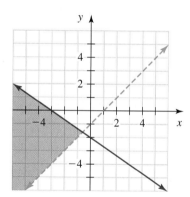

Figure 41 Graph of the solution set of the system

Figure 42 Verify our work (where we imagine that the border $y = x - 1$ is dashed)

We can use a graphing calculator to draw a graph of the solution set of the system (see Fig. 42), where we imagine that the border $y = x - 1$ is drawn with a dashed line. The graph of the solution set is the region shaded by vertical lines (in blue) and horizontal lines (in red). ∎

Example 6 Graphing the Solution Set of a System of Inequalities

Graph the solution set of the system

$$y > -\frac{3}{2}x$$
$$-2x + 5y < 5$$

Solution

First, we sketch the graph of $y > -\frac{3}{2}x$ (see Fig. 43, blue region) and the graph of $-2x + 5y < 5$ (see Fig. 43, red region). The graph of the solution set of the system is the intersection of the graphs of the inequalities, which is shown in Fig. 44.

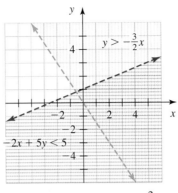

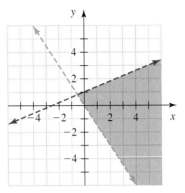

Figure 43 Graphs of $y > -\dfrac{3}{2}x$ (in blue) and $-2x + 5y < 5$ (in red)

Figure 44 Graph of the solution set of the system

Example 7 Graphing the Solution Set of a System of Inequalities

Graph the solution set of the system

$$-x + 3y \leq 9$$
$$-x + 3y \geq 3$$
$$x \geq 1$$
$$x \leq 4$$

Solution

In Fig. 45, we use arrows to indicate the graphs of each of the four inequalities. The solution set of the system is the intersection of the four graphs.

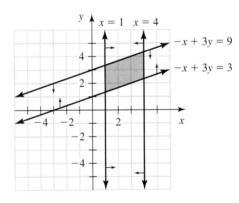

Figure 45 Graph of the solution set of the system

Modeling with Systems of Inequalities

We can use a system of inequalities to help us make estimates about authentic situations.

Example 8 Using a System of Inequalities to Make Estimates

For you to get the most benefits, yet be safe, while exercising, your heart rate should be within a certain range, called the *target heart-rate zone*. Table 33 shows the lower and upper limits of the target heart-rate zones for various ages.

Let h be the heart rate (in beats per minute) for a person who is a years of age. Linear models for the lower and upper limits of the target heart-rate zones are

$$h = -0.75a + 165 \quad \text{Upper limit}$$
$$h = -0.5a + 110 \quad \text{Lower limit}$$

1. Find a system of inequalities that describes the target heart-rate zones for people between the ages of 10 years and 100 years.
2. Graph the solution set of the system of inequalities that you found in Problem 1.
3. What is the target heart-rate zone for a person who is 23 years old?

Table 33 Target Heart-Rate Zone

| Age | Target Heart-Rate Zone (beats per minute) | |
	Lower Limit	Upper Limit
20	100	150
30	95	142.5
40	90	135
50	85	127.5
60	80	120
70	75	112.5
80	70	105
90	65	97.5

Source: *American Heart Association*

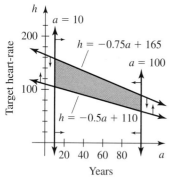

h = −0.75a + 165

a = 100

h = −0.5a + 110

Figure 46 Graph of the solution set of the system

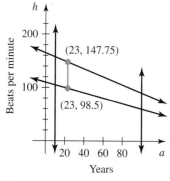

(23, 147.75)

(23, 98.5)

Figure 47 The target heart-rate zone for a 23-year-old person

Solution

1. To be in the target heart-rate zone, a person's heart rate must be less than or equal to the upper limit ($h \leq -0.75a + 165$) and greater than or equal to the lower limit ($h \geq -0.5a + 110$). We are seeking the zones for people between the ages of 10 years and 100 years, inclusive: $a \geq 10$ and $a \leq 100$.

   So, the target heart-rate zones can be described by the system

$$h \leq -0.75a + 165$$
$$h \geq -0.5a + 110$$
$$a \geq 10$$
$$a \leq 100$$

2. In Fig. 46, we use arrows to indicate the graphs of each of the four inequalities. The solution set of the system is the intersection of the four graphs.

3. The target heart-rate zone for a 23-year-old person is represented by the blue vertical line segment above $a = 23$ on the a-axis in Fig. 47. From this line segment, we see that the target heart-rate zone for a 23-year-old person is between 98.5 beats per minute and 147.75 beats per minute, inclusive.

   We can also find the target heart-rate zone for a 23-year-old person by substituting 23 for a in the upper limit and lower limit models:

$$h = -0.75(23) + 165 = 147.75 \qquad \text{Upper limit}$$
$$h = -0.5(23) + 110 = 98.5 \qquad \text{Lower limit}$$

This means that the target heart-rate zone is between 98.5 beats per minute and 147.75 beats per minute, inclusive, which checks. ∎

group exploration

Looking ahead: Graphing quadratic equations in two variables

Graph the equation by hand. To begin, substitute the values $-3, -2, -1, 0, 1, 2$, and 3 for x. Make other substitutions as necessary. Then, use a graphing calculator to verify your work.

1. $y = x^2$ 2. $y = x^2 + 1$ 3. $y = x^2 - 3$
4. $y = 2x^2$ 5. $y = -x^2$

TIPS FOR SUCCESS: The Value of Learning Mathematics

Imagine someone who is a math expert. Are you imagining a person wearing broken glasses that are taped together, a pocket protector, and wrinkled, unfashionable clothing—in other words, a math nerd? Of course, this stereotype does not accurately describe most mathematicians. But perhaps it is because of the stereotype of a math nerd that many students don't want to become "too good" at mathematics.

Learning mathematics will not transform you into a nerd. Rather, it will transform you into a more educated, well-rounded person. Learning mathematics will equip you to be more effective in many lines of work, as well as when you do math-intensive activities such as investing or filling out a tax return. Most people have a high respect for mathematicians due to their commitment to a useful and challenging subject.

HOMEWORK 6.6   FOR EXTRA HELP ▶

Student Solutions Manual   PH Math/Tutor Center   MathXL   MathXL®   MyMathLab   MyMathLab

Which of the given ordered pairs satisfy the given inequality?

1. $y < 2x - 3$ $(4, 1), (-2, 3), (-3, -1)$

2. $y > -3x + 2$ $(4, -1), (-4, 3), (-2, -5)$

3. $2x - 5y \geq 10$ $(-3, -1), (-1, -4), (5, 0)$

4. $3y \leq -4x$ $(2, 1), (-3, 4), (-2, -1)$

Graph the inequality by hand.

5. $y > x - 3$

6. $y < 2x - 4$

7. $y \leq -2x + 3$

8. $y \geq -3x + 4$

9. $y < \dfrac{1}{3}x + 1$

10. $y > \dfrac{2}{3}x - 5$

11. $y \geq -\dfrac{3}{5}x - 1$

12. $y \leq -\dfrac{1}{2}x + 3$

13. $y > x$

14. $y < -x$

15. $y - 2x < 0$

16. $y + 3x \leq 0$

17. $3x + y \leq 2$

18. $2x + y \geq 3$

19. $4x - y > 1$

20. $x - y < 3$

21. $4x - 3y \geq 0$

22. $5x - 2y \leq 0$

23. $4x + 5y \geq 10$

24. $2x + 4y > 4$

25. $2x - 5y < 5$

26. $3x - 4y > 12$

27. $y > 3$

28. $y < -2$

29. $x \geq -3$

30. $x \leq 4$

31. $x < -2$

32. $y > 1$

33. $y \leq 0$

34. $x \geq 0$

Graph the solution set of the system of inequalities by hand.

35. $y \geq \dfrac{1}{3}x - 3$
$y \leq -\dfrac{3}{2}x + 2$

36. $y \leq \dfrac{2}{3}x + 1$
$y \geq -\dfrac{4}{3}x - 2$

37. $y > 2x - 3$
$y > -\dfrac{3}{4}x + 1$

38. $y < -3x + 5$
$y < \dfrac{2}{5}x - 3$

39. $y \leq -\dfrac{2}{3}x - 3$
$y > 2x + 1$

40. $y \leq \dfrac{4}{3}x - 2$
$y > -x + 4$

41. $y \geq -2x - 1$
$y > \dfrac{1}{3}x + 2$

42. $y \leq 3x - 3$
$y < -\dfrac{1}{4}x + 1$

43. $y > -3x + 4$
$y \leq 2x + 3$

44. $y > -4x + 5$
$y \leq -x + 1$

45. $y \leq \dfrac{2}{3}x$
$y < -\dfrac{2}{5}x$

46. $y \leq -\dfrac{1}{2}x$
$y < \dfrac{4}{3}x$

47. $5x - 3y \leq 12$
$-2y < x$

48. $x - 2y < 6$
$3y \geq 6x$

49. $x - y \leq 2$
$2x + y < 1$

50. $x + y < 1$
$3x - y \geq 2$

51. $2x - 3y > 3$
$3x + 5y \geq 10$

52. $-3x + 4y \leq 4$
$5x - 2y < 2$

53. $y < 3$
$x \geq -2$

54. $y \geq 2$
$x < -1$

55. $y \leq 1$
$y \geq -2$
$x \geq -3$
$x \leq 4$

56. $y \leq -1$
$y \geq -4$
$x \leq 2$
$x \geq -5$

57. $2x - 5y \geq -5$
$2x - 5y \leq 15$
$x \geq -1$
$x \leq 3$

58. $3x + 4y \leq 16$
$3x + 4y \geq 8$
$x \geq 1$
$x \leq 4$

59. The lower and upper limits of ideal weights of men with a medium frame are listed in Table 34 for various heights. Assume that the men are wearing 5 pounds of clothing, including shoes with 1-inch heels.

Table 34 Ideal Weights of Men with a Medium Frame

Height (inches)	Ideal Weight (pounds)	
	Lower Limit	Upper Limit
65	137	148
67	142	154
69	148	160
71	154	166
73	160	174
76	171	187

Source: *Metropolitan Life Insurance Company*

Let w be the ideal weight (in pounds, including clothes) of a man who has a medium frame and a height of h inches (including shoes). Linear models for the lower and upper limits of the ideal weights are

$$w = 3.50h - 80.97 \quad \text{Upper limit}$$
$$w = 3.08h - 64.14 \quad \text{Lower limit}$$

a. Find a system of inequalities that describes the ideal weights of men who have a medium frame and are between the heights of 63 inches and 78 inches, inclusive.

b. Graph by hand the solution set of the system of inequalities that you found in part (a).

c. What is the ideal weight range of men who have a medium frame and a height of 68 inches?

60. In Exercise 19 of Homework 5.3, you found an equation close to $H = 4.5d + 15.3$, where H is the number of "human years" that is equivalent to d "dog years" for dogs that weigh between 20 pounds and 50 pounds, inclusive (see Table 35). A reasonable model for 50–90-pound dogs is $H = 5.4d + 12.3$.

Assume that the model $H = 4.5d + 15.5$ is best suited for dogs that weigh 35 pounds (the mean of 20 pounds and 50 pounds) and that the model $H = 5.4d + 12.3$ is best suited for dogs that weigh 70 pounds (the mean of 50 pounds and 90 pounds).

Table 35 Dog Years Compared with Human Years		
	Human Years	
Dog Years	**20-50-Pound Dogs**	**50-90-Pound Dogs**
1	15	14
2	24	22
3	29	29
5	38	40
7	47	50
9	56	61
11	65	72
13	74	82
15	83	93

Source: *Fred Metzger, State College, PA*

a. Find a system of inequalities that describes the numbers of human years that are equivalent to dog years between 10 dog years and 20 dog years, inclusive, for dog weights between 35 pounds and 70 pounds, inclusive.
b. Graph by hand the solution set of the system of inequalities that you found in part (a).
c. What are the numbers of human years that are equivalent to 15 dog years for dogs that weigh between 35 pounds and 70 pounds, inclusive?

61. Find an inequality in two variables whose graph is shown in Fig. 48.

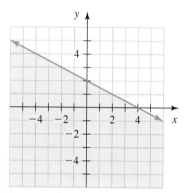

Figure 48 Exercise 61

62. Find a system of two linear inequalities whose solution set is shown in Fig. 49.

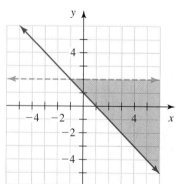

Figure 49 Exercise 62

63. A student believes that the graph of $5x - 2y < 6$ is the region below the line $5x - 2y = 6$. Is the student correct? Explain.

64. A student believes that the graph of $y \leq 3x - 1$ is the region below the line $y = 3x - 1$. Is the student correct? Explain.

65. The graphs of $y = ax + b$ and $y = cx + d$ are sketched in Fig. 50.

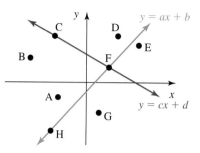

Figure 50 Exercise 65

Which of the points A, B, C, D, E, F, G, and H represent an ordered pair that
a. satisfies the inequality $y > ax + b$?
b. satisfies the inequality $y \leq cx + d$?
c. satisfies both $y > ax + b$ and $y \leq cx + d$?
d. satisfies neither $y > ax + b$ nor $y \leq cx + d$?

66. Consider the system of inequalities

$$y \geq 2x + 1$$
$$y < -x - 2$$

Find an ordered pair that
a. satisfies $y \geq 2x + 1$, but does not satisfy $y < -x - 2$.
b. satisfies $y < -x - 2$, but does not satisfy $y \geq 2x + 1$.
c. satisfies both inequalities.
d. does not satisfy either inequality.

67. Find three ordered pairs that are solutions of the inequality $y \geq 2x - 3$. Also, find three ordered pairs that are not solutions.

68. Find three ordered pairs that are solutions of the inequality $2x - y < 4$. Also, find three ordered pairs that are not solutions.

69. Explain how to graph an inequality in two variables by hand.

70. Explain how to solve a system of inequalities in two variables.

Related Review

71. Graph $y = 2x - 1$ by hand.

72. Graph $3x - 4y = 4$ by hand.

73. Graph the inequality $y \leq 2x - 1$ by hand.

74. Graph the inequality $3x - 4y > 4$ by hand.

75. Solve the inequality $3 \leq 2x - 1$, an inequality in *one* variable. Describe the solution set as an inequality, in interval notation, and in a graph.

76. Solve the inequality $3x - 4 > 4$, an inequality in *one* variable. Describe the solution set as an inequality, in interval notation, and in a graph.

Graph the solution set of the system of linear inequalities by hand.

77. $y \leq 2x - 1$
$y \geq -x + 5$

78. $3x - 4y > 4$
$x + 2y < 8$

Solve the system of linear equations.

79. $y = 2x - 1$
$y = -x + 5$

80. $3x - 4y = 4$
$x + 2y = 8$

Expressions, Equations, and Graphs

Give an example of the following, and then solve, simplify, or graph, as appropriate:

81. linear equation in one variable

82. linear equation in two variables

83. linear inequality in one variable

84. linear inequality in two variables

85. system of two linear equations in two variables

86. system of two linear inequalities in two variables

analyze this

GLOBAL WARMING LAB (continued from Chapter 5)

Recall from the Global Warming Lab in Chapter 5 that, to avoid a climate catastrophe, the IPCC has recommended that carbon emissions be lowered to 0.3 metric ton per person per year by 2050. Developed countries' carbon emissions are about 2.7 metric tons per person per year. So, these countries will have to reduce their per-person emissions significantly while trying to sustain their relatively strong economies. The United States, with carbon emissions of 5.4 metric tons per person per year, will have to reduce its emissions drastically.[*]

Developing countries' carbon emissions are about 0.4 metric ton per person per year. Due to their already low per-person emissions, it seems that these countries will have an easy time meeting IPCC's goal. However, by 2050 many such countries will have become *developed* countries, and their economies will have become much stronger. As a result of this growth, their carbon emissions will greatly increase without intervention.[†]

For example, China's carbon emissions were only 0.4 metric ton per person in 1980. Due to China's booming economy, however, the country's carbon emissions in 2050 may reach 3.2 metric tons per person per year—eight times the 1980 per-person level. This large increase in per-person carbon emissions will be amplified by a population increase of 500 million people in those 70 years (from 0.98 billion to 1.48 billion).[‡]

An analysis of the data in Table 36 shows that developing countries' carbon emissions are growing at a greater rate than developed countries' carbon emissions. This is occurring not only because developing countries are becoming more industrialized, but also because their populations are increasing significantly. Developed countries' economies and populations are growing at a much slower rate; in fact, most developed countries' populations will begin to decline slowly after 2020.[§]

Many challenges lie ahead for developed and developing countries, both of which will need to develop efficient systems that rely on alternative energy sources whenever possible. Citizens of developed countries will have the extra challenges of foregoing certain conveniences that up until now have been taken for granted. Developing countries will have the extra challenge of harnessing large population growths.

Table 36 Carbon Emissions

Year	Carbon Emissions (million metric tons carbon)	
	Developed Countries	Developing Countries
1994	3602	2480
1995	3628	2589
1996	3660	2700
1997	3698	2759
1998	3708	2729

Source: *Carbon Dioxide Information Analysis Center (CDIAC)*

Analyzing the Situation

1. Let c be carbon emissions (in million metric tons) of developed countries in the year that is t years since 1990. Use a graphing calculator to draw a scattergram, and then find a model of the situation.

2. Let c be carbon emissions (in million metric tons) of developing countries in the year that is t years since 1990. Use a graphing calculator to draw a scattergram, and then find a model of the situation.

3. Compare the slopes of both models. What does the comparison tell you about the situation?

4. Explain why developing countries' carbon emissions are increasing at a greater rate than developed countries' carbon emissions.

5. Use substitution or elimination to predict when developing countries' carbon emissions will equal developed countries' carbon emissions. What is that carbon emission value?

6. Use your result from Problem 5 to predict the per-person carbon emissions in the year when developing countries' carbon emissions will equal developed countries' carbon emissions. Assume that world population will be about 7.7 billion in that year.[¶] Given that the carbon emissions in 2000 were 1.1 metric tons per person, will per-person carbon emissions have increased or decreased by that year? Explain.

7. What are the challenges that lie ahead in trying to reduce carbon emissions? What challenges are unique to developed countries? To developing countries?

[*]World Development Report 1999/2000, The World Bank Group.

[†]Carbon Dioxide Information Analysis Center (CDIAC).

[‡]United Nations Population Division.

[§]U.S. Census Bureau.

[¶]United Nations Population Division.

Chapter Summary

Key Points
OF CHAPTER 6

Solution, solution set, and solve for a system (Section 6.1)

We say that an ordered pair (a, b) is a **solution** of a system of two equations in two variables if it satisfies both equations. The **solution set** of a system is the set of all solutions of the system. We **solve** a system by finding its solution set.

Using graphing to solve a system (Section 6.1)

The solution set of a system of two linear equations can be found by locating any intersection point(s) of the graphs of the two equations.

Intersection point of the graphs of two models (Section 6.1)

If the independent variable of two models represents time, then an intersection point of the graphs of the two models indicates when the quantities represented by the dependent variables were or will be equal.

Types of linear systems (Section 6.1)

There are three types of linear systems of two equations:

1. *One-solution system:* The lines intersect in one point. The solution of the system is the ordered pair that corresponds to that point.
2. **Inconsistent system:** The lines are parallel. The solution set of the system is the empty set.
3. **Dependent system:** The lines are identical. The solution set of the system is the set of the infinite number of solutions represented by points on the same line.

Using tables to solve a linear system (Section 6.1)

If an ordered pair is listed in both tables of solutions of two linear equations, then that ordered pair is a solution of that system.

Using substitution to solve a linear system (Section 6.2)

To use **substitution** to solve a system of two linear equations,

1. Isolate a variable to one side of either equation.
2. Substitute the expression for the variable found in step 1 into the other equation.
3. Solve the equation in one variable found in step 2.
4. Substitute the solution found in step 3 into one of the original equations, and solve for the other variable.

Using elimination to solve a linear system (Section 6.3)

To use **elimination** to solve a system of two linear equations,

1. Use the multiplication property of equality to get the coefficients of one variable to be equal in absolute value and opposite in sign.
2. Add the left sides and add the right sides of the equations to eliminate one of the variables.
3. Solve the equation in one variable found in step 2.
4. Substitute the solution found in step 3 into one of the original equations and solve for the other variable.

Solving inconsistent systems and dependent systems by substitution or elimination (Sections 6.2 and 6.3)

If the result of applying substitution or elimination to a linear system of two equations is

- a false statement, then the system is inconsistent; that is, the solution set is the empty set.
- a true statement that can be put into the form $a = a$, then the system is dependent; that is, the solution set is the set of ordered pairs represented by every point on the (same) line.

Methods of solving a linear system (Section 6.3)

Any linear system of two equations can be solved by graphing, substitution, or elimination. All three methods give the same result.

Methods of finding an intersection point of two linear models (Section 6.4)

We can find an intersection point of two linear models by graphing, substitution, or elimination.

Five-step problem-solving method (Section 6.5)

To solve some problems in which we want to find two quantities, it is useful to perform the following five steps:

- *Step 1: Define each variable.*
- *Step 2: Write a system of two equations.*
- *Step 3: Solve the system.*
- *Step 4: Describe each result.*
- *Step 5: Check.*

Using arithmetic to suggest how to form an equation (Section 6.5)	If you have difficulty forming an equation, try making up numbers for unknown quantities and performing arithmetic to compute a related quantity. Your computation may suggest how to form the desired equation.
Annual simple interest rate (Section 6.5)	The **annual simple interest rate** is the percentage of the **principal** that equals the **interest** earned per year.
Satisfy, solution, solution set, and solve for an inequality in two variables (Section 6.6)	If an inequality in the two variables x and y becomes a true statement when a is substituted for x and b is substituted for y, we say that the ordered pair (a, b) **satisfies** the inequality and call (a, b) a **solution** of the inequality. The **solution set** of an inequality is the set of all solutions of the inequality. We **solve** the inequality by finding its solution set.
Graph of an inequality in two variables (Section 6.6)	The graph of an inequality of the form $y > mx + b$ is the region above the line $y = mx + b$. The graph of an inequality of the form $y < mx + b$ is the region below the line $y = mx + b$. For either inequality, we use a dashed line to show that $y = mx + b$ is not part of the graph. The graph of an inequality of the form $y \geq mx + b$ is the line $y = mx + b$, as well as the region above that line. The graph of an inequality of the form $y \leq mx + b$ is the line $y = mx + b$, as well as the region below that line.
Solution, solution set, and solve for a system of inequalities in two variables (Section 6.6)	We say that an ordered pair (a, b) is a **solution** of a system of inequalities in two variables if it satisfies all of the inequalities of the system. The **solution set** of a system is the set of all solutions of the system. We **solve** a system by finding its solution set.
Finding the solution set of a system of inequalities (Section 6.6)	The solution set of a system of inequalities in two variables can be found by locating the intersection of the graphs of all of the inequalities.

CHAPTER 6 REVIEW EXERCISES

Find the solution of the system by graphing the equations by hand. Use "intersect" on a graphing calculator to verify your work.

1. $y = 2x - 3$
$y = -3x + 7$

2. $y = \dfrac{3}{2}x + 4$
$y = -\dfrac{1}{2}x - 4$

3. $y = \dfrac{2}{5}x$
$y = -2x$

4. $4x + y = 3$
$-3x + y = -4$

5. $-x + y = 4$
$2x + y = -5$

6. $x - 3y = 3$
$2x + 3y = -12$

Solve the system by substitution. Verify your solution by checking that it satisfies both equations of the system.

7. $3x - 2y = 11$
$y = 5x - 16$

8. $4x - 3y - 5 = 0$
$x = 4 - 2y$

9. $y = -5x$
$y = 2x$

10. $y = -3(x + 2)$
$y = 4(x - 5)$

11. $x + y = -1$
$2x - y = 4$

12. $x + 2y = 5$
$4x + 2y = -4$

Use substitution to solve the system, with coordinates of solutions rounded to the second decimal place. Verify your work by using "intersect" on a graphing calculator.

13. $y = -2.19x + 3.51$
$y = 1.54x - 6.22$

14. $y = -4.98x - 1.18$
$y = -0.57x + 4.08$

Solve the system by elimination. Verify your work by using "intersect" on a graphing calculator or by checking that your result satisfies both equations of the system.

15. $x - 2y = -1$
$3x + 5y = 19$

16. $2x - 5y = -3$
$4x + 3y = -19$

17. $3x + 8y = 2$
$5x - 2y = -12$

18. $4x - 3y = -6$
$-7x + 5y = 11$

19. $-2x - 5y = 2$
$3x + 6y = 0$

20. $y = 3x - 5$
$y = -2x + 5$

21. $2(x + 3) + y = 6$
$x - 3(y - 2) = -1$

22. $\dfrac{1}{2}x - \dfrac{2}{3}y = -\dfrac{5}{3}$
$\dfrac{1}{3}x - \dfrac{3}{2}y = -\dfrac{13}{6}$

Use elimination to solve the system, with coordinates of solutions rounded to the second decimal place. Verify your work by using "intersect" on a graphing calculator.

23. $y = 4.59x + 1.25$
$y = 0.52x + 4.39$

24. $y = 0.91x - 3.57$
$y = -3.58x + 6.05$

Solve the system by either substitution or elimination. Verify your work by using "intersect" on a graphing calculator or by checking that your result satisfies both equations of the system.

25. $2x - 7y = -13$
$5x + 3y = -12$

26. $4x + 7y = 8$
$x = 3 - 2y$

27. $-3x + 7y = 6$
$6x + 2y = -12$

28. $y = -x + 7$
$y = 2x - 5$

29. $4x + 5y = -6$
$2y = -3x - 8$

30. $y = x - 2$
$3x + 5y - 30 = 0$

31. $2x - 3y = 0$
$5x - 7y = -1$

32. $2(4x - 3) - 5y = 12$
$5(3x - 1) + 2y = 6$

Solve the system by elimination. If the system is inconsistent or dependent, say so.

33. $2x - 6y = 4$
$-3x + 9y = -3$

34. $y = -4x + 3$
$8x + 2y = 6$

35. Consider the system

$$y = -2x + 7$$
$$y = 3x - 3$$

Find an ordered pair that
 a. satisfies $y = -2x + 7$, but does not satisfy $y = 3x - 3$.
 b. satisfies $y = 3x - 3$, but does not satisfy $y = -2x + 7$.
 c. does not satisfy either equation.
 d. satisfies both equations.

36. Figure 51 shows the graphs of two linear equations. To the first decimal place, estimate the coordinates of the solution of the system.

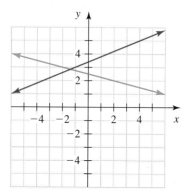

Figure 51 Exercise 36

37. Some solutions of two linear equations are shown in Tables 37 and 38. Find the solution of the system of the two equations.

Table 37 Solutions of One Equation	
x	y
0	75
1	73
2	71
3	69
4	67

Table 38 Solutions of the Other Equation	
x	y
0	5
1	8
2	11
3	14
4	17

38. Solve the system of equations three times, once by each of the three methods: graphing by hand, substitution, and elimination. Decide which method you prefer for this system.

$$3x + 4y = 15$$
$$2y = -5x + 11$$

39. In recent years, airlines have been increasing their use of small "regional jets" in place of turboprops or large jets (see Table 39).

Table 39 Regional Jets versus Turboprops or Large Jets		
	Percent	
Year	**Regional Jets**	**Turboprops or Large Jets**
2000	8	86
2001	11	83
2002	16	78
2003	21	72

Source: *Department of Transportation; Federal Aviation Administration*

 a. Let p be the percentage of airplanes that are regional jets at t years since 2000. Find an equation of a model to describe the data.

 b. Let p be the percentage of airplanes that are turboprops or large jets at t years since 2000. Find an equation of a model to describe the data.
 c. Use substitution or elimination to predict when the percentage of airplanes that are regional jets will be equal to the percentage of airplanes that are turboprops or large jets. What is that percentage? Use "intersect" on a graphing calculator to verify your work.

40. In 1995, 58% of car and light-truck sales were car sales, and that value has decreased by 1.25 percentage points per year since then. In 1995, 42% of car and light-truck sales were light-truck sales, and that value has increased by 1.25 percentage points per year (Source: *Autodata*).
 a. Let p be the percentage of car and light-truck sales that were car sales at t years since 1995. Find an equation of a model to describe the data.
 b. Let p be the percentage of car and light-truck sales that were light-truck sales at t years since 1995. Find an equation of a model to describe the data.
 c. Estimate when sales of cars and light trucks were equal by solving a system of equations.
 d. Estimate when sales of cars and light trucks were equal by substituting an appropriate value for p in one of the two models and then solving the resulting equation for t.
 e. Compare the slope of the car model with the slope of the light-truck model. What do you notice about these slopes? Why does this make sense?

41. The length of a rectangle is 2 feet more than three times the width. If the perimeter is 44 feet, find the dimensions of the rectangle.

42. An 8000-seat amphitheater has tickets for sale at $22 and $39. How many tickets should be sold at each price for a sellout performance to generate a total revenue of $201,500?

Graph the inequality in two variables by hand.

43. $y \le 3x - 5$ **44.** $y \ge -2x + 4$ **45.** $3x - 2y > 4$

46. $2y - 5x < 0$ **47.** $x \ge 3$ **48.** $y < -2$

Graph the solution set of the system of inequalities by hand.

49. $y > x + 1$
$y \le -2x + 5$

50. $y \ge \frac{3}{5}x + 1$
$x < -1$

51. $3x - 4y \ge 12$
$5y \le -3x$

52. $x > 2$
$y \le -1$

53. The lower and upper limits of ideal weights of women with a medium frame are listed in Table 40 for various heights. Assume that the women are wearing 5 pounds of clothing, including shoes with 1-inch heels.

Table 40 Ideal Weights of Women with a Medium Frame		
Height (inches)	**Ideal Weight (pounds)**	
	Lower Limit	**Upper Limit**
61	115	129
63	121	135
65	127	141
67	133	147
69	139	153
72	148	162

Source: *Metropolitan Life Insurance Company*

Let w be the ideal weight (in pounds, including clothes) of a woman who has a medium frame and a height of h inches (including shoes). Linear models for the lower and upper limits of the ideal weights are

$$w = 3h - 54 \qquad \text{Upper limit}$$
$$w = 3h - 68 \qquad \text{Lower limit}$$

a. Find a system of inequalities that describes the ideal weights of women who have a medium frame and

are between the heights of 58 inches and 74 inches, inclusive.

b. Graph by hand the solution set of the system of inequalities that you found in part (a).

c. What is the ideal weight range of women who have a medium frame and a height of 70 inches?

54. Find three ordered pairs that are solutions of the inequality $4x - 3y > 9$. Also, find three ordered pairs that are not solutions.

CHAPTER 6 TEST

1. Find the solution of this system by graphing the equations by hand:
$$y = -\frac{2}{5}x - 1$$
$$y = -2x + 7$$

2. Use "intersect" on a graphing calculator to solve this system, with the coordinates of the solution rounded to the second decimal place:
$$y = \frac{2}{3}x + 4$$
$$3x + 4y = -2$$

Solve the system by substitution.

3. $5x - 2y = 4$
     $y = 3x - 1$

4. $3x + 4y = 9$
     $x - 2y = -7$

Solve the system by elimination.

5. $-7x - 2y = -8$
     $5x + 4y = -2$

6. $2x - 5y = -18$
     $3x + 4y = -4$

Solve the system by substitution or elimination. If the system is inconsistent or dependent, say so.

7. $3x - 5y = -21$
     $x = 2(2 - y)$

8. $2x - 3y = 4$
     $-4x + 6y = -8$

9. $x = 2y - 3$
     $3x - 6y = 12$

10. $4x - 7y = 6$
      $-5x + 2y = -21$

11. Use substitution or elimination to solve this system, with coordinates of the solution rounded to the second decimal place:
$$y = -1.94x + 8.62$$
$$y = 1.25x - 2.38$$

12. The graphs of $y = ax + b$ and $y = cx + d$ are sketched in Fig. 52.

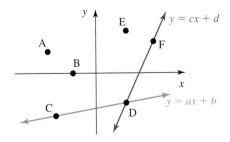

Figure 52
Exercise 12

Which of the points A, B, C, D, E, and F

a. satisfy the equation $y = ax + b$?

b. satisfy the equation $y = cx + d$?

c. satisfy both equations?

d. do not satisfy either equation?

13. Figure 53 shows the graphs of two linear equations. Find the solution of the system.

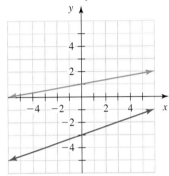

Figure 53 Exercise 13

14. Create a system of two linear equations whose solution is $(2, 4)$.

15. Find the coordinates of the points A, B, C, and D as shown in Fig. 54. The equations of the lines l_1 and l_2 are provided, but no attempt has been made to sketch the lines accurately except for showing the intersection points.

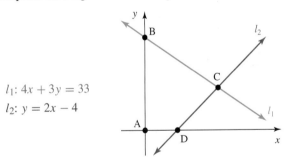

l_1: $4x + 3y = 33$

l_2: $y = 2x - 4$

Figure 54 Exercise 15

16. Supreme Court Justice Sandra Day O'Connor increasingly sided with the liberal justices in 5–4 decisions (see Table 41).

Table 41 Supreme Court Justice O'Connor's Votes in 5-4 Rulings

Year	Number of 5-4 Rulings	Number of Rulings Voted with Conservatives	Voted with Liberals
2000	18	11 (61.1%)	1 (5.6%)
2001	27	15 (55.6%)	5 (18.5%)
2002	21	10 (47.6%)	4 (19.1%)
2003	14	5 (35.7%)	4 (28.6%)

Source: *Harvard Law Review*

a. Let p be the percentage of 5–4 rulings in which O'Connor voted with conservative justices in the year that is t years since 2000. Find an equation of a model to describe the data.

b. Let p be the percentage of 5–4 rulings in which O'Connor voted with liberal justices in the year that is t years since 2000. Find an equation of a model to describe the data.

c. Use substitution or elimination to estimate when Justice O'Connor voted with liberal justices as often as she did with conservative justices. What is that percentage?

17. The number of guns in circulation in the United States was 230 million in 2000 and has since increased by about 4.9 million per year (Source: *Bureau of Alcohol, Tobacco, and Firearms*). The U.S. population was 280 million in 2000 and has since increased by about 2.9 million per year (Source: *U.S. Census Bureau*).

a. Let n be the number of guns in circulation (in millions) at t years since 2000. Find an equation of a model to describe the data.

b. Let n be the U.S. population (in millions) at t years since 2000. Find an equation of a model to describe the data.

c. Predict when there will be, on average, one gun per person in the United States. Will everyone own a gun in that year? How many guns will be in circulation?

18. A person plans to invest a total of $7000, part of it in an account at 3% annual interest and the rest in an account at 7% annual interest. How much should the person invest in each account so that the total interest in one year will be $410?

Graph the inequality in two variables by hand.

19. $5x - 2y \le 6$ **20.** $y < -3$

Graph the solution set of the system of inequalities by hand.

21. $y \le -3x + 4$ **22.** $y > 2$
$\quad\; x - 3y > 6$ $\quad\;\; x \ge -3$

CUMULATIVE REVIEW OF CHAPTERS 1-6

1. The low temperatures in New York City for the first four days of March are $4°F$, $-2°F$, $-1°F$, and $3°F$. Let F be the temperature in Fahrenheit degrees. Use points on a number line to describe these low temperatures.

2. The rate at which a cricket chirps depends on the temperature. Let n be the number of times a cricket chirps per minute when the temperature is t degrees Fahrenheit. What does the ordered pair (70, 129) represent?

Perform the indicated operations. Use a calculator to verify your work.

3. $-\dfrac{26}{27} \cdot \dfrac{12}{13}$ **4.** $\dfrac{5}{7} - \left(-\dfrac{3}{5}\right)$

5. Evaluate $a^2 - bc + b^2$, where $a = -3$, $b = -2$, and $c = 4$.

6. Find the slope of the line that contains the points $(-3, -2)$ and $(5, -8)$. State whether the line is increasing, decreasing, horizontal, or vertical.

7. Graphs of four equations are shown in Fig. 55. State whether m and b are positive, negative, zero, or undefined for the $y = mx + b$ form of each equation.

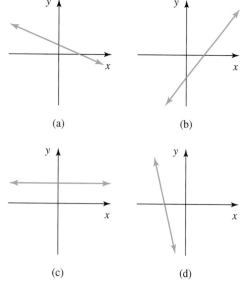

(a)   (b)

(c)   (d)

Figure 55 Exercise 7

8. Some solutions of four linear equations are provided in Table 42. Find an equation of each of the four lines.

Table 42 Solutions of Four Linear Equations

Equation 1		Equation 2		Equation 3		Equation 4	
x	**y**	**x**	**y**	**x**	**y**	**x**	**y**
0	49	0	11	3	39	2	14
1	41	1	15	4	37	4	20
2	33	2	19	5	35	5	23
3	25	3	23	6	33	8	32
4	17	4	27	7	31	9	35

For Exercises 9–12, simplify the expression or solve the equation, as appropriate. Use a graphing calculator to verify your work.

9. $2 - 5(4x + 8) = 3(2x - 7) + 3$

10. $-3(2p - w) - (7p + 2w) + 5$

11. $\dfrac{2}{3}(6w + 9y - 15)$ **12.** $\dfrac{3m}{5} - \dfrac{2}{3} = \dfrac{4}{5}$

13. Solve $ax + by = c$ for y.

Let x be a number. Translate the following to an expression or an equation, as appropriate, and then simplify the expression or solve the equation.

14. Six, plus 3 times the sum of 4 and the number

15. Four minus the quotient of the number and 3 is 2.

Determine the slope and the y-intercept, and use them to graph the equation by hand. Use a graphing calculator to verify your work.

16. $y = 2x - 4$ **17.** $x - 2y = 6$

18. $5x + 2y - 12 = 0$ **19.** $y = -3$

For Exercises 20 and 21, find the x-intercept and y-intercept. Then graph the equation by hand.

20. $2x - 5y = 10$ **21.** $y = -2x + 4$

22. Graph the equation $x = 4$.

23. Determine whether the lines $2x + 5y = 7$ and $y = \dfrac{2}{3}x - 3$ are parallel.

24. Find an equation of the line that has slope $-\dfrac{2}{5}$ and contains the point $(3, -2)$. Write your equation in slope–intercept form. Check that the point $(3, -2)$ satisfies your equation.

Find an equation of the line that passes through the two given points. If possible, write your equation in slope–intercept form. Verify your equation by using a graphing calculator.

25. $(-5, 1)$ and $(-2, -3)$ **26.** $(-2, 8)$ and $(-2, 1)$

For Exercises 27–30, refer to Fig. 56.

27. Find y when $x = -3$.   **28.** Find x when $y = -1$.

29. Find the x-intercept.   **30.** Find an equation of the line.

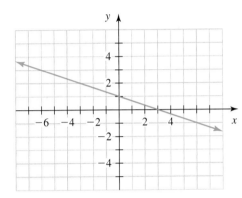

Figure 56 Exercises 27–30

31. Solve the inequality $-3(x - 5) > 18$, an inequality in one variable. Describe the solution set as an inequality, in interval notation, and in a graph.

32. Find the solution of this system by graphing the equations by hand:

$$y = 2x - 3$$
$$x + y = 3$$

33. Use "intersect" on a graphing calculator to solve this system, with the coordinates of the solution rounded to the second decimal place:

$$y = -2.9x + 7.8$$
$$y = 1.3x - 6.1$$

Solve the system by substitution or elimination. Use "intersect" on a graphing calculator to verify your work.

34. $3x + 5y = -1$ **35.** $4x - y - 9 = 0$
$\quad\;\, 2x - 3y = 12$ $\qquad\;\; y = 5 - 3x$

36. Use elimination or substitution to solve the system, with coordinates of solutions rounded to the second decimal place. Verify your work by using "intersect" on a graphing calculator.

$$y = -2.9x + 97.8$$
$$y = 3.1x - 45.6$$

37. Solve the system three times, once by each of the three methods: graphing by hand, substitution, and elimination. Decide which method you prefer for this system.

$$2x - 3y = 19$$
$$y = 4x - 13$$

38. Some solutions of two linear equations are shown in Tables 43 and 44. Find the solution of the system of the two equations.

Table 43 Solutions of One Equation	
x	**y**
0	97
1	93
2	89
3	85
4	81

Table 44 Solutions of the Other Equation	
x	**y**
0	7
1	9
2	11
3	13
4	15

39. Graph the inequality $y < -\dfrac{2}{3}x + 3$.

40. Graph the solution set of the system

$$3x - 5y \le 10$$
$$x > -4$$

41. Traffic safety is usually measured by the fatality rate, which is the number of traffic-related deaths per 100 million miles traveled. Fatality rates are shown in Table 45 for various years.

Table 45 Traffic Fatality Rates		
Year	**Fatalities (thousands)**	**Fatality Rate (number of deaths per 100 million miles)**
1998	41.5	1.58
1999	41.7	1.55
2000	41.9	1.53
2001	42.2	1.51
2002	43.0	1.51
2003	42.6	1.48
2004	42.6	1.46

Source: *National Center for Statistics & Analysis*

Let r be the traffic fatality rate (in number of deaths per 100 million miles traveled) at t years since 1990.
  a. Explain how it is possible for the fatality rate to be generally decreasing when the number of fatalities is generally increasing.
  b. Use a graphing calculator to draw a scattergram of the data that describes the values of t and r.
  c. Find an equation of a linear model that describes the relationship between t and r.
  d. Find the r-intercept. What does it mean in this situation?
  e. Transportation Secretary Norman Mineta proposed a plan to lower the traffic fatality rate to 1.0 death per 100 million vehicle miles traveled in 2008. Predict when the fatality rate will reach 1.0 death per 100 million miles traveled. Is Mineta's plan an ambitious one? Explain.
  f. Predict the fatality rate in 2008.

42. Since 1990, the divorce rate in the United States has decreased, but the divorce rate in Japan has nearly doubled (see Table 46).
  a. Let r be the divorce rate (number of divorces per 1000 people) in Japan at t years since 1990. Find an equation of a model to describe the data.
  b. Let r be the divorce rate (number of divorces per 1000 people) in the United States at t years since 1990. Find an equation of a model to describe the data.

Table 46 Divorce Rates in Japan and the United States

| Year | Divorce Rate (number of divorces per 1000 people) | |
	Japan	United States
1990	1.3	4.7
1992	1.5	4.8
1994	1.6	4.6
1996	1.7	4.3
1998	2.0	4.2
2000	2.1	4.2
2002	2.3	4.0
2003	2.3	4.0

Source: *Japanese Ministry of Health, Labor and Welfare; U.S. National Center for Health Statistics*

c. Predict when the divorce rates in Japan and the United States will be equal. What is that divorce rate?

43. The enrollment of a college was 25,700 students in 2000, and since then enrollment has decreased by 375 students per year. Let E be the enrollment (in number of students) at t years since 2000.
   a. Find the slope of the linear model that describes the situation. What does it mean in this situation?
   b. Find an equation of the model.
   c. Predict when enrollment will be 21,500 students.

d. Predict in which years enrollment will be less than 20,000 students.

44. The Telecommunications Act of 1996 forced regional Bell® companies to lease local phone lines at a discounted price to competing companies. Bell companies controlled 181.3 million telephone lines in 2000, and their number of lines has since decreased by 6.8 million lines per year. Other companies controlled 9.4 million lines in 2000, and their number of lines has since increased by 6.8 million lines per year (Source: *Federal Communications Commission*).
   a. Let n be the number of lines (in millions) controlled by Bell companies. Find an equation of a model to describe the data.
   b. Now let n be the number of lines (in millions) controlled by other companies. Find an equation of a model to describe the data.
   c. Explain why it makes sense that the slopes of the two models are equal in absolute value and opposite in sign.
   d. Predict when Bell companies will control the same number of local lines as other companies do.
   e. After November 2004, Bell companies no longer had to lease telephone lines at a discounted price to other companies. Do you think Bell companies will control the same number of local lines as other companies before or after your prediction in part (d)? Explain.

45. How many quarts of a 16% acid solution and a 28% acid solution need to be mixed to form 12 gallons of a 20% acid solution?

Polynomials

Teachers open the door, but you must enter by yourself.

—*Chinese Proverb*

How much time do you spend driving each day? Table 1 lists the average times that Americans of various age groups spend driving each day. In Example 1 of Section 7.2, we will estimate the age of drivers who spend the most time driving.

Table 1 Average Times Spent Driving Each Day		
Age Group	Age Used to Represent Age Group	Average Daily Driving Time (minutes)
15–19	17.0	25
20–24	22.0	52
25–54	39.5	64
55–64	59.5	58
over 64	70.0	39

Source: *U.S. Department of Transportation*

In Chapters 1–6, we worked with linear expressions and equations. In Chapters 7–9, we will work with new expressions and equations called *polynomial* expressions and equations. This work will enable us to make estimates and predictions about a new type of authentic situation, such as the driving data just described.

7.1 GRAPHING QUADRATIC EQUATIONS

Objectives

▶ Graph *quadratic equations in two variables.*

▶ Find any intercepts of a parabola.

In this section, we will graph *quadratic equations in two variables*.

Graphing Quadratic Equations in Two Variables

So far, we have graphed linear equations in two variables. These equations are one type of *polynomial equation in two variables*. Here are some other examples of polynomial equations in two variables:

$$y = 2x^2 - 18x + 36 \qquad y = x^3 - x^2 - 6x + 2 \qquad y = x^4 - 4x^2 - x$$

The graphs of these equations are shown in Fig. 1.

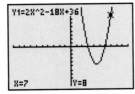 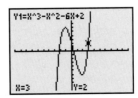 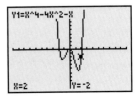

Figure 1 Graphs of three polynomial equations in two variables

Graphing most types of polynomial equations is beyond the scope of this course. After having graphed linear equations in two variables, we will now focus on graphing polynomial equations such as $y = 2x^2 - 18x + 36$, which is called a *quadratic equation in two variables*. Here are some more equations of this type:

$$y = 3x^2 - 5x + 4 \qquad y = -2x^2 + 9 \qquad y = 4x^2 \qquad y = x^2 - 6x$$

DEFINITION Quadratic equation in two variables

A **quadratic equation in two variables** is an equation that can be put into the form

$$y = ax^2 + bx + c,$$

where a, b, and c are constants and $a \neq 0$. This form is called **standard form**.

The equation $y = 3x^2 - 5x + 4$ is in the form $y = ax^2 + bx + c$, with $a = 3$, $b = -5$, and $c = 4$. The equation $y = -2x^2 + 9$ is also in the form $y = ax^2 + bx + c$, with $a = -2$, $b = 0$, and $c = 9$.

Table 2 Solutions of $y = x^2$

x	y
−3	$(-3)^2 = 9$
−2	$(-2)^2 = 4$
−1	$(-1)^2 = 1$
0	$0^2 = 0$
1	$1^2 = 1$
2	$2^2 = 4$
3	$3^2 = 9$

Example 1 Sketching a Graph

Sketch the graph of $y = x^2$.

Solution

First, we list some solutions of $y = x^2$ in Table 2. Then, in Fig. 2, we sketch a curve that contains the points corresponding to the solutions. ∎

Recall from Section 3.1 that every point on the graph of an equation represents a solution of the equation. Every point *not* on the graph represents an ordered pair that is *not* a solution.

The graph of a quadratic equation in two variables is called a **parabola**. So, the curve sketched in Fig. 2 is a parabola. Two more examples of parabolas are sketched in Figs. 3 and 4.

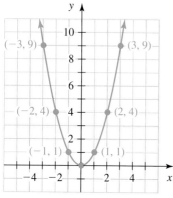

Figure 2 Graph of $y = x^2$

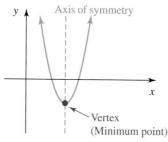

Figure 3 A parabola that opens upward

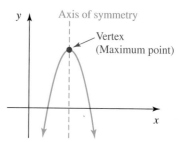

Figure 4 A parabola that opens downward

The lowest point of a parabola that *opens upward* (see Fig. 3) is called the **minimum point**. The highest point of a parabola that *opens downward* (see Fig. 4) is called the **maximum point**. The minimum point or maximum point of a parabola is called the **vertex** of the parabola.

The vertical line that passes through a parabola's vertex is called the **axis of symmetry** (see Figs. 3 and 4). The part of the parabola that lies to the left of the axis of symmetry is the mirror reflection of the part that lies to the right.

Example 2 Sketching a Graph and Finding the Vertex

Sketch the graph of $y = 2x^2 - 8x + 6$, and give the coordinates of the vertex.

Solution

First, we list some solutions of $y = 2x^2 - 8x + 6$ in Table 3. Then, in Fig. 5, we sketch a curve that contains the points corresponding to the solutions.

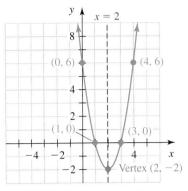

Table 3 Solutions of $y = 2x^2 - 8x + 6$

x	y	
0	$2(0)^2 - 8(0) + 6 = 6$	
1	$2(1)^2 - 8(1) + 6 = 0$	
2	$2(2)^2 - 8(2) + 6 = -2$	← Vertex
3	$2(3)^2 - 8(3) + 6 = 0$	
4	$2(4)^2 - 8(4) + 6 = 6$	

Figure 5 Graph of $y = 2x^2 - 8x + 6$

The vertex is $(2, -2)$, the lowest point of this parabola. The axis of symmetry is the line $x = 2$, the vertical line that contains the vertex.

To verify our work, we check that the part of the parabola which lies to the left of the axis of symmetry is the mirror reflection of the part which lies to the right. In particular, we check that the point $(3, 0)$ is the mirror reflection of the point $(1, 0)$ and that the point $(4, 6)$ is the mirror reflection of the point $(0, 6)$. We can also use a graphing calculator to verify our work (see Fig. 6).

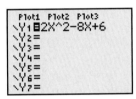 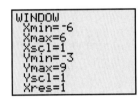 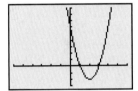

Figure 6 Verify the work

Intercepts of Curves

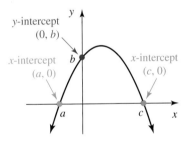

Figure 7 Intercepts of a curve

The definition of an intercept of a line is similar to the definition of an intercept of any type of curve. An **x-intercept of a curve** is a point where the curve and the x-axis intersect; a **y-intercept of a curve** is a point where the curve and the y-axis intersect (see Fig. 7).

Example 3   Sketching a Graph and Finding the Intercepts and the Vertex

Sketch the graph of $y = -x^2 - 2x + 3$, and give the coordinates of the intercepts and the vertex.

Solution

First, we list some solutions of $y = -x^2 - 2x + 3$ in Table 4. Then, in Fig. 8, we sketch a curve that contains the points corresponding to the solutions.

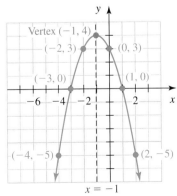

Table 4 Solutions of $y = -x^2 - 2x + 3$

x	y	
-4	$-(-4)^2 - 2(-4) + 3 = -5$	
-3	$-(-3)^2 - 2(-3) + 3 = 0$	
-2	$-(-2)^2 - 2(-2) + 3 = 3$	
-1	$-(-1)^2 - 2(-1) + 3 = 4$	← Vertex
0	$-(0)^2 - 2(0) + 3 = 3$	
1	$-(1)^2 - 2(1) + 3 = 0$	
2	$-(2)^2 - 2(2) + 3 = -5$	

Figure 8 Graph of $y = -x^2 - 2x + 3$

The x-intercepts are $(-3, 0)$ and $(1, 0)$. The y-intercept is $(0, 3)$. The vertex is $(-1, 4)$, the highest point of the parabola. The axis of symmetry is the line $x = -1$, the vertical line that contains the vertex.

We use a graphing calculator to verify our work (see Fig. 9).

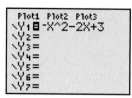 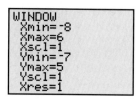 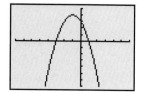

Figure 9 Verify the work

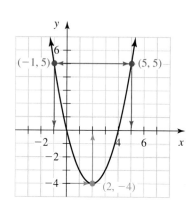

Figure 10 Graph of a parabola

Example 4　Finding Coordinates of Points on a Parabola

A parabola is sketched in Fig. 10.

1. Find x when $y = 5$.
2. Find x when $y = -4$.
3. Find x when $y = -5$.

Solution

1. The red arrows in Fig. 10 show that the output $y = 5$ originates from the two inputs $x = -1$ and $x = 5$. So, the values of x are -1 and 5 when $y = 5$.
2. The blue arrows in Fig. 10 show that the output $y = -4$ originates from the single input $x = 2$. So, the value of x is 2 when $y = -4$. (There is a single input because the vertex $(2, -4)$ is the only point on the parabola that has a y-coordinate equal to -4.)
3. No point of the upward-opening parabola is below the vertex, which has a y-coordinate of -4. So, there is no point on the parabola with a y-coordinate of -5.

By considering the shape of any parabola that opens upward or downward, we see that each (output) value of y originates from either two, one, or no (input) values of x.

Note that this situation is different from that for lines. For a nonhorizontal line, each value of y originates from *exactly* one value of x.

group exploration

Significance of a, h, and k for $y = a(x - h)^2 + k$

It is possible to write any quadratic equation in two variables in the form $y = a(x - h)^2 + k$, where a, h, and k are constants and $a \neq 0$. You will explore the graphical significance of a, h, and k.

1. Use ZStandard followed by ZSquare on a graphing calculator to draw a graph of $y = x^2$.
2. Graph these equations of the form $y = x^2 + k$ in order: $y = x^2 + 1$, $y = x^2 + 2$, $y = x^2 + 3$, and $y = x^2 + 4$. State in terms of k how you could "move" the graph of $y = x^2$ to produce each graph. Do the same with the equations $y = x^2 - 1$, $y = x^2 - 2$, $y = x^2 - 3$, and $y = x^2 - 4$.
3. Graph these equations of the form $y = (x - h)^2$ in order: $y = (x - 1)^2$, $y = (x - 2)^2$, $y = (x - 3)^2$, and $y = (x - 4)^2$. State in terms of h how you could "move" the graph of $y = x^2$ to produce each graph. Do the same with the equations $y = (x + 1)^2$, $y = (x + 2)^2$, $y = (x + 3)^2$, and $y = (x + 4)^2$.
4. Graph these equations of the form $y = ax^2$ in order: $y = 0.1x^2$, $y = 0.4x^2$, $y = x^2$, $y = 2x^2$, and $y = 5x^2$. State in terms of a how you could adjust the graph of $y = x^2$ to produce each graph. Do the same with the equations $y = -0.1x^2$, $y = -0.4x^2$, $y = -x^2$, $y = -2x^2$, and $y = -5x^2$.

5. a. Graph these equations in order: $y = x^2$, $y = 2x^2$, $y = -2x^2$, $y = -2(x+1)^2$, and $y = -2(x+1)^2 - 6$. Explain how these graphs relate to your observations in Problems 2, 3, and 4.

b. Referring to your graph of $y = -2(x + 1)^2 - 6$ from part (a), find the coordinates of the vertex. Compare these coordinates with the equation $y = -2(x + 1)^2 - 6$. What do you notice?

6. Summarize your findings about a, h, and k in terms of how you could move or adjust the graph of $y = x^2$ to produce the graph of $y = a(x - h)^2 + k$. Also, discuss how the coordinates of the vertex are related to a, h, and k. If you are unsure, continue exploring.

TIPS FOR SUCCESS: Examine Your Mistakes

When you make mistakes on a quiz or an exam, it is tempting to disregard these errors by calling them silly mistakes. Or perhaps you know you've made significant errors, but don't want to face them fully. This is understandable; no one enjoys examining his or her mistakes! However, your mistakes can be the gateway to your success. If you examine your errors closely, you can learn a lot about related concepts and perform better next time.

For each error that you made on a quiz or an exam, think carefully about what went wrong. It is a good idea to read about the related concepts and then do many related exercises to build your understanding and confidence in using those concepts. Discussing the concepts with other students or your instructor can help, too.

Ask your instructor or tutor to look for a pattern in your errors. It's possible that you made one kind of error five times rather than five different errors. Learning to correct that one type of error could save you many points on the next test.

HOMEWORK 7.1

FOR EXTRA HELP ▶

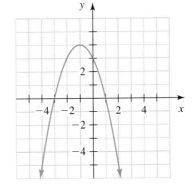

Student Solutions Manual · PH Math/Tutor Center · MathXL® · MyMathLab

Graph the equation by hand. To begin, substitute the values -3, $-2, -1, 0, 1, 2,$ *and 3 for x. Make other substitutions as necessary. Then use a graphing calculator to verify your work.*

1. $y = 2x^2$

2. $y = 3x^2$

3. $y = -x^2$

4. $y = -2x^2$

5. $y = x^2 + 1$

6. $y = x^2 - 4$

7. $y = 3x^2 - 5$

8. $y = -2x^2 + 4$

9. $y = x^2 + 4x$

10. $y = x^2 - 4x$

11. $y = -3x^2 - 6x$

12. $y = 2x^2 + 4x$

13. $y = x^2 - 4x + 3$

14. $y = x^2 - 2x - 1$

15. $y = -2x^2 + 4x - 2$

16. $y = -2x^2 + 8x - 3$

Figure 11 Exercises 17–28

For Exercises 17–28, refer to the graph sketched in Fig. 11.

17. Find y when $x = 2$.

18. Find y when $x = -2$.

19. Find x when $y = 3$.

20. Find x when $y = -5$.

21. Find x when $y = 4$.

22. Find x when $y = 0$.

23. Find x when $y = 5$.

24. Find x when $y = 6$.

25. Find the x-intercept(s).

26. Find the y-intercept(s).

27. Find the vertex.

28. Find the maximum point.

For Exercises 29–38, refer to the graph sketched in Fig. 12.

29. Find y when $x = 0$.

30. Find y when $x = 4$.

31. Find x when $y = 0$.

32. Find x when $y = 3$.

33. Find x when $y = -3$.

34. Find x when $y = -1$.

35. Find the y-intercept(s).

36. Find the x-intercept(s).

37. Find the minimum point.

38. Find the vertex.

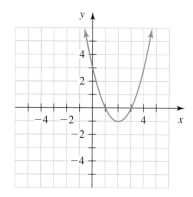

Figure 12 Exercises 29–38

39. Describe the Rule of Four as it applies to the equation $y = x^2 - 3$:
  a. Describe the solutions of the equation by using a graph.
  b. Describe five solutions of the equation by using a table.
  c. Describe the solutions of the equation by using words.

40. Describe the Rule of Four as it applies to the equation $y = x^2 + 2x$:
  a. Describe the solutions of the equation by using a graph.
  b. Describe five solutions of the equation by using a table.
  c. Describe the solutions of the equation by using words.

For Exercises 41–44, refer to Table 5.

41. Find y when $x = 1$.   **42.** Find y when $x = 2$.

43. Find x when $y = 3$.   **44.** Find x when $y = 4$.

Table 5 Values of x and y (Exercises 41-44)	
x	**y**
0	0
1	3
2	4
3	3
4	0

45. a. What is the smallest possible value of x^2? Explain.
  b. What is the smallest possible value of $x^2 + 3$? Explain.
  c. What is the vertex of $y = x^2 + 3$? Explain how you can determine this without graphing the equation. Then graph the equation either by hand or by graphing calculator to verify your result.

46. Solve the system
$$y = x^2 + 2$$
$$y = -x^2 + 2$$
  [**Hint:** Use graphing.]

47. a. For the equation $y = x^2 + 1$, a quadratic equation in two variables, find all outputs for the given input. State how many outputs there are for that single input.
    i. the input $x = 2$
    ii. the input $x = 4$
    iii. the input $x = -3$
  b. For $y = x^2 + 1$, how many outputs originate from any single input? Explain.
  c. Give an example of a quadratic equation in two variables. Using your equation, find all outputs for the given input. State how many outputs there are for that single input.
    i. the input $x = 1$
    ii. the input $x = 3$
    iii. the input $x = -2$
  d. For your equation, how many outputs originate from any single input? Explain.
  e. For *any* quadratic equation in two variables, how many outputs originate from a single input? Explain.

48. Use Fig. 13 to find x when
  a. $y = -1$.   **b.** $y = 2$.   **c.** $y = 3$.
  d. $y = 4$.   **e.** $y = 5$.
  f. For each of parts (a)–(e), you found two, one, or no values for x. Now describe all value(s) of y for which there are
    i. two values of x.
    ii. one value of x.
    iii. no values of x.

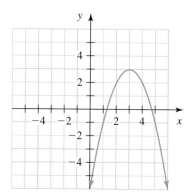

Figure 13 Exercise 48

Related Review

Graph the equation by hand. Then use a graphing calculator to verify your work.

49. $y = -2x - 1$ **50.** $y = 3x - 2$

51. $y = -2x^2 - 1$ **52.** $y = 3x^2 - 2$

53. a. Graph $y = 2x$.
  b. Graph $y = x^2$.
  c. Compare the y-intercepts of the graphs of $y = 2x$ and $y = x^2$.
  d. Which of the curves $y = 2x$ and $y = x^2$ appears to be "steeper" for large values of x? Explain.

54. a. Create a table of values of x and y for the equation $y = 2x$. Use the values 1, 100, 200, 300, 400, and 500 for x.
  b. Create a table of values of x and y for the equation $y = x^2$. Use the values 1, 100, 200, 300, 400, and 500 for x.
  c. As the values of x increase beyond 1, do the values of y increase at a greater rate for the equation $y = 2x$ or for the equation $y = x^2$? Explain.

Expressions, Equations, and Graphs

Perform the indicated instruction. Then use words such as linear, quadratic, one variable, *and* two variables *to describe the expression, equation, or system.*

55. Solve $8(3x - 2) = 4(x - 5)$.

56. Graph $y = 2x^2 - 4$ by hand.

57. Simplify $8(3x - 2) - 4(x - 5)$.

58. Graph $y = 2x - 4$ by hand.

7.2 QUADRATIC MODELS

Objectives

▷ Know the meaning of *quadratic model*.

▷ Use a quadratic model to make estimates and predictions.

▷ Determine whether to model a situation by using a linear model, a quadratic model, or neither.

▷ Use a graphing calculator to make predictions with a quadratic model.

In Chapters 1–6, we used linear models to describe authentic situations. In Chapters 7–9, we will use *quadratic models* to describe other types of authentic situations.

Using a Quadratic Model to Make Estimates and Predictions

We begin by defining a quadratic model.

> **DEFINITION** Quadratic model
>
> A **quadratic model** is a quadratic equation in two variables that describes the relationship between two quantities for an authentic situation. We also refer to the quadratic equation's graph (a parabola) as a quadratic model.

Example 1 Using a Parabola to Make Estimates

Table 6 lists the average times spent driving each day by various age groups.

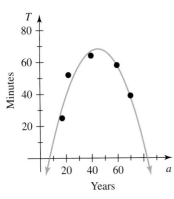

Gotta go, honey, my toast just popped up.

Table 6	Average Times Spent Driving Each Day	
Age Group	Age Used to Represent Age Group	Average Daily Driving Time (minutes)
15–19	17.0	25
20–24	22.0	52
25–54	39.5	64
55–64	59.5	58
over 64	70.0	39

Source: *U.S. Department of Transportation*

Let T be the average daily driving time (in minutes) for a person of age a years. A reasonable model is drawn from a scattergram of the data in Fig. 14.

1. Estimate the average daily driving time of 22-year-old drivers.
2. What was the actual average daily driving time of 22-year-old drivers? Compute the error in the estimate from Problem 1.
3. Estimate the age(s) of drivers whose average daily driving time is 56 minutes.
4. Estimate the age of drivers who spend the most time driving.
5. Find the a-intercepts. What do they mean in this situation?

Figure 14 Scattergram and model of daily driving times

Solution

1. The blue arrows in Fig. 15 show that the input $a = 22$ leads to the approximate output $T = 43$. So, according to the model, the average daily driving time for 22-year-old drivers is 43 minutes.
2. The estimated average daily driving time of 43 minutes for 22-year-old drivers is less than the actual time of 52 minutes (see Table 6 or Fig. 15). Since $52 - 43 = 9$ and we underestimated the time, the error is -9 minutes.

3. The red arrows in Fig. 16 show that the output $T = 56$ originates from the two approximate inputs $a = 29$ and $a = 61$. So, according to the model, both 29-year-old drivers and 61-year-old drivers have an average daily driving time of 56 minutes.

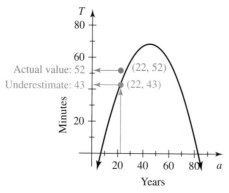

Figure 15 Estimated and actual daily driving times for 22-year-old drivers

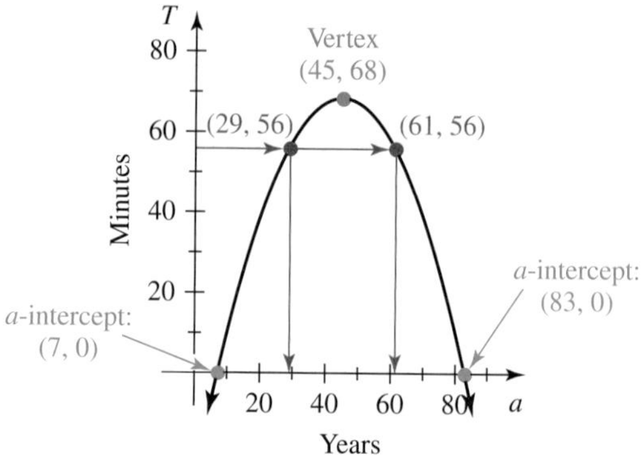

Figure 16 Using the model to make more estimates

4. The maximum point on the parabola is the vertex, which is about $(45, 68)$. See Fig. 16. So, according to the model, 45-year-old drivers have an average daily driving time of 68 minutes, which is greater than the average for any other age.

5. From Fig. 16, it appears that the a-intercepts are $(7, 0)$ and $(83, 0)$. According to the model, 7-year-old children do not drive, which is true. However, there is model breakdown for values of a near (but not equal to) 7; for example, the model suggests that 6-year-old children drive for negative amounts of time, which doesn't make sense, and that 8-year-old children drive briefly, which is not true.

The a-intercept $(83, 0)$ means that 83-year-old adults do not drive. Model breakdown has occurred, because some 83-year-old adults drive. ∎

In Example 1, we found that $T = 56$ corresponds to *two* values of a. Recall from Section 7.1 that, for a parabola, each value of the dependent variable corresponds to two, one, or zero values of the independent variable.

Deciding Which Type of Model to Use

By graphing a scattergram of some data, we can determine whether to use a linear model, a quadratic model, or neither.

Example 2 Selecting a Model

Consider the scattergrams of data shown in Figs. 17, 18, and 19 for situations 1, 2, and 3, respectively. For each situation, determine whether a linear model, quadratic model, or neither would describe the situation well.

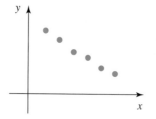

Figure 17 Use a linear model

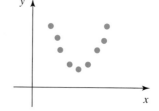

Figure 18 Use a quadratic model

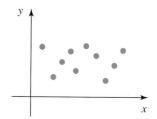

Figure 19 Do not use a linear or a quadratic model

Solution

A linear model would describe the data for situation 1 well. It appears that a quadratic model would fit the data for situation 2 well. Neither a linear model nor a quadratic model would fit the data well for situation 3. ∎

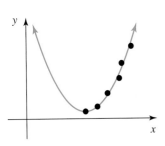

Figure 20 Scattergram and a quadratic model

Throughout this course, we have seen that usually only a part of a model describes a situation well. For the scattergram shown in Fig. 20, the sketched quadratic model fits the data well. At least part of the model that lies to the right of the vertex can be used to make reasonable estimates or predictions. The part of the model that lies to the left of the vertex may or may not describe the situation well. More data points could reveal whether this part of the parabola gives good estimates.

Using a Graphing Calculator to Make Predictions

When modeling a situation by using a quadratic equation, we can use TRACE on a graphing calculator to help us make estimates and predictions. We can also use "minimum" or "maximum" to locate the vertex of the model.

Example 3 Using a Graphing Calculator

Table 7 shows the U.S. school enrollment (kindergarten through grade 12) for various years. Let E be the school enrollment (in millions of students) at t years since 1990. A linear model of the situation is

$$E = 0.58t + 41.64$$

A quadratic model of the situation is

$$E = -0.021t^2 + 0.83t + 41.21$$

1. Which of the two models comes closer to the points in the scattergram of the data?
2. Substitute a value for one of the variables in the quadratic model to predict the enrollment in 2012.
3. Use "maximum" on a graphing calculator to find the vertex of the quadratic model. What does it mean in this situation?
4. Which model predicts that enrollments will eventually decrease?
5. Use TRACE on a graphing calculator together with the quadratic model to estimate when the enrollment was 49 million students and when the enrollment will be 49 million students.

Table 7 Kindergarten through Grade 12 Enrollments

Year	Enrollment (millions)
1990	41.2
1992	42.8
1994	44.1
1996	45.6
1998	46.5
2000	47.2
2002	48.2

Source: *U.S. Department of Education*

Solution

1. We use ZoomStat to get the window settings shown in Fig. 21. Although the linear model comes close to the points (see Fig. 22), the quadratic model comes even closer (see Fig. 23).

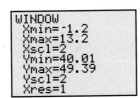

Figure 21 Window settings

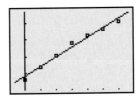

Figure 22 Linear model and scattergram

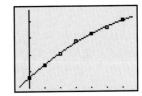

Figure 23 Quadratic model and scattergram

2. Since 2012 is 22 years since 1990, we substitute the input 22 for t into the quadratic model and solve for E:

$$E = -0.021(22)^2 + 0.83(22) + 41.21 = 49.31$$

The enrollment will be 49.3 million students in 2012, according to the quadratic model.

3. The screen in Fig. 23 does not show the vertex. We use Zoom Out on a graphing calculator to adjust the window settings so that we can see the vertex (see Fig. 24). We also turn off the plotter so that the scattergram is no longer displayed. Then we use "maximum" on a graphing calculator to find the approximate vertex (see Fig. 24). See Sections A.6 and A.18 for graphing calculator instructions.

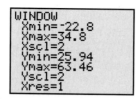 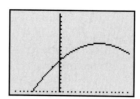 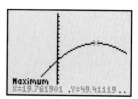

Figure 24 Use Zoom Out and "maximum" to find the vertex

The vertex is about (19.76, 49.41). So, according to the quadratic model, the enrollment in 2010 will be about 49.4 million students, the largest enrollment ever.

4. The part of the quadratic model to the right of its vertex predicts that enrollments will decrease (after 2010).

5. From Fig. 25, we see that $E \approx 49$ when $t = 15.3$ and when $t = 24.2$. So, the enrollment was about 49 million students in 2005 and will be about 49 million students in 2014, according to the quadratic model.

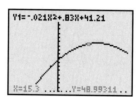 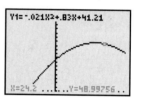

Figure 25 Find when the enrollment was or will be 49 million students

In Example 3, we used "maximum" on a graphing calculator to find the vertex of a quadratic model. In general, we can use "minimum" to find the vertex of a parabola that opens upward (see Fig. 26) and "maximum" to find the vertex of a parabola that opens downward (see Fig. 27). See Section A.18 for calculator instructions.

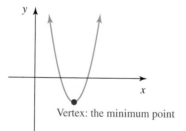

Figure 26 A parabola that opens upward

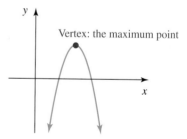

Figure 27 A parabola that opens downward

group exploration

Looking ahead: Combining like terms

Recall from Section 4.1 that we can write $2x + 3x = 5x$ because of the distributive law:

$$2x + 3x = (2 + 3)x = 5x$$

1. Use the distributive law to simplify $2x^2 + 3x^2$.

2. Can the distributive law be used to simplify $2x^2 + 3x$? Explain.

3. Use the distributive law twice to simplify $3x^2 + 5x^2 + 2x + 7x$.

4. Simplify $8x^2 - 5x + 2 - 3x^2 + x - 7$.

5. Simplify $5x^2 - 3(x^2 - 4x) - 2x + 3$.

group exploration

Looking ahead: Combining expressions that represent authentic quantities

In Example 1 of Section 6.4, we used the linear expression $11.75t + 101.77$ to model the number of part-time college instructors (in thousands) and the linear expression $7.38t + 376.15$ to model the number of full-time college instructors (in thousands), both at t years since 1970 (see Table 8).

Table 8 Numbers of Part-Time and Full-Time Instructors

| Year | Number of Instructors (thousands) | |
	Part Time	Full Time
1970	104	369
1975	188	440
1980	236	450
1985	256	459
1991	291	536
1995	381	551
1999	437	591
2003	543	632

Source: *U.S. National Center for Education Statistics*

1. Find the sum of the expressions $7.38t + 376.15$ and $11.75t + 101.77$. What does the result represent?

2. Evaluate the result of Problem 1 for $t = 40$. What does your result mean in this situation?

3. Find the difference of the expressions $7.38t + 376.15$ and $11.75t + 101.77$. What does the result represent?

4. Evaluate the result of Problem 3 for $t = 40$. What does your result mean in this situation? [**Hint:** The difference $7 - 2 = 5$ means that 7 is 5 more than 2.]

HOMEWORK 7.2

FOR EXTRA HELP ▶

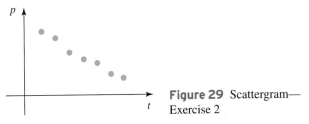

Student Solutions Manual   PH Math/Tutor Center   MathXL®   MyMathLab

---

1. A scattergram of a given situation is graphed in Fig. 28. Determine whether a linear model, a quadratic model, or neither type of model would be reasonable for modeling the data. Also, sketch a curve that you would use to model the data.

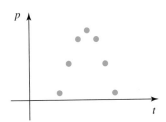

Figure 28 Scattergram—Exercise 1

2. A scattergram of a given situation is graphed in Fig. 29. Determine whether a linear model, a quadratic model, or neither type of model would be reasonable for modeling the data. Also, sketch a curve that you would use to model the data.

Figure 29 Scattergram—Exercise 2

3. A student believes that the data listed in Table 9 suggest a quadratic relationship, since the values of y increase and then decrease. Is the student correct? Explain. [**Hint:** Use a graphing calculator to draw a scattergram of the data.]

Table 9 Is There a Quadratic Relationship?

x	y	x	y
0	1	6	18
1	2	7	7
2	3	8	3
3	7	9	2
4	18	10	1
5	30		

4. A student believes that the data listed in Table 10 suggest a quadratic relationship, since the values of *y* decrease and then increase. Is the student correct? Explain. [**Hint:** Use a graphing calculator to draw a scattergram of the data.]

Table 10 Is There a Quadratic Relationship?	
x	**y**
0	34
1	33
2	32
3	28
4	17
5	5
6	17
7	28
8	32
9	33
10	34

5. CD album sales are shown in Table 11 for various years.

Table 11 CD Album Sales	
Year	**CD Album Sales (millions)**
1998	711
1999	800
2000	813
2001	775
2002	675
2003	550

Source: *Nielsen SoundScan*

Let *s* be the CD album sales (in millions) for the year that is *t* years since 1990. A quadratic model is graphed in Fig. 30.

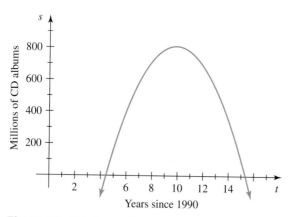

Figure 30 CD album sales model

a. Trace the graph in Fig. 30 carefully onto a piece of paper. Then, on the same graph, draw a scattergram of the data. Does the model fit the data well?
b. Estimate the sales in 2005.
c. Estimate when the sales were 340 million CD albums.
d. What is the vertex of the model? What does this result mean in this situation, and how might it be related to free online music swapping?
e. What are the *t*-intercepts? What do they mean in this situation?

6. Table 12 lists the percentages of Americans of various age groups who listen to talk radio.

Table 12 Percentages of Americans Who Listen to Talk Radio		
Age Group (years)	**Age Used to Represent Age Group (years)**	**Percent**
12–17	14.5	6
18–34	26.0	16
35–44	39.5	25
45–54	49.5	26
55–64	59.5	21
over 64	75.0	6

Source: Talkers Magazine; *Mediamark*

Let *p* be the percentage of Americans of age *a* years who listen to talk radio. A quadratic model of the situation is graphed in Fig. 31.

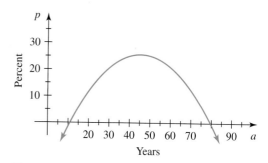

Figure 31 Talk radio model

a. Trace the graph in Fig. 31 carefully onto a piece of paper. Then, on the same graph, draw a scattergram of the data. Does the model fit the data well?
b. Estimate the percentage of 20-year-old Americans who listen to talk radio.
c. Estimate the age(s) at which 20% of Americans listen to talk radio.
d. What is the vertex of the model? What does it mean in this situation?
e. What are the *a*-intercepts? What do they mean in this situation?

7. The numbers of Supreme Court cases are shown in Table 13 for various years.

Table 13 Numbers of U.S. Supreme Court Cases	
Year	**Number of Cases (thousands)**
1994	8.1
1995	7.6
1996	7.6
1997	7.7
1998	8.0
1999	8.4
2000	9.0

Source: *Office of the Clerk, Supreme Court of the United States*

Let n be the number (in thousands) of cases in the year that is t years since 1990. A quadratic model is graphed in Fig. 32.

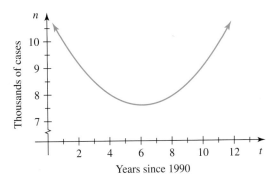

Figure 32 Supreme Court case model

a. Trace the graph in Fig. 32 carefully onto a piece of paper. Then, on the same graph, draw a scattergram of the data. Does the model fit the data well?

b. Estimate the number of cases in 2001. Find the average number of cases per workday. Assume that the justices work five days a week for 48 weeks each year.

c. Estimate when 9 thousand cases were heard.

d. Estimate when the least number of cases were heard. Does that match what really happened? Explain.

8. The numbers of people killed while riding motorcycles are shown in Table 14 for various years.

Table 14 Numbers of Motorcycle Deaths	
Year	Number of People Killed while Riding Motorcycles (thousands)
1985	4.4
1988	3.5
1991	2.7
1994	2.2
1997	2.1
2000	2.8
2003	3.6

Source: *National Highway Traffic Safety Administration*

Let n be the number of people (in thousands) killed while riding motorcycles in the year that is t years since 1985. A quadratic model is graphed in Fig. 33.

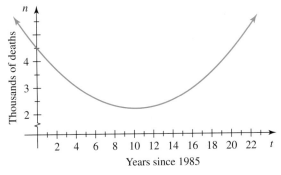

Figure 33 Motorcycle deaths model

a. Trace the graph in Fig. 33 carefully onto a piece of paper. Then, on the same graph, draw a scattergram of the data. Does the model fit the data well?

b. Estimate the number of motorcycle deaths in 2006.

c. Estimate in which year(s) 3.3 thousand motorcycle deaths occurred or will occur.

d. What is the n-intercept? What does it mean in this situation?

e. Estimate when the least number of motorcycle deaths occurred. What was that number of deaths?

9. The percentages of married people who reach major anniversaries are listed in Table 15.

Table 15 Percentages of Married People Who Reach Major Anniversaries	
Anniversary	Percent
5th	82
10th	65
15th	52
25th	33
35th	20
50th	5

Source: *U.S. Census Bureau*

Let p be the percentage of married people who reach their ath anniversary. A quadratic model is graphed in Fig. 34.

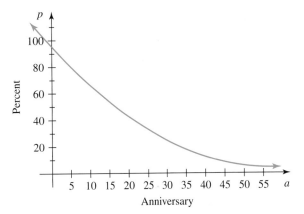

Figure 34 Anniversary model

a. Trace the graph in Fig. 34 carefully onto a piece of paper. Then, on the same graph, draw a scattergram of the data. Does the model fit the data well?

b. What is the p-intercept? What does it mean in this situation?

c. What percentage of married people reach their 20th anniversary?

d. What anniversary does exactly half of marriages reach?

e. Does the part of the parabola that lies to the right of the vertex describe the situation well? Explain.

10. The average prices of flat-panel plasma televisions are shown in Table 16 for various years.

Table 16 Average Prices of Flat-Panel Plasma Televisions	
Year	Average Price (thousands of dollars)
2000	9.8
2001	6.8
2003	4.6
2005	2.5
2006	1.7

Source: *DisplaySearch*

Let p be the average price (in thousands of dollars) of flat-panel plasma televisions at t years since 2000. A quadratic model is graphed in Fig. 35.

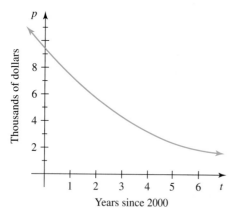

Figure 35 Flat-panel plasma television model

a. Trace the graph in Fig. 35 carefully onto a piece of paper. Then, on the same graph, draw a scattergram of the data. Does the model fit the data well?

b. Estimate when the average price of a flat-panel plasma television was $5.7 thousand.

c. Estimate the average price of a flat-panel plasma television in 2004.

d. Does the part of the parabola that lies to the right of the vertex describe the situation well? Explain.

11. American companies' investment in other countries increased during the economic boom of the 1990s, but has declined since the economic slump that began in 2000 (see Table 17).

Table 17 Investment by U.S. Companies in Other Countries

Year	Investment in Other Countries (billions of dollars)
1997	105
1998	143
1999	189
2000	178
2001	128

Source: *U.S. Bureau of Economic Analysis*

Let I be U.S. companies' investment (in billions of dollars) in other countries in the year that is t years since 1995. A scattergram of the data and a quadratic model are graphed in Fig. 36.

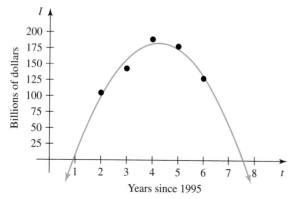

Figure 36 Foreign investment model

a. Use the quadratic model to estimate foreign investment in 1998.

b. What was the actual foreign investment in 1998?

c. Is your result in part (a) an underestimate or an overestimate? Explain how you can tell this from the graph of the scattergram and the model. Compute the error in your estimate.

12. For this exercise, refer to Exercise 11, including Table 17 and Fig. 36.

a. Use the quadratic model to estimate foreign investment in 1999.

b. What was the actual foreign investment in 1999?

c. Is your result in part (a) an underestimate or an overestimate? Explain how you can tell this from the graph of the scattergram and the model. Compute the error in your estimate.

13. The numbers of electric cars in use are shown in Table 18 for various years.

Table 18 Numbers of Electric Cars in Use

Year	Number of Electric Cars in Use (thousands)
1996	3.3
1998	5.2
2000	11.8
2002	33.0
2004	56.0

Source: *Department of Energy*

Let n be the number of electric cars (in thousands) in use at t years since 1995. A model of the situation is $n = t^2 - 3.48t + 5.8$.

a. Use a graphing calculator to draw the graph of the model and a scattergram of the data in the same viewing window. Does the model fit the data well?

b. Substitute a value for one of the variables in the model's equation to predict the number of electric cars that will be in use in 2011.

c. Use TRACE on a graphing calculator to predict when exactly 150 thousand electric cars will be in use.

d. Does the part of the parabola that lies to the left of the vertex describe the situation well? Explain.

14. The percentages of the U.S. population that are foreign born are shown in Table 19 for various years.

Table 19 Percentages of U.S. Population That Is Foreign Born

Year	Percent
1920	13.2
1930	11.6
1940	8.8
1950	6.9
1960	5.4
1970	4.7
1980	6.2
1990	8.0
2000	10.4
2004	12.0

Source: *U.S. Census Bureau*

Let p be the percentage of the U.S. population that is foreign born at t years since 1900. A model of the situation is $p = 0.0042t^2 - 0.55t + 23.44$.

a. Use a graphing calculator to draw the graph of the model and a scattergram of the data in the same viewing window. Does the model fit the data well?

b. Substitute a value for one of the variables in the model's equation to predict the percentage of the U.S. population that will be foreign born in 2010.

c. Use "minimum" on a graphing calculator to estimate when the percentage of the U.S. population that was foreign born was the lowest.

d. Use TRACE on a graphing calculator to predict when 14% of the U.S. population will be foreign born.

15. For nine straight seasons before Mark Cuban bought the Dallas Mavericks basketball team during the 1999–2000 season, the team had won less than half of their 82 season games. From 2001 through 2004, the team won more than half of their games every season (see Table 20).

Table 20 Number of Games Won and Lost by the Mavericks

End of Season (Year)	Wins	Losses
2000	40	42
2001	53	29
2002	57	25
2003	60	22
2004	52	30

Source: *National Basketball Association*

Let w be the number of Maverick wins for the season that ends t years since 2000. A quadratic model of the situation is $w = -3.07t^2 + 15.39t + 40.1$. A linear model of the situation is $w = 3.1t + 46.2$.

a. Use a graphing calculator to draw graphs of the quadratic and linear models and a scattergram of the data in the same viewing window. Which model comes closer to the points in the scattergram?

b. Which of the two models would Mark Cuban and the Mavericks want to be the most accurate for future seasons? Explain.

c. Use "maximum" on a graphing calculator, along with the quadratic model, to estimate in which season the Mavericks had the greatest number of wins or to predict in which season they will do so. What is that number of wins, according to the model?

d. Use TRACE on a graphing calculator, along with the quadratic model, to estimate in which season the Mavericks won 40 games and lost 42 games or to predict in which season they will do so.

16. The numbers of prisoners cleared and released due to DNA testing are shown in Table 21 for various years. Let n be the number of prisoners cleared and released due to DNA testing in the year that is t years since 1980. A quadratic model of the situation is $p = 0.074t^2 - 0.93t + 3.75$. A linear model of the situation is $p = 1.45t - 13.74$.

a. Use a graphing calculator to draw graphs of the quadratic and linear models and a scattergram of the data in the same viewing window. Which model comes closer to the points in the scattergram?

Table 21 Numbers of Prisoners Cleared and Released Due to DNA Testing

Year	Prisoners Cleared and Released Due to DNA Testing
1989	2
1991	2
1993	4
1995	7
1997	9
1999	10
2001	23
2003	19

Source: *University of Michigan Law School*

b. For each model, substitute a value for one of the variables in the model's equation to predict the number of prisoners who will be cleared and released due to DNA testing in 2011. Refer to the models' graphs to explain why the quadratic model's prediction is so much larger than the linear model's prediction.

c. Which of the two models estimates that the number of prisoners cleared and released due to DNA testing was decreasing for years before 1986? Has model breakdown likely occurred for that model for those years?

d. Which of the two models estimates that DNA testing led to a negative number of prisoners being cleared and released in each year before 1989? Has model breakdown occurred for that model for those years? Explain.

e. Use TRACE on a graphing calculator, along with the quadratic model, to predict in which year 40 prisoners will be cleared and released due to DNA testing.

17. Suppose that a situation can be modeled well by a parabola, where t is the independent variable and p is the dependent variable. If a data point (c, d) is below the parabola, does the model underestimate or overestimate the value of p when $t = c$? Explain.

18. Suppose that a situation can be modeled well by a parabola, where t is the independent variable and p is the dependent variable. If a data point (c, d) is above the parabola, does the model underestimate or overestimate the value of p when $t = c$? Explain.

19. Which is more desirable, finding a quadratic model whose graph contains several (but not all) data points or finding a quadratic model whose graph does not contain any data points but comes close to all data points? Include in your discussion some sketches of scattergrams and quadratic models.

20. If a quantity increases from one year to the next, explain how in some cases it might be better to model the situation by a quadratic equation rather than a linear equation. Include an example of a scattergram in your explanation.

21. Describe the meaning of *quadratic model* in your own words. (See page 9 for guidelines on writing a good response.)

22. Explain why the vertex and any intercepts of a quadratic model are often of special interest. (See page 9 for guidelines on writing a good response.)

Related Review

23. Because of economic pressures, the male birthrate (number of male births per 100 female births) of many countries has increased. In particular, the male birthrate has increased due to an increase in inexpensive prenatal scans, followed by abortions to prevent the birth of unwanted daughters. Male birthrates in China are shown in Table 22 for various years.

Table 22 Male Birthrates in China	
Year	Male Birthrate (number of male births per 100 female births)
1982	108
1985	110
1990	112
1995	116
2000	117

Source: *China News Agency; Population Reference Bureau*

Let M be the male birthrate (number of male births per 100 female births) in China at t years since 1980.

a. Use a graphing calculator to draw a scattergram of the data.
b. Find an equation of a model to describe the data.
c. Predict the male birthrate in China in 2010.
d. Estimate when the number of male and female births was equal.

24. The average personal incomes of Americans are shown in Table 23 for various years. Let I be the personal income (in thousands of dollars) of an American for the year that is t years since 1980.

a. Use a graphing calculator to draw a scattergram of the data.
b. Find an equation of a model to describe the data.
c. What is the slope? What does it mean in this situation?
d. Predict the average personal income of Americans in 2011.
e. Predict the average personal income of Americans *per work hour* in 2011. Assume that a typical work schedule consists of 8 hours per day, 5 days per week, and 50 weeks per year.

Table 23 Average Personal Income of Americans	
Year	Average Personal Income (thousands of dollars)
1983	13
1985	15
1990	20
1995	24
2000	30
2004	33

Source: *U.S. Bureau of Economic Analysis*

25. Explain how creating a scattergram can help you determine whether to describe some data by a linear model, a quadratic model, or neither.

26. Suppose that both a linear model and a quadratic model fit the points in a scattergram of data well and that the independent variable is time (in years). Will using the two models to make predictions for a certain year give similar results?

Expressions, Equations, and Graphs

Perform the indicated instruction. Then use words such as linear, quadratic, one variable, *and* two variables *to describe the expression, equation, or system.*

27. Solve the following:

$$3x - 2y = -4$$
$$4x + 5y = 33$$

28. Simplify $4x + 9(2x - 3)$.

29. Graph $3x - 2y = -4$ by hand.

30. Solve $4x + 9(2x - 3) = 0$.

7.3 ADDING AND SUBTRACTING POLYNOMIALS

Objectives

▷ Know the meaning of *term, monomial, polynomial, degree, coefficient,* and *like terms.*
▷ Combine like terms.
▷ Add and subtract polynomials.
▷ Use a sum or a difference of two expressions to model an authentic situation.

In Sections 7.1 and 7.2, we worked with quadratic equations in two variables, such as $y = 2x^2 - 8x + 6$, which is a special type of polynomial equation. We will now begin working with *polynomial expressions,* or *polynomials* for short. For example, $2x^2 - 8x + 6$ is a polynomial.

Polynomials

Recall from Section 4.1 that a **term** is a constant, a variable, or a product of a constant and one or more variables raised to powers. Here are some examples of terms: $5x^4$, x, 17, $-6x^8y^2$, and x^{-3}.

A **monomial** is a constant, a variable, or a product of a constant and one or more variables raised to *counting-number* powers. Here are some examples of monomials:

$$-2x^7, \ y, \ -3, \ \frac{2}{7}x^4y^9, \ \text{and } x^3$$

A **polynomial,** or **polynomial expression,** is a monomial or a sum of monomials. Here are some examples of polynomials:

$$3x^2 + 5x + 2 \qquad 9x^5 y^3 - 6x^3 y^2 - xy + 5 \qquad -2x + 5 \qquad 7 \qquad 3x^4$$

The polynomial $7x^3 - 5x^2 + 8x - 1$ is a *polynomial in one variable*. It has four terms: $7x^3$, $-5x^2$, $8x$, and -1. We usually write polynomials in one variable so that the exponents of the terms decrease from left to right, an arrangement called **decreasing order.** If a polynomial contains more than one variable, we usually write the polynomial so that the exponents of one of the variables decrease from left to right.

The **degree of a term** in one variable is the exponent on the variable. For example, the degree of the term $2x^7$ is 7. The degree of a term in two or more variables is the sum of the exponents on the variables. For example, the term $5x^4 y^2$ has degree $4 + 2 = 6$. The **degree of a polynomial** is the largest degree of any nonzero term of the polynomial. For example, the polynomial $7x^4 - 2x^2 + 5$ has degree 4. A constant polynomial such as 5 has degree 0.

Polynomials with degrees 1, 2, or 3 have special names:

Degree	Name	Examples
1	linear (first-degree) polynomial	$-3x + 7$, $2x$
2	quadratic (second-degree) polynomial	$8x^2 - 5x + 2$, $x^2 - 4$
3	cubic (third-degree) polynomial	$4x^3 + 2x^2 - 6x + 5$, $-6x^3 + x$

Example 1 Describing Polynomials

Use words such as *linear, quadratic, cubic, polynomial, degree, one variable,* and *two variables* to describe the expression.

1. $5x^2 - 3x + 7$ **2.** $-7x^3 + 8x^2 - 1$ **3.** $4a^3 b^2 - 2a^2 b^2 + 9ab^2$

Solution

1. The term $5x^2$ has degree 2, which is larger than the degrees of the other terms. So, $5x^2 - 3x + 7$ is a quadratic (or second-degree) polynomial in one variable.
2. The term $-7x^3$ has degree 3, which is larger than the degrees of the other terms. So, $-7x^3 + 8x^2 - 1$ is a cubic (or third-degree) polynomial in one variable.
3. The term $4a^3 b^2$ has degree $3 + 2 = 5$, which is larger than the degrees of the other terms. So, $4a^3 b^2 - 2a^2 b^2 + 9ab^2$ is a fifth-degree polynomial in two variables.

    ■

Combining Like Terms

Recall from Section 4.2 that the **coefficient** of a term is the constant factor of the term. For the term $-3x^2$, the coefficient is -3. For the term x^3, the coefficient is 1, because $x^3 = 1x^3$. The coefficients of a polynomial are the coefficients of the variable terms. For example, the coefficients of the polynomial $7x^3 - 5x^2 + 8x - 1$ are 7, -5, and 8. The **leading coefficient** of a polynomial is the coefficient of the term with the largest degree. For $7x^3 - 5x^2 + 8x - 1$, the leading coefficient is 7.

Recall from Section 4.2 that **like terms** are either constant terms or variable terms that contain the same variable(s) raised to exactly the same power(s). For example, the terms $9x^2 y^7$ and $4x^2 y^7$ are like terms, because both terms have an x with the exponent 2 and a y with the exponent 7. Recall from Section 4.2 that if terms are not like terms, we say that they are **unlike terms.** For example, $3x^3$ and $5x^2$ are unlike terms, because the exponents of x are different.

Recall also from Section 4.2 that we can combine like terms of linear polynomials by using the distributive law. For example,

$$3x + 5x = (3 + 5)x = 8x$$

Likewise, we can *combine like terms* of polynomials of higher degree, such as $4x^2 + 6x^2$, by using the distributive law:

$$4x^2 + 6x^2 = (4 + 6)x^2 = 10x^2$$

Note that we can find $4x^2 + 6x^2$ in one step by adding the coefficients:

$$4x^2 + 6x^2 = 10x^2$$

$$4 + 6 = 10$$

We can also find $6x^3 - 2x^3$ by adding the coefficients:

$$6x^3 - 2x^3 = 4x^3$$

$$6 + (-2) = 4$$

However, we can't add the coefficients in $8x^4 + 3x^2$, because $8x^4$ and $3x^2$ are not like terms. There is no helpful way to use the distributive law for unlike terms.

Combining Like Terms

To combine like terms, add the coefficients of the terms.

Example 2   Combining Like Terms

Combine like terms when possible.

1. $7x^2 + 5x^2$
2. $-4x^3y^2 - 6x^3y^2$
3. $5a^2b^4 + 3a^4b^2$
4. $-6x^2 + x^2$

Solution

1. $7x^2 + 5x^2 = 12x^2$ Add coefficients: $7 + 5 = 12$
2. $-4x^3y^2 - 6x^3y^2 = -10x^3y^2$ Add coefficients: $-4 + (-6) = -10$
3. Although $5a^2b^4$ and $3a^4b^2$ have the same variables, the variables do not have the same exponents. They are not like terms, so we cannot combine them.
4. $-6x^2 + x^2 = -6x^2 + 1x^2 = -5x^2$ Add coefficients: $-6 + 1 = -5$ ∎

Example 3   Combining Like Terms

Combine like terms: $8x - 7x^2 - 2x + 3x^2$.

Solution

$$8x - 7x^2 - 2x + 3x^2 = -7x^2 + 3x^2 + 8x - 2x \qquad \text{Rearrange terms.}$$
$$= -4x^2 + 6x \qquad \text{Combine like terms.} \quad ∎$$

Addition of Polynomials

Now that we know how to combine like terms, we can add polynomials.

Adding Polynomials

To add polynomials, combine like terms.

Example 4   Adding Polynomials

Find the sum.

1. $(4x^2 + 3x) + (6x^2 - 2x)$
2. $(5x^3 - 3x^2 + 8x - 1) + (7x^3 - x^2 - 3x + 4)$
3. $(2x^3 - 5x) + (7x^2 - 4x)$
4. $(4x^2 + 3xy - 5y^2) + (3x^2 - 7xy + 9y^2)$

Solution

1. $(4x^2 + 3x) + (6x^2 - 2x) = 4x^2 + 6x^2 + 3x - 2x \qquad \text{Rearrange terms.}$
$$= 10x^2 + x \qquad \text{Combine like terms.}$$

2. $\left(5x^3 - 3x^2 + 8x - 1\right) + \left(7x^3 - x^2 - 3x + 4\right)$
$$= 5x^3 + 7x^3 - 3x^2 - x^2 + 8x - 3x - 1 + 4 \qquad \text{Rearrange terms.}$$
$$= 12x^3 - 4x^2 + 5x + 3 \qquad \text{Combine like terms.}$$

3. $\left(2x^3 - 5x\right) + \left(7x^2 - 4x\right) = 2x^3 + 7x^2 - 5x - 4x \qquad \text{Rearrange terms.}$
$$= 2x^3 + 7x^2 - 9x \qquad \text{Combine like terms.}$$

We use a graphing calculator table to verify the work (see Fig. 37).

Figure 37 Verify the work

4. $\left(4x^2 + 3xy - 5y^2\right) + \left(3x^2 - 7xy + 9y^2\right)$
$$= 4x^2 + 3x^2 + 3xy - 7xy - 5y^2 + 9y^2 \qquad \text{Rearrange terms.}$$
$$= 7x^2 - 4xy + 4y^2 \qquad \text{Combine like terms.} \qquad \blacksquare$$

Subtraction of Polynomials

In Section 4.1, we discussed how to subtract expressions; those expressions were actually linear polynomials. In this section, we take similar steps to subtract polynomials of higher degree.

Example 5 Subtracting Polynomials

Find the difference.
  1. $\left(8x^3 + 5x^2\right) - \left(6x^3 + 2x^2\right)$
  2. $\left(2a^2 - 5ab + 7b^2\right) - \left(6a^2 - 4ab + 3b^2\right)$

Solution

1. To begin, we write $\left(8x^3 + 5x^2\right) - \left(6x^3 + 2x^2\right)$ as $\left(8x^3 + 5x^2\right) - 1\left(6x^3 + 2x^2\right)$ and then distribute the -1:

$$\left(8x^3 + 5x^2\right) - \left(6x^3 + 2x^2\right) = \left(8x^3 + 5x^2\right) - 1\left(6x^3 + 2x^2\right) \qquad a - b = a - 1b$$
$$= 8x^3 + 5x^2 - 6x^3 - 2x^2 \qquad \text{Distributive law}$$
$$= 8x^3 - 6x^3 + 5x^2 - 2x^2 \qquad \text{Rearrange terms.}$$
$$= 2x^3 + 3x^2 \qquad \text{Combine like terms.}$$

We use a graphing calculator table to verify the work (see Fig. 38).

2. $\left(2a^2 - 5ab + 7b^2\right) - \left(6a^2 - 4ab + 3b^2\right)$
$$= \left(2a^2 - 5ab + 7b^2\right) - 1\left(6a^2 - 4ab + 3b^2\right) \qquad a - b = a - 1b$$
$$= 2a^2 - 5ab + 7b^2 - 6a^2 + 4ab - 3b^2 \qquad \text{Distributive law}$$
$$= 2a^2 - 6a^2 - 5ab + 4ab + 7b^2 - 3b^2 \qquad \text{Rearrange terms.}$$
$$= -4a^2 - ab + 4b^2 \qquad \text{Combine like terms.} \qquad \blacksquare$$

Figure 38 Verify the work

In Problem 1 of Example 5, a key step in finding $\left(8x^3 + 5x^2\right) - \left(6x^3 + 2x^2\right)$ is to write this difference as $\left(8x^3 + 5x^2\right) - 1\left(6x^3 + 2x^2\right)$, so that we can distribute the -1.

Subtracting Polynomials
To subtract polynomials, first distribute -1 and then combine like terms.

Using a Sum or Difference to Model an Authentic Situation

Suppose that A and B represent quantities. Then $A + B$ represents the sum of the two quantities.

The difference $A - B$ tells us how much more there is of one quantity than the other. For example, suppose that $A = 7$ and $B = 2$. Then $A - B = 7 - 2 = 5$ tells us that A is 5 more than B. Now suppose that $A = 2$ and $B = 7$. Then $A - B = 2 - 7 = -5$ tells us that A is 5 less than B.

The Meaning of the Sign of a Difference

If a difference $A - B$ is positive, then A is more than B. If a difference $A - B$ is negative, then A is less than B.

In Example 6, we will add and subtract expressions that represent authentic quantities and will interpret the meanings of the sum and difference.

Example 6 Describing Quantities by Combining Expressions

The numbers of subscriptions to cable TV and satellite TV in the United States are shown in Table 24 for various years.

Table 24 Subscriptions to Cable TV and Satellite TV

Year	Number of Subscriptions (millions)	
	Cable TV	Satellite TV
2000	66	14
2001	66	17
2002	65	19
2003	64	22
2004	63	24

Source: *J. D. Power & Associates*

Let n be the number of subscriptions (in millions) at t years since 2000. Reasonable models of the subscriptions to cable TV and satellite TV are

$$n = -0.8t + 66.4 \quad \text{Cable TV}$$
$$n = 2.5t + 14.2 \quad \text{Satellite TV}$$

1. Check that the models fit the data well.
2. Find the sum of the expressions $-0.8t + 66.4$ and $2.5t + 14.2$. What does the result represent?
3. Evaluate the result in Problem 2 for $t = 10$. What does your result mean in this situation?
4. Find the difference of the expressions $-0.8t + 66.4$ and $2.5t + 14.2$. What does the result represent?
5. Evaluate the result in Problem 4 for $t = 10$. What does your result mean in this situation?

Solution

1. To keep track of the two types of data points, we use square "Marks" for the cable TV scattergram and plus-sign "Marks" for the satellite TV scattergram (see Fig. 39). The models appear to fit the data quite well.
2. $$(-0.8t + 66.4) + (2.5t + 14.2) = 1.7t + 80.6$$

The expression $1.7t + 80.6$ represents the *total* number of subscriptions to cable TV and satellite TV at t years since 2000.
3. We substitute 10 for t in the expression $1.7t + 80.6$:

$$1.7(10) + 80.6 = 97.6$$

The total number of subscriptions in 2010 will be 97.6 million subscriptions.

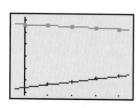

Figure 39 The models fit the data well

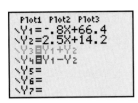

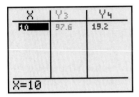

Figure 40 Verify the work

4. $(-0.8t + 66.4) - (2.5t + 14.2)$

$$= (-0.8t + 66.4) - 1(2.5t + 14.2) \qquad a - b = a - 1b$$
$$= -0.8t + 66.4 - 2.5t - 14.2 \qquad \text{Distributive law}$$
$$= -3.3t + 52.2 \qquad \text{Combine like terms.}$$

The expression $-3.3t + 52.2$ represents the *difference* in subscriptions (in millions) to cable TV and satellite TV at t years since 2000.

5. We substitute 10 for t in the expression $-3.3t + 52.2$:

$$-3.3(10) + 52.2 = 19.2$$

In 2010, the difference in subscriptions to cable TV and satellite TV will be 19.2 million subscriptions. In other words, there will be 19.2 million more cable TV subscriptions than satellite TV subscriptions.

We can use Y_n references on a graphing calculator to verify our work in Problems 3 and 5 (see Fig. 40). See Section A.22 for graphing calculator instructions. ∎

group exploration

The degree of a sum of polynomials

In this exploration, you will explore the degree of the sum of two polynomials.

1. Find the degrees of the given expressions. Next, add the expressions. Then determine the degree of the sum.
   a. $2x + 6$ and $4x^2 + 5x + 4$
   b. $4x + 2$ and $3x^3 + 5x^2 + x + 6$
   c. $3x^2 + 6x + 1$ and $4x^3 + 2x^2 + 5x + 7$

2. What is the degree of the sum of a polynomial of degree 1 and a polynomial of degree 2?

3. What is the degree of the sum of a polynomial of degree 1 and a polynomial of degree 3?

4. What is the degree of the sum of a polynomial of degree 2 and a polynomial of degree 3?

5. Find the degrees of the given expressions. Next, add the expressions. Then determine the degree of the sum.
   a. $3x^2 + 2x + 6$ and $5x^2 + 7x + 4$
   b. $2x^2 + 3x + 5$ and $-2x^2 + 4x + 1$

6. If two polynomials have the same degree, what can you say about the degree of the sum of the two polynomials?

HOMEWORK 7.3   FOR EXTRA HELP ▶

Student Solutions Manual   PH Math/Tutor Center   *Math* XL   **MyMathLab**
                                                    MathXL®   MyMathLab

Use words such as linear, quadratic, cubic, polynomial, degree, one variable, *and* two variables *to describe the expression.*

1. $3x^2 - 4x + 2$
2. $9x^2 + 5x^3 - 1$
3. $-7x^3 - 9x - 4$
4. $-4x^3 + 6x^2 - 9x + 8$
5. $3p^5q^2 - 5p^3q^3 + 7pq^4$
6. $8m^7n + 2m^4n^3 + 7m^2n^4$

Combine like terms when possible. Use a graphing calculator table to verify your work when possible.

7. $2x + 4x$
8. $3x + 7x$
9. $-4x - 9x$
10. $6x - 8x$
11. $3t^2 + 5t^2$
12. $-4a^2 + 9a^2$
13. $-8a^4b^3 - 3a^4b^3$
14. $9x^2y^5 - 2x^2y^5$

15. $4x^2 + x^2$
16. $-8x^2 - x^2$
17. $7x^2 - 3x$
18. $6x^2 + 2x$
19. $5b^3 - 8b^3$
20. $-4m^5 + 2m^5$
21. $-x^6 + 7x^6$
22. $-x^4 + 6x^4$
23. $2t^3w^5 + 4t^5w^3$
24. $4a^7b^2 - 7a^2b^7$
25. $-2.5p^4 + 9.9p^4$
26. $-3.05y^3 - 7.38y^3$

Perform the indicated operations. Use a graphing calculator table to verify your work.

27. $8x^2 + 2x - 3x^2 - 5x$
28. $-7x + 9x^2 + 4x - x^2$
29. $9x - 4x^2 + 5x - 2 + 3x^2 - 6$

30. $6x^2 - 5x - x^2 - 2x - 1$

31. $5x + 8x^3 - x^2 + 4x^3 + 2x - x^3$

32. $-5x^2 - 9x + 8x^3 - x + 1 - 7x^3$

33. $20.3t^2 - 5.4t - 45.1t^3 - 3.6t + 93.8t^2$

34. $-5.99k^2 + 2.35 - 6.911k^3 + 7.91k^2 - 1.99k^3$

Perform the addition. Use a graphing calculator table to verify your work when possible.

35. $\left(5x^2 - 4x - 2\right) + \left(-9x^2 - 3x + 8\right)$

36. $\left(-6x^2 + 2x - 5\right) + \left(4x^2 + 3x - 2\right)$

37. $\left(4x^2 - 3x + 2\right) + \left(-4x^2 + 3x - 2\right)$

38. $\left(-5x^2 + 2x - 7\right) + \left(5x^2 - 2x + 7\right)$

39. $\left(4x^3 - 7x^2 + 2x - 9\right) + \left(-7x^3 - 3x^2 + 5x - 2\right)$

40. $\left(-5x^3 + 4x^2 - 6x + 4\right) + \left(-2x^3 - 8x^2 + x - 5\right)$

41. $\left(5a^3 - a\right) + \left(8a^3 - 3a^2 - 1\right)$

42. $\left(2t^2 - 8t + 4\right) + \left(-7t^3 + 2t^2\right)$

43. $\left(5a^2 - 3ab + 7b^2\right) + \left(-4a^2 - 6ab + 2b^2\right)$

44. $\left(3m^2 + 6mn - 4n^2\right) + \left(6m^2 - 7mn - 2n^2\right)$

45. $\left(14.1x^3 - 7.9x^2 - 4.8x + 31.9\right) + \left(-8.2x^3 + 28.8x^2 - 9.5x + 32.2\right)$

46. $\left(-76.3x^3 + 8.2x^2 - 14.4x - 57.1\right) + \left(-19.3x^3 - 3.8x^2 - 3.2x + 80.9\right)$

For Exercises 47–58, find the difference. Use a graphing calculator table to verify your work when possible.

47. $(-3x + 8) - (4x - 3)$

48. $(2x - 4) - (-3x - 1)$

49. $\left(2x^2 + 5x - 1\right) - \left(4x^2 + 9x - 7\right)$

50. $\left(3x^2 - 7x + 2\right) - \left(2x^2 - x + 1\right)$

51. $\left(6y^3 - 2y^2 - 4y + 5\right) - \left(5y^3 - 8y^2 - 5y + 3\right)$

52. $\left(-3t^3 + 7t^2 - 2t + 4\right) - \left(4t^3 + 6t^2 - t - 4\right)$

53. $\left(5x^3 - 9x^2 + 2x - 4\right) - \left(5x^3 - 9x^2 + 2x - 4\right)$

54. $\left(-3x^3 + 2x^2 + 7x - 1\right) - \left(-3x^3 + 2x^2 + 7x - 1\right)$

55. $\left(3x^2 + 7xy - 2y^2\right) - \left(5x^2 - 4xy - 3y^2\right)$

56. $\left(-7a^2 - 4ab + 3b^2\right) - \left(-5a^2 + ab - 6b^2\right)$

57. $\left(2.54x^2 + 6.29x - 7.99\right) - \left(-4.21x^2 - 8.45x + 9.29\right)$

58. $\left(-7.72x^2 - 5.38x + 6.13\right) - \left(2.49x^2 - 3.36x - 8.83\right)$

59. Find the sum of the expressions $\left(2x^2 + 5\right)$ and $(3x + 9)$.

60. Find the sum of the expressions $\left(-4x^2 + x\right)$ and $\left(x^2 - 1\right)$.

61. Find the difference of the expressions $\left(5x^3 - 5x + 8\right)$ and $\left(8x^2 - 3x - 3\right)$.

62. Find the difference of the expressions $\left(4x^3 + 6x^2 - x\right)$ and $\left(7x^3 - 4x - 2\right)$.

63. Subtract $\left(3x^2 - 5x + 4\right)$ from $\left(2x^2 + 6x - 1\right)$.

64. Subtract $\left(4x^2 + 2x - 4\right)$ from $\left(7x^2 - 9x + 5\right)$.

Perform the indicated operations. Use a graphing calculator to verify your work.

65. $\left(6x^3 - 2x\right) - \left(4x^2 - 7x + 2\right) + \left(3x^3 - x^2\right)$

66. $\left(4x^3 - 7x^2\right) + \left(5x^2 - 6x\right) - \left(7x^3 + 4x^2 - 5\right)$

67. In Exercise 1 of Homework 6.4, the number n (in thousands) of people who have earned a bachelor's degree is modeled by the system

$$n = 13.28t + 440.09 \quad \text{Women}$$
$$n = 3.42t + 468.14 \quad \text{Men}$$

where t is the number of years since 1980 (see Table 25).

Table 25 Women and Men Who Have Earned a Bachelor's Degree		
	Number of People Who Have Earned a Bachelor's Degree (thousands)	
Year	Women	Men
1980	456	474
1985	497	483
1990	560	492
1995	634	526
2000	708	530
2002	742	550

Source: *U.S. National Center for Education Statistics*

a. Find the sum of the expressions $13.28t + 440.09$ and $3.42t + 468.14$. What does the result represent?

b. Evaluate the result in part (a) for $t = 31$. What does your result mean in this situation?

c. Find the difference of the expressions $13.28t + 440.09$ and $3.42t + 468.14$. What does the result represent?

d. Evaluate the result in part (c) for $t = 31$. What does your result mean in this situation?

68. In Exercise 2 of Homework 6.4, the number n of daily newspapers is modeled by the system

$$n = 18.13t + 386.86 \quad \text{Morning newspapers}$$
$$n = -31.52t + 1382.53 \quad \text{Evening newspapers}$$

where t is the number of years since 1980 (see Table 26).

Table 26 Numbers of Morning Dailies and Evening Dailies		
	Number of Daily Newspapers	
Year	Morning	Evening
1980	387	1388
1985	482	1220
1990	559	1084
1995	656	891
2000	766	727
2004	813	653

Source: *Editor & Publisher Co.*

a. Find the sum of the expressions $18.13t + 386.86$ and $-31.52t + 1382.53$. What does the result represent?

b. Evaluate the result in part (a) for $t = 30$. What does your result mean in this situation?

c. Find the difference of the expressions $18.13t + 386.86$ and $-31.52t + 1382.53$. What does the result represent?

d. Evaluate the result in part (c) for $t = 30$. What does your result mean in this situation?

69. Crossover utility vehicles are sport utility vehicles (SUVs) with car designs. Many consumers are choosing crossover utility vehicles because they have better gasoline mileage, smoother rides, and better handling than ordinary SUVs do (see Table 27).

Table 27 Crossover Utility Vehicle and SUV Market Shares

| | Percentage of All U.S. Vehicle Sales | |
| | Crossover Utility Vehicles | SUVs |
Year		
1996	0.5	13.2
1998	1.7	15.0
2000	3.1	17.2
2002	7.4	17.7
2004	11.5	16.5

Source: *Ward's Auto Infobank*

Let p be the percentage of U.S. vehicle sales at t years since 1990. Models of crossover utility vehicle and SUV sales are

$p = 0.155t^2 - 1.72t + 5.28$ *Crossover utility vehicles*
$p = -0.138t^2 + 3.22t - 1.38$ *SUVs*

a. Use a graphing calculator to check that the models fit the data well.

b. Find the sum of the expressions $0.155t^2 - 1.72t + 5.28$ and $-0.138t^2 + 3.22t - 1.38$. What does the result represent?

c. Evaluate the result in part (b) for $t = 18$. What does your result mean in this situation?

d. Find the difference of the expressions $0.155t^2 - 1.72t + 5.28$ and $-0.138t^2 + 3.22t - 1.38$. What does the result represent?

e. Evaluate the result in part (d) for $t = 18$. What does your result mean in this situation?

70. Per person annual health care costs and the portions paid by employers with more than 5000 employees are shown in Table 28 for various years.

Table 28 Health Care Costs for Companies with More than 500 Employees

| | Annual Health Care Costs (thousands of dollars) | |
Year	Per Person Cost	Portion of per Person Cost Paid by Employers
1998	4.0	3.0
1999	4.2	3.2
2000	4.7	3.5
2001	5.1	3.8
2002	6.0	4.3
2003	7.0	4.9

Source: *Hewitt Associates*

Let H be the per person annual health care cost (in thousands of dollars) at t years since 1990. Models of the per person cost and the portion of the per person cost paid by employers are

$H = 0.10t^2 - 1.51t + 9.66$ *Per person cost*

$H = 0.05t^2 - 0.68t + 5.22$ *Portion of per person cost paid by employers*

a. Use a graphing calculator to check that the models fit the data well.

b. Find the difference of the expressions $0.10t^2 - 1.51t + 9.66$ and $0.05t^2 - 0.68t + 5.22$. What does the result represent?

c. Evaluate the result in part (b) for $t = 20$. What does your result mean in this situation?

d. Predict the *percentage* of per person annual health care costs that will be paid by *employees* in 2010. [**Hint:** First find the per person annual cost in 2010. Then use your result in part (c).]

71. A student tries to find the difference $(5x^2 + 3x + 7) - (3x^2 + 2x + 1)$:

$$(5x^2 + 3x + 7) - (3x^2 + 2x + 1)$$
$$= 5x^2 + 3x + 7 - 3x^2 + 2x + 1$$
$$= 2x^2 + 5x + 8$$

Describe any errors. Then find the difference correctly.

72. A student tries to find the sum $3x^2 + 2x$:

$$3x^2 + 2x = 5x$$

Describe any errors. Then find the sum correctly.

73. Find two polynomials of degree 3 whose sum is the polynomial $6x^3 - 4x^2 + x - 2$.

74. Find two polynomials of degree 3 whose difference is $6x^3 - 4x^2 + x - 2$.

75. Find two polynomials of degree 3 whose sum is $3x^2 - 5x + 4$.

76. Find two polynomials of degree 3 whose sum is 0.

77. Use the distributive law to explain why $2x^3 + 5x^3 = 7x^3$.

78. Use the distributive law to explain why $9x^4 - 3x^4 = 6x^4$.

Related Review

For Exercises 79–82, let x be a number. Translate the English phrase to a mathematical expression. Then simplify the expression.

79. The number squared, minus 3 times the number, plus 5 times the number squared

80. The number squared plus twice the number squared minus 3

81. Four times the number squared, minus the sum of the number squared and the number

82. Six times the number squared, plus the difference of the number squared and 5

83. The amounts of domestic and imported petroleum consumed in the United States are shown in Table 29 for various years.

Table 29 Amounts of Domestic and Imported Petroleum Consumed in the United States

| | Sources of Petroleum (millions of barrels per day) | |
Year	Domestic	Imported
1989	9.1	7.9
1994	8.4	9.0
1999	7.6	10.5
2004	7.1	12.8

Source: *Energy Information Administration*

Let A be the amount of petroleum (in millions of barrels of petroleum per day) consumed in the United States at t years since 1980. Models of domestic petroleum and imported petroleum are

$$A = -0.14t + 10.29 \quad \text{Domestic}$$
$$A = 0.32t + 4.70 \quad \text{Imported}$$

a. Use a graphing calculator to check that the models fit the data well.
b. Find the sum of the expressions $-0.14t + 10.29$ and $0.32t + 4.70$. What does the result represent?
c. Evaluate the result in part (b) for $t = 31$. What does your result mean in this situation?
d. Use elimination or substitution to estimate in what year the United States consumed the same amount of domestic petroleum as imported petroleum.

84. U.S. revenues from pacemakers and defibrillators, devices that help maintain heart rhythms, are shown in Table 30 for various years. Let r be the revenue (in billions of dollars) from heart devices in the year that is t years since 2000. Models of pacemaker and defibrillator revenues are

$$r = 0.1t + 2.4 \quad \text{Pacemakers}$$
$$r = 0.82t + 1.34 \quad \text{Defibrillators}$$

a. Use a graphing calculator to check that the models fit the data well.
b. Find the sum of the expressions $0.1t + 2.4$ and $0.82t + 1.34$. What does the result represent?

Table 30 Revenues from Pacemakers and Defibrillators

| Year | Revenues from Heart Devices (billions of dollars) | |
	Pacemakers	Defibrillators
2000	2.4	1.7
2001	2.5	1.8
2002	2.6	2.8
2003	2.7	3.8
2004	2.8	4.8

Source: *Merrill Lynch*

c. Evaluate the result in part (b) for $t = 10$. What does your result mean in this situation?
d. Use elimination or substitution to estimate in what year the revenue from pacemakers was equal to the revenue from defibrillators.

Expressions, Equations, and Graphs

Perform the indicated instruction. Then use words such as linear, quadratic, cubic, polynomial, degree, one variable, and two variables to describe the expression, equation, or system.

85. Simplify $3x^2 - 1 - 2x^2$.
86. Evaluate $3t^2 - 1 - 2t^2$ for $t = -2$.
87. Graph $y = 3x^2 - 1 - 2x^2$ by hand.
88. Solve $3p - 1 = -2p$.

7.4 MULTIPLYING POLYNOMIALS

I think and think for months and years. Ninety-nine times, the conclusion is false. The hundredth time I am right.

—Albert Einstein

Objectives

▸ Know the meaning of *binomial* and *trinomial*.
▸ Know the product property for exponents.
▸ Multiply polynomials.
▸ Use a product of two expressions to model an authentic situation.

In Section 7.3, we added and subtracted polynomials. How do we multiply polynomials?

Monomials, Binomials, and Trinomials

We refer to a polynomial as a monomial, a **binomial,** or a **trinomial,** depending on whether it has one, two, or three nonzero terms, respectively:

Name	Examples	Meaning
monomial	$8x^3$, x^2, $-4xy^2$, 9	one nonzero term
binomial	$7x^3 - 2x$, $-2x^2 + x$, $8y^2 + 5$	two nonzero terms
trinomial	$-x^3 + 8x - 7$, $4x^2 + 6x - 1$	three nonzero terms

Product Property for Exponents

Consider the product $x^2 \cdot x^3$. We can write this product as a single power:

$$x^2 \cdot x^3 = (x \cdot x)(x \cdot x \cdot x) \quad \text{Write factors without exponents.}$$
$$= x \cdot x \cdot x \cdot x \cdot x \quad \text{Remove parentheses.}$$
$$= x^5 \quad \text{Simplify.}$$

Note that we can find the same product by adding the exponents:

$$x^2 \cdot x^3 = x^{2+3} = x^5$$

Product Property for Exponents

If n and m are counting numbers, then

$$x^m x^n = x^{m+n}$$

In words: To multiply two powers of x, keep the base and add the exponents.

For example, $x^4 x^8 = x^{4+8} = x^{12}$. Also, $x^5 x = x^5 x^1 = x^{5+1} = x^6$.

Multiplication of Monomials

We can use the product property to help us multiply monomials.

Example 1 Finding Products of Monomials

Find the product.

1. $2x(7x^3)$ **2.** $3x^4(-5x^2)$

Solution

1. We rearrange factors so that the coefficients are adjacent and the powers of x are adjacent:

$$2x(7x^3) = (2 \cdot 7)(x^1 x^3) \quad \text{Rearrange factors; } x = x^1$$
$$= 14x^4 \quad \text{Add exponents: } x^m x^n = x^{m+n}$$

2. $3x^4(-5x^2) = 3(-5)(x^4 \cdot x^2) \quad \text{Rearrange factors.}$
$$= -15x^6 \quad \text{Add exponents: } x^m x^n = x^{m+n} \quad \blacksquare$$

WARNING   The expression $3x^4(-5x^2)$ is *not* the same as the expression $3x^4 - 5x^2$. The expression $3x^4(-5x^2)$ is a product, whereas $3x^4 - 5x^2$ is a difference. We can tell that $3x^4(-5x^2)$ is a product because there is no operation symbol between the expressions $3x^4$ and $(-5x^2)$.

Example 2 Finding the Product of Two Monomials

Find the product $3a^2b(2a^3b^2)$.

Solution

We rearrange factors so that the coefficients are adjacent, the powers of a are adjacent, and the powers of b are adjacent:

$$3a^2b(2a^3b^2) = (3 \cdot 2)(a^2 a^3)(b^1 b^2) \quad \text{Rearrange factors.}$$
$$= 6a^5b^3 \quad \text{Add exponents: } x^m x^n = x^{m+n} \quad \blacksquare$$

Multiplication of a Monomial and a Polynomial

We can use the distributive law to help us multiply a monomial and a polynomial.

Example 3   Finding the Product of a Monomial and a Polynomial

Find the product.

1. $2x(3x + 5)$ **2.** $-4t\left(8t^2 - 3t\right)$ **3.** $3xy\left(2x^2 - 4xy + y^2\right)$

Solution

1. $2x(3x + 5) = 2x \cdot 3x + 2x \cdot 5$ Distributive law

$\qquad\qquad\quad = 6x^2 + 10x$ $x \cdot x = x^2$

2. $-4t\left(8t^2 - 3t\right) = -4t \cdot 8t^2 + 4t \cdot 3t$ Distributive law

$\qquad\qquad\qquad = -32t^3 + 12t^2$ Add exponents: $x^m x^n = x^{m+n}$

3. $3xy\left(2x^2 - 4xy + y^2\right) = 3xy \cdot 2x^2 - 3xy \cdot 4xy + 3xy \cdot y^2$ Distributive law

$\qquad\qquad\qquad\qquad = 6x^3y - 12x^2y^2 + 3xy^3$ Add exponents: $x^m x^n = x^{m+n}$

◼

Multiplication of Two Polynomials

We can also use the distributive law to help us find the product of two binomials. For instance, we can find the product $(a + b)(c + d)$ by using the distributive law three times:

$$(a + b)(c + d) = a(c + d) + b(c + d) \qquad \text{Distribute } (c + d).$$
$$= ac + ad + bc + bd \qquad \text{Distribute } a \text{ and distribute } b.$$

By examining the expression $ac + ad + bc + bd$, we can find the product $(a + b)(c + d)$ more directly by adding the four products formed by multiplying each term in the first sum by each term in the second sum:

$$(a + b)(c + d) = ac + ad + bc + bd$$

After using this technique, we combine like terms if possible.

Multiplying Two Polynomials

To multiply two polynomials, multiply each term in the first polynomial by each term in the second polynomial. Then combine like terms if possible.

Example 4   Finding Products of Binomials

Find the product.

1. $(x + 3)(x + 5)$
2. $(3x + 4y)(2x - 5y)$
3. $(-2.8x + 3.5)(4.1x + 9.7)$
4. $(x - 4)\left(x^2 - 3\right)$

Solution

1. $(x + 3)(x + 5) = x \cdot x + x \cdot 5 + 3 \cdot x + 3 \cdot 5$ Multiply pairs of terms.

$\qquad\qquad\qquad = x^2 + 5x + 3x + 15$ $x \cdot x = x^2$

$\qquad\qquad\qquad = x^2 + 8x + 15$ Combine like terms.

2. $(3x + 4y)(2x - 5y) = 3x \cdot 2x - 3x \cdot 5y + 4y \cdot 2x - 4y \cdot 5y$ Multiply pairs of terms.

$$= 6x^2 - 15xy + 8xy - 20y^2$$ $x \cdot x = x^2$

$$= 6x^2 - 7xy - 20y^2$$ Combine like terms.

3. $(-2.8x + 3.5)(4.1x + 9.7)$ Multiply pairs of terms.

$$= -2.8x(4.1x) - 2.8x(9.7) + 3.5(4.1x) + 3.5(9.7)$$

$$= -11.48x^2 - 27.16x + 14.35x + 33.95$$ $x \cdot x = x^2$

$$= -11.48x^2 - 12.81x + 33.95$$ Combine like terms.

4. $(x - 4)(x^2 - 3) = x \cdot x^2 - x \cdot 3 - 4 \cdot x^2 + 4 \cdot 3$ Multiply pairs of terms.

$$= x^3 - 3x - 4x^2 + 12$$ Add exponents: $x^m x^n = x^{m+n}$

$$= x^3 - 4x^2 - 3x + 12$$ Rearrange terms.

We use a graphing calculator table to verify our work (see Fig. 41).

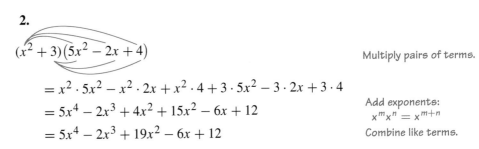

Figure 41 Verify the work

We can find the product of two polynomials of any degree in a similar fashion. The key idea is to multiply each term in the first polynomial by each term in the second polynomial. Then we combine like terms.

Example 5   Finding the Products of Polynomials

Find the product.

1. $(3a - 2)(a^2 + 4a + 5)$ **2.** $(x^2 + 3)(5x^2 - 2x + 4)$

Solution

1.

$(3a - 2)(a^2 + 4a + 5)$ Multiply pairs of terms.

$$= 3a \cdot a^2 + 3a \cdot 4a + 3a \cdot 5 - 2 \cdot a^2 - 2 \cdot 4a - 2 \cdot 5$$ Add exponents: $x^m x^n = x^{m+n}$

$$= 3a^3 + 12a^2 + 15a - 2a^2 - 8a - 10$$

$$= 3a^3 + 12a^2 - 2a^2 + 15a - 8a - 10$$ Rearrange terms.

$$= 3a^3 + 10a^2 + 7a - 10$$ Combine like terms.

2.

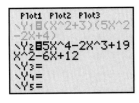

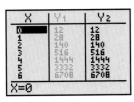

Figure 42 Verify the work

$(x^2 + 3)(5x^2 - 2x + 4)$ Multiply pairs of terms.

$$= x^2 \cdot 5x^2 - x^2 \cdot 2x + x^2 \cdot 4 + 3 \cdot 5x^2 - 3 \cdot 2x + 3 \cdot 4$$

$$= 5x^4 - 2x^3 + 4x^2 + 15x^2 - 6x + 12$$ Add exponents: $x^m x^n = x^{m+n}$

$$= 5x^4 - 2x^3 + 19x^2 - 6x + 12$$ Combine like terms.

We use a graphing calculator table to verify our work (see Fig. 42).

Using a Product of Two Expressions to Model a Situation

In Section 7.3, we modeled authentic situations by using sums and differences of expressions. Sometimes it is meaningful to find the product of two expressions that represent authentic situations.

Example 6 Using a Product to Model a Situation

The average annual cost C of tuition (in dollars per student) at a community college can be modeled by the equation

$$C = 55t + 832$$

where t is the number of years since 1990 (see Table 31). The total enrollment E at all community colleges (in millions of students) can be modeled by the equation

$$E = 0.15t + 3.84$$

where t is the number of years since 1990 (see Table 31).

Table 31 Average Tuition and Total Enrollment at All Community Colleges			
Year*	Average Annual Tuition (dollars)	Year	Total Enrollment (millions)
1990	756	1998	5.0
1993	1025	1999	5.3
1995	1192	2000	5.3
1997	1276	2001	5.4
2000	1338	2002	5.7
2002	1380		
2004	1670		

Source: *U.S. National Center for Education Statistics*
*For school years ending in year shown

1. Check that the models fit the data well.
2. Perform a unit analysis of the expression $(55t + 832)(0.15t + 3.84)$.
3. Find the product of the expressions $55t + 832$ and $0.15t + 3.84$. What does the result represent?
4. Evaluate the result in Problem 3 for $t = 21$. What does your result mean in this situation?

Solution

1. We check the fit of the tuition model in Fig. 43 and the fit of the enrollment model in Fig. 44. The models appear to fit the data fairly well.

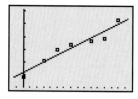   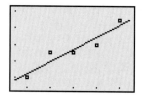

Figure 43 Tuition scattergram and model

Figure 44 Enrollment scattergram and model

2. Here is a unit analysis of the expression $(55t + 832)(0.15t + 3.84)$:

$$\underbrace{(55t + 832)}_{\frac{\text{dollars}}{\text{student}}} \quad \underbrace{(0.15t + 3.84)}_{\text{millions of students}}$$

The units of the expression are millions of dollars.

3. $(55t + 832)(0.15t + 3.84)$

Multiply pairs of terms.

$$= 55t(0.15t) + 55t(3.84) + 832(0.15t) + 832(3.84)$$
$$= 8.25t^2 + 211.2t + 124.8t + 3194.88 \qquad x \cdot x = x^2$$
$$= 8.25t^2 + 336t + 3194.88 \qquad \text{Combine like terms.}$$

The expression represents the total money (in millions of dollars) paid by all community college students for annual tuition in the year that is t years since 1990.

4. We substitute 21 for t in the expression $8.25t^2 + 336t + 3194.88$:

$$8.25(21)^2 + 336(21) + 3194.88 = 13{,}889.13$$

In 2011, the total money paid by all community college students for tuition will be about \$13,889 million dollars, or about \$13.9 billion. ■

group exploration

Looking ahead: Product of binomial conjugates

1. Find the products.
 a. $(x + 2)(x - 2)$ **b.** $(x + 3)(x - 3)$ **c.** $(x + 4)(x - 4)$

2. What pattern do you notice in your work from Problem 1?

3. Use the pattern you described in Problem 2 to help you find each of the following products:
 a. $(x + 5)(x - 5)$ **b.** $(x + 9)(x - 9)$ **c.** $(x + k)(x - k)$

4. Compare the results of finding the product $(x + k)(x - k)$ with the results of finding the product $(x - k)(x + k)$. Explain in terms of one of the laws of operations why the results are the same.

TIPS FOR SUCCESS: "Work Out" by Solving Problems

Although there are many things you can do to enhance your learning, there is no substitute for solving problems. Your mathematical ability will respond to solving problems in much the same way that your muscles respond to lifting weights. Muscles increase greatly in strength when you work out intensely, frequently, and consistently.

Just as with building muscles, to learn math, you must be an *active* participant. No amount of watching weight lifters lift, reading about weight-lifting techniques, or conditioning yourself psychologically can replace working out by lifting weights. The same is true of learning math: No amount of watching your instructor do problems, reading your text, or listening to a tutor can replace "working out" by solving problems.

HOMEWORK 7.4

FOR EXTRA HELP ▶

Student Solutions Manual   PH Math/Tutor Center   Math XL
MathXL®   MyMathLab
MyMathLab

Find the product. Use a graphing calculator table to verify your work when possible.

1. $x^4 x^3$ **2.** $x^3 x^5$ **3.** $w^8 w$ **4.** pp^8

5. $-5x^4(-6x^3)$ **6.** $-6x^9(-2x^2)$

7. $4p^2 t(-9p^3 t^2)$ **8.** $2a^3 b^2(-7ab^2)$

9. $\frac{4}{5}x^3\left(-\frac{7}{2}x^2\right)$ **10.** $-\frac{3}{8}x^5\left(-\frac{4}{5}x\right)$

11. $3w(w - 2)$ **12.** $6y(y + 1)$

13. $2.8x(9.4x - 7.3)$ **14.** $4.6x(5.7x - 1.4)$

15. $-4x(2x^2 + 3)$ **16.** $-5x(3x^2 - 7)$

17. $2mn^2(3m^2 + 5n)$ **18.** $4pq(3p + 2q^2)$

19. $2x(3x^2 - 2x + 7)$ **20.** $4x(x^2 + 3x - 1)$

21. $-3t^2(2t^2 + 4t - 2)$ **22.** $-2a^2(3a^2 - a - 4)$

23. $2xy(3x^2 - 4xy + 5y^2)$ **24.** $3p^2q(4p^2 - pq + 2q^2)$

Find the product. Use a graphing calculator table to verify your work when possible.

25. $(x + 2)(x + 4)$ **26.** $(x + 3)(x + 7)$

27. $(x - 2)(x + 5)$ **28.** $(x + 4)(x - 6)$

29. $(a - 3)(a - 2)$ **30.** $(p - 6)(p - 5)$

31. $(x + 6)(x - 6)$ **32.** $(x - 8)(x + 8)$

33. $(x - 5.3)(x - 9.2)$ **34.** $(x - 1.7)(x + 4.3)$

35. $(5y - 2)(3y + 4)$ **36.** $(6a + 3)(2a - 4)$

37. $(2x + 4)(2x + 4)$ **38.** $(5x + 2)(5x + 2)$

39. $(3x - 1)(3x - 1)$ **40.** $(4x - 3)(4x - 3)$

41. $(3x - 5y)(4x + y)$ **42.** $(2x + 7y)(3x - 2y)$

43. $(2a - 3b)(3a - 2b)$ **44.** $(3m - 7n)(2m - 4n)$

45. $(3x - 4)(3x + 4)$ **46.** $(2x + 7)(2x - 7)$

47. $(9x + 4y)(9x - 4y)$ **48.** $(6a - 5b)(6a + 5b)$

49. $(2.5x + 9.1)(4.6x - 7.7)$

50. $(4.8x - 2.5)(3.1x - 8.9)$

51. $(0.37x + 20.45)(-1.7x + 50.8)$

52. $(-8.2x + 143.7)(0.23x - 8.29)$

53. $(x + 6)(x^2 - 3)$ **54.** $(x - 4)(x^2 + 2)$

55. $(2t^2 - 5)(3t - 2)$ **56.** $(3a^2 - 3)(4a - 5)$

57. $(3a^2 + 5b^2)(2a^2 - 3b^2)$ **58.** $(4m^2 - 7n)(3m^2 - 2n)$

Find the product. Use a graphing calculator table to verify your work when possible.

59. $(x + 2)(x^2 + 3x + 5)$ **60.** $(x + 3)(x^2 + 2x + 4)$

61. $(x + 2)(x^2 - 2x + 4)$ **62.** $(x - 4)(x^2 + 4x + 16)$

63. $(2b^2 - 3b + 2)(b - 4)$ **64.** $(3y^2 - y + 5)(y - 2)$

65. $(2x^2 + 3)(3x^2 - x + 4)$

66. $(5x^2 + 2)(2x^2 - 2x - 3)$

67. $(4w^2 - 2w + 1)(3w^2 - 2)$

68. $(3t^2 + 3t - 4)(2t^2 - 1)$

69. $(2x^2 + 4x - 1)(3x^2 - x + 2)$

70. $(4x^2 - x + 3)(3x^2 + 2x + 1)$

71. $(a + b + c)(a + b - c)$ **72.** $(a - b - c)(a + b + c)$

73. The average monthly cost c (in dollars) of cable TV per subscription in the United States can be modeled by the equation $c = 1.5t + 30.4$, where t is the number of years since 2000 (see Table 32). In Example 6 of Section 7.3, we used the model $n = -0.8t + 66.4$, where n is the number of cable TV subscribers (in millions) in the United States at t years since 2000.

Table 32 Costs of Cable TV per Subscription

Year	Average Monthly Cost per Subscription (dollars)	Number of Subscriptions (millions)
2000	30	66
2001	32	66
2002	34	65
2003	35	64
2004	36	63

Sources: *Veronis Suhler Stevenson Communications Industry Forecast; J. D. Power & Associates*

a. Perform a unit analysis of the quadratic expression $(1.5t + 30.4)(-0.8t + 66.4)$.

b. Find the product of the expressions $1.5t + 30.4$ and $-0.8t + 66.4$. What does the result represent?

c. Evaluate the result in part (b) for $t = 11$. What does your result mean in this situation?

74. In Exercise 18 of Homework 5.3, you found an equation close to $C = 0.14t + 0.8$, where C is the annual consumption of sports drinks (in gallons consumed per person) in the United States at t years since 1990 (see Table 33). The U.S. population can be modeled by the equation $P = 3.3t + 248$, where P is the population (in millions of people) at t years since 1990.

Table 33 U.S. per Person Consumption of Sports Drinks; U.S. Population

Year	Consumption (gallons per person)	Year	Population (millions)
1998	1.9	1990	248.8
1999	2.1	1993	257.8
2000	2.2	1996	265.2
2001	2.3	1999	279.0
2002	2.5	2002	287.9
		2004	293.7

Sources: Beverage Digest; U.S. Census Bureau

a. Perform a unit analysis of the quadratic expression $(0.14t + 0.8)(3.3t + 248)$.

b. Find the product of the expressions $0.14t + 0.8$ and $3.3t + 248$. What does the result represent?

c. Evaluate the result in part (b) for $t = 21$. What does your result mean in this situation?

75. In Exercise 11 of Homework 1.4, you modeled average salaries for public school teachers. A reasonable model is $s = 1.15t + 19$, where s is the average salary (in thousands of dollars) at t years since 1980 (see Table 34). The number n (in millions) of teachers can be modeled well by the equation $n = 0.044t + 1.97$, where t is the number of years since 1980.

a. Perform a unit analysis of the quadratic expression $(1.15t + 19)(0.044t + 1.97)$. [**Hint:** What is one thousand times one million?]

b. Find the product of the expressions $1.15t + 19$ and $0.044t + 1.97$. Round the coefficients of your result to the second decimal place. What does the result represent?

Table 34 Average Salaries and Numbers of Public School Teachers

Year	Average Salary (thousands of dollars)	Number of Teachers (millions)
1985	23.6	2.2
1990	31.3	2.4
1995	37.3	2.6
2000	42.2	2.9
2004	45.6	3.0

Sources: *National Center for Education Statistics; National Education Association*

c. Evaluate the result in part (b) for $t = 30$. What does your result mean in this situation?

76. In Exercise 10 of Homework 1.4, you modeled the death rate from heart disease in the United States. A reasonable model is $r = -7.7t + 563$, where r is the death rate (number of deaths per 100,000 people) from heart disease for the year that is t years since 1960 (see Table 35). The U.S. population can be modeled by the equation $P = 33t + 1493$, where P is the population (in 100,000s of people) at t years since 1960.

Table 35 Death Rates from Heart Disease; U.S. Population

Year	Death Rate (number of deaths per 100,000 people)	Year	Population (in 100,000s)
1960	559	1990	2488
1970	493	1993	2578
1980	412	1996	2652
1990	322	1999	2790
2000	258	2002	2879
2003	236	2004	2937

Sources: *U.S. Center for Health Statistics; U.S. Census Bureau*

a. Perform a unit analysis of the quadratic expression $(-7.7t + 563)(33t + 1493)$.
b. Find the product of the expressions $-7.7t + 563$ and $33t + 1493$. What does the result represent?
c. Evaluate the result in part (b) for $t = 51$. What does your result mean in this situation?

77. A student tries to find the product $6x(-4x)$:

$$6x(-4x) = 6x - 4x = 2x$$

Describe any errors. Then find the product correctly.

78. A student tries to find the product $(x + 3)(x + 5)$:

$$(x + 3)(x + 5) = 2x + 5x + 3x + 8$$
$$= 10x + 8$$

Describe any errors. Then find the product correctly.

79. a. Find the given product. Then decide whether the result is a linear, quadratic, or cubic polynomial.
  i. $(2x + 3)(4x + 5)$
  ii. $(3x - 7)(5x + 2)$
  b. Create two linear polynomials and find their product. Is the result a linear, quadratic, or cubic polynomial?
  c. In general, is the product of two linear polynomials a linear, quadratic, or cubic polynomial? Explain.

80. a. Find the given product. Then decide whether the result is a linear, quadratic, or cubic polynomial.
  i. $(x + 2)(x^2 + 3x + 5)$
  ii. $(2x - 3)(3x^2 + 4x - 1)$
  b. Create a linear polynomial and a quadratic polynomial, and find their product. Is the result a linear, quadratic, or cubic polynomial?
  c. In general, is the product of a linear polynomial and a quadratic polynomial a linear, quadratic, or cubic polynomial? Explain.

81. a. Find the product $(x + 4)(x + 7)$.
  b. Find the product $(x + 7)(x + 4)$.
  c. Explain in terms of the laws of operations why it makes sense that $(x + 4)(x + 7)$ and $(x + 7)(x + 4)$ are equivalent expressions.

82. Use the distributive law three times to help you show that $(x + 2)(x + 7)$ and $x^2 + 9x + 14$ are equivalent expressions.

83. Which of the following expressions are equivalent?

$$(x - 5)(x + 2) \quad x^2 - 3x + 10 \quad x(x - 3) - 10$$
$$(x - 2)(x + 5) \quad x^2 - 3x - 10 \quad (x + 2)(x - 5)$$

84. Which of the following expressions are equivalent?

$$(x + 2)(x + 6) \quad x^2 + 8x + 8 \quad (x + 6)(x + 2)$$
$$x^2 + 8x + 12 \quad (x - 2)(x - 6) \quad x^2 + 4(2x + 3)$$

Related Review

Perform the indicated operation. Use a graphing calculator table to verify your work.

85. $(3x - 5)(2x^2 - 4x + 2)$
86. $(2x^2 - x - 8)(4x - 1)$
87. $(3x - 5) - (2x^2 - 4x + 2)$
88. $(2x^2 - x - 8) - (4x - 1)$

Simplify the right-hand side of the equation to help you decide whether the equation is linear or quadratic. State whether the graph of the equation is a line or a parabola. Use a graphing calculator graph to verify your decision.

89. $y = 3x(x - 2)$
90. $y = -4x(2x - 1)$
91. $y = (x - 3)(x - 4)$
92. $y = (x - 4)(x + 6)$
93. $y = (x + 2) - (3x + 5)$
94. $y = (4x - 3) + (-9x + 7)$
95. $y = (2x + 1)(5x - 2)$
96. $y = (3x - 5)(6x - 4)$

Expressions, Equations, and Graphs

Perform the indicated instruction. Then use words such as linear, quadratic, cubic, polynomial, degree, one variable, and two variables to describe the expression, equation, or system.

97. Solve $2w - 5 = 7w + 5$.
98. Graph $y = 2x - 5$ by hand.
99. Find the product $(2w - 5)(7w + 5)$.
100. Solve the following:

$$y = 2x - 5$$
$$y = 7x + 5$$

7.5 POWERS OF POLYNOMIALS; PRODUCT OF BINOMIAL CONJUGATES

Objectives

▸ Know how to raise a product to a power.

▸ Find the power of a monomial.

▸ Simplify the square of a binomial.

▸ Put a quadratic equation into standard form, $y = ax^2 + bx + c$.

▸ Find the product of two binomial conjugates.

Recall from Section 2.6 that we refer to x^n as "the nth power of x." For most of this section, we will discuss how to find powers of polynomials.

Raising a Product to a Power

Consider the power $(xy)^3$. We can write this power as a product of monomials:

$$(xy)^3 = (xy)(xy)(xy) \qquad \text{Write power without exponents.}$$
$$= (x \cdot x \cdot x)(y \cdot y \cdot y) \qquad \text{Rearrange factors.}$$
$$= x^3 y^3 \qquad \text{Simplify.}$$

Note that we can get this same result by distributing the exponent 3 to both factors of the base xy:

$$(xy)^3 = x^3 y^3$$

Raising a Product to a Power

If n is a counting number, then

$$(xy)^n = x^n y^n$$

In words: To raise a product to a power, raise each factor to the power.

For example, $(xy)^6 = x^6 y^6$.

Power of a Monomial

We can use the property of raising a product to a power to help us find the power of a monomial.

Example 1 Finding Powers of Monomials

Perform the indicated operation.

1. $(xy)^7$ **2.** $(7x)^2$ **3.** $(-2x)^4$

Solution

1. $(xy)^7 = x^7 y^7$
2. $(7x)^2 = 7^2 x^2 = 49x^2$
3. $(-2x)^4 = (-2)^4 x^4 = 16x^4$ ■

WARNING   The expressions $5x^2$ and $(5x)^2$ are *not* equivalent expressions:

$$5x^2 = 5x \cdot x$$
$$(5x)^2 = (5x)(5x) = 25x \cdot x$$

For $5x^2$, the base is the variable x. For $(5x)^2$, the base is the product $5x$.

Here we show a typical error and the correct work performed to find the power $(5x)^2$:

$$(5x)^2 = 5x^2 \qquad \text{Incorrect}$$
$$(5x)^2 = 5^2 x^2 = 25x^2 \qquad \text{Correct}$$

Since the base $5x$ is a product, we need to distribute the exponent 2 to *both* factors, 5 and x.

In general, when finding a power of the form $(AB)^n$, don't forget to distribute the exponent n to both of the factors A and B.

Square of a Binomial

Next, we will discuss how to square a sum. The square of the binomial $x+4$ is $(x+4)^2$:

$$\begin{aligned}(x+4)^2 &= (x+4)(x+4) && y^2 = yy\\ &= x^2 + 4x + 4x + 16 && \text{Multiply pairs of terms.}\\ &= x^2 + 8x + 16 && \text{Combine like terms.}\end{aligned}$$

We have simplified $(x+4)^2$ by writing it as $x^2 + 8x + 16$. We **simplify the square of a binomial** by writing it as an expression that does not have parentheses.

Now we generalize and simplify $(A+B)^2$:

$$\begin{aligned}(A+B)^2 &= (A+B)(A+B) && y^2 = yy\\ &= A^2 + AB + BA + B^2 && \text{Multiply pairs of terms.}\\ &= A^2 + 2AB + B^2 && \text{Combine like terms.}\end{aligned}$$

So, $(A+B)^2 = A^2 + 2AB + B^2$. We can use similar steps to find the property $(A-B)^2 = A^2 - 2AB + B^2$.

Squaring a Binomial

$$(A+B)^2 = A^2 + 2AB + B^2 \qquad \text{Square of a sum}$$
$$(A-B)^2 = A^2 - 2AB + B^2 \qquad \text{Square of a difference}$$

In words: The square of a binomial equals the first term squared, plus (or minus) twice the product of the two terms, plus the second term squared.

We can instead simplify $(x+4)^2$ by substituting x for A and 4 for B in the formula for the square of a sum:

$$(A+B)^2 = A^2 + 2AB + B^2$$
$$(x+4)^2 = x^2 + 2 \cdot x \cdot 4 + 4^2 \qquad \text{Substitute.}$$
$$= x^2 + 8x + 16 \qquad \text{Simplify.}$$

The result is the same as our result from writing $(x+4)^2$ as $(x+4)(x+4)$ and then multiplying. So, there are two ways to simplify $(x+4)^2$. Similarly, there are two ways to simplify the square of any binomial. If you experiment with both methods on several exercises in the homework, you will be able to make an informed choice of methods for solving future problems.

Example 2   Simplifying Squares of Binomials

Simplify.

  1. $(x+3)^2$ **2.** $(x-7)^2$

Solution

  1. We substitute x for A and 3 for B:

$$(A+B)^2 = A^2 + 2AB + B^2$$
$$(x+3)^2 = x^2 + 2 \cdot x \cdot 3 + 3^2 \qquad \text{Substitute.}$$
$$= x^2 + 6x + 9 \qquad \text{Simplify.}$$

Another way to simplify $(x + 3)^2$ is to write it as $(x + 3)(x + 3)$ and then multiply each term in the first binomial by each term in the second binomial:

$$\begin{aligned}(x + 3)^2 &= (x + 3)(x + 3) &&b^2 = bb\\ &= x^2 + 3x + 3x + 9 &&\text{Multiply pairs of terms.}\\ &= x^2 + 6x + 9 &&\text{Combine like terms.}\end{aligned}$$

2. We substitute x for A and 7 for B:

$$(A - B)^2 = A^2 - 2AB + B^2$$
$$(x - 7)^2 = x^2 - 2 \cdot x \cdot 7 + 7^2 \qquad \text{Substitute.}$$
$$= x^2 - 14x + 49 \qquad \text{Simplify.}$$

Another way to simplify $(x - 7)^2$ is to write it as $(x - 7)(x - 7)$ and then multiply each term in the first binomial by each term in the second binomial:

$$\begin{aligned}(x - 7)^2 &= (x - 7)(x - 7) &&b^2 = bb\\ &= x^2 - 7x - 7x + 49 &&\text{Multiply pairs of terms.}\\ &= x^2 - 14x + 49 &&\text{Combine like terms.} \qquad \blacksquare\end{aligned}$$

In Example 2, we found that $x^2 + 6x + 9$ and $x^2 - 14x + 49$ are *squares* of binomials. Both of these *trinomials* are called perfect-square trinomials. A **perfect-square trinomial** is a trinomial that is equivalent to the square of a binomial.

Example 3   Simplifying Squares of Binomials

Simplify.

1. $(5x - 4)^2$

2. $(3x + 5y)^2$

Solution

1. We substitute $5x$ for A and 4 for B:

$$(A - B)^2 = A^2 - 2AB + B^2$$
$$(5x - 4)^2 = (5x)^2 - 2 \cdot 5x \cdot 4 + 4^2 \qquad \text{Substitute.}$$
$$= 25x^2 - 40x + 16 \qquad \text{Simplify.}$$

We use a graphing calculator table to verify our work (see Fig. 45).

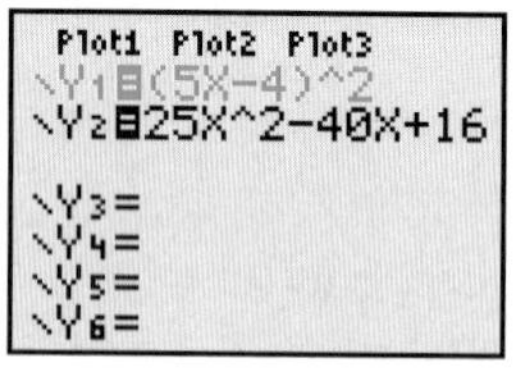

2. We substitute $3x$ for A and $5y$ for B:

$$(A + B)^2 = A^2 + 2AB + B^2$$
$$(3x + 5y)^2 = (3x)^2 + 2 \cdot 3x \cdot 5y + (5y)^2 \qquad \text{Substitute.}$$
$$= 9x^2 + 30xy + 25y^2 \qquad \text{Simplify.} \qquad \blacksquare$$

Here we compare squaring a product with squaring a sum:

$$(AB)^2 = A^2 B^2 \qquad \text{Square of a product}$$
$$(A + B)^2 = A^2 + 2AB + B^2 \qquad \text{Square of a sum}$$

WARNING   It is a common error to omit the middle term in squaring a binomial. Here we show not only a typical error made in simplifying $(x + 6)^2$, but also the correct result:

$$(x + 6)^2 = x^2 + 36 \qquad \text{Incorrect}$$
$$(x + 6)^2 = x^2 + 12x + 36 \qquad \text{Correct}$$

When simplifying $(A + B)^2$, don't omit the middle term, $2AB$, of $A^2 + 2AB + B^2$. Likewise, when simplifying $(A - B)^2$, don't omit the middle term, $-2AB$, of $A^2 - 2AB + B^2$.

Plot1 Plot2 Plot3
\Y1◼(5X-4)^2
\Y2◼25X^2-40X+16
\Y3=
\Y4=
\Y5=
\Y6=

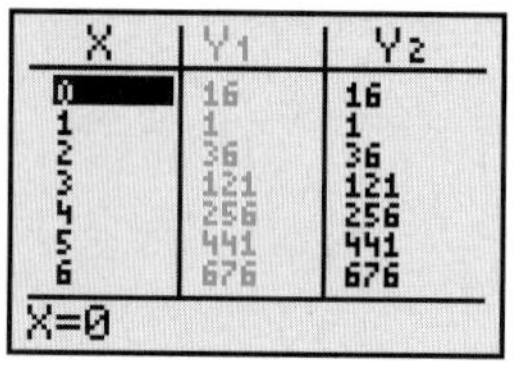

X	Y1	Y2
0	16	16
1	1	1
2	36	36
3	121	121
4	256	256
5	441	441
6	676	676

X=0

Figure 45 Verify the work

Quadratic Equations Written in Standard Form

We can sometimes use our knowledge of squaring a binomial to help us write a quadratic equation in standard form, $y = ax^2 + bx + c$.

Example 4 Writing a Quadratic Equation in Standard Form

Write $y = -2(x - 5)^2 + 3$ in standard form.

Solution

We begin by simplifying $(x - 5)^2$, because we work with exponents before multiplying or adding:

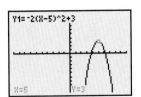

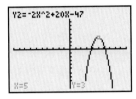

Figure 46 Verify the work

$$
\begin{aligned}
y &= -2(x - 5)^2 + 3 & &\text{Original equation} \\
&= -2(x^2 - 10x + 25) + 3 & &(A - B)^2 = A^2 - 2AB + B^2 \\
&= -2x^2 + 20x - 50 + 3 & &\text{Distributive law} \\
&= -2x^2 + 20x - 47 & &\text{Combine like terms.}
\end{aligned}
$$

We use graphing calculator graphs to verify our work (see Fig. 46). ∎

Multiplication of Binomial Conjugates

We say that the sum of two terms and the difference of the same two terms are **binomial conjugates** of each other. For instance, $5x + 7$ and $5x - 7$ are binomial conjugates. To find the binomial conjugate of a binomial, we change the plus symbol to a minus symbol or vice versa.

How do we find the product of two binomial conjugates? To begin, we find the product of the binomial conjugates $x + 3$ and $x - 3$:

$$
\begin{aligned}
(x + 3)(x - 3) &= x^2 - 3x + 3x - 9 & &\text{Multiply pairs of terms.} \\
&= x^2 - 9 & &\text{Combine like terms.}
\end{aligned}
$$

Now we generalize and find the product of the binomial conjugates $(A+B)$ and $(A-B)$:

$$
\begin{aligned}
(A + B)(A - B) &= A^2 - AB + AB - B^2 & &\text{Multiply pairs of terms.} \\
&= A^2 - B^2 & &\text{Combine like terms.}
\end{aligned}
$$

We see that the product of $A + B$ and $A - B$ is the binomial $A^2 - B^2$, which is called a **difference of two squares.**

Product of Binomial Conjugates

$$(A + B)(A - B) = A^2 - B^2$$

In words: The product of two binomial conjugates is the difference of the square of the first term and the square of the second term.

By the commutative law of multiplication, we have

$$(A - B)(A + B) = (A + B)(A - B) = A^2 - B^2.$$

So, it is also true that

$$(A - B)(A + B) = A^2 - B^2$$

Example 5 Finding the Product of Binomial Conjugates

Find the product.

1. $(x + 5)(x - 5)$
2. $(3x - 8)(3x + 8)$
3. $(2a - 7b)(2a + 7b)$

Figure 47 Verify the work

Solution

1. We substitute x for A and 5 for B:

$$(A + B)(A - B) = A^2 - B^2$$
$$(x + 5)(x - 5) = x^2 - 5^2 \quad \text{Substitute.}$$
$$= x^2 - 25 \quad \text{Simplify.}$$

2. We substitute $3x$ for A and 8 for B:

$$(A - B)(A + B) = A^2 - B^2$$
$$(3x - 8)(3x + 8) = (3x)^2 - 8^2 \quad \text{Substitute.}$$
$$= 9x^2 - 64 \quad \text{Simplify.}$$

We use a graphing calculator table to verify our work (see Fig. 47).

3. We substitute $2a$ for A and $7b$ for B:

$$(A - B)(A + B) = A^2 - B^2$$
$$(2a - 7b)(2a + 7b) = (2a)^2 - (7b)^2 \quad \text{Substitute.}$$
$$= 4a^2 - 49b^2 \quad \text{Simplify.}$$

■

group exploration

Looking ahead: Properties of exponents

1. **a.** Write each expression as a product of fractions. Then simplify so that the result is a single fraction.

   i. $\left(\dfrac{x}{y}\right)^2$ **ii.** $\left(\dfrac{x}{y}\right)^3$ **iii.** $\left(\dfrac{x}{y}\right)^4$

   b. Write $\left(\dfrac{x}{y}\right)^n$ in another form. [**Hint:** Refer to your results in part (a).]

   c. Use your result in part (b) to write $\left(\dfrac{x}{y}\right)^{28}$ in another form in one step.

2. **a.** Write each expression as a product of powers. Then simplify so that the result is a single power.

   i. $\left(x^2\right)^3$ [**Hint:** Write as a product of three x^2 factors.]
   ii. $\left(x^5\right)^2$
   iii. $\left(x^3\right)^4$

   b. Write $\left(x^m\right)^n$ in another form. [**Hint:** Refer to part (a) and determine whether we add, subtract, multiply, or divide the exponents m and n.]

   c. Use your result in part (b) to write $\left(x^6\right)^9$ in another form in one step.

3. **a.** We can write the power x^3 as $x \cdot x \cdot x$ without using exponents. Write each expression without using exponents. Then simplify the result.

   i. $\dfrac{x^5}{x^3}$ **ii.** $\dfrac{x^6}{x^2}$ **iii.** $\dfrac{x^8}{x^5}$

   b. Write $\dfrac{x^m}{x^n}$ in another form. [**Hint:** Refer to part (a) and determine whether we add, subtract, multiply, or divide the exponents m and n.]

   c. Use your result in part (b) to write $\dfrac{x^{25}}{x^{21}}$ in another form in one step.

HOMEWORK 7.5 FOR EXTRA HELP ▶

 Student Solutions Manual   PH Math/Tutor Center   Math XL MathXL®   MyMathLab MyMathLab

Perform the indicated operation. Use a graphing calculator table to verify your work when possible.

1. $(xy)^8$ **2.** $(xy)^5$ **3.** $(6x)^2$ **4.** $(9x)^2$

5. $(4x)^3$ **6.** $(2w)^5$ **7.** $(-8x)^2$ **8.** $(-5x)^2$

9. $(-3x)^3$ **10.** $(-2x)^5$ **11.** $(-a)^5$ **12.** $(-w)^4$

Simplify. Use a graphing calculator table to verify your work when possible.

13. $(x + 5)^2$ **14.** $(x + 2)^2$ **15.** $(x - 4)^2$

16. $(x - 3)^2$ **17.** $(2x + 3)^2$ **18.** $(3x + 7)^2$

19. $(5y - 2)^2$ **20.** $(4a - 9)^2$ **21.** $(3x + 6)^2$

22. $(4x + 3)^2$ **23.** $(9x - 2)^2$ **24.** $(6x - 3)^2$

25. $(2a + 5b)^2$ **26.** $(3t + 7w)^2$ **27.** $(8x - 3y)^2$

28. $(4m - 3n)^2$

Write the equation in standard form. Use graphing calculator graphs to verify your work.

29. $y = (x + 6)^2$ **30.** $y = (x - 8)^2$

31. $y = (x - 3)^2 + 1$ **32.** $y = (x + 4)^2 - 5$

33. $y = 2(x + 4)^2 - 3$ **34.** $y = 3(x - 1)^2 + 2$

35. $y = -3(x - 1)^2 - 2$ **36.** $y = -2(x - 2)^2 - 1$

Find the product. Use a graphing calculator table to verify your work when possible.

37. $(x + 4)(x - 4)$ **38.** $(x + 6)(x - 6)$

39. $(t - 7)(t + 7)$ **40.** $(a - 8)(a + 8)$

41. $(x + 1)(x - 1)$ **42.** $(x + 9)(x - 9)$

43. $(3a + 7)(3a - 7)$ **44.** $(5y + 4)(5y - 4)$

45. $(2x - 3)(2x + 3)$ **46.** $(4x - 7)(4x + 7)$

47. $(3x + 6y)(3x - 6y)$ **48.** $(5m + 8n)(5m - 8n)$

49. $(3t - 7w)(3t + 7w)$ **50.** $(2x - 5y)(2x + 5y)$

Perform the indicated operation. Use a graphing calculator table to verify your work when possible.

51. $(x + 11)(x - 11)$ **52.** $(x - 7)(x + 7)$

53. $\left(4x^2 - 5x\right) - \left(2x^3 - 8x\right)$

54. $\left(7x^3 - 2x\right) - \left(3x^2 - 6x\right)$

55. $(6m - 2n)^2$ **56.** $(2x - 7y)^2$ **57.** $5t\left(-2t^2\right)$

58. $-3w^2(-4w)$ **59.** $(3x + 4)\left(x^2 - x + 2\right)$

60. $(2x - 5)\left(x^2 + 4x - 1\right)$ **61.** $(2t - 3p)(2t + 3p)$

62. $(5a - 2b)(5a + 2b)$ **63.** $2xy^2\left(4x^2 - 8x - 5\right)$

64. $-3xy\left(8x^2 - 2xy + 9y^2\right)$

65. $\left(-6x^2 - 4x + 5\right) + \left(-2x^2 + 3x - 8\right)$

66. $\left(5x^3 - 2x + 1\right) + \left(-9x^3 - 5x^2 - 8x\right)$

67. $(x + 3)(x - 7)$ **68.** $(x - 2)(x - 4)$

69. $(4w - 8)^2$ **70.** $(6c - 4)^2$

71. $\left(x^2 + 2x - 5\right)(x - 3)$ **72.** $\left(x^2 - 2x + 3\right)(x + 2)$

73. $(3x - 7y)(2x + 3y)$ **74.** $(2a + 3b)(4a - 5b)$

75. $(6x - 7)(6x + 7)$ **76.** $(5x - 2)(5x + 2)$

77. $(2t + 7)^2$ **78.** $(5k + 4)^2$

79. A student tries to find the power $(4x)^2$:

$$(4x)^2 = 4x^2$$

Describe any errors. Then find the power correctly.

80. A student tries to find the power $(-3x)^2$:

$$(-3x)^2 = -9x^2$$

Describe any errors. Then find the power correctly.

81. A student tries to simplify $(x + 7)^2$:

$$(x + 7)^2 = x^2 + 7^2$$
$$= x^2 + 49$$

Describe any errors. Then simplify the expression correctly.

82. A student tries to simplify $(x - 9)^2$:

$$(x - 9)^2 = x^2 - 9^2$$
$$= x^2 - 81$$

Describe any errors. Then simplify the expression correctly.

83. Use graphing calculator tables to show that $(x - 5)^2$ is not equivalent to $x^2 - 5^2$ and to show that $(x - 5)^2$ is not equivalent to $x^2 + 5^2$. Then simplify $(x - 5)^2$, and verify your work by using a graphing calculator table.

84. Use a graphing calculator table to show that $(x + 5)^2$ is not equivalent to $x^2 + 5^2$. Then simplify $(x + 5)^2$, and verify your work by using a graphing calculator table.

85. a. Use a graphing calculator table to show that $(x + 4)^2$ and $x^2 + 4^2$ are not equivalent.
   b. Simplify $(x + 4)^2$.
   c. Use a graphing calculator table to show that $(x + 4)^2$ and your result in part (b) are equivalent.

86. a. Use a graphing calculator table to show that $(x - 3)^2$ and $x^2 - 3^2$ are not equivalent.
   b. Simplify $(x - 3)^2$.
   c. Use a graphing calculator table to show that $(x - 3)^2$ and your result in part (b) are equivalent.

87. Show that the statement $(A + B)^2 = A^2 + B^2$ is false by substituting 2 for A and 3 for B. Explain why your work shows that the statement is false. Then write an important true statement that has the expression $(A + B)^2$ on the left side.

88. Show that the statement $(A - B)^2 = A^2 - B^2$ is false by substituting 5 for A and 3 for B. Explain why your work shows that the statement is false. Then write an important true statement that has the expression $(A - B)^2$ on the left side.

89. Which of the following expressions are equivalent?

$(x - 2)^2$ $(x + 2)^2 - 4x$ $x(x - 4) + 4$ $(x - 2)(x + 2)$

$x^2 - 4x + 4$ $(x - 1)(x - 4)$

90. Which of the following expressions are equivalent?

$(x + 3)^2$ $x^2 + 6x + 9$ $(x + 3)^2 + 6x$ $(x + 3)(x + 3)$

$(x - 3)^2 + 12x$ $(x + 3)(x - 3) + 6x$

91. Suppose that a student asks you, "Where did the term $14x$ come from?" in the equation $(x + 7)^2 = x^2 + 14x + 49$. How would you respond?

92. A student tries to simplify the expression $2(x + 4)^2$:

$$2(x + 4)^2 = (2x + 8)^2$$
$$= (2x)^2 + 2(2x)(8) + 8^2$$
$$= 4x^2 + 32x + 64$$

Describe any errors. Then simplify the expression correctly.

93. In this exercise, you will explore a reasonable definition of b^0, where b is a nonzero real number.
  a. i. Perform the exponentiation for $2^5, 2^4, 2^3, 2^2$, and 2^1.
     ii. What pattern do you notice in your results in part (i)?
     iii. Based on your results in part (i), what would be a reasonable value for 2^0?
  b. Perform the exponentiations for $3^4, 3^3, 3^2$, and 3^1. Based on these values, what would be a reasonable value for 3^0?
  c. If b is a nonzero real number, what would be a reasonable value for b^0?

94. a. Perform the exponentiations for $(-2)^2, (-2)^4, (-2)^6$, and $(-2)^8$. State whether each result is positive, negative, or 0.
  b. Perform the exponentiations for $(-2)^3, (-2)^5, (-2)^7$, and $(-2)^9$. State whether each result is positive, negative, or 0.

  c. For which counting numbers n is $(-2)^n$ positive? negative? Explain.
  d. Let k be a negative number. For which counting numbers n is k^n positive? negative? Explain.

Related Review

Simplify the right-hand side of the equation to help you decide whether the equation is linear or quadratic. State whether the graph of the equation is a line or a parabola. Use a graphing calculator graph to verify your work.

95. $y = 2x(5x - 3)$ **96.** $y = -3(7x + 2)$
97. $y = (4x - 3)(6x - 5)$ **98.** $y = 2x(x + 5) - 2x^2$
99. $y = x^2 - (x - 3)^2$ **100.** $y = x - (x + 2)^2$

Expressions, Equations, and Graphs

Perform the indicated instruction. Then use words such as linear, quadratic, cubic, polynomial, degree, one variable, and *two variables to describe the expression, equation, or system.*

101. Solve the following:

$$2x - 5y = 15$$
$$y = 3x - 16$$

102. Simplify $(3a - 8)^2$.

103. Graph $2x - 5y = 15$ by hand.

104. Solve $2(3a - 8) = 2a + 7$.

7.6 PROPERTIES OF EXPONENTS

Many of life's failures are people who did not realize how close they were to success when they gave up.

—*Thomas Edison*

Objectives

▸ Know properties of exponents.

▸ Recognize whether a power expression is simplified.

▸ Use the properties of exponents to simplify power expressions.

▸ Know the meaning of the exponent zero.

In Sections 7.4 and 7.5, we discussed two properties of exponents. In this section, we will discuss several more properties of exponents.

Product Property and Raising a Product to a Power

We begin by reviewing the product property for exponents, from Section 7.4, and how to raise a product to a power, from Section 7.5. If n and m are counting numbers, then

$$x^m x^n = x^{m+n} \quad \text{Product property for exponents}$$
$$(xy)^n = x^n y^n \quad \text{Raising a product to a power}$$

To multiply two powers of x, we keep the base and add the exponents. For example, $x^2 x^5 = x^{2+5} = x^7$. To raise a product to a power, we raise each factor to the power. For example, $(xy)^5 = x^5 y^5$.

Example 1 Performing Operations

Perform the indicated operation.

1. $(2x^3 y^6)(3x^8 y^7)$ **2.** $(-2xy)^3$

Solution

1. We rearrange the factors so that the coefficients are adjacent, the powers of x are adjacent, and the powers of y are adjacent:

$$(2x^3 y^6)(3x^8 y^7) = (2 \cdot 3)(x^3 x^8)(y^6 y^7) \quad \text{Rearrange factors.}$$
$$= 6x^{11} y^{13} \quad \text{Add exponents: } x^m x^n = x^{m+n}$$

2. $(-2xy)^3 = (-2)^3 x^3 y^3$ Raise each factor to third power: $(xy)^n = x^n y^n$
$$= -8x^3 y^3 \quad \text{Simplify.} \qquad \blacksquare$$

Recognizing whether a Power Expression Is Simplified

In Example 1, we worked with the power expressions $(2x^3 y^6)(3x^8 y^7)$ and $(-2xy)^3$. A **power expression** is an expression that contains one or more powers. Here are some more power expressions:

$$(9x^7)(5x^8) \qquad 3x^5(2xy)^3 \qquad \frac{8x^6 y^9}{6x^2 y^4} \qquad \left(\frac{7x^8}{x^3}\right)^2 \qquad x^3 y + \frac{x}{y^2}$$

We can use the two properties about exponents that we have discussed, as well as three more properties that we will discuss in this section, to *simplify a power expression.*

Simplifying a Power Expression

A power expression is simplified if

1. It includes no parentheses.

2. In any monomial, each variable or constant appears as a base at most once. For example, for nonzero x, we write $x^3 x^5$ as x^8.

3. Each numerical expression (such as 5^2) has been calculated, and each numerical fraction has been simplified.

Quotient Property for Exponents

Consider the quotient $\dfrac{x^5}{x^2}$. We can simplify this quotient by first writing the expression without exponents:

$$\frac{x^5}{x^2} = \frac{x \cdot x \cdot x \cdot x \cdot x}{x \cdot x} \quad \text{Write quotient without exponents.}$$
$$= \frac{x \cdot x}{x \cdot x} \cdot \frac{x \cdot x \cdot x}{1} \quad \frac{ac}{bd} = \frac{a}{b} \cdot \frac{c}{d}$$
$$= 1 \cdot x \cdot x \cdot x \quad \text{Simplify: } \frac{x \cdot x}{x \cdot x} = 1$$
$$= x^3 \quad \text{Simplify.}$$

Note that we can simplify the quotient of two powers of x by subtracting the exponents:

$$\frac{x^5}{x^2} = x^{5-2} = x^3$$

> ### Quotient Property for Exponents
> If m and n are counting numbers and x is nonzero, then
> $$\frac{x^m}{x^n} = x^{m-n}$$
> In words: To divide two powers of x, keep the base and subtract the exponents.

For example, $\dfrac{x^9}{x^4} = x^{9-4} = x^5$.

Example 2   Quotient Property

Simplify.

1. $\dfrac{6x^9}{4x^5}$

2. $\dfrac{12x^6y^8}{4xy^7}$

Solution

1. $\dfrac{6x^9}{4x^5} = \dfrac{6}{4} \cdot \dfrac{x^9}{x^5}$ $\dfrac{ac}{bd} = \dfrac{a}{b} \cdot \dfrac{c}{d}$

$= \dfrac{3}{2} \cdot x^{9-5}$ Simplify; subtract exponents: $\dfrac{x^m}{x^n} = x^{m-n}$

$= \dfrac{3}{2} \cdot \dfrac{x^4}{1}$ Simplify.

$= \dfrac{3x^4}{2}$ Multiply numerators; multiply denominators.

2. $\dfrac{12x^6y^8}{4xy^7} = \dfrac{12}{4} \cdot \dfrac{x^6}{x^1} \cdot \dfrac{y^8}{y^7}$ $\dfrac{ac}{bd} = \dfrac{a}{b} \cdot \dfrac{c}{d}$

$= 3 \cdot x^{6-1} \cdot y^{8-7}$ Simplify; subtract exponents: $\dfrac{x^m}{x^n} = x^{m-n}$

$= 3x^5y^1$ Simplify.

$= 3x^5y$ $y^1 = y$

Zero as an Exponent

What is the meaning of 2^0? Computing powers of 2 can suggest the meaning:

The exponent decreases by 1.

$2^4 = 16$
$2^3 = 8$
$2^2 = 4$ The value is divided by 2.
$2^1 = 2$
$2^0 = 1$

Each time we decrease the exponent by 1, the value is divided by 2. This pattern suggests that $2^0 = 1$.

Similar work with any other nonzero base x would suggest that $x^0 = 1$.

> ### DEFINITION   Zero exponent
> For nonzero x,
> $$x^0 = 1$$

So, for nonzero x and y, $4^0 = 1$, $(xy)^0 = 1$ and $\left(\dfrac{5x^2}{y^4}\right)^0 = 1$.

Raising a Quotient to a Power

Consider the power expression $\left(\dfrac{x}{y}\right)^4$. We begin to write this expression in another form by writing it without exponents:

$$\left(\frac{x}{y}\right)^4 = \frac{x}{y} \cdot \frac{x}{y} \cdot \frac{x}{y} \cdot \frac{x}{y} \qquad \text{Write power expression without exponents.}$$

$$= \frac{x \cdot x \cdot x \cdot x}{y \cdot y \cdot y \cdot y} \qquad \text{Multiply numerators; multiply denominators.}$$

$$= \frac{x^4}{y^4} \qquad \text{Simplify.}$$

Note that we can get the same result by distributing the exponent 4 to both the numerator and the denominator of the base $\dfrac{x}{y}$:

$$\left(\frac{x}{y}\right)^4 = \frac{x^4}{y^4}$$

This method is called *raising a quotient to a power.*

Raising a Quotient to a Power

If n is a counting number and y is nonzero, then

$$\left(\frac{x}{y}\right)^n = \frac{x^n}{y^n}$$

In words: To raise a quotient to a power, raise both the numerator and the denominator to the power.

Example 3 Raising a Quotient to a Power

Simplify.

1. $\left(\dfrac{x}{y}\right)^8$ **2.** $\left(\dfrac{x}{2}\right)^3$

Solution

1. $\left(\dfrac{x}{y}\right)^8 = \dfrac{x^8}{y^8}$ Raise numerator and denominator to eighth power: $\left(\dfrac{x}{y}\right)^n = \dfrac{x^n}{y^n}$

2. $\left(\dfrac{x}{2}\right)^3 = \dfrac{x^3}{2^3}$ Raise numerator and denominator to third power: $\left(\dfrac{x}{y}\right)^n = \dfrac{x^n}{y^n}$

                $= \dfrac{x^3}{8}$ Simplify.   ■

Raising a Power to a Power

Consider the power expression $\left(x^2\right)^3$. We can write this expression as the power x^6:

$$\left(x^2\right)^3 = x^2 \cdot x^2 \cdot x^2 \qquad \text{Write expression without exponent 3.}$$

$$= x^{2+2+2} \qquad \text{Add exponents: } x^m x^n = x^{m+n}$$

$$= x^6 \qquad \text{Simplify.}$$

Note that we can get the same result by multiplying the exponents 2 and 3:

$$\left(x^2\right)^3 = x^{2 \cdot 3} = x^6$$

This method is called *raising a power to a power.*

Raising a Power to a Power

If m and n are counting numbers, then

$$(x^m)^n = x^{mn}$$

In words: To raise a power to a power, keep the base and multiply the exponents.

Example 4   Raising a Power to a Power

Simplify.

1. $(x^2)^7$ **2.** $(x^5)^8$

Solution

1. $(x^2)^7 = x^{2 \cdot 7}$ Multiply exponents: $(x^m)^n = x^{mn}$

$\quad\quad\quad = x^{14}$ Simplify.

2. $(x^5)^8 = x^{5 \cdot 8}$ Multiply exponents: $(x^m)^n = x^{mn}$

$\quad\quad\quad = x^{40}$ Simplify.

■

Using Combinations of Properties of Exponents

Next, we use more than one property of exponents to simplify an expression.

Example 5   Simplifying Power Expressions

Simplify.

1. $(2x^8 y^5)^3$ **2.** $2x^7(4x^4)^2$

3. $\dfrac{8x^3 x^4}{2x^7}$

Solution

1. $(2x^8 y^5)^3 = 2^3 (x^8)^3 (y^5)^3$ Raise each factor to third power: $(xy)^n = x^n y^n$

$\quad\quad\quad\quad = 8x^{24} y^{15}$ Multiply exponents: $(x^m)^n = x^{mn}$

2. $2x^7(4x^4)^2 = 2x^7[4^2(x^4)^2]$ Raise each factor to second power: $(xy)^n = x^n y^n$

$\quad\quad\quad\quad\quad = 2x^7[16x^8]$ Multiply exponents: $(x^m)^n = x^{mn}$

$\quad\quad\quad\quad\quad = (2 \cdot 16)(x^7 x^8)$ Rearrange factors.

$\quad\quad\quad\quad\quad = 32x^{15}$ Add exponents: $x^m x^n = x^{m+n}$

3. $\dfrac{8x^3 x^4}{2x^7} = \dfrac{4x^7}{x^7}$ Simplify; add exponents: $x^m x^n = x^{m+n}$

$\quad\quad\quad = 4x^{7-7}$ Subtract exponents: $\dfrac{x^m}{x^n} = x^{m-n}$

$\quad\quad\quad = 4x^0$ Simplify.

$\quad\quad\quad = 4$ $x^0 = 1$.

■

Example 6   Simplifying Power Expressions

Simplify.

1. $\left(\dfrac{2x^4}{3y^7}\right)^3$ **2.** $\dfrac{(2x^3 y)^5}{x^8 y}$

Solution

1. $\left(\dfrac{2x^4}{3y^7}\right)^3 = \dfrac{(2x^4)^3}{(3y^7)^3}$ Raise numerator and denominator to third power: $\left(\dfrac{x}{y}\right)^n = \dfrac{x^n}{y^n}$

$= \dfrac{2^3(x^4)^3}{3^3(y^7)^3}$ Raise factors to third power: $(xy)^n = x^n y^n$

$= \dfrac{8x^{12}}{27y^{21}}$ Multiply exponents: $(x^m)^n = x^{mn}$

2. $\dfrac{(2x^3 y)^5}{x^8 y} = \dfrac{2^5(x^3)^5 y^5}{x^8 y}$ Raise factors to fifth power: $(xy)^n = x^n y^n$

$= \dfrac{32x^{15} y^5}{x^8 y^1}$ Multiply exponents: $(x^m)^n = x^{mn}$; $y = y^1$

$= 32x^{15-8} y^{5-1}$ Subtract exponents: $\dfrac{x^m}{x^n} = x^{m-n}$

$= 32x^7 y^4$ Simplify. ∎

group exploration

Properties of exponents

1. For the statement $x^2 x^3 = x^5$, a student wants to know why there are two x's on the left-hand side of the equation and only one x on the right-hand side. What would you tell the student?

2. A student tries to simplify the expression $(3x^3)^2$:

$$(3x^3)^2 = (3 \cdot 2)x^{3\cdot 2} = 6x^6$$

Describe any errors. Then simplify the expression correctly.

3. In simplifying power expressions, a student is confused about when to add exponents and when to multiply exponents. What would you tell the student?

4. A student tries to simplify $\dfrac{(2x^5)^4}{x^2}$:

$$\dfrac{(2x^5)^4}{x^2} = (2x^{5-2})^4 = (2x^3)^4 = 2^4(x^3)^4 = 16x^{12}$$

Describe any errors. Then simplify the expression correctly.

5. Simplify $x^3 + x^3$. Then simplify $x^3 x^3$. Explain why your two results have different exponents.

group exploration

Looking ahead: Negative-integer exponents

In this exploration, you will explore the meaning of a negative-integer exponent.

1. Use your graphing calculator to compute 2^{-1}. Then press $\boxed{\text{MATH}}$ $\boxed{1}$ $\boxed{\text{ENTER}}$ to convert your result into a fraction.

2. Use your graphing calculator to compute 3^{-1}. Then press $\boxed{\text{MATH}}$ $\boxed{1}$ $\boxed{\text{ENTER}}$ to convert your result into a fraction.

3. For each part, use your graphing calculator to compute the expression and then convert the result into a fraction.
   a. 4^{-1} **b.** 5^{-1} **c.** 6^{-1}

4. Review your work in Problems 1–3. What pattern do you notice? If you do not see a pattern, then keep working with more powers of the form x^{-1} until you do.

5. Without using any type of calculator, write 13^{-1} as a fraction.

6. Write x^{-1} as a fraction.

7. For each part, use your graphing calculator to compute the expression and then convert the result into a fraction.

    a. 3^{-2} **b.** 4^{-2} **c.** 5^{-2}

8. Review your work in Problem 7. What pattern do you notice? If you do not see a pattern, then keep working with more powers of the form x^{-2} until you do.

9. Without using any type of calculator, write 9^{-2} as a fraction.

10. Write x^{-2} as a fraction.

11. Without using any type of calculator, write each expression as a fraction.

    a. 8^{-1} **b.** 7^{-2} **c.** 4^{-3} **d.** 2^{-5}

12. Write each expression as a fraction.

    a. x^{-3} **b.** x^{-4} **c.** x^{-5} **d.** x^{-6}

13. Write the expression x^{-n} as a fraction.

HOMEWORK 7.6

FOR EXTRA HELP ▶

Student Solutions Manual   PH Math/Tutor Center   Math XL MathXL®   MyMathLab MyMathLab

Simplify.

1. $x^3 x^5$

2. $x^2 x^7$

3. $r^5 r$

4. $y y^8$

5. $(5x^4)(3x^5)$

6. $(-2x^7)(6x^8)$

7. $(-4b^3)(-8b^5)$

8. $(3t^5)(-7t^4)$

9. $(6a^2 b^5)(9a^4 b^3)$

10. $(10c^7 d^2)(3c^5 d)$

11. $(rt)^7$

12. $(ab)^4$

13. $(8x)^2$

14. $(2x)^4$

15. $(2xy)^5$

16. $(3xy)^3$

17. $(-2a)^4$

18. $(-3t)^3$

19. $(9xy)^0$

20. $(-3x)^0$

21. $\dfrac{a^5}{a^2}$

22. $\dfrac{w^8}{w^6}$

23. $\dfrac{6x^7}{3x^3}$

24. $\dfrac{8x^6}{4x^5}$

25. $\dfrac{15x^6 y^8}{12x^3 y}$

26. $\dfrac{14x^5 y^9}{21x^3 y^6}$

27. $\left(\dfrac{t}{w}\right)^7$

28. $\left(\dfrac{a}{b}\right)^4$

29. $\left(\dfrac{3}{t}\right)^3$

30. $\left(\dfrac{b}{7}\right)^2$

31. $\left(\dfrac{x}{3}\right)^0$

32. $\left(\dfrac{2}{x}\right)^0$

33. $(r^2)^4$

34. $(a^3)^8$

35. $(x^4)^9$

36. $(x^6)^9$

Simplify.

37. $(6x^3)^2$

38. $(4x^5)^3$

39. $(-t^3)^4$

40. $(-a^2)^3$

41. $(2a^2 a^7)^3$

42. $(2w^3 w^4)^5$

43. $(x^2 y^3)^4 x^5 y^8$

44. $(xy^6)^3 x^2 y^4$

45. $5x^4(3x^6)^2$

46. $4x^2(2x^5)^3$

47. $-3c^6(c^4)^5$

48. $-7w^4(w^3)^6$

49. $(xy^3)^5 (xy)^4$

50. $(x^2 y)^3 (xy^3)^2$

51. $\dfrac{10t^5 t^7}{8t^4}$

52. $\dfrac{15y^9 y^2}{20y^8}$

53. $\dfrac{18x^{10}}{24x^4 x^6}$

54. $\dfrac{12x^{15}}{21x^9 x^6}$

55. $\left(\dfrac{y}{2x}\right)^3$

56. $\left(\dfrac{8x}{y}\right)^2$

57. $\left(\dfrac{x^2}{y^5}\right)^4$

58. $\left(\dfrac{x^8}{y^2}\right)^5$

59. $\left(\dfrac{r^6}{6}\right)^2$

60. $\left(\dfrac{2}{a^4}\right)^3$

61. $\left(\dfrac{2a^4}{3b^2}\right)^3$

62. $\left(\dfrac{4w^6}{7t^3}\right)^2$

63. $\left(\dfrac{3x^4}{5y^7}\right)^0$

64. $\left(\dfrac{-5y^2}{9x^4}\right)^0$

65. $\left(\dfrac{2a^6 b}{3c^5}\right)^3$

66. $\left(\dfrac{5xy^4}{7w^3}\right)^2$

67. $\dfrac{(x^4 y)^4}{x^5}$

68. $\dfrac{(x^3 y^7)^5}{y^{10}}$

69. $\dfrac{(w^3)^4}{(2w)^5}$

70. $\dfrac{(3t)^3}{(t^2)^5}$

71. $\dfrac{(4x^5 y^8)^2}{8x^8 y^9}$

72. $\dfrac{(2x^4 y^3)^3}{12x^5 y^4}$

73. If an object falls from rest for t seconds, the distance d (in feet) the object will fall is described by the equation $d = 16t^2$, assuming no air resistance. Estimate how far a sky diver will have fallen 5 seconds after jumping out of an airplane.

74. The weight w (in ounces) of a thick-crust pizza with three toppings at Papa Del's in Urbana, Illinois, is described by the equation $w = 0.41d^2$, where d is the diameter (in inches) of the pizza. Estimate the weight of such a pizza with diameter 12 inches.

75. The power P (in watts) generated by a windmill is modeled by the equation $P = 0.8r^2 v^3$, where r is the radius (in meters)

of the windmill and v is the wind speed (in meters per second). Estimate the power a windmill with radius 0.57 meter can generate if the wind speed is 12.5 meters per second.

76. The intensity I (in watts per square meter) of a television signal at a distance of d kilometers is described by the equation $I = \dfrac{180}{d^2}$. Find the intensity of the signal at a distance of 5 kilometers.

77. A student tries to simplify $x^3 x^5$:

$$x^3 x^5 = x^{3 \cdot 5} = x^{15}$$

Describe any errors. Then simplify the expression correctly.

78. A student tries to simplify $\left(x^4\right)^6$:

$$\left(x^4\right)^6 = x^{4+6} = x^{10}$$

Describe any errors. Then simplify the expression correctly.

79. A student tries to simplify $\left(5x^3\right)^2$:

$$\left(5x^3\right)^2 = 5\left(x^3\right)^2 = 5x^6$$

Describe any errors. Then simplify the expression correctly.

80. A student tries to simplify $2(3x)^2$:

$$2(3x)^2 = (6x)^2 = 36x^2$$

Describe any errors. Then simplify the expression correctly.

81. A student tries to simplify $\left(2x^2\right)^4$:

$$\left(2x^2\right)^4 = (2 \cdot 4)x^{2 \cdot 4} = 8x^8$$

Describe any errors. Then simplify the expression correctly.

82. Describe what it means to use properties of exponents to simplify a power expression. Include several examples.

Related Review

Simplify.

83. $x^3 x^2$ **84.** $x^5 x^3$ **85.** $x^3 + x^2$

86. $x^5 + x^3$ **87.** $2x^4 + 3x^4$ **88.** $5x^3 - 2x^3$

89. $\left(2x^4\right)\left(3x^4\right)$ **90.** $\left(5x^3\right)\left(-2x^3\right)$ **91.** $(3x)^2$

92. $(6x)^2$ **93.** $(3+x)^2$ **94.** $(6+x)^2$

Expressions, Equations, and Graphs

Perform the indicated instruction. Then use words such as linear, quadratic, cubic, power, polynomial, degree, one variable, *and* two variables *to describe the expression, equation, or system.*

95. Solve $\dfrac{2}{3}x - \dfrac{5}{6} = \dfrac{1}{2}x$.

96. Find the product $(2p - 1)\left(3p^2 + p - 2\right)$.

97. Combine like terms: $\dfrac{2}{3}x - \dfrac{5}{6} - \dfrac{1}{2}x$.

98. Graph $y = x^2 - 4x + 3$ by hand.

7.7 NEGATIVE-INTEGER EXPONENTS

Objectives

▹ Know the meaning of a negative-integer exponent.

▹ Simplify power expressions containing negative-integer exponents.

▹ Work with models whose equations contain negative-integer exponents.

▹ Use scientific notation.

In Section 7.6, we worked with power expressions with nonnegative-integer exponents. In this section, we will extend our work with power expressions to include negative-integer exponents.

Definition of a Negative-Integer Exponent

We can begin to see the meaning of a negative-integer exponent by considering the expression $\dfrac{x^3}{x^5}$. We write this expression in two different forms:

$$\frac{x^3}{x^5} = x^{3-5} = x^{-2}$$

$$\frac{x^3}{x^5} = \frac{x \cdot x \cdot x}{x \cdot x \cdot x \cdot x \cdot x} = \frac{1}{x^2}$$

Equating the two results, we have

$$x^{-2} = \frac{1}{x^2}$$

Likewise, we could write $\dfrac{x^2}{x^5}$ in two different forms to show that

$$x^{-3} = \frac{1}{x^3}$$

Next, we describe this pattern in general.

DEFINITION Negative-integer exponent

If n is a counting number and x is nonzero, then

$$x^{-n} = \frac{1}{x^n}$$

In words: To find x^{-n}, take its reciprocal and change the sign of the exponent.

Simplifying Power Expressions Containing Negative-Integer Exponents

Simplifying a power expression includes writing it so that each exponent is positive.

Example 1   Simplifying Power Expressions

Simplify.

1. 5^{-2} **2.** x^{-6}

Solution

1. $5^{-2} = \dfrac{1}{5^2}$ Write power so exponent is positive: $x^{-n} = \dfrac{1}{x^n}$

         $= \dfrac{1}{25}$ Simplify.

2. $x^{-6} = \dfrac{1}{x^6}$ Write power so exponent is positive: $x^{-n} = \dfrac{1}{x^n}$ ∎

Next, we write $\dfrac{1}{x^{-n}}$ in another form, where x is nonzero and n is a counting number:

$$\frac{1}{x^{-n}} = 1 \div x^{-n} \qquad \frac{a}{b} = a \div b$$

$$= 1 \div \frac{1}{x^n} \qquad \text{Write power so exponent is positive: } x^{-n} = \frac{1}{x^n}$$

$$= 1 \cdot \frac{x^n}{1} \qquad \text{Multiply by reciprocal of } \frac{1}{x^n}, \text{ which is } \frac{x^n}{1}.$$

$$= x^n \qquad \text{Simplify.}$$

So, $\dfrac{1}{x^{-n}} = x^n$.

Negative-Integer Exponent in a Denominator

If x is nonzero and n is a counting number, then

$$\frac{1}{x^{-n}} = x^n$$

In words: To find $\dfrac{1}{x^{-n}}$, find its reciprocal and change the sign of the exponent.

Example 2   Simplifying Power Expressions

Simplify.

1. $\dfrac{1}{2^{-3}}$

2. $\dfrac{1}{x^{-7}}$

Solution

1. $\dfrac{1}{2^{-3}} = 2^3$ Write power so exponent is positive: $\dfrac{1}{x^{-n}} = x^n$

 $\phantom{\dfrac{1}{2^{-3}}} = 8$ Simplify.

2. $\dfrac{1}{x^{-7}} = x^7$ Write power so exponent is positive: $\dfrac{1}{x^{-n}} = x^n$ ∎

In Example 3, we will simplify some quotients of two powers.

Example 3   Simplifying Power Expressions

Simplify.

1. $\dfrac{x^{-4}}{y^2}$

2. $\dfrac{x^3}{y^{-7}}$

Solution

1. $\dfrac{x^{-4}}{y^2} = x^{-4} \cdot \dfrac{1}{y^2}$ Write quotient as a product: $\dfrac{A}{B} = A \cdot \dfrac{1}{B}$

 $\phantom{\dfrac{x^{-4}}{y^2}} = \dfrac{1}{x^4} \cdot \dfrac{1}{y^2}$ Write powers so exponents are positive: $x^{-n} = \dfrac{1}{x^n}$

 $\phantom{\dfrac{x^{-4}}{y^2}} = \dfrac{1}{x^4 y^2}$ Multiply numerators; multiply denominators: $\dfrac{A}{B} \cdot \dfrac{C}{D} = \dfrac{AC}{BD}$

2. $\dfrac{x^3}{y^{-7}} = x^3 \cdot \dfrac{1}{y^{-7}}$ Write quotient as a product: $\dfrac{A}{B} = A \cdot \dfrac{1}{B}$

 $\phantom{\dfrac{x^3}{y^{-7}}} = x^3 y^7$ Write powers so exponents are positive: $\dfrac{1}{x^{-n}} = x^n$ ∎

For Problem 1 of Example 3, we simplified $\dfrac{x^{-4}}{y^2}$. When simplifying such expressions, we usually skip the first two steps shown in the example and write

$$\dfrac{x^{-4}}{y^2} = \dfrac{1}{x^4 y^2}$$

Likewise, we can skip the first step of our work in Problem 2 of Example 3 by simplifying $\dfrac{x^3}{y^{-7}}$ as follows:

$$\dfrac{x^3}{y^{-7}} = x^3 y^7$$

Example 4   Simplifying Power Expressions

Simplify $\dfrac{4a^{-2}b^4}{7c^{-5}}$.

Solution

$$\dfrac{4a^{-2}b^4}{7c^{-5}} = \dfrac{4c^5 b^4}{7a^2}$$ Write powers so exponents are positive: $x^{-n} = \dfrac{1}{x^n}$; $\dfrac{1}{x^{-n}} = x^n$ ∎

In Section 7.6, we discussed various properties of counting-number exponents. It turns out that these properties are also true for all negative-integer exponents and the zero exponent.

Properties of Exponents

If m and n are integers and x and y are nonzero, then

1. $x^m x^n = x^{m+n}$ Product property for exponents

2. $\dfrac{x^m}{x^n} = x^{m-n}$ Quotient property for exponents

3. $(xy)^n = x^n y^n$ Raising a product to a power

4. $\left(\dfrac{x}{y}\right)^n = \dfrac{x^n}{y^n}$ Raising a quotient to a power

5. $\left(x^m\right)^n = x^{mn}$ Raising a power to a power

We can use these properties to expand our rules for simplifying power expressions to include those which contain negative-integer exponents.

Simplifying Power Expressions

A power expression is simplified if

1. It includes no parentheses.
2. In any monomial, each variable or constant appears as a base at most once.
3. Each numerical expression (such as 7^2) has been calculated, and each numerical fraction has been simplified.
4. Each exponent is positive.

Example 5   Simplifying Power Expressions

Simplify.

1. $\left(x^{-4}\right)^3$

2. $\left(8t^6\right)\left(-3t^{-10}\right)$

Solution

1. $\left(x^{-4}\right)^3 = x^{-12}$ Multiply exponents: $(x^m)^n = x^{mn}$

$\qquad = \dfrac{1}{x^{12}}$ Write power so exponent is positive: $x^{-n} = \dfrac{1}{x^n}$

2. $\left(8t^6\right)\left(-3t^{-10}\right) = 8(-3)t^6 t^{-10}$ Rearrange factors.

$\qquad = -24t^{-4}$ Add exponents: $x^m x^n = x^{m+n}$

$\qquad = -\dfrac{24}{t^4}$ Write power so exponent is positive: $x^{-n} = \dfrac{1}{x^n}$ ∎

In Example 6, we will discuss how to simplify the quotient of two powers in which the bases are the same.

Example 6   Simplifying Power Expressions

Simplify.

1. $\dfrac{x^4}{x^9}$

2. $\dfrac{4x^7}{3x^{-2}}$

3. $\dfrac{6b^{-3}c^{-2}}{9b^4 c^{-5}}$

Solution

1. $\dfrac{x^4}{x^9} = x^{4-9}$ Subtract exponents: $\dfrac{x^m}{x^n} = x^{m-n}$

     $= x^{-5}$ Subtract.

     $= \dfrac{1}{x^5}$ Write power so exponent is positive: $x^{-n} = \dfrac{1}{x^n}$

2. $\dfrac{4x^7}{3x^{-2}} = \dfrac{4x^{7-(-2)}}{3}$ Subtract exponents: $\dfrac{x^m}{x^n} = x^{m-n}$

     $= \dfrac{4x^9}{3}$ Simplify.

3. $\dfrac{6b^{-3}c^{-2}}{9b^4c^{-5}} = \dfrac{2b^{-3-4}c^{-2-(-5)}}{3}$ Simplify; subtract exponents: $\dfrac{x^m}{x^n} = x^{m-n}$

     $= \dfrac{2b^{-7}c^3}{3}$ Simplify.

     $= \dfrac{2c^3}{3b^7}$ Write powers so exponents are positive: $x^{-n} = \dfrac{1}{x^n}$ ∎

In the first step of Problem 2 of Example 6, we found that

$$\frac{4x^7}{3x^{-2}} = \frac{4x^{7-(-2)}}{3}$$

WARNING Note that we need a subtraction symbol *and* a negative symbol in the expression on the right-hand side. It is a common error to omit writing one of these two symbols in problems of this type.

Example 7 Simplifying Power Expressions

Simplify.

1. $\left(2x^{-5}\right)^{-3}$ **2.** $\left(\dfrac{2x^{-4}}{y^6}\right)^{-5}$

Solution

1. $\left(2x^{-5}\right)^{-3} = 2^{-3}\left(x^{-5}\right)^{-3}$ Raise factors to power -3: $(xy)^n = x^n y^n$

     $= 2^{-3}x^{15}$ Multiply exponents: $(x^m)^n = x^{mn}$

     $= \dfrac{x^{15}}{2^3}$ Write powers so exponents are positive: $x^{-n} = \dfrac{1}{x^n}$

     $= \dfrac{x^{15}}{8}$ Simplify.

2. $\left(\dfrac{2x^{-4}}{y^6}\right)^{-5} = \dfrac{\left(2x^{-4}\right)^{-5}}{\left(y^6\right)^{-5}}$ Raise numerator and denominator to power -5: $\left(\dfrac{x}{y}\right)^n = \dfrac{x^n}{y^n}$

     $= \dfrac{2^{-5}\left(x^{-4}\right)^{-5}}{\left(y^6\right)^{-5}}$ Raise each factor to power -5: $(xy)^n = x^n y^n$

     $= \dfrac{2^{-5}x^{20}}{y^{-30}}$ Multiply exponents: $(x^m)^n = x^{mn}$

     $= \dfrac{x^{20}y^{30}}{2^5}$ Write powers so exponents are positive: $x^{-n} = \dfrac{1}{x^n}; \dfrac{1}{x^{-n}} = x^n$

     $= \dfrac{x^{20}y^{30}}{32}$ Simplify. ∎

Models Whose Equations Contain Negative-Integer Exponents

Many authentic situations can be modeled by equations that contain negative-integer exponents. Some examples of such situations are the sound level of a guitar, the intensity of illumination by a light bulb, and the gravitational force of the Sun acting on Earth.

Example 8 A Model Whose Equation Contains Negative-Integer Exponents

The intensity of a television signal I (in watts per square meter) at a distance d kilometers from the transmitter is described by the equation

$$I = 250d^{-2}$$

1. Simplify the right-hand side of the equation.
2. Substitute 5 for d in the equation found in Problem 1, and solve for I. What does the result mean in this situation?

Solution

1. We use the definition of a negative-integer exponent to write the expression on the right-hand side without any negative exponents:

   $I = 250d^{-2}$ Original equation

   $I = \dfrac{250}{d^2}$ Write power so exponent is positive: $x^{-n} = \dfrac{1}{x^n}$

2. We substitute 5 for d in the equation $I = \dfrac{250}{d^2}$:

   $$I = \frac{250}{5^2} = \frac{250}{25} = 10$$

   So, the intensity is 10 watts per square meter at a distance of 5 kilometers from the transmitter. ∎

Scientific Notation

Now that we know how to work with negative-integer exponents, we can use exponents to describe numbers in *scientific notation*. This will enable us to describe compactly a number whose absolute value is very large or very small. For example, the distance to Proxima Centauri, the nearest star other than the Sun, is 40,100,000,000,000 kilometers. Here, we write 40,100,000,000,000 in scientific notation:

$$4.01 \times 10^{13}$$

The symbol "$\times$" stands for multiplication.

As another example, 1 square inch is approximately 0.00000016 acre. Here, we write 0.00000016 in scientific notation:

$$1.6 \times 10^{-7}$$

DEFINITION Scientific notation

A number is written in **scientific notation** if it has the form $N \times 10^k$, where k is an integer and the absolute value of N is between 1 and 10 or is equal to 1.

Here are more examples of numbers in scientific notation:

$$8.6 \times 10^{19} \qquad 2.159 \times 10^{8} \qquad -4.23 \times 10^{-14} \qquad 7.94 \times 10^{-97}$$

The problems in Example 9 suggest how to convert a number from scientific notation $N \times 10^k$ to standard decimal notation.

Example 9 Writing Numbers in Standard Decimal Notation

Simplify.

1. 7×10^3

2. 7×10^{-3}

3. 9.48×10^{-4}

Solution

1. We simplify $7 \times 10^3 = 7.0 \times 10^3$ by *multiplying* 7.0 by 10 three times and hence moving the decimal point three places to the *right:*

$$7.0 \times 10^3 = 7000.0 = 7000$$

three places to the right

2. Since

$$7 \times 10^{-3} = 7 \times \frac{1}{10^3} = \frac{7}{1} \times \frac{1}{10^3} = \frac{7}{10^3}$$

we see that we can simplify 7.0×10^{-3} by *dividing* 7.0 by 10 three times and hence moving the decimal point three places to the *left:*

$$7.0 \times 10^{-3} = 0.007$$

three places to the left

3. We *divide* 9.48 by 10 four times and hence move the decimal point of 9.48 four places to the *left:*

$$9.48 \times 10^{-4} = 0.000948$$

four places to the left ∎

Converting from Scientific Notation to Standard Decimal Notation

To write the scientific notation $N \times 10^k$ in standard decimal notation, we move the decimal point of the number N as follows:

- If k is *positive,* we multiply N by 10 k times and hence move the decimal point k places to the *right.*
- If k is *negative,* we divide N by 10 k times and hence move the decimal point k places to the *left.*

The problems in Example 10 suggest how to convert a number from standard decimal notation to scientific notation.

Example 10 Writing Numbers in Scientific Notation

Write the number in scientific notation.

1. 845,000,000

2. 0.0000382

Solution

1. In scientific notation, we would have 8.45×10^k, but what is k? If we move the decimal point of 8.45 eight places to the right, the result is 845,000,000. So, $k = 8$ and the scientific notation is 8.45×10^8.

2. In scientific notation, we would have 3.82×10^k, but what is k? If we move the decimal point of 3.82 five places to the left, the result is 0.0000382. So, $k = -5$ and the scientific notation is 3.82×10^{-5}. ∎

> **Converting from Standard Decimal Notation to Scientific Notation**
>
> To write a number in scientific notation, count the number k of places that the decimal point needs to be moved so that the absolute value of the new number N is between 1 and 10 or is equal to 1.
>
> - If the decimal point is moved to the left, then the scientific notation is written as $N \times 10^k$.
> - If the decimal point is moved to the right, then the scientific notation is written as $N \times 10^{-k}$.

Example 11 Writing Numbers in Scientific Notation

Write the number in scientific notation.

1. 778,000,000   (Jupiter's average distance, in kilometers, from the Sun)
2. 0.000012   (the diameter, in meters, of a white blood cell)

Solution

1. For 778,000,000, the decimal point needs to be moved eight places to the left so that the new number is between 1 and 10. Therefore, the scientific notation is 7.78×10^8.
2. For 0.000012, the decimal point needs to be moved five places to the right so that the new number is between 1 and 10. Therefore, the scientific notation is 1.2×10^{-5}.

∎

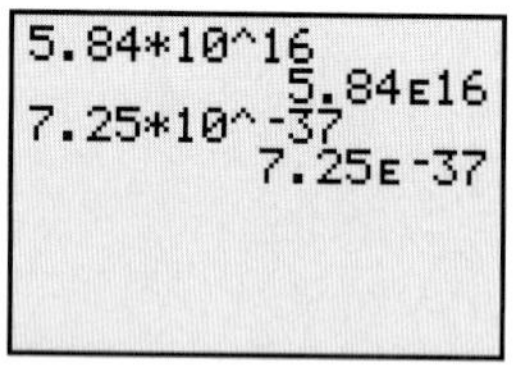

```
5.84*10^16
           5.84E16
7.25*10^-37
           7.25E-37
```

Figure 48 The numbers 5.84×10^{16} and 7.25×10^{-37}

Calculators express numbers in scientific notation so that the numbers "fit" on the screen. To represent 5.84×10^{16}, most calculators use the notation 5.84 E 16, where E stands for <u>e</u>xponent (of 10). Calculators represent 7.25×10^{-37} as 7.25 E −37 (see Fig. 48).

group exploration

Properties of exponents

1. Since $x^{-5} = \dfrac{1}{x^5}$ for nonzero x, does it follow that $-5 = \dfrac{1}{5}$? Explain.

2. A student tries to simplify $7x^{-2}$:

$$7x^{-2} = \frac{1}{7x^2}$$

   Describe any errors. Then simplify the expression correctly.

3. A student tries to simplify $\dfrac{x^8}{x^{-5}}$:

$$\frac{x^8}{x^{-5}} = x^{8-5} = x^3$$

   Describe any errors. Then simplify the expression correctly.

4. A student tries to simplify $\left(5x^3\right)^{-2}$:

$$\left(5x^3\right)^{-2} = 5\left(x^3\right)^{-2} = 5x^{-6} = \frac{5}{x^6}$$

   Describe any errors. Then simplify the expression correctly.

TIPS FOR SUCCESS: Complete the Rest of the Assignment

If you have spent a good amount of time trying to solve an exercise but can't, consider going on to the next exercise in the assignment. You may find that the next exercise involves a different concept or involves a more familiar situation. You may even find that, after completing the rest of the assignment, you are able to complete the exercise(s) you skipped. One explanation of this phenomenon is that you may have learned or remembered some concept in a later exercise that relates to the exercise with which you were struggling.

HOMEWORK 7.7

FOR EXTRA HELP ▶

 Student Solutions Manual   PH Math/Tutor Center   *Math XL* MathXL®   *MyMathLab* MyMathLab

Simplify.

1. 6^{-2}

2. 9^{-2}

3. x^{-4}

4. x^{-2}

5. b^{-1}

6. t^{-8}

7. $\dfrac{1}{2^{-4}}$

8. $\dfrac{1}{9^{-2}}$

9. $\dfrac{1}{w^{-2}}$

10. $\dfrac{1}{a^{-5}}$

11. $\dfrac{x^{-3}}{y^5}$

12. $\dfrac{x^2}{y^{-7}}$

13. $\dfrac{a^{-4}}{b^{-2}}$

14. $\dfrac{r^{-4}}{t^{-8}}$

15. $\dfrac{2a^3b^{-5}}{5c^{-8}}$

16. $\dfrac{3x^{-1}y^6}{7w^{-2}}$

17. $\dfrac{4x^{-9}}{-6y^4w^{-1}}$

18. $\dfrac{-9x^{-1}}{6y^{-5}w^7}$

Simplify.

19. $\left(x^{-2}\right)^7$

20. $\left(x^5\right)^{-8}$

21. $\left(t^{-4}\right)^{-3}$

22. $\left(a^{-6}\right)^{-1}$

23. $\left(6t^{-4}\right)\left(5t^2\right)$

24. $\left(9w^4\right)\left(2w^{-6}\right)$

25. $\left(-4x^{-1}\right)\left(3x^{-8}\right)$

26. $\left(-5x^{-3}\right)\left(-8x^{-5}\right)$

27. $\left(-4x^3y^{-7}\right)\left(-x^{-5}y^4\right)$

28. $\left(-3b^{-2}c^{-6}\right)\left(-b^9c^{-1}\right)$

29. $\dfrac{x^2}{x^6}$

30. $\dfrac{x^5}{x^8}$

31. $\dfrac{a^{-3}}{a^5}$

32. $\dfrac{r^{-4}}{r^7}$

33. $\dfrac{a^3}{a^{-2}}$

34. $\dfrac{y^6}{y^{-1}}$

35. $\dfrac{7x^{-3}}{4x^{-9}}$

36. $\dfrac{4x^{-2}}{9x^{-5}}$

37. $\dfrac{2^{-1}}{2^4}$

38. $\dfrac{3^{-2}}{3^2}$

39. $\dfrac{5^{-6}}{5^{-4}}$

40. $\dfrac{7^{-5}}{7^{-6}}$

41. $\dfrac{3^4w^{-8}}{3^2w^{-3}}$

42. $\dfrac{2^5b^{-7}}{2^2b^{-2}}$

43. $\left(8c^{-3}\right)^2$

44. $\left(4t^{-7}\right)^3$

45. $\left(2x^{-1}\right)^{-5}$

46. $\left(2x^{-4}\right)^{-4}$

47. $\left(x^{-2}y^5\right)^{-6}$

48. $\left(x^4y^{-7}\right)^{-5}$

49. $\left(2a^{-6}b\right)^{-3}$

50. $\left(4r^3t^{-8}\right)^{-2}$

51. $\left(ab^2\right)^3\left(a^3\right)^{-2}$

52. $\left(t^3w\right)^5\left(w^2\right)^{-4}$

53. $\dfrac{1}{(xy)^{-3}}$

54. $\dfrac{1}{(xy)^{-5}}$

55. $\left(\dfrac{x^{-3}}{y^2}\right)^4$

56. $\left(\dfrac{x^5}{y^{-8}}\right)^3$

57. $\left(\dfrac{2}{c^{-7}}\right)^{-3}$

58. $\left(\dfrac{w^{-3}}{7}\right)^{-1}$

59. $\left(\dfrac{3r^{-5}}{t^9}\right)^3$

60. $\left(\dfrac{a^3}{5b^{-6}}\right)^2$

61. $\left(\dfrac{8a^{-3}b}{c^{-5}d^4}\right)^{-2}$

62. $\left(\dfrac{x^{-4}y^2}{tw^{-7}}\right)^{-3}$

63. $\dfrac{\left(2a^{-2}b\right)^{-3}}{\left(3cd^{-3}\right)^2}$

64. $\dfrac{\left(5xy^{-3}\right)^{-2}}{\left(2t^2w^{-1}\right)^{-3}}$

65. $\dfrac{6b^{-3}c^4}{8b^2c^{-3}}$

66. $\dfrac{9t^4w^{-6}}{3t^{-2}w^3}$

67. If an object is moving at a constant speed s (in miles per hour), then $s = dt^{-1}$, where d is the distance traveled (in miles) and t is the time (in hours).
   a. Simplify the right-hand side of the equation.
   b. Substitute 186 for d and 3 for t in the equation you found in part (a), and solve for s. What does your result mean in this situation?

68. The force F (in pounds) you must exert on a wrench handle of length L inches to loosen a bolt is described by the equation $F = 720L^{-1}$.
   a. Simplify the right-hand side of the equation.
   b. Substitute 12 for L in the equation you found in part (a), and solve for F. What does your result mean in this situation?

69. If a person plays an electric guitar outside, the sound level L (in decibels) at a distance d yards from the amplifier is described by the equation $L = 5760d^{-2}$.
   a. Simplify the right-hand side of the equation.
   b. Substitute 8 for d in the equation you found in part (a), and solve for L. What does your result mean in this situation?

70. The amount of light I (in milliwatts per square centimeter) of a 50-watt light bulb at a distance d centimeters is described by the equation $I = 8910d^{-2}$.

a. Simplify the right-hand side of the equation.
b. Substitute 80 for d in the equation you found in part (a), and solve for I. What does your result mean in this situation?

Write the number in standard decimal form.

71. 4.9×10^4

72. 8.31×10^6

73. 8.59×10^{-3}

74. 6.488×10^{-5}

75. 2.95×10^{-4}

76. 8.7×10^{-2}

77. -4.512×10^8

78. -9.46×10^{10}

Write the number in scientific notation.

79. 45,700,000

80. 280,000

81. 0.0000659

82. 0.000023

83. $-5,987,000,000,000$

84. $-308,000,000$

85. 0.000001

86. 0.0004

For each of Exercises 87 and 88, numbers are displayed in scientific notation in a graphing calculator table shown in the given figure. Write each of these numbers in the Y_1 column in standard decimal form.

87. See Fig. 49.

88. See Fig. 50.

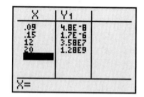

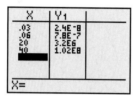

Figure 49 Exercise 87   **Figure 50** Exercise 88

For Exercises 89–94, the given sentence contains a number written in scientific notation. Write the number in standard decimal form.

89. The Milky Way galaxy contains about 4×10^{11} stars.

90. Light travels at 2.998×10^8 meters per second.

91. Bill Gates's net worth was 4.66×10^{10} dollars in January 2005 (Source: *Bloomberg*).

92. The intensity of the sound inside a running car is about 1×10^{-8} watt per square meter.

93. Orange juice has a hydrogen ion concentration of 6.3×10^{-4} mole per liter.

94. One second is 3.17×10^{-8} year.

For Exercises 95–100, the given sentence contains a number (other than a date) written in standard decimal form. Write the number in scientific notation.

95. The area of the United States is about 3,720,000 square miles.

96. The diameter of Earth is about 12,700 kilometers.

97. The temperature of the Sun's core is approximately 27,000,000°F.

98. The diameter of a human hair is about 0.00254 centimeter.

99. A human red blood cell has a diameter of about 0.0000075 meter.

100. One teaspoon is approximately 0.0013 gallon.

101. A student tries to simplify $\dfrac{5x^{-3}y^2}{w}$:

$$\frac{5x^{-3}y^2}{w} = \frac{y^2}{5x^3w}$$

Describe any errors. Then simplify the expression correctly.

102. A student tries to simplify $\dfrac{-8x^5}{y^3}$:

$$\frac{-8x^5}{y^3} = \frac{x^5}{8y^3}$$

Describe any errors. Then simplify the expression correctly.

103. A student tries to simplify $\dfrac{x^6}{x^{-4}}$:

$$\frac{x^6}{x^{-4}} = x^{6-4} = x^2$$

Describe any errors. Then simplify the expression correctly.

104. A student thinks that 4^{-3} is a negative number. Is the student correct? Explain.

105. a. Find 1^5.
   b. Find 1^0.
   c. Find 1^{-3}.
   d. Find 1^n, where n is an integer.

106. Explain how you can write a power expression with negative-integer exponents as an equivalent expression with only positive exponents. (See page 9 for guidelines on writing a good response.)

Related Review

Simplify.

107. $-3x^{-3} + \left(-3x^{-3}\right)$

108. $3x^3 + 3x^3$

109. $-3x^{-3}\left(-3x^{-3}\right)$

110. $3x^3\left(3x^3\right)$

111. $\left(-3x^{-3}\right)^3$

112. $\left(3x^3\right)^3$

113. $\left(-3x^3\right)^{-3}$

114. $\left(-3x^{-3}\right)^{-3}$

115. a. Simplify $\dfrac{x^4}{x^{-3}}$ by each of the following methods:

   i. Use the quotient property $\dfrac{x^m}{x^n} = x^{m-n}$.

   ii. Use the property $\dfrac{1}{x^{-n}} = x^n$.

   b. Compare the results that you found in part (a).

116. a. Simplify $\dfrac{x^6}{x^2}$ by each of the following methods:

   i. Use the quotient property $\dfrac{x^m}{x^n} = x^{m-n}$.

   ii. Write x^6 and x^2 without exponents.

   b. Compare the results that you found in part (a).

Expressions, Equations, and Graphs

Give an example of the following. Then solve, simplify, or graph, as appropriate.

117. A quadratic equation in two variables

118. A linear equation in two variables

119. A quadratic polynomial in one variable and with four terms

120. A linear equation in one variable

121. A cubic polynomial in one variable and with five terms

122. A system of two linear equations in two variables

analyze this

GLOBAL WARMING LAB (continued from Chapter 6)

Throughout our study of global warming, we have used the IPCC's yardstick that, by 2050, carbon emissions should be 0.3 metric ton per person per year. Yet, it is difficult to imagine developed countries, with average annual carbon emissions of 2.7 metric tons per person today, reducing their emissions by 89% to meet the IPCC's recommendation. In particular, it is difficult to imagine that the United States will reduce its annual carbon emissions of 5.4 metric tons per person by 94% to reach that desired level.

Recall from the Global Warming Lab in Chapter 5 that the IPCC's yardstick is equivalent to recommending that 2050 carbon emissions be 60% less than the carbon emissions in 1990. We can reach this goal without reducing per-person carbon emissions at all—if we reduce world population.

Reducing world population seems impossible, because it is currently growing by leaps and bounds. The United Nations (U.N.) predicts that, from 2000 to 2050, the world population will increase from 6.07 billion to 8.92 billion—a 47% increase (see Table 36).

Table 36 World Population

Year	Population (billions)	Year	Population (billions)
A.D. 1	0.17	1600	0.55
200	0.19	1800	0.81
400	0.19	2000	6.07
600	0.20	2025	7.85
800	0.22	2050	8.92
1000	0.25	2075	9.22
1200	0.36	2100	9.06
1400	0.35		

Source: *The Cambridge Factfinder*

Even though world population is increasing rapidly, the U.N. has predicted that world population will reach a maximum around 2075, due to markedly lower family sizes in recent years and continuing fatalities from AIDS.*

The U.N. uses a complicated model to make these predictions. A simpler model that matches well with the U.N. model for the years 2000 to 2100 is the quadratic equation $p = -0.00053t^2 + 0.0825t + 6.09$, where p is world population at t years since 2000. Both models predict that world population will slightly decline from 2075 to 2100. The complicated U.N. model predicts that, after 2100, world population will dip a bit below the 2050 level of 8.92 billion and then slowly increase, returning to 8.97 billion in 2300.

Using the yardstick of 2050 carbon emissions at 60% less than the carbon emissions in 1990, we can calculate the largest carbon emissions that Earth can handle each year:

$$0.40(6.1) = 2.44 \text{ billion metric tons}$$

Next, we will consider two scenarios that stay within this limit.

For our first scenario, let's assume that in the future each person in the world emits 5.4 metric tons per year, the current per-person annual emissions rate for Americans. Then we could still be at the yearly limit of 2.44 billion metric tons of carbon if the world population were a mere 452 million. By modeling the data in Table 37, we could show that U.S. population alone will reach 452 million before 2080. Common sense dictates that countries would never voluntarily lower their populations enough to help reach this level.

Table 37 United States Population

Year	U.S. Population (billions)
1960	0.18
1970	0.21
1980	0.23
1990	0.25
2000	0.28

Source: *U.S. Census Bureau; The Cambridge Factfinder*

For our second scenario, let's make an assumption about how low carbon emissions could be. Although 0.3 metric ton per person per year seems unreachable, perhaps 1.4 metric tons per person per year is attainable. After all, Sweden, Switzerland, and France already have per-person annual carbon emissions at or near 1.4 metric tons (see Table 38). If all countries annually produced 1.4 metric tons of carbon per person, Earth could handle 1.7 billion people.

Table 38 Gross National Product (GNP) Ranks and per-Person Annual Carbon Emissions

Country	1998 GNP Rank	1998 GNP per-Person Rank	2000 per-Person Carbon Emissions (metric tons)
Sweden	20	14	1.4
Switzerland	18	3	1.5
France	4	17	1.7
Italy	6	25	2.0
Austria	21	12	2.1
Denmark	23	6	2.3
Japan	2	7	2.6
Germany	3	13	2.6
United Kingdom	5	22	2.6
Belgium	19	15	2.7
Norway	25	4	3.0
Australia	14	24	4.9
United States	1	10	5.4
Netherlands	12	18	12.6

Source: *The World Bank Group*

*United Nations Population Division.

The first scenario reveals that we will not be able to get carbon emissions under control merely by reducing world population. Earlier, we observed that simply reducing per-person emissions will not be acceptable. So, as the second scenario suggests, it is only by reducing both world population and per-person emissions that we can reach the IPCC's goal.

Analyzing the Situation

1. Let w be world population (in billions of people) in the year t. For example, $t = 2005$ represents the year 2005.
   a. Use a graphing calculator to draw a scattergram of world population data.
   b. Calculate the changes in world population for every 200 (or 199) years up until 2000.
   c. When has world population grown most? Explain why it is surprising that world population will reach its maximum in 2075.

2. Now let w be world population (in billions of people) at t years *since 2000*.
   a. Create a scattergram for the years 2000–2100 that are listed in Table 36. In the same viewing window, draw a graph of the scattergram and the quadratic world population model provided earlier.
   b. Use the quadratic model to predict world population for the years 2000, 2025, 2050, 2075, and 2100.
   c. Use "maximum" on your graphing calculator to predict the maximum world population. In what year will this happen?
   d. Does the quadratic model make predictions that are similar to the U.N. predictions listed in Table 36? Explain.

3. In Problem 4 of the Global Warming Lab in Chapter 5, you found a model of the U.S. population (in billions) at t years since 1950. If you didn't find such an equation, find it now. Then use it to predict when the U.S. population will reach 452 million. Finally, compare your result with the claim that this will happen before 2080.

4. Perform a unit analysis for the estimate of the amount of carbon emissions (in billions of tons of carbon) that Earth can withstand each year:

$$0.40(6.1) = 2.44 \text{ billion metric tons}$$

Explain why this computation gives the amount of carbon emissions that Earth can withstand each year.

5. Show the work for the claim that if everyone in the world emitted carbon as Americans do, then Earth could support only 452 million people.

6. Verify the claim that if worldwide annual carbon emissions were 1.4 tons of carbon per person, then Earth could support 1.7 billion people.

PROJECTILE LAB

In this lab, you will estimate your vertical throwing speed.

Materials

You will need at least three people and the following items:

1. A baseball or other ball (a solid, heavy ball works best)
2. A digital stopwatch or some other timing device

Recording of Data

The first person should throw the ball straight up and say "Mark" at the moment the ball is released. The second person should hold his or her hand at the position of the ball's release and should say "Mark" when the ball returns to its initial height. The third person should begin timing when the first person says "Mark" and stop timing when the second person says "Mark."

Theory

Throughout this lab, we will use "height" to mean distance above the ground. Let h_0 be the ball's height (in feet) at the moment of its release. Let v_0 be the ball's speed (in feet per second) at that moment. We call h_0 the *initial height* and v_0 the *initial velocity*. The ball's height h (in feet) is given by

$$h = -16t^2 + v_0t + h_0$$

where t is the time (in seconds) since the ball was released. So, t and h are variables and v_0 and h_0 are constants.

Analyzing the Situation

1. Substitute 0 for t in the equation $h = -16t^2 + v_0t + h_0$ and solve for h. Explain why your work shows that h_0 represents the initial height.

2. What was the actual initial height of the ball in your experiment? Substitute this value for h_0 in the equation $h = -16t^2 + v_0t + h_0$.

3. How long did it take for the ball to return to its initial height? Substitute that time for t and the initial height for h in the equation you found in Problem 2. Then solve for v_0. What does your value for v_0 mean in this situation?

4. Substitute the value for v_0 that you found in Problem 3 into the equation you found in Problem 2. [**Hint:** Your equation should still have the variables t and h in it.]

5. Use a graphing calculator to draw a graph of your model. Use a pencil and paper to copy the graph.

6. What is the h-intercept of the model? What does it mean in this situation?

7. Use the model to estimate the height of the ball at 1 second.

Chapter Summary

Key Points

OF CHAPTER 7

Quadratic equation in two variables (Section 7.1)

A **quadratic equation in two variables** is an equation that can be put into the form $y = ax^2 + bx + c$, where a, b, and c are constants and $a \neq 0$. This form is called **standard form.**

Parabola, minimum point, maximum point, and vertex (Section 7.1)

A **parabola** is the graph of a quadratic equation in two variables. The lowest point of a parabola that *opens upward* is called the **minimum point.** The highest point of a parabola that *opens downward* is called the **maximum point.** The minimum point or maximum point of a parabola is called the **vertex** of the parabola.

Intercepts of curves (Section 7.1)

An **x-intercept of a curve** is a point where the curve and the x-axis intersect; a **y-intercept of a curve** is a point where the curve and the y-axis intersect.

Quadratic model (Section 7.2)

A **quadratic model** is a quadratic equation in two variables that describes the relationship between two quantities for an authentic situation. We also refer to the quadratic equation's graph (a parabola) as a quadratic model.

Deciding which type of model to use (Section 7.2)

By graphing a scattergram of some data, we can determine whether to use a linear model, a quadratic model, or neither.

Term (Section 7.3)

A **term** is a constant, a variable, or a product of a constant and one or more variables raised to powers.

Monomial (Section 7.3)

A **monomial** is a constant, a variable, or a product of a constant and one or more variables raised to counting-number powers.

Polynomial or polynomial expression (Section 7.3)

A **polynomial,** or **polynomial expression,** is a monomial or a sum of monomials.

Like terms (Section 7.3)

Like terms are either constant terms or variable terms that contain the same variable(s) raised to exactly the same power(s).

Combining like terms (Section 7.3)

To combine like terms, add the coefficients of the terms.

Adding polynomials (Section 7.3)

To add polynomials, combine like terms.

Subtracting polynomials (Section 7.3)

To subtract polynomials, first distribute -1 and then combine like terms.

The meaning of the sign of a difference (Section 7.3)

If a difference $A - B$ is positive, then A is more than B. If a difference $A - B$ is negative, then A is less than B.

Binomial and trinomial (Section 7.4)

We refer to a polynomial as a **binomial** or a **trinomial,** depending on whether it has two or three nonzero terms, respectively.

Product property for exponents (Section 7.4)

If n and m are counting numbers, then $x^m x^n = x^{m+n}$.

Multiplying two polynomials (Section 7.4)

To multiply two polynomials, multiply each term in the first polynomial by each term in the second polynomial. Then combine like terms if possible.

Raising a product to a power (Section 7.5)

If n is a counting number, then $(xy)^n = x^n y^n$.

Square of a sum (Section 7.5)

$$(A + B)^2 = A^2 + 2AB + B^2$$

Square of a difference (Section 7.5)

$$(A - B)^2 = A^2 - 2AB + B^2$$

Product of binomial conjugates (Section 7.5)

$$(A + B)(A - B) = A^2 - B^2$$

Simplifying power expressions (Section 7.6)

A power expression is simplified if

1. It includes no parentheses.

2. In any monomial, each variable or constant appears as a base at most once.

3. Each numerical expression (such as 7^2) has been calculated, and each numerical fraction has been simplified.

Quotient property for exponents (Section 7.6)

If n and m are counting numbers and x is nonzero, then $\dfrac{x^m}{x^n} = x^{m-n}$.

Zero exponent (Section 7.6)

For nonzero x, $x^0 = 1$.

Raising a quotient to a power (Section 7.6)

If n is a counting number and y is nonzero, then $\left(\dfrac{x}{y}\right)^n = \dfrac{x^n}{y^n}$.

Raising a power to a power (Section 7.6)

If m and n are counting numbers, then $\left(x^m\right)^n = x^{mn}$.

Negative-integer exponent (Section 7.7)

If n is a counting number and x is nonzero, then $x^{-n} = \dfrac{1}{x^n}$.

Negative-integer exponent in a denominator (Section 7.7)

If x is nonzero and n is a counting number, then $\dfrac{1}{x^{-n}} = x^n$.

Properties of integer exponents (Section 7.7)

If m and n are integers and x and y are nonzero, then

- $x^m x^n = x^{m+n}$ **Product property for exponents**
- $\dfrac{x^m}{x^n} = x^{m-n}$ **Quotient property for exponents**
- $(xy)^n = x^n y^n$ **Raising a product to a power**
- $\left(\dfrac{x}{y}\right)^n = \dfrac{x^n}{y^n}$ **Raising a quotient to a power**
- $\left(x^m\right)^n = x^{mn}$ **Raising a power to a power**

Simplifying power expressions (Section 7.7)

A power expression is simplified if
1. It includes no parentheses.
2. In any monomial, each variable or constant appears as a base at most once.
3. Each numerical expression has been calculated, and each numerical fraction has been simplified.
4. Each exponent is positive.

Scientific notation (Section 7.7)

A number is written in **scientific notation** if it has the form $N \times 10^k$, where k is an integer and the absolute value of N is between 1 and 10 or is equal to 1.

Converting from scientific notation to standard decimal notation (Section 7.7)

To write the scientific notation $N \times 10^k$ in standard decimal notation, we move the decimal point of the number N as follows:
- If k is *positive*, we multiply N by 10 k times and hence move the decimal point k places to the *right*.
- If k is *negative*, we divide N by 10 k times and hence move the decimal point k places to the *left*.

Converting from standard decimal notation to scientific notation (Section 7.7)

To write a number in scientific notation, count the number k of places that the decimal point needs to be moved so that the absolute value of the new number N is either between 1 and 10 or equal to 1.
- If the decimal point is moved to the left, then the scientific notation is written as $N \times 10^k$.
- If the decimal point is moved to the right, then the scientific notation is written as $N \times 10^{-k}$.

CHAPTER 7 REVIEW EXERCISES

Graph the equation by hand. To begin, substitute the values -3, $-2, -1, 0, 1, 2,$ and 3 for x. Make other substitutions as necessary. Then use a graphing calculator to verify your work.

1. $y = -2x^2$ **2.** $y = 3x^2 - 4$ **3.** $y = 2x^2 - 8x + 4$

For Exercises 4–7, refer to Table 39.

4. Find y when $x = 3$.   **5.** Find y when $x = 2$.

6. Find x when $y = 2$.   **7.** Find x when $y = 1$.

Table 39 Values of x and y (Exercises 4-7)

x	y
0	5
1	2
2	1
3	2
4	5

8. Average annual expenditures by Americans are shown in Table 40 for various age groups.

Table 40 Average Annual Expenditures

Age Group (years)	Age Used to Represent Age Group (years)	Average Annual Expenditure (thousands of dollars)
18–24	21.0	24.3
25–34	29.5	40.3
35–44	39.5	48.3
45–54	49.5	48.7
55–64	59.5	44.3
65–74	69.5	32.2

Source: *Consumer Expenditure Survey*

Let E be the average annual expenditure (in thousands of dollars) by Americans at age a years. A quadratic model is graphed in Fig. 51.

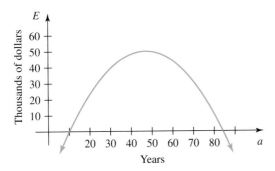

Figure 51 Expenditure model

a. Trace the graph in Fig. 51 carefully onto a piece of paper. Then, on the same graph, draw a scattergram of the data. Does the model fit the data well?

b. Estimate the average annual expenditure for 19-year-old Americans.

c. Estimate the age(s) at which the average annual expenditure is $35 thousand.

d. What is the vertex of the model? What does it mean in this situation?

e. Find the a-intercepts. What do they mean in this situation?

9. Americans' faith in executives who are running major corporations has been declining (see Table 41).

Table 41 Americans' Faith in Executives at Major Corporations

Year	Percent
2000	28
2001	20
2002	16
2003	13
2004	12

Source: *Harris Interactive*

Let p be the percentage of Americans who have faith in executives at major corporations at t years since 2000.

a. A quadratic model of the situation is $p = 1.07t^2 - 8.19t + 27.7$. A linear model of the situation is $p = -3.9t + 25.6$. Use a graphing calculator to draw graphs of the linear and quadratic models and, in the same viewing window, the scattergram of the data. Which model comes closer to the points in the scattergram?

b. Which of the two models predicts that Americans' confidence in executives at major corporations will eventually increase? Explain.

c. Use "minimum" on a graphing calculator to find the vertex of the quadratic model. What does the vertex represent in this situation?

d. Use the quadratic model to predict what percentage of Americans will have faith in executives at major corporations in 2006.

e. Use the quadratic model to predict when in the future 27.7% of Americans will again have confidence in executives at major corporations.

Perform the indicated operation. Use a graphing calculator table to check your work.

10. $\left(-4x^3 + 7x^2 - x\right) + \left(-5x^3 - 9x^2 + 3\right)$

11. $\left(6x^3 - 2x^2 + 5\right) - \left(8x^3 - 4x^2 + 3x\right)$

12. Perform the indicated operations: $4x - 9x^2 - 6x + 2 - 7x^3 + x^2 - 8x$.

13. In Exercise 39 of the Review Exercises in Chapter 6, the percentage of airplanes that are regional jets or that are turboprops or large jets is modeled by the system

$$p = 4.4t + 7.4 \quad \text{Regional jets}$$
$$p = -4.7t + 86.8 \quad \text{Turboprops or large jets}$$

where p is the percentage of airplanes and t is the number of years since 2000 (see Table 42).

Table 42 Regional Jets versus Turboprops or Large Jets

	Percent	
Year	**Regional Jets**	**Turboprops or Large Jets**
2000	8	86
2001	11	83
2002	16	78
2003	21	72

Source: *Department of Transportation; Federal Aviation Administration*

a. Find the sum of the linear expressions $4.4t + 7.4$ and $-4.7t + 86.8$. What does the result represent?

b. Evaluate the expression in part (a) for $t = 10$. What does your result mean in this situation?

c. Find the difference of the expressions $4.4t + 7.4$ and $-4.7t + 86.8$. What does the result represent?

d. Evaluate the expression in part (c) for $t = 10$. What does your result mean in this situation?

Perform the indicated operation. Use a graphing calculator table to verify your work when possible.

14. $-3x^2\left(7x^5\right)$

15. $5x^3\left(2x^2 - 7x + 4\right)$

16. $(w - 3)(w - 9)$

17. $(2a + 5b)(3a - 8b)$

18. $\left(3x^2 - 4\right)(5x + 6)$

19. $(x + 4)\left(x^2 - 3x + 5\right)$

20. $\left(4b^2 - b + 3\right)(2b - 7)$

21. $(-2t)^3$

22. $(x + 7)^2$

23. $(x - 4)^2$

24. $(2p + 5)^2$

25. $-5(c + 2)^2$

26. $(x + 6)(x - 6)$

27. $(4m - 7n)(4m + 7n)$

28. $(3p - 2t)^2$

29. $(2a^2 - a + 3)(a^2 + 2a - 1)$

30. Is the product of a linear polynomial and a quadratic polynomial a linear, quadratic, or cubic polynomial? Give an example.

Write the equation in standard form. Use graphing calculator graphs to verify your work.

31. $y = (x - 5)^2 + 3$

32. $y = -2(x + 3)^2 - 6$

Simplify.

33. $(-x^5)^2$

34. $(2x^3)(6x^4)$

35. $(8a^2b^3)(-5a^4b^9)$

36. $\dfrac{8x^4y^8}{16x^3y^5}$

37. $\left(\dfrac{x}{2}\right)^3$

38. $(2x^9y^3)^5$

39. $3x^6(5x^4)^2$

40. $\dfrac{15c^2c^7}{10c^4}$

41. $\left(\dfrac{a^4}{9}\right)^2$

42. $\left(\dfrac{-9x^5}{5y^7}\right)^0$

43. $\dfrac{(3x^5y^4)^2}{6x^7y^3}$

44. $\left(\dfrac{3x^5}{4x^2}\right)^3$

Simplify.

45. 4^{-3}

46. $\dfrac{1}{x^{-5}}$

47. $\dfrac{r^{-7}}{r^3}$

48. $\dfrac{t^{-5}}{t^{-2}}$

49. $\dfrac{x^{-5}y^2}{w^{-7}}$

50. $(c^6)^{-4}$

51. $(-4x^3)(5x^{-9})$

52. $(2x^{-3})^{-5}$

53. $(3x^{-4}y^6)^{-3}$

54. $\dfrac{6a^{-2}b^3}{8a^{-5}b^6}$

55. $\dfrac{(3a^{-2}b)^2}{(2ab^{-3})^3}$

56. $\left(\dfrac{x^3y^{-2}}{w^{-4}}\right)^{-7}$

57. A student tries to simplify $\dfrac{x^9}{x^{-6}}$:

$$\dfrac{x^9}{x^{-6}} = x^{9-6}$$
$$= x^3$$

Describe any errors. Then simplify the expression correctly.

Write the number in standard decimal form.

58. 5.832×10^8

59. 3.17×10^{-4}

Write the number in scientific notation.

60. 74,200,000

61. 0.00008

62. Saturn is an average distance of about 1,426,000,000 kilometers from the Sun. Write 1,426,000,000 in scientific notation.

CHAPTER 7 TEST

1. Graph $y = -2x^2 + 4x + 1$ by hand. To begin, substitute the values $-3, -2, -1, 0, 1, 2,$ and 3 for x. Make other substitutions as necessary. Then use a graphing calculator to verify your work.

For Exercises 2–7, refer to the graph sketched in Fig. 52.

2. Find y when $x = -5$.

3. Find x when $y = -3$.

4. Find x when $y = -4$.

5. Find x when $y = -5$.

6. Find the x-intercepts.

7. Find the minimum point.

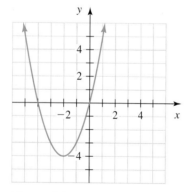

Figure 52 Exercises 2–7

8. The sales of stiletto pumps are shown in Table 43 for various years.

Table 43 Sales of Stiletto Pumps	
Year	Sales (millions of pairs)
1989	24
1991	15
1993	11
1995	16
1997	24
1999	42
2000	55
2001	67

Source: Fashion Weekly

Let s be sales of stiletto pumps (in millions of pairs) at t years since 1980. A quadratic model is graphed in Fig. 53.

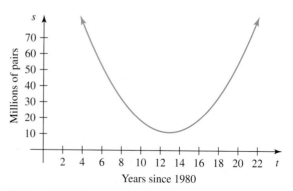

Figure 53 Stiletto pumps model

a. Trace the graph in Fig. 53 carefully onto a piece of paper. Then, on the same graph, draw a scattergram of the data. Does the model fit the data well?

b. Estimate sales in 1996.

c. Estimate when sales were 30 million pairs.

d. What is the vertex of the model? What does it mean in this situation?

Perform the indicated operations.

9. $\left(2x^3 - 4x^2 + 7x\right) - \left(6x^3 - 3x^2 + 9x\right)$

10. $-5x^2\left(3x^2 - 8x + 2\right)$ **11.** $(3p - 2)(4p + 6)$

12. $(2k - 5)\left(3k^2 - 4k - 2\right)$ **13.** $(x - 6)^2$

14. $(4a + 7b)^2$ **15.** $\left(3w^2 - 2w + 3\right)\left(w^2 + w - 2\right)$

16. In Exercise 7 of Homework 6.4, you modeled the prime-time household viewing shares for both broadcast TV and ad-supported cable TV (see Table 44). Two models of the situation are

$$s = -2.17t + 97.74 \qquad \text{Broadcast TV}$$
$$s = 2.02t - 4.84 \qquad \text{Ad-supported cable TV}$$

where s is the prime-time household viewing share at t years since 1980.

a. Find the sum of the expressions $-2.17t + 97.74$ and $2.02t - 4.84$. What does the result represent?

b. Evaluate the expression in part (a) for $t = 31$. What does your result mean in this situation?

c. Find the difference of the expressions $-2.17t + 97.74$ and $2.02t - 4.84$. What does the result represent?

d. Evaluate the expression in part (c) for $t = 31$. What does your result mean in this situation?

Table 44 Broadcast and Ad-Supported Cable TV

	Prime-time Household Viewing Shares (Percent)	
Year	Broadcast TV	Ad-Supported Cable TV
1988	78.8	11.6
1990	74.8	16.4
1992	72.1	19.8
1994	70.6	21.3
1996	64.9	26.7
1998	58.2	32.5
2000	54.0	35.2
2003	46.5	42.4

Source: *Cabletelevision Advertising Bureau (analysis of Nielsen data)*

17. Write $y = -3(x - 1)^2 + 5$ in standard form.

18. A student tries to simplify $(x + 4)^2$:

$$(x + 4)^2 = x^2 + 4^2$$
$$= x^2 + 16$$

Describe any errors. Then simplify the expression correctly.

Simplify.

19. $\dfrac{6x^7 y^4}{8x^3 y^9}$ **20.** $\left(4a^3 b^5\right)^3 a^6 b$ **21.** $\left(\dfrac{x^3}{y^4}\right)^6$

22. $\left(7x^{-3}\right)^{-2}$ **23.** $\left(\dfrac{x^2 y^{-6}}{w^{-3}}\right)^4$ **24.** $\dfrac{2p^{-5} t^2}{4p^{-2} t^{-3}}$

25. Write 0.000468 in scientific notation.

26. In 2002, college enrollment in the United States was about 1.65×10^7 students (Source: *U.S. Census Bureau*). Write 1.65×10^7 in standard decimal form.

Chapter 8

Factoring Polynomials and Solving Polynomial Equations

"I can't do it" never yet accomplished anything; "I will try" has performed wonders.

—George P. Burnham

Table 1 Revenues from Organic Food Sales

Year	Revenue (billions of dollars)
1997	3.6
1999	5.1
2001	7.1
2003	10.4
2005	13.9

Source: *Organic Trade Association*

Are you aware of the health benefits of organic foods? Certified organic foods are untouched by pesticides, preservatives, hormones, and antibiotics. The annual revenue from organic food sales has more than tripled in the United States since 1997 (see Table 1). In Exercise 8 of Homework 8.6, you will predict the revenue from organic food sales in 2011.

In Chapter 7, we discussed how to multiply polynomials. In this chapter, we will discuss how to do the reverse process, *factoring,* which will help us solve new types of equations. In turn, solving these equations will help us use quadratic models to make predictions about authentic situations, such as the organic food data just described.

8.1 FACTORING TRINOMIALS OF THE FORM $x^2 + bx + c$ AND DIFFERENCES OF TWO SQUARES

Objectives

▷ Know that multiplying and *factoring* are reverse processes.

▷ Factor a trinomial of the form $x^2 + bx + c$.

▷ Know the meaning of a *prime* polynomial.

▷ Factor a difference of two squares.

We multiply 3 and 7 as follows: $3 \cdot 7 = 21$. We factor the number 21 as follows: $21 = 3 \cdot 7$. Thus, factoring 21 is the reverse process of multiplying 3 and 7. Next, we will make similar observations about polynomials.

Multiplying Polynomials versus Factoring Polynomials

In Chapter 7, we found products of polynomials—for example,

$$(x + 3)(x + 5) = x^2 + 5x + 3x + 15 \quad \text{Multiply pairs of terms.}$$
$$= x^2 + 8x + 15 \quad \text{Combine like terms.}$$

In this section, we will learn how to go backward. That is, we will learn how to write $x^2 + 8x + 15$ as a product. This process is called *factoring*.

We **factor** a polynomial by writing it as a product. We say that $(x + 3)(x + 5)$ is a **factored polynomial** and that both $(x + 3)$ and $(x + 5)$ are factors of the polynomial.

> **Multiplying versus Factoring**
>
> Multiplying and factoring are reverse processes. For example,
>
> Multiplying
>
> $$(x + 3)(x + 5) = x^2 + 8x + 15$$
>
> Factoring

Factoring a Trinomial of the Form $x^2 + bx + c$

To see how to factor $x^2 + 8x + 15$, let's take another look at how we find the product $(x + 3)(x + 5)$:

last terms

$$(x + 3)(x + 5) = x^2 + 5x + 3x + 3 \cdot 5$$
$$= x^2 + 8x + 15$$

sum of   product of
last terms   last terms
$3 + 5 = 8$ $3 \cdot 5 = 15$

For $x^2 + 8x + 15$, notice that the coefficient of x is 8, which is the sum of the *last terms* of $(x + 3)(x + 5)$, 3 and 5. Also, the constant term of $x^2 + 8x + 15$ is 15, which is the product of 3 and 5. Now we find the product $(x + p)(x + q)$:

$$(x + p)(x + q) = x^2 + qx + px + pq \qquad \text{Multiply pairs of terms.}$$
$$= x^2 + px + qx + pq \qquad \text{Rearrange terms.}$$
$$= x^2 + (p + q)x + pq \qquad \text{Distributive law}$$

In the result, we see that the coefficient of x is the sum of the last terms p and q and that the constant term is the product of the last terms p and q. This observation can help us factor some quadratic trinomials.

Trinomials with Positive Constant Terms

For the trinomial $x^2 + 8x + 15$, the constant term is positive. In Examples 1–3, we will factor three more trinomials whose constant term is positive.

Example 1   Factoring Trinomials of the Form $x^2 + bx + c$

Factor $x^2 + 6x + 8$.

Solution

To factor $x^2 + 6x + 8$, we need two integers whose product is 8 and whose sum is 6. We try only positive integers, since both their product and their sum have to be positive. The only pairs of factors of 8 whose product is 8 are 1 and 8 or 2 and 4:

Product = 8	**Sum = 6?**
$1(8) = 8$	$1 + 8 = 9$
$2(4) = 8$	$2 + 4 = 6 \leftarrow$ Success!

Since $2(4) = 8$ and $2 + 4 = 6$, we conclude that the last terms of the factors are 2 and 4:

$$x^2 + 6x + 8 = (x + 2)(x + 4)$$

We check the result by finding the product $(x + 2)(x + 4)$:

$$(x + 2)(x + 4) = x^2 + 4x + 2x + 8 = x^2 + 6x + 8$$

By the commutative law, $(x+2)(x+4) = (x+4)(x+2)$, so we can write the factors $x+2$ and $x+4$ in either order. ∎

We now summarize how to factor a trinomial of the form $x^2 + bx + c$.

Factoring $x^2 + bx + c$

To factor $x^2 + bx + c$, look for two integers p and q whose product is c and whose sum is b. That is, $pq = c$ and $p + q = b$. If such integers exist, the factored polynomial is

$$(x + p)(x + q)$$

Example 2 Factoring Trinomials of the Form $x^2 + bx + c$

Factor $x^2 + 8x + 12$.

Solution

To factor $x^2 + 8x + 12$, we need two integers whose product is 12 and whose sum is 8. We try only positive integers, since both their product and their sum have to be positive. Here are the possibilities:

Product = 12	**Sum = 8?**
$1(12) = 12$	$1 + 12 = 13$
$2(6) = 12$	$2 + 6 = 8$ ⟵ Success!
$3(4) = 12$	$3 + 4 = 7$

Since $2(6) = 12$ and $2 + 6 = 8$, we conclude that the last terms of the factors are 2 and 6:

$$x^2 + 8x + 12 = (x + 2)(x + 6)$$

We check the result by finding the product $(x + 2)(x + 6)$:

$$(x + 2)(x + 6) = x^2 + 6x + 2x + 12 = x^2 + 8x + 12$$ ∎

Example 3 Factoring a Trinomial of the Form $x^2 + bx + c$

Factor $x^2 - 12x + 36$.

Solution

To factor $x^2 - 12x + 36$, we need two integers whose product is 36 and whose sum is -12. Since the product 36 is positive, the two integers must have the same sign. Therefore, both integers must be negative, because the sum -12 is negative. Here are the possibilities:

Product = 36	**Sum = -12?**
$-1(-36) = 36$	$-1 + (-36) = -37$
$-2(-18) = 36$	$-2 + (-18) = -20$
$-3(-12) = 36$	$-3 + (-12) = -15$
$-4(-9) = 36$	$-4 + (-9) = -13$
$-6(-6) = 36$	$-6 + (-6) = -12$ ⟵ Success!

Since $-6(-6) = 36$ and $-6 + (-6) = -12$, we conclude that the last terms of the factors are -6 and -6:

$$x^2 - 12x + 36 = (x - 6)(x - 6) = (x - 6)^2$$

Note that our result, $(x - 6)^2$, is the square of a binomial. So, the original expression $x^2 - 12x + 36$ is a perfect-square trinomial (Section 7.5).

We use a graphing calculator table to verify our work (see Fig. 1). ∎

Figure 1 Verify the work

When the constant term of a trinomial is positive, we need to consider only certain possibilities for the factors of that constant term. In Examples 1 and 2, we worked with only positive factors of the positive constant term, because the coefficient of the middle term was positive. In Example 3, we worked with only negative factors of the positive constant term, because the coefficient of the middle term was negative.

Factoring $x^2 + bx + c$ with c Positive

To factor a trinomial of the form $x^2 + bx + c$ with a positive constant term c,

- If b is positive, look for two *positive* integers whose product is c and whose sum is b. For example,

$$x^2 + 10x + 21 = (x + 3)(x + 7)$$

Positive   Positive   Both last
b c terms are positive.

- If b is negative, look for two *negative* integers whose product is c and whose sum is b. For example,

$$x^2 - 11x + 28 = (x - 7)(x - 4)$$

Negative   Positive   Both last
b c terms are negative.

Trinomials with Negative Constant Terms

How do we factor a quadratic trinomial whose constant term is negative?

Example 4   Factoring Trinomials of the Form $x^2 + bx + c$

Factor $w^2 - w - 20$.

Solution

To factor $w^2 - w - 20$, we need two integers whose product is -20 and whose sum is -1. Since the product -20 is negative, the two integers must have different signs. Here are the possibilities:

Product $= -20$	Sum $= -1$?
$1(-20) = -20$	$1 + (-20) = -19$
$2(-10) = -20$	$2 + (-10) = -8$
$4(-5) = -20$	$4 + (-5) = -1$ ⟵ Success!
$5(-4) = -20$	$5 + (-4) = 1$
$10(-2) = -20$	$10 + (-2) = 8$
$20(-1) = -20$	$20 + (-1) = 19$

Since $4(-5) = -20$ and $4 + (-5) = -1$, we conclude that the last terms of the factors are 4 and -5:

$$w^2 - w - 20 = (w + 4)(w - 5)$$

Figure 2 Verify the work

We use a graphing calculator table to verify our work (see Fig. 2). ■

When the constant term of a trinomial is negative, any two integers whose product equals that negative constant term have different signs. For instance, in Example 4 we factored the trinomial $w^2 - w - 20$ by working with integers with different signs whose product is -20.

> **Factoring $x^2 + bx + c$ with c Negative**
>
> To factor a trinomial of the form $x^2 + bx + c$ with a negative constant term c, look for two integers with *different* signs whose product is c and whose sum is b. For example,
>
> $$x^2 + 2x - 24 = (x - 4)(x + 6)$$
>
> Negative
> c
>
> The last terms have different signs.

We can use a similar method to factor trinomials that have two variables.

Example 5 Factoring Trinomials with Two Variables

Factor $x^2 + 5xy + 6y^2$.

Solution

To help us find the last terms, we write the trinomial in the form $x^2 + (5y)x + 6y^2$. We need two monomials whose product is $6y^2$ and whose sum is $5y$. So, the last terms are $2y$ and $3y$:

$$x^2 + 5xy + 6y^2 = (x + 2y)(x + 3y)$$

We check by finding the product $(x + 2y)(x + 3y)$:

$$(x + 2y)(x + 3y) = x^2 + 3xy + 2xy + 6y^2 = x^2 + 5xy + 6y^2 \qquad \blacksquare$$

Prime Polynomials

Just as a prime number has no positive factors other than itself and 1, a polynomial that cannot be factored is called **prime.**

For example, consider the polynomial $x^2 + 4x + 6$. To factor this polynomial, we need two integers whose product is 6 and whose sum is 4. We try only positive integers, since both their product and sum have to be positive. Here are the possibilities:

Product = 6	**Sum = 4?**
$1(6) = 6$	$1 + 6 = 7$
$2(3) = 6$	$2 + 3 = 5$

None of the possible sums equal 4, so we conclude that the trinomial $x^2 + 4x + 6$ is prime.

Example 6 Identifying a Prime Polynomial

Factor $3x - 15 + x^2$.

Solution

First, we write $3x - 15 + x^2$ in descending order to avoid confusion about the coefficients:

$$x^2 + 3x - 15$$

To factor $x^2 + 3x - 15$, we need two integers whose product is -15 and whose sum is 3. Since the product -15 is negative, the two integers must have different signs. Here are the possibilities:

Product = −15	**Sum = 3?**
$1(-15) = -15$	$1 + (-15) = -14$
$3(-5) = -15$	$3 + (-5) = -2$
$5(-3) = -15$	$5 + (-3) = 2$
$15(-1) = -15$	$15 + (-1) = 14$

Since none of the sums equal 3, we conclude that the trinomial $x^2 + 3x - 15$ is prime. So, the original trinomial, $3x - 15 + x^2$, is prime. $\qquad \blacksquare$

Factoring the Difference of Two Squares

In Section 7.5, we found the product of two binomial conjugates by using the property $(A + B)(A - B) = A^2 - B^2$. The expression $A^2 - B^2$ is the difference of two squares. To factor a difference of two squares, we can use that property in reverse.

Difference of Two Squares

$$A^2 - B^2 = (A + B)(A - B)$$

In words: The difference of the squares of two terms is the product of the sum of the terms and the difference of the terms.

Example 7 Factoring Differences of Two Squares

Factor.

1. $x^2 - 16$ **2.** $9m^2 - 4$ **3.** $25p^2 - 49q^2$

Solution

1. Since $x^2 - 16 = (x)^2 - (4)^2$, we substitute x for A and 4 for B:

$$A^2 - B^2 = (A + B)(A - B)$$
$$x^2 - 16 = x^2 - 4^2 = (x + 4)(x - 4)$$

2. Since $9m^2 - 4 = (3m)^2 - (2)^2$, we substitute $3m$ for A and 2 for B:

$$A^2 - B^2 = (A + B)(A - B)$$
$$9m^2 - 4 = (3m)^2 - 2^2 = (3m + 2)(3m - 2)$$

3. Since $25p^2 - 49q^2 = (5p)^2 - (7q)^2$, we substitute $5p$ for A and $7q$ for B:

$$A^2 - B^2 = (A + B)(A - B)$$
$$25p^2 - 49q^2 = (5p)^2 - (7q)^2 = (5p + 7q)(5p - 7q)$$ ■

WARNING   The binomial $x^2 + 16$ is prime. Some students think that this polynomial can be factored as $(x + 4)^2$, but simplify $(x + 4)^2$; you'll see that the result is $x^2 + 8x + 16$, not $x^2 + 16$. In general, **a polynomial of the form $x^2 + k^2$, where $k \neq 0$, is prime.**

group exploration

Looking ahead: Factoring out the greatest common factor

1. Simplify the following.
   a. $2(x + 3)$ **b.** $5(x - 4)$ **c.** $3(x^2 - 5x + 4)$

2. Factor the following. [**Hint:** Refer to your work in Problem 1.]
   a. $2x + 6$ **b.** $5x - 20$ **c.** $3x^2 - 15x + 12$

3. Factor the following.
   a. $3x + 12$ **b.** $2x - 10$ **c.** $5x^2 - 20x + 15$

4. Find the product of the following.
   a. $x(3x + 9)$ **b.** $2x^2(5x - 3)$ **c.** $7x(2x^2 + 5x + 3)$

5. Factor the following. [**Hint:** Refer to your work in Problem 4.]
   a. $3x^2 + 9x$ **b.** $10x^3 - 6x^2$ **c.** $14x^3 + 35x^2 + 21x$

6. Factor the following.
   a. $x^2 + 5x$ **b.** $6x^3 - 15x^2$ **c.** $4x^3 - 6x^2 - 10x$

HOMEWORK 8.1

FOR EXTRA HELP ▶

Student Solutions Manual   PH Math/Tutor Center   Math XL   MyMathLab
MathXL®   MyMathLab

Factor when possible. If a polynomial is prime, say so. Verify that you have factored correctly by finding the product of your factored polynomial.

1. $x^2 + 5x + 6$ **2.** $x^2 + 9x + 8$ **3.** $t^2 + 9t + 20$

4. $w^2 + 10w + 16$ **5.** $x^2 + 8x + 16$ **6.** $x^2 + 14x + 49$

7. $x^2 - 2x - 8$ **8.** $x^2 - 3x - 10$ **9.** $a^2 - 6a - 16$

10. $w^2 - 5w - 24$ **11.** $x^2 + 5x - 24$ **12.** $x^2 + 7x - 30$

13. $x^2 + 8x - 12$ **14.** $x^2 + 9x - 20$ **15.** $3t - 28 + t^2$

16. $y^2 - 14 + 5y$ **17.** $x^2 - 10x + 16$ **18.** $x^2 - 11x + 28$

19. $24 - 11x + x^2$ **20.** $36 + x^2 - 15x$ **21.** $x^2 - 3x + 10$

22. $x^2 - 2x + 8$ **23.** $r^2 - 10r + 25$ **24.** $a^2 - 6a + 9$

25. $x^2 + 36 - 12x$ **26.** $1 - 2x + x^2$

27. $x^2 + 10xy + 9y^2$ **28.** $a^2 + 8ab + 12b^2$

29. $m^2 - mn - 6n^2$ **30.** $r^2 + rt - 20t^2$

31. $a^2 - 7ab + 6b^2$ **32.** $m^2 - 7mn + 10n^2$

Factor when possible. Use a graphing calculator table to verify your work when possible.

33. $x^2 - 25$ **34.** $x^2 - 49$ **35.** $x^2 - 81$

36. $x^2 - 36$ **37.** $t^2 - 1$ **38.** $a^2 - 64$

39. $x^2 + 36$ **40.** $x^2 + 4$ **41.** $4x^2 - 25$

42. $9x^2 - 49$ **43.** $81r^2 - 1$ **44.** $25p^2 - 64$

45. $36x^2 + 49$ **46.** $25x^2 + 9$ **47.** $49p^2 - 100q^2$

48. $4a^2 - 9b^2$ **49.** $64m^2 - 9n^2$ **50.** $25b^2 - 9c^2$

Factor when possible. Use a graphing calculator table to verify your work when possible.

51. $x^2 - 3x - 18$ **52.** $x^2 - 8x - 20$

53. $x^2 + 14x + 49$ **54.** $x^2 + 10x + 25$

55. $a^2 - 4$ **56.** $m^2 - 100$

57. $x^2 + 4x + 12$ **58.** $x^2 + 12x + 24$

59. $x^2 - 8x + 12$ **60.** $x^2 - 9x + 18$

61. $-2w - 48 + w^2$ **62.** $-4t - 60 + t^2$

63. $x^2 - 8x + 16$ **64.** $x^2 - 2x + 1$

65. $w^2 + 49$ **66.** $b^2 + 25$

67. $m^2 - 6mn - 27n^2$ **68.** $a^2 - 19ab - 20b^2$

69. $32 - 18x + x^2$ **70.** $36 - 13x + x^2$

71. $100p^2 - 9t^2$ **72.** $64a^2 - 49b^2$

73. $p^2 + 12p + 36$ **74.** $m^2 + 18m + 81$

75. A student tries to factor the polynomial $x^2 + 9$:

$$x^2 + 9 = (x + 3)(x + 3) = (x + 3)^2$$

Describe any errors. Then factor the polynomial correctly.

76. Two students try to factor the polynomial $x^2 + 14x + 48$:

Student A

$$x^2 + 14x + 48 = (x + 6)(x + 8)$$

Student B

$$x^2 + 14x + 48 = (x + 8)(x + 6)$$

Are both students, one student, or neither student correct? Explain.

77. Which of the following expressions are equivalent?

$$(x - 3)(x + 7) \qquad x^2 - 21 \qquad x^2 - 4x - 21$$
$$x^2 + 4x - 21 \qquad (x + 7)(x - 3)$$

78. Which of the following expressions are equivalent?

$$x^2 - 2x - 24 \qquad (x - 6)(x + 4) \qquad x^2 - 10x - 24$$
$$(x + 4)(x - 6) \qquad x^2 + 2x - 24$$

79. Factor the expression $x^2 - 5x - 24$. Then find the product of the result. What do you observe?

80. Find the product $(x - 2)(x + 8)$. Then factor the result. What do you observe?

81. Give three examples of quadratic polynomials in which $x + 5$ is a factor.

82. Give three examples of quadratic polynomials in which $x - 3$ is a factor.

83. Find all possible values of k so that $x^2 + kx + 12$ can be factored.

84. Find all possible values of k so that $x^2 + kx - 20$ can be factored.

85. Compare the process of factoring an expression with that of finding the product of an expression. (See page 9 for guidelines on writing a good response.)

86. Describe how to factor a difference of two squares. (See page 9 for guidelines on writing a good response.)

Related Review

If the expression is not factored, then factor it. If the expression is factored, then find the product.

87. $(x - 9)(x + 2)$ **88.** $(x - 1)(x - 8)$

89. $x^2 - 15x + 50$ **90.** $x^2 + 19x - 20$

91. $(3x - 7)(3x + 7)$ **92.** $(4x + 1)(4x - 1)$

93. $25x^2 - 36$ **94.** $49x^2 - 81$

Expressions, Equations, and Graphs

Perform the indicated instruction. Then use words such as lin-ear, quadratic, cubic, polynomial, degree, one variable, and two variables to describe the expression, equation, or system.

95. Find the difference $(5p + 7w) - (2p - 4w)$.

96. Solve $6 = 3(x - 5)$.

97. Find the product $(5p + 7w)(2p - 4w)$.

98. Graph $y = 3(x - 5)$ by hand.

99. Factor $p^2 - 11pw + 18w^2$.

100. Solve the following:

$$y = 3(x - 5)$$
$$7x + 4y = -3$$

8.2 FACTORING OUT THE GCF; FACTORING BY GROUPING

Objectives

▶ Factor out the *greatest common factor (GCF)* of a polynomial.

▶ Completely factor polynomials.

▶ Factor out the opposite of the GCF of a polynomial.

▶ Factor a polynomial by grouping.

In Section 8.1, we discussed some ways to factor polynomials. In this section, we will discuss more factoring techniques.

Factoring Out the GCF

Consider the polynomial $2x + 10$. Note that 2 is a common factor of both $2x = 2 \cdot x$ and $10 = 2 \cdot 5$:

$$2x + 10 = 2 \cdot x + 2 \cdot 5$$

We use the distributive law to "factor out" the common factor 2:

$$2x + 10 = 2 \cdot x + 2 \cdot 5 = 2(x + 5)$$

So, we have factored $2x + 10$ as $2(x + 5)$. We check the result by finding the product $2(x + 5)$:

$$2(x + 5) = 2x + 10$$

Example 1   Factoring Out a Common Factor

Factor.

1. $3x + 21$ **2.** $8x^2 - 6x$ **3.** $6x^3 + 12x^2$

Solution

1. The number 3 is a common factor of $3x = 3 \cdot x$ and $21 = 3 \cdot 7$. So, we use the distributive law to factor out 3:

$$3x + 21 = 3 \cdot x + 3 \cdot 7 \qquad \text{3 is a common factor.}$$
$$= 3(x + 7) \qquad \text{Factor out 3.}$$

We can verify our work by finding the product $3(x + 7)$:

$$3(x + 7) = 3x + 21$$

2. The expression $2x$ is a common factor of $8x^2 = 2x \cdot 4x$ and $6x = 2x \cdot 3$. So, we use the distributive law to factor out $2x$:

$$8x^2 - 6x = 2x \cdot 4x - 2x \cdot 3 \qquad \text{2x is a common factor.}$$
$$= 2x(4x - 3) \qquad \text{Factor out 2x.}$$

3. The expression $6x^2$ is a common factor of $6x^3 = 6x^2 \cdot x$ and $12x^2 = 6x^2 \cdot 2$. So, we use the distributive law to factor out $6x^2$:

$$6x^3 + 12x^2 = 6x^2 \cdot x + 6x^2 \cdot 2 \qquad \text{6x}^2 \text{ is a common factor.}$$
$$= 6x^2(x + 2) \qquad \text{Factor out 6x}^2. \qquad \blacksquare$$

In Problem 3 of Example 1, notice that $2x$ is also a common factor of $6x^3 = 2x \cdot 3x^2$ and $12x^2 = 2x \cdot 6x$. So, we could have factored $6x^3 + 12x^2$ by factoring out $2x$ rather than $2x^2$:

$$6x^3 + 12x^2 = 2x(3x^2 + 6x)$$

However, the resulting expression is not completely factored: We can still factor $3x^2 + 6x$ by factoring out $3x$:

$$6x^3 + 12x^2 = 2x(3x^2 + 6x) = 2x \cdot 3x(x + 2) = 6x^2(x + 2)$$

Although we have found the same final result, it was more efficient to factor out $6x^2$, which has a larger coefficient and a higher degree than $2x$. We call $6x^2$ the *greatest common factor* of $6x^3$ and $12x^2$.

> **DEFINITION** Greatest common factor
>
> The **greatest common factor (GCF)** of two or more terms is the monomial with the largest coefficient and the largest degree that is a factor of all the terms.

For each polynomial in Example 1, the common factor that we factored out of the polynomial was the GCF. In Example 2, we factor some more polynomials by factoring out the GCF.

Example 2   Factoring Out the GCF

Factor.

1. $20x^2 + 35x$

2. $14p^3 - 21p^2$

Solution

1. We begin by factoring $20x^2$ and $35x$:

$$20x^2 = 2 \cdot 2 \cdot 5 \cdot x \cdot x$$
$$35x = 5 \cdot 7 \cdot x$$

Both 5 and x are common factors. So, the GCF is $5x$:

$$20x^2 + 35x = 5x \cdot 4x + 5x \cdot 7 \qquad \text{5x is the GCF.}$$
$$= 5x(4x + 7) \qquad \text{Factor out 5x.}$$

2. We begin by factoring $14p^3$ and $21p^2$:

$$14p^3 = 2 \cdot 7 \cdot p \cdot p \cdot p$$
$$21p^2 = 3 \cdot 7 \cdot p \cdot p$$

There are three common factors: 7, p, and p. So, the GCF is $7p^2$:

$$14p^3 - 21p^2 = 7p^2 \cdot 2p - 7p^2 \cdot 3 \qquad \text{7p}^2 \text{ is the GCF.}$$
$$= 7p^2(2p - 3) \qquad \text{Factor out 7p}^2.$$

We use a graphing calculator table to verify our work (see Fig. 3).   ∎

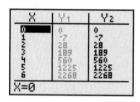

Figure 3 Verify the work

After you factor a polynomial, verify your work by finding the product of your result or by using a graphing calculator table.

So far, we have factored out the GCF for some binomials with one variable. We can also factor out the GCF for some polynomials with more than two terms and more than one variable.

Example 3   Factoring Out the GCF

Factor $12x^4y^2 - 6x^2y^3 + 15xy^2$.

Solution

We begin by factoring $12x^4y^2$, $6x^2y^3$, and $15xy^2$:

$$12x^4y^2 = 2 \cdot 2 \cdot 3 \cdot x \cdot x \cdot x \cdot x \cdot y \cdot y$$
$$6x^2y^3 = 2 \cdot 3 \cdot x \cdot x \cdot y \cdot y \cdot y$$
$$15xy^2 = 3 \cdot 5 \cdot x \cdot y \cdot y$$

There are four common factors: 3, x, y, and y. So, the GCF is $3xy^2$:

$$12x^4y^2 - 6x^2y^3 + 15xy^2 = 3xy^2 \cdot 4x^3 - 3xy^2 \cdot 2xy + 3xy^2 \cdot 5 \qquad \text{3xy}^2 \text{ is the GCF.}$$
$$= 3xy^2(4x^3 - 2xy + 5) \qquad \text{Factor out 3xy}^2.$$
∎

Completely Factoring Polynomials

After we factor the GCF out of a polynomial, we must check whether the result can be further factored by using factoring techniques discussed in Section 8.1. If a result cannot be further factored, it is said to be **completely factored.**

Example 4 Completely Factoring Polynomials

Factor $4x^2 - 36$.

Solution

The GCF of $4x^2$ and 36 is 4:

$$\begin{aligned} 4x^2 - 36 &= 4(x^2 - 9) && \text{Factor out GCF, 4.} \\ &= 4(x + 3)(x - 3) && A^2 - B^2 = (A + B)(A - B) \quad\blacksquare \end{aligned}$$

To factor $4x^2 - 36$ in Example 4, we first factored out the GCF, 4, and then factored the resulting difference of two squares, $x^2 - 9$. These steps require less work than first using the property for the difference of two squares:

$$\begin{aligned} 4x^2 - 36 &= (2x + 6)(2x - 6) && A^2 - B^2 = (A + B)(A - B) \\ &= 2(x + 3)(2)(x - 3) && \text{Factor out GCF, 2.} \\ &= 4(x + 3)(x - 3) && \text{Simplify.} \end{aligned}$$

Not only are there fewer steps in Example 4, but it is easier to use the property for the difference of squares to factor $x^2 - 9$ than $4x^2 - 36$. In general, **when the leading coefficient of a polynomial is positive and the GCF is not 1, we first factor out the GCF.** (We will soon discuss what to do when the leading coefficient of a polynomial is negative.)

Example 5 Completely Factoring Polynomials

Factor $2x^4y - 6x^3y - 20x^2y$.

Solution

The GCF of $2x^4y$, $6x^3y$, and $20x^2y$ is $2x^2y$:

$$\begin{aligned} 2x^4y - 6x^3y - 20x^2y &= 2x^2y(x^2 - 3x - 10) && \text{Factor out GCF, } 2x^2y. \\ &= 2x^2y(x - 5)(x + 2) && \begin{array}{l}\text{Find two integers whose product} \\ \text{is } -10 \text{ and whose sum is } -3.\end{array} \quad\blacksquare \end{aligned}$$

WARNING   It is a common error when factoring a polynomial to forget to *completely* factor it. In Example 5, we factored the GCF, $2x^2y$, out of $2x^4y - 6x^3y - 20x^2y$:

$$2x^4y - 6x^3y - 20x^2y = 2x^2y(x^2 - 3x - 10) \quad \text{Not completely factored}$$

However, we were not done factoring, because we could still factor $x^2 - 3x - 10$:

$$2x^2y(x^2 - 3x - 10) = 2x^2y(x - 5)(x + 2) \quad \text{Completely factored}$$

When factoring a polynomial, always *completely* factor it.

Example 6 Completely Factoring Polynomials

Factor.

1. $12x^3 - 75x$
2. $36x + 4x^3 - 24x^2$

Solution

1. The GCF of $12x^3$ and $75x$ is $3x$:

$$\begin{aligned} 12x^3 - 75x &= 3x(4x^2 - 25) && \text{Factor out GCF, } 3x. \\ &= 3x(2x + 5)(2x - 5) && A^2 - B^2 = (A + B)(A - B) \end{aligned}$$

2. We begin by writing $36x + 4x^3 - 24x^2$ in descending order:

$$36x + 4x^3 - 24x^2 = 4x^3 - 24x^2 + 36x \qquad \text{Write in descending order.}$$
$$= 4x(x^2 - 6x + 9) \qquad \text{Factor out GCF, } 4x.$$
$$= 4x(x-3)(x-3) \qquad \text{Find two integers whose product is 9 and whose sum is } -6.$$
$$= 4x(x-3)^2 \qquad bb = b^2 \qquad ∎$$

Factoring Out the Opposite of the GCF of a Polynomial

How do we factor a polynomial in which the leading coefficient is negative? For example, consider the polynomial $-5x^3 + 30x^2 - 40x$, which has a negative leading coefficient, -5.

> **How to Factor when the Leading Coefficient Is Negative**
>
> When the leading coefficient of a polynomial is negative, we first factor out the opposite of the GCF.

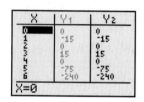

Figure 4 Verify the work

Example 7   Factoring Out the Opposite of the GCF

Factor $-5x^3 + 30x^2 - 40x$.

Solution

For $-5x^3 + 30x^2 - 40x$, the GCF is $5x$. Since the leading coefficient, -5, is negative, we factor out the opposite of the GCF:

$$-5x^3 + 30x^2 - 40x = -5x(x^2 - 6x + 8) \qquad \text{Factor out } -5x, \text{ opposite of GCF.}$$
$$= -5x(x-2)(x-4) \qquad \text{Find two integers whose product is 8 and whose sum is } -6.$$

We use a graphing calculator table to verify our work (see Fig. 4). ∎

Example 8   Factoring Out the Opposite of the GCF

Factor $49 - w^2$.

Solution

First we write $49 - w^2$ in descending order: $-w^2 + 49$. The GCF of $-w^2 + 49$ is 1. Since the leading coefficient, -1, is negative, we factor out the opposite of the GCF:

$$-w^2 + 49 = -1(w^2 - 49) \qquad \text{Factor out } -1, \text{ opposite of GCF.}$$
$$= -1(w+7)(w-7) \qquad A^2 - B^2 = (A+B)(A-B)$$
$$= -(w+7)(w-7) \qquad -1a = -a \qquad ∎$$

Factoring by Grouping

So far, we have factored out a GCF when the GCF is a monomial. For example, here we factor out the monomial p from the polynomial $x^2(p) + 2(p)$:

$$x^2(p) + 2(p) = (x^2 + 2)(p)$$

We can also factor out a GCF that is a binomial. For example, here we factor out the binomial $x + 3$ from the polynomial $x^2(x+3) + 2(x+3)$:

$$x^2(x+3) + 2(x+3) = (x^2 + 2)(x+3)$$

We can factor some polynomials that contain four terms by first factoring the first two terms and the last two terms—for example,

$$\underbrace{x^3 + 3x^2}_{\text{factor}} + \underbrace{2x + 6}_{\text{factor}} = x^2(x + 3) + 2(x + 3) \qquad \textit{Factor both pairs of terms.}$$

$$= (x^2 + 2)(x + 3) \qquad \textit{Factor out GCF, x + 3.}$$

We call this method *factoring by grouping.*

Example 9   Factoring by Grouping

Factor $2x^3 - 8x^2 - 3x + 12$.

Solution

$$2x^3 - 8x^2 - 3x + 12 = 2x^2(x - 4) - 3(x - 4) \qquad \textit{Factor both pairs of terms.}$$

$$= (2x^2 - 3)(x - 4) \qquad \textit{Factor out GCF, x - 4.} ■$$

WARNING   It is a common error to think that a polynomial such as $2x^2(x - 4) - 3(x - 4)$ in Example 9 is factored. Even though both of the terms $2x^2(x - 4)$ and $3(x - 4)$ are factored, the entire expression $2x^2(x - 4) - 3(x - 4)$ is a difference, not a product. The polynomial $(2x^2 - 3)(x - 4)$ in Example 9 *is* factored, because it is a product.

We now describe in general how to factor a polynomial by grouping.

Factoring by Grouping

For a polynomial with four terms, we **factor by grouping** (if it can be done) by

1. Factoring the first two terms and the last two terms.
2. Factoring out the binomial GCF.

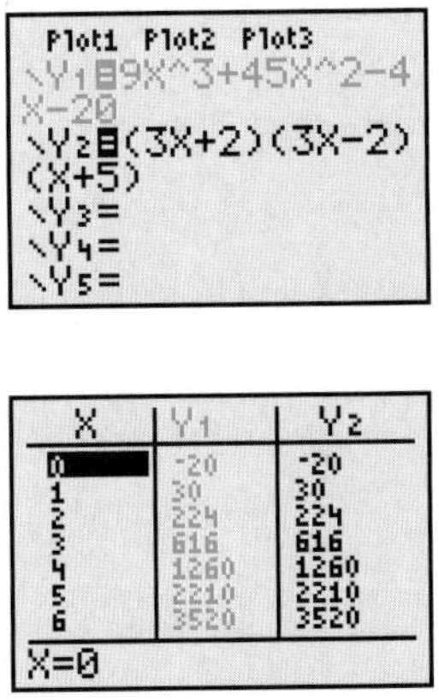

When trying to factor a polynomial with four terms, consider trying to factor it by grouping.

Example 10   Factoring by Grouping

Factor $9x^3 + 45x^2 - 4x - 20$.

Solution

$$9x^3 + 45x^2 - 4x - 20 = 9x^2(x + 5) - 4(x + 5) \qquad \textit{Factor both pairs of terms.}$$

$$= (9x^2 - 4)(x + 5) \qquad \textit{Factor out GCF, x + 5.}$$

$$= (3x + 2)(3x - 2)(x + 5) \qquad \textit{A}^2 - \textit{B}^2 = (\textit{A} + \textit{B})(\textit{A} - \textit{B})$$

We use a graphing calculator table to verify our work (see Fig. 5).   ■

Figure 5 Verify the work

Example 11   Factoring by Grouping

Factor $2a - 2b + xa - xb$.

Solution

$$2a - 2b + xa - xb = 2(a - b) + x(a - b) \qquad \textit{Factor both pairs of terms.}$$

$$= (2 + x)(a - b) \qquad \textit{Factor out GCF, a - b.} ■$$

In summary, there are two aspects to factoring that are important to remember:

1. If the leading coefficient of a polynomial is positive and the GCF is not 1, first factor out the GCF. If the leading coefficient is negative, first factor out the opposite of the GCF.
2. Always *completely* factor a polynomial.

group exploration
Equivalent expressions and locating errors

1. Consider the following correct work for factoring $3x^3 + 2x^2 - 27x - 18$:

$$3x^3 + 2x^2 - 27x - 18 = x^2(3x + 2) - 9(3x + 2)$$
$$= (x^2 - 9)(3x + 2)$$
$$= (x + 3)(x - 3)(3x + 2)$$

Use graphing calculator tables to show that all of the preceding four expressions are equivalent.

2. Consider the following incorrect work for factoring $4x^3 - 3x^2 - 16x + 12$:

$$4x^3 - 3x^2 - 16x + 12 = x^2(4x - 3) - 4(4x + 3)$$
$$= (x^2 - 4)(4x - 3)(4x + 3)$$
$$= (x + 2)(x - 2)(4x - 3)(4x + 3)$$

Use graphing calculator tables to find the step(s) in which an error was made. Then do the factoring correctly.

HOMEWORK 8.2

FOR EXTRA HELP ▶

 Student Solutions Manual

 PH Math/Tutor Center

Math XL MathXL®

MyMathLab MyMathLab

Factor. Use a graphing calculator table to verify your work when possible.

1. $6x + 8$
2. $3x + 15$
3. $20w^2 + 35w$
4. $28p^2 + 21p$
5. $12x^2 - 30x$
6. $18x^2 - 12x$
7. $6a^2b - 9ab$
8. $8pq^3 - 6p^2q$
9. $8x^3y^2 + 12x^2y^3$
10. $27x^3y - 45x^2y^3$
11. $15x^3 - 10x - 30$
12. $28x^3 + 12x - 20$
13. $12t^4 + 8t^3 - 16t$
14. $6r^4 - 9r^2 - 12r$
15. $10a^4b - 15a^3b + 25ab$
16. $8p^4q + 6p^2q - 4pq$

Factor. Verify that you have factored correctly by finding the product of your factored polynomial.

17. $2x^2 - 18$
18. $3x^2 - 75$
19. $3m^2 + 21m + 30$
20. $5p^2 - 5p - 60$
21. $2x^2 - 18x + 36$
22. $4x^2 - 44x + 72$
23. $3x^3 - 27x$
24. $5x^3 - 20x$
25. $4r^3 - 16r^2 - 20r$
26. $2t^3 - 12t^2 - 32t$
27. $6x^4 - 24x^2$
28. $2x^4 - 50x^2$
29. $8m^4n - 18m^2n$
30. $75p^4y - 27p^2y$
31. $5x^4 + 10x^3 - 120x^2$
32. $8x^4 - 24x^3 + 16x^2$
33. $8x - 2x^3$
34. $24x - 6x^3$
35. $36t^2 + 32t + 4t^3$
36. $60y - 40y^2 + 5y^3$
37. $-12x^3 + 27x$
38. $-4x^4 + 4x^2$
39. $-3x^3 - 18x^2 + 48x$
40. $-4x^3 + 20x^2 - 24x$
41. $6a^4b + 36a^3b + 54a^2b$
42. $4m^4n - 40m^3n + 100m^2n$

Factor. Use a graphing calculator table to verify your work when possible.

43. $5x^2(x - 3) + 2(x - 3)$
44. $8x^2(x + 4) + 3(x + 4)$
45. $6x^2(2x + 5) - 7(2x + 5)$
46. $5x^2(4x - 1) - 3(4x - 1)$
47. $2p^3 + 6p^2 + 5p + 15$
48. $12r^3 + 3r^2 + 8r + 2$
49. $6x^3 - 2x^2 + 21x - 7$
50. $4x^3 - 16x^2 + 3x - 12$
51. $15w^3 + 5w^2 - 6w - 2$
52. $6b^3 - 10b^2 - 21b + 35$
53. $16x^3 - 12x^2 - 36x + 27$
54. $50x^3 + 125x^2 - 8x - 20$
55. $2b^3 - 5b^2 - 18b + 45$
56. $4t^3 - 7t^2 - 16t + 28$
57. $x^3 - x^2 - x + 1$
58. $x^3 + x^2 - x - 1$
59. $3x + 3y + ax + ay$
60. $5a + 5b + xa + xb$
61. $2xy - 8x + 3y - 12$
62. $5ab - 10a + 3b - 6$

Factor. Use a graphing calculator table to verify your work when possible.

63. $81x^2 - 25$
64. $9x^2 - 100$
65. $w^2 - 10w + 16$
66. $p^2 + 3p - 40$
67. $24 - 10x + x^2$
68. $-14 - 5x + x^2$
69. $20a^2b - 15ab^3$
70. $14xy^2 + 21y^2x$
71. $3r^2 + 30r + 75$
72. $2y^2 + 16y + 32$
73. $64x^3 - 49x$
74. $2x^3 - 162x$
75. $-m^2 + 6m - 9$
76. $-w^2 + 16w - 64$
77. $x^3 + 9x^2 - 4x - 36$
78. $28x^3 - 35x^2 + 8x - 10$
79. $2m^3n - 10m^2n^2 + 12mn^3$
80. $4p^3y + 8p^2y^2 - 12py^3$

81. A student tries to factor $6x^3 + 8x^2 + 15x + 20$:

$$6x^3 + 8x^2 + 15x + 20 = 2x^2(3x + 4) + 5(3x + 4)$$

Describe any errors. Then factor the polynomial correctly.

82. Two students try to factor $15x^3 + 3x^2 - 35x - 7$:

Student A

$$15x^3 + 3x^2 - 35x - 7 = 3x^2(5x + 1) - 7(5x + 1)$$
$$= (3x^2 - 7)(5x + 1)$$

Student B

$$15x^3 + 3x^2 - 35x - 7 = 15x^3 - 35x + 3x^2 - 7$$
$$= 5x(3x^2 - 7) + 1(3x^2 - 7)$$
$$= (5x + 1)(3x^2 - 7)$$

Are both students, one student, or neither student correct?

83. A student tries to factor $4x^3 + 28x^2 + 40x$:

$$4x^3 + 28x^2 + 40x = 4x(x^2 + 7x + 10)$$

Describe any errors. Then factor the polynomial correctly.

84. A student tries to factor $5x^3 - 45x$:

$$5x^3 - 45x = 5x(x^2 - 9)$$

Describe any errors. Then factor the polynomial correctly.

85. A student tries to factor $4x^2 - 100$:

$$4x^2 - 100 = (2x + 10)(2x - 10)$$
$$= 2(x + 5)(2)(x - 5)$$
$$= 4(x + 5)(x - 5)$$

What would you tell the student?

86. A student tries to factor $64x^2 - 36$:

$$64x^2 - 36 = (8x + 6)(8x - 6)$$
$$= 2(4x + 3)(2)(4x - 3)$$
$$= 4(4x + 3)(4x - 3)$$

What would you tell the student?

87. Give three examples of cubic polynomials in which $2x$ is a factor.

88. Give three examples of quadratic polynomials in which $-5x$ is a factor.

89. A student tries to factor $2x^2 + 10x + 12$:

$$2x^2 + 10x + 12 = 2(x^2 + 5x + 6)$$

The student then checks that tables for $y = 2x^2 + 10x + 12$ and $y = 2(x^2 + 5x + 6)$ are the same. Explain why the student's work is incorrect even though the tables for the two equations are the same.

90. Explain why, when factoring a polynomial, it is a good idea to factor out the GCF first if it is not 1 and the leading coefficient of the polynomial is positive.

Related Review

If the expression is not factored, then factor it. If the expression is factored, then find the product.

91. $2x(x - 3)(x + 4)$

92. $-5(x - 2)(x - 7)$

93. $5x^2 - 40x + 80$

94. $3x^2 + 12x + 12$

95. $6x^3 - 9x^2 - 4x + 6$

96. $8x^3 + 6x^2 - 20x - 15$

97. $(x - 3)(x^2 + 5)$

98. $(x^2 - 2)(x - 8)$

Expressions, Equations, and Graphs

Perform the indicated instruction. Then use words such as linear, quadratic, cubic, polynomial, degree, one variable, *and* two variables *to describe the expression, equation, or system.*

99. Graph $y = -4x + 1$ by hand.

100. Find the product $(3x - 2)(4x^2 - x + 5)$.

101. Solve $-4x + 1 = 2x - 5$.

102. Find the sum $(3x - 2) + (4x^2 - x + 5)$.

103. Solve the following:

$$y = -4x + 1$$
$$y = 2x - 5$$

104. Factor $36x^2 - 81$.

FACTORING TRINOMIALS OF THE FORM
8.3 $ax^2 + bx + c$

The work will teach you how to do it.

—Estonian proverb

Objectives

▹ Factor a trinomial by trial and error.

▹ Know how to rule out possibilities when factoring by trial and error.

▹ Factor a trinomial by grouping.

In Section 8.1, we factored trinomials of the form $ax^2 + bx + c$, where $a = 1$. In this section, we will factor trinomials of the form $ax^2 + bx + c$, where $a \neq 1$. We will discuss two methods: factoring by trial and error and factoring by grouping. These two methods give equivalent results.

Method 1: Factoring Trinomials by Trial and Error

One way to factor trinomials of the form $ax^2 + bx + c$, where $a \neq 1$, is to make educated guesses at the factorization and then find the product of these guesses to see if any of them work. This method is called **factoring by trial and error.**

Example 1   Factoring by Trial and Error

Factor $3x^2 + 17x + 10$.

Solution

If we can factor $3x^2 + 17x + 10$, the result will be of the form

$$(3x + ?)(x + ?)$$

The product of the last terms has to be 10, so the last terms must be 1 and 10 or 2 and 5, where each pair of terms can be written in either order. We can rule out negative last terms in the factors, because the middle term, $17x$, of $3x^2 + 17x + 10$ has the positive coefficient 17. We decide between the two pairs of possible last terms by multiplying:

$$(3x + 1)(x + 10) = 3x^2 + 30x + x + 10 \quad = 3x^2 + 31x + 10$$
$$(3x + 10)(x + 1) = 3x^2 + 3x + 10x + 10 = 3x^2 + 13x + 10$$
$$(3x + 2)(x + 5) = 3x^2 + 15x + 2x + 10 = 3x^2 + 17x + 10 \quad \longleftarrow \text{ Success!}$$
$$(3x + 5)(x + 2) = 3x^2 + 6x + 5x + 10 \quad = 3x^2 + 11x + 10$$

So, $3x^2 + 17x + 10 = (3x + 2)(x + 5)$. We use a graphing calculator table to verify our work (see Fig. 6).   ∎

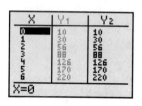

Figure 6 Verify the work

In trying to factor a polynomial, once we find the factored polynomial, there is no need to multiply the other possibilities. In Example 1, we multiplied all possible factorizations of $3x^2 + 17x + 10$ only to show how to organize the work in case the last possibility is the correct one.

In order to use the method shown in Example 1, it is helpful to be able to multiply two binomials in one step. Consider the product of $2x + 3$ and $4x + 5$:

$$(2x + 3)(4x + 5) = 8x^2 + 10x + 12x + 15$$
$$= 8x^2 + 22x + 15$$

To find the product in one step, we must combine the like terms $10x$ and $12x$ mentally. Note that these like terms come from the product of the two *outer terms* and the product of the two *inner terms* of $(2x + 3)(4x + 5)$:

$$(2x + 3)(4x + 5) = 8x^2 + 10x + 12x + 15$$
$$= 8x^2 + 22x + 15$$

Example 2   Factoring by Trial and Error

Factor $2x^2 - 3x - 9$.

Solution

If we can factor $2x^2 - 3x - 9$, the result will be of the form

$$(2x + ?)(x + ?)$$

The product of the last terms is -9, so the last terms must be 1 and -9, 3 and -3, or -1 and 9, where each pair can be written in either order. We decide among the three pairs of possible last terms by multiplying:

$$(2x + 1)(x - 9) = 2x^2 - 17x - 9$$
$$(2x - 9)(x + 1) = 2x^2 - 7x - 9$$
$$(2x + 3)(x - 3) = 2x^2 - 3x - 9 \longleftarrow \text{Success!}$$
$$(2x - 3)(x + 3) = 2x^2 + 3x - 9$$
$$(2x - 1)(x + 9) = 2x^2 + 17x - 9$$
$$(2x + 9)(x - 1) = 2x^2 + 7x - 9$$

Therefore, $2x^2 - 3x - 9 = (2x + 3)(x - 3)$. ■

Factoring $ax^2 + bx + c$ by Trial and Error

To **factor a trinomial** of the form $ax^2 + bx + c$ **by trial and error,** if the trinomial can be factored as a product of two binomials, then the product of the coefficients of the first terms of the binomials is equal to a and the product of the last terms of the binomials is equal to c. For example,

$$\underset{\substack{\uparrow \quad \uparrow \quad \uparrow \\ a = 6 \quad b = 23 \quad c = 20}}{6x^2 + 23x + 20} = \underset{\substack{\uparrow \qquad \uparrow \\ \text{Last terms:} \\ 4 \cdot 5 = 20 = c}}{(3x + 4)(2x + 5)}$$

Coefficients of first terms:
$3 \cdot 2 = 6 = a$

To find the correct factored expression, multiply the possible products and identify the one for which the coefficient of x is b.

FACTORING OUT THE GCF AND THEN FACTORING BY TRIAL AND ERROR

When factoring a polynomial, recall from Section 8.2 that if the GCF is not 1, then we first factor out the GCF (or its opposite) and continue factoring if possible. Always completely factor a polynomial.

Example 3 Completely Factoring a Polynomial

Factor $15x^4 - 39x^3 + 18x^2$.

Solution

To factor $15x^4 - 39x^3 + 18x^2$, we first factor out the GCF, $3x^2$:

$$3x^2(5x^2 - 13x + 6)$$

If we can factor further, the desired result is of the form

$$3x^2(5x + ?)(x + ?)$$

The product of the last terms has to be 6, so the last terms must be -1 and -6 or -2 and -3, where each pair can be written in either order. We decide by multiplying:

(We have temporarily put aside the GCF, $3x^2$.)

$$(5x - 1)(x - 6) = 5x^2 - 31x + 6$$
$$(5x - 6)(x - 1) = 5x^2 - 11x + 6$$
$$(5x - 2)(x - 3) = 5x^2 - 17x + 6$$
$$(5x - 3)(x - 2) = 5x^2 - 13x + 6 \longleftarrow \text{Success!}$$

So, $15x^4 - 39x^3 + 18x^2 = 3x^2(5x^2 - 13x + 6) = 3x^2(5x - 3)(x - 2)$. ■

WARNING   When we factor a polynomial by trial and error, we can easily forget about the GCF by the time we have found the other factors. **If there is more factoring to be done after factoring out the GCF, write a note several lines down that reminds you to include the GCF in your result.**

RULING OUT POSSIBILITIES WHILE FACTORING BY TRIAL AND ERROR

Example 4 shows how to rule out possible factorizations to help speed up the process of factoring.

Example 4   Ruling Out Possibilities

Factor $6x^2 - 19x + 8$.

Solution

If we can factor $6x^2 - 19x + 8$, the result is of one of the following two forms:

$$(6x + ?)(x + ?) \qquad (3x + ?)(2x + ?)$$

The product of the last terms has to be 8, so the last terms must be -1 and -8, or -2 and -4, where each pair can be written in either order. We can rule out positive last terms, because the middle term, $-19x$, has a negative coefficient, -19.

Since the terms of $6x^2 - 19x + 8$ do not have a common factor of 2, we can also rule out products that have a factor of 2. For example, we can rule out $(6x - 8)(x - 1)$, because it has a factor of 2:

$$(6x - 8)(x - 1) = 2(3x - 4)(x - 1)$$

We decide among the remaining possible last terms by multiplying:

$$(6x - 1)(x - 8) = 6x^2 - 49x + 8$$

Contains a factor of 2, rule out: $(6x - 2)(x - 4)$

Contains a factor of 2, rule out: $(6x - 4)(x - 2)$

Contains a factor of 2, rule out: $(3x - 1)(2x - 8)$

$(3x - 8)(2x - 1) = 6x^2 - 19x + 8$ ← Success!

Contains a factor of 2, rule out: $(3x - 2)(2x - 4)$

Contains a factor of 2, rule out: $(3x - 4)(2x - 2)$

Figure 7 Verify the work

So, $6x^2 - 19x + 8 = (3x - 8)(2x - 1)$. We use a graphing calculator table to verify our work (see Fig. 7). ∎

We can use a similar method to factor trinomials that have two variables.

Example 5   Factoring a Trinomial That Has Two Variables

Factor $10p^2 + 19pw + 6w^2$.

Solution

If we can factor $10p^2 + 19pw + 6w^2$, the result is of one of the following two forms:

$$(10p + ?)(p + ?) \qquad (5p + ?)(2p + ?)$$

The product of the last terms has to be $6w^2$, so the last terms must be $6w$ and w or $3w$ and $2w$, where each pair can be written in either order.

Since the terms of $10p^2 + 19pw + 6w^2$ do not have a common factor of 2, we can rule out products that have a factor of 2. We decide among the remaining possible last terms by multiplying:

Contains a factor of 2, rule out: $(10p + 6w)(p + w)$
$$(10p + w)(p + 6w) = 10p^2 + 61pw + 6w^2$$
$$(10p + 3w)(p + 2w) = 10p^2 + 23pw + 6w^2$$

Contains a factor of 2, rule out: $(10p + 2w)(p + 3w)$
$$(5p + 6w)(2p + w) = 10p^2 + 17pw + 6w^2$$

Contains a factor of 2, rule out: $(5p + w)(2p + 6w)$

Contains a factor of 2, rule out: $(5p + 3w)(2p + 2w)$
$$(5p + 2w)(2p + 3w) = 10p^2 + 19pw + 6w^2 \quad \longleftarrow \text{Success!}$$

So, $10p^2 + 19pw + 6w^2 = (5p + 2w)(2p + 3w)$. ■

Method 2: Factoring Trinomials by Grouping

Instead of using trial and error to factor a trinomial, we can factor by grouping.

To factor a trinomial of the form $x^2 + bx + c$, recall from Section 8.1 that we look for two integers whose product is c and whose sum is b. To factor a trinomial of the form $ax^2 + bx + c$, we must look for two integers whose product is ac and whose sum is b.

Factoring $ax^2 + bx + c$ by Grouping

To **factor a trinomial** of the form $ax^2 + bx + c$ **by grouping** (if it can be done),

1. Find pairs of numbers whose product is ac.
2. Determine which of the pairs of numbers from step 1 has the sum b. Call this pair of numbers m and n.
3. Write the bx term as $mx + nx$:
$$ax^2 + bx + c = ax^2 + mx + nx + c$$
4. Factor $ax^2 + mx + nx + c$ by grouping.

Another name for this technique is the *ac* **method.**

Example 6 Factoring a Trinomial by Grouping

Factor $2x^2 + 13x + 15$.

Solution

Here, $a = 2$, $b = 13$, and $c = 15$.

Step 1. Find the product ac: $ac = 2(15) = 30$.

Step 2. We want to find two numbers m and n that have the product $ac = 30$ and the sum $b = 13$:

Product = 30	**Sum = 13?**
$1(30) = 30$	$1 + 30 = 31$
$2(15) = 30$	$2 + 15 = 17$
$3(10) = 30$	$3 + 10 = 13 \quad \longleftarrow$ Success!
$5(6) = 30$	$5 + 6 = 11$

Since $3(10) = 30$ and $3 + 10 = 13$, we conclude that the two numbers m and n are 3 and 10.

Step 3. We write $2x^2 + 13x + 15 = 2x^2 + 3x + 10x + 15$.

Step 4. We factor $2x^2 + 3x + 10x + 15$ by grouping:

$$2x^2 + 3x + 10x + 15 = x(2x + 3) + 5(2x + 3) \qquad \text{Factor both pairs of terms.}$$
$$= (x + 5)(2x + 3) \qquad \text{Factor out GCF, } (2x + 3).$$

So, $2x^2 + 13x + 15 = (x + 5)(2x + 3)$. We check the result by finding the product $(x + 5)(2x + 3)$:

$$(x + 5)(2x + 3) = 2x^2 + 3x + 10x + 15 = 2x^2 + 13x + 15 \qquad \blacksquare$$

Example 7 Factoring Out the GCF and Then Factoring by Grouping

Factor $18x^3 + 70x^2 - 8x$.

Solution

First, we factor out the GCF, $2x$:

$$18x^3 + 70x^2 - 8x = 2x(9x^2 + 35x - 4)$$

Next, we use grouping to try to factor $9x^2 + 35x - 4$. Here, $a = 9$, $b = 35$, and $c = -4$.

Step 1. Find the product ac: $ac = 9(-4) = -36$.

Step 2. We want to find two numbers m and n that have the product $ac = -36$ and the sum $b = 35$:

	Product $= -36$	Sum $= 35$?
	$1(-36) = -36$	$1 + (-36) = -35$
(We have	$2(-18) = -36$	$2 + (-18) = -16$
temporarily put	$3(-12) = -36$	$3 + (-12) = -9$
aside the GCF, $2x$.)	$4(-9) = -36$	$4 + (-9) = -5$
	$6(-6) = -36$	$6 + (-6) = 0$
	$9(-4) = -36$	$9 + (-4) = 5$
	$12(-3) = -36$	$12 + (-3) = 9$
	$18(-2) = -36$	$18 + (-2) = 16$
	$36(-1) = -36$	$36 + (-1) = 35 \longleftarrow$ Success!

Since $36(-1) = -36$ and $36 + (-1) = 35$, we conclude that the two numbers m and n are 36 and -1.

Step 3. We write $9x^2 + 35x - 4 = 9x^2 + 36x - 1x - 4$.

Step 4. We factor $9x^2 + 36x - 1x - 4$ by grouping:

$$9x^2 + 36x - 1x - 4 = 9x(x + 4) - 1(x + 4) \qquad \text{Factor both pairs of terms.}$$
$$= (9x - 1)(x + 4) \qquad \text{Factor out GCF, } (x + 4).$$

So, $18x^3 + 70x^2 - 8x = 2x(9x^2 + 35x - 4) = 2x(9x - 1)(x + 4)$. We use a graphing calculator table to verify our work (see Fig. 8).

Figure 8 Verify the work

group exploration

Looking ahead: Developing a factoring strategy

In this exploration, you will summarize what you have learned about factoring.

1. When factoring a polynomial, what should you try to do first? Give an example.
2. Describe various techniques that you can use to factor a polynomial with the given number of terms. For each technique, give an example of factoring a polynomial.
   - **a.** two terms
   - **b.** three terms
   - **c.** four terms
3. Explain how you know when you are done factoring a polynomial.

HOMEWORK 8.3 FOR EXTRA HELP ▶

Student Solutions Manual   PH Math/Tutor Center   MathXL®   MyMathLab

Factor if possible. Verify that you have factored correctly by finding the product of your factored polynomial.

1. $2x^2 + 7x + 3$

2. $3x^2 + 16x + 5$

3. $5x^2 + 11x + 2$

4. $4x^2 + 13x + 3$

5. $3x^2 + 8x + 4$

6. $5x^2 + 13x + 6$

7. $2t^2 + t - 6$

8. $3b^2 - 10b + 8$

9. $6x^2 - 13x + 6$

10. $8x^2 - 14x - 15$

11. $4x^2 + 20x + 25$

12. $9x^2 + 12x + 4$

13. $6x^2 - 37x + 6$

14. $4x^2 + 15x - 4$

15. $2r^2 + 5r + 4$

16. $3a^2 - 2a - 6$

17. $18x^2 + 21x - 4$

18. $12x^2 - 17x + 6$

19. $3m^2 - 22m + 24$

20. $10t^2 - 27t + 18$

21. $2x^2 - 21x + 40$

22. $4x^2 + 81x + 20$

23. $2a^2 + 5ab + 3b^2$

24. $3m^2 + 8mn + 4n^2$

25. $5x^2 + 18xy - 8y^2$

26. $2p^2 - 5pt - 12t^2$

27. $6b^2 - 15bc + 6c^2$

28. $8r^2 - 33rw + 4w^2$

Factor. Use a graphing calculator table to verify your work when possible.

29. $4x^2 + 26x + 30$

30. $6x^2 + 27x + 30$

31. $20a^2 - 40a + 15$

32. $12b^2 - 56b + 32$

33. $24x^2 + 15x - 9$

34. $12x^2 - 27x + 15$

35. $-20x^2 + 22x + 12$

36. $-24x^2 + 54x - 27$

37. $4w^4 - 6w^3 - 12w^2$

38. $9t^4 - 21t^3 - 12t^2$

39. $10x^4 - 5x^3 - 50x^2$

40. $36x^4 - 30x^3 + 6x^2$

41. $6a^2 - 34ab - 12b^2$

42. $6m^2 + 38mn + 12n^2$

43. $12r^3 + 40r^2w + 32rw^2$

44. $15x^3 + 36x^2y + 12xy^2$

Factor when possible. Use a graphing calculator table to verify your work when possible.

45. $x^2 - 6x - 27$

46. $x^2 - 4x - 45$

47. $-48x^2 + 40x$

48. $-35x^2 - 42x$

49. $5a^3 + 2a^2 - 15a - 6$

50. $10r^3 + 25r^2 - 6r - 15$

51. $x^2 + 9$

52. $4x^2 + 25$

53. $4x^2 - 12x + 9$

54. $25x^2 - 10x + 1$

55. $-17p^2 + 17$

56. $-5t^2 + 20$

57. $24 + 10x + x^2$

58. $-12 + x + x^2$

59. $b^2 - 3bc - 28c^2$

60. $p^2 + 4pt - 32t^2$

61. $8t^2 - 10t + 3$

62. $6a^2 - 19a + 15$

63. $7x^4 - 28x^2$

64. $2x^4 - 50x^2$

65. $12p^3 - 4p^2 - 27p + 9$

66. $50m^3 - 75m^2 - 2m + 3$

67. $x^2 - 6x + 12$

68. $x^2 - 3x - 8$

69. $3x^4 - 21x^3 + 30x^2$

70. $4x^4 - 12x^3 + 8x^2$

71. $20x^2 + 16x^4 - 42x^3$

72. $-14x^3 + 6x^4 - 40x^2$

73. $36a^2 - 49b^2$

74. $64m^2 - 25n^2$

75. $-2x^2y + 8xy + 24y$

76. $-6x^2y + 24xy - 24y$

77. $4y^3 - 9y^2 - 9y$

78. $12t^3 + 7t^2 - 5t$

79. A student tries to factor the polynomial $2x^2 + 7x + 10$. Since the integers 2 and 5 have the product $2(5) = 10$ and the sum $2 + 5 = 7$, the student writes

$$2x^2 + 7x + 10 = (2x + 5)(x + 2)$$

Is the student correct? Explain.

80. A student tries to factor the polynomial $3x^2 + 12$:

$$3x^2 + 12 = 3(x^2 + 4) = 3(x + 2)(x + 2) = 3(x + 2)^2$$

Describe any errors. Then factor the polynomial correctly.

81. A student tries to factor the polynomial $8x^2 + 28x + 12$:

$$8x^2 + 28x + 12 = (4x + 2)(2x + 6)$$

The student then uses a graphing calculator table to verify that the expressions $8x^2 + 28x + 12$ and $(4x + 2)(2x + 6)$ are equivalent. Explain why the student's work is incorrect even though the expressions are equivalent.

82. A student tries to factor the polynomial $6x^2 + 28x - 10$. First, the student divides the polynomial by 2: $3x^2 + 14x - 5$. Then the student factors $3x^2 + 14x - 5$:

$$3x^2 + 14x - 5 = (3x - 1)(x + 5)$$

The student thinks that the factorization of $6x^2 + 28x - 10$ is $(3x - 1)(x + 5)$. What would you tell the student? Also, explain how using a graphing calculator table to check the work can help the student identify the error.

83. Give three examples of quadratic polynomials in which $2x - 3$ is a factor.

84. Give three examples of cubic polynomials in which $3x + 1$ is a factor.

Related Review

If the expression is not factored, then factor it. If the expression is factored, then find the product.

85. $3x^2 + 16x - 12$

86. $5x^2 - 29x + 20$

87. $(4x - 7)(3x - 1)$

88. $(2x - 9)(5x + 3)$

89. $(x - 3)(2x^2 + 3x - 5)$

90. $(3x^2 - x - 2)(2x + 3)$

91. $6x^3 + 10x^2 - 4x$

92. $6x^4 - 33x^3 + 36x^2$

Expressions, Equations, and Graphs

Perform the indicated instruction. Then use words such as lin-ear, quadratic, cubic, polynomial, degree, one variable, *and* two variables *to describe the expression, equation, or system.*

93. Graph $y = x^2 - 3$ by hand.

94. Solve the following:

$$7x + 2y = -6$$
$$3x - 4y = -22$$

95. Factor $x^2 - 2x - 3$.

96. Graph $7x + 2y = -6$ by hand.

97. Evaluate $x^2 - 2x - 3$ for $x = -5$.

98. Find the product $(7x + 2y)(3x - 4y)$.

8.4 SUMS AND DIFFERENCES OF CUBES; A FACTORING STRATEGY

Objectives

▸ Factor a sum or difference of two cubes.

▸ Know a factoring strategy.

In Section 8.1, we discussed how to factor a difference of two squares. Here, first we will discuss how to factor a sum or difference of two cubes. Then we will discuss how to sift through the many factoring techniques we have discussed in this chapter and select the best ones to completely factor a given polynomial.

The Sum or Difference of Two Cubes

To see how to factor the sum of two cubes, we begin by multiplying the expressions $A + B$ and $A^2 - AB + B^2$:

$$(A + B)(A^2 - AB + B^2) = A \cdot A^2 - A \cdot AB + A \cdot B^2 + B \cdot A^2 - B \cdot AB + B \cdot B^2$$
$$= A^3 - A^2B + AB^2 + A^2B - AB^2 + B^3$$
$$= A^3 + B^3$$

So, $(A + B)(A^2 - AB + B^2) = A^3 + B^3$. Note that the right-hand side of the equation, $A^3 + B^3$, is a sum of two cubes. By similar work, we can also find a property for the difference of two cubes.

Sum or Difference of Two Cubes

$$A^3 + B^3 = (A + B)(A^2 - AB + B^2) \qquad \text{Sum of two cubes}$$
$$A^3 - B^3 = (A - B)(A^2 + AB + B^2) \qquad \text{Difference of two cubes}$$

We can use these two properties to factor any polynomial that is a sum or difference of two cubes. In order to use the properties, it will help to memorize the following cubes:

$$2^3 = 8 \qquad 3^3 = 27 \qquad 4^3 = 64 \qquad 5^3 = 125 \qquad 10^3 = 1000$$

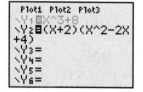

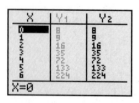

Figure 9 Verify the work

Example 1   Factoring a Sum and a Difference of Two Cubes

Factor.

1. $x^3 + 8$

2. $x^3 - 64$

Solution

1.

$$A^3 + B^3 = (A + B)(A^2 - AB + B^2)$$
$$\downarrow \quad \downarrow \qquad \downarrow \quad \downarrow \ \downarrow \quad \downarrow \downarrow \quad \downarrow$$
$$x^3 + 8 = x^3 + 2^3 = (x + 2)(x^2 - x \cdot 2 + 2^2) \qquad \text{Factor.}$$
$$= (x + 2)(x^2 - 2x + 4) \qquad \text{Simplify.}$$

The trinomial $x^2 - 2x + 4$ is prime, so we have completely factored $x^3 + 8$. We use a graphing calculator table to verify our work (see Fig. 9).

2.

$$A^3 - B^3 = (A - B)(A^2 + AB + B^2)$$
$$\downarrow \quad \downarrow \qquad \downarrow \quad \downarrow \ \downarrow \quad \downarrow \downarrow \quad \downarrow$$
$$x^3 - 64 = x^3 - 4^3 = (x - 4)(x^2 + x \cdot 4 + 4^2) \qquad \text{Factor.}$$
$$= (x - 4)(x^2 + 4x + 16) \qquad \text{Simplify.}$$

The trinomial $x^2 + 4x + 16$ is prime, so we have completely factored $x^3 - 64$.

Example 2   Factoring a Sum and a Difference of Two Cubes

Factor.

1. $27x^3 + 125$

2. $16p^3 - 2$

Solution

1.

$$27x^3 + 125 = (3x)^3 + 5^3 \qquad \text{Write as a sum of cubes.}$$
$$= (3x + 5)\big((3x)^2 - 3x \cdot 5 + 5^2\big) \qquad A^3 + B^3 = (A + B)(A^2 - AB + B^2)$$
$$= (3x + 5)(9x^2 - 15x + 25) \qquad \text{Simplify.}$$

The trinomial $9x^2 - 15x + 25$ is prime, so we have completely factored $27x^3 + 125$.

2. We begin by factoring out the GCF, 2:

$$16p^3 - 2 = 2(8p^3 - 1) \qquad \text{Factor out GCF, 2.}$$
$$= 2\big((2p)^3 - 1^3\big) \qquad \text{Write as a difference of cubes.}$$
$$= 2(2p - 1)\big((2p)^2 + 2p \cdot 1 + 1^2\big) \qquad A^3 - B^3 = (A - B)(A^2 + AB + B^2)$$
$$= 2(2p - 1)(4p^2 + 2p + 1) \qquad \text{Simplify.}$$

The trinomial $4p^2 + 2p + 1$ is prime, so we have completely factored $16p^3 - 2$. ■

Consider the properties for the sum of cubes and difference of cubes:

$$A^3 + B^3 = (A + B)\big(A^2 - AB + B^2\big) \qquad A^3 - B^3 = (A - B)\big(A^2 + AB + B^2\big)$$

Provided that we have first factored out the GCF (or its opposite), we can assume that the trinomials $A^2 - AB + B^2$ and $A^2 + AB + B^2$ are prime.

A Factoring Strategy

We will now discuss a five-step factoring strategy that will help us determine the best factoring techniques to use to completely factor a given polynomial.

Five-Step Factoring Strategy

The five steps that follow can be used to factor many polynomials. Steps 2–4 can be applied to the entire polynomial or to a factor of the polynomial.

1. If the leading coefficient is positive and the GCF is not 1, we factor out the GCF. If the leading coefficient is negative, we factor out the opposite of the GCF.

2. For a binomial, try using one of the properties for the difference of two squares, the sum of two cubes, or the difference of two cubes.

3. For a trinomial of the form $ax^2 + bx + c$,
   a. If $a = 1$, try to find two integers whose product is c and whose sum is b.
   b. If $a \neq 1$, try to factor by trial and error or by grouping.

4. For an expression with four terms, try factoring by grouping.

5. Continue applying steps 2–4 until the polynomial is completely factored.

Example 3   Factoring a Polynomial

Factor $x^3 - 7x^2y + 12xy^2$.

Solution

For $x^3 - 7x^2y + 12xy^2$, the GCF is x. First, we factor out x:

$$x^3 - 7x^2y + 12xy^2 = x\big(x^2 - 7xy + 12y^2\big)$$

Since the factor $x^2 - 7xy + 12y^2$ is a trinomial with a leading coefficient that is 1, we try to find two monomials whose product is $12y^2$ and whose sum is $-7y$. The monomials are $-3y$ and $-4y$, so we have

$$x^3 - 7x^2y + 12xy^2 = x(x^2 - 7xy + 12y^2) = x(x - 3y)(x - 4y)$$ ∎

Example 4   Factoring a Polynomial

Factor $32w^3 - 98w$.

Solution

For $32w^3 - 98w$, the GCF is $2w$. First, we factor out $2w$:

$$32w^3 - 98w = 2w(16w^2 - 49)$$

Since the factor $16w^2 - 49$ has two terms, we check to see whether it is the difference of two squares, which it is. So, we have

$$32w^3 - 98w = 2w(16w^2 - 49) = 2w(4w + 7)(4w - 7)$$ ∎

Example 5   Factoring a Polynomial

Factor $3x - 10 + 4x^2$.

Solution

First, we write $3x - 10 + 4x^2$ in descending order:

$$4x^2 + 3x - 10$$

Since $4x^2 + 3x - 10$ is a trinomial with a leading coefficient that is not 1, we try to factor it by grouping. Here, $a = 4$, $b = 3$, and $c = -10$.

Step 1. Find the product ac: $ac = 4(-10) = -40$.

Step 2. We want to find two numbers m and n that have the product $ac = -40$ and the sum $b = 3$. The integers are 8 and -5.

Step 3. We write $4x^2 + 3x - 10 = 4x^2 + 8x - 5x - 10$.

Step 4. We factor $4x^2 + 8x - 5x - 10$ by grouping:

$$4x^2 + 8x - 5x - 10 = 4x(x + 2) - 5(x + 2) \qquad \text{Factor both pairs of terms.}$$
$$= (4x - 5)(x + 2) \qquad \text{Factor out GCF, } x + 2.$$

So, $3x - 10 + 4x^2 = 4x^2 + 3x - 10 = (4x - 5)(x + 2)$.
Instead of using grouping, we could have used trial and error. ∎

Example 6   Factoring a Polynomial

Factor $12x^3 - 8x^2 - 27x + 18$.

Solution

Since $12x^3 - 8x^2 - 27x + 18$ has four terms, we try to factor it by grouping:

$$12x^3 - 8x^2 - 27x + 18 = 4x^2(3x - 2) - 9(3x - 2) \qquad \text{Factor both pairs of terms.}$$
$$= (4x^2 - 9)(3x - 2) \qquad \text{Factor out GCF, } 3x - 2.$$
$$= (2x + 3)(2x - 3)(3x - 2) \qquad A^2 - B^2 = (A + B)(A - B)$$

We use a graphing calculator table to verify our work (see Fig. 10). ∎

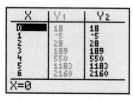

Figure 10 Verify the work

Example 7 Factoring a Polynomial

Factor $-5x^4 + 20x^3 - 30x^2$.

Solution

For $-5x^4 + 20x^3 - 30x^2$, the GCF is $5x^2$. Since the leading coefficient of the polynomial, -5, is negative, we factor out $-5x^2$:

$$-5x^4 + 20x^3 - 30x^2 = -5x^2(x^2 - 4x + 6)$$

Because the factor $x^2 - 4x + 6$ is a trinomial with a leading coefficient that is 1, we try to find two integers whose product is 6 and whose sum is -4. However, there is no such pair; the factor $x^2 - 4x + 6$ is prime. So, the completely factored result is $-5x^2(x^2 - 4x + 6)$. ∎

group exploration

Looking ahead: Zero factor property

1. What can you say about A or B if $AB = 0$?
2. What can you say about A or B if $A(B - 3) = 0$?
3. What can you say about x if $x(x - 3) = 0$?
4. Solve the equation $x^2 - 3x = 0$. [**Hint:** Does this have something to do with Problem 3?]
5. Solve $3x^2 - 6x = 0$.
6. Solve $x^2 - 7x + 10 = 0$.

TIPS FOR SUCCESS: Solutions Manual

If you are having trouble solving a homework exercise, it can help to refer to the Solutions Manual, which provides step-by-step solutions to the exercises. It is a nice complement to other forms of support, such as your instructor, a tutor, or friends, because you can refer to it at any time, day or night.

However, do not use the manual as a crutch. It is a common error to consult the manual before trying to complete an exercise without it. But if you have been trying to solve an exercise for a while and begin to feel frustrated, reach for the manual.

Once you have completed your homework assignment, assess how often you sought help from the manual. If you did so frequently, then continue to do more exercises until you can do any type of problem in the homework without referring to the manual. The ultimate goal is for you to understand the material, not just complete the assignment.

HOMEWORK 8.4   FOR EXTRA HELP ▶

Student Solutions Manual   PH Math/Tutor Center   *Math*XL MathXL®   *MyMathLab* MyMathLab

Factor. Use a graphing calculator table to verify your work when possible.

1. $x^3 + 27$
2. $x^3 + 64$
3. $x^3 + 125$
4. $x^3 + 1000$
5. $x^3 - 8$
6. $x^3 - 27$
7. $x^3 - 1$
8. $x^3 - 125$
9. $8t^3 + 27$
10. $27p^3 + 64$
11. $27x^3 - 8$
12. $8x^3 - 125$
13. $5x^3 + 40$
14. $2x^3 + 2$
15. $2x^3 - 54$

16. $3x^3 - 24$
17. $8x^3 + 27y^3$
18. $27m^3 + 64n^2$
19. $64a^3 - 27b^3$
20. $125p^3 - 8t^3$

Factor when possible. Use a graphing calculator table to verify your work when possible.

21. $x^2 - 64$
22. $x^2 - 1$
23. $m^2 + 11m + 28$
24. $a^2 + 5a - 14$

25. $2t^2 + 2t - 24$

26. $3a^2 - 21a + 30$

27. $x^2 + 49$

28. $x^2 + 4$

29. $-3x^2 + 24x - 45$

30. $-2x^2 + 22x - 60$

31. $1 + 15p^2 - 8p$

32. $-6t - 8 + 5t^2$

33. $x^2 - 2x + 1$

34. $x^2 - 8x + 16$

35. $24r^2 + 4r - 4$

36. $20y^3 - 12y^2 - 8y$

37. $-4ab^3 + 6a^2b^2$

38. $-6a^2b + 9ab^3$

39. $a^2 - ab - 20b^2$

40. $m^2 - 12mn + 32n^2$

41. $8x^3 - 20x^2 - 2x + 5$

42. $5x^3 - 2x^2 - 5x + 2$

43. $-12x^4 - 4x^3$

44. $-25x^4 + 35x^3$

45. $15a^4 + 25a^3 + 10a^2$

46. $6t^4 - 3t^3 - 30t^2$

47. $24 - 14x + x^2$

48. $9 - 6x + x^2$

49. $2w^4 + 4w^3 - 8w^2$

50. $3p^4 - 9p^3 + 24p^2$

51. $12x^4 - 27x^2$

52. $98x^4 - 18x^2$

53. $x^2 + 10x + 25$

54. $x^2 + 4x + 4$

55. $2x^2 + x - 21$

56. $4x^2 + 20x + 21$

57. $m^3 - 13m^2n + 36mn^2$

58. $p^2q + 5pq^2 - 12q^3$

59. $4x - 5 + 2x^2$

60. $7x - 4 + 3x^2$

61. $100x^2 - 9y^2$

62. $49b^2 - 36c^2$

63. $4x^2 + 12x + 9$

64. $16x^2 - 8x + 1$

65. $18x^3 + 27x^2 - 8x - 12$

66. $12x^3 + 9x^2 + 8x + 6$

67. $3a^3 - 10a^2b + 8ab^2$

68. $5p^2t + 18pt^2 - 8t^3$

69. $x^2 - 9x - 20$

70. $x^2 - 13x - 30$

71. $x^3 - 1000$

72. $x^3 + 1$

73. $64a^3 + 27$

74. $27r^3 + 8$

75. $3x^3 + 24$

76. $7x^3 - 7$

For Exercises 77–80, discuss which technique(s) you should consider using when factoring the given type of polynomial. Also, refer to an exercise in Homework 8.4 that illustrates this technique.

77. binomial

78. trinomial of the form $x^2 + bx + c$

79. trinomial of the form $ax^2 + bx + c$, where $a \neq 1$

80. polynomial with four terms

81. A student tries to factor the polynomial $x^3 - 27$:

$$x^3 - 27 = (x - 3)(x^2 + 6x + 9)$$

Describe any errors. Then factor the polynomial correctly.

82. A student tries to factor the polynomial $x^3 + 8$:

$$x^3 + 8 = (x + 2)(x^2 + 2x + 4)$$

Describe any errors. Then factor the polynomial correctly.

83. Is the following statement true? Explain.

$$A^3 - B^3 = (A - B)(A^2 + 2AB + B^2)$$

84. When factoring a polynomial with a negative leading coefficient, what should you do? Give an example.

85. When factoring a polynomial, does it matter whether you factor out the GCF in the first step or in the last step? Explain.

86. Describe a factoring strategy in your own words.

Related Review

If the expression is not factored, then factor it. If the expression is factored, then find the product.

87. $(5x - 7)(5x + 7)$

88. $(2x + 5)(x^2 - x + 4)$

89. $81x^2 - 16$

90. $x^2 - 5x - 36$

91. $3x^3 + 9x^2 - 12x$

92. $30x^2 - 24$

93. $-2x(7x^2 - 5x + 1)$

94. $(4x - 3)(2x + 9)$

Expressions, Equations, and Graphs

Perform the indicated instruction. Then use words such as linear, quadratic, cubic, polynomial, degree, one variable, and two variables to describe the expression, equation, or system.

95. Graph $y = 4x - 5$ by hand.

96. Factor $x^3 - 3x^2 - 4x + 12$.

97. Solve $4x - 5 = 3x + 2$.

98. Simplify $(2x - 7)^2$.

99. Find the product $(4x - 5)(3x + 2)$.

100. Graph $y = 3x^2$ by hand.

8.5 SOLVING POLYNOMIAL EQUATIONS

Objectives

▷ Know the *zero factor property*.

▷ Use factoring to solve *quadratic equations in one variable*.

▷ Use graphing to solve a polynomial equation in one variable.

▷ Know how many real-number solutions a quadratic equation in one variable can have.

▷ Combine like terms to help solve quadratic equations in one variable.

▷ Solve quadratic equations in one variable that contain fractions.

▷ Find the *x*-intercept(s) of the graph of a polynomial equation in two variables.

▷ Use factoring to solve *cubic equations in one variable*.

▷ Compare solving polynomial equations with factoring polynomials.

In this section, we will discuss how to use factoring to help solve equations. In Section 8.6, we will use this skill to make efficient predictions with some quadratic models.

Zero Factor Property

In Section 7.1, we graphed quadratic equations in *two* variables. In this section, we will solve quadratic equations in *one* variable, such as

$$x^2 - 4x - 21 = 0 \qquad 8x^2 - 50 = 0 \qquad 4x^2 = 20x - 25$$

A **quadratic equation in one variable** is an equation that can be put into the form

$$ax^2 + bx + c = 0$$

where a, b, and c are constants and $a \neq 0$. The connection between solving a quadratic equation and factoring an expression lies in the *zero factor property*.

Zero Factor Property

Let A and B be real numbers:

$$\text{If } AB = 0, \text{ then } A = 0 \text{ or } B = 0.$$

In words: If the product of two numbers is zero, then at least one of the numbers must be zero.

Solving Quadratic Equations in One Variable

In the next several examples, we will use the zero factor property to solve some quadratic equations in one variable.

Example 1 Solving a Quadratic Equation

Solve $(x - 3)(x - 8) = 0$.

Solution

$$
\begin{aligned}
(x - 3)(x - 8) &= 0 && \text{Original equation} \\
x - 3 = 0 \quad &\text{or} \quad x - 8 = 0 && \text{Zero factor property} \\
x = 3 \quad &\text{or} \quad x = 8 && \text{Add 3 to both sides./Add 8 to both sides.}
\end{aligned}
$$

So, the solutions are 3 and 8. We check that both 3 and 8 satisfy the original equation:

Check $x = 3$	**Check $x = 8$**
$(x - 3)(x - 8) = 0$	$(x - 3)(x - 8) = 0$
$(3 - 3)(3 - 8) \overset{?}{=} 0$	$(8 - 3)(8 - 8) \overset{?}{=} 0$
$0(-5) \overset{?}{=} 0$	$5(0) \overset{?}{=} 0$
$0 \overset{?}{=} 0$	$0 \overset{?}{=} 0$
true	true

Example 2 Solving a Quadratic Equation by Factoring

Solve $x^2 - 4x - 21 = 0$.

Solution

The zero factor property tells us about the *product* of two numbers that equals zero. So, to solve $x^2 - 4x - 21 = 0$ we begin by factoring the left side:

$$
\begin{aligned}
x^2 - 4x - 21 &= 0 && \text{Original equation} \\
(x + 3)(x - 7) &= 0 && \text{Factor left side.} \\
x + 3 = 0 \quad &\text{or} \quad x - 7 = 0 && \text{Zero factor property} \\
x = -3 \quad &\text{or} \quad x = 7 && \text{Subtract 3 from both sides./Add 7 to both sides.}
\end{aligned}
$$

So, the solutions are -3 and 7. We check that both -3 and 7 satisfy the original equation:

$$\textbf{Check } x = -3 \qquad\qquad \textbf{Check } x = 7$$

$$x^2 - 4x - 21 = 0 \qquad\qquad x^2 - 4x - 21 = 0$$

$$(-3)^2 - 4(-3) - 21 \stackrel{?}{=} 0 \qquad\qquad 7^2 - 4(7) - 21 \stackrel{?}{=} 0$$

$$9 + 12 - 21 \stackrel{?}{=} 0 \qquad\qquad 49 - 28 - 21 \stackrel{?}{=} 0$$

$$0 \stackrel{?}{=} 0 \qquad\qquad 0 \stackrel{?}{=} 0$$

$$\text{true} \qquad\qquad\qquad \text{true} \qquad\qquad \blacksquare$$

In Examples 1 and 2, each original equation has one side that is 0. If neither side of a quadratic equation in one variable is 0, we must first use properties of equality to get one side to be 0 and then apply the zero factor property.

Example 3   Solving a Quadratic Equation by Factoring

Solve $4x^2 = 20x - 25$.

Solution

$$\begin{array}{ll}
4x^2 = 20x - 25 & \text{Original equation} \\
4x^2 - 20x + 25 = 0 & \text{Write in } ax^2 + bx + c = 0 \text{ form.} \\
(2x - 5)(2x - 5) = 0 & \text{Factor left side.} \\
2x - 5 = 0 & \text{Zero factor property} \\
2x = 5 & \text{Add 5 to both sides.} \\
x = \dfrac{5}{2} & \text{Divide both sides by 2.}
\end{array}$$

So, the solution is $\dfrac{5}{2}$. We use a graphing calculator table to check that if $\dfrac{5}{2}$ is substituted for x in the expressions $4x^2$ and $20x - 25$, the two results are equal (see Fig. 11).

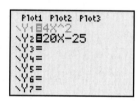

Figure 11 Verify the work $\blacksquare$

Example 4   Solving a Quadratic Equation by Factoring

Solve $12w^2 - 75 = 0$.

Solution

$$\begin{array}{ll}
12w^2 - 75 = 0 & \text{Original equation} \\
3(4w^2 - 25) = 0 & \text{Factor out GCF, 3.} \\
3(2w + 5)(2w - 5) = 0 & \text{Completely factor left side.}
\end{array}$$

Now we can apply a variation on the zero factor property: If $3AB = 0$, then $A = 0$ or $B = 0$. Here we take $(2w + 5)$ to be A and $(2w - 5)$ to be B:

$$\begin{array}{lll}
2w + 5 = 0 & \text{or} \quad 2w - 5 = 0 & \text{Zero factor property} \\
2w = -5 & \text{or} \quad 2w = 5 & \\
w = -\dfrac{5}{2} & \text{or} \quad w = \dfrac{5}{2} &
\end{array}$$

So, the solutions are $-\dfrac{5}{2}$ and $\dfrac{5}{2}$. We can check that the solutions satisfy the original equation. $\blacksquare$

Example 5 Solving a Quadratic Equation by Factoring

Solve $6x^2 = 18x$.

Solution

$$6x^2 = 18x \qquad \text{Original equation}$$
$$6x^2 - 18x = 0 \qquad \text{Write in } ax^2 + bx = 0 \text{ form.}$$
$$6x(x - 3) = 0 \qquad \text{Factor left side.}$$
$$6x = 0 \quad \text{or} \quad x - 3 = 0 \qquad \text{Zero factor property}$$
$$x = 0 \quad \text{or} \quad x = 3$$

So, the solutions are 0 and 3. We can check that the solutions satisfy the original equation. (Try it.) ∎

WARNING   In Example 5, we solved the equation $6x^2 = 18x$ by first writing it in the form $ax^2 + bx = 0$ (there is no c term) and then factoring the left side of the equation. It is a common error to try to solve the equation $6x^2 = 18x$ by dividing both sides by $6x$ to get the result $x = 3$. But note that this incorrect method gives only one of the two solutions 0 and 3.

In general, when solving quadratic equations, we can't divide by x because we don't know if the value of x is zero. Recall from Section 2.2 that division by zero is undefined.

Solving a Quadratic Equation by Graphing

We can solve quadratic equations by graphing in much the same way that we solved linear equations by graphing in Sections 4.3 and 4.4.

Example 6 Solving a Quadratic Equation in One Variable by Graphing

The graphs of $y = x^2 - 10x + 28$, $y = 7$, $y = 3$, and $y = 1$ are shown in Fig. 12. Use these graphs to solve the given equation.

1. $x^2 - 10x + 28 = 7$ **2.** $x^2 - 10x + 28 = 3$ **3.** $x^2 - 10x + 28 = 1$

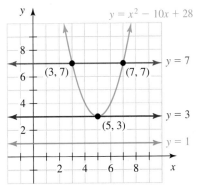

Figure 12 Solving three quadratic equations in one variable

Solution

1. The graphs of $y = x^2 - 10x + 28$ and $y = 7$ intersect only at the points $(3, 7)$ and $(7, 7)$, which have x-coordinates 3 and 7. So, 3 and 7 are the solutions of $x^2 - 10x + 28 = 7$.
2. The graphs of $y = x^2 - 10x + 28$ and $y = 3$ intersect only at the point $(5, 3)$, which has x-coordinate 5. So, 5 is the solution of $x^2 - 10x + 28 = 3$.
3. The graphs of $y = x^2 - 10x + 28$ and $y = 1$ do not intersect. So, no real number is a solution of $x^2 - 10x + 28 = 1$. ∎

Our results in Example 6 suggest how many real-number solutions a quadratic equation in one variable can have.

Number of Solutions of a Quadratic Equation in One Variable

A quadratic equation in one variable has either two real-number solutions, one real-number solution, or no real-number solutions.

In this section, we will not solve quadratic equations that have no real-number solutions.

Combining Like Terms to Help Solve a Quadratic Equation in One Variable

After getting one side of an equation equal to zero, it is sometimes necessary to combine like terms to write the equation in $ax^2 + bx + c = 0$ form.

Example 7 Solving a Quadratic Equation by Factoring

Solve $12x^2 - 31x = 3x - 24$.

Solution

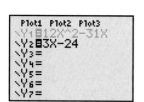

$$12x^2 - 31x = 3x - 24 \qquad \text{Original equation}$$
$$12x^2 - 31x - 3x + 24 = 3x - 24 - 3x + 24 \qquad \begin{array}{l}\text{Subtract } 3x \text{ from both sides;}\\ \text{add } 24 \text{ to both sides.}\end{array}$$
$$12x^2 - 34x + 24 = 0 \qquad \text{Combine like terms.}$$
$$6x^2 - 17x + 12 = 0 \qquad \text{Divide both sides by 2.}$$
$$(2x - 3)(3x - 4) = 0 \qquad \text{Factor left side.}$$
$$2x - 3 = 0 \quad \text{or} \quad 3x - 4 = 0 \qquad \text{Zero factor property}$$
$$2x = 3 \quad \text{or} \quad 3x = 4$$
$$x = \frac{3}{2} \quad \text{or} \quad x = \frac{4}{3}$$

Figure 13 Verify the work

So, the solutions are $\frac{3}{2}$ and $\frac{4}{3}$. For each solution, we check that if it is substituted for x in the expressions $12x^2 - 31x$ and $3x - 24$, the two results are equal (see Fig. 13). ■

Example 8 Solving a Quadratic Equation by Factoring

Solve $(x - 2)(x + 5) = 8$.

Solution

Although the left-hand side of $(x - 2)(x + 5) = 8$ is factored, the right-hand side is not zero. First, we multiply the left-hand side and then write the equation in the form $ax^2 + bx + c = 0$:

$$(x - 2)(x + 5) = 8 \qquad \text{Original equation}$$
$$x^2 + 3x - 10 = 8 \qquad \text{Multiply left-hand side.}$$
$$x^2 + 3x - 18 = 0 \qquad \text{Write in } ax^2 + bx + c = 0 \text{ form.}$$
$$(x + 6)(x - 3) = 0 \qquad \text{Factor left-hand side.}$$
$$x + 6 = 0 \quad \text{or} \quad x - 3 = 0 \qquad \text{Zero factor property}$$
$$x = -6 \quad \text{or} \quad x = 3$$

The solutions are -6 and 3. Recall from Section 4.3 that we can use "intersect" on a graphing calculator to solve an equation in one variable. In Fig. 14, we graph the equations $y = (x - 2)(x + 5)$ and $y = 8$ and find the intersection points $(-6, 8)$ and $(3, 8)$, which have x-coordinates -6 and 3, respectively. So, the solutions of the original equation are -6 and 3, which checks.

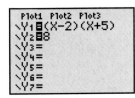
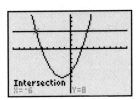
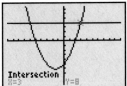

Figure 14 Verify the work ■

Quadratic Equations That Contain Fractions

Recall from Section 4.4 that if an equation contains fractions, we multiply both sides of the equation by the least common denominator (LCD) of all the fractions.

Example 9 Solving a Quadratic Equation That Contains Fractions

Solve $\dfrac{1}{2}x^2 + 4 = \dfrac{11}{3}x$.

Solution

To find the LCD of the two fractions in the equation, we list the multiples of 2 and the multiples of 3:

$$\text{Multiples of 2: } 2, 4, 6, 8, 10, \ldots$$
$$\text{Multiples of 3: } 3, 6, 9, 12, 15, \ldots$$

The LCD is 6. To clear the equation of fractions, we multiply both sides of the equation by the LCD, 6:

$$\frac{1}{2}x^2 + 4 = \frac{11}{3}x \qquad \text{Original equation}$$
$$6 \cdot \frac{1}{2}x^2 + 6 \cdot 4 = 6 \cdot \frac{11}{3}x \qquad \text{Multiply both sides by LCD, 6.}$$
$$3x^2 + 24 = 22x \qquad \text{Simplify.}$$
$$3x^2 - 22x + 24 = 0 \qquad \text{Write in } ax^2 + bx + c = 0 \text{ form.}$$
$$(3x - 4)(x - 6) = 0 \qquad \text{Factor left side.}$$
$$3x - 4 = 0 \quad \text{or} \quad x - 6 = 0 \qquad \text{Zero factor property}$$
$$3x = 4 \quad \text{or} \quad x = 6$$
$$x = \frac{4}{3} \quad \text{or} \quad x = 6$$

So, the solutions are $\dfrac{4}{3}$ and 6. To verify our work, we use ZStandard followed by ZoomFit to graph the equations $y = \dfrac{1}{2}x^2 + 4$ and $y = \dfrac{11}{3}x$ and then use "intersect" to find the approximate intersection points $(1.33, 4.89)$ and $(6, 22)$, which have x-coordinates 1.33 and 6 (see Fig. 15). So, the approximate solutions of the original equation are 1.33 and 6, which checks.

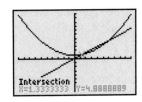
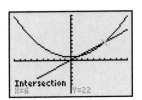

Figure 15 Verify the work ■

Finding *x*-intercept(s) of a Parabola

Recall from Section 5.1 that for an equation containing the variables x and y, we can find the x-intercept(s) of the graph by substituting 0 for y and then solving for x. In Example 10, we use this strategy to find the x-intercepts of a parabola.

Example 10 Finding *x*-Intercepts

Find the x-intercepts of the parabola $y = x^2 - 2x - 8$.

Solution

We substitute 0 for y and solve for x:

$$0 = x^2 - 2x - 8 \qquad \text{Substitute 0 for } y.$$
$$0 = (x + 2)(x - 4) \qquad \text{Factor right-hand side.}$$
$$x + 2 = 0 \quad \text{or} \quad x - 4 = 0 \qquad \text{Zero factor property}$$
$$x = -2 \quad \text{or} \quad x = 4$$

So, the x-intercepts are $(-2, 0)$ and $(4, 0)$. We use "zero" on a graphing calculator to verify our work (see Fig. 16). See Section A.19 for graphing calculator instructions. ■

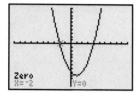

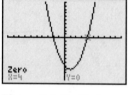

Figure 16 Verify the work

Cubic Equations in One Variable

So far in this section, we have solved quadratic equations. We will now solve some cubic equations, such as

$$3x^3 - 21x^2 + 30x = 0 \qquad x^3 - 9x = 4x^2 - 36$$

A **cubic equation in one variable** is an equation that can be put into the form

$$ax^3 + bx^2 + cx + d = 0$$

where a, b, c, and d are constants and $a \neq 0$. We can solve some cubic equations by using the zero factor property for three factors:

$$\text{If } ABC = 0, \text{ then } A = 0, B = 0, \text{ or } C = 0.$$

Example 11 Solving a Cubic Equation by Factoring

Solve $3x^3 - 21x^2 + 30x = 0$.

Solution

$$3x^3 - 21x^2 + 30x = 0 \qquad \text{Original equation}$$
$$3x(x^2 - 7x + 10) = 0 \qquad \text{Factor out GCF, } 3x.$$
$$3x(x - 2)(x - 5) = 0 \qquad \text{Factor } x^2 - 7x + 10.$$
$$3x = 0 \quad \text{or} \quad x - 2 = 0 \quad \text{or} \quad x - 5 = 0 \qquad \text{Zero factor property}$$
$$x = 0 \quad \text{or} \quad x = 2 \quad \text{or} \quad x = 5$$

So, the solutions are 0, 2, and 5. We can check that the solutions satisfy the original equation. (Try it.) ■

Note that solving a cubic equation in one variable is similar to solving a quadratic equation in one variable. We now summarize the steps used to solve either type of equation.

Solving Quadratic and Cubic Equations in One Variable

If an equation can be solved by factoring, we solve it by the following steps:

1. Write the equation so that one side of it is equal to zero.
2. Factor the nonzero side of the equation.
3. Apply the zero factor property.
4. Solve each equation that results from using the zero factor property.

Example 12   Solving a Cubic Equation by Factoring

Solve $x^3 - 9x = 4x^2 - 36$.

Solution

$$
\begin{aligned}
x^3 - 9x &= 4x^2 - 36 && \text{Original equation}\\
x^3 - 4x^2 - 9x + 36 &= 0 && \text{Write in } ax^3 + bx^2 + cx + d = 0 \text{ form.}\\
x^2(x-4) - 9(x-4) &= 0 && \text{Factor both pairs of terms.}\\
(x^2 - 9)(x-4) &= 0 && \text{Factor out GCF, } (x-4).\\
(x+3)(x-3)(x-4) &= 0 && A^2 - B^2 = (A+B)(A-B)
\end{aligned}
$$

$x+3=0$ or $x-3=0$ or $x-4=0$ Zero factor property

$x=-3$ or $x=3$ or $x=4$

So, the solutions are -3, 3, and 4. We can check that the solutions satisfy the original equation. (Try it.) ■

WARNING   It is a common error to try to apply the zero factor property to an equation such as $x^2(x-4) - 9(x-4) = 0$ and incorrectly conclude that the solutions are 0 and 4. However, the expression $x^2(x-4) - 9(x-4)$ is *not* factored, because it is a difference, not a product. Only after we factor the left side of the equation $x^2(x-4) - 9(x-4) = 0$ can we apply the zero factor property.

Solving Polynomial Equations versus Factoring Polynomials

In Section 4.5, we compared solving a linear equation with simplifying a linear expression. **When we solve any equation, our objective is to find all *numbers* that satisfy the equation. When we factor a polynomial, our objective is to write the polynomial as a product of polynomials, which is an *expression*.**

Here we compare solving a polynomial equation with factoring a polynomial:

Solving the Equation $x^2 - 5x + 6 = 0$

$$
\begin{aligned}
x^2 - 5x + 6 &= 0 && \text{Original equation}\\
(x-2)(x-3) &= 0 && \text{Factor left side.}\\
x-2=0 \ \text{ or } \ x-3 &= 0 && \text{Zero factor property}\\
x=2 \ \text{ or } \ x &= 3
\end{aligned}
$$

The solutions, 2 and 3, are numbers.

Factoring the Expression $x^2 - 5x + 6$

$$x^2 - 5x + 6 = (x-2)(x-3)$$

The result, $(x-2)(x-3)$, is an expression.

group exploration

Solving systems of quadratic equations

1. Use "intersect" on a graphing calculator to solve the system

$$y = x^2 - 6x$$
$$y = x - 10$$

2. Now solve the system by substitution.

3. Solve the equation $x^2 - 6x = x - 10$.

4. Compare your results in Problem 3 with the x-coordinates of the results in Problem 2.

5. Use "intersect" to help you solve $x^2 - 2x = x + 4$. Then solve this equation without using graphing. Compare your results.

6. Explain why you cannot solve the equation $x^2 + 3x = x + 5$ by using factoring. Then use "intersect" to find approximate solutions of the equation.

7. Summarize what you have learned in this exploration. Explain how you can use "intersect" to verify work, as well as to solve equations that you cannot solve by factoring.

HOMEWORK 8.5

FOR EXTRA HELP ▶

Student Solutions Manual   PH Math/Tutor Center   Math XL   MyMathLab
MathXL®   MyMathLab

Solve. Verify any results by checking that they satisfy the equation.

1. $(x - 2)(x - 7) = 0$ **2.** $(x + 3)(x - 1) = 0$

3. $x^2 + 7x + 10 = 0$ **4.** $x^2 + 5x + 6 = 0$

5. $w^2 + 3w - 28 = 0$ **6.** $t^2 - 4t - 45 = 0$

7. $x^2 - 14x + 49 = 0$ **8.** $x^2 - 2x + 1 = 0$

9. $r^2 + 3r = 4$ **10.** $p^2 + 2p = 24$

11. $5x = x^2 + 6$ **12.** $9x = x^2 + 14$

13. $k^2 - 64 = 0$ **14.** $y^2 - 9 = 0$

15. $36x^2 - 49 = 0$ **16.** $4x^2 - 25 = 0$

17. $x^2 = 49$ **18.** $x^2 = 4$

19. $w^2 = w$ **20.** $r^2 = -r$

Solve. Use a graphing calculator to verify your work.

21. $(3x - 7)(2x + 5) = 0$ **22.** $(4x + 3)(9x - 1) = 0$

23. $4x^2 - 8x + 3 = 0$ **24.** $3x^2 - 11x + 10 = 0$

25. $10k^2 + 3k - 4 = 0$ **26.** $6p^2 - 5p - 6 = 0$

27. $25x^2 + 20x + 4 = 0$ **28.** $9p^2 - 6p + 1 = 0$

29. $x^2 - 2x = 12 - 6x$ **30.** $x^2 - 10x = 6x - 15$

31. $3t^2 - 4t = 6t - 8$ **32.** $2r^2 + 5r = -15 - 6r$

33. $(x - 2)(x - 5) = 4$ **34.** $(x - 3)(x - 1) = 8$

35. $4x(x + 1) = 15$ **36.** $-9x(x - 1) = 2$

37. $(2w - 3)(w - 2) = 3$ **38.** $(3y + 4)(y - 1) = -2$

39. $(x - 2)^2 - 3x = -6$ **40.** $(x + 3)^2 + 2x = -3$

Solve. Use a graphing calculator to verify your work.

41. $\frac{1}{4}x^2 + 2x + 3 = 0$ **42.** $\frac{1}{5}x^2 + \frac{6}{5}x + 1 = 0$

43. $\frac{3}{8}m^2 + m - 2 = 0$ **44.** $\frac{5}{2}t^2 + 9t - 4 = 0$

45. $-\frac{1}{3}x^2 + \frac{1}{3}x + 10 = 6$ **46.** $-\frac{1}{2}x^2 - \frac{5}{2}x + 12 = 5$

For Exercises 47–50, use the graph of $y = x^2 - 6x + 5$ shown in Fig. 17 to solve the given equation.

47. $x^2 - 6x + 5 = -3$ **48.** $x^2 - 6x + 5 = 0$

49. $x^2 - 6x + 5 = -4$ **50.** $x^2 - 6x + 5 = -5$

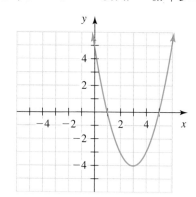

Figure 17 Exercises 47–50

For Exercises 51–54, use the graph of $y = -x^2 - 4x - 1$ shown in Fig. 18 to solve the given equation.

51. $-x^2 - 4x - 1 = 2$

52. $-x^2 - 4x - 1 = -1$

53. $-x^2 - 4x - 1 = 4$

54. $-x^2 - 4x - 1 = 3$

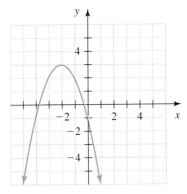

Figure 18 Exercises 51–54

Use "intersect" on a graphing calculator to solve the equation. Round any solutions to the second decimal place.

55. $x^2 - 5x = -2x + 3$ **56.** $x^2 + 3x = x + 5$

57. $-x^2 - 2x + 4 = -x - 5$

58. $-2x^2 + 4x + 8 = 2x - 1$

For Exercises 59–62, solve the given equation by referring to the solutions of $y = 2x^2 - 8x + 10$ shown in Table 2.

59. $2x^2 - 8x + 10 = 4$ **60.** $2x^2 - 8x + 10 = 10$

61. $2x^2 - 8x + 10 = 2$ **62.** $2x^2 - 8x + 10 = -3$

Table 2 Exercises 59-62

x	y
−1	20
0	10
1	4
2	2
3	4
4	10
5	20

Find all x-intercepts. Verify your work by using a graphing calculator.

63. $y = x^2 - 3x - 28$ **64.** $y = x^2 - 9x + 18$

65. $y = x^2 - 8x + 16$ **66.** $y = x^2 + 6x + 9$

67. $y = 2x^2 - x - 3$ **68.** $y = 3x^2 + x - 10$

Solve. Use a graphing calculator to verify your work.

69. $2x^3 - 5x^2 - 18x + 45 = 0$

70. $x^3 - x^2 - x + 1 = 0$

71. $3x^3 - 4x^2 - 12x + 16 = 0$

72. $8x^3 + 12x^2 - 18x - 27 = 0$

73. $18y^3 - 27y^2 = 8y - 12$

74. $12w^3 - 75w = 50 - 8w^2$

Solve. Use a graphing calculator to verify your work.

75. $5x^2 + 10x - 40 = 0$

76. $4x^2 + 16x - 20 = 0$

77. $2x^3 - 50x = 0$

78. $3x^3 - 12x = 0$

79. $3p^3 + 15p^2 = -18p$

80. $3r^3 + 27r^2 = -60r$

81. $2x^3 = 4x^2 + 16x$

82. $2x^3 = 4x - 2x^2$

83. A student tries to solve the equation $4x^2 = 8x$:

$$4x^2 = 8x$$
$$\frac{4x^2}{4x} = \frac{8x}{4x}$$
$$x = 2$$

Describe any errors. Then solve the equation correctly.

84. A student tries to solve the equation $x^2 = 25$:

$$x^2 = 25$$
$$x = 5$$

Describe any errors. Then solve the equation correctly.

85. A student tries to solve the equation $2x^2 - 26x + 80 = 0$:

$$2x^2 - 26x + 80 = 0$$
$$2(x^2 - 13x + 40) = 0$$
$$2(x - 5)(x - 8) = 0$$
$$x = 2, \ x = 5, \text{ or } x = 8$$

Describe any errors. Then solve the equation correctly.

86. A student tries to solve the equation $x^3 + 3x^2 - 4x - 12 = 0$:

$$x^3 + 3x^2 - 4x - 12 = 0$$
$$x^2(x + 3) - 4(x + 3) = 0$$
$$x + 3 = 0$$
$$x = -3$$

Describe any errors. Then solve the equation correctly.

87. Give an example of a quadratic equation in one variable whose solutions are 2 and 5.

88. Give an example of a quadratic equation in one variable whose solutions are −4 and 1.

89. Consider the equation $y = x^2 - 6x + 8$.
  a. Find x when $y = 3$.
  b. Find x when $y = -1$.
  c. Use your results from parts (a) and (b) to help you graph the equation $y = x^2 - 6x + 8$ by hand. [**Hint:** Your work in part (b) should help you determine the vertex of the graph. (Why?)]

90. Here is the work done in solving the equation $x^2 = 5x - 6$:

$$x^2 = 5x - 6$$
$$x^2 - 5x + 6 = 0$$
$$(x - 2)(x - 3) = 0$$
$$x - 2 = 0 \quad \text{or} \quad x - 3 = 0$$
$$x = 2 \quad \text{or} \quad x = 3$$

Show that *each line* of the work becomes a true statement when 2 or 3 is substituted for x.

91. Compare factoring a quadratic polynomial with solving a quadratic equation. (See page 9 for guidelines on writing a good response.)

92. Explain in your own words how to solve a quadratic equation in one variable. (See page 9 for guidelines on writing a good response.)

Related Review

Solve.

93. $x^2 + 3x = 7x + 12$ **94.** $3x(x - 2) = 24$

95. $x + 3 = 7x + 12$ **96.** $3(x - 2) = 24$

97. Solve the formula $A = P + PRT$ for P. [**Hint:** Begin by factoring the right-hand side of the formula.]

98. Solve the formula $S = 2WL + 2WH + 2LH$ for L. [**Hint:** You will need to use factoring eventually.]

Expressions, Equations, and Graphs

Perform the indicated instruction. Then use words such as linear, quadratic, cubic, polynomial, degree, one variable, *and* two variables *to describe the expression, equation, or system.*

99. Factor $x^2 - 4x - 5$. **100.** Factor $x^2 - 4$.

101. Solve $x^2 - 4x - 5 = 0$. **102.** Solve $x^2 - 4 = 0$.

103. Graph $y = x^2 - 4x - 5$ by hand.

104. Graph $y = x^2 - 4$ by hand.

8.6 USING FACTORING TO MAKE PREDICTIONS WITH QUADRATIC MODELS

Objectives

▷ Use an equation of a quadratic model to make estimates and predictions.

▷ Model projectile motion and the area of rectangular objects.

In Section 7.2, we used the *graph* of a quadratic model to make predictions. In this section, we will use an *equation* of such a model to make similar predictions.

Using Quadratic Equations to Make Predictions

In Example 1, we will check that a specific quadratic equation models a situation well and then use the model to make estimates and a prediction.

Example 1 Making Predictions

Table 3 National Defense Funding for Research and Development

Year	Defense Funding for Research and Development (billions of dollars)
1989	47
1993	42
1997	38
1999	39
2001	40

Source: *Congressional Budget Office*

Defense funding for research and development (to create innovations for future wars) in the United States declined in the early 1990s, but has increased since 1997 (see Table 3). Let D be U.S. defense funding (in billions of dollars) for the year that is t years since 1989. A model of the situation is

$$D = \frac{1}{9}t^2 - 2t + 48$$

1. Use a graphing calculator to draw the graph of the model and, in the same viewing window, the scattergram of the data. Does the model fit the data well?
2. Estimate when U.S. defense funding for research and development was $43 billion.
3. Use a graphing calculator to find the vertex of the model. What does the vertex mean in this situation?
4. Estimate U.S. defense funding for research and development for 2005. The actual amount was approximately $66 billion. Compute the error in the estimate.

Solution

1. The graph of the model and the scattergram of the data are shown in Fig. 19. The model appears to fit the data fairly well.

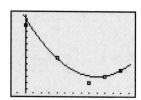

Figure 19 Check the fit

2. To estimate when the funding was \$43 billion, we substitute the output 43 for D in the equation $D = \frac{1}{9}t^2 - 2t + 48$ and solve for t:

$$43 = \frac{1}{9}t^2 - 2t + 48 \qquad \text{Substitute 43 for } D.$$

$$0 = \frac{1}{9}t^2 - 2t + 5 \qquad \text{Write in } 0 = ax^2 + bx + c \text{ form.}$$

$$0 = t^2 - 18t + 45 \qquad \text{Multiply both sides by LCD, 9.}$$

$$0 = (t - 3)(t - 15) \qquad \text{Factor right-hand side.}$$

$$t - 3 = 0 \quad \text{or} \quad t - 15 = 0 \qquad \text{Zero factor property}$$

$$t = 3 \quad \text{or} \qquad t = 15$$

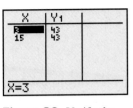

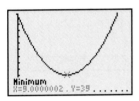

Figure 20 Verify the work

So, U.S. defense funding for research and development was \$43 billion in both 1992 and 2004, according to the model. We can verify our work by using a graphing calculator table (see Fig. 20).

3. Since the parabola opens upward, we use "minimum" on a graphing calculator to find the approximate vertex (see Fig. 21). See Section A.18 for graphing calculator instructions.

The vertex is about $(9, 39)$. This means that U.S. defense funding for research and development was a minimum in 1998 (\$39 billion), according to the model. However, in reality, \$39 billion was not the minimum funding; we see from Table 3 that defense funding in 1997 was lower (\$38 billion).

Figure 21 Finding the approximate vertex

4. We represent the year 2005 as $t = 2005 - 1989 = 16$. We substitute the input 16 for t in $D = \frac{1}{9}t^2 - 2t + 48$:

$$D = \frac{1}{9}(16)^2 - 2(16) + 48 \approx 44$$

So, U.S. defense funding for research and development in 2005 was \$44 billion, according to the model. The error is $66 - 44 = 22$ billion dollars. It is likely that defense funding for research and development has increased considerably due to fighting the Iraq War, fighting terrorism, and having greater resources because of an improving economy.   ■

In making a prediction about the dependent variable of a quadratic model, substitute a chosen value for the independent variable in the model. Then solve for the dependent variable.

In making a prediction about the independent variable of a quadratic model, substitute a chosen value for the dependent variable in the model. Then solve for the independent variable, which involves writing the equation so that one side of it is zero and then factoring the nonzero side.

On the basis of my assumption of the initial speed of the ball, I've got 6 seconds to get into position...

Projectile Motion

Any thrown object is called a *projectile*. If you launch a projectile such as a baseball or stone into the air, the relationship between time and the height of the projectile can be described well by a quadratic model. Such a model will not work well for an object that encounters a lot of air resistance—a feather, for example.

Example 2   Making Estimates

A batter hits a baseball into the air. The height h (in feet) of the baseball at t seconds after the ball is hit is given by

$$h = -16t^2 + 96t + 4$$

1. When is the baseball at a height of 84 feet? Explain why there are two times.
2. When is the baseball at a height of 148 feet? Explain why there is one time.
3. When is the baseball at a height of 150 feet?

Solution

1. We substitute the output 84 for h in the equation $h = -16t^2 + 96t + 4$ and solve for t:

$$84 = -16t^2 + 96t + 4 \qquad \text{Substitute 84 for } h.$$
$$0 = -16t^2 + 96t - 80 \qquad \text{Write in } 0 = ax^2 + bx + c \text{ form.}$$
$$0 = -16(t^2 - 6t + 5) \qquad \text{Factor out } -16 \text{ (on right-hand side).}$$
$$0 = -16(t - 1)(t - 5) \qquad \text{Completely factor right-hand side.}$$
$$t - 1 = 0 \quad \text{or} \quad t - 5 = 0 \qquad \text{Zero factor property}$$
$$t = 1 \quad \text{or} \quad t = 5$$

So, 1 second after the baseball is hit, and again 5 seconds after it is hit, the baseball is at a height of 84 feet. It makes sense that there are two times, because the baseball reaches 84 feet both on its way up and on its way down (see Fig. 22).

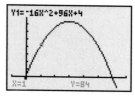 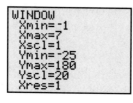 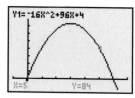

Figure 22 Finding when the baseball is at a height of 84 feet

2. We substitute the output 148 for h in the equation $h = -16t^2 + 96t + 4$ and solve for t:

$$148 = -16t^2 + 96t + 4 \qquad \text{Substitute 148 for } h.$$
$$0 = -16t^2 + 96t - 144 \qquad \text{Write in } 0 = ax^2 + bx + c \text{ form.}$$
$$0 = -16(t^2 - 6t + 9) \qquad \text{Factor out } -16 \text{ (on right-hand side).}$$
$$0 = -16(t - 3)(t - 3) \qquad \text{Completely factor right-hand side.}$$
$$t - 3 = 0 \qquad \text{Zero factor property}$$
$$t = 3$$

So, 3 seconds after the baseball is hit, it is at a height of 148 feet. It makes sense that there is only one time *if* the baseball reaches a height of 148 feet only at the top of its climb—in other words, if the vertex of the parabola is (3, 148). Figure 23 confirms the situation.

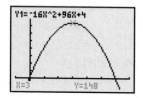

Figure 23 Finding when the baseball is at a height of 148 feet

3. Since the maximum height of the baseball is 148 feet, the baseball never reaches a height of 150 feet (see Fig. 23). ∎

Consider the following observations about a quadratic model, a parabola, and a quadratic equation in one variable:

- Our work in Example 2 suggests that the baseball reaches each height either two times, one time, or never.

- In Sections 7.1 and 7.2, we found that, for a parabola opening upward or downward, each value of y corresponds to two, one, or no values of x.

- In Section 8.5, we learned that a quadratic equation in one variable has either two, one, or no real-number solutions.

These observations suggest an important common thread between quadratic models, parabolas, and quadratic equations in one variable. For the purposes of this section, that common thread means that when we use a quadratic model to predict the value of an independent variable for a specific value of the dependent variable, there will be either two, one, or no real-number values.

Area of Rectangular Objects

Recall from Section 4.6 that the area A of a rectangle is given by the formula $A = LW$, where L is the length and W is the width.

Example 3   Using the Area of a Rectangular Object

A rectangular garden has an area of 40 square feet. If the length is 3 feet more than the width, find the dimensions of the garden.

Solution

We use the five-step problem-solving method of Section 6.5.

Step 1: Define each variable.   Let L be the length (in feet) and W be the width (in feet). See Fig. 24.

Step 2: Write a system of two equations.   Since the length is 3 feet more than the width, our first equation is $L = W + 3$. Because the area is 40 square feet, we find our second equation by substituting 40 for A in the area formula $A = LW$: $40 = LW$. The system is

$$L = W + 3 \qquad \text{Equation (1)}$$
$$LW = 40 \qquad \text{Equation (2)}$$

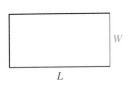

Figure 24 The length L and width W of a rectangular garden

Step 3: Solve the system.   We can solve the system by substitution. We substitute $W + 3$ for L in equation (2) and then solve for W:

$$
\begin{aligned}
LW &= 40 & &\text{Equation (2)} \\
(W + 3)W &= 40 & &\text{Substitute } W + 3 \text{ for } L. \\
W^2 + 3W &= 40 & &\text{Distributive law} \\
W^2 + 3W - 40 &= 0 & &\text{Subtract 40 from both sides.} \\
(W + 8)(W - 5) &= 0 & &\text{Factor left side.} \\
W + 8 = 0 \quad \text{or} \quad W - 5 &= 0 & &\text{Zero factor property} \\
W = -8 \quad \text{or} \quad W &= 5
\end{aligned}
$$

The result $W = -8$ means that the width is negative; model breakdown has occurred, because a width must be positive. So, the width is 5 feet (our other result). To find the length, we substitute 5 for W in equation (1):

$$L = W + 3 = 5 + 3 = 8$$

Step 4: Describe each result.   The width is 5 feet and the length is 8 feet.

Step 5: Check.   The product of 5 and 8 is 40, which checks for an area of 40 square feet. Also, 8 is indeed 3 more than 5. ∎

group exploration

Looking ahead: Product property for square roots

1. If a is a nonnegative number, then $\sqrt{a}$ is the nonnegative number we square to get a. We call $\sqrt{a}$ the *principal square root* of a. So, $\sqrt{9} = 3$ because $3^2 = 9$. Find the principal square root or the product of principal square roots.

   a. $\sqrt{25}$ b. $\sqrt{2 \cdot 8}$

   c. $\sqrt{25} \cdot \sqrt{49}$ d. $\sqrt{13}$ [**Hint:** Use a calculator.]

2. Find the principal square root $\sqrt{4 \cdot 9}$ and the product $\sqrt{4} \cdot \sqrt{9}$. Then compare your results.

3. Find the principal square root $\sqrt{4 \cdot 25}$ and the product $\sqrt{4} \cdot \sqrt{25}$. Then compare your results.

4. Write $\sqrt{ab}$ as a product of two principal square roots. [**Hint:** See Problems 2 and 3.] This is called the *product property for square roots.*

5. Use a calculator to estimate $\sqrt{4+9}$. Then find the sum $\sqrt{4} + \sqrt{9}$. Finally, compare your results.

6. Use a calculator to estimate $\sqrt{16+25}$. Then find the sum $\sqrt{16} + \sqrt{25}$. Finally, compare your results.

7. Is the statement $\sqrt{a+b} = \sqrt{a} + \sqrt{b}$ true for all real numbers a and b? Explain.

TIPS FOR SUCCESS: Create an Example

When learning a definition or property, try to create an example. For example, while studying the material in Sections 8.1–8.4, you could create polynomials that might be factored by the following methods:

- Factor out the GCF.
- Factor a cubic polynomial of the form $ax^3 + bx^2 + cx + d$ by grouping.
- Factor a difference of two squares.
- Factor a sum of two cubes.
- Factor a difference of two cubes.
- Factor a quadratic trinomial of the form $x^2 + bx + c$.
- Factor a quadratic trinomial of the form $ax^2 + bx + c$, where $a \neq 1$, by using trial and error or by grouping.
- Completely factor a polynomial by using a combination of methods.

You could create these examples by finding products of polynomials. For instance, create a trinomial of the form $ax^2 + bx + c$ by finding the product $(2x+5)(3x-4)$. After creating many polynomials, let time pass so that you won't remember their factored forms. Then practice factoring them. While factoring each polynomial, reflect on the polynomial's type and on why you chose to use a particular method to factor that type.

Creating examples will shed light on many details of a concept and will personalize the information. It will also help you understand the similarities and differences between related concepts.

HOMEWORK 8.6

FOR EXTRA HELP ▶

Student Solutions Manual · PH Math/Tutor Center · Math XL MathXL® · MyMathLab MyMathLab

---

1. The number of children with autism who are enrolled in special-education classes has increased greatly in recent years (see Table 4), possibly due to refined diagnoses and public awareness.

Table 4 Numbers of Special-Education Students with Autism

Year	Number of Special-Education Students with Autism (thousands)
1995	23
1997	34
1999	53
2001	80
2003	120

Source: *Individuals with Disabilities Education Act*

Let n be the number (in thousands) of special-education students with autism at t years since 1995. A model of the situation is $n = t^2 + 4t + 22$.

a. Use a graphing calculator to draw the graph of the model and, in the same viewing window, the scattergram of the data. Does the model fit the data well?

b. Predict the number of special-education students with autism in 2010.

c. Estimate when there were 43 thousand special-education students with autism.

2. Sales of hot tubs are shown in Table 5 for various years. Let n be the number (in thousands) of hot tubs sold in the year that is t years since 1995.

a. Use a graphing calculator to draw a scattergram of the data. Can the data be modeled better by a linear equation or a quadratic equation? Explain.

Table 5 Numbers of Hot Tubs Sold

Year	Hot-Tub Sales (thousands)
1995	260
1996	235
1997	263
1998	275
1999	305
2000	375

Source: *National Spa & Pool Institute*

b. Use a graphing calculator to draw the graph of the model $n = 8t^2 - 16t + 250$ and, in the same viewing window, the scattergram of the data. Does the model fit the data well?

c. Predict hot-tub sales in 2010.

d. Estimate in what year(s) 890 thousand hot tubs were or will be sold.

e. Use "minimum" on a graphing calculator to find the vertex of the model. What does the vertex mean in this situation?

3. Total spending by domestic and foreign travelers in the United States are shown in Table 6 for various years.

Table 6 Total Spending by Travelers in the United States

Year	Spending (billions of dollars)
2000	580
2001	550
2002	537
2003	552
2004	585

Source: *Travel Association of America*

Let s be the total spending (in billions of dollars) by domestic and foreign travelers in the United States in the year that is t years since 2000.

a. Use a graphing calculator to draw a scattergram of the data. Can the data be modeled better by a linear equation or a quadratic equation? Explain.

b. Use a graphing calculator to draw the graph of the model $s = 11t^2 - 44t + 582$ and, in the same viewing window, the scattergram of the data. Does the model fit the data well?

c. Estimate when total spending by travelers was or will be $637 billion.

d. Use "minimum" on a graphing calculator to find the vertex of the model. What does the vertex mean in this situation?

4. In 1992, Edison schools, which are public schools run by a private company, opened. Although the company has not yet earned a profit, test scores are higher in its schools than in other types of schools, and the number of Edison students has increased greatly (see Table 7). Let n be the number of students (in thousands) in Edison schools at t years since 1995.

a. Use a graphing calculator to draw a scattergram of the data. Can the data be modeled better by a linear equation or a quadratic equation? Explain.

Table 7 Numbers of Students in Edison Schools

Year	Number of Students (thousands)
1996	2
1998	13
1999	23
2000	37
2001	57
2002	76

Source: *Edison Schools, Inc.*

b. Use a graphing calculator to draw the graph of the model $n = \frac{7}{4}t^2 - 2t + 4$ and, in the same viewing window, the scattergram of the data. Does the model fit the data well?

c. Predict the number of students that will attend Edison schools in 2011.

d. Estimate when there were 7 thousand students in Edison schools.

5. US Airways filed for bankruptcy in 2003, recovered, and filed for bankruptcy again in 2004. Table 8 shows US Airways revenues for various years.

Table 8 US Airways Revenues

Year	Revenue (billions of dollars)
1999	8.6
2000	9.3
2001	8.3
2002	7.0
2003	5.3

Source: *Hoover's*

Let r be US Airways revenue (in billions of dollars) in the year that is t years since 1999.

a. Use a graphing calculator to draw a scattergram of the data. Can the data be modeled better by a linear equation or a quadratic equation? Explain.

b. Use a graphing calculator to draw the graph of the model $r = -\frac{1}{3}t^2 + \frac{1}{2}t + 9$ and, in the same viewing window, the scattergram of the data. Does the model fit the data well?

c. Estimate the revenue in 2004.

d. What are the t-intercepts? What do they mean in this situation?

6. The revenues from snack-vending machines in the United States are shown in Table 9 for various years.

Table 9 Revenues from Snack-Vending Machines

Year	Revenue (billions of dollars)
1997	22.1
1998	23.3
1999	24.5
2000	25.6
2001	24.3
2002	23.1

Source: *Automatic Merchandiser*

Let r be the revenue (in billions of dollars) from snack-vending machines in the United States in the year that is t years since 1997.

a. Use a graphing calculator to draw a scattergram of the data. Can the data be modeled better by a linear equation or a quadratic equation? Explain.

b. Use your graphing calculator to draw the graph of the model $r = -\frac{2}{5}t^2 + \frac{11}{5}t + 22$ and, in the same viewing window, the scattergram of the data. Does the model fit the data well?

c. Estimate the revenue in 2005.

d. Estimate when the revenue was or will be $14 billion.

e. Use "maximum" on a graphing calculator to estimate when the revenue was the highest. What was that revenue?

7. Firestone® tire sales are shown in Table 10 for various years.

Table 10 Firestone Tire Sales

Year	Tire Sales (millions of tires)
1999	23.0
2000	19.2
2001	16.1
2002	16.1
2003	16.4

Source: *Modern Tire Dealer*

Let s be Firestone tire sales (in millions of tires) in the year that is t years since 1999. A linear model of the situation is $s = -1.63t + 21.42$. A quadratic model of the situation is $s = \frac{4}{5}t^2 - \frac{24}{5}t + 23$.

a. Use a graphing calculator to draw a scattergram of the data and both models in the same viewing window. Which model comes closer to the points in the scattergram?

b. As a result of an investigation led by the National Highway Traffic Safety Administration in May 2000 and 500 lawsuits involving 270 deaths and 800 injuries, Firestone recalled 6.5 million unsafe tires. Most economists felt that the brand would never recover from the stigma of selling unsafe tires. Which of the two models do you think economists believed would give better predictions of future sales?

c. To most economists' surprise, sales held constant from 2001 to 2002 and increased from 2002 to 2003. Consumers' confidence in the tires has increased. As a result, Firestone launched its first major ad campaign in four years, which is costing $18 million. Which of the two models do you think Firestone believes will give better predictions of future sales? Explain.

d. Use the quadratic model to estimate in which years sales were 19 million tires.

8. Revenues from organic-food sales are shown in Table 11 for various years. Let r be the revenue (in billions of dollars) from organic-food sales in the year that is t years since 1997. A linear model of the situation is $r = 1.30t + 2.84$. A quadratic model of the situation is $r = \frac{1}{10}t^2 + \frac{1}{2}t + \frac{37}{10}$.

Table 11 Revenues from Organic-Food Sales

Year	Revenue (billions of dollars)
1997	3.6
1999	5.1
2001	7.1
2003	10.4
2005	13.9

Source: *Organic Trade Association*

a. Use a graphing calculator to draw a scattergram of the data and both models in the same viewing window. Which model comes closer to the points in the scattergram?

b. Use first the linear model and then the quadratic model to estimate the revenue from organic-food sales in 2001. Which is the better estimate? Explain.

c. Which of the two models would farmers who grow organic foods hope will describe future revenues from organic food sales better? Explain.

d. Use the quadratic model to estimate in which year the revenue from organic-food sales were $12.1 billion.

e. Use the quadratic model to predict the revenue from organic-food sales in 2011.

9. A company's profit can be modeled by $p = t^2 - 3t + 5$, where p is the profit (in thousands of dollars) for the year that is t years since 2005. Predict when the profit was or will be $23 thousand.

10. A company's profit can be modeled by $p = t^2 - 6t + 17$, where p is the profit (in thousands of dollars) for the year that is t years since 2005. Predict when the profit was or will be $24 thousand.

11. A company's revenue can be modeled by $r = 2t^2 - 13t + 25$, where r is the revenue (in millions of dollars) for the year that is t years since 2005. Predict when the revenue was or will be $10 million.

12. A company's revenue can be modeled by $r = 3t^2 - 19t + 35$, where r is the revenue (in millions of dollars) for the year that is t years since 2005. Predict when the revenue was or will be $26 million.

13. A batter hits a baseball into the air. The height h (in feet) of the baseball after t seconds is given by the quadratic equation $h = -16t^2 + 64t + 3$.

a. When is the baseball at a height of 3 feet? Explain why there are two such times. [**Hint:** Use a graphing calculator to graph the model.]

b. When is the baseball at a height of 51 feet? Explain why there are two such times.

c. When is the baseball at a height of 67 feet? Explain why there is just one such time.

14. A person throws a ball into the air. The height h (in feet) of the ball after t seconds is given by $h = -16t^2 + 96t + 4$.

a. When is the baseball at a height of 4 feet? Explain why there are two such times. [**Hint:** Use a graphing calculator to graph the model.]

b. When is the baseball at a height of 84 feet? Explain why there are two such times.

c. When is the baseball at a height of 148 feet? Explain why there is just one such time.

15. A rectangular boardroom table has an area of 60 square feet. If the length is 7 feet more than the width, find the dimensions of the table.

16. A rectangular rug has an area of 80 square feet. If the length is 2 feet more than the width, find the dimensions of the rug.

17. A rectangular office floor has an area of 98 square feet. If the length is twice the width, find the dimensions of the floor.

18. A rectangular garden has an area of 162 square feet. If the length is twice the width, find the dimensions of the garden.

19. For a given set of data, describe how you can determine whether to model the situation by a linear model, a quadratic model, or neither.

20. Explain how to make a prediction for the dependent variable of a quadratic model. Explain how to make a prediction for the independent variable.

Related Review

21. The average tax refund has increased since 1997 (see Table 12).

Table 12 Average Tax Refunds

Year	Average Tax Refund (dollars)
1997	1295
1998	1347
1999	1542
2000	1624
2001	1714
2002	1939
2003	1973

Source: *Internal Revenue Service*

Let T be the average tax refund (in dollars) in the year that is t years since 1990.

a. Use a graphing calculator to draw a scattergram of the data. Can the data be modeled better by a linear equation or a quadratic equation? Explain.

b. Find an equation of a model to describe the data.

c. Predict when the average tax refund will be $2800.

22. The revenues of Wal-Mart®, the largest private employer in the United States, are shown in Table 13 for various years.

Table 13 Wal-Mart's Revenues

Year	Revenues (billions of dollars)
1999	138
2000	165
2001	191
2002	218
2003	245

Sources: *Wal-Mart; Redbook*

Let r be the revenue (in billions of dollars) in the year that is t years since 1990.

a. Use a graphing calculator to draw a scattergram of the data. Can the data be modeled better by a linear equation or a quadratic equation? Explain.

b. Find an equation of a model to describe the data.

c. Predict when the revenue will be $400 billion.

23. The U.S. annual per person consumption of seafood was 10.3 pounds in 1960 and has increased by about 0.14 pound per year since then (Source: *National Marine Fisheries Service*). Let c be the U.S. annual per person consumption of seafood (in pounds) at t years since 1960.

a. Find an equation of a model to describe the data.

b. Predict when the U.S. annual per person consumption of seafood will be 17.4 pounds.

24. In 2000, 3059 schools were on year-round schedules. The number of such schools has increased by about 118 schools per year since then (Source: *National Association for Year-Round Education*). Let n be the number of schools that have been on year-round schedules at t years since 2000.

a. Find an equation of a model to describe the data.

b. Predict the number of schools that will be on year-round schedules in 2010.

25. The average cost of campus housing has increased approximately linearly from $4340 for the school year ending in 1999 to $5475 for the school year ending in 2004 (Source: *The College Board*). Predict the average cost of campus housing for the school year ending in 2011.

26. The percentage of boys ages 3 to 17 who have been diagnosed as having attention deficit hyperactivity disorder (ADHD) has increased approximately linearly from 8.3% in 1997 to 10.3% in 2002 (Source: *Child Trends DataBank*). Predict the percentage of boys who will have ADHD in 2010.

Expressions, Equations, and Graphs

Give an example of each. Then solve, simplify, or graph, as appropriate.

27. cubic expression in one variable with five terms

28. quadratic expression in one variable with four terms

29. linear expression in one variable with four terms

30. quadratic equation in one variable

31. system of two linear equations in two variables

32. linear equation in one variable

33. quadratic equation in two variables

34. linear equation in two variables

Chapter Summary

Key Points
OF CHAPTER 8

Factor (Section 8.1)

We **factor** a polynomial by writing it as a product.

Multiplying versus factoring (Section 8.1)

Multiplying and factoring are reverse processes.

Factoring $x^2 + bx + c$ (Section 8.1)

To factor $x^2 + bx + c$, look for two integers p and q whose product is c and whose sum is b. That is, $pq = c$ and $p + q = b$. If such integers exist, the factored polynomial is $(x + p)(x + q)$.

Factoring $x^2 + bx + c$ with c positive (Section 8.1)

To factor a trinomial of the form $x^2 + bx + c$ with a positive constant term c,

  • If b is positive, look for two *positive* integers whose product is c and whose sum is b.

  • If b is negative, look for two *negative* integers whose product is c and whose sum is b.

Factoring $x^2 + bx + c$ with c negative (Section 8.1)

To factor a trinomial of the form $x^2 + bx + c$ with a negative constant term c, look for two integers with *different* signs whose product is c and whose sum is b.

Prime polynomial (Section 8.1)

A polynomial that cannot be factored is called **prime.**

Difference of two squares (Section 8.1)

$A^2 - B^2 = (A + B)(A - B)$.

Polynomial $x^2 + k^2$, where $k \neq 0$ (Section 8.1)

A polynomial of the form $x^2 + k^2$, where $k \neq 0$, is prime.

GCF (Section 8.2)

The **greatest common factor (GCF)** of two or more terms is the monomial with the largest coefficient and the largest degree that is a factor of all the terms.

Factor out GCF (Section 8.2)

When the leading coefficient of a polynomial is positive and the GCF is not 1, we first factor out the GCF.

Completely factor (Section 8.2)

When factoring a polynomial, always *completely* factor it.

Factoring when the leading coefficient is negative (Section 8.2)

When the leading coefficient of a polynomial is negative, we first factor out the opposite of the GCF.

Factor by grouping (Section 8.2)

For a polynomial with four terms, **factor by grouping** (if it can be done) by

1. Factoring the first two terms and the last two terms.

2. Factoring out the binomial GCF.

Factor $ax^2 + bx + c$ by trial and error (Section 8.3)

To **factor a trinomial** of the form $ax^2 + bx + c$ **by trial and error,** if the trinomial can be factored as a product of two binomials, then the product of the coefficients of the first terms of the binomials is equal to a and the product of the last terms of the binomials is equal to c. To find the correct factored expression, multiply the possible products and identify the one for which the coefficient of x is b.

Factoring $ax^2 + bx + c$ by grouping (ac method) (Section 8.3)

To **factor a trinomial** of the form $ax^2 + bx + c$ **by grouping** (if it can be done),

1. Find pairs of numbers whose product is ac.

2. Determine which of the pairs of numbers from step 1 has the sum b. Call this pair of numbers m and n.

3. Write the bx term as $mx + nx$:

$$ax^2 + bx + c = ax^2 + mx + nx + c$$

4. Factor $ax^2 + mx + nx + c$ by grouping.

Another name for this technique is the *ac* **method.**

Sum of two cubes (Section 8.4)

$A^3 + B^3 = (A + B)\left(A^2 - AB + B^2\right)$

Difference of two cubes (Section 8.4)

$A^3 - B^3 = (A - B)\left(A^2 + AB + B^2\right)$

Factoring strategy (Section 8.4)

The five steps that follow can be used to factor many polynomials. Steps 2–4 can be applied to the entire polynomial or to a factor of the polynomial.

1. If the leading coefficient is positive and the GCF is not 1, we factor out the GCF. If the leading coefficient is negative, we factor out the opposite of the GCF.

2. For a binomial, try using one of the properties for the difference of two squares, the sum of two cubes, or the difference of two cubes.

3. For a trinomial of the form $ax^2 + bx + c$,
   a. If $a = 1$, try to find two integers whose product is c and whose sum is b.
   b. If $a \neq 1$, try to factor by trial and error or by grouping.

4. For an expression with four terms, try factoring by grouping.

5. Continue applying steps 2–4 until the polynomial is completely factored.

Quadratic equation in one variable (Section 8.5)

A **quadratic equation in one variable** is an equation that can be put into the form $ax^2 + bx + c = 0$, where a, b, and c are constants and $a \neq 0$.

Zero factor property (Section 8.5)

Let A and B be real numbers: If $AB = 0$, then $A = 0$ or $B = 0$.

Number of solutions: quadratic equation (Section 8.5)

A quadratic equation in one variable has either two real-number solutions, one real-number solution, or no real-number solutions.

Cubic equation in one variable (Section 8.5)

A **cubic equation in one variable** is an equation that can be put into the form $ax^3 + bx^2 + cx + d = 0$, where a, b, c, and d are constants and $a \neq 0$.

Solving quadratic and cubic equations in one variable (Section 8.5)

If an equation can be solved by factoring, we solve it by the following steps:

1. Write the equation so that one side of it is equal to zero.

2. Factor the nonzero side of the equation.

3. Apply the zero factor property.

4. Solve each equation that results from using the zero factor property.

Making a prediction about the dependent variable (Section 8.6)

In making a prediction about the dependent variable of a quadratic model, substitute a chosen value for the independent variable in the model. Then solve for the dependent variable.

Making a prediction about the independent variable (Section 8.6)

In making a prediction about the independent variable of a quadratic model, substitute a chosen value for the dependent variable in the model. Then solve for the independent variable, which involves writing the equation so that one side of it is zero and then factoring the nonzero side.

CHAPTER 8 REVIEW EXERCISES

For Exercises 1–32, factor when possible. Use a graphing calculator table to verify your work when possible.

1. $x^2 + 9x + 20$
2. $6x^2 - 2x - 8$
3. $81x^2 - 49$
4. $x^2 + 14x + 49$
5. $-18t^4 - 33t^3 + 30t^2$
6. $p^2 - 3pq - 54q^2$
7. $32 - 12x + x^2$
8. $-9x^2 + 4$
9. $4w^2 + 25$
10. $20m^2n - 45mn^3$
11. $16x^2 + 14x + 2x^3$
12. $16x^3 - 32x^2 + 16x$
13. $24x^3 - 32x^2$
14. $5x^4y - 35x^3y + 60x^2y$
15. $-m^2 - 2m + 35$
16. $4r^3 - 10r^2 + 6r - 15$
17. $2p^2 + 7p + 3$
18. $x^2 - 9x + 20$
19. $6t^2 + 11ty - 10y^2$
20. $2x^3 - 50x$
21. $x^2 - 10x + 25$
22. $p^2 - 81$
23. $8w^2 - 12w + 3$
24. $x^2 + 4$
25. $4x^3 - 4x^2 - 9x + 9$
26. $4x^2 + 20x + 25$
27. $12w^3 - 50w^2 + 8w$
28. $49a^2 - 9b^2$
29. $x^2 - 7x - 12$
30. $x^3 + 3x^2 - 4x - 12$
31. $r^3 + 8$
32. $8t^3 - 27$

33. A student tries to factor $x^2 + 25$:

$$x^2 + 25 = (x + 5)(x + 5) = (x + 5)^2$$

Describe any errors. Then factor the polynomial correctly.

34. A student tries to factor $5x^3 + 35x^2 + 60x$:

$$5x^3 + 35x^2 + 60x = 5x\left(x^2 + 7x + 12\right)$$

Describe any errors. Then solve the equation correctly.

35. Find all possible values of k so that $x^2 + kx + 24$ can be factored.

For Exercises 36–51, solve the given equation. Verify your results by checking that they satisfy the equation.

36. $x^2 + 9x + 14 = 0$
37. $2x^3 + 16x^2 = -24x$
38. $(m - 3)(m + 2) = -4$
39. $t^2 - 6t + 9 = 0$
40. $x^2 - 3x = 5x - 15$
41. $25x^2 - 81 = 0$
42. $2x^3 - 7x^2 - 2x + 7 = 0$
43. $6x^2 + x - 2 = 0$
44. $3x^2 = 15x$
45. $8r^2 - 18r + 9 = 0$
46. $a^2 = 2a + 35$
47. $x(x - 4) = 12$
48. $\frac{1}{3}x^2 - \frac{1}{3}x - 4 = 6$
49. $3x^3 - 2x^2 = 27x - 18$
50. $a^2 = 4$
51. $5p^2 + 20p - 60 = 0$

52. Use "intersect" on a graphing calculator to solve the equation $2x^2 + 4x - 5 = 2x + 3$. Round the solution(s) to the second decimal place.

For Exercises 53–56, solve the given equation by referring to the solutions of $y = -3x^2 + 6x + 20$ shown in Table 14.

53. $-3x^2 + 6x + 20 = 23$ **54.** $-3x^2 + 6x + 20 = -4$

55. $-3x^2 + 6x + 20 = 27$ **56.** $-3x^2 + 6x + 20 = 20$

Table 14 Exercises 53–56

x	y
−2	−4
−1	11
0	20
1	23
2	20
3	11
4	−4

For Exercises 57 and 58, find all x-intercepts. Verify your work by using your graphing calculator.

57. $y = x^2 - 49$ **58.** $y = 8x^2 - 14x - 15$

59. Give an example of a quadratic equation whose solutions are 3 and −6.

60. Give an example of a cubic equation whose solutions are −2, 0, and 1.

61. The asthma episode rates are shown in Table 15 for various years.

Table 15 Asthma Episode Rates

Year	Number of Episodes per 1000 People
1999	38.6
2000	40.0
2001	43.2
2002	42.6
2003	38.7

Source: *National Center for Health Statistics*

Let r be the annual asthma episode rate (number of episodes per 1000 people) at t years since 1999.

a. Use a graphing calculator to draw a scattergram of the data. Can the data be modeled better by a linear equation or a quadratic equation? Explain.

b. Use a graphing calculator to draw the graph of the model $r = -t^2 + 4t + 39$ and, in the same viewing window, the scattergram of the data. Does the model fit the data well?

c. Estimate the annual asthma episode rate in 2004.

d. Estimate when the annual asthma episode rate was or will be 27 episodes per 1000 people.

e. Use "maximum" on a graphing calculator to find the vertex of the model. What does the vertex mean in this situation?

62. A rectangular banner has area 30 square feet. If the length is 4 feet more than twice the width, find the dimensions of the banner.

CHAPTER 8 TEST

For Exercises 1–9, factor.

1. $x^2 - 3x - 40$ **2.** $24 + x^2 - 10x$

3. $8m^2n^3 - 10m^3n$ **4.** $p^2 - 14pq + 40q^2$

5. $25p^2 - 36y^2$ **6.** $3x^4y - 21x^3y + 36x^2y$

7. $8x^3 + 20x^2 - 18x - 45$ **8.** $8x^2 - 26x + 15$

9. $64x^3 - 1$

10. Which of the following expressions are equivalent?

$$x^2 - 7x - 10 \qquad (x - 5)(x + 2) \qquad x^2 - 3x - 10$$
$$x^2 + 3x - 10 \qquad (x + 2)(x - 5)$$

11. A student tries to factor $5x^3 + 3x^2 - 20x - 12$:

$$5x^3 + 3x^2 - 20x - 12 = x^2(5x + 3) - 4(5x + 3)$$

Describe any errors. Then factor the polynomial correctly.

Solve.

12. $x^2 - 13x + 36 = 0$ **13.** $49x^2 - 9 = 0$

14. $t(t + 14) = 2(t - 18)$ **15.** $\frac{1}{4}p^2 - \frac{1}{2}p - 6 = 0$

16. $3x^3 - 12x = 8 - 2x^2$ **17.** $2x^3 = 8x^2 + 10x$

For Exercises 18–21, use the graph of $y = x^2 + 6x + 7$ shown in Fig. 25 to solve the given equation.

18. $x^2 + 6x + 7 = 2$

19. $x^2 + 6x + 7 = -1$

20. $x^2 + 6x + 7 = -2$

21. $x^2 + 6x + 7 = -4$

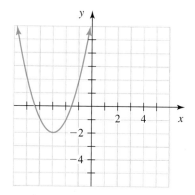

Figure 25 Exercises 18–21

22. Find the x-intercepts of the graph of $y = 10x^2 - 11x - 6$.

23. Give an example of a quadratic equation whose solutions are −3 and 8.

24. The percentages of Americans who think that the protection of the environment should be given priority even at the risk of curbing economic growth are shown in Table 16 for various years. Let p be the percentage of Americans at t years since 2000 who think that the protection of the environment should be given priority even at the risk of curbing economic growth.

a. Use a graphing calculator to draw a scattergram of the data. Can the data be modeled better by a linear equation or a quadratic equation? Explain.

Table 16 Percentages of Americans Who Think That the Environment Should Be Given Priority Even at the Risk of Curbing Economic Growth

Year	Percent
2000	69
2001	57
2002	54
2003	47
2004	49
2005	53

Source: *The Gallup Organization*

b. Use a graphing calculator to draw the graph of the model $p = 2t^2 - 13t + 69$ and, in the same viewing window, the scattergram of the data. Does the model fit the data well?

c. Predict what percentage of Americans in 2007 will think that the environment should be given top priority.

d. Estimate when 63% of Americans thought that the environment should be given top priority.

e. Use "minimum" on a graphing calculator to find the vertex of the quadratic model. What does the vertex mean in this situation?

25. A company's revenue can be modeled by $r = t^2 - 4t + 34$, where r is the revenue (in millions of dollars) for the year that is t years since 2005. Estimate when the revenue was or will be $66 million.

CUMULATIVE REVIEW OF CHAPTERS 1-8

1. Let p be the percentage of Americans of age a years who work. What is the independent variable? What is the dependent variable?

2. Suppose that an independent variable t and a dependent variable p are approximately linearly related. If a data point (c, d) lies above a linear model, does the model underestimate or overestimate the value of p when $t = c$? Explain.

3. An hour ago, the temperature was $4°F$. The temperature is now $-2°F$. What is the change in temperature?

4. Find the slope of the line that passes through the points $(-3, -1)$ and $(5, 3)$. State whether the line is increasing, decreasing, horizontal, or vertical.

5. For the equation $y = -4x + 9$, describe the change in the value of y as the value of x is increased by 1.

6. Find an equation of the line that has slope $\frac{2}{3}$ and contains the point $(-2, -6)$. Write your equation in slope–intercept form.

7. Find an equation of the line that contains the points $(-4, 7)$ and $(2, -3)$. Write your equation in slope–intercept form.

8. Find an equation of the line sketched in Fig. 26.

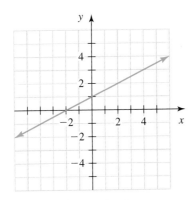

Figure 26 Exercise 8

Let x be a number. Translate each statement into an expression or an equation. Then simplify the expression or solve the equation, as appropriate.

9. Three times the sum of the number and 5 is 9.

10. Seven, minus 4 times the quotient of the number and 2

For Exercises 11 and 12, solve the system by either substitution or elimination. Verify your work by using "intersect" on a graphing calculator or by checking that your result satisfies both equations in the system.

11. $2x - 3y = 7$
    $y = 4x - 9$

12. $3x + 4y = 4$
    $7x - 5y = 38$

13. Solve the system of equations three times, once by each of the three methods: elimination, substitution, and graphing by hand. Decide which method you prefer for this system.

$$4x - 5y = -22$$
$$x + 2y = 1$$

14. Figure 27 shows the graphs of two linear equations. Find the solution of the system.

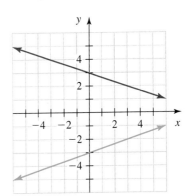

Figure 27 Exercise 14

Perform the indicated operations.

15. $5(-2) - (-6)^2 + 4$

16. $9 - (6 - 8)^3 + 4 \div (-2)$

Evaluate the given expression for $a = -4$, $b = -2$, and $c = 5$.

17. $b^2 - 4ac$

18. $\dfrac{c + a^2}{a - b^2}$

For Exercises 19–24, refer to the graph sketched in Fig. 28.

19. Find y when $x = 4$.

20. Find x when $y = 4$.

21. Find x when $y = 3$.

22. Find x when $y = 5$.

23. Find the x-intercepts.

24. Find the maximum point.

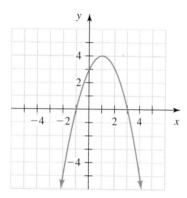

Figure 28 Exercises 19–24

25. Graph $x = 2$ by hand.

For Exercises 26–43, perform the indicated instruction. Then use words such as linear, quadratic, cubic, polynomial, degree, one variable, *and* two variables *to describe the expression, equation, or system.*

26. Graph $y = 2x^2$ by hand.

27. Factor $4m^2 - 49n^2$.

28. Simplify $(3p - 7q)^2$.

29. Factor $w^2 + 5w - 14$.

30. Solve $5x^2 + 18x - 8 = 0$.

31. Solve $x^2 = 2x + 35$.

32. Factor $6x^3y + x^2y - 15xy$.

33. Find the product $(8p - 3)(2p + 3)$.

34. Find the product $-4xy(5x^2 + 2xy - 3y^2)$.

35. Factor $a^2 - 3ab - 40b^2$.

36. Solve $5(t - 2) = 4 - 7t$.

37. Find the sum $(3x^2 - 5x) + (-7x^2 - 2x + 9)$.

38. Factor $3x^2 - 33x + 54$.

39. Graph $2x - 4y = 8$ by hand.

40. Find the product $(2m - 5)(3m^2 - 2m + 4)$.

41. Find the difference $(x^2 - x) - (2x^3 - 4x^2 + 5x)$.

42. Factor $4r^3 + 8r^2 - 9r - 18$.

43. Factor $8x^3 - 27$.

44. Solve the formula $S = 2\pi r^2 + rh$ for h.

45. Solve the inequality $2(x - 1) \le 5(x + 2)$. Describe the solution set as an inequality, in interval notation, and in a graph.

46. Graph the inequality $4x - 5y \ge 20$ by hand.

47. Graph the solution set of the system.

$$y > -\frac{1}{2}x + 2$$
$$y < 3$$

48. Which of the following expressions are equivalent?

$$(x - 2)(x + 4) \qquad 8 - 2x + x^2 \qquad x^2 + 2x - 8$$
$$(x + 2)(x - 4) \qquad (x + 4)(x - 2) \qquad x(x + 2) - 8$$

49. Write the equation $y = -3(x - 2)^2$ in standard form.

Find all x-intercepts. Verify your work by using a graphing calculator.

50. $y = x^2 + 4x - 21$

51. $2x - 5y = 20$

For Exercises 52–55, simplify.

52. $\left(-2ab^2\right)^3$

53. $\dfrac{2^2 r^{-2}}{2^5 r^{-7}}$

54. $\left(4x^{-8}y^5\right)\left(2x^3y^{-2}\right)$

55. $\left(\dfrac{x^2 y^{-6}}{w^{-4}}\right)^3$

56. A 6000-seat theater has tickets for sale at $20 and $35. How many tickets should be sold at each price for a sellout performance to generate a total revenue of $147,000?

57. American sprinter Marion Jones won a record five medals at the 2000 Sydney Olympics. Her best times in the 100-meter run are shown in Table 17 for various years.

Table 17 Marion Jones's Best Times in the 100-Meter Run	
Year	**100-Meter Time (seconds)**
1998	10.65
1999	10.70
2000	10.75
2001	10.84
2002	10.84
2004	11.04

Source: USA Today

Let B be Jones's best time (in seconds) in the 100-meter run at t years since 1990.

a. Use a graphing calculator to draw a scattergram of the data.

b. Find an equation of a model to describe the data.

c. What is the slope? What does it mean in this situation?

d. Jones did not compete in 2003. If she had continued to compete, estimate what her best time for the 100-meter run would have been in 2003.

e. For the period 1998–2002, what was the smallest change in Jones's best time in the 100-meter run from one year to the next? When did this occur?

f. The U.S. Anti-Doping Agency (USADA) has been investigating whether Jones used drugs to enhance her performance. Explain why the data suggest that Jones did not *begin* using drugs anytime after 1998.

58. For decades, the life expectancy (at birth) for Japanese who live in Okinawa has been higher than the life expectancy (at birth) for Japanese who live elsewhere in Japan. However, this has changed in recent years, because Okinawans have largely adopted an American diet consisting of foods such as hamburgers, fried chicken, and pizza (Source: New York Times). The life expectancy for Okinawans was 77.2 years in 1995 and has increased by 0.10 year each year since. The life expectancy for all Japanese who live anywhere in Japan was 76.7 years in 1995 and has increased by 0.20 year each year since (Source: *Ministry of Health*).

a. Let L be the life expectancy (in years) for Okinawans at t years since 1995. Find an equation of a model to describe the data.

b. Let L be the life expectancy (in years) for all Japanese living in Japan at t years since 1995. Find an equation of a model to describe the data.

c. Estimate when the life expectancy for Okinawans was equal to the life expectancy for all Japanese living in Japan. What was that life expectancy?

59. Table 18 lists the percentages of Americans of various age groups who visit online trading sites.

Table 18 Americans Who Visit Online Trading Sites		
Age Group (years)	Age Used to Represent Age Group (years)	Percent
2–17	9.5	4.7
18–24	21.0	9.7
25–34	29.5	22.9
35–44	39.5	22.2
45–54	49.5	20.6
55–64	59.5	14.6
65+	70.0	5.3

Source: *comScore Media Matrix*

Let p be the percentage of Americans of age t years who visit online trading sites. A quadratic model is sketched in Fig. 29.

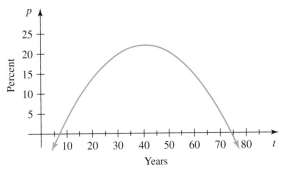

Figure 29 Online trading site model

a. Carefully trace the graph in Fig. 29 onto a piece of paper. Then, on the same graph, draw a scattergram of the data. Does the model fit the data well?

b. What are the t-intercepts? What do they mean in this situation?

c. What is the vertex of the model? What does it mean in this situation?

d. Estimate the percentage of 19-year-old Americans who visit online trading sites.

e. Estimate the age(s) at which 18% of Americans visit online trading sites.

60. A company's profit can be modeled by $p = t^2 - 2t + 8$, where p is the profit (in millions of dollars) for the year that is t years since 2005. Estimate when the profit was or will be $32 million.

61. A rectangular rug has an area of 84 square feet. If the length is 8 feet more than the width, find the dimensions of the rug.

Chapter 9

Solving Quadratic Equations

Obstacles are those frightful things you see when you take your eyes off your goal.

—*Henry Ford*

	Total Value of Goods Sold (billions of dollars)
Year	
1998	1.0
1999	2.6
2000	4.7
2001	9.1
2002	14.5
2003	22.8
2004	34.2

Table 1 Total Value of Goods Sold on eBay

Source: *eBay*

Have you ever bought or sold an item on eBay®? The total value of goods sold on eBay has increased greatly since 1998 (see Table 1). In Exercise 7 of Homework 9.6, you will predict in which year the total value of goods sold on eBay will be $115 billion.

In Chapter 8, we used factoring to solve quadratic equations in one variable. However, most quadratic equations cannot be solved by factoring. In this chapter, we will discuss other ways to solve quadratic equations; some of those ways can be used to solve *any* quadratic equation. This work will enable us to make estimates and predictions about any situation that can be described well by a quadratic model, including the eBay data just presented.

9.1 SIMPLIFYING RADICAL EXPRESSIONS

Objectives

▸ Know the meaning of *square root* and *principal square root*.

▸ Approximate a principal square root.

▸ Know the *product property for square roots*.

▸ Use the product property for square roots to simplify radicals.

So far, we have worked with polynomial expressions. In this section and Section 9.2, we will work with a new type of expression, called a *radical expression*. This work will prepare us to solve more quadratic equations in Sections 9.3–9.6.

Principal Square Roots

What number squared is equal to 9? The numbers that "work" are −3 and 3:

$$(-3)^2 = 9 \qquad 3^2 = 9$$

We say that −3 and 3 are *square roots* of 9. A **square root** of a number a is the number we square to get a.

Of the square roots of 9, only the nonnegative square root, 3, is the *principal square root* of 9. Using symbols, we write $\sqrt{9} = 3$.

DEFINITION Principal square root

If a is a nonnegative number, then $\sqrt{a}$ is the nonnegative number we square to get a. We call $\sqrt{a}$ the **principal square root** of a.

For example, $\sqrt{25} = 5$, since $5^2 = 25$. Also, $\sqrt{64} = 8$, because $8^2 = 64$.

The symbol "$\sqrt{}$" is called a **radical sign.** An expression under a radical sign is called a **radicand.** For $\sqrt{4x-7}$, the radicand is $4x-7$. A radical sign together with a radicand is called a **radical.** Here, we label the radical sign and radicand of the radical $\sqrt{2x+5}$:

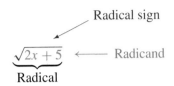

Here are some more radicals: $\sqrt{5}$, $\sqrt{x}$, $\sqrt{9x+4}$.

An expression that contains a radical is called a **radical expression.** Here are some radical expressions:

$$\sqrt{5}, \quad \sqrt{x}, \quad \sqrt{2x-5}, \quad 4\sqrt{x+8}-8x, \quad \left(3\sqrt{x}-2\right)\left(\sqrt{x}+7\right)$$

WARNING Is a square root of a negative number a real number? Consider $\sqrt{-9}$. Note that $\sqrt{-9} \neq -3$, because $(-3)^2$ is equal to 9, not -9. Since any number squared is nonnegative, we see that $\sqrt{-9}$ is not a real number. In general, a square root of a negative number is not a real number.

Example 1 Finding Square Roots

Find the square root.

1. $\sqrt{49}$
2. $\sqrt{-49}$
3. $-\sqrt{49}$
4. $-\sqrt{-49}$

Solution

1. $\sqrt{49} = 7$, because $7^2 = 49$.
2. $\sqrt{-49}$ is not a real number, because the radicand -49 is negative.
3. $-\sqrt{49} = -7$.
4. $-\sqrt{-49}$ is not a real number, because the radicand is negative. ∎

Rational and Irrational Square Roots

Recall from Section 1.1 that a rational number is a number that can be written in the form $\frac{n}{d}$, where n and d are integers and d is nonzero. A **perfect square** is a number whose principal square root is rational. For example, 25 is a perfect square, because $\sqrt{25} = 5 = \frac{5}{1}$ is rational. By squaring the integers from 0 to 15, we can find the integer perfect squares between 0 and 225, inclusive (see Table 2). You should memorize the perfect squares shown in this table, because you will work with them again and again.

For a number that is not a perfect square, any square root of the number is not rational. Recall from Section 1.1 that we call such a number *irrational*. For example, $\sqrt{7}$ is irrational. We know that $\sqrt{7}$ is a number between 2 and 3, because $2^2 = 4$ and $3^2 = 9$. To use a calculator to get the estimate $\sqrt{7} \approx 2.645751311$, press $\boxed{\text{2nd}}$ $\boxed{\sqrt{}}$ $\boxed{7}$ $\boxed{)}$ $\boxed{\text{ENTER}}$ (see Fig. 1).

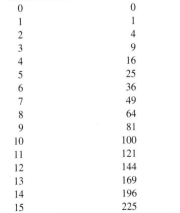

Table 2 Perfect Squares

x	Perfect Square x^2
0	0
1	1
2	4
3	9
4	16
5	25
6	36
7	49
8	64
9	81
10	100
11	121
12	144
13	169
14	196
15	225

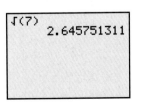

Figure 1 Estimating $\sqrt{7}$

Example 2   Approximating Square Roots

State whether the square root is rational or irrational. If the square root is rational, find the (exact) value. If the square root is irrational, estimate its value by rounding to the second decimal place.

1. $\sqrt{19}$ **2.** $\sqrt{169}$

```
√(19)
       4.358898944
```

Figure 2 Estimating $\sqrt{19}$

Solution

1. The number 19 is not a perfect square, so $\sqrt{19}$ is irrational. We use a calculator to compute $\sqrt{19} \approx 4.36$ (see Fig. 2).

2. The number 169 is a perfect square, so $\sqrt{169}$ is rational. In fact, $\sqrt{169} = 13$, because $13^2 = 169$. ∎

Product Property for Square Roots

Consider the following computations with radicals:

$$\sqrt{4 \cdot 9} = \sqrt{36} = 6$$
$$\sqrt{4}\sqrt{9} = 2 \cdot 3 = 6$$

Since both results are equal, we can write

$$\sqrt{4 \cdot 9} = \sqrt{4} \cdot \sqrt{9}$$

This equation suggests the following **product property for square roots,** which can help us work with the principal square root of a product.

Product Property for Square Roots

For nonnegative numbers a and b,

$$\sqrt{ab} = \sqrt{a}\sqrt{b}$$

In words: The square root of a product is the product of the square roots.

For example, $\sqrt{16 \cdot 49} = \sqrt{16} \cdot \sqrt{49} = 4 \cdot 7 = 28$.

Using the Product Property for Square Roots to Simplify Radicals

We can sometimes write a radical as an expression with a smaller radicand. For example, consider $\sqrt{24}$. Note that the perfect square 4 is a factor of 24: $4 \cdot 6 = 24$. We can use the product property for square roots to write

$$\sqrt{24} = \sqrt{4 \cdot 6} = \sqrt{4}\sqrt{6} = 2\sqrt{6}$$

When writing an expression such as $2\sqrt{6}$, we write the "2" to the left of "$\sqrt{6}$." If we were to write the expression as $\sqrt{62}$, the radicand might appear to be 62, not 6.

A square root is **simplified** when the radicand does not have any perfect-square factors other than 1.

Example 3   Simplifying Radicals

Simplify.

1. $\sqrt{12}$ **2.** $\sqrt{32x}$ **3.** $7\sqrt{45x}$

$\sqrt{(12)}$
     3.464101615
$2\sqrt{(3)}$
     3.464101615

Figure 3 Verify the work

Solution

1. Note that 4 is the largest perfect-square factor of 12. We write 12 as a product of 4 and 3 and then apply the product property for square roots:

$$\sqrt{12} = \sqrt{4 \cdot 3} \qquad \text{4 is the largest perfect-square factor.}$$
$$= \sqrt{4}\sqrt{3} \qquad \text{Product property: } \sqrt{ab} = \sqrt{a}\sqrt{b}$$
$$= 2\sqrt{3} \qquad \sqrt{4} = 2$$

We verify our work by using a calculator to compare approximations for $\sqrt{12}$ and $2\sqrt{3}$ (see Fig. 3).

2. Both 4 and 16 are perfect-square factors of $32x$. Since 16 is the largest perfect-square factor, we write $32x$ as the product of 16 and $2x$ and then apply the product property for square roots:

$$\sqrt{32x} = \sqrt{16 \cdot 2x} \qquad \text{16 is the largest perfect-square factor.}$$
$$= \sqrt{16}\sqrt{2x} \qquad \text{Product property: } \sqrt{ab} = \sqrt{a}\sqrt{b}$$
$$= 4\sqrt{2x} \qquad \sqrt{16} = 4$$

We use a graphing calculator table with nonnegative values of x to verify our work (see Fig. 4).

 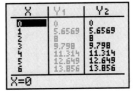

Figure 4 Verify the work

3.
$$7\sqrt{45x} = 7\sqrt{9 \cdot 5x} \qquad \text{9 is the largest perfect-square factor.}$$
$$= 7\sqrt{9}\sqrt{5x} \qquad \text{Product property: } \sqrt{ab} = \sqrt{a}\sqrt{b}$$
$$= 7 \cdot 3 \cdot \sqrt{5x} \qquad \sqrt{9} = 3$$
$$= 21\sqrt{5x} \qquad \text{Multiply.} \qquad ■$$

In Problem 2 of Example 3, we used the largest perfect-square factor, 16, of the radicand $32x$ to simplify $\sqrt{32x}$. What would happen if we had used the smaller perfect-square factor, 4?

$$\sqrt{32x} = \sqrt{4 \cdot 8x} = \sqrt{4}\sqrt{8x} = 2\sqrt{8x}$$

The radicand, $8x$, has a perfect-square factor of 4, so we continue to simplify:

$$\sqrt{32x} = 2\sqrt{8x} = 2\sqrt{4 \cdot 2x} = 2\sqrt{4}\sqrt{2x} = 2 \cdot 2\sqrt{2x} = 4\sqrt{2x}$$

The result is the same as our result in Example 3. This exploration suggests two things:

- The most efficient way to simplify a square root is to use the *largest* perfect-square factor of the radicand.
- If we don't use the largest perfect-square factor of the radicand, we can simplify the square root by continuing to use perfect-square factors until the radicand has no perfect-square factors other than 1.

Simplifying a Square Root with a Small Radicand

To simplify a square root in which it is easy to determine the largest perfect-square factor of the radicand,

1. Write the radicand as the product of the *largest* perfect-square factor and another number.

2. Apply the product property for square roots.

In Example 4, we will discuss how to simplify a radical that has a large radicand.

Example 4 Simplifying a Radical with a Large Radicand

Simplify $\sqrt{72}$.

Solution

We begin by finding the prime factorization of the radicand, 72:

$$\sqrt{72} = \sqrt{2 \cdot 2 \cdot 2 \cdot 3 \cdot 3} \qquad \text{Find prime factorization of 72.}$$

$$= \sqrt{2 \cdot 2 \cdot 3 \cdot 3 \cdot 2} \qquad \text{Rearrange factors to highlight pairs of identical factors.}$$

$$= \sqrt{2 \cdot 2} \cdot \sqrt{3 \cdot 3} \cdot \sqrt{2} \qquad \text{Product property: } \sqrt{ab} = \sqrt{a}\sqrt{b}$$

$$= 2 \cdot 3 \cdot \sqrt{2} \qquad \sqrt{2 \cdot 2} = \sqrt{4} = 2;\ \sqrt{3 \cdot 3} = \sqrt{9} = 3$$

$$= 6\sqrt{2} \qquad \text{Multiply.} \qquad ■$$

Simplifying a Square Root with a Large Radicand

To simplify a square root in which it is difficult to determine the largest perfect-square factor of the radicand,

1. Find the prime factorization of the radicand.
2. Look for pairs of identical factors of the radicand and rearrange the factors to highlight them.
3. Use the product property for square roots.

Simplifying a Square Root Whose Radicand Contains x^2

If x is nonnegative, then $\sqrt{x^2} = x$, because squaring x gives x^2. In this course, we will not discuss the case in which x is negative.

Simplifying $\sqrt{x^2}$

If x is nonnegative, then

$$\sqrt{x^2} = x$$

Example 5 Simplifying a Radical

Simplify. Assume that any variables are nonnegative.

1. $\sqrt{21t^2 w^2}$
2. $3\sqrt{20x^2 y}$

Solution

1.
$$\sqrt{21t^2 w^2} = \sqrt{21}\sqrt{t^2}\sqrt{w^2} \qquad \text{Product property: } \sqrt{ab} = \sqrt{a}\sqrt{b}$$

$$= \left(\sqrt{21}\right)tw \qquad \sqrt{x^2} = x \text{ for nonnegative } x$$

$$= tw\sqrt{21} \qquad \text{Rearrange factors.}$$

2.
$$3\sqrt{20x^2 y} = 3\sqrt{4 \cdot 5 \cdot x^2 \cdot y} \qquad 4 \text{ and } x^2 \text{ are perfect squares.}$$

$$= 3\sqrt{4 \cdot x^2 \cdot 5y} \qquad \text{Rearrange factors.}$$

$$= 3\sqrt{4}\sqrt{x^2}\sqrt{5y} \qquad \text{Product property: } \sqrt{ab} = \sqrt{a}\sqrt{b}$$

$$= 3 \cdot 2 \cdot x \cdot \sqrt{5y} \qquad \sqrt{4} = 2,\ \sqrt{x^2} = x \text{ for nonnegative } x$$

$$= 6x\sqrt{5y} \qquad \text{Multiply.} \qquad ■$$

group exploration

Looking ahead: Quotient property for square roots

1. Find $\sqrt{\dfrac{4}{9}}$ and $\dfrac{\sqrt{4}}{\sqrt{9}}$. Then compare your results.

2. Find $\sqrt{\dfrac{16}{25}}$ and $\dfrac{\sqrt{16}}{\sqrt{25}}$. Then compare your results.

3. Find $\sqrt{\dfrac{49}{81}}$ and $\dfrac{\sqrt{49}}{\sqrt{81}}$. Then compare your results.

4. Write $\sqrt{\dfrac{a}{b}}$ as a quotient of two square roots. [**Hint:** See Problems 1, 2, and 3.]

5. Write each expression as a radical expression in which no radicand is a fraction. Assume that x is positive.

   a. $\sqrt{\dfrac{5}{16}}$

   b. $\sqrt{\dfrac{20}{x^2}}$

TIPS FOR SUCCESS: Use 3-by-5 Cards

Do you have trouble memorizing definitions and properties? If so, try writing a word or phrase on one side of a 3-by-5 card, and, on the other side, put its definition or state its property and how it can be applied. For example, you could write "product property for square roots" on one side of a card and "For nonnegative numbers a and b, $\sqrt{ab} = \sqrt{a}\sqrt{b}$" on the other side. You could also describe, in your own words, the meaning of the property and how you can apply it.

Once you have completed a card for each definition and property, shuffle the cards and quiz yourself until you are confident that you know the definitions and properties and how to apply them. Quiz yourself again later to make sure that you have retained the information.

In addition to memorizing definitions and properties, it is important that you strive to understand their meanings and how to apply them.

HOMEWORK 9.1

FOR EXTRA HELP ▶

 Student Solutions Manual

 PH Math/Tutor Center

Math XL MathXL®

MyMathLab MyMathLab

Find the square root.

1. $\sqrt{4}$
2. $\sqrt{16}$
3. $\sqrt{81}$
4. $\sqrt{36}$
5. $\sqrt{121}$
6. $\sqrt{100}$
7. $\sqrt{144}$
8. $\sqrt{225}$
9. $-\sqrt{16}$
10. $-\sqrt{9}$
11. $-\sqrt{81}$
12. $-\sqrt{169}$
13. $\sqrt{-9}$
14. $\sqrt{-4}$
15. $-\sqrt{-25}$
16. $-\sqrt{-16}$

State whether the square root is rational or irrational. If the square root is rational, find the (exact) value. If the square root is irrational, estimate its value by rounding to the second decimal place.

17. $\sqrt{30}$
18. $\sqrt{62}$
19. $\sqrt{78}$
20. $\sqrt{256}$
21. $\sqrt{196}$
22. $\sqrt{95}$

Simplify. Use a calculator to verify your work.

23. $\sqrt{20}$
24. $\sqrt{28}$
25. $\sqrt{45}$
26. $\sqrt{18}$
27. $\sqrt{27}$
28. $\sqrt{8}$
29. $\sqrt{50}$
30. $\sqrt{32}$

31. $\sqrt{300}$
32. $\sqrt{75}$
33. $-\sqrt{98}$
34. $-\sqrt{80}$
35. $4\sqrt{72}$
36. $5\sqrt{52}$
37. $3\sqrt{120}$
38. $2\sqrt{135}$

Simplify. Assume that any variables are nonnegative.

39. $\sqrt{9x}$
40. $\sqrt{25x}$
41. $\sqrt{64t}$
42. $\sqrt{49a}$
43. $\sqrt{81x^2}$
44. $\sqrt{36x^2}$
45. $\sqrt{225t^2w^2}$
46. $\sqrt{169a^2b^2}$
47. $\sqrt{5x^2}$
48. $\sqrt{17x^2}$
49. $7\sqrt{39x^2y}$
50. $4\sqrt{23x^2y}$
51. $\sqrt{12p}$
52. $\sqrt{24t}$
53. $2\sqrt{63x}$
54. $5\sqrt{40x}$
55. $\sqrt{60a^2b^2}$
56. $\sqrt{72t^2w^2}$
57. $3\sqrt{125xy^2}$
58. $5\sqrt{48xy^2}$

59. Suppose that, while traveling on a dry road, a driver slams on the brakes. Let D be the distance (in feet) that the car will skid and S be the speed (in miles per hour) before braking. The relationship between D and S is described by the model

$S = \sqrt{30FD}$, where F is a drag factor, which is a measure of the roughness of the surface of the road. The drag factor on new concrete is 0.95, and the drag factor on polished concrete or asphalt is 0.75.

A motorist who was involved in an accident claims that he was driving at the posted speed limit of 60 miles per hour. A police officer measures the car's skid marks on the asphalt road to be 210 feet long. Assuming that the motorist applied the brakes suddenly, estimate the speed at which the motorist was traveling before braking.

60. The size of a rectangular television screen is usually described by the length of its diagonal, given (in inches) by $d = \sqrt{w^2 + h^2}$, where w is the screen's width and h is the screen's height, both in inches. Find the length of the diagonal of a television screen whose width is 17 inches and height is 11 inches.

61. A student thinks that $\sqrt{-25} = -5$. Is the student correct? Explain.

62. A student thinks that $-\sqrt{-9} = 3$. Is the student correct? Explain.

Without using a calculator, find the two consecutive integers nearest to the given radical. Verify your result by using a calculator.

63. $\sqrt{22}$ **64.** $\sqrt{15}$ **65.** $\sqrt{71}$ **66.** $\sqrt{43}$

67. a. Which of the two numbers is larger?
    i. 2, $\sqrt{2}$ **ii.** 5, $\sqrt{5}$ **iii.** 8, $\sqrt{8}$
    b. Describe any patterns in your results for part (a).
    c. Which of the two numbers is larger? [**Hint:** Use a calculator to be sure.]
    i. 0.2, $\sqrt{0.2}$ **ii.** 0.5, $\sqrt{0.5}$ **iii.** 0.8, $\sqrt{0.8}$
    d. Describe any patterns in your results for part (c).
    e. For which type of number does finding the principal square root give a smaller result? A larger result?

68. a. Square the given number. Then find the principal square root of the result.
    i. 3   **ii.** 5   **iii.** 6
    b. Find the square root.
    i. $\sqrt{3^2}$ **ii.** $\sqrt{5^2}$ **iii.** $\sqrt{6^2}$
    c. Describe any patterns from your work in parts (a) and (b).
    d. Explain why $\sqrt{x^2} = x$, where x is nonnegative.

69. Is $\sqrt{10}$ twice as big as $\sqrt{5}$? Explain. Include the product rule for square roots in your discussion.

70. Is $2\sqrt{3}$ equal to $\sqrt{6}$? Explain. Include the product rule for square roots in your discussion.

Sketch the graph of the equation. Verify your result by using a graphing calculator.

71. $y = \sqrt{x}$ **72.** $y = -\sqrt{x}$

Related Review

Simplify. Assume that x is nonnegative.

73. $(7x)^2$ **74.** $(5x)^2$ **75.** $\sqrt{49x^2}$ **76.** $\sqrt{25x^2}$

Expressions, Equations, and Graphs

Perform the indicated instruction. Then use words such as linear, quadratic, cubic, power, radical, polynomial, degree, one variable, *and* two variables *to describe the expression, equation, or system.*

77. Graph $y = -2x^2$ by hand.

78. Factor $4r^3 - 12r^2 - r + 3$.

79. Solve $x^2 = 6x - 8$.

80. Simplify $(m + 2n)^2$.

81. Simplify $\sqrt{68x}$.

82. Solve the following:

$$6x + 5y = 8$$
$$2x - 3y = -16$$

9.2 SIMPLIFYING MORE RADICAL EXPRESSIONS

Objectives

▷ Know the *quotient property for square roots*.

▷ Simplify a radical expression.

▷ *Rationalize the denominator* of a fraction.

▷ Know how to completely simplify a radical expression.

In Section 9.1, we discussed how to use the product property for square roots to help us simplify the principal square root of a product. In this section, we will discuss a property that will help us simplify the principal square root of a quotient.

Quotient Property for Square Roots

Here, we compute the square root of a quotient and a quotient of square roots:

$$\text{Square root of a quotient:} \quad \sqrt{\frac{16}{49}} = \frac{4}{7} \quad \text{because} \quad \left(\frac{4}{7}\right)^2 = \frac{16}{49}$$

$$\text{Quotient of square roots:} \quad \frac{\sqrt{16}}{\sqrt{49}} = \frac{4}{7}$$

Since the two results are equal, we can write

$$\sqrt{\frac{16}{49}} = \frac{\sqrt{16}}{\sqrt{49}}$$

This equation suggests the **quotient property for square roots.**

Quotient Property for Square Roots

For a nonnegative number a and a positive number b,

$$\sqrt{\frac{a}{b}} = \frac{\sqrt{a}}{\sqrt{b}}$$

In words: The square root of a quotient is the quotient of the square roots.

Using the Quotient Property for Square Roots to Simplify Radicals

If a radical has a fractional radicand, we **simplify the radical** by writing it as an expression whose radicand is not a fraction.

Example 1   Simplifying Radicals

Simplify. Assume that x and y are positive.

1. $\sqrt{\dfrac{5}{x^2}}$
2. $\sqrt{\dfrac{50}{81}}$
3. $\sqrt{\dfrac{3x^2 y}{49}}$

Solution

1. $\sqrt{\dfrac{5}{x^2}} = \dfrac{\sqrt{5}}{\sqrt{x^2}}$ Quotient property: $\sqrt{\dfrac{a}{b}} = \dfrac{\sqrt{a}}{\sqrt{b}}$

$\phantom{\sqrt{\dfrac{5}{x^2}}} = \dfrac{\sqrt{5}}{x}$ $\sqrt{x^2} = x$ for nonnegative x

We use a graphing calculator table with positive values of x to verify our work (see Fig. 5).

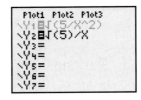 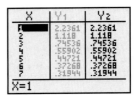

Figure 5 Verify the work

2. $\sqrt{\dfrac{50}{81}} = \dfrac{\sqrt{50}}{\sqrt{81}}$ Quotient property: $\sqrt{\dfrac{a}{b}} = \dfrac{\sqrt{a}}{\sqrt{b}}$

$\phantom{\sqrt{\dfrac{50}{81}}} = \dfrac{\sqrt{25 \cdot 2}}{9}$ 25 is a perfect square; $\sqrt{81} = 9$.

$\phantom{\sqrt{\dfrac{50}{81}}} = \dfrac{\sqrt{25}\sqrt{2}}{9}$ Product property: $\sqrt{ab} = \sqrt{a}\sqrt{b}$

$\phantom{\sqrt{\dfrac{50}{81}}} = \dfrac{5\sqrt{2}}{9}$ $\sqrt{25} = 5$

3. $\sqrt{\dfrac{3x^2 y}{49}} = \dfrac{\sqrt{3x^2 y}}{\sqrt{49}}$ Quotient property: $\sqrt{\dfrac{a}{b}} = \dfrac{\sqrt{a}}{\sqrt{b}}$

$\phantom{\sqrt{\dfrac{3x^2 y}{49}}} = \dfrac{\sqrt{x^2}\sqrt{3y}}{7}$ Product property: $\sqrt{ab} = \sqrt{a}\sqrt{b}$; $\sqrt{49} = 7$

$\phantom{\sqrt{\dfrac{3x^2 y}{49}}} = \dfrac{x\sqrt{3y}}{7}$ $\sqrt{x^2} = x$ for nonnegative x

■

Rationalizing the Denominator of a Radical Expression

We simplify an expression of the form $\dfrac{p}{\sqrt{q}}$ by leaving no denominator as a radical expression. We call this process **rationalizing the denominator.**

Example 2 Rationalizing the Denominator

Simplify.

1. $\dfrac{5}{\sqrt{7}}$ **2.** $\dfrac{2}{\sqrt{x}}$

Solution

1. $\dfrac{5}{\sqrt{7}} = \dfrac{5}{\sqrt{7}} \cdot 1$ $a = a \cdot 1$

     $= \dfrac{5}{\sqrt{7}} \cdot \dfrac{\sqrt{7}}{\sqrt{7}}$ Rationalize denominator: $\dfrac{\sqrt{7}}{\sqrt{7}} = 1$

     $= \dfrac{5\sqrt{7}}{\sqrt{49}}$ Product property: $\sqrt{a}\sqrt{b} = \sqrt{ab}$

     $= \dfrac{5\sqrt{7}}{7}$ $\sqrt{49} = 7$

2. $\dfrac{2}{\sqrt{x}} = \dfrac{2}{\sqrt{x}} \cdot \dfrac{\sqrt{x}}{\sqrt{x}}$ Rationalize denominator: $\dfrac{\sqrt{x}}{\sqrt{x}} = 1$

     $= \dfrac{2\sqrt{x}}{\sqrt{x^2}}$ Product property: $\sqrt{a}\sqrt{b} = \sqrt{ab}$

     $= \dfrac{2\sqrt{x}}{x}$ $\sqrt{x^2} = x$ for nonnegative x   ■

As shown in Example 2, **to rationalize the denominator of a fraction of the form** $\dfrac{p}{\sqrt{q}}$**, where** q **is positive, we multiply the fraction by 1 in the form** $\dfrac{\sqrt{q}}{\sqrt{q}}$**.**

Example 3 Rationalizing the Denominator

Simplify.

1. $\sqrt{\dfrac{7}{3}}$ **2.** $\sqrt{\dfrac{3t}{20}}$

Solution

1. We first use the quotient property for square roots and then rationalize the denominator.

$$\sqrt{\dfrac{7}{3}} = \dfrac{\sqrt{7}}{\sqrt{3}} \qquad \text{Quotient property: } \sqrt{\dfrac{a}{b}} = \dfrac{\sqrt{a}}{\sqrt{b}}$$

$$= \dfrac{\sqrt{7}}{\sqrt{3}} \cdot \dfrac{\sqrt{3}}{\sqrt{3}} \qquad \text{Rationalize denominator.}$$

$$= \dfrac{\sqrt{21}}{\sqrt{9}} \qquad \text{Product property: } \sqrt{a}\sqrt{b} = \sqrt{ab}$$

$$= \dfrac{\sqrt{21}}{3} \qquad \sqrt{9} = 3$$

2. $\sqrt{\dfrac{3t}{20}} = \dfrac{\sqrt{3t}}{\sqrt{20}}$ Quotient property: $\sqrt{\dfrac{a}{b}} = \dfrac{\sqrt{a}}{\sqrt{b}}$

$\qquad = \dfrac{\sqrt{3t}}{2\sqrt{5}}$ $\sqrt{20} = \sqrt{4 \cdot 5} = \sqrt{4}\sqrt{5} = 2\sqrt{5}$

$\qquad = \dfrac{\sqrt{3t}}{2\sqrt{5}} \cdot \dfrac{\sqrt{5}}{\sqrt{5}}$ Rationalize denominator.

$\qquad = \dfrac{\sqrt{15t}}{2\sqrt{25}}$ Product property: $\sqrt{a}\sqrt{b} = \sqrt{ab}$

$\qquad = \dfrac{\sqrt{15t}}{10}$ $\sqrt{25} = 5$ ∎

Simplifying a Radical Quotient

Next we will **simplify a radical quotient.** We will use this skill when solving quadratic equations in Section 9.5.

Example 4   Simplifying a Radical Quotient

Simplify $\dfrac{6 + 3\sqrt{2}}{12}$.

Solution

We first factor the numerator and then simplify the expression:

$$\dfrac{6 + 3\sqrt{2}}{12} = \dfrac{3(2 + \sqrt{2})}{3(4)} \qquad \text{Factor out 3.}$$

$$= \dfrac{3}{3} \cdot \dfrac{2 + \sqrt{2}}{4} \qquad \dfrac{ac}{bd} = \dfrac{a}{b} \cdot \dfrac{c}{d}$$

$$= \dfrac{2 + \sqrt{2}}{4} \qquad \dfrac{3}{3} = 1; \text{ simplify.}$$ ∎

Example 5   Simplifying a Radical Quotient

Simplify $\dfrac{8 - \sqrt{28}}{10}$.

Solution

We begin by simplifying the radical $\sqrt{28}$:

$$\dfrac{8 - \sqrt{28}}{10} = \dfrac{8 - 2\sqrt{7}}{10} \qquad \sqrt{28} = \sqrt{4 \cdot 7} = \sqrt{4}\sqrt{7} = 2\sqrt{7}$$

$$= \dfrac{2(4 - \sqrt{7})}{2(5)} \qquad \text{Factor out 2.}$$

$$= \dfrac{2}{2} \cdot \dfrac{4 - \sqrt{7}}{5} \qquad \dfrac{ac}{bd} = \dfrac{a}{b} \cdot \dfrac{c}{d}$$

$$= \dfrac{4 - \sqrt{7}}{5} \qquad \dfrac{2}{2} = 1; \text{ simplify.}$$

Figure 6 Verify the work

We verify our work by using a calculator to compare approximations for $\dfrac{8 - \sqrt{28}}{10}$ and $\dfrac{4 - \sqrt{7}}{5}$ (see Fig. 6). When we perform such a check, it is important that all parentheses be entered as shown. ∎

In Example 5, we obtained the result $\dfrac{4 - \sqrt{7}}{5}$. Keep in mind that this radical expression is a number. From our check in Fig. 6, we see that $\dfrac{4 - \sqrt{7}}{5} \approx 0.27$.

Summary of Simplifying a Radical Expression

We have discussed various ways to simplify a radical expression. What follows is a summary of these methods.

Simplifying a Radical Expression

To simplify a radical expression,

1. Use the quotient property for square roots so that no radicand is a fraction.
2. Use the product property for square roots so that no radicands have perfect-square factors other than 1.
3. Rationalize denominators so that no denominator is a radical expression.
4. Use the property $\dfrac{a}{a} = 1$, where a is nonzero, to simplify.
5. Continue applying steps 1–4 until the radical expression is completely simplified.

---

group exploration

Estimating the value of a radical expression

1. For parts (a) through (c), do not use a calculator.
   a. Between what two consecutive counting numbers is $\sqrt{19}$? Explain.
   b. Between what two consecutive counting numbers is $3 + \sqrt{19}$? Explain. [**Hint:** Build on your result from part (a).]
   c. Between what two rational numbers is $\dfrac{3 + \sqrt{19}}{2}$? Explain. Your results should be no more than 0.5 apart. [**Hint:** Build on your result from part (b).]
   d. Use a calculator to estimate $\dfrac{3 + \sqrt{19}}{2}$. Is your work in part (c) correct? Explain.

2. a. Without using a calculator, determine between what two rational numbers is $\dfrac{9 - \sqrt{29}}{2}$. Your results should be no more than 0.5 apart.
   b. Use a calculator to estimate $\dfrac{9 - \sqrt{29}}{2}$. Is your work in part (a) correct? Explain.

HOMEWORK 9.2   FOR EXTRA HELP ▶

Student Solutions Manual   PH Math/Tutor Center   MathXL®   MyMathLab

---

Simplify. Assume that any variables are positive. Use a graphing calculator table to verify your work when possible.

1. $\sqrt{\dfrac{25}{36}}$
2. $\sqrt{\dfrac{9}{49}}$
3. $\sqrt{\dfrac{121}{x^2}}$
4. $\sqrt{\dfrac{196}{x^2}}$

5. $\sqrt{\dfrac{7x}{25}}$
6. $\sqrt{\dfrac{3x}{16}}$
7. $\sqrt{\dfrac{19}{a^2}}$
8. $\sqrt{\dfrac{23}{t^2}}$

9. $\sqrt{\dfrac{5}{x^2 y^2}}$
10. $\sqrt{\dfrac{3}{a^2 b^2}}$
11. $-\sqrt{\dfrac{8}{49}}$
12. $-\sqrt{\dfrac{27}{16}}$

13. $\sqrt{\dfrac{20}{81}}$
14. $\sqrt{\dfrac{45}{4}}$
15. $\sqrt{\dfrac{75w}{36}}$
16. $\sqrt{\dfrac{32y}{9}}$

17. $\sqrt{\dfrac{4a}{b^2}}$
18. $\sqrt{\dfrac{25x}{y^2}}$
19. $\sqrt{\dfrac{80}{x^2}}$
20. $\sqrt{\dfrac{200}{x^2}}$

21. $\sqrt{\dfrac{7r^2 t}{81}}$
22. $\sqrt{\dfrac{5ab^2}{49}}$

Simplify. Assume that any variables are positive. Use a graphing calculator table to verify your work when possible.

23. $\dfrac{2}{\sqrt{3}}$
24. $\dfrac{5}{\sqrt{2}}$
25. $\dfrac{6}{\sqrt{5}}$
26. $\dfrac{4}{\sqrt{7}}$

27. $\dfrac{a}{\sqrt{13}}$
28. $\dfrac{t}{\sqrt{17}}$
29. $\dfrac{7}{\sqrt{x}}$
30. $\dfrac{9}{\sqrt{x}}$

31. $\sqrt{\dfrac{2}{7}}$
32. $\sqrt{\dfrac{3}{5}}$
33. $\sqrt{\dfrac{11}{2}}$
34. $\sqrt{\dfrac{5}{6}}$

35. $\sqrt{\dfrac{7}{p}}$
36. $\sqrt{\dfrac{2}{b}}$
37. $\sqrt{\dfrac{3}{8}}$
38. $\sqrt{\dfrac{5}{18}}$

39. $\sqrt{\dfrac{3x}{50}}$ **40.** $\sqrt{\dfrac{7x}{32}}$ **41.** $\sqrt{\dfrac{5w^2}{12}}$ **42.** $\sqrt{\dfrac{3c^2}{28}}$

43. $\sqrt{\dfrac{7x^2 y}{5}}$ **44.** $\sqrt{\dfrac{3xy^2}{7}}$

Simplify. Use a calculator to verify your work.

45. $\dfrac{9 + 3\sqrt{2}}{6}$ **46.** $\dfrac{10 + 6\sqrt{5}}{8}$ **47.** $\dfrac{8 - 4\sqrt{7}}{4}$

48. $\dfrac{15 - 5\sqrt{2}}{5}$ **49.** $\dfrac{8 + 12\sqrt{13}}{6}$ **50.** $\dfrac{21 + 6\sqrt{17}}{9}$

51. $\dfrac{4 + \sqrt{12}}{8}$ **52.** $\dfrac{6 + \sqrt{20}}{4}$ **53.** $\dfrac{10 - \sqrt{50}}{20}$

54. $\dfrac{6 - \sqrt{27}}{12}$ **55.** $\dfrac{9 - \sqrt{45}}{6}$ **56.** $\dfrac{25 - \sqrt{75}}{10}$

57. Let T be the amount of time (in seconds) it takes a baseball to fall to the ground when the ball has dropped from h feet above the ground. A good model is $T = \sqrt{\dfrac{h}{16}}$.
  a. Simplify the right-hand side of the equation.
  b. How long would it take for a baseball to reach the ground if it were to be dropped from the top of Chicago's Sears Tower, which is 1450 feet tall?

58. The distance d (in miles) to the horizon at an altitude of h feet above sea level is given by the equation $d = \sqrt{\dfrac{3h}{2}}$.
  a. Simplify the right-hand side of the equation.
  b. New York City's Empire State Building is 1250 feet tall. What would be the distance from the top of the building to the horizon? Assume that the base of the building is at sea level.
  c. If an airplane flies at an altitude of 34,000 feet over the skyscraper, what is the distance from the airplane to the horizon?

59. A student tries to simplify $\dfrac{5}{\sqrt{3}}$:

$$\frac{5}{\sqrt{3}} = \left(\frac{5}{\sqrt{3}}\right)^2$$

$$= \frac{5^2}{\left(\sqrt{3}\right)^2}$$

$$= \frac{25}{3}$$

Describe any errors. Then simplify the expression correctly.

60. A student thinks that $\dfrac{\sqrt{14}}{7} = 2$. Is the student correct? Explain.

61. A student tries to simplify $\dfrac{3}{\sqrt{20}}$:

$$\frac{3}{\sqrt{20}} = \frac{3}{\sqrt{20}} \cdot \frac{\sqrt{20}}{\sqrt{20}}$$

$$= \frac{3\sqrt{20}}{20}$$

$$= \frac{3\sqrt{4 \cdot 5}}{20}$$

$$= \frac{3(2)\sqrt{5}}{2(10)}$$

$$= \frac{3\sqrt{5}}{10}$$

Is the work correct? Is there an easier approach? Explain.

62. Describe how to simplify a radical expression.

Related Review

Factor.

63. $12 + 8x$ **64.** $10 - 35x$

65. $12 + 8\sqrt{3}$ **66.** $10 - 35\sqrt{2}$

67. $14 - 21x$ **68.** $18 + 27x$

69. $14 - 21\sqrt{7}$ **70.** $18 + 27\sqrt{5}$

Expressions, Equations, and Graphs

Perform the indicated instruction. Then use words such as linear, quadratic, cubic, power, radical, polynomial, degree, one variable, *and* two variables *to describe the expression, equation, or system.*

71. Solve $x^2 - 4x = 0$.

72. Find the difference $(2x + 4) - (5x - 3)$.

73. Factor $x^2 - 4x$.

74. Find the product $(2x + 4)(5x - 3)$.

75. Graph $y = x^2 - 4x$ by hand.

76. Solve $(2x + 4)(5x - 3) = 0$.

SOLVING QUADRATIC EQUATIONS BY THE SQUARE ROOT PROPERTY; 9.3 THE PYTHAGOREAN THEOREM

Objectives

▷ Use the *square root property* to solve quadratic equations.

▷ Know the *Pythagorean theorem*.

▷ Use the Pythagorean theorem to find the length of a side of a right triangle.

▷ Use the Pythagorean theorem to model a situation.

In Section 8.5, we discussed how to use factoring to help solve some quadratic equations in one variable. In this section, we will discuss another way to solve some quadratic equations.

Solving Quadratic Equations of the Form $x^2 = k$

Consider the equation $x^2 = 9$. We first use factoring to solve this equation:

$$x^2 = 9 \qquad \text{Original equation}$$
$$x^2 - 9 = 0 \qquad \text{Subtract 9 from both sides.}$$
$$(x + 3)(x - 3) = 0 \qquad \text{Factor left side.}$$
$$x + 3 = 0 \quad \text{or} \quad x - 3 = 0 \qquad \text{Zero factor property}$$
$$x = -3 \quad \text{or} \quad x = 3$$

So, the solutions are -3 and 3. We can use the notation ± 3 to stand for the numbers -3 and 3.

It is not necessary to use factoring to solve the equation $x^2 = 9$, however. Notice that a solution of this equation is any number that, when squared, is equal to 9. So, the solutions are the square roots of 9; in symbols, we write $x = \pm\sqrt{9} = \pm 3$.

Here, we solve $x^2 = 16$:

$$x^2 = 16$$
$$x = \pm\sqrt{16}$$
$$x = \pm 4$$

This work suggests the *square root property*.

Square Root Property

Let k be a nonnegative constant. Then $x^2 = k$ is equivalent to

$$x = \pm\sqrt{k}$$

Example 1 Using the Square Root Property to Solve Equations

Solve.

1. $x^2 = 25$ **2.** $x^2 = 3$ **3.** $x^2 = -9$

Solution

1. $x^2 = 25$ Original equation

     $x = \pm\sqrt{25}$ Square root property

     $x = \pm 5$ Simplify.

2. $x^2 = 3$ Original equation

     $x = \pm\sqrt{3}$ Square root property

3. Since the square of a number is nonnegative, we conclude that $x^2 = -9$ has no real-number solutions. ∎

WARNING   It is a common error to confuse solving an equation such as $x^2 = 25$ with computing a principal square root such as $\sqrt{25}$. In Problem 1 of Example 1, we found that the solutions of $x^2 = 25$ are the *two* numbers -5 and 5. Recall from Section 9.1 that $\sqrt{25}$ is the *one* number 5.

Example 2 Using the Square Root Property to Solve Equations

Solve.

1. $x^2 - 24 = 0$ **2.** $3x^2 - 4 = 3$

Solution

1. First, we isolate x^2 to (get x^2 alone on) one side of the equation:

$$x^2 - 24 = 0 \qquad \text{Original equation}$$
$$x^2 = 24 \qquad \text{Add 24 to both sides.}$$
$$x = \pm\sqrt{24} \qquad \text{Square root property}$$
$$x = \pm 2\sqrt{6} \qquad \text{Simplify.}$$

2. First, we isolate x^2 to one side of the equation:

$$3x^2 - 4 = 3 \qquad \text{Original equation}$$
$$3x^2 = 7 \qquad \text{Add 4 to both sides.}$$
$$x^2 = \frac{7}{3} \qquad \text{Divide both sides by 3.}$$
$$x = \pm\sqrt{\frac{7}{3}} \qquad \text{Square root property}$$
$$x = \pm\frac{\sqrt{7}}{\sqrt{3}} \qquad \text{Quotient property: } \sqrt{\frac{a}{b}} = \frac{\sqrt{a}}{\sqrt{b}}$$
$$x = \pm\frac{\sqrt{7}}{\sqrt{3}} \cdot \frac{\sqrt{3}}{\sqrt{3}} \qquad \text{Rationalize denominator.}$$
$$x = \pm\frac{\sqrt{21}}{3} \qquad \text{Simplify.}$$

To verify our work, we use a graphing calculator to graph $y = 3x^2 - 4$ and $y = 3$ in the same coordinate system and then use "intersect" to find the approximate intersection points $(-1.53, 3)$ and $(1.53, 3)$, which have x-coordinates -1.53 and 1.53 (see Fig. 7). So, the approximate solutions of the original equation are -1.53 and 1.53, which check.

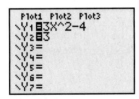 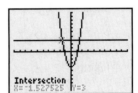 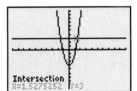

Figure 7 Verify the work

From Example 2, we see that **to solve an equation of the form $ax^2 + c = k$, we isolate x^2 to one side of the equation before using the square root property.**

Solving Quadratic Equations of the Form $(x + p)^2 = k$

We can also use the square root property to solve equations of the form $(x + p)^2 = k$.

Example 3   Using the Square Root Property to Solve Equations

Solve.

1. $(x + 4)^2 = 36$
2. $(x - 7)^2 = 50$

Solution

1. Note that the base of $(x + 4)^2$ is $x + 4$. We can still use the square root property to solve the equation $(x + 4)^2 = 36$.

$$(x + 4)^2 = 36 \qquad \text{Original equation}$$
$$x + 4 = \pm\sqrt{36} \qquad \text{Square root property}$$
$$x + 4 = \pm 6 \qquad \text{Simplify.}$$
$$x + 4 = -6 \quad \text{or} \quad x + 4 = 6 \qquad \text{Write as two equations}$$
$$x = -10 \quad \text{or} \quad x = 2$$

2. $(x - 7)^2 = 50$ Original equation

$x - 7 = \pm\sqrt{50}$ Square root property

$x - 7 = \pm 5\sqrt{2}$ Simplify.

$x = 7 \pm 5\sqrt{2}$ Add 7 to both sides.

So, the solutions are $7 - 5\sqrt{2}$ and $7 + 5\sqrt{2}$. We use graphing calculator tables to check that if $7 - 5\sqrt{2}$ or $7 + 5\sqrt{2}$ is substituted for x in the expression $(x - 7)^2$, the result is 50 (see Fig. 8).

 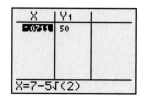 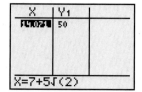

Figure 8 Verify the work ■

The Pythagorean Theorem

Now that we know the square root property, we can use it to help us work with a special type of triangle called a *right triangle*. An angle of 90° is called a *right angle*. If one angle of a triangle measures 90°, the triangle is a **right triangle** (see Fig. 9).

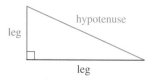

Figure 9 A right triangle

The side opposite the right angle is the longest side. We call that side the **hypotenuse,** and we call the two shorter sides the **legs.** The **Pythagorean theorem** describes the relationship between the lengths of the legs and the hypotenuse of a right triangle.

Figure 10 The Pythagorean theorem: $a^2 + b^2 = c^2$

Pythagorean Theorem

If a and b are the lengths of the legs of a right triangle and c is the length of the hypotenuse, then

$$a^2 + b^2 = c^2$$

In words: The sum of the squares of the lengths of the legs is equal to the square of the length of the hypotenuse (see Fig. 10).

Finding the Length of a Side of a Right Triangle

If we know the lengths of two of the three sides of a right triangle, how can we use the Pythagorean theorem to find the length of the third side?

Example 4   Finding the Length of a Side of a Right Triangle

The lengths of two sides of a right triangle are given. Find the length of the third side.

1.

2.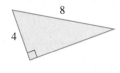

Solution

1. Since the lengths of the legs are given, we are to find the length of the hypotenuse. We substitute $a = 3$ and $b = 4$ into the equation $c^2 = a^2 + b^2$ and solve for c:

$$c^2 = a^2 + b^2 \qquad \text{Pythagorean theorem}$$
$$c^2 = 3^2 + 4^2 \qquad \text{Substitute 3 for } a \text{ and 4 for } b.$$
$$c^2 = 9 + 16 \qquad \text{Find the squares.}$$
$$c^2 = 25 \qquad \text{Add.}$$
$$c = \sqrt{25} \qquad c \text{ is nonnegative, so disregard } -\sqrt{25}.$$
$$c = 5 \qquad \text{Simplify.}$$

The length of the hypotenuse is 5 units.

2. The length of the hypotenuse is 8 and the length of one of the legs is 4. We substitute $a = 4$ and $c = 8$ into the equation $a^2 + b^2 = c^2$ and solve for b:

$$a^2 + b^2 = c^2 \qquad \text{Pythagorean theorem}$$
$$4^2 + b^2 = 8^2 \qquad \text{Substitute 4 for } a \text{ and 8 for } c.$$
$$16 + b^2 = 64 \qquad \text{Find the squares.}$$
$$b^2 = 48 \qquad \text{Isolate } b^2: \text{ Subtract 16 from both sides.}$$
$$b = \sqrt{48} \qquad b \text{ is nonnegative, so disregard } -\sqrt{48}.$$
$$b = 4\sqrt{3} \qquad \text{Simplify.}$$

The length of the second leg is $4\sqrt{3}$ units (about 6.93 units). ■

Applying the Pythagorean Theorem to Model a Situation

The Pythagorean theorem can be used to model many situations. It is used by surveyors, architects, pilots, navigators, engineers, and many other professionals. Several applications are included in the Homework exercises and in the next example.

Example 5 Modeling a Situation

A person is flying a kite. If the person has let out 102.9 feet of string and the horizontal distance from the person to the kite is 61.5 feet, find the vertical distance between the bottom of the kite and the ground.

Solution

First, we define h to be the vertical distance (in feet) between the bottom of the kite and the ground. Then, we draw a sketch that describes the situation (see Fig. 11).

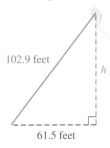

102.9 feet

h

61.5 feet   **Figure 11** Flying a kite

The triangle in Fig. 11 is a right triangle in which the hypotenuse has length 102.9 feet and one of its legs has length 61.5 feet. We use the Pythagorean theorem to find h.

$$61.5^2 + h^2 = 102.9^2 \qquad \text{Pythagorean theorem}$$
$$3782.25 + h^2 = 10{,}588.41 \qquad \text{Find the squares.}$$
$$h^2 = 6806.16 \qquad \text{Isolate } h^2: \text{ Subtract 3782.25 from both sides.}$$
$$h = \sqrt{6806.16} \qquad h \text{ is nonnegative, so disregard } -\sqrt{6806.16}.$$
$$h \approx 82.50 \qquad \text{Find approximate square root.}$$

So, the bottom of the kite is approximately 82.5 feet above the ground. ■

group exploration

Pythagorean theorem and its converse

For this exploration, you will need a ruler, scissors, and paper to help you work with right triangles and other triangles. It will be helpful to have a tool for drawing right angles, such as a protractor or graph paper, although a corner of a piece of paper will suffice. For each triangle, assume that c is the length of the *longest* side (in some triangles, more than one side can be equally the longest) and a and b are the lengths of the other sides.

1. Sketch three right triangles of different sizes. Measure the sides. Show that, for each right triangle, $a^2 + b^2 = c^2$.

2. Sketch three triangles of different sizes that are *not* right triangles. For these triangles, check whether $a^2 + b^2 = c^2$.

3. Sketch a triangle that has an angle close, but not equal, to 90°. Check whether $a^2 + b^2 \approx c^2$. If you cannot show this for your triangle, repeat the problem with a triangle that has an angle even closer to 90°.

4. Note that if $a = 3$, $b = 5$, and $c = \sqrt{34}$, then $a^2 + b^2 = c^2$. Cut three thin strips of paper that are about 3, 5, and $\sqrt{34}$ inches in length. Form a triangle with the three strips of paper. Is the triangle a right triangle?

5. Find values of a, b, and c, other than the ones in Problem 4, such that $a^2 + b^2 = c^2$. Then repeat Problem 4, using your values.

6. Find three more values of a, b, and c, other than those in Problems 4 and 5, such that $a^2 + b^2 = c^2$. Then repeat Problem 4, using your values.

7. Summarize at least three concepts addressed in this exploration.

TIPS FOR SUCCESS: Plan for Final Exams

Don't wait until the last minute to begin studying for your final exams. Look at your finals schedule and decide how you will allocate your time to prepare for each final.

It is important that you be well rested during your finals so that you can concentrate fully during your exams. Plan to do some fun activities that involve exercise, as that is a great way to neutralize stress.

HOMEWORK 9.3

FOR EXTRA HELP ▶

Student Solutions Manual

PH Math/Tutor Center

Math XL
MathXL®

MyMathLab
MyMathLab

Solve. Use a graphing calculator to verify your work.

1. $x^2 = 4$
2. $x^2 = 16$
3. $x^2 = 196$
4. $x^2 = 169$
5. $x^2 = 0$
6. $x^2 = 1$
7. $x^2 = 15$
8. $x^2 = 17$
9. $x^2 = 20$
10. $x^2 = 18$
11. $x^2 = 27$
12. $x^2 = 50$
13. $x^2 = -49$
14. $x^2 = -25$
15. $x^2 - 28 = 0$
16. $x^2 - 40 = 0$
17. $x^2 + 17 = 0$
18. $x^2 + 8 = 0$
19. $4x^2 = 5$
20. $9x^2 = 11$
21. $5x^2 = 7$
22. $7x^2 = 3$
23. $8m^2 = 5$
24. $12y^2 = 11$
25. $2x^2 + 4 = 7$
26. $3x^2 + 1 = 8$
27. $5x^2 - 3 = 11$
28. $7x^2 - 8 = -3$

Solve. Use a graphing calculator to verify your work.

29. $(x + 2)^2 = 16$
30. $(x + 4)^2 = 25$
31. $(p - 3)^2 = 36$
32. $(t - 6)^2 = 49$
33. $(x - 7)^2 = 13$
34. $(x - 4)^2 = 35$
35. $(x + 3)^2 = -16$
36. $(x - 1)^2 = -9$
37. $(x + 2)^2 = 18$
38. $(x + 7)^2 = 45$
39. $(r - 6)^2 = 24$
40. $(p - 1)^2 = 75$
41. $(x - 5)^2 = 0$
42. $(x + 9)^2 = 0$

Use factoring or the square root property to solve the equation. Use a graphing calculator to verify your work.

43. $x^2 = 4x + 12$
44. $x^2 = -12x - 20$
45. $y^2 - 81 = 0$
46. $m^2 - 49 = 0$
47. $(x - 2)^2 = 24$
48. $(x + 8)^2 = 32$

49. $3x^2 + 4 = 15$

50. $25x^2 - 1 = 16$

51. $2x^2 - 15 = -7x$

52. $3x^2 + 2 = 5x$

53. $(t - 1)^2 = 1$

54. $(y - 7)^2 = 36$

The lengths of two sides of a right triangle are given. Find the length of the third side. (The triangle is not drawn to scale.)

55. See Fig. 12.

56. See Fig. 13.

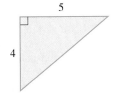

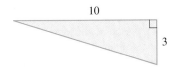

Figure 12 Exercise 60

Figure 13 Exercise 56

57. See Fig. 14.

58. See Fig. 15.

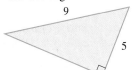

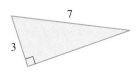

Figure 14 Exercise 57

Figure 15 Exercise 58

59. See Fig. 16.

60. See Fig. 17.

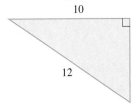

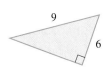

Figure 16 Exercise 59

Figure 17 Exercise 60

Let a and b be the lengths of the legs of a right triangle, and let c be the length of the hypotenuse. Values for two of the three lengths are given. Find the third length.

61. $a = 6$ and $b = 8$

62. $a = 5$ and $b = 12$

63. $a = 2$ and $b = 3$

64. $a = 2$ and $b = 7$

65. $a = 4$ and $b = 8$

66. $a = 3$ and $b = 6$

67. $a = 6$ and $c = 10$

68. $a = 12$ and $c = 13$

69. $a = 2$ and $c = 5$

70. $a = 8$ and $c = 11$

71. $b = 2$ and $c = 7$

72. $b = 6$ and $c = 8$

For Exercises 73–86, round approximate results to the first decimal place.

73. A student drives 8 miles west from home and then 17 miles north to get to school. What would be the length of the trip if it were possible to drive along a straight line from home to school?

74. A commuter drives 5 miles south from home and then 9 miles east to get to work. What would be the length of the trip if it were possible to drive along a straight line from home to work?

75. A 12-foot-long ladder is leaning against a building. The bottom of the ladder is 4 feet from the base of the building. Will the ladder reach the bottom of a window that is 11 feet above the ground?

76. A 20-foot-long ladder is leaning against a building. The bottom of the ladder is 5 feet from the base of the building. Will the ladder reach the bottom of a window that is 19 feet above the ground?

77. The size of a rectangular television is usually described in terms of the length of its diagonal. If a 19-inch television screen has a height of 12 inches, what is the width of the screen?

78. The size of a rectangular television is usually described in terms of the length of its diagonal. If a 26-inch television has a height of 15 inches, what is the width of the screen?

79. A rectangular painting has a width of 24 inches. The length of the diagonal is 37 inches. Find the length of the painting.

80. A rectangular painting has a length of 42 inches. The length of the diagonal is 53 inches. Find the width of the painting.

81. A surveyor wants to estimate the distance (in miles) across a lake from point A to point B, as shown in Fig. 18. Find that distance.

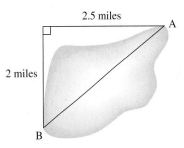

Figure 18 Exercise 81

82. A surveyor wants to estimate the distance (in miles) across a lake from point A to point B, as shown in Fig. 19. Find that distance.

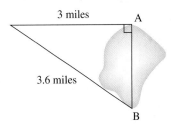

Figure 19 Exercise 82

83. Not including the end zones, an official football field is rectangular with a length of 100 yards and a width of 53 yards, 1 foot. If a player runs diagonally across this rectangle, how far will he run? [**Hint:** Use units of feet to perform your calculations.]

84. A soccer field is rectangular. For international matches, the length must be between 110 and 120 yards, inclusive, and the width must be between 70 and 80 yards, inclusive.
   a. Find the length of the diagonal for the largest permitted field for international matches.
   b. Find the length of the diagonal for the smallest permitted field for international matches.

85. Sioux Falls, South Dakota, is about 266 miles almost directly south of Fargo, North Dakota, and is almost directly west of Madison, Wisconsin. The distance between Fargo and Madison is about 514 miles. What is the approximate distance between Sioux Falls and Madison?

86. Jackson, Mississippi, is about 406 miles almost directly east of Dallas, Texas, and is almost directly south of Memphis, Tennessee. The distance between Dallas and Memphis is about 460 miles. What is the approximate distance between Jackson and Memphis?

87. a. Use factoring to solve the equation $x^2 - 36 = 0$.
   b. Use the square root property to solve the equation $x^2 - 36 = 0$.
   c. Compare your results for parts (a) and (b).
   d. In your opinion, is it easier to solve $x^2 - 36 = 0$ by factoring or by using the square root property? Explain.

88. a. Use factoring to solve the equation $(x - 3)^2 = 25$.
   b. Use the square root property to solve the equation $(x - 3)^2 = 25$.
   c. Compare your results for parts (a) and (b).
   d. In your opinion, is it easier to solve $(x - 3)^2 = 25$ by factoring or by using the square root property? Explain.

89. a. Can the equation $(x - 2)^2 = 7$ be solved by means of the square root property? If so, solve the equation by this method.
   b. Can the equation $(x - 2)^2 = 7$ be solved by factoring? If so, solve the equation by this method.
   c. Can all equations that can be solved by means of the square root property be solved by factoring? Explain.

90. A student tries to solve the equation $x^2 - 2x - 24 = 0$:

$$x^2 - 2x - 24 = 0$$
$$x^2 = 2x + 24$$
$$x = \pm\sqrt{2x + 24}$$

Describe any errors. Then solve the equation correctly.

91. Explain how to solve a quadratic equation by using the square root property.

92. Describe the Pythagorean theorem in your own words. Give an example of how to use it.

Related Review

93. a. Solve $x^2 = 36$.
   b. Find $\sqrt{36}$.
   c. Explain why your results for parts (a) and (b) are different.

94. A student says that 3 is a solution of the equation $-x^2 = 9$ because $-3^2 = 9$. Is the student correct? Explain.

95. A rectangle has an area of 16 square inches. Let W be the width, L be the length, D be the length of the diagonal (all in inches), and A be the area (in inches squared).
   a. Sketch three possible rectangles, including one that is a square.
   b. For each of your three rectangles, find the approximate length D of the diagonal. Round each result to the second decimal place.
   c. Which of your three rectangles has the smallest diagonal length D?
   d. Which of the symbols W, L, D, and A are variables? Explain.
   e. Which of the symbols W, L, D, and A are constants? Explain.

96. A rectangle has a perimeter of 20 inches. Let W be the width, L be the length, D be the length of the diagonal, and P be the perimeter, all with units in feet.
   a. Sketch three possible rectangles, including one that is a square.
   b. For each of your three rectangles, find the approximate length D of the diagonal. Round each result to the second decimal place.
   c. Which of your three rectangles has the smallest diagonal length D?
   d. Which of the symbols W, L, D, and P are variables? Explain.
   e. Which of the symbols W, L, D, and P are constants? Explain.

Expressions, Equations, and Graphs

Perform the indicated instruction. Then use words such as linear, quadratic, cubic, power, radical, polynomial, degree, one variable, *and* two variables *to describe the expression, equation, or system.*

97. Find the product $(3x^2 - 2)(4x^2 + x - 5)$.

98. Solve $5 - 2(3w - 7) = 4 + 2w$.

99. Factor $3x^4 - 12x^3 - 63x^2$.

100. Solve $4x^2 - 13 = 0$.

101. Graph $3x + 5y = 15$ by hand.

102. Simplify $\sqrt{\dfrac{5}{x}}$.

9.4 SOLVING QUADRATIC EQUATIONS BY COMPLETING THE SQUARE

The man who makes no mistakes does not usually make anything.

—*Bishop W. C. Magee*

Objectives

▶ Know the relationship between b and c for a perfect-square trinomial of the form $x^2 + bx + c$.

▶ Solve quadratic equations by completing the square.

▶ Know that any quadratic equation can be solved by completing the square.

So far, we have discussed how to solve quadratic equations both by factoring and by using the square root property. In this section, we will discuss yet another way, called *completing the square*.

Perfect-Square Trinomials

To begin our study of completing the square, we simplify the square of a sum. For example,

$$(x + 5)^2 = x^2 + 10x + 25$$

So, $x^2 + 10x + 25$ is a perfect-square trinomial. Recall from Section 7.5 that a *perfect-square trinomial* is a trinomial equivalent to the square of a binomial.

For $x^2 + 10x + 25$, there is a special connection between the 10 and the 25. If we divide the 10 by 2 and then square the result, we get 25:

$$x^2 + 10x + 25$$

$$\left(\frac{10}{2}\right)^2 = 5^2 = 25$$

This is no coincidence. Consider simplifying $(x - 3)^2$: $(x - 3)^2 = x^2 - 6x + 9$. If we divide -6 by 2 and then square the result, we get 9:

$$x^2 - 6x + 9$$

$$\left(\frac{-6}{2}\right)^2 = (-3)^2 = 9$$

For the general case, we simplify $(x + k)^2$, where k is a constant:

$$(x + k)^2 = x^2 + (2k)x + k^2$$

$$\left(\frac{2k}{2}\right)^2 = k^2$$

For $x^2 + (2k)x + k^2$, we see that if we divide the coefficient of x by 2 and then square the result, we get the constant term k^2.

Perfect-Square Trinomial Property

For a perfect-square trinomial of the form $x^2 + bx + c$, dividing b by 2 and then squaring the result gives c:

$$x^2 + bx + c$$

$$\left(\frac{b}{2}\right)^2 = c$$

Note that this property is for a perfect-square trinomial $ax^2 + bx + c$, where $a = 1$.

Example 1   Factoring Perfect-Square Trinomials

Find the value of c such that the expression is a perfect-square trinomial. Then factor the perfect-square trinomial.

  1. $x^2 + 12x + c$ **2.** $x^2 - 18x + c$

Solution

  1. We divide $b = 12$ by 2 and then square the result:

$$\left(\frac{12}{2}\right)^2 = 6^2 = 36 = c$$

The expression is $x^2 + 12x + 36$, with factored form $(x + k)^2$, for some positive integer k. So, we have

$$x^2 + 12x + 36 = (x + k)^2 = x^2 + 2kx + k^2$$

The constant terms
36 and k^2 are equal.

Here, $k^2 = 36$, or $k = 6$ (k is positive). So, the factored form of $x^2 + 12x + 36$ is $(x + 6)^2$.

2. We divide $b = -18$ by 2 and then square the result:

$$\left(\frac{-18}{2}\right)^2 = (-9)^2 = 81 = c$$

The expression is $x^2 - 18x + 81$, with factored form $(x - k)^2$, for some positive integer k. So, we have

$$x^2 - 18x + 81 = (x - k)^2 = x^2 - 2kx + k^2$$

The constant terms
81 and k^2 are equal.

Here, $k^2 = 81$, or $k = 9$ (k is positive). So, the factored form of $x^2 - 18x + 81$ is $(x - 9)^2$.   ∎

Solving $x^2 + bx + c = 0$ by Completing the Square

In Example 2, we will begin to solve a quadratic equation by forming a perfect-square trinomial on one side of the equation.

Example 2   Solving by Completing the Square

Solve $x^2 + 8x = 3$.

Solution

Since $\left(\dfrac{8}{2}\right)^2 = 4^2 = 16$, we add 16 to both sides of $x^2 + 8x = 3$ so that the left side will be a perfect-square trinomial:

$$\begin{aligned}
x^2 + 8x &= 3 & &\text{Original equation} \\
x^2 + 8x + 16 &= 3 + 16 & &\text{Add 16 to both sides.} \\
(x + 4)^2 &= 19 & &\text{Factor the left side.} \\
x + 4 &= \pm\sqrt{19} & &\text{Square root property} \\
x &= -4 \pm \sqrt{19} & &\text{Subtract 4 from both sides.}
\end{aligned}$$

We use graphing calculator tables to check that if $-4 - \sqrt{19}$ or $-4 + \sqrt{19}$ is substituted for x in the expression $x^2 + 8x$, the result is 3 (see Fig. 20).

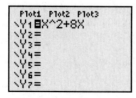
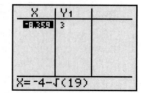
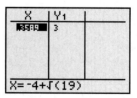

Figure 20 Verify the work   ∎

In Example 2, we added 16 to both sides of $x^2 + 8x = 3$ and then factored the left side to get $(x + 4)^2$. By adding 16 to both sides of the equation, we say that we *completed the square* for $x^2 + 8x$.

WARNING In solving a quadratic equation such as $x^2 + 8x = 3$ by completing the square, it is a common error to add a number (in this case 16) to only the left side of the equation. Remember to add the number to *both* sides of the equation.

Example 3 Solving by Completing the Square

Solve $p^2 - 14p + 4 = 0$.

Solution

We're trying to get a perfect-square trinomial on the left side of the equation $p^2 - 14p + 4 = 0$, so we subtract 4 from both sides:

$$p^2 - 14p = -4$$

Since $\left(\dfrac{-14}{2}\right)^2 = (-7)^2 = 49$, we add 49 to both sides of $p^2 - 14p = -4$ to complete the square for $p^2 - 14p$:

$$
\begin{array}{ll}
p^2 - 14p + 49 = -4 + 49 & \text{Add 49 to both sides.} \\
(p - 7)^2 = 45 & \text{Factor left side.} \\
p - 7 = \pm\sqrt{45} & \text{Square root property} \\
p - 7 = \pm 3\sqrt{5} & \text{Simplify.} \\
p = 7 \pm 3\sqrt{5} & \text{Add 7 to both sides.} \quad\blacksquare
\end{array}
$$

Recall from Section 8.5 that some quadratic equations do not have any real-number solutions. We will see such an equation in Example 4.

Example 4 A Quadratic Equation with No Real-Number Solutions

Solve $x^2 + 2x = -7$.

Solution

Since $\left(\dfrac{2}{2}\right)^2 = 1^2 = 1$, we add 1 to both sides of $x^2 + 2x = -7$ to complete the square for $x^2 + 2x$:

$$
\begin{array}{ll}
x^2 + 2x = -7 & \text{Original equation} \\
x^2 + 2x + 1 = -7 + 1 & \text{Add 1 to both sides.} \\
(x + 1)^2 = -6 & \text{Factor left side.}
\end{array}
$$

Since the square of a number is nonnegative, we conclude that $x^2 + 2x = -7$ has no real-number solutions. $\quad\blacksquare$

Solving $ax^2 + bx + c = 0$ by Completing the Square

So far, we have worked with trinomials only of the form $ax^2 + bx + c$, where $a = 1$. Consider simplifying $(2x + 3)^2$:

$$(2x + 3)^2 = 4x^2 + 12x + 9$$

Dividing $b = 12$ by 2 and then squaring the result does *not* give 9:

$$\left(\dfrac{12}{2}\right)^2 = 6^2 = 36 \neq 9$$

So, the perfect-square trinomial property given for $ax^2 + bx + c$, where $a = 1$, does not extend to trinomials with $a \neq 1$. However, **when solving a quadratic equation of the form $ax^2 + bx + c = 0$ with $a \neq 1$, we can first divide both sides by a to obtain an equation involving $1x^2$, one to which we can apply the property.**

Example 5 Solving by Completing the Square

Solve $2x^2 + 12x - 22 = 0$.

Solution

$$2x^2 + 12x - 22 = 0 \qquad \text{Original equation}$$

$$2x^2 + 12x = 22 \qquad \text{Add 22 to both sides.}$$

$$x^2 + 6x = 11 \qquad \text{Divide both sides by 2 so that coefficient of } x^2 \text{ will be 1.}$$

$$x^2 + 6x + 9 = 11 + 9 \qquad \left(\frac{6}{2}\right)^2 = 3^2 = 9; \text{ add 9 to both sides.}$$

$$(x + 3)^2 = 20 \qquad \text{Factor left side.}$$

$$x + 3 = \pm\sqrt{20} \qquad \text{Square root property}$$

$$x + 3 = \pm 2\sqrt{5} \qquad \text{Simplify.}$$

$$x = -3 \pm 2\sqrt{5} \qquad \text{Subtract 3 from both sides.} \qquad \blacksquare$$

Which Equations Can Be Solved by Completing the Square?

Any quadratic equation can be solved by completing the square. Here is a summary of this method:

Solving a Quadratic Equation by Completing the Square

To solve a quadratic equation by **completing the square,**

1. Write the equation in the form $ax^2 + bx = k$, where a, b, and k are constants.
2. If $a \neq 1$, divide both sides of the equation by a.
3. Complete the square for the expression on the left side of the equation.
4. Solve the equation by using the square root property.

WARNING   To solve an equation of the form $ax^2 + bx = k$ with $a \neq 1$, we must divide both sides of the equation by a before completing the square for the left side of the equation.

group exploration

Identifying errors in solving by completing the square

1. A student tries to solve the equation $x^2 + 6x - 5 = 0$:

$$x^2 + 6x - 5 = 0$$
$$x^2 + 6x = 5$$
$$x^2 + 6x + 9 = 5$$
$$(x + 3)^2 = 5$$
$$x + 3 = \pm\sqrt{5}$$
$$x = -3 \pm \sqrt{5}$$

Describe any errors. Then solve the equation correctly.

2. A student tries to solve the equation $4x^2 - 8x = 12$:

$$4x^2 - 8x = 12$$
$$4x^2 - 8x + 16 = 12 + 16$$
$$(2x - 4)^2 = 28$$
$$2x - 4 = \pm\sqrt{28}$$
$$2x - 4 = \pm 2\sqrt{7}$$
$$2x = 4 \pm 2\sqrt{7}$$
$$x = 2 \pm \sqrt{7}$$

Describe any errors. Then solve the equation correctly.

Find the value of c for which the expression is a perfect-square trinomial. Then factor the perfect-square trinomial.

1. $x^2 + 6x + c$

2. $x^2 + 16x + c$

3. $x^2 + 14x + c$

4. $x^2 + 4x + c$

5. $x^2 + 2x + c$

6. $x^2 + 20x + c$

7. $x^2 - 8x + c$

8. $x^2 - 12x + c$

9. $x^2 - 10x + c$

10. $x^2 - 2x + c$

11. $x^2 - 20x + c$

12. $x^2 - 16x + c$

Solve by completing the square. Use a graphing calculator to verify your work.

13. $x^2 + 6x = 5$

14. $x^2 + 8x = 1$

15. $x^2 + 14x = -20$

16. $x^2 + 4x = 13$

17. $x^2 - 10x = -4$

18. $x^2 - 16x = 3$

19. $t^2 - 2t = 5$

20. $p^2 - 12p = -3$

21. $x^2 + 12x = -4$

22. $x^2 + 10x = 3$

23. $x^2 - 4x = 14$

24. $x^2 - 14x = -9$

25. $x^2 - 8x = 8$

26. $x^2 - 2x = 19$

27. $x^2 - 16x = -70$

28. $x^2 - 6x = -10$

29. $y^2 + 20y = -40$

30. $r^2 + 18r = -6$

31. $x^2 + 6x + 1 = 0$

32. $x^2 + 12x - 3 = 0$

33. $x^2 - 4x + 1 = 0$

34. $x^2 - 16x + 9 = 0$

35. $x^2 - 10x - 7 = 0$

36. $x^2 - 14x - 11 = 0$

37. $y^2 + 20y + 120 = 0$

38. $m^2 + 18m + 100 = 0$

39. $x^2 + 8x + 4 = 0$

40. $x^2 + 2x - 17 = 0$

Solve by completing the square. Use a graphing calculator to verify your work.

41. $2x^2 - 16x = 6$

42. $3x^2 - 18x = 6$

43. $5x^2 + 10x = 35$

44. $2x^2 + 8x = 16$

45. $3w^2 - 30w = -21$

46. $5t^2 - 60t = -20$

47. $6x^2 + 12x - 6 = 0$

48. $2x^2 + 20x + 6 = 0$

49. $4x^2 - 24x + 4 = 0$

50. $3x^2 - 6x - 33 = 0$

51. $5x^2 + 20x - 20 = 0$

52. $2x^2 + 16x - 8 = 0$

Solve by the method of your choice. Use a graphing calculator to verify your work.

53. $x^2 - 9 = 0$

54. $x^2 - 16 = 0$

55. $r^2 = 11r - 30$

56. $p^2 = 6p + 27$

57. $(x - 5)^2 = 32$

58. $(x + 3)^2 = 45$

59. $3x^2 + 5x = 12$

60. $2x^2 - 3x = 5$

61. $x^2 = 13$

62. $x^2 = 21$

63. $t^2 - 6t - 3 = 0$

64. $m^2 + 8m - 4 = 0$

65. a. Solve the equation $x^2 - 6x + 8 = 0$ by factoring.

   b. Solve the equation $x^2 - 6x + 8 = 0$ by completing the square.

   c. In your opinion, is it easier to solve $x^2 - 6x + 8 = 0$ by factoring or by completing the square? Explain.

66. a. Solve the equation $x^2 + 4x = 12$ by factoring.

   b. Solve the equation $x^2 + 4x = 12$ by completing the square.

   c. In your opinion, is it easier to solve $x^2 + 4x = 12$ by factoring or by completing the square? Explain.

67. a. Can the equation $x^2 + 4x = 7$ be solved by completing the square? If so, solve the equation by using this method.

   b. Can the equation $x^2 + 4x = 7$ be solved by factoring? If so, solve the equation by using this method.

   c. Explain how to decide whether to solve a quadratic equation by completing the square or by factoring.

68. Compare the methods of solving a quadratic equation by factoring, by using the square root property, and by completing the square. Describe the methods, as well as their advantages and disadvantages. (See page 9 for guidelines on writing a good response.)

69. Describe how to solve a quadratic equation by completing the square. (See page 9 for guidelines on writing a good response.)

70. For an equation of the form $ax^2 + bx = c$, where $a \neq 1$, explain why it is necessary to divide both sides of the equation by a when we solve by completing the square.

Related Review

71. Simplify $(x + 7)^2$.

72. Simplify $(x - 5)^2$.

73. Factor $x^2 - 16x + 64$.

74. Factor $x^2 + 12x + 36$.

Expressions, Equations, and Graphs

Perform the indicated instruction. Then use words such as linear, quadratic, cubic, power, radical, polynomial, degree, one variable, *and* two variables *to describe the expression, equation, or system.*

75. Simplify $(x - 4)^2$.

76. Solve the following:
$$y = -3x + 1$$
$$5x + 2y = 1$$

77. Solve $(x - 4)^2 = 3$.

78. Solve $4(2x - 5) = -3x + 1$.

79. Factor $p^2 - 8pq + 16q^2$.

80. Graph $y = -3x + 1$ by hand.

SOLVING QUADRATIC EQUATIONS
9.5 BY THE QUADRATIC FORMULA

Objectives

▷ Solve quadratic equations by using the *quadratic formula*.

▷ Find approximate solutions of a quadratic equation.

▷ Find approximate *x*-intercepts of the graph of a quadratic equation in two variables.

▷ Determine whether to solve a quadratic equation by factoring, by using the square root property, by completing the square, or by using the quadratic formula.

Any quadratic equation can be solved by completing the square. However, it is difficult to use this method with most quadratic equations. An easier option is to use an important equation called the *quadratic formula*, which can also be used to solve *any* quadratic equation.

The Quadratic Formula

We will find the quadratic formula by solving the general quadratic equation

$$ax^2 + bx + c = 0$$

by completing the square. For now, we assume that *a* is positive:

$$ax^2 + bx + c = 0 \qquad \text{General quadratic equation}$$

$$ax^2 + bx = -c \qquad \text{Subtract } c \text{ from both sides.}$$

$$x^2 + \frac{b}{a}x = -\frac{c}{a} \qquad \text{Divide both sides by (nonzero) } a \text{ so}$$

that the coefficient of x^2 will be 1.

To begin completing the square, we divide $\dfrac{b}{a}$ by 2:

$$\frac{b}{a} \div 2 = \frac{b}{a} \cdot \frac{1}{2} = \frac{b}{2a}$$

Then we square the result:

$$\left(\frac{b}{2a}\right)^2 = \frac{b^2}{4a^2}$$

Next, we add $\dfrac{b^2}{4a^2}$ to both sides of the equation $x^2 + \dfrac{b}{a}x = -\dfrac{c}{a}$:

$$x^2 + \frac{b}{a}x + \frac{b^2}{4a^2} = -\frac{c}{a} + \frac{b^2}{4a^2} \qquad \text{Add } \frac{b^2}{4a^2} \text{ to both sides.}$$

$$\left(x + \frac{b}{2a}\right)^2 = -\frac{c}{a} \cdot \frac{4a}{4a} + \frac{b^2}{4a^2} \qquad \begin{array}{l}\text{Factor left-hand side; find a common}\\ \text{denominator, } 4a^2, \text{ for right-hand side.}\end{array}$$

$$\left(x + \frac{b}{2a}\right)^2 = \frac{b^2 - 4ac}{4a^2} \qquad \text{Add fractions.}$$

$$x + \frac{b}{2a} = \pm\sqrt{\frac{b^2 - 4ac}{4a^2}} \qquad \text{Square root property}$$

$$x + \frac{b}{2a} = \pm\frac{\sqrt{b^2 - 4ac}}{\sqrt{4a^2}} \qquad \sqrt{\frac{A}{B}} = \frac{\sqrt{A}}{\sqrt{B}}$$

$$x = -\frac{b}{2a} \pm \frac{\sqrt{b^2 - 4ac}}{2a} \qquad \begin{array}{l}\text{Subtract } \frac{b}{2a} \text{ from both sides;}\\ \sqrt{4a^2} = \sqrt{4}\sqrt{a^2} = 2a \text{ for } a \text{ nonnegative.}\end{array}$$

$$x = \frac{-b \pm \sqrt{b^2 - 4ac}}{2a} \qquad \text{Add the fractions.}$$

So, we have found a formula (the last line) for the solutions of a quadratic equation $ax^2 + bx + c = 0$, where a is positive. In a similar way, we could derive the same formula for a quadratic equation in which a is negative.

> ### Quadratic Formula
>
> The solutions of a quadratic equation $ax^2 + bx + c = 0$ are given by the **quadratic formula:**
>
> $$x = \frac{-b \pm \sqrt{b^2 - 4ac}}{2a}$$

WARNING   For the fraction in the quadratic formula, notice that the term $-b$ is part of the numerator:

$$\frac{-b \pm \sqrt{b^2 - 4ac}}{2a} \qquad \text{Correct}$$

$$-b \pm \frac{\sqrt{b^2 - 4ac}}{2a} \qquad \text{Incorrect}$$

Example 1   Solving by Using the Quadratic Formula

Solve $x^2 - 7x + 10 = 0$ by using the quadratic formula.

Solution

Comparing $x^2 - 7x + 10$ with $ax^2 + bx + c$, we see that $a = 1$, $b = -7$, and $c = 10$. We substitute these values for a, b, and c into the quadratic formula:

$$x = \frac{-b \pm \sqrt{b^2 - 4ac}}{2a} \qquad \text{Quadratic formula}$$

$$x = \frac{-(-7) \pm \sqrt{(-7)^2 - 4(1)(10)}}{2(1)} \qquad \text{Substitute into quadratic formula.}$$

$$x = \frac{7 \pm \sqrt{49 - 40}}{2} \qquad \text{Simplify.}$$

$$x = \frac{7 \pm \sqrt{9}}{2} \qquad \text{Subtract.}$$

$$x = \frac{7 \pm 3}{2} \qquad \sqrt{9} = 3$$

$$x = \frac{7 - 3}{2} \quad \text{or} \quad x = \frac{7 + 3}{2} \qquad \text{Write as two equations.}$$

$$x = \frac{4}{2} \quad \text{or} \quad x = \frac{10}{2} \qquad \text{Subtract./Add.}$$

$$x = 2 \quad \text{or} \quad x = 5$$

The solutions are 2 and 5. We can check that the solutions satisfy the original equation. (Try it.)   ∎

Note that instead of using the quadratic formula, we can solve $x^2 - 7x + 10 = 0$ by factoring:

$$x^2 - 7x + 10 = 0 \qquad \text{Original equation}$$

$$(x - 2)(x - 5) = 0 \qquad \text{Factor left side.}$$

$$x - 2 = 0 \quad \text{or} \quad x - 5 = 0 \qquad \text{Zero factor property}$$

$$x = 2 \quad \text{or} \quad x = 5$$

A benefit of the quadratic formula is that we can use it to solve equations that are difficult or impossible to solve by factoring. In Example 2, we solve an equation that would be impossible to solve by factoring.

Example 2 Solving by Using the Quadratic Formula

Solve $5x^2 - x = 2$.

Solution

First, we write the equation in the form $ax^2 + bx + c = 0$:

$$5x^2 - x = 2 \qquad \text{Original equation}$$

$$5x^2 - x - 2 = 0 \qquad \text{Subtract 2 from both sides.}$$

Then we substitute $a = 5$, $b = -1$, and $c = -2$ into the quadratic formula:

$$x = \frac{-b \pm \sqrt{b^2 - 4ac}}{2a} \qquad \text{Quadratic formula}$$

$$x = \frac{-(-1) \pm \sqrt{(-1)^2 - 4(5)(-2)}}{2(5)} \qquad \text{Substitute into quadratic formula.}$$

$$x = \frac{1 \pm \sqrt{1 + 40}}{10} \qquad \text{Simplify.}$$

$$x = \frac{1 \pm \sqrt{41}}{10} \qquad \text{Add.}$$

We use graphing calculator tables to check that if $\dfrac{1 - \sqrt{41}}{10}$ or $\dfrac{1 + \sqrt{41}}{10}$ is substituted for x in the expression $5x^2 - x$, the result is 2 (see Fig. 21).

 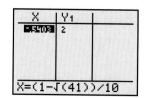 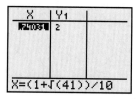

Figure 21 Verify the work ■

By the check in Fig. 21, we see that $\dfrac{1 - \sqrt{41}}{10}$ is about -0.54. So, $\dfrac{1 - \sqrt{41}}{10}$ is simply a number. Likewise, $\dfrac{1 + \sqrt{41}}{10}$ is a number (about 0.74).

Example 3 Solving by Using the Quadratic Formula

Solve $-2x^2 - 6x + 3 = 0$.

Solution

First, we multiply both sides of the equation by -1 so that we can avoid having a negative denominator after we use the quadratic formula:

$$-2x^2 - 6x + 3 = 0 \qquad \text{Original equation}$$

$$2x^2 + 6x - 3 = 0 \qquad \text{Multiply both sides by } -1.$$

(Another benefit is that we will have fewer negative numbers to substitute into the quadratic formula.)

Then we substitute $a = 2$, $b = 6$, and $c = -3$ into the quadratic formula:

$$x = \frac{-b \pm \sqrt{b^2 - 4ac}}{2a}$$ Quadratic formula

$$x = \frac{-(6) \pm \sqrt{(6)^2 - 4(2)(-3)}}{2(2)}$$ Substitute into quadratic formula.

$$x = \frac{-6 \pm \sqrt{36 + 24}}{4}$$ Simplify.

$$x = \frac{-6 \pm \sqrt{60}}{4}$$ Add.

$$x = \frac{-6 \pm 2\sqrt{15}}{4}$$ $\sqrt{60} = \sqrt{4 \cdot 15} = \sqrt{4}\sqrt{15} = 2\sqrt{15}$

$$x = \frac{2(-3 \pm \sqrt{15})}{2(2)}$$ Factor out 2.

$$x = \frac{2}{2} \cdot \frac{-3 \pm \sqrt{15}}{2}$$ $\dfrac{ab}{cd} = \dfrac{a}{c} \cdot \dfrac{b}{d}$

$$x = \frac{-3 \pm \sqrt{15}}{2}$$ $\dfrac{2}{2} = 1$; simplify.   ∎

Recall from Section 9.1 that the square root of a negative number is not a real number. For instance, $\sqrt{-23}$ is not a real number. We will deal with such a situation in Example 4.

Example 4   Solving by Using the Quadratic Formula

Solve $\dfrac{3}{2}x^2 = x - \dfrac{5}{2}$.

Solution

We clear the equation of fractions by multiplying by the LCD, 2, and then write the equation in the form $ax^2 + bx + c = 0$:

$$\frac{3}{2}x^2 = x - \frac{5}{2}$$ Original equation

$$2 \cdot \frac{3}{2}x^2 = 2 \cdot x - 2 \cdot \frac{5}{2}$$ Multiply both sides by the LCD, 2.

$$3x^2 = 2x - 5$$ Simplify.

$$3x^2 - 2x + 5 = 0$$ Write in $ax^2 + bx + c = 0$ form.

Next we substitute $a = 3$, $b = -2$, and $c = 5$ into the quadratic formula:

$$x = \frac{-(-2) \pm \sqrt{(-2)^2 - 4(3)(5)}}{2(3)}$$ Substitute into quadratic formula.

$$x = \frac{2 \pm \sqrt{4 - 60}}{6}$$ Simplify.

$$x = \frac{2 \pm \sqrt{-56}}{6}$$ Subtract.

Since the square root of a negative number is not a real number, we conclude that there are no real-number solutions. To verify our work, we use a graphing calculator to graph $y = \dfrac{3}{2}x^2$ and $y = x - \dfrac{5}{2}$ (see Fig. 22). Because the curves do not intersect, the original equation has no real-number solutions.   ∎

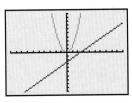

Figure 22 Verify the work

If the result of solving a quadratic equation is a radical expression in which a radicand is negative, such as $\dfrac{2 \pm \sqrt{-56}}{6}$, then there are no real-number solutions.

Finding Approximate Solutions of a Quadratic Equation

When we use a model to make predictions, it is not necessary to find exact solutions of an equation. To prepare us for working with quadratic models in Section 9.6, we use a calculator to help us find approximate solutions in Example 5.

Example 5   Finding Approximate Solutions

Find approximate solutions of $7.2w^2 + 4.8w + 2.3 = 3.7$. Round the results to the second decimal place.

Solution

$$7.2w^2 + 4.8w + 2.3 = 3.7 \qquad \text{Original equation}$$

$$7.2w^2 + 4.8w - 1.4 = 0 \qquad \text{Subtract 3.7 from both sides.}$$

$$w = \frac{-4.8 \pm \sqrt{4.8^2 - 4(7.2)(-1.4)}}{2(7.2)} \qquad \begin{array}{l}\text{Substitute } a = 7.2, b = 4.8, \text{ and} \\ c = -1.4 \text{ into quadratic formula.}\end{array}$$

$$w = \frac{-4.8 \pm \sqrt{23.04 + 40.32}}{14.4} \qquad \text{Simplify.}$$

$$w = \frac{-4.8 \pm \sqrt{63.36}}{14.4} \qquad \text{Add.}$$

$$w \approx \frac{-4.8 \pm 7.9599}{14.4} \qquad \text{Approximate square root.}$$

$$w \approx \frac{-4.8 - 7.9599}{14.4} \quad \text{or} \quad w \approx \frac{-4.8 + 7.9599}{14.4} \qquad \text{Write as two equations.}$$

$$w \approx \frac{-12.7599}{14.4} \quad \text{or} \quad w \approx \frac{3.1599}{14.4} \qquad \text{Subtract./Add.}$$

$$w \approx -0.89 \quad \text{or} \quad w \approx 0.22$$

Finding Approximate x-Intercepts of a Parabola

Recall from Section 8.5 that we can find the x-intercepts of a quadratic equation $y = ax^2 + bx + c$ by substituting 0 for y and then solving for x.

Example 6   Finding Approximate x-Intercepts

Find approximate x-intercepts of the parabola $y = -1.9x^2 + 25.7x - 68.2$. Round the coordinates of the results to the second decimal place.

Solution

First, we substitute 0 for y. Then, we use the quadratic formula to solve for x:

$$0 = -1.9x^2 + 25.7x - 68.2 \qquad \text{Substitute 0 for } y.$$

$$x = \frac{-25.7 \pm \sqrt{25.7^2 - 4(-1.9)(-68.2)}}{2(-1.9)} \qquad \begin{array}{l}\text{Substitute } a = -1.9, b = 25.7, \text{ and} \\ c = -68.2 \text{ into quadratic formula.}\end{array}$$

$$x = \frac{-25.7 \pm \sqrt{660.49 - 518.32}}{-3.8} \qquad \text{Simplify.}$$

$$x = \frac{-25.7 \pm \sqrt{142.17}}{-3.8} \qquad \text{Subtract.}$$

$$x \approx \frac{-25.7 \pm 11.9235}{-3.8} \qquad \text{Approximate square root.}$$

$$x \approx 9.90 \quad \text{or} \quad x \approx 3.63 \qquad \text{Calculate.}$$

So, the x-intercepts are about $(3.63, 0)$ and $(9.90, 0)$. We use "zero" on a graphing calculator to verify our work (see Fig. 23).

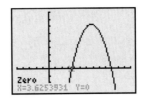 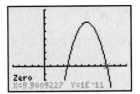

Figure 23 Verify the work ∎

Deciding Which Method to Use to Solve a Quadratic Equation

In Chapter 8 and in this chapter, we have discussed four ways to solve quadratic equations: factoring, the square root property, completing the square, and the quadratic formula. How do we know which method to use?

Remember that **any quadratic equation can be solved by the quadratic formula.** Although any equation can also be solved by completing the square, that method is much more difficult to use than the quadratic formula.

Not many quadratic equations can be solved by factoring, because almost all polynomials are prime. However, if an equation can be solved by simple factoring techniques, then it is easier to solve the equation by factoring than by any of the other three methods.

Here are some guidelines on deciding which method to use to solve a quadratic equation:

Method:	When to Use:
Factoring	For equations that can easily be put into the form $ax^2 + bx + c = 0$ and where $ax^2 + bx + c$ can easily be factored.
Square root property	For equations that can easily be put into the form $x^2 = k$ or $(x + p)^2 = k$.
Completing the square	When the directions require it.
Quadratic formula	For all equations except those which can easily be solved by factoring or by the square root property.

Example 7   Deciding Which Method to Use

Solve.

1. $(x - 5)^2 = 7$ **2.** $x^2 + 5x + 3 = 0$ **3.** $x^2 + 4x - 21 = 0$

Solution

1. The equation is of the form $(x + p)^2 = k$, so we solve it by using the square root property:

$$(x - 5)^2 = 7 \qquad \text{Original equation}$$
$$x - 5 = \pm\sqrt{7} \qquad \text{Square root property}$$
$$x = 5 \pm \sqrt{7} \qquad \text{Add 5 to both sides.}$$

2. The trinomial $x^2 + 5x + 3$ is prime, so we can't solve the equation $x^2 + 5x + 3 = 0$ by factoring. The equation $x^2 + 5x + 3 = 0$ can't be put into the form $x^2 = k$, and it can't easily be put into the form $(x + p)^2 = k$, so we don't try to solve it by using the square root property. Instead, we substitute $a = 1$, $b = 5$, and $c = 3$ into the quadratic formula:

$$x = \frac{-5 \pm \sqrt{5^2 - 4(1)(3)}}{2(1)} \qquad \text{Quadratic formula}$$
$$x = \frac{-5 \pm \sqrt{25 - 12}}{2} \qquad \text{Simplify.}$$
$$x = \frac{-5 \pm \sqrt{13}}{2} \qquad \text{Subtract.}$$

3. The trinomial $x^2 + 4x - 21$ is factorable, so we solve the equation $x^2 + 4x - 21 = 0$ by factoring:

$$x^2 + 4x - 21 = 0 \qquad \text{Original equation}$$
$$(x + 7)(x - 3) = 0 \qquad \text{Factor left side.}$$
$$x + 7 = 0 \quad \text{or} \quad x - 3 = 0 \qquad \text{Zero factor property}$$
$$x = -7 \quad \text{or} \quad x = 3$$

group exploration

Comparing methods of solving quadratic equations

1. Solve the equation $x^2 + 6x + 8 = 0$ by factoring.

2. Solve the equation $x^2 + 6x + 8 = 0$ by completing the square.

3. Solve the equation $x^2 + 6x + 8 = 0$ by using the quadratic formula.

4. Compare your results from Problems 1, 2, and 3. In your opinion, which method was easiest to use?

5. Try solving the equation $x^2 + 2x - 5 = 0$ by each of the three methods. What did you find?

6. Compare the methods of solving quadratic equations by factoring, by completing the square, and by using the quadratic formula. What are the advantages and disadvantages of each method?

HOMEWORK 9.5

FOR EXTRA HELP ▶

Student Solutions Manual   PH Math/Tutor Center   MathXL®   MyMathLab

Solve by using the quadratic formula.

1. $2x^2 + 5x + 3 = 0$
2. $2x^2 + 3x + 1 = 0$
3. $4x^2 + 7x + 2 = 0$
4. $3x^2 + 5x + 1 = 0$
5. $x^2 + 3x - 5 = 0$
6. $x^2 + 7x - 2 = 0$
7. $3w^2 - 5w - 3 = 0$
8. $2p^2 - 7p - 2 = 0$
9. $5x^2 + 2x - 1 = 0$
10. $2x^2 + 4x - 3 = 0$
11. $-3m^2 + 6m - 2 = 0$
12. $-5t^2 + 8t - 2 = 0$
13. $4x^2 + 2x + 3 = 0$
14. $5x^2 + 3x + 2 = 0$
15. $2x^2 + 5x = 4$
16. $3x^2 - 2x = 7$
17. $-r^2 = r - 1$
18. $-2y^2 = 3y - 4$
19. $5x^2 + 3 = 2x$
20. $7x^2 + 4 = x$
21. $x^2 = \dfrac{7}{3}x - 1$
22. $2x^2 = \dfrac{5}{2}x + 1$
23. $3x(3x - 1) = 1$
24. $5x(5x - 1) = 1$
25. $(t + 4)(t - 2) = 3$
26. $(p - 3)(p - 5) = 2$

Find approximate solutions of the equation. Round your results to the second decimal place. Use a graphing calculator table to verify your work.

27. $4x^2 - 3x - 9 = 0$
28. $8x^2 + 2x - 5 = 0$
29. $2.1r^2 + 6.8r - 17.1 = 0$
30. $0.5m^2 - 3.5m - 14.5 = 0$
31. $-1.2x^2 = 2.8x - 12.9$
32. $-1.8x^2 = 3.8x - 52.9$

33. $0.4x^2 - 3.4x + 17.4 = 54.9$
34. $2.7x^2 + 3.8x - 39.7 = 5.6$

Find approximate x-intercept(s). Round the coordinates of your results to the second decimal place. Use a graphing calculator to verify your work.

35. $y = 2x^2 - 5x + 1$
36. $y = 7x^2 + 4x - 3$
37. $y = -5x^2 + 3x + 4$
38. $y = -4x^2 - 2x + 5$
39. $y = 3.7x^2 + 5.2x - 7.5$
40. $y = 1.2x^2 - 5.4x - 3.9$
41. $y = -2.9x^2 - 1.9x + 8.4$
42. $y = -4.1x^2 + 9.8x + 6.6$

Solve by the method of your choice.

43. $x^2 + 11x = -18$
44. $x^2 - 3x = 28$
45. $(x + 4)^2 = 13$
46. $(x - 7)^2 = 5$
47. $3r^2 + 5r = 1$
48. $5t^2 - 6t = 7$
49. $36x^2 - 49 = 0$
50. $25x^2 - 9 = 0$
51. $14x^2 = 21x$
52. $4x^2 = 6x$
53. $7x^2 - 3 = 2$
54. $4x^2 + 5 = 8$
55. $6x^2 = 7x - 2$
56. $4x^2 = -4x - 3$
57. $6m^2 - 2m + 5 = 0$
58. $3p^2 + 4p + 6 = 0$
59. $\dfrac{7}{2}x^2 = x + \dfrac{3}{2}$
60. $\dfrac{2}{7}x^2 = x + \dfrac{3}{7}$
61. $(w + 3)^2 - 2w = 7$
62. $(r - 2)^2 + 3r = 7$

For Exercises 63–66, use the graph of $y = -x^2 + 2x + 3$ shown in Fig. 24 to solve the given equation.

63. $-x^2 + 2x + 3 = 0$ **64.** $-x^2 + 2x + 3 = 3$

65. $-x^2 + 2x + 3 = 4$ **66.** $-x^2 + 2x + 3 = 5$

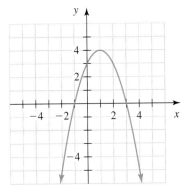

Figure 24 Exercises 63–66

For Exercises 67–70, use the graph of $y = x^2 + 6x + 7$ shown in Fig. 25 to solve the given equation.

67. $x^2 + 6x + 7 = 2$ **68.** $x^2 + 6x + 7 = -1$

69. $x^2 + 6x + 7 = -4$ **70.** $x^2 + 6x + 7 = -2$

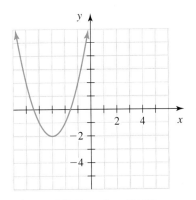

Figure 25 Exercises 67–70

Use "intersect" on a graphing calculator to solve the equation. Round the solution(s) to the second decimal place.

71. $x^2 - 5 = 2x - 1$ **72.** $7x - 2x^2 = -2x + 5$

73. $2x^2 - 5x + 1 = -x^2 + 3x + 3$

74. $-2x^2 + 6x - 4 = x^2 - 5x + 1$

For Exercises 75–78, solve the given equation by referring to the solutions of $y = -2x^2 + 4x + 4$ shown in Table 3.

75. $-2x^2 + 4x + 4 = -12$ **76.** $-2x^2 + 4x + 4 = -2$

77. $-2x^2 + 4x + 4 = 6$ **78.** $-2x^2 + 4x + 4 = 7$

Table 3 Exercises 75-78	
x	**y**
-2	-12
-1	-2
0	4
1	6
2	4
3	-2
4	-12

79. A student tries to solve the equation $x^2 + 5x = 3$:

$$x^2 + 5x = 3$$
$$x = \frac{-5 \pm \sqrt{5^2 - 4(1)(3)}}{2(1)}$$
$$x = \frac{-5 \pm \sqrt{25 - 12}}{2}$$
$$x = \frac{-5 \pm \sqrt{13}}{2}$$

Describe any errors. Then solve the equation correctly.

80. A student tries to solve the equation $3x^2 + 2x - 7 = 0$:

$$3x^2 + 2x - 7 = 0$$
$$x = \frac{-2 \pm \sqrt{2^2 - 4(3)(-7)}}{2(3)}$$
$$x = \frac{-2 \pm \sqrt{4 + 84}}{6}$$
$$x = \frac{-2 \pm \sqrt{88}}{6}$$
$$x = \frac{-2 \pm 2\sqrt{22}}{6}$$
$$x = \frac{-1 \pm 2\sqrt{22}}{3}$$

Describe any errors. Then solve the equation correctly.

81. a. Solve $x^2 + 2x - 3 = 0$ by factoring.
  b. Solve $x^2 + 2x - 3 = 0$ by using the quadratic formula.
  c. Solve $x^2 + 2x - 3 = 0$ by completing the square.
  d. Compare your results from parts (a), (b), and (c).
  e. In your opinion, which method was easiest to use? Explain.

82. A student tries to solve the equation $7x^2 + 3x - 2 = 0$:

$$7x^2 + 3x - 2 = 0$$
$$x = -3 \pm \frac{\sqrt{3^2 - 4(7)(-2)}}{2(7)}$$
$$x = -3 \pm \frac{\sqrt{9 + 56}}{14}$$
$$x = -3 \pm \frac{\sqrt{65}}{14}$$

Describe any errors. Then solve the equation correctly.

83. Describe how to solve a quadratic equation by using the quadratic formula.

84. Compare the methods of solving a quadratic equation by factoring, by the square root property, by completing the square, and by using the quadratic formula. Describe the methods, as well as their advantages and disadvantages.

Related Review

Solve.

85. $2x - 6 = 1$ **86.** $3x - 8 = -2$ **87.** $2x^2 - 6 = 1$

88. $3x^2 - 8 = -2$ **89.** $2x^2 - 6x = 1$ **90.** $3x^2 - 8x = -2$

Find approximate x-intercept(s). Round the coordinates of your result(s) to the second decimal place. Use a graphing calculator to verify your work.

91. $y = 3x^2 - 2x - 4$ **92.** $y = -2x^2 + 4x + 7$

93. $y = 3x - 2$ **94.** $y = -2x + 4$

Expressions, Equations, and Graphs

Perform the indicated instruction. Then use words such as linear, quadratic, cubic, power, radical, polynomial, degree, one variable, *and* two variables *to describe the expression, equation, or system.*

95. Solve $x^2 = 5 - 3x$.

96. Graph $y = x^2 - 3$ by hand.

97. Factor $6x^2 - x - 1$.

98. Solve $x^2 - 3 = 0$.

99. Simplify $3x - 4(7x - 5) + 3$.

100. Find the product $(x^2 - 3)(x^3 + 2)$.

9.6 MORE QUADRATIC MODELS

Objectives

▶ Use quadratic models to make estimates and predictions.

In Section 8.6, we used quadratic models to describe authentic situations. The situations and models were carefully chosen so that we would be able to use factoring to help make estimates and predictions about the independent variable. Recall from Section 9.5 that factoring is rarely helpful, because almost all polynomials are prime. *Any* quadratic equation can be solved by means of the quadratic formula. So, now that we know it, **we can use the quadratic formula to make estimates and predictions about the independent variable for any situation that is described well by a quadratic model.**

Example 1 Using a Quadratic Model to Make Estimates

Table 4 lists the percentages of Americans of various age groups who are Internet users.

Table 4 Percentages of Americans Who Are Internet Users		
Age Group (years)	Age Used to Represent Age Group (years)	Percent
18–24	21	75.7
25–34	29.5	71.1
35–44	39.5	65.4
45–54	49.5	53.4
55–64	59.5	48.6
over 64	70	19.4

Source: *Pew Internet & American Life Project*

Let p be the percentage of Americans at age a years who are Internet users. A model of the situation is

$$p = -0.02a^2 + 0.8a + 65.7$$

1. Use a graphing calculator to draw the graph of the model and, in the same viewing window, the scattergram of the data. Does the model fit the data well?
2. Estimate the percentage of 19-year-old Americans who are Internet users.
3. Estimate the age at which half of Americans are Internet users.
4. Use "maximum" on a graphing calculator to estimate the age of Americans who are most likely to use the Internet. According to the model, what percentage of Americans at this age use the Internet?
5. Find the p-intercept. What does it mean in this situation?
6. Find the a-intercepts. What do they mean in this situation?

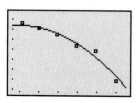

Figure 26 Check how well the model fits the data

Solution

1. The graph of the model and the scattergram of the data are shown in Fig. 26. The model appears to fit the data well.

2. We substitute the input 19 for a in the equation $p = -0.02a^2 + 0.8a + 65.7$ and solve for p:

$$p = -0.02(19)^2 + 0.8(19) + 65.7 = 73.68$$

So, about 73.7% of 19-year-old Americans are Internet users.

3. Half of all Americans is 50%, so, to find the age, we substitute the output 50 for p in the equation $p = -0.02a^2 + 0.8a + 65.7$ and solve for a:

$50 = -0.02a^2 + 0.8a + 65.7$	Substitute 50 for p.
$0 = -0.02a^2 + 0.8a + 15.7$	Subtract 50 from both sides.
$a = \dfrac{-0.8 \pm \sqrt{0.8^2 - 4(-0.02)(15.7)}}{2(-0.02)}$	Substitute into quadratic formula.
$a = \dfrac{-0.8 \pm \sqrt{1.896}}{-0.04}$	Simplify.
$a \approx \dfrac{-0.8 \pm 1.3770}{-0.04}$	Approximate square root.
$a \approx 54.43$ or $a \approx -14.43$	Calculate.

The values of a that we found represent the approximate ages -14 years and 54 years. A negative age does not make sense, so model breakdown has occurred for the estimate -14 years. The model does make a reasonable estimate that half of 54-year-old Americans are Internet users.

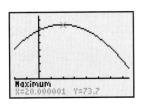

Figure 27 Finding the approximate vertex

4. We use "maximum" on a graphing calculator to approximate the vertex (see Fig. 27). The vertex is about $(20.00, 73.70)$. So, according to the model, about 73.7% of 20-year-old Americans use the Internet, the highest percentage for any age group. However, the actual maximum percentage is larger than 73.7%. From Table 4, we see that 75.7% of Americans between the ages of 18 and 24, inclusive, use the Internet.

5. To find the p-intercept, we substitute 0 for a in the equation $p = -0.02a^2 + 0.8a + 65.7$ and solve for p:

$$p = -0.02(0)^2 + 0.8(0) + 65.7 = 65.7$$

The p-intercept is $(0, 65.7)$. So, the model estimates that 65.7% of newborns are Internet users; model breakdown has occurred.

6. To find the a-intercepts, we substitute 0 for p in the equation $p = -0.02a^2 + 0.8a + 65.7$ and solve for a:

$0 = -0.02a^2 + 0.8a + 65.7$	Substitute 0 for p.
$a = \dfrac{-0.8 \pm \sqrt{0.8^2 - 4(-0.02)(65.7)}}{2(-0.02)}$	Substitute into quadratic formula.
$a = \dfrac{-0.8 \pm \sqrt{5.896}}{-0.04}$	Simplify.
$a \approx \dfrac{-0.8 \pm 2.4282}{-0.04}$	Approximate square root.
$a \approx 80.71$ or $a \approx -40.71$	Calculate.

The a-intercepts are approximately $(-40.71, 0)$ and $(80.71, 0)$. For the a-intercept $(-40.71, 0)$, note that the age -41 does not make sense. For the a-intercept $(80.71, 0)$, the model estimates that no 81-year-old Americans are Internet users. However, a little research would show that some 81-year-old Americans are Internet users. So, model breakdown occurs at both a-intercepts.

We use a graphing calculator to verify our work in Problems 2, 3, 5, and 6 (see Fig. 28).

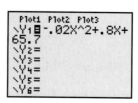
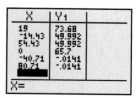

Figure 28 Verify the work ■

Recall from Section 8.6 that, to make a prediction about the dependent variable of a quadratic model, we substitute a value for the independent variable and then solve for the dependent variable. To make a prediction about the independent variable, we substitute a value for the dependent variable and then solve for the independent variable, usually by using the quadratic formula.

Example 2 Comparing a Linear Model and a Quadratic Model

Sales of gasoline–electric hybrid cars in the United States have increased greatly since 2000 (see Table 5).

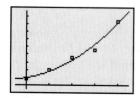

Once there's definite proof that carbon emissions are linked to global warming, I'll trade my Hummer in for a hybrid.

Table 5 Sales of Gasoline-Electric Hybrid Cars	
Year	**Sales (thousands of cars)**
2000	7.8
2001	20.0
2002	35.9
2003	45.9
2004	83.2

Source: *R. L. Polk & Co.*

Let s be hybrid-car sales (in thousands of cars) in the year that is t years since 2000. A linear model of the situation is

$$s = 17.67t + 3.22$$

A quadratic model of the situation is

$$s = 3.16t^2 + 5.01t + 9.55$$

1. Which model describes the situation better?
2. Use first the linear model and then the quadratic model to estimate hybrid-car sales in 2004. Which is the better estimate? Explain.
3. Honda and Toyota have dominated the hybrid-car market. Which mathematical model would those two companies hope describes the situation better? Explain.
4. Use the quadratic model to estimate in which year hybrid-car sales were or will be 400 thousand cars.

Solution

1. We see how well each model fits the data (see Figs. 29 and 30). It appears that the quadratic model fits the data a little better than the linear model does.
2. First we substitute the input 4 for t in the linear model:

$$s = 17.67(4) + 3.22 = 73.90$$

The linear model estimates that 73.9 thousand hybrid cars were sold in 2004. Now we substitute the input 4 for t in the quadratic model to make another estimate of hybrid-car sales in 2004:

$$s = 3.16(4)^2 + 5.01(4) + 9.55 = 80.15$$

The quadratic model estimates that 80.2 thousand hybrid cars were sold in 2004. The actual sales in 2004 were 83.2 thousand hybrid cars (see Table 5). So, the quadratic model's estimate is better than the linear model's estimate.

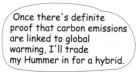

Figure 29 Linear model

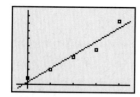

Figure 30 Quadratic model

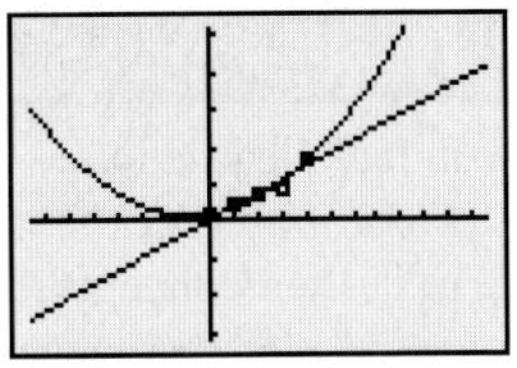

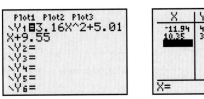

Figure 31 Using Zoom Out to see more of the models

3. We use Zoom Out to display the screen shown in Fig. 31. The quadratic model predicts that, as time passes, hybrid-car sales will be much higher than what the linear model predicts. So, Honda and Toyota would hope that the quadratic model describes the situation better than the linear model does.

4. To find when the hybrid-car sales were or will be 400 thousand cars, we substitute the output 400 for s in the equation $s = 3.16t^2 + 5.01t + 9.55$ and solve for t:

$$400 = 3.16t^2 + 5.01t + 9.55 \qquad \text{Substitute 400 for } s.$$

$$0 = 3.16t^2 + 5.01t - 390.45 \qquad \text{Subtract 400 from both sides.}$$

$$t = \frac{-5.01 \pm \sqrt{5.01^2 - 4(3.16)(-390.45)}}{2(3.16)} \qquad \text{Substitute into quadratic formula.}$$

$$t = \frac{-5.01 \pm \sqrt{4960.3881}}{6.32} \qquad \text{Simplify.}$$

$$t \approx \frac{-5.01 \pm 70.43}{6.32} \qquad \text{Approximate square root.}$$

$$t \approx -11.94 \quad \text{or} \quad t \approx 10.35 \qquad \text{Calculate.}$$

The solutions represent the years 1988 and 2010. Model breakdown occurs for 1988, because hybrid cars were not available for sale then. So, we predict that 400 thousand hybrid cars will be sold in 2010. We verify our work in Fig. 32.

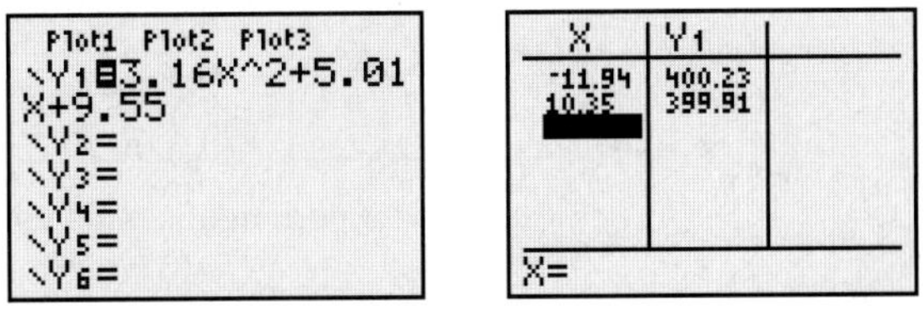

```
Plot1 Plot2 Plot3
\Y1■3.16X^2+5.01
X+9.55
\Y2=
\Y3=
\Y4=
\Y5=
\Y6=
```

```
   X    │ Y1
 -11.94 │400.23
  10.35 │399.91



X=
```

Figure 32 Verify the work ■

group exploration

Looking ahead: Simplifying rational expressions

In Sections 9.2 and 9.5, we simplified radical quotients by factoring the numerator and the denominator and then using the property $\frac{a}{a} = 1$, where $a \neq 0$. Here, we use similar steps to simplify $\frac{14x}{21x}$:

$$\frac{14x}{21x} = \frac{2(7x)}{3(7x)} = \frac{2}{3} \cdot \frac{7x}{7x} = \frac{2}{3} \cdot 1 = \frac{2}{3}$$

Simplify the following expressions:

1. $\dfrac{3x}{5x}$

2. $\dfrac{15x}{25x}$

3. $\dfrac{(x+2)(x-5)}{(x-3)(x-5)}$

4. $\dfrac{x^2 - 9}{x^2 + 7x + 12}$ [**Hint:** First factor the numerator and the denominator.]

5. $\dfrac{3x^2 - 12x}{x^2 - 8x + 16}$

TIPS FOR SUCCESS: Create a Mind Map for the Final Exam

In preparing for a final exam, consider how all the concepts you have learned are interconnected. One way to help yourself do this is to make a *mind map*. Put the main topic in the middle. Around it, attach concepts that relate to it, then concepts that relate to those concepts, and so on.

A portion of a mind map that describes this course is illustrated in Fig. 33. Many more "concept rectangles" could be added to it. You could make one mind map showing an overview of the course and several mind maps for the components of the overview map.

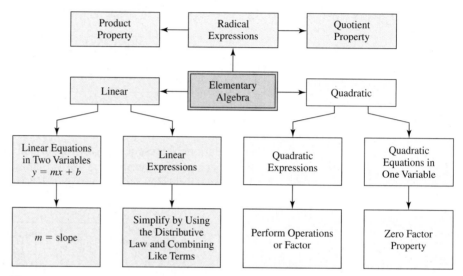

Figure 33 A portion of a mind map describing the course

HOMEWORK 9.6 FOR EXTRA HELP ▶

Student Solutions Manual   PH Math/Tutor Center   *Math* XL MathXL®   MyMathLab MyMathLab

---

1. Prime-time rankings of the television program *20/20* are shown in Table 6 for various years. For example, a ranking of 11 means that, for programs running at prime times, the show had the 11th-largest viewing audience.

Table 6 Prime-Time Rankings of *20/20*

Year	Prime-Time Ranking
1988	57
1990	44
1992	22
1994	14
1996	11
1998	18
2000	36
2002	54
2003	71

Source: *Nielsen Media Research*

Let r be the prime-time ranking for the show at t years since 1980. A model of the situation is $r = 0.94t^2 - 28.5t + 228.8$.

a. Use a graphing calculator to draw the graph of the model and, in the same viewing window, the scattergram of the data. Does the model fit the data well?

b. Estimate when the show's ranking was 30.

c. Use "minimum" on a graphing calculator to estimate when the program had its best ranking. According to the model, what was that ranking?

d. After conducting interviews for 25 years, Barbara Walters left the program in 2004. Estimate the show's ranking in that year. Give a possible reason she may have left the show.

2. The percentages of people working at age 65 or older are shown in Table 7 for various years.

Table 7 Percentages of People Working at Age 65 or Older

Year	Percent
1970	17
1975	14
1980	13
1985	11
1990	12
1995	12
2000	13
2003	14

Source: *Bureau of Labor Statistics*

Let p be the percentage of people working at age 65 or older at t years since 1970. A model of the situation is $p = 0.0143t^2 - 0.55t + 16.8$.

a. Use a graphing calculator to draw the graph of the model and, in the same viewing window, the scattergram of the data. Does the model fit the data well?

b. Predict the percentage of people working at age 65 or older in 2011.

c. Estimate when the percentage of people working at age 65 or older was or will be 16%.

d. Use "minimum" on a graphing calculator to estimate when the lowest percentage of people worked at age 65 or older. According to the model, what was that percentage?

3. The numbers of bank robberies in New York City are shown in Table 8 for various years.

Table 8 Bank Robberies in New York City

Year	Number of Bank Robberies
1998	254
1999	176
2000	136
2001	223
2002	249
2003	408

Source: *New York Police Department*

Let n be the number of bank robberies in New York City at t years since 1990. A model of the situation is the equation $n = 25.9t^2 - 513t + 2695$.

a. Use a graphing calculator to draw the graph of the model and, in the same viewing window, the scattergram of the data. Does the model fit the data well?

b. Use "minimum" on a graphing calculator to estimate in which year the number of bank robberies was at its lowest. According to the model, what was that number of bank robberies?

c. Estimate when there were or will be 600 bank robberies in New York City.

4. The percentages of drivers who admit to running red lights are shown in Table 9 for various age groups.

Table 9 Percentages of Drivers Who Admit to Running Red Lights

Age Group (years)	Age Used to Represent Age Group (years)	Percent
18–25	21.5	75
26–35	30.5	73
36–45	40.5	63
46–55	50.5	56
over 55	65.0	35

Source: *The Social Science Research Center*

Let p be the percentage of drivers at age a years who admit to running red lights. A model of the situation is the equation $p = -0.015a^2 + 0.38a + 74$.

a. Use a graphing calculator to draw the graph of the model and, in the same viewing window, the scattergram of the data. Does the model fit the data well?

b. Estimate the percentage of 37-year-old drivers who admit to running red lights.

c. At what age do half of all drivers admit to running red lights?

d. Find the a-intercepts. What do they mean in this situation?

e. Use "minimum" on a graphing calculator to estimate the age at which drivers are least likely to admit to running red lights. According to the model, what was that percentage?

5. Table 10 lists the percentages of U.S. Internet users, by household income groups, who use e-mail/instant messaging.

Table 10 Percentages of U.S. Internet Users Who Use E-mail/Instant Messaging

Household Income Group (in thousands of dollars)	Income Used to Represent Household Income Group (in thousands of dollars)	Percent
0–15	7.5	72.0
15–25	20	75.5
25–35	30	78.7
35–50	42.5	81.2
50–75	62.5	85.0
over 75	100	89.1

Source: *U.S. Census Bureau*

Let p be the percentage of U.S. Internet users with a household income of d thousand dollars who use e-mail/instant messaging. A model of the situation is the quadratic equation $p = -0.00141d^2 + 0.34d + 69.5$.

a. Use a graphing calculator to draw the graph of the model and, in the same viewing window, the scattergram of the data. Does the model fit the data well?

b. Estimate the percentage of Internet users whose household income is $23,000 and who use e-mail/instant messaging.

c. Estimate the income(s) at which 80% of Internet users use e-mail/instant messaging.

6. The average sizes of U.S. households are shown in Table 11 for various years.

Table 11 Average Sizes of U.S. Households

Year	Average Size (number of people)
1930	4.1
1940	3.7
1950	3.4
1960	3.4
1970	3.1
1980	2.8
1990	2.6
2000	2.6
2003	2.6

Source: *U.S. Census Bureau*

Let s be the average size (in number of people) of a U.S. household at t years since 1900. A model of the situation is $s = 0.00015t^2 - 0.041t + 5.15$.

a. Use a graphing calculator to draw the graph of the model and, in the same viewing window, the scattergram of the data. Does the model fit the data well?

b. Predict the average size of a U.S. household in 2010.

c. Explain why the part of the parabola that lies to the right of the vertex will probably not describe future average household sizes well. Predict when the average U.S. household size will be 2.4 people.

7. eBay® has experienced incredible growth since it began in 1996. The total values of goods sold on eBay are shown in Table 12 for various years.

Table 12 Total Values of Goods Sold on eBay	
Year	Total Value of Goods Sold (billions of dollars)
1998	1.0
1999	2.6
2000	4.7
2001	9.1
2002	14.5
2003	22.8
2004	34.2

Source: *eBay*

Let v be the total value (in billions of dollars) of goods sold on eBay in the year that is t years since 1990. A model of the situation is $v = 0.98t^2 - 16.1t + 68$.

a. Use a graphing calculator to draw the graph of the model and, in the same viewing window, the scattergram of the data. Does the model fit the data well?

b. Predict the total value of goods sold in 2011.

c. Predict in which year the total value of goods sold will be $115 billion.

8. DVD revenues (sales and rentals) are shown in Table 13 for various years.

Table 13 DVD Revenues	
Year	DVD Revenue (billions of dollars)
1998	0.4
1999	1.5
2000	3.9
2001	7.1
2002	11.9
2003	17.5

Source: *Adams Media Research*

Let r be the revenue (in billions of dollars) for the year that is t years since 1990. A model of the situation is $r = 0.57t^2 - 8.6t + 32.6$.

a. Use a graphing calculator to draw the graph of the model and, in the same viewing window, the scattergram of the data. Does the model fit the data well?

b. Predict the revenue in 2010.

c. Predict when the revenue will be $60 billion.

9. The percentage of law degrees earned by women is increasing (see Table 14).

Table 14 Percentages of Law Degrees Earned by Women	
Year	Percent
1990	42.2
1992	42.7
1994	43.0
1996	43.5
1998	44.4
2000	45.9
2002	48.0

Source: *U.S. National Center for Education Statistics*

Let p be the percentage of law degrees earned by women in the year that is t years since 1990. A linear model of the situation is $p = 0.45t + 41.54$. A quadratic model of the situation is $p = 0.044t^2 - 0.08t + 42.4$.

a. Use a graphing calculator to draw the graphs of both models and, in the same viewing window, the scattergram of the data. Which model describes the situation better?

b. Use first the linear model and then the quadratic model to estimate the percentage in 1996. Which estimate is better? Explain.

c. Find the p-intercept of the linear model first and then that of the quadratic model. Which p-intercept describes the situation better? Explain.

d. Estimate in what year women earned half of all law degrees.

10. The average per-minute cost of cell phones has declined greatly since 1998 (see Table 15).

Table 15 Average per-Minute Cost of Cell Phones	
Year	Average per-Minute Cost (cents)
1998	34
1999	27
2000	22
2001	18
2002	15
2003	13
2004	11

Source: *J. D. Power & Associates*

Let c be the average per-minute cost (in cents) at t years since 1995. A linear model of the situation is $c = -3.71t + 42.29$. A quadratic model of the situation is $c = 0.5t^2 - 9.7t + 58$.

a. Use a graphing calculator to draw the graphs of both models and, in the same viewing window, the scattergram of the data. Which model describes the situation better?

b. Use first the linear model and then the quadratic model to estimate the average per-minute cost of cell phones in 2001. Which is the better estimate? Explain.

c. In 1995, the average cost of a cell phone was 64 cents per minute. Find the c-intercept first of the linear model and then of the quadratic model. Which intercept describes the situation better? Explain.

d. For the period 2004–2006, which model would cell-phone subscribers hope would describe the situation better? Explain.

11. The amount of donations from corporations are shown in Table 16 for various years.

Table 16 Donations from Corporations	
Year	Amount (billions of dollars)
1990	5.5
1992	5.9
1994	7.0
1996	7.5
1998	9.7
2000	12.5

Source: *Giving USA*

Let A be the amount (in billions of dollars) of donations from corporations in the year that is t years since 1990. A model of the situation is $A = 0.073t^2 - 0.06t + 5.6$.

  a. Use a graphing calculator to draw the graph of the model and, in the same viewing window, the scattergram of the data. Does the model fit the data well?

  b. Estimate the amount of donations from corporations in 1999. The total 1999 donations from individuals, foundations, bequests (money left in wills), and corporations was $190.2 billion. What percent of all donations were from corporations?

  c. Predict when corporations will donate $35 billion.

12. Internet gambling (e-gambling) is illegal in the United States. Estimates of e-gambling revenues are shown in Table 17 for various years.

Table 17 Global Internet Gambling Revenue Estimates	
Year	Revenue (billions of U.S. dollars)
1999	1.2
2000	2.2
2001	3.1
2002	4.5
2003	6.4

Source: *Christiansen Capitol Advisors, LLC*

Let r be the e-gambling revenue (in billions of U.S. dollars) in the year that is t years since 1990. A model of the situation is $r = 0.16t^2 - 2.3t + 9.1$.

  a. Use a graphing calculator to draw the graph of the model and, in the same viewing window, the scattergram of the data. Does the model fit the data well?

  b. Predict the e-gambling revenue in 2010. Gaming industry leaders predict that the e-gambling revenue in that year will be $20 billion. Is your result larger or smaller than this prediction?

  c. Predict when the revenue will reach $25 billion.

13. Table 18 shows that people with more education tend to earn more money.

Let I be the median annual income (in thousands of dollars) for people with a full-time equivalent of t years of education. A model for the situation is $I = 0.175t^2 - 1.1t + 3.8$.

  a. Use a graphing calculator to draw the graph of the model and, in the same viewing window, the scattergram of the data. Does the model fit the data well?

Table 18 Education versus Median Annual Income in 2001		
Grade-Level Completion or Degree	Full-Time Equivalent Years in School	Median Annual Income (thousands of dollars)
9th to 12th Grade, no diploma	10	10.3
High school	12	15.7
Some college	13	20.1
Associate's	14	22.6
Bachelor's	16	31.0
Master's	18	40.7
Doctoral	20	52.2

Source: *U.S. Census Bureau*

  b. Estimate the full-time equivalent number of years of education for people whose median annual income is $45 thousand.

  c. In August 2001, a high school graduate is trying to decide whether to get a job or to go to a public college to get a bachelor's degree. For the 2001–2002 academic year, average tuition and fees for a public 4-year college are $4281. In parts (i)–(iv), assume that tuition and fees, as well as incomes, are constant.

   i. If the student doesn't go to college, estimate the student's total earnings in four years.

   ii. If the student goes to college, estimate the total cost of four years of college.

   iii. If the student goes to college, estimate after how many years of work will the student be in the same financial position (total earnings minus total costs) as if the student hadn't gone to college and had gotten a job instead.

   iv. Assume that the student earns a bachelor's degree in 4 years and then works for 33 years until retirement. How much more will the student's lifetime earnings be than if the person had not gone to college (and had worked for 37 years)?

   v. During the past 20 years, those with larger incomes have had larger growth in incomes. Is your estimate in part (iv) an underestimate or an overestimate? Explain.

14. The numbers of women who have run in the New York City (NYC) Marathon are shown in Table 19 for various years.

Table 19 Numbers of Women Who Have Run in the NYC Marathon	
Year	Number of Women
1970	1
1973	12
1983	2355
1993	6151
2003	11,927

Source: *New York Road Runners Club*

Let n be the number of women who have run in the NYC Marathon at t years since 1970. A model of the situation is $n = 9.1t^2 + 64.2t - 102.6$.

  a. Use a graphing calculator to draw the graph of the model and, in the same viewing window, the scattergram of the data. Does the model fit the data well?

b. Predict the number of women who will run in the marathon in 2010.

c. Predict in which year 18,000 women will run in the marathon—when about half of the runners will be women.

15. Discuss how to make predictions about the dependent or independent variable of a quadratic model.

16. Discuss how to find intercepts of a quadratic model.

Related Review

17. Of the more than two million students who take the SAT® each year, a small, but growing, number of students are exempted from the three-hour testing limit (see Table 20).

Table 20 Numbers of Students Exempted from the Three-Hour SAT Limit

Year	Number of Students Given Extra Time (thousands)
1993	12
1995	16
1997	22
1999	24
2001	26
2003	33

Source: *The College Board*

Let n be the number of students (in thousands) who are given extra time on the SAT in the year that is t years since 1990.

a. Find an equation of a model to describe the data. [**Hint:** Begin by using your graphing calculator to draw a scattergram of the data.]

b. Predict when 40 thousand students will be given extra time on the SAT.

c. Predict the number of students who will be given extra time on the SAT in 2011.

18. The rate of illnesses and injuries suffered in the workplace has declined since 1997 (see Table 21).

Table 21 Illness and Injury Rate in the Workplace

Year	Illness and Injury Rate (cases per 100 full-time workers per year)
1997	7.1
1998	6.7
1999	6.3
2000	6.0
2001	5.7
2002	5.3
2003	5.0

Source: *Bureau of Labor Statistics Surveys*

Let r be the rate of illnesses and injuries (cases per 100 full-time workers per year) suffered in the workplace at t years since 1990.

a. Find an equation of a model to describe the data.

b. Predict when the illness and injury rate will be 3.5 cases per 100 full-time workers.

c. Use your model to estimate the illness and injury rate in 2004.

d. In 2004, Wal-Mart® was the largest U.S. company, with 1.2 million employees. Estimate the number of illness and injury cases at Wal-Mart in 2004.

19. In 2003, 54% of farms owned or leased computers. The percentage has increased about 4 percentage points per year since then (Source: *National Agricultural Statistics Service*). Let p be the percentage of farms that own or lease computers at t years since 2003.

a. Find an equation of a model to describe the data.

b. Predict the percentage of farms that will own or lease computers in 2010.

c. Predict when all farms will own or lease computers.

20. In 1999, 31% of college students smoked cigarettes. The percentage has increased about 0.89 percentage point per year since then (Source: *University of Michigan Institute for Social Research*). Let p be the percentage of college students who smoke at t years since 1999.

a. Find an equation of a model to describe the data.

b. Predict what percentage of college students will smoke cigarettes in 2011.

c. Predict when 40% of college students will smoke cigarettes.

21. The percentage of radio listeners tuned in to public radio has increased approximately linearly from 2.0% in 1986 to 5.2% in 2003 (Source: *Corporation for Public Broadcasting*). Predict the percentage of radio listeners tuned in to public radio in 2012.

22. The total revenue (adjusted for inflation to 2004 dollars) from Broadway shows increased approximately linearly from $450 in 1994 to $735 in 2003 (Source: *The League of American Theatres and Producers, Inc.*). Predict the total revenue from Broadway shows in 2010.

Expressions, Equations, and Graphs

Give an example of each. Then solve, simplify, or graph, as appropriate.

23. quadratic equation in one variable that can be solved by factoring

24. linear equation in one variable

25. radical expression in which a radicand is a fraction

26. quadratic expression with four terms

27. linear equation in two variables

28. system of two linear equations in two variables

29. quadratic equation in one variable that can't be solved by factoring

30. cubic expression with five terms

31. linear expression with four terms

32. quadratic equation in two variables

analyze this

GLOBAL WARMING LAB (continued from Chapter 7)

Some developing countries' carbon emissions are increasing greatly as their economies continue to grow. In particular, China's future carbon emissions will affect the planet considerably, because of that country's large population, large population growth, and increasing per-person carbon emissions (see Table 22). China's per-person carbon emissions are projected to be 3.2 metric tons per person in 2050—7.8 times the 1980 per-person emissions.

Table 22 China's Population and Carbon Emissions

Year	Population (billions)	Per-Person Carbon Emissions (metric tons per person)
1980	0.98	0.41
1985	1.05	0.50
1990	1.14	0.54
1995	1.23	0.65
2000	1.30	0.78

Source: *U.S. Census Bureau; U.S. Energy Information Administration*

Currently, the United States has the largest gross national product (GNP) and carbon emissions in the world. China's economy is growing so rapidly that its GNP is likely to surpass U.S. GNP by as early as 2020, although GNP per person will still be larger in the United States.* China's carbon emissions will overtake U.S. carbon emissions even sooner (see Table 23).

Table 23 U.S. Population and Carbon Emissions

Year	Population (billions)	Per-Person Carbon Emissions (metric tons per person)
1960	0.18	4.4
1970	0.21	5.7
1980	0.23	5.7
1990	0.25	5.6
2000	0.28	5.4

Source: *U.S. Census Bureau;* The Cambridge Factfinder

Recall from the Global Warming Lab in Chapter 7 that if world population were to decline to 1.7 billion, then the IPCC's 2050 carbon emissions goal could be met if carbon emissions were limited to 1.4 metric tons per person each year. This per-person limit would create room for developing countries such as China to expand their economies.

Such a population–emissions strategy would require developing countries to learn to expand their economies in new ways that would not increase per-person carbon emissions so much. It would also require developed countries to learn to maintain their economies while reducing their per-person carbon emissions. Finally, it would require that world population be reduced by 72% from its 2000 level

*The New York Times.

of 6.07 billion. All three requirements would be great challenges.

Analyzing the Situation

1. Let p be China's population (in billions) at t years since 1950. Create a scattergram for China's population data. Then find an equation of a model that describes the situation well.

2. Let y be China's *per-person* carbon emissions (in metric tons per person) in the year that is t years since 1950. A reasonable model is $c = 0.00043t^2 - 0.0165t + 0.53$. Verify the claim that China's per-person emissions in 2050 will be 3.2 metric tons of carbon per person.

3. Let c be China's carbon emissions (in billions of metric tons) in the year that is t years since 1950. Find an equation that describes the relationship between t and c. [**Hint:** Add, subtract, multiply, or divide the right-hand sides of your result from Problem 1 and the China per-person carbon emissions model.]

4. Perform a unit analysis of the model you found in Problem 3.

5. In Problem 5 of the Global Warming Lab in Chapter 5, you found a model of U.S. carbon emissions c (in billions of metric tons) for the year that is t years since 1950. If, however, you didn't find such an equation, find it now. Then use "intersect" on a graphing calculator to predict when China and the United States will have the same annual amount of carbon emissions. What is that annual amount of carbon emissions?

6. Predict when China's *per-person* annual carbon emissions will reach the United States' level of 5.4 metric tons of carbon per person each year.

7. Explain why the year you found in Problem 5 is so much earlier than the year you found in Problem 6.

PROJECTILE LAB (continued from Chapter 7)

In the Projectile Lab in Chapter 7, you found an equation of a quadratic model that describes the times and heights of a ball that you threw straight up. In this lab, you will analyze that situation further.

Analyzing the Situation

1. What model did you work with in the Projectile Lab in Chapter 7? Provide the equation and define the variables.

2. Use the model to estimate both times when the ball was at a height of 10 feet.

3. Find the average of the two times you found in Problem 2.

4. Use the model to estimate the height of the ball at the time you found in Problem 3.

5. Use a graphing calculator to draw a graph of the model. Copy the graph and indicate on the graph the two points that represent when the ball was at a height of 10 feet and the point that represents when the ball was at the height you found in Problem 4.

6. By referring to your graph in Problem 5, explain why your result in Problem 4 is the maximum height of the ball.

7. Use "maximum" on a graphing calculator to estimate the maximum height of the ball. Compare this result with your result from Problem 6.

8. Find the average speed of the ball on its way up. That is, find the ball's average speed from the moment it was released to the moment it reached its maximum height.

9. Explain why it makes sense that your result in Problem 8 is less than the initial speed you found in the Projectile Lab in Chapter 7.

Chapter Summary

Key Points
OF CHAPTER 9

Principal square root (Section 9.1)	If a is a nonnegative number, then $\sqrt{a}$ is the nonnegative number we square to get a. We call $\sqrt{a}$ the **principal square root** of a.
Perfect square (Section 9.1)	A **perfect square** is a number whose principal square root is rational.
Product property for square roots (Section 9.1)	For nonnegative numbers a and b, $$\sqrt{ab} = \sqrt{a}\sqrt{b}$$
Simplified square root (Section 9.1)	A square root is **simplified** when the radicand does not have any perfect-square factors other than 1.

Simplifying a square root with a small radicand (Section 9.1)

To simplify a square root in which it is easy to determine the largest perfect-square factor of the radicand,

1. Write the radicand as the product of the *largest* perfect-square factor and another number.

2. Apply the product property for square roots.

Simplifying a square root with a large radicand (Section 9.1)

To simplify a square root in which it is difficult to determine the largest perfect-square factor of the radicand,

1. Find the prime factorization of the radicand.

2. Look for pairs of identical factors of the radicand and rearrange the factors to highlight them.

3. Use the product property for square roots.

Simplifying $\sqrt{x^2}$ (Section 9.1)	If x is nonnegative, then $\sqrt{x^2} = x$.
Quotient property for square roots (Section 9.2)	For a nonnegative number a and a positive number b, $$\sqrt{\frac{a}{b}} = \frac{\sqrt{a}}{\sqrt{b}}$$
Rationalize the denominator (Section 9.2)	To **rationalize the denominator** of a fraction of the form $\dfrac{p}{\sqrt{q}}$, where q is positive, we multiply the fraction by 1 in the form $\dfrac{\sqrt{q}}{\sqrt{q}}$.

Simplifying a radical expression (Section 9.2)

To simplify a radical expression,

1. Use the quotient property for square roots so that no radicand is a fraction.

2. Use the product property for square roots so that no radicands have perfect-square factors other than 1.

3. Rationalize denominators so that no denominator is a radical expression.

4. Use the property $\dfrac{a}{a} = 1$, where a is nonzero, to simplify.

5. Continue applying steps 1–4 until the radical expression is completely simplified.

Square root property (Section 9.3)

Let k be a nonnegative constant. Then $x^2 = k$ is equivalent to $x = \pm\sqrt{k}$.

Pythagorean theorem (Section 9.3)

If a and b are the lengths of the legs of a right triangle and c is the length of the hypotenuse, then $a^2 + b^2 = c^2$.

Perfect-square trinomial property (Section 9.4)

For a perfect-square trinomial of the form $x^2 + bx + c$, dividing b by 2 and then squaring the result gives c:

$$x^2 + bx + c$$

$$\left(\frac{b}{2}\right)^2 = c$$

When completing the square can be used (Section 9.4)

Any quadratic equation can be solved by completing the square.

Solving a quadratic equation by completing the square (Section 9.4)

To solve a quadratic equation by **completing the square,**
1. Write the equation in the form $ax^2 + bx = k$, where a, b, and k are constants.
2. If $a \neq 1$, divide both sides of the equation by a.
3. Complete the square for the expression on the left side of the equation.
4. Solve the equation by using the square root property.

Quadratic formula (Section 9.5)

The solutions of a quadratic equation $ax^2 + bx + c = 0$ are given by the **quadratic formula:**

$$x = \frac{-b \pm \sqrt{b^2 - 4ac}}{2a}$$

When the quadratic formula can be used (Section 9.5)

Any quadratic equation can be solved by the quadratic formula.

Guidelines on solving quadratic equations (Section 9.5)

Here are some guidelines on deciding which method to use to solve a quadratic equation:

Method:	When to Use:
Factoring	For equations that can easily be put into the form $ax^2 + bx + c = 0$ and where $ax^2 + bx + c$ can easily be factored.
Square root property	For equations that can easily be put into the form $x^2 = k$ or $(x + p)^2 = k$.
Completing the square	When the directions require it.
Quadratic formula	For all equations except those which can easily be solved by factoring or by the square root property.

Using the quadratic formula to make estimates and predictions (Section 9.6)

We can use the quadratic formula to make estimates and predictions about the independent variable for any situation that is described well by a quadratic model.

CHAPTER 9 REVIEW EXERCISES

Simplify.

1. $\sqrt{196}$ **2.** $-\sqrt{64}$ **3.** $\sqrt{-25}$ **4.** $-\sqrt{-81}$

Find an approximate value of the radical. Round your result to the second decimal place.

5. $\sqrt{95}$ **6.** $-7.29\sqrt{38.36}$

Simplify. Assume that any variables are nonnegative.

7. $\sqrt{84}$ **8.** $\sqrt{36x}$

9. $\sqrt{18ab^2}$ **10.** $\sqrt{98m^2n^2}$

11. Without using a calculator, find the two nearest consecutive integers to $\sqrt{39}$.

Simplify. Assume that any variables are positive.

12. $\sqrt{\dfrac{5}{9}}$ **13.** $\sqrt{\dfrac{50y}{x^2}}$ **14.** $\dfrac{4}{\sqrt{x}}$ **15.** $\sqrt{\dfrac{5b^2}{32}}$

16. A student tries to simplify $\dfrac{3}{\sqrt{7}}$:

$$\frac{3}{\sqrt{7}} = \left(\frac{3}{\sqrt{7}}\right)^2$$

$$= \frac{3^2}{\left(\sqrt{7}\right)^2}$$

$$= \frac{9}{7}$$

Describe any errors. Then simplify the expression correctly.

Solve. Use a graphing calculator to verify your work.

17. $p^2 = 45$

18. $5t^2 = 7$

19. $5x^2 + 4 = 12$

20. $(x + 4)^2 = 27$

21. $(w - 6)^2 = -15$

22. $(m - 8)^2 = 0$

Let a and b be the lengths of the legs of a right triangle, and let c be the length of the hypotenuse. Values for two of the three lengths are given. Find the third length.

23. $a = 4$ and $b = 8$

24. $b = 3$ and $c = 6$

The lengths of two sides of a right triangle are given. Find the length of the third side. (The triangle is not drawn to scale.)

25. See Fig. 34.

26. See Fig. 35.

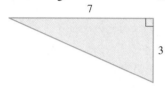

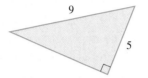

Figure 34 Exercise 25   **Figure 35** Exercise 26

27. A 12-foot-long ladder is leaning against a building. The bottom of the ladder is 3.5 feet from the base of the building. Will the ladder reach the bottom of a window that is 11 feet above the ground?

Solve by completing the square.

28. $x^2 + 6x = 2$

29. $x^2 + 2x = 17$

30. $w^2 - 8w - 4 = 0$

31. $t^2 - 12t + 40 = 0$

32. $2x^2 + 8x = 12$

33. $3x^2 - 18x - 27 = 0$

Solve by using the quadratic formula.

34. $3x^2 + 7x + 1 = 0$

35. $2x^2 - 5x - 4 = 0$

36. $-5y^2 = 3 - 2y$

37. $-p^2 + 3p = -5$

38. $2x^2 - x = \dfrac{3}{2}$

39. $2x(x - 1) = 5$

Find approximate solutions of the equation. Round your results to the second decimal place. Use a graphing calculator table to verify your work.

40. $6w^2 = 8w + 5$

41. $-1.9t^2 - 5.4t + 27.9 = 14.1$

42. A student tries to solve the equation $2x^2 + 3x - 7 = 0$:

$$2x^2 + 3x - 7 = 0$$

$$x = -3 \pm \frac{\sqrt{3^2 - 4(2)(-7)}}{2(2)}$$

$$= -3 \pm \frac{\sqrt{65}}{4}$$

Describe any errors. Then solve the equation correctly.

Solve by the method of your choice.

43. $x^2 = 5x + 14$

44. $3x^2 - 4x = 1$

45. $r^2 + 13 = 0$

46. $(t - 4)^2 = 24$

47. $x^2 = 2x + 35$

48. $5x^2 + 8 = 20$

49. $(x - 2)^2 - 3x = 1$

50. $3x^2 + 2x - 8 = 0$

51. $(w + 7)^2 = 27$

52. $-2p^2 = 7p - 3$

53. $x^2 - 2x = \dfrac{5}{2}$

54. $4x(x - 2) = -3$

For Exercises 55–58, use the graph of $y = \dfrac{1}{2}x^2 - 2x - 1$ shown in Fig. 36 to solve the given equation.

55. $\dfrac{1}{2}x^2 - 2x - 1 = -3$

56. $\dfrac{1}{2}x^2 - 2x - 1 = -4$

57. $\dfrac{1}{2}x^2 - 2x - 1 = -1$

58. $\dfrac{1}{2}x^2 - 2x - 1 = 5$

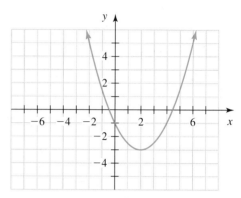

Figure 36 Exercises 55–58

59. Use "intersect" on a graphing calculator to solve the equation $7 - 2x^2 = x^2 - x - 8$. Round the solution(s) to the second decimal place.

Find approximate x-intercepts. Round the coordinates of your results to the second decimal place. Use a graphing calculator to verify your results.

60. $y = 3x^2 - 8x + 2$

61. $y = -3.9x^2 + 7.1x + 54.9$

62. a. Solve $x^2 - 2x - 15 = 0$ by factoring.

    b. Solve $x^2 - 2x - 15 = 0$ by using the quadratic formula.

    c. Solve $x^2 - 2x - 15 = 0$ by completing the square.

    d. Compare your results from parts (a), (b), and (c).

63. A person is said to have moderate or severe memory impairment if the person can recall 4 or fewer words out of 20 on combined immediate and delayed recall tests. The percentages of Americans with moderate or severe memory impairment are shown in Table 24 for various age groups.

Table 24 Percentages of Americans with Moderate or Severe Memory Impairment		
Age Group (years)	**Age Used to Represent Age Group (years)**	**Percent**
65–69	67	1.1
70–74	72	2.5
75–79	77	4.5
80–84	82	6.4
over 84	90	12.9

Source: *Health and Retirement Study*

Let p be the percentage of Americans with moderate or severe memory impairment at age a years. A model of the situation is $p = 0.016a^2 - 2a + 63$.

a. Use a graphing calculator to draw the graph of the model and, in the same viewing window, the scattergram of the data. Does the model fit the data well?

b. Estimate the percentage of Americans at age 70 that have moderate or severe memory impairment.

c. Estimate at what age 10% of Americans have moderate or severe memory impairment.

64. The percentages of medical degrees earned by women has increased greatly since 1970 (see Table 25). Let p be the percentage of medical degrees earned by women at t years since 1970. A model of the situation is the equation $p = -0.0176t^2 + 1.71t + 7.3$.

   a. Use a graphing calculator to draw the graph of the model and, in the same viewing window, the scattergram of the data. Does the model fit the data well?

   b. Predict the percentage of medical degrees that will be earned by women in 2010.

   c. Predict when women will earn 48% of medical degrees.

Table 25 Medical Degrees Earned by Women	
Year	**Percent**
1970	8.4
1975	13.1
1980	23.4
1985	30.4
1990	34.2
1995	38.8
2000	42.7
2002	44.4

Source: *U.S. National Center for Education Statistics*

CHAPTER 9 TEST

Simplify.

1. $-\sqrt{121}$

2. $-\sqrt{-9}$

Simplify. Assume that any variables are positive.

3. $\sqrt{48}$

4. $\sqrt{64x}$

5. $\sqrt{45a^2b}$

6. $\dfrac{3}{\sqrt{7}}$

7. $\sqrt{\dfrac{2m^2}{n}}$

8. $\sqrt{\dfrac{3}{20}}$

9. The lengths of two sides of the triangle shown in Fig. 37 are given. Find the length of the third side. (The triangle is not drawn to scale.)

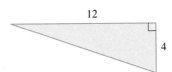

Figure 37 Exercise 9

10. The width of a picture frame is 11 inches, and the diagonal has a length of 17 inches. Estimate the length of the picture frame to the second decimal place.

Solve by the method of your choice.

11. $3x^2 = 5$

12. $-5x^2 + 4 = -28$

13. $(t - 5)^2 = 18$

14. $2w^2 + 3w - 6 = 0$

15. $x(x - 3) = 40$

16. $\dfrac{3}{2}x^2 = x + 2$

17. Solve $5p^2 - 20p - 35 = 0$ by completing the square.

18. Find approximate solutions of the quadratic equation $1.4x^2 - 2.3x - 38.5 = 7.4$. Round your results to the second decimal place.

19. Find approximate x-intercepts of $y = -1.2x^2 + 37.9x - 50.4$. Round the coordinates of your results to the second decimal place.

For Exercises 20–23, solve the given equation by referring to the solutions of $y = -x^2 + 6x + 5$ shown in Table 26.

20. $-x^2 + 6x + 5 = 13$

21. $-x^2 + 6x + 5 = 10$

22. $-x^2 + 6x + 5 = 14$

23. $-x^2 + 6x + 5 = 15$

Table 26 Exercises 20-23	
x	**y**
0	5
1	10
2	13
3	14
4	13
5	10
6	5

24. A student tries to solve the equation $2x^2 + 4x - 3 = 0$:

$$2x^2 + 4x - 3 = 0$$
$$x = \frac{-4 \pm \sqrt{4^2 - 4(2)(-3)}}{2(2)}$$
$$x = \frac{-4 \pm \sqrt{16 + 24}}{4}$$
$$x = \frac{-4 \pm \sqrt{40}}{4}$$
$$x = \frac{-4 \pm 2\sqrt{10}}{4}$$
$$x = -1 \pm 2\sqrt{10}$$

Describe any errors. Then solve the equation correctly.

25. The Walton family, Wal-Mart's founders, have given hundreds of millions of dollars to help reform education (see Table 27).

Table 27 Walton Family's Donations to Education Charities	
Year	**Donations (millions of dollars)**
1998	140
1999	55
2000	35
2001	95
2002	364

Source: *Walton Family Foundation*

Let d be the total donations (in millions of dollars) from the Walton family to education charities in the year that is t years since 1995. A model of the situation is the equation $d = 56.3t^2 - 514t + 1188$.

a. Use a graphing calculator to draw the graph of the model and, in the same viewing window, the scattergram of the data. Does the model fit the data well?

b. Use "minimum" on a graphing calculator to estimate in what year total donations were at a minimum. According to the model, what were total donations in that year? Is your result an underestimate or an overestimate?

c. Estimate total donations in 2003.

d. The family expects to donate as much as 20% of its $100 billion in Wal-Mart stock. Predict in which year the Waltons will donate one billion dollars to education charities. [**Hint:** One billion is how many millions?]

10

Rational Expressions and Equations

Twenty years from now you will be more disappointed by the things you didn't do than by the ones you did do.

—Mark Twain

Do you pay your bills online? The numbers of households in the United States that pay bills online are shown in Table 1 for various years. In Example 6 of Section 10.5, we will predict when 20% of households will pay their bills online.

Year	Table 1 Households That Pay Bills Online	
	Households That Pay Bills Online (millions)	**Total Number of Households (millions)**
1998	1.7	102.5
1999	4.1	103.9
2000	5.2	104.7
2001	7.3	108.2
2002	9.9	109.3

Sources: *Yankee Group; U.S. Census Bureau*

In Chapters 1–6, we worked with linear expressions and equations. In Chapters 7–9, we worked with polynomial expressions and equations. In this chapter, we will work with yet another type of expressions and equations called *rational* expressions and equations. This work will enable us to make estimates and predictions about authentic situations, such as the percentage of households in the United States that pay bills online.

10.1 SIMPLIFYING RATIONAL EXPRESSIONS

Objectives

▷ Know the meaning of *rational expression*.

▷ Find all *excluded values* of a rational expression.

▷ Simplify a rational expression.

▷ Make estimates with a *rational model*.

In this section, we will discuss the meaning of a *rational expression* and how to simplify it. We will also discuss how to use a *rational model* to make estimates about an authentic situation.

Rational Expressions

Throughout this course, we have worked with polynomials. If P and Q are polynomials with Q nonzero, we call the ratio $\dfrac{P}{Q}$ a **rational expression.** The name "rational" refers to a ratio. Here are some examples:

$$\frac{x^2 + 5x - 3}{x - 7} \qquad \frac{4x^2 - 7}{x^3} \qquad \frac{5x^4}{x^2 - 9}$$

Excluded Values

Consider the rational expression $\dfrac{3}{x}$. Note that substituting 0 for x in $\dfrac{3}{x}$ leads to $\dfrac{3}{0}$, which is undefined because we cannot divide by 0 (Section 2.2). We say that 0 is an excluded value of $\dfrac{3}{x}$.

DEFINITION Excluded value

A number is an **excluded value** of a rational expression if substituting the number into the expression leads to a division by 0.

Example 1 Finding Excluded Values

Find any excluded values of the given expression.

1. $\dfrac{5}{x-2}$ 　　　　　　　　　　　　　　　　　**2.** $\dfrac{x}{3}$

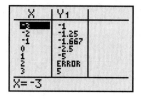

Figure 1 Verify the work

Solution

1. The number 2 is an excluded value, because $\dfrac{5}{2-2}$ has a division by 0. No other value of x leads to a division by 0, so 2 is the only excluded value. To verify our work, we create a graphing calculator table for $y = \dfrac{5}{x-2}$. See Fig. 1. The "ERROR" message across from $x = 2$ in the table supports the idea that the value 2 for x leads to a division by 0. (The TI-83 and TI-84 display either "ERROR" or "ERR:.")

2. The denominator is the constant 3, so no substitution for x will lead to a division by 0. Thus, there are no excluded values. ∎

WARNING　For the expression $\dfrac{5}{x-2}$, 0 itself is *not* an excluded value, because $\dfrac{5}{0-2} = \dfrac{5}{-2}$ does not contain a division by 0.

For the expression $\dfrac{x}{3}$, again 0 is not an excluded value, because $\dfrac{0}{3}$ does not have a division by 0. In fact, $\dfrac{0}{3}$ is defined and $\dfrac{0}{3} = 0$.

Example 2 Finding Excluded Values

Find any excluded values of the given expression.

1. $\dfrac{x+1}{4x-7}$ 　　　　　　　　　　　　　　　**2.** $\dfrac{x-4}{2x^2-5x-3}$

Solution

1. To find which values of x lead to a division by 0, we set the denominator of $\dfrac{x+1}{4x-7}$ equal to 0 and solve for x:

$$4x - 7 = 0 \qquad \text{Set denominator of } \tfrac{x+1}{4x-7} \text{ equal to 0.}$$
$$4x = 7 \qquad \text{Add 7 to both sides.}$$
$$x = \frac{7}{4} \qquad \text{Divide both sides by 4.}$$

So, the only excluded value is $\dfrac{7}{4}$.

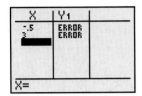

Figure 2 Verify the work

2. We set the denominator of $\dfrac{x-4}{2x^2-5x-3}$ equal to 0 and solve for x:

$$2x^2 - 5x - 3 = 0 \qquad \text{Set denominator of } \dfrac{x-4}{2x^2-5x-3} \text{ equal to 0.}$$

$$(2x+1)(x-3) = 0 \qquad \text{Factor left side.}$$

$$2x+1 = 0 \quad \text{or} \quad x-3 = 0 \qquad \text{Zero factor property}$$

$$2x = -1 \quad \text{or} \quad x = 3$$

$$x = -\frac{1}{2} \quad \text{or} \quad x = 3$$

So, the excluded values are $-\dfrac{1}{2}$ and 3. We use a graphing calculator table to verify our work (see Fig. 2). ■

Finding Excluded Values

To find any excluded values of a rational expression $\dfrac{P}{Q}$,

1. Set the denominator Q equal to 0.
2. Solve the equation $Q = 0$.

Simplifying a Rational Expression

In Sections 9.2 and 9.5, we simplified radical quotients by factoring the numerator and the denominator and then using the property $\dfrac{a}{a} = 1$, where $a \neq 0$. Here, we use similar steps to simplify $\dfrac{15}{35}$:

$$\frac{15}{35} = \frac{5 \cdot 3}{5 \cdot 7} = \frac{5}{5} \cdot \frac{3}{7} = 1 \cdot \frac{3}{7} = \frac{3}{7}$$

We simplify the rational expression $\dfrac{3x+6}{x^2-4}$ in a similar manner:

$$\frac{3x+6}{x^2-4} = \frac{3(x+2)}{(x-2)(x+2)} \qquad \text{Factor numerator and denominator.}$$

$$= \frac{3}{x-2} \cdot \frac{x+2}{x+2} \qquad \frac{AC}{BD} = \frac{A}{B} \cdot \frac{C}{D}$$

$$= \frac{3}{x-2} \cdot 1 \qquad \text{Simplify: } \frac{x+2}{x+2} = 1$$

$$= \frac{3}{x-2} \qquad a \cdot 1 = a$$

Figure 3 Verify the work

The number 2 is the only excluded value of our result, $\dfrac{3}{x-2}$. The numbers -2 and 2 are the only excluded values of the original expression, $\dfrac{3x+6}{x^2-4}$. (Try it.) When we write

$$\frac{3x+6}{x^2-4} = \frac{3}{x-2}$$

we mean that the two expressions yield the same number for each real number substituted for x, except for any excluded values of either expression (-2 and 2). In Fig. 3, we use a graphing calculator table to verify our work.

A rational expression is in **lowest terms** if the numerator and denominator have no common factors other than 1 or -1. We **simplify a rational expression** by writing it in lowest terms. We also write the numerator and the denominator in factored form.

Simplifying a Rational Expression

To simplify a rational expression,

1. Factor the numerator and the denominator.
2. Use the property

$$\frac{AB}{AC} = \frac{A}{A} \cdot \frac{B}{C} = 1 \cdot \frac{B}{C} = \frac{B}{C}$$

where A and C are nonzero, so that the expression is in lowest terms.

Throughout the rest of this chapter, you may assume that the form $\frac{A}{B}$ represents a rational expression.

For the expression $\frac{AB}{AC}$, note that the polynomial A is a factor of both the numerator and the denominator. The expression $\frac{2x}{5x}$ can be simplified to $\frac{2}{5}$ when $x \neq 0$, because x is a factor of both the numerator and the denominator.

WARNING   The expression

$$\frac{2x + 7}{5x}$$

is in lowest terms already. Although x is a factor of the term $2x$, it is not a factor of $2x + 7$, the (entire) *numerator*.

Likewise, the expression

$$\frac{(x + 3)(x + 5) + 4}{7(x + 3)}$$

is in lowest terms already. The expression $x + 3$ is not a factor of $(x + 3)(x + 5) + 4$, the (entire) numerator. To see that the rational expression is in lowest terms, we simplify the numerator to get $\frac{x^2 + 8x + 19}{7(x + 3)}$. (Try it.)

Example 3   Simplifying Rational Expressions

Simplify.

1. $\dfrac{20x^2}{12x^4}$

2. $\dfrac{5t + 15}{t^2 + 5t + 6}$

Solution

1.
$$\frac{20x^2}{12x^4} = \frac{2 \cdot 2 \cdot 5 \cdot x \cdot x}{2 \cdot 2 \cdot 3 \cdot x \cdot x \cdot x \cdot x} \qquad \text{Factor numerator and denominator.}$$

$$= \frac{5}{3x^2} \qquad \text{Simplify: } \frac{2 \cdot 2 \cdot x \cdot x}{2 \cdot 2 \cdot x \cdot x} = 1$$

2.
$$\frac{5t + 15}{t^2 + 5t + 6} = \frac{5(t + 3)}{(t + 2)(t + 3)} \qquad \text{Factor numerator and denominator.}$$

$$= \frac{5}{t + 2} \qquad \text{Simplify: } \frac{t + 3}{t + 3} = 1$$

We use a graphing calculator table to verify our work (see Fig. 4). There are "ERROR" messages across from $x = -3$ and $x = -2$ in the table, because both -3 and -2 are excluded values of the original expression and -2 is an excluded value of our result.   ■

Example 4   Simplifying Rational Expressions

Simplify.

1. $\dfrac{2a^2 - 50}{2a^2 - 6a - 20}$

2. $\dfrac{x^2 - 5x + 6}{x^3 + 3x^2 - 4x - 12}$

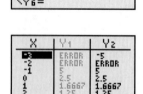

Figure 4 Verify the work

Solution

1.
$$\frac{2a^2 - 50}{2a^2 - 6a - 20} = \frac{2(a^2 - 25)}{2(a^2 - 3a - 10)}$$

$$= \frac{2(a - 5)(a + 5)}{2(a - 5)(a + 2)}$$ Factor numerator and denominator.

$$= \frac{a + 5}{a + 2}$$ Simplify: $\frac{2(a - 5)}{2(a - 5)} = 1$

2.
$$\frac{x^2 - 5x + 6}{x^3 + 3x^2 - 4x - 12} = \frac{(x - 2)(x - 3)}{x^2(x + 3) - 4(x + 3)}$$

$$= \frac{(x - 2)(x - 3)}{(x^2 - 4)(x + 3)}$$ Factor numerator and denominator.

$$= \frac{(x - 2)(x - 3)}{(x - 2)(x + 2)(x + 3)}$$

$$= \frac{x - 3}{(x + 2)(x + 3)}$$ Simplify: $\frac{x - 2}{x - 2} = 1$ ∎

Graphs of Rational Equations in Two Variables

A **rational equation in two variables** is an equation in two variables in which each side can be written as a rational expression. Here are some examples of such equations:

$$y = \frac{1}{x}, \quad y = \frac{x + 2}{x - 4}, \quad y = \frac{3x}{x^2 - x - 6}$$

We show graphing calculator graphs of these equations in Fig. 5.

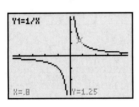 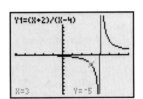 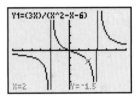

Figure 5 Graphs of three rational equations in two variables; the vertical lines (in red) displayed in the middle and right-hand screens are not part of the graph

Graphing rational equations in two variables by hand is beyond the scope of this course.

Rational Models

A **rational model** is a rational equation in two variables that describes the relationship between two quantities for an authentic situation. Rational models can help us describe equal shares of a quantity. For example, a rational model can describe the relationship between the number of people who attend a party and the equal per-person cost of the party.

Example 5 Finding a Rational Model

For a ski trip, the ski club at a college plans to charter a bus for a total cost of $350. Let p be the per-person bus cost (in dollars per person) if n students go on the trip.

1. Use a table to help find an equation for n and p. Show the arithmetic to help you see a pattern.
2. Perform a unit analysis of the equation you found in Problem 1.
3. Substitute 16 for n in the equation. What does the result mean in this situation?
4. Use a graphing calculator graph to verify the result in Problem 3.
5. Is the graph of the model in Quadrant I increasing, decreasing, or neither? What does that mean in this situation?

Table 2 Number of Students and per-Person Bus Cost	
Number of Students n	**Per-Person Bus Cost (dollars per person)** p
1	$\dfrac{350}{1}$
2	$\dfrac{350}{2}$
3	$\dfrac{350}{3}$
4	$\dfrac{350}{4}$
n	$\dfrac{350}{n}$

Solution

1. We create Table 2. From the last row of the table, we see that the per-person cost (in dollars per person) can be represented by $\dfrac{350}{n}$. So, $p = \dfrac{350}{n}$.

2. Here is a unit analysis of the equation $p = \dfrac{350}{n}$:

$$\underbrace{p}_{\text{dollars per person}} = \underbrace{\dfrac{350 \;\leftarrow\; \text{dollars}}{n \;\leftarrow\; \text{number of students}}}_{\text{dollars per person}}$$

The units of the expressions on both sides of the equation are dollars per person, which suggests that our equation is correct.

3. We substitute 16 for n in the equation $p = \dfrac{350}{n}$:

$$p = \dfrac{350}{16} = 21.875$$

So, the per-person cost is about \$21.88 if 16 students go on the trip. For the ski club to pay exactly \$350, eight members could pay \$21.87 each and the other eight members could pay \$21.88 each.

4. In Fig. 6, we verify our work in Problem 3.

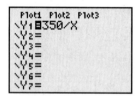 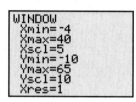 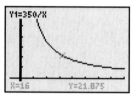

Figure 6 Verify the work

5. The graph appears to be decreasing in Quadrant I. That means that the more people who go on the trip, the smaller the per-person cost will be, which makes sense. ∎

In Example 5, we used a rational equation in two variables to model equal shares of a quantity. Rational equations in two variables can also be helpful in modeling percentages of a population or the average value of some quantity. For instance, a rational model can describe the relationship between the year and the percentage of college students who work. A rational model can also describe the relationship between the year and the average price of a textbook.

Example 6 Using a Rational Model to Make Estimates

The number of people between the ages of 16 and 24 years who graduated from high school and the number of those students who then enrolled in college within 12 months are shown in Table 3 for various years.

Table 3 College Enrollment of Recent High School Graduates		
Year	**Number of High School Graduates (millions)**	**Number of High School Graduates Who Enrolled in College within 12 Months (millions)**
1997	2.77	1.86
1998	2.81	1.84
1999	2.90	1.82
2000	2.76	1.75
2001	2.55	1.57

Source: *U.S. National Center for Education Statistics,* Digest of Education Statistics *(annual)*

Let p be the percentage of high school graduates who enrolled in college at t years since 1990 and within 12 months of receiving their high school diploma. A reasonable rational model is

$$p = \frac{-2.64t^2 + 41t + 28}{-0.0521t^2 + 0.89t - 0.92}$$

1. Use the model to estimate the percentage of recent high school graduates who enrolled in college in 2001.
2. Find the *actual* percentage of recent high school graduates who enrolled in college in 2001.
3. Is your result in Problem 1 an underestimate or an overestimate? Explain.
4. Use the model to estimate the percentage of recent high school graduates who enrolled in college in 2002.

Solution

1. Since $2001 - 1990 = 11$, we substitute 11 for t in the equation
$p = \dfrac{-2.64t^2 + 41t + 28}{-0.0521t^2 + 0.89t - 0.92}$ and solve for p:

$$p = \frac{-2.64(11)^2 + 41(11) + 28}{-0.0521(11)^2 + 0.89(11) - 0.92} = \frac{159.56}{2.5659} \approx 62.18$$

According to the model, 62.2% of recent high school graduates enrolled in college in 2001.

2. From Table 3 we see that, of the 2.55 million recent high school graduates, 1.57 million enrolled in college. We begin to find the percentage by dividing the number enrolled in college by the number of recent high school graduates:

$$\frac{1.57}{2.55} \approx 0.616$$

By moving the decimal point of 0.616 two places to the right, we see that 61.6% of recent high school graduates enrolled in college in 2001.

3. Our result in Problem 1 is an overestimate: $62.2\% > 61.6\%$.

4. Since $2002 - 1990 = 12$, we substitute 12 for t in the equation
$p = \dfrac{-2.64t^2 + 41t + 28}{-0.0521t^2 + 0.89t - 0.92}$ and solve for p:

$$p = \frac{-2.64(12)^2 + 41(12) + 28}{-0.0521(12)^2 + 0.89(12) - 0.92} = \frac{139.84}{2.2576} \approx 61.94$$

According to the model, 61.9% of recent high school graduates enrolled in college in 2002. ∎

group exploration

Simplifying rational expressions

We can write $\dfrac{2x}{5x} = \dfrac{2}{5}$ because x is a common factor of the numerator and the denominator. For each problem, use this concept to help you determine whether the work is correct. If you are unsure, it may help to decide whether you can show work similar to the following:

$$\frac{2x}{5x} = \frac{2}{5} \cdot \frac{x}{x} = \frac{2}{5} \cdot 1 = \frac{2}{5}$$

1. a. $\dfrac{4x}{9x} = \dfrac{4}{9}$

b. $\dfrac{4+x}{9x} = \dfrac{4+1}{9} = \dfrac{5}{9}$

3. a. $\dfrac{8+x^5}{6+x^2} = \dfrac{4+x^3}{3}$

b. $\dfrac{8x^5}{6x^2} = \dfrac{4x^3}{3}$

4. a. $\dfrac{(x+2)(x-4)}{5(x-4)} = \dfrac{x+2}{5}$

b. $\dfrac{(x+2)(x-4)+1}{5(x-4)} = \dfrac{(x+2)+1}{5} = \dfrac{x+3}{5}$

2. a. $\dfrac{3x \cdot 5}{7x} = \dfrac{3 \cdot 5}{7} = \dfrac{15}{7}$

b. $\dfrac{3x+5}{7x} = \dfrac{3+5}{7} = \dfrac{8}{7}$

HOMEWORK 10.1

FOR EXTRA HELP ▶

 Student Solutions Manual

PH Math/Tutor Center

 Math XL
MathXL®

MyMathLab
MyMathLab

Find all excluded values.

1. $\dfrac{7}{x}$

2. $\dfrac{3}{x}$

3. $\dfrac{x}{4}$

4. $\dfrac{x}{6}$

5. $\dfrac{3}{x-4}$

6. $\dfrac{8}{x-1}$

7. $-\dfrac{x}{x+9}$

8. $-\dfrac{x}{x+2}$

9. $\dfrac{2p}{3p-12}$

10. $\dfrac{8t}{5t-20}$

11. $\dfrac{x-4}{6x+8}$

12. $\dfrac{x+1}{12x+9}$

13. $\dfrac{2}{x^2+5x+6}$

14. $\dfrac{5}{x^2+8x+15}$

15. $\dfrac{r}{r^2-2r-35}$

16. $\dfrac{2y}{y^2-10y+24}$

17. $\dfrac{x-9}{x^2-10x+25}$

18. $\dfrac{x+7}{x^2+6x+9}$

19. $\dfrac{3x-1}{x^2-16}$

20. $\dfrac{x+8}{x^2-25}$

21. $\dfrac{c+4}{25c^2-49}$

22. $\dfrac{k-3}{36k^2-1}$

23. $\dfrac{2x-5}{2x^2+13x+15}$

24. $\dfrac{7x-4}{3x^2-7x+2}$

25. $\dfrac{3x}{6x^2-13x+6}$

26. $\dfrac{x+1}{4x^2+4x-3}$

27. $\dfrac{w-5}{w^3+2w^2-4w-8}$

28. $\dfrac{t+2}{t^3+3t^2-9t-27}$

Simplify.

29. $\dfrac{4x}{6}$

30. $\dfrac{12}{15x}$

31. $\dfrac{12t^3}{15t}$

32. $\dfrac{21y^4}{14y^2}$

33. $\dfrac{18x^3y}{27x^2y^4}$

34. $\dfrac{35x^2y^3}{25x^5y}$

35. $\dfrac{3x-6}{5x-10}$

36. $\dfrac{7x+21}{4x+12}$

37. $\dfrac{2x+12}{3x+18}$

38. $\dfrac{5x-25}{2x-10}$

39. $\dfrac{a^3+4a^2}{7a^2+28a}$

40. $\dfrac{4r^2-24r}{r^3-6r^2}$

41. $\dfrac{x^2-y^2}{3x+3y}$

42. $\dfrac{x^2-xy}{2x-2y}$

43. $\dfrac{4x+8}{x^2+7x+10}$

44. $\dfrac{3x+12}{x^2+7x+12}$

45. $\dfrac{5x-35}{x^2-9x+14}$

46. $\dfrac{4x-4}{x^2-9x+8}$

47. $\dfrac{t^2+5t+4}{t^2+9t+20}$

48. $\dfrac{b^2+3b+2}{b^2+8b+12}$

49. $\dfrac{x^2-9x+14}{x^2-8x+7}$

50. $\dfrac{x^2-x-30}{x^2-4x-12}$

51. $\dfrac{x^2-4}{x^2-3x-10}$

52. $\dfrac{x^2-9}{x^2+5x-24}$

53. $\dfrac{x^3+8x^2+16x}{x^3-16x}$

54. $\dfrac{x^3-14x^2+49x}{x^3-49x}$

55. $\dfrac{6x-16}{9x^2-64}$

56. $\dfrac{27x+21}{81x^2-49}$

57. $\dfrac{-4w+8}{w^2+2w-8}$

58. $\dfrac{-6a+30}{a^2-6a+5}$

59. $\dfrac{4x^2-25}{2x^2+x-15}$

60. $\dfrac{9x^2-4}{3x^2-8x+4}$

61. $\dfrac{3x^2+9x+6}{6x^2+5x-1}$

62. $\dfrac{2x^2+6x-20}{4x^2-5x-6}$

63. $\dfrac{a^2+2ab+b^2}{a^2-b^2}$

64. $\dfrac{a^2+3ab+2b^2}{a^2+4ab+3b^2}$

65. $\dfrac{5x+10}{x^3+2x^2-3x-6}$

66. $\dfrac{5x-15}{x^3-3x^2-4x+12}$

67. $\dfrac{t^2+2t+1}{t^3+t^2-t-1}$

68. $\dfrac{a^2+4a+4}{a^3+2a^2-4a-8}$

69. $\dfrac{x^3+8}{x^2+7x+10}$

70. $\dfrac{x^3+27}{x^2-2x-15}$

71. $\dfrac{x^3-64}{x^2-16}$

72. $\dfrac{x^3-8}{x^2-4}$

73. Some students have a party and share equally in the expense, which is \$60 for both soft drinks and snacks. Let p be the per-person cost (in dollars per person) if n students go to the party.
   a. Complete Table 4 to help find an equation for n and p. Show the arithmetic to help you see a pattern.

Table 4 Per-Person Cost for the Party

Number of Students n	Cost per Student (dollars per student) p
10	
20	
30	
40	
n	

b. Perform a unit analysis of the equation you found in part (a).

c. Substitute 25 for n in the equation and solve for p. What does the result mean in this situation?

d. For positive values of n, what happens to the value of p as the value of n is increased? What does that trend mean in this situation?

74. A student has taken a total of 70 units (hours or credits) of courses at a college. Let A be the average number of units the student has taken per semester and n be the number of semesters the student has been enrolled at the college.

a. Complete Table 5 to help find an equation for n and A. Show the arithmetic to help you see a pattern.

Table 5 Average Number of Units per Semester

Number of Semesters n	Average Number of Units per Semester A
4	
5	
6	
7	
n	

b. Perform a unit analysis of the equation you found in part (a).

c. Substitute 8 for n in the equation and solve for A. What does the result mean in this situation?

d. For positive values of n, what happens to the value of A as the value of n is increased? What does that trend mean in this situation?

75. The numbers of prisoners and the numbers of releases from prisons are shown in Table 6 for various years.

Table 6 Prisoners and Releases from Prisons

Year	Number of Prisoners Released from Prison (thousands)	Year	Total Number of Prisoners (thousands)
1980	170	1980	316
1985	234	1985	481
1990	420	1990	740
1995	492	1995	1085
2000	585	2000	1331
2004	630	2003	1409

Source: *Bureau of Justice Statistics*

Let p be the percentage of prisoners who are released in the year that is t years since 1980. A reasonable model of the situation is

$$p = \frac{-28t^2 + 2694t + 15{,}122}{50.8t + 275}$$

a. Use the model to estimate the percentage of prisoners that were released in 2000.

b. Now find the *actual* percentage of prisoners that were released in 2000.

c. Is your result in part (a) an underestimate or an overestimate? Explain.

d. Predict the percentage of prisoners that will be released in 2010.

76. The numbers of households that have cable television are shown in Table 7 for various years.

Table 7 Numbers of Households with Cable Television

Year	Number of Households with Cable Television (millions)	Total Number of Households (millions)
1980	17.5	80.8
1985	35.4	86.8
1990	50.5	93.3
1995	60.6	99.0
2000	66.3	104.7
2004	65.9	112.0

Sources: *Paul Kagan Associates, Inc.; U.S. Census Bureau*

Let p be the percentage of households that have cable television at t years since 1980. A reasonable model is

$$p = \frac{-9.3t^2 + 428t + 1706}{1.26t + 80.51}$$

a. Use the model to estimate the percentage of households that had cable television in 2004.

b. Now find the *actual* percentage of households that had cable television in 2004.

c. Is your result in part (a) an underestimate or an overestimate? Explain.

d. Estimate the percentage of households that had cable television in 2002.

77. A student thinks that 0 is an excluded value of the expression $\frac{x}{2}$. Is the student correct? Explain.

78. A student thinks that the numbers 3 and 5 are excluded values for the expression $\frac{x-3}{x-5}$. Is the student correct? Explain.

79. It is a common error to think that 0 is an excluded value for the expression $\frac{5}{x+2}$. Substitute 0 for x in the expression $\frac{5}{x+2}$ to show that 0 is not an excluded value.

80. It is a common error to think that 4 is an excluded value of the expression $\frac{x-4}{x+6}$. Substitute 4 for x in the expression $\frac{x-4}{x+6}$ to show that 4 is not an excluded value.

81. A student tries to simplify the expression $\frac{4x+3}{2x}$:

$$\frac{4x+3}{2x} = \frac{4+3}{2} = \frac{7}{2}$$

Describe any errors. Then simplify the expression correctly.

82. A student tries to simplify the expression
$$\frac{(x-2)(x+4)}{x^2(x-2)+5(x-2)}:$$

$$\frac{(x-2)(x+4)}{x^2(x-2)+5(x-2)} = \frac{x+4}{x^2+5(x-2)}$$
$$= \frac{x+4}{x^2+5x-10}$$

Describe any errors. Then simplify the expression correctly.

83. Give three examples of rational expressions, each of whose only excluded value is 7.

84. Give three examples of rational expressions, each of whose only excluded values are -3 and 5.

85. Give three examples of rational expressions, each equivalent to the expression $\frac{x}{x-4}$.

86. Give three examples of rational expressions, each equivalent to the expression $\frac{5}{x+2}$.

87. Describe how to find all excluded values of a rational expression. (See page 9 for guidelines on writing a good response.)

88. Describe how to simplify a rational expression. (See page 9 for guidelines on writing a good response.)

Related Review

Simplify.

89. $\frac{x^{-4}y^9}{5w^{-2}}$

90. $\frac{2w^{-3}}{x^{-5}y^{-1}}$

91. $\frac{4x-12}{x^2-7x+12}$

92. $\frac{x^2-4}{x^2+5x-14}$

93. $\frac{6-\sqrt{32}}{8}$

94. $\frac{10+\sqrt{50}}{35}$

Expressions, Equations, and Graphs

Perform the indicated instruction. Then use words such as linear, quadratic, cubic, power, radical, rational, polynomial, degree, one variable, *and* two variables *to describe the expression, equation, or system.*

95. Factor $2p^3 - 3p^2 - 18p + 27$.

96. Simplify $3(2x-5) - (5x+2) + 2(x-3)$.

97. Graph $y = x^2 - 2x$ by hand.

98. Simplify $-2(4x-6)^2$.

99. Solve $x(3x-2) = 4$.

100. Graph $y = 3x^2$ by hand.

10.2 MULTIPLYING AND DIVIDING RATIONAL EXPRESSIONS

Objectives

▶ Multiply and divide rational expressions.

▶ Convert units of quantities.

In Section 10.1, we simplified rational expressions. In this section, we will multiply and divide rational expressions.

Multiplication of Rational Expressions

Recall from Section 2.2 that we multiply fractions by using the property $\frac{a}{b} \cdot \frac{c}{d} = \frac{ac}{bd}$, where b and d are nonzero. For example,

$$\frac{2}{5} \cdot \frac{3}{7} = \frac{2 \cdot 3}{5 \cdot 7} = \frac{6}{35}$$

We multiply more complicated rational expressions in a similar way.

Multiplying Rational Expressions

If $\frac{A}{B}$ and $\frac{C}{D}$ are rational expressions and B and D are nonzero, then

$$\frac{A}{B} \cdot \frac{C}{D} = \frac{AC}{BD}$$

In words: To multiply two rational expressions, write the numerators as a product and write the denominators as a product.

Example 1 Finding the Product of Two Rational Expressions

Find the product $\frac{4x}{5} \cdot \frac{7}{6x^4}$. Simplify the result.

Solution

$$\frac{4x}{5} \cdot \frac{7}{6x^4} = \frac{4x \cdot 7}{5 \cdot 6x^4}$$ Write numerators as a product and denominators as a product: $\frac{A}{B} \cdot \frac{C}{D} = \frac{AC}{BD}$

$$= \frac{2 \cdot 2 \cdot x \cdot 7}{5 \cdot 2 \cdot 3 \cdot x \cdot x \cdot x \cdot x}$$ Factor numerator and denominator.

$$= \frac{14}{15x^3}$$ Simplify: $\frac{2 \cdot x}{2 \cdot x} = 1$ ∎

To find the product of rational expressions, we usually begin by factoring the numerators and the denominators if possible. That will put us in a good position to simplify after we have found the product.

Example 2   Finding the Product of Two Rational Expressions

Find the product $\dfrac{2x + 6}{3} \cdot \dfrac{6x}{5x + 15}$. Simplify the result.

Solution

$$\frac{2x + 6}{3} \cdot \frac{6x}{5x + 15} = \frac{2(x + 3)}{3} \cdot \frac{2 \cdot 3 \cdot x}{5(x + 3)}$$ Factor numerators and denominators.

$$= \frac{2(x + 3) \cdot 2 \cdot 3 \cdot x}{3 \cdot 5(x + 3)}$$ Write both numerators and denominators as a product: $\frac{A}{B} \cdot \frac{C}{D} = \frac{AC}{BD}$

$$= \frac{4x}{5}$$ Simplify: $\frac{(x + 3) \cdot 3}{(x + 3) \cdot 3} = 1$

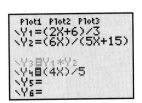

We verify our work by creating a graphing calculator table for $y = \dfrac{2x + 6}{3} \cdot \dfrac{6x}{5x + 15}$ and $y = \dfrac{4x}{5}$. See Fig. 7. To enter an equation by using Y_n references, see Section A.22. There is an "ERROR" message across from $x = -3$ in the table, because -3 is an excluded value of the original expression.   ∎

Figure 7 Verify the work

Example 3   Finding the Product of Two Rational Expressions

Find the product $\dfrac{x^2 - 49}{5x - 10} \cdot \dfrac{x^2 - 2x}{x^2 - 9x + 14}$. Simplify the result.

Solution

$$\frac{x^2 - 49}{5x - 10} \cdot \frac{x^2 - 2x}{x^2 - 9x + 14} = \frac{(x - 7)(x + 7)}{5(x - 2)} \cdot \frac{x(x - 2)}{(x - 7)(x - 2)}$$ Factor numerators and denominators.

$$= \frac{(x - 7)(x + 7) \cdot x(x - 2)}{5(x - 2)(x - 7)(x - 2)}$$ Multiply numerators; multiply denominators.

$$= \frac{x(x + 7)}{5(x - 2)}$$ Simplify: $\frac{(x - 7)(x - 2)}{(x - 7)(x - 2)} = 1$ ∎

How to Multiply Rational Expressions

To multiply two rational expressions,

1. Factor the numerators and the denominators.
2. Multiply by using the property $\dfrac{A}{B} \cdot \dfrac{C}{D} = \dfrac{AC}{BD}$, where B and D are nonzero.
3. Simplify the result.

Example 4 Finding the Product of Two Rational Expressions

Find the product $\dfrac{4x^2 - 9}{2x - 10} \cdot \dfrac{x^2 - 10x + 25}{2x^2 - 7x + 6}$. Simplify the result.

Solution

$$\dfrac{4x^2 - 9}{2x - 10} \cdot \dfrac{x^2 - 10x + 25}{2x^2 - 7x + 6} = \dfrac{(2x - 3)(2x + 3)}{2(x - 5)} \cdot \dfrac{(x - 5)(x - 5)}{(2x - 3)(x - 2)} \qquad \text{Factor numerators and denominators.}$$

$$= \dfrac{(2x - 3)(2x + 3)(x - 5)(x - 5)}{2(x - 5)(2x - 3)(x - 2)} \qquad \text{Multiply numerators; multiply denominators.}$$

$$= \dfrac{(x - 5)(2x + 3)}{2(x - 2)} \qquad \text{Simplify: } \dfrac{(2x - 3)(x - 5)}{(2x - 3)(x - 5)} = 1 \quad \blacksquare$$

Division of Rational Expressions

Recall from Section 2.2 that, to divide two fractions, we use the property $\dfrac{a}{b} \div \dfrac{c}{d} = \dfrac{a}{b} \cdot \dfrac{d}{c}$, where b, c, and d are nonzero. For example,

$$\dfrac{2}{5} \div \dfrac{3}{7} = \dfrac{2}{5} \cdot \dfrac{7}{3} = \dfrac{14}{15}$$

Note that dividing by $\dfrac{3}{7}$ is the same as multiplying by the reciprocal of $\dfrac{3}{7}$, which is $\dfrac{7}{3}$. We divide more complicated rational expressions in a similar way.

Dividing Rational Expressions

If $\dfrac{A}{B}$ and $\dfrac{C}{D}$ are rational expressions and B, C, and D are nonzero, then

$$\dfrac{A}{B} \div \dfrac{C}{D} = \dfrac{A}{B} \cdot \dfrac{D}{C}$$

In words: To divide by a rational expression, multiply by its reciprocal.

Example 5 Finding the Quotient of Two Rational Expressions

Find the quotient $\dfrac{15x^3}{2} \div \dfrac{35x^7}{4}$.

Solution

$$\dfrac{15x^3}{2} \div \dfrac{35x^7}{4} = \dfrac{15x^3}{2} \cdot \dfrac{4}{35x^7} \qquad \text{Multiply by reciprocal of } \dfrac{35x^7}{4}, \text{ which is } \dfrac{4}{35x^7}.$$

$$= \dfrac{3 \cdot 5x^3}{2} \cdot \dfrac{2 \cdot 2}{5 \cdot 7x^7} \qquad \text{Factor numerators and denominators.}$$

$$= \dfrac{3 \cdot 5 \cdot x^3 \cdot 2 \cdot 2}{2 \cdot 5 \cdot 7 \cdot x^7} \qquad \text{Multiply numerators; multiply denominators.}$$

$$= \dfrac{6}{7x^4} \qquad \text{Simplify: } \dfrac{2 \cdot 5 \cdot x^3}{2 \cdot 5 \cdot x^3} = 1 \quad \blacksquare$$

How to Divide Rational Expressions

To divide two rational expressions,

1. Write the quotient as a product by using the property $\dfrac{A}{B} \div \dfrac{C}{D} = \dfrac{A}{B} \cdot \dfrac{D}{C}$, where B, C, and D are nonzero.
2. Find the product.
3. Simplify.

Example 6   Finding the Quotient of Two Rational Expressions

Find the quotient.

1. $\dfrac{x+1}{x^2 - x - 6} \div \dfrac{3x+3}{x^2 - 9}$

2. $\dfrac{3x^2 + 11x + 10}{x^2 + 2x - 24} \div \dfrac{x^2 - 4}{3x^2 + 18x}$

Solution

1. $\dfrac{x+1}{x^2 - x - 6} \div \dfrac{3x+3}{x^2 - 9}$

$= \dfrac{x+1}{x^2 - x - 6} \cdot \dfrac{x^2 - 9}{3x + 3}$ Multiply by reciprocal of $\frac{3x+3}{x^2-9}$.

$= \dfrac{x+1}{(x-3)(x+2)} \cdot \dfrac{(x-3)(x+3)}{3(x+1)}$ Factor numerators and denominators.

$= \dfrac{(x+1)(x-3)(x+3)}{(x-3)(x+2) \cdot 3(x+1)}$ Multiply numerators; multiply denominators.

$= \dfrac{x+3}{3(x+2)}$ Simplify: $\frac{(x+1)(x-3)}{(x+1)(x-3)} = 1$

We verify our work by creating a graphing calculator table for the equations $y = \dfrac{x+1}{x^2 - x - 6} \div \dfrac{3x+3}{x^2 - 9}$ and $y = \dfrac{x+3}{3(x+2)}$. See Fig. 8. There are "ERROR" messages across from the values $x = -3$, $x = -2$, $x = -1$, and $x = 3$, because each of these values is an excluded value of either the original expression or our result (why?).

2. $\dfrac{3x^2 + 11x + 10}{x^2 + 2x - 24} \div \dfrac{x^2 - 4}{3x^2 + 18x}$

$= \dfrac{3x^2 + 11x + 10}{x^2 + 2x - 24} \cdot \dfrac{3x^2 + 18x}{x^2 - 4}$ Multiply by reciprocal of $\frac{x^2-4}{3x^2+18x}$.

$= \dfrac{(3x+5)(x+2)}{(x-4)(x+6)} \cdot \dfrac{3x(x+6)}{(x-2)(x+2)}$ Factor numerators and denominators.

$= \dfrac{(3x+5)(x+2) \cdot 3x(x+6)}{(x-4)(x+6)(x-2)(x+2)}$ Multiply numerators; multiply denominators.

$= \dfrac{3x(3x+5)}{(x-4)(x-2)}$ Simplify: $\frac{(x+2)(x+6)}{(x+2)(x+6)} = 1$ ■

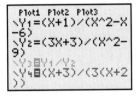

Figure 8 Verify the work

Example 7   Finding the Quotient of Two Rational Expressions

Find the quotient $\dfrac{x^2 + 6x + 9}{x - 4} \div (x^2 + x - 6)$.

Solution

$$\frac{x^2 + 6x + 9}{x - 4} \div (x^2 + x - 6) = \frac{x^2 + 6x + 9}{x - 4} \cdot \frac{1}{x^2 + x - 6}$$

Multiply by reciprocal of $\frac{x^2 + x - 6}{1}$.

$$= \frac{(x + 3)(x + 3)}{x - 4} \cdot \frac{1}{(x + 3)(x - 2)}$$

Factor numerators and denominators.

$$= \frac{(x + 3)(x + 3)}{(x - 4)(x + 3)(x - 2)}$$

Multiply numerators; multiply denominators.

$$= \frac{x + 3}{(x - 4)(x - 2)}$$

Simplify: $\frac{x + 3}{x + 3} = 1$ ∎

Converting Units of Quantities

Suppose that we're ordering a rug measured in feet, but we measured the width and the length of the floor in inches. There is no need to remeasure: We can convert the units of a quantity to equivalent units by multiplying by ratios of units where each ratio is equal to 1. For example, since there are 12 inches in 1 foot, the following ratio is equal to 1:

$$\frac{12 \text{ inches}}{1 \text{ foot}} = 1$$

The reciprocal is also equal to 1:

$$\frac{1 \text{ foot}}{12 \text{ inches}} = 1$$

Suppose that the width of a floor is 110 inches. Here, we convert from units of inches to feet:

$$\frac{110 \text{ inches}}{1} \cdot \frac{1 \text{ foot}}{12 \text{ inches}} \approx 9.2 \text{ feet}$$

When converting, we can eliminate "inches," because $\frac{\text{inches}}{\text{inches}} = 1$.

We can eliminate a pair of the same units even if one is in singular form and the other is in plural form. For example, $\frac{\text{feet}}{\text{foot}} = 1$. However, $\frac{\text{inch}^2}{\text{inch}} = \text{inch}$, not 1.

Some equivalent units are shown in the margin.

Equivalent Units

Length

1 inch = 2.54 centimeters
1 foot = 12 inches
1 yard = 3 feet
1 mile = 5280 feet
1 mile ≈ 1.61 kilometers

Volume

1 cup = 8 ounces
1 quart = 4 cups
1 quart ≈ 0.946 liter
1 gallon = 4 quarts

Weight

1 gram = 1000 milligrams
1 pound = 16 ounces

Time

1 year ≈ 365 days

Example 8 Converting Units

Make the indicated unit conversions. Round the results to two decimal places for Problems 2–4.

1. The official height of a basketball hoop is 10 feet. What is its height in yards?
2. An electric 1974 Fender® Jazz Bass® is 46.25 inches long. What is the length in centimeters?
3. Americans consume an average of 199.3 pounds of meat per year. What is an American's average consumption in ounces per day?
4. A person walks at a speed of 4 miles per hour. What is the person's speed in feet per second?

Solution

1. Since there are 3 feet in 1 yard, we can multiply 10 feet by $\frac{1 \text{ yard}}{3 \text{ feet}} = 1$. By doing so, we can eliminate "feet," because $\frac{\text{feet}}{\text{feet}} = 1$:

$$\frac{10 \text{ feet}}{1} \cdot \frac{1 \text{ yard}}{3 \text{ feet}} = \frac{10}{3} \text{ yards}$$

The official height of the hoop is $3\frac{1}{3}$ yards.

2. Since there are 2.54 centimeters in 1 inch, we multiply 46.25 inches by $\dfrac{2.54 \text{ centimeters}}{1 \text{ inch}} = 1$ so that the inches are eliminated:

$$\frac{46.25 \text{ inches}}{1} \cdot \frac{2.54 \text{ centimeters}}{1 \text{ inch}} \approx 117.48 \text{ centimeters}$$

The bass is approximately 117.48 centimeters long.

3. There are 16 ounces in 1 pound and approximately 365 days in 1 year. To convert, we multiply by ratios equal to 1 or approximately equal to 1. We arrange the ratios so that the units we want to eliminate appear in one numerator and one denominator:

$$\frac{199.3 \text{ pounds}}{1 \text{ year}} \cdot \frac{16 \text{ ounces}}{1 \text{ pound}} \cdot \frac{1 \text{ year}}{365 \text{ days}} \approx 8.74 \frac{\text{ounces}}{\text{day}}$$

On average, Americans consume about 8.74 ounces of meat per day.

4. There are 5280 feet in 1 mile, 60 minutes in 1 hour, and 60 seconds in 1 minute. To convert, we multiply by 1's. Again, we arrange the ratios so that the units we want to eliminate appear in one numerator and one denominator:

$$\frac{4 \text{ miles}}{1 \text{ hour}} \cdot \frac{5280 \text{ feet}}{1 \text{ mile}} \cdot \frac{1 \text{ hour}}{60 \text{ minutes}} \cdot \frac{1 \text{ minute}}{60 \text{ seconds}} \approx 5.87 \frac{\text{feet}}{\text{second}}$$

So, the person is walking at a speed of about 5.87 feet per second. ∎

Converting Units

To convert the units of a quantity,

1. Write the quantity in the original units.

2. Multiply by fractions equal to 1 so that the units you want to eliminate appear in one numerator and one denominator.

---

group exploration

Looking ahead: Adding rational expressions

Perform each addition. Simplify the result.

1. $\dfrac{2}{7} + \dfrac{3}{7}$

2. $\dfrac{4}{x} + \dfrac{5}{x}$

3. $\dfrac{3}{x+2} + \dfrac{6}{x+2}$

4. $\dfrac{2x+5}{3x} + \dfrac{4x+1}{3x}$

5. $\dfrac{x^2}{x+3} + \dfrac{5x+6}{x+3}$ [**Hint:** Remember to simplify the result.]

6. $\dfrac{x^2+x}{x^2-4} + \dfrac{2x-10}{x^2-4}$

HOMEWORK 10.2 FOR EXTRA HELP ▸

Student Solutions Manual   PH Math/Tutor Center   Math XL — MathXL®   MyMathLab — MyMathLab

Perform the indicated operation. Simplify the result.

1. $\dfrac{3}{x} \cdot \dfrac{5}{x}$

2. $\dfrac{x}{2} \cdot \dfrac{x}{7}$

3. $\dfrac{x}{6} \div \dfrac{3}{2x}$

4. $\dfrac{8}{x} \div \dfrac{10x}{7}$

5. $\dfrac{6a^2}{7} \cdot \dfrac{21}{5a^8}$

6. $\dfrac{15}{3t^4} \cdot \dfrac{2t^9}{40}$

7. $\dfrac{2}{x-3} \cdot \dfrac{x-4}{x+5}$

8. $\dfrac{x-1}{x+7} \cdot \dfrac{x-6}{5}$

9. $\dfrac{k-2}{k-6} \div \dfrac{k+6}{k+4}$

10. $\dfrac{y-4}{y+8} \div \dfrac{y+3}{y+4}$

11. $\dfrac{6}{7x-14} \cdot \dfrac{5x-10}{9}$

12. $\dfrac{3x+15}{8} \cdot \dfrac{3}{4x+20}$

13. $\dfrac{3x+18}{x-6} \div \dfrac{x+6}{2x-12}$

14. $\dfrac{x+4}{7x-7} \div \dfrac{3x+12}{x-1}$

15. $\dfrac{4w^6}{w+3} \cdot \dfrac{w+5}{2w^2}$

16. $\dfrac{b-1}{6b^3} \cdot \dfrac{9b^7}{b+4}$

17. $\dfrac{(x-4)(x+1)}{(x-7)(x+2)} \cdot \dfrac{5(x+2)}{3(x-4)}$

18. $\dfrac{(x+3)(x+9)}{(x-6)(x-1)} \cdot \dfrac{x(x-6)}{6(x+3)}$

19. $\dfrac{4(x-4)^2}{(x+5)^2} \div \dfrac{14(x-4)}{15(x+5)}$

20. $\dfrac{9(x+2)^2}{8(x-1)^2} \div \dfrac{3(x+2)}{10(x-1)}$

21. $\dfrac{4t^7}{3t-9} \cdot \dfrac{5t-15}{8t^3}$

22. $\dfrac{10w+10}{18w^5} \cdot \dfrac{27w^3}{5w+5}$

23. $\dfrac{8x^2}{x^2-49} \div \dfrac{4x^5}{3x+21}$

24. $\dfrac{33x^6}{5x-30} \div \dfrac{11x^{10}}{x^2-36}$

25. $\dfrac{15a^4b}{8ab^5} \div \dfrac{25ab^3}{4a}$

26. $\dfrac{4xy^3}{9x^3y} \div \dfrac{10y}{3xy^5}$

27. $\dfrac{x^2-3x-28}{x^2+4x-45} \cdot \dfrac{x+9}{x-7}$

28. $\dfrac{x^2+x-42}{x^2-x-20} \cdot \dfrac{x-5}{x-6}$

29. $\dfrac{t^2-36}{t^2-81} \div \dfrac{t+6}{t-9}$

30. $\dfrac{r^2-1}{r^2-49} \div \dfrac{r-1}{r+7}$

31. $\dfrac{x^2+6x+8}{x^2-5x} \cdot \dfrac{4x-20}{3x+6}$

32. $\dfrac{6x-12}{5x+35} \cdot \dfrac{x^2+10x+21}{2x^2-4x}$

33. $\dfrac{x^2-4}{x^2-7x+12} \div \dfrac{x^2-8x+12}{x^2-2x-3}$

34. $\dfrac{x^2-11x+28}{x^2-x} \div \dfrac{x^2-5x-14}{x^2-1}$

35. $\dfrac{a^2+6a+9}{a^2+6a} \cdot \dfrac{a^2+11a+30}{4a^2-36}$

36. $\dfrac{c^2-10c+25}{3c^2-12c} \cdot \dfrac{c^2-6c+8}{2c^2-50}$

37. $\dfrac{x^2+x-6}{x^2+9x+8} \cdot \dfrac{x^2+5x-24}{x^2-7x+10}$

38. $\dfrac{x^2-x-30}{x^2-x-12} \cdot \dfrac{x^2-6x+8}{x^2+3x-10}$

39. $\dfrac{x^2+16}{x^2+8x+16} \div \dfrac{4x}{x^2-16}$

40. $\dfrac{3x}{x^2-6x+9} \div \dfrac{x^2+9}{x^2-9}$

41. $\dfrac{9x^2-25}{x^2-8x+16} \cdot \dfrac{x^2-4x}{3x^2-2x-5}$

42. $\dfrac{x^2+5x}{2x^2-x-21} \cdot \dfrac{4x^2-49}{x^2+10x+25}$

43. $\dfrac{3p^2+15p+12}{4p^2} \div \dfrac{9p+36}{6p^4}$

44. $\dfrac{15w^6}{2w^2+20w+48} \div \dfrac{5w^2}{8w+48}$

45. $\dfrac{x^2-25}{x^3-3x^2-4x} \div \dfrac{x^2-10x+25}{x^2-4x}$

46. $\dfrac{3x^3+18x^2+24x}{x^2-36} \div \dfrac{6x^2+24x}{5x-30}$

47. $\dfrac{-6x+12}{x^2+10x+21} \cdot \dfrac{x^2-9}{-3x+6}$

48. $\dfrac{x^2-2x+1}{-10x+40} \cdot \dfrac{-6x+24}{x^2-3x+2}$

49. $(b^2-25) \cdot \dfrac{2b}{b^2-6b+5}$

50. $(t^2-11t+24) \cdot \dfrac{t+1}{4t-32}$

51. $\dfrac{x^2+8x}{x-4} \div (x^2+16x+64)$

52. $\dfrac{4x+28}{x+1} \div (x^2-49)$

53. $\dfrac{a^2-b^2}{4a^2-9b^2} \cdot \dfrac{2a-3b}{a+b}$

54. $\dfrac{a^2b-ab^2}{a+3b} \cdot \dfrac{3a^2+9ab}{a^2-b^2}$

55. $\dfrac{x^2-y^2}{3x+6y} \div \dfrac{x^2+2xy+y^2}{x^2+2xy}$

56. $\dfrac{2x^2+3xy+y^2}{5x-10y} \div \dfrac{4x^2-y^2}{xy-2y^2}$

57. $\dfrac{x^3-2x^2+3x-6}{x^2-3x} \cdot \dfrac{3x-9}{x^2-4}$

58. $\dfrac{x^3+3x^2-2x-6}{x^2+2x} \cdot \dfrac{5x+10}{x^2-9}$

59. $\dfrac{x^3+27}{6x-3} \div \dfrac{2x^2-6x+18}{2x^2-x}$

60. $\dfrac{x^3-8}{x^2-9} \cdot \dfrac{5x+15}{3x-6}$

61. Perform the indicated operation. Simplify the result.

a. $\dfrac{x^2-1}{x^2-4} \cdot \dfrac{x^2+3x+2}{x^2-3x+2}$

b. $\dfrac{x^2-1}{x^2-4} \div \dfrac{x^2+3x+2}{x^2-3x+2}$

62. Perform the indicated operation. Simplify the result.

a. $\dfrac{x^2+x-6}{x^2-4x-5} \cdot \dfrac{x^2-x-2}{x^2-2x-15}$

b. $\dfrac{x^2+x-6}{x^2-4x-5} \div \dfrac{x^2-x-2}{x^2-2x-15}$

For Exercises 63–72, round approximate results to two decimal places. Refer to the list of equivalent units in the margin of p. 510 as needed.

63. On an official tennis court, the height of the net is 3 feet at the center. What is that height in inches?

64. An NBA basketball court is 94 feet long. What is its length in yards?

65. A person races in a 10-kilometer run. How long is the race in miles?

66. A person buys 13.9 gallons of gasoline. How many liters of gasoline is that?

67. In 2003, Americans consumed an average of 16.3 pounds of fish and shellfish per year (Source: *U.S. Department of Commerce*). What is this average in ounces per day?

68. A person drives at a speed of 68 miles per hour. What is the driving speed in feet per second?

69. A Porsche® 911 Turbo® Coupe has gas mileage equal to 6.29 kilometers per liter. What is the car's gas mileage in miles per gallon?

70. There are 2250 milligrams of salt in a 10-ounce bag of Snyder's® pretzels. How many grams of salt are there in a pound of Snyder's pretzels?

71. A 12-ounce can of Jolt® has 71.2 milligrams of caffeine (Source: *National Soft Drink Association*). How many grams of caffeine are in 1 gallon of Jolt?

72. A 1-cup serving of Lucky Charms® has 13 grams of sugar. How many milligrams of sugar are in 1 ounce of the cereal?

73. Find the product of $\dfrac{x^3}{12}$ and $\dfrac{3}{x}$.

74. Find the product of $\dfrac{15}{x^2}$ and $\dfrac{x^6}{20}$.

75. Find the quotient of $\dfrac{x-2}{x^2-3x-18}$ and $\dfrac{x+4}{x^2+4x+3}$.

76. Find the quotient of $\dfrac{x^2-4}{6x-42}$ and $\dfrac{5x-10}{3x-21}$.

77. A student tries to find the product $\dfrac{x+2}{x+4}\cdot\dfrac{x+4}{x+6}$:

$$\frac{x+2}{x+4}\cdot\frac{x+4}{x+6}=\frac{(x+2)(x+4)}{(x+4)(x+6)}=\frac{x^2+6x+8}{x^2+10x+24}$$

Describe any errors. Then find the product correctly.

78. A student tries to find the quotient $\dfrac{x+4}{x-5}\div\dfrac{x-5}{x+8}$:

$$\frac{x+4}{x-5}\div\frac{x-5}{x+8}=\frac{x+4}{x+8}$$

Describe any errors. Then find the quotient correctly.

79. A student tries to find the product $\dfrac{x}{3}\cdot\dfrac{x+4}{x-7}$:

$$\frac{x}{3}\cdot\frac{x+4}{x-7}=\frac{x^2+4}{3x-7}$$

Find a value to substitute for x to show that the student's work is incorrect. Then perform the multiplication correctly. Use a graphing calculator table to verify your result.

80. A student tries to find the product $\dfrac{x+3}{x+7}\cdot\dfrac{x+3}{x-7}$:

$$\frac{x+3}{x+7}\cdot\frac{x+3}{x-7}=\frac{x^2+9}{x^2-49}$$

Find a value to substitute for x to show that the student's work is incorrect. Then perform the multiplication correctly. Use a graphing calculator table to verify your result.

81. Give an example of two rational expressions whose product is 1.

82. Give an example of two rational expressions whose quotient is 1.

83. Describe how to multiply two rational expressions. Include an example.

84. Describe how to divide two rational expressions. Include an example.

Related Review

Perform the indicated operation. Simplify the result.

85. $(2x-6)\left(x^2-7x+12\right)$

86. $\left(2x^2+3x-2\right)(3x+6)$

87. $(2x-6)\div\left(x^2-7x+12\right)$

88. $\left(2x^2+3x-2\right)\div(3x+6)$

Expressions, Equations, and Graphs

Perform the indicated instruction. Then use words such as linear, quadratic, cubic, power, radical, rational, polynomial, degree, one variable, *and* two variables *to describe the expression, equation, or system.*

89. Simplify $-3(x-5)^2$.

90. Graph $y=3x^2$ by hand.

91. Solve $-3(x-5)^2=-24$.

92. Find the difference $\left(3x^2-2x\right)-\left(7x^2+4x\right)$.

93. Evaluate $-3(x-5)^2$ for $x=2$.

94. Solve $3x^2-2x=4$.

10.3 ADDING RATIONAL EXPRESSIONS

Objectives

▸ Add rational expressions with a common denominator.

▸ Add rational expressions with different denominators.

In Section 10.2, we multiplied and divided rational expressions. How do we add rational expressions?

Addition of Rational Expressions with a Common Denominator

Recall from Section 2.2 that we add fractions with a common denominator by using the property $\dfrac{a}{b}+\dfrac{c}{b}=\dfrac{a+c}{b}$, where b is nonzero. For example,

$$\frac{2}{7}+\frac{3}{7}=\frac{5}{7}$$

We add more complicated rational expressions with a common denominator in a similar way.

> ### Adding Rational Expressions with a Common Denominator
>
> If $\dfrac{A}{B}$ and $\dfrac{C}{B}$ are rational expressions, where B is nonzero, then
>
> $$\frac{A}{B} + \frac{C}{B} = \frac{A+C}{B}$$
>
> In words: To add two rational expressions with a common denominator, add the numerators and keep the common denominator.

Example 1 Adding Two Rational Expressions with a Common Denominator

Perform the addition $\dfrac{4x+3}{x} + \dfrac{2}{x}$.

Solution

$$\frac{4x+3}{x} + \frac{2}{x} = \frac{4x+3+2}{x}$$

Add numerators and keep common denominator: $\dfrac{A}{B} + \dfrac{C}{B} = \dfrac{A+C}{B}$

$$= \frac{4x+5}{x}$$

Combine like terms. ∎

After we add two rational expressions, we may be able to simplify the result.

Example 2 Adding Two Rational Expressions with a Common Denominator

Perform the addition $\dfrac{x^2 - 10x}{x - 5} + \dfrac{2x + 15}{x - 5}$.

Solution

$$\frac{x^2 - 10x}{x-5} + \frac{2x+15}{x-5} = \frac{x^2 - 10x + 2x + 15}{x-5}$$

Add numerators and keep common denominator: $\dfrac{A}{B} + \dfrac{C}{B} = \dfrac{A+C}{B}$

$$= \frac{x^2 - 8x + 15}{x - 5}$$

Combine like terms.

$$= \frac{(x-5)(x-3)}{x-5}$$

Factor numerator.

$$= x - 3$$

Simplify: $\dfrac{x-5}{x-5} = 1$ ∎

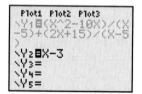

Figure 9 Verify the work

We use a graphing calculator table to verify our work (see Fig. 9).

Addition of Rational Expressions with Different Denominators

When adding two fractions with different denominators in Section 2.2, we listed multiples of the denominators to help us find the least common denominator (LCD) of the fractions. In this section, we discuss another way to find the LCD of two fractions.

Suppose that two brothers, John and Paul, own the following numbers and types of musical instruments:

John	Paul
3 guitars	1 guitar
1 bass guitar	2 bass guitars
1 sitar	1 sitar

The brothers will not give their instruments to each other, but each brother wants to own the same numbers and types of musical instruments that the other brother has.

Since Paul has 1 more bass than John, John wants 1 more bass guitar. Since John has 2 more guitars than Paul, Paul wants 2 more guitars. John and Paul have the same number of sitars, so neither brother wants another sitar.

Let's compare this situation with finding the LCD of the sum $\frac{7}{24} + \frac{5}{18}$. First, we find the prime factorization of each denominator:

$$24 = 2 \cdot 2 \cdot 2 \cdot 3$$
$$18 = 2 \cdot 3 \cdot 3$$

What factors does each of the denominators need so that the denominators can become the same? Since 18 has one more 3 factor than 24, the denominator 24 needs one 3 factor. Since 24 has two more 2 factors than 18, the denominator 18 needs two 2 factors.

We use the fact that $\frac{A}{A} = 1$ when $A \neq 0$ to introduce the one 3 factor for 24 and two 2 factors for 18:

$$\frac{7}{24} + \frac{5}{18} = \frac{7}{2 \cdot 2 \cdot 2 \cdot 3} + \frac{5}{2 \cdot 3 \cdot 3} \qquad \text{Find prime factorization of each denominator.}$$

$$= \frac{7}{2 \cdot 2 \cdot 2 \cdot 3} \cdot \frac{3}{3} + \frac{5}{2 \cdot 3 \cdot 3} \cdot \frac{2 \cdot 2}{2 \cdot 2} \qquad \text{Introduce missing factors.}$$

$$= \frac{7 \cdot 3}{2 \cdot 2 \cdot 2 \cdot 3 \cdot 3} + \frac{5 \cdot 2 \cdot 2}{2 \cdot 3 \cdot 3 \cdot 2 \cdot 2} \qquad \text{Find products.}$$

$$= \frac{21}{72} + \frac{20}{72} \qquad \text{Simplify.}$$

$$= \frac{41}{72} \qquad \text{Add numerators and keep common denominator: } \frac{A}{B} + \frac{C}{B} = \frac{A+C}{B}$$

Note that this method not only helps us find the LCD, but also suggests what forms of $\frac{A}{A}$ to use to introduce missing factors.

Example 3 Adding Two Rational Expressions with Different Denominators

Find the sum $\frac{1}{6} + \frac{9}{2w}$.

Solution

We begin by factoring the denominators:

$$6 = 2 \cdot 3$$
$$2w = 2 \cdot w$$

Since $2w$ has a w factor and 6 has no w factors, the denominator 6 needs a w factor. Since 6 has a 3 factor and $2w$ has no 3 factors, the denominator $2w$ needs a 3 factor.

We use the fact that $\frac{A}{A} = 1$, where A is nonzero, to introduce the missing factors:

$$\frac{1}{6} + \frac{9}{2w} = \frac{1}{2 \cdot 3} + \frac{9}{2 \cdot w} \qquad \text{Factor denominators.}$$

$$= \frac{1}{2 \cdot 3} \cdot \frac{w}{w} + \frac{9}{2 \cdot w} \cdot \frac{3}{3} \qquad \text{Introduce missing factors.}$$

$$= \frac{w}{6w} + \frac{27}{6w} \qquad \text{Find products.}$$

$$= \frac{w + 27}{6w} \qquad \text{Add numerators and keep common denominator: } \frac{A}{B} + \frac{C}{B} = \frac{A+C}{B}$$

Notice that the result is in lowest terms. ∎

Example 4   Adding Two Rational Expressions with Different Denominators

Find the sum $\dfrac{7}{12x} + \dfrac{5}{8x^3}$.

Solution

We begin by factoring the denominators:

$$12x = 2 \cdot 2 \cdot 3 \cdot x$$
$$8x^3 = 2 \cdot 2 \cdot 2 \cdot x \cdot x \cdot x$$

Since $8x^3$ has one more 2 factor and two more x factors than $12x$, the denominator $12x$ needs one more 2 factor and two more x factors. Since $12x$ has a 3 factor and $8x^3$ does not have any 3 factors, the denominator $8x^3$ needs a 3 factor.

We use the fact that $\dfrac{A}{A} = 1$, where A is nonzero, to introduce the missing factors:

$$\frac{7}{12x} + \frac{5}{8x^3} = \frac{7}{2 \cdot 2 \cdot 3 \cdot x} + \frac{5}{2 \cdot 2 \cdot 2 \cdot x \cdot x \cdot x} \qquad \text{Factor denominators.}$$

$$= \frac{7}{2 \cdot 2 \cdot 3 \cdot x} \cdot \frac{2 \cdot x \cdot x}{2 \cdot x \cdot x} + \frac{5}{2 \cdot 2 \cdot 2 \cdot x \cdot x \cdot x} \cdot \frac{3}{3} \qquad \text{Introduce missing factors.}$$

$$= \frac{14x^2}{24x^3} + \frac{15}{24x^3} \qquad \text{Find products.}$$

$$= \frac{14x^2 + 15}{24x^3} \qquad \begin{array}{l}\text{Add numerators and keep}\\\text{common denominator.}\end{array}$$

Notice that the result is in lowest terms.   ■

Example 5   Adding Two Rational Expressions with Different Denominators

Find the sum $\dfrac{2}{x-4} + \dfrac{6}{x+7}$.

Solution

The denominator $x - 4$ needs an $x + 7$ factor, and the denominator $x + 7$ needs an $x - 4$ factor. We use the fact that $\dfrac{A}{A} = 1$, where A is nonzero, to introduce the missing factors:

$$\frac{2}{x-4} + \frac{6}{x+7} = \frac{2}{x-4} \cdot \frac{x+7}{x+7} + \frac{6}{x+7} \cdot \frac{x-4}{x-4} \qquad \text{Introduce missing factors.}$$

$$= \frac{2(x+7)}{(x-4)(x+7)} + \frac{6(x-4)}{(x+7)(x-4)} \qquad \text{Find products.}$$

$$= \frac{2(x+7) + 6(x-4)}{(x-4)(x+7)} \qquad \begin{array}{l}\text{Add numerators and keep common}\\\text{denominator: } \dfrac{A}{B} + \dfrac{C}{B} = \dfrac{A+C}{B}\end{array}$$

$$= \frac{2x + 14 + 6x - 24}{(x-4)(x+7)} \qquad \text{Distributive law}$$

$$= \frac{8x - 10}{(x-4)(x+7)} \qquad \text{Combine like terms.}$$

Notice that the result is in lowest terms.   ■

For the sum of two expressions with different denominators, we first factor the denominators, if possible, to help us find the LCD.

Example 6 Adding Two Rational Expressions with Different Denominators

Find the sum $\dfrac{5}{6} + \dfrac{7}{2x - 8}$.

Solution

First, we factor the denominators:

$$6 = 2 \cdot 3$$
$$2x - 8 = 2(x - 4)$$

Since $2x - 8$ has an $x - 4$ factor and 6 does not, the denominator 6 needs an $x - 4$ factor. Since 6 has a 3 factor and $2x - 8$ does not, the denominator $2x - 8$ needs a 3 factor:

$$\dfrac{5}{6} + \dfrac{7}{2x - 8} = \dfrac{5}{2 \cdot 3} + \dfrac{7}{2(x - 4)} \qquad \text{Factor denominators.}$$

$$= \dfrac{5}{2 \cdot 3} \cdot \dfrac{x - 4}{x - 4} + \dfrac{7}{2(x - 4)} \cdot \dfrac{3}{3} \qquad \text{Introduce missing factors.}$$

$$= \dfrac{5(x - 4)}{6(x - 4)} + \dfrac{21}{6(x - 4)} \qquad \text{Find products.}$$

$$= \dfrac{5(x - 4) + 21}{6(x - 4)} \qquad \text{Add numerators and keep common denominator: } \dfrac{A}{B} + \dfrac{C}{B} = \dfrac{A + C}{B}$$

$$= \dfrac{5x - 20 + 21}{6(x - 4)} \qquad \text{Distributive law}$$

$$= \dfrac{5x + 1}{6(x - 4)} \qquad \text{Combine like terms.} \qquad \blacksquare$$

Example 7 Adding Two Rational Expressions with Different Denominators

Find the sum $\dfrac{x}{2x - 4} + \dfrac{-4}{x^2 - 4}$.

Solution

First, we factor the denominators:

$$2x - 4 = 2(x - 2)$$
$$x^2 - 4 = (x - 2)(x + 2)$$

Since $x^2 - 4$ has an $x + 2$ factor and $2x - 4$ does not, the denominator $2x - 4$ needs an $x + 2$ factor. Since $2x - 4$ has a 2 factor and $x^2 - 4$ does not, the denominator $x^2 - 4$ needs a 2 factor:

$$\dfrac{x}{2x - 4} + \dfrac{-4}{x^2 - 4} = \dfrac{x}{2(x - 2)} + \dfrac{-4}{(x - 2)(x + 2)} \qquad \text{Factor denominators.}$$

$$= \dfrac{x}{2(x - 2)} \cdot \dfrac{x + 2}{x + 2} + \dfrac{-4}{(x - 2)(x + 2)} \cdot \dfrac{2}{2} \qquad \text{Introduce missing factors.}$$

$$= \dfrac{x(x + 2)}{2(x - 2)(x + 2)} + \dfrac{-8}{2(x - 2)(x + 2)} \qquad \text{Find products.}$$

$$= \dfrac{x(x + 2) - 8}{2(x - 2)(x + 2)} \qquad \text{Add numerators and keep common denominator.}$$

$$= \dfrac{x^2 + 2x - 8}{2(x - 2)(x + 2)} \qquad \text{Distributive law}$$

$$= \dfrac{(x - 2)(x + 4)}{2(x - 2)(x + 2)} \qquad \text{Factor numerator.}$$

$$= \dfrac{x + 4}{2(x + 2)} \qquad \text{Simplify: } \dfrac{x - 2}{x - 2} = 1$$

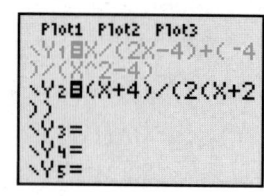

Figure 10 Verify the work

We use a graphing calculator table to verify our work (see Fig. 10). There are "ERROR" messages across from $x = -2$ and $x = 2$ in the table, because -2 and 2 are excluded values of either the original expression or our result (why?). ∎

How to Add Two Rational Expressions with Different Denominators

To add two rational expressions with different denominators,

1. Factor the denominators of the expressions if possible. Determine which factors are missing.

2. Use the property $\dfrac{A}{A} = 1$, where A is nonzero, to introduce missing factors.

3. Add the expressions by using the property $\dfrac{A}{B} + \dfrac{C}{B} = \dfrac{A + C}{B}$, where B is nonzero.

4. Simplify.

Example 8   Adding Two Rational Expressions with Different Denominators

Find the sum $\dfrac{2x}{x^2 - x - 2} + \dfrac{x + 4}{x + 1}$.

Solution

First, we factor the denominators:

$$x^2 - x - 2 = (x - 2)(x + 1)$$
$$x + 1 = x + 1$$

The denominator $x^2 - x - 2$ does not need any factors. Since $x^2 - x - 2$ has an $x - 2$ factor and $x + 1$ does not, the denominator $x + 1$ needs an $x - 2$ factor:

$$\frac{2x}{x^2 - x - 2} + \frac{x + 4}{x + 1} = \frac{2x}{(x - 2)(x + 1)} + \frac{x + 4}{x + 1} \qquad \text{Factor denominator.}$$

$$= \frac{2x}{(x - 2)(x + 1)} + \frac{x + 4}{x + 1} \cdot \frac{x - 2}{x - 2} \qquad \text{Introduce missing factor.}$$

$$= \frac{2x}{(x - 2)(x + 1)} + \frac{(x + 4)(x - 2)}{(x + 1)(x - 2)} \qquad \text{Find product.}$$

$$= \frac{2x + (x + 4)(x - 2)}{(x - 2)(x + 1)} \qquad \begin{array}{l}\text{Add numerators and keep} \\ \text{common denominator.}\end{array}$$

$$= \frac{2x + x^2 + 2x - 8}{(x - 2)(x + 1)} \qquad \text{Find product.}$$

$$= \frac{x^2 + 4x - 8}{(x - 2)(x + 1)} \qquad \text{Combine like terms.}$$

Notice that the result is in lowest terms. ∎

group exploration

Performing operations with rational expressions

For Problems 1–3, a student tries to perform an operation and then simplify the result. If the result is correct, decide whether there is a more efficient way to do the problem. If the result is incorrect, describe any errors and do the problem correctly.

1. $\dfrac{x + 2}{x + 5} + \dfrac{6}{(x + 2)(x + 5)} = \dfrac{1}{x + 5} + \dfrac{6}{x + 5}$

$$= \dfrac{7}{x + 5}$$

2. $\dfrac{1}{x^2+6x+8}+\dfrac{1}{x^2+5x+6}$

$=\dfrac{1}{x^2+6x+8}\cdot\dfrac{x^2+5x+6}{x^2+5x+6}+\dfrac{1}{x^2+5x+6}\cdot\dfrac{x^2+6x+8}{x^2+6x+8}$

$=\dfrac{(x^2+5x+6)+(x^2+6x+8)}{(x^2+5x+6)(x^2+6x+8)}$

$=\dfrac{2x^2+11x+14}{(x^2+5x+6)(x^2+6x+8)}$

3. $\dfrac{4}{x+2}\cdot\dfrac{3}{x+1}=\left(\dfrac{4}{x+2}\cdot\dfrac{x+1}{x+1}\right)\cdot\left(\dfrac{3}{x+1}\cdot\dfrac{x+2}{x+2}\right)$

$=\dfrac{4(x+1)}{(x+2)(x+1)}\cdot\dfrac{3(x+2)}{(x+1)(x+2)}$

$=\dfrac{12(x+1)(x+2)}{(x+1)^2(x+2)^2}$

$=\dfrac{12}{(x+1)(x+2)}$

group exploration

Looking ahead: Subtracting rational expressions

1. Find the difference. Simplify the result.

  a. $\dfrac{7}{5}-\dfrac{4}{5}$

  b. $\dfrac{x^2-2x}{x-5}-\dfrac{15}{x-5}$

  c. $\dfrac{4}{2x}-\dfrac{7}{x^2}$

  d. $\dfrac{5}{x+3}-\dfrac{2}{x^2-9}$

2. For finding the difference $\dfrac{x^2}{x+1}-\dfrac{3x+7}{x+1}$, which of the methods that follow yields the correct answer? Use graphing calculator tables to help you decide.

Method A

$\dfrac{x^2}{x+1}-\dfrac{3x+7}{x+1}=\dfrac{x^2-(3x+7)}{x+1}$

$=\dfrac{x^2-3x-7}{x+1}$

Method B

$\dfrac{x^2}{x+1}-\dfrac{3x+7}{x+1}=\dfrac{x^2-3x+7}{x+1}$

3. Find the difference. Simplify the result.

  a. $\dfrac{x^2}{x-5}-\dfrac{4x-3}{x-5}$

  b. $\dfrac{3x}{x^2+6x+8}-\dfrac{5}{x^2+7x+12}$

HOMEWORK 10.3 FOR EXTRA HELP ▶

Student Solutions Manual

PH Math/Tutor Center

Math XL MathXL®

MyMathLab MyMathLab

Find the sum. Simplify the result.

1. $\dfrac{7}{x} + \dfrac{2}{x}$

2. $\dfrac{3}{x^2} + \dfrac{5}{x^2}$

3. $\dfrac{2x}{x-1} + \dfrac{6x}{x-1}$

4. $\dfrac{-4x}{x+5} + \dfrac{x}{x+5}$

5. $\dfrac{3x-2}{x+3} + \dfrac{5x+4}{x+3}$

6. $\dfrac{6x+3}{x-4} + \dfrac{9x-8}{x-4}$

7. $\dfrac{t^2}{t+5} + \dfrac{7t+10}{t+5}$

8. $\dfrac{a^2+9a}{a+3} + \dfrac{18}{a+3}$

9. $\dfrac{x}{x^2-4} + \dfrac{2}{x^2-4}$

10. $\dfrac{x}{x^2-16} + \dfrac{-4}{x^2-16}$

11. $\dfrac{x^2-5x}{x^2+5x+6} + \dfrac{4x-12}{x^2+5x+6}$

12. $\dfrac{x^2-4x}{x^2+3x-4} + \dfrac{6x-8}{x^2+3x-4}$

13. $\dfrac{3r^2-5r}{r^2+6r+9} + \dfrac{-2r^2+r-21}{r^2+6r+9}$

14. $\dfrac{5w^2-2w+12}{w^2-12w+36} + \dfrac{-4w^2-6w}{w^2-12w+36}$

Find the sum. Simplify the result.

15. $\dfrac{3}{x} + \dfrac{5}{2x}$

16. $\dfrac{8}{3x} + \dfrac{2}{x}$

17. $\dfrac{3}{2w} + \dfrac{5}{6}$

18. $\dfrac{3}{8} + \dfrac{7}{4a}$

19. $\dfrac{5x}{6} + \dfrac{3}{4x}$

20. $\dfrac{4}{9x} + \dfrac{7x}{6}$

21. $\dfrac{5}{8x^3} + \dfrac{3}{10x}$

22. $\dfrac{7}{6x} + \dfrac{1}{3x^3}$

23. $\dfrac{a}{b} + \dfrac{b}{a}$

24. $\dfrac{a}{3b} + \dfrac{b}{2a}$

25. $\dfrac{5}{4x} + \dfrac{2}{x+3}$

26. $\dfrac{7}{x-5} + \dfrac{4}{3x}$

27. $\dfrac{3}{r+2} + \dfrac{4}{r-5}$

28. $\dfrac{2}{b-1} + \dfrac{6}{b+7}$

29. $\dfrac{6}{x-1} + \dfrac{3}{5x}$

30. $\dfrac{7}{2x} + \dfrac{4}{x+6}$

31. $\dfrac{5x}{x-4} + \dfrac{2}{x+2}$

32. $\dfrac{3}{x+5} + \dfrac{4x}{x-3}$

33. $\dfrac{1}{a+b} + \dfrac{1}{a-b}$

34. $\dfrac{1}{x+y} + \dfrac{1}{x+2y}$

35. $\dfrac{x}{x+4} + \dfrac{2}{5x+20}$

36. $\dfrac{3}{x-6} + \dfrac{7x}{4x-24}$

37. $\dfrac{p}{3p-9} + \dfrac{-1}{p-3}$

38. $\dfrac{c}{2c-4} + \dfrac{-1}{c-2}$

39. $\dfrac{2x}{5x-25} + \dfrac{4}{3x-15}$

40. $\dfrac{3x}{4x+12} + \dfrac{2}{6x+18}$

41. $\dfrac{6}{x^2-1} + \dfrac{3}{x+1}$

42. $\dfrac{1}{x+2} + \dfrac{4}{x^2-4}$

43. $\dfrac{t^2+2t}{t^2+11t+18} + \dfrac{4}{t+9}$

44. $\dfrac{r^2+3r}{r^2+7r+12} + \dfrac{2}{r+4}$

45. $\dfrac{x}{2x-6} + \dfrac{3}{x^2-9}$

46. $\dfrac{2x}{x^2-4} + \dfrac{5}{3x+6}$

47. $\dfrac{x}{x^2-4} + \dfrac{1}{x^2+2x}$

48. $\dfrac{1}{x^2+x} + \dfrac{2x}{x^2-1}$

49. $\dfrac{4}{(a-3)(a+1)} + \dfrac{2}{(a+1)(a+4)}$

50. $\dfrac{7}{(w-5)(w-2)} + \dfrac{4}{(w-3)(w-2)}$

51. $\dfrac{3}{x^2-16} + \dfrac{5}{x^2+5x+4}$

52. $\dfrac{2}{x^2+5x-14} + \dfrac{6}{x^2-49}$

53. $\dfrac{4x}{x^2+3x-18} + \dfrac{2}{x^2+10x+24}$

54. $\dfrac{5}{x^2+3x+2} + \dfrac{7x}{x^2-x-2}$

55. $\dfrac{x}{x^2+9x+20} + \dfrac{-4}{x^2+8x+15}$

56. $\dfrac{6}{x^2-9} + \dfrac{-5}{x^2-x-6}$

57. $3 + \dfrac{w-2}{w+5}$

58. $2 + \dfrac{a+3}{a-4}$

Find the sum. Simplify the result.

59. $\dfrac{x-3}{x+5} + \dfrac{x+2}{x-4}$

60. $\dfrac{x+7}{x-2} + \dfrac{x-5}{x+1}$

61. $\dfrac{y-2}{y-3} + \dfrac{y+3}{y+2}$

62. $\dfrac{a+5}{a+1} + \dfrac{a-1}{a-5}$

63. $\dfrac{x-6}{x-5} + \dfrac{2x}{x^2-2x-15}$

64. $\dfrac{x+4}{x+3} + \dfrac{7}{x^2+5x+6}$

65. $\dfrac{5}{b^2-7b+12} + \dfrac{b+2}{b-4}$

66. $\dfrac{3t}{t^2-4t-5} + \dfrac{t-3}{t+1}$

67. $\dfrac{x-2}{4x+12} + \dfrac{5}{x^2-9}$

68. $\dfrac{3}{x^2-4} + \dfrac{x-3}{5x-10}$

69. $\dfrac{2x}{x-y} + \dfrac{2xy}{x^2-2xy+y^2}$

70. $\dfrac{3y}{x+y} + \dfrac{-9xy}{3x^2+4xy+y^2}$

71. $\dfrac{2x}{x^3 - 4x^2 + 2x - 8} + \dfrac{5}{3x^2 + 6}$

72. $\dfrac{2}{x^3 + 2x^2 - 5x - 10} + \dfrac{3x}{2x^2 - 10}$

73. $\dfrac{5x}{x^3 - 8} + \dfrac{4}{x^2 - 4}$

74. $\dfrac{3x}{x^3 + 27} + \dfrac{6}{x^2 - 9}$

75. There are two 100-watt lights in a room. A person is twice as far from one light as from the other. The illumination (the brightness of the light) can be measured in watts per square meter (W/m²). The illumination (in W/m²) from the closer light source is

$$\dfrac{18}{d^2}$$

where d is the distance (in meters) to that light. The illumination (in W/m²) from the other light source is

$$\dfrac{18}{(2d)^2}$$

where $2d$ is the distance (in meters) to that light.
 a. Find an expression for the total illumination (in W/m²) from the two light sources.
 b. Write your result from part (a) as a single fraction.
 c. Evaluate the expression you found in part (b) for $d = 1.2$. What does your result mean in this situation?

76. A student plans to drive from El Camino College in Torrance, California, to Chandler-Gilbert Community College in Sun Lakes, Arizona. The trip involves 231 miles in California, followed by 174 miles in Arizona. The speed limit is 70 mph in California and 75 mph in Arizona. The student plans to drive at a mph above the speed limits. The driving time (in hours) in California is

$$\dfrac{231}{a + 70}$$

The driving time (in hours) in Arizona is

$$\dfrac{174}{a + 75}$$

 a. Find an expression for the total driving time (in hours).
 b. Write your result from part (a) as a single fraction.
 c. Evaluate the expression you found in part (b) for $a = 5$. What does your result mean in this situation?

77. A student tries to find the sum $\dfrac{3}{x + 2} + \dfrac{5}{x + 3}$:

$$\dfrac{3}{x+2} + \dfrac{5}{x+3} = \dfrac{3}{x+2} \cdot \dfrac{1}{x+3} + \dfrac{5}{x+3} \cdot \dfrac{1}{x+2}$$

$$= \dfrac{3}{(x+2)(x+3)} + \dfrac{5}{(x+3)(x+2)}$$

$$= \dfrac{8}{(x+2)(x+3)}$$

Describe any errors. Then find the sum correctly.

78. A student tries to find the sum $\dfrac{x}{3} + \dfrac{2}{x}$:

$$\dfrac{x}{3} + \dfrac{2}{x} = \dfrac{x+2}{3+x} = \dfrac{x+2}{x+3}$$

Describe any errors. Then find the sum correctly.

79. A student tries to find the sum $\dfrac{2}{x^2 + 2x} + \dfrac{3}{x + 2}$:

$$\dfrac{2}{x^2+2x} + \dfrac{3}{x+2} = \dfrac{2}{x^2+2x} \cdot \dfrac{x+2}{x+2} + \dfrac{3}{x+2} \cdot \dfrac{x^2+2x}{x^2+2x}$$

$$= \dfrac{2x+4}{(x^2+2x)(x+2)} + \dfrac{3x^2+6x}{(x+2)(x^2+2x)}$$

$$= \dfrac{3x^2+8x+4}{(x^2+2x)(x+2)}$$

$$= \dfrac{(3x+2)(x+2)}{x(x+2)(x+2)}$$

$$= \dfrac{3x+2}{x(x+2)}$$

Is the work correct? Is there a better way to find the sum? Explain.

80. a. Add $\dfrac{2}{5} + \dfrac{4}{7}$.
 b. Add $\dfrac{7}{2} + \dfrac{5}{3}$.
 c. Add $\dfrac{a}{b} + \dfrac{c}{d}$, where b and d are nonzero.
 d. Use your result from part (c) to help you find the sum $\dfrac{2}{5} + \dfrac{4}{7}$. Compare your result with your result from part (a).

81. Describe how to add two rational expressions that have a common denominator. Include an example.

82. Describe how to add two rational expressions that have different denominators. Include an example.

Related Review

Find the sum. Simplify the result.

83. $(4x^2 - 7x + 2) + (-3x^2 + 2x + 4)$

84. $(-5x^2 + 8x - 17) + (6x^2 - 5x - 23)$

85. $\dfrac{4x^2 - 7x + 2}{x - 3} + \dfrac{-3x^2 + 2x + 4}{x - 3}$

86. $\dfrac{-5x^2 + 8x - 17}{x - 5} + \dfrac{6x^2 - 5x - 23}{x - 5}$

Expressions, Equations, and Graphs

Perform the indicated instruction. Then use words such as linear, quadratic, cubic, power, radical, rational, polynomial, degree, one variable, *and* two variables *to describe the expression, equation, or system.*

87. Solve the following:

$$y = 3x - 2$$
$$y = 5x + 4$$

88. Factor $t^2 - t - 20$.

89. Graph $y = 3x - 2$ by hand.

90. Solve $t^2 - t - 20 = 0$.

91. Solve $3x - 2 = 5x + 4$.

92. Find the sum $\dfrac{t}{t^2 - t - 20} + \dfrac{3}{t^2 + 7t + 12}$.

10.4 SUBTRACTING RATIONAL EXPRESSIONS

Being defeated is often a temporary condition. Giving up is what makes it permanent.
—*Marilyn vos Savant*

Objectives

▹ Subtract rational expressions with a common denominator.

▹ Subtract rational expressions with different denominators.

In Sections 10.2 and 10.3, we multiplied, divided, and added rational expressions. In this section, we subtract rational expressions.

Subtraction of Rational Expressions with a Common Denominator

Recall from Section 2.2 that we subtract fractions with a common denominator by using the property $\frac{a}{b} - \frac{c}{b} = \frac{a-c}{b}$, where b is nonzero. For example,

$$\frac{5}{7} - \frac{2}{7} = \frac{3}{7}$$

We subtract more complicated rational expressions with a common denominator in a similar way.

Subtracting Rational Expressions with a Common Denominator

If $\frac{A}{B}$ and $\frac{C}{B}$ are rational expressions, where B is nonzero, then

$$\frac{A}{B} - \frac{C}{B} = \frac{A-C}{B}$$

In words: To subtract two rational expressions with a common denominator, subtract the numerators and keep the common denominator.

After subtracting two rational expressions, it may be possible to simplify the result.

Example 1 Subtracting Two Rational Expressions with a Common Denominator

Find the difference $\dfrac{x}{x^2-4} - \dfrac{2}{x^2-4}$.

Solution

$$\frac{x}{x^2-4} - \frac{2}{x^2-4} = \frac{x-2}{x^2-4} \qquad \text{Subtract numerators and keep common denominator: } \frac{A}{B} - \frac{C}{B} = \frac{A-C}{B}$$

$$= \frac{x-2}{(x-2)(x+2)} \qquad \text{Factor denominator.}$$

$$= \frac{1}{x+2} \qquad \text{Simplify: } \frac{x-2}{x-2} = 1$$

■

Example 2 Subtracting Two Rational Expressions with a Common Denominator

Find the difference $\dfrac{x^2}{x+4} - \dfrac{3x+28}{x+4}$.

Solution

$$\frac{x^2}{x+4} - \frac{3x+28}{x+4} = \frac{x^2 - (3x+28)}{x+4}$$ Subtract numerators and keep common denominator: $\frac{A}{B} - \frac{C}{B} = \frac{A-C}{B}$

$$= \frac{x^2 - 3x - 28}{x+4}$$ Simplify.

$$= \frac{(x-7)(x+4)}{x+4}$$ Factor numerator.

$$= x - 7$$ Simplify: $\frac{x+4}{x+4} = 1$ ∎

WARNING   It is a common error to write

$$\frac{x^2}{x+4} - \frac{3x+28}{x+4} = \frac{x^2 - 3x + 28}{x+4}$$ Incorrect

This work is incorrect. **When subtracting rational expressions, be sure to subtract the *entire* numerator:**

$$\frac{x^2}{x+4} - \frac{3x+28}{x+4} = \frac{x^2 - (3x+28)}{x+4} = \frac{x^2 - 3x - 28}{x+4}$$

See Example 2 for the rest of the work.

Subtraction of Rational Expressions with Different Denominators

When subtracting rational expressions with different denominators, we use the method discussed in Section 10.3 to find the LCD.

Example 3   Subtracting Two Rational Expressions with Different Denominators

Find the difference $\dfrac{2}{9a} - \dfrac{5}{6a^2}$.

Solution

$$\frac{2}{9a} - \frac{5}{6a^2} = \frac{2}{3 \cdot 3 \cdot a} - \frac{5}{2 \cdot 3 \cdot a \cdot a}$$ Factor denominators.

$$= \frac{2}{3 \cdot 3 \cdot a} \cdot \frac{2 \cdot a}{2 \cdot a} - \frac{5}{2 \cdot 3 \cdot a \cdot a} \cdot \frac{3}{3}$$ Introduce missing factors.

$$= \frac{4a}{18a^2} - \frac{15}{18a^2}$$ Find products.

$$= \frac{4a - 15}{18a^2}$$ Subtract numerators and keep common denominator: $\frac{A}{B} - \frac{C}{B} = \frac{A-C}{B}$ ∎

Example 4   Subtracting Two Rational Expressions with Different Denominators

Find the difference $\dfrac{x}{x^2 - 9} - \dfrac{2}{5x + 15}$.

Solution

$$\frac{x}{x^2-9} - \frac{2}{5x+15} = \frac{x}{(x-3)(x+3)} - \frac{2}{5(x+3)}$$ Factor denominators.

$$= \frac{x}{(x-3)(x+3)} \cdot \frac{5}{5} - \frac{2}{5(x+3)} \cdot \frac{x-3}{x-3}$$ Introduce missing factors.

$$= \frac{5x}{5(x-3)(x+3)} - \frac{2(x-3)}{5(x+3)(x-3)}$$ Find products.

$$= \frac{5x - 2(x-3)}{5(x-3)(x+3)}$$ $\dfrac{A}{B} - \dfrac{C}{B} = \dfrac{A-C}{B}$

$$= \frac{5x - 2x + 6}{5(x-3)(x+3)}$$ Distributive law

$$= \frac{3x + 6}{5(x-3)(x+3)}$$ Combine like terms.

$$= \frac{3(x+2)}{5(x-3)(x+3)}$$ Factor numerator.

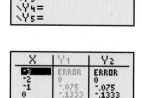

Figure 11 Verify the work

Note that $\dfrac{3(x+2)}{5(x-3)(x+3)}$ is in lowest terms. We use a graphing calculator table to verify our work (see Fig. 11). There are "ERROR" messages across from $x = -3$ and $x = 3$ in the table, because -3 and 3 are excluded values of either the original expression or our result (why?). ∎

How to Subtract Rational Expressions with Different Denominators

To subtract two rational expressions with different denominators,

1. Factor the denominators of the expressions if possible. Determine which factors are missing.

2. Use the property $\dfrac{A}{A} = 1$, where A is nonzero, to introduce missing factors.

3. Subtract the expressions by using the property $\dfrac{A}{B} - \dfrac{C}{B} = \dfrac{A-C}{B}$, where B is nonzero.

4. Simplify.

Example 5   Subtracting Two Rational Expressions with Different Denominators

Find the difference $\dfrac{x+2}{x-3} - \dfrac{x-4}{x+1}$.

Solution

$$\frac{x+2}{x-3} - \frac{x-4}{x+1} = \frac{x+2}{x-3} \cdot \frac{x+1}{x+1} - \frac{x-4}{x+1} \cdot \frac{x-3}{x-3}$$ Introduce missing factors.

$$= \frac{(x+2)(x+1)}{(x-3)(x+1)} - \frac{(x-4)(x-3)}{(x+1)(x-3)}$$ Find products.

$$= \frac{(x+2)(x+1) - (x-4)(x-3)}{(x-3)(x+1)}$$ Subtract numerators and keep common denominator: $\dfrac{A}{B} - \dfrac{C}{B} = \dfrac{A-C}{B}$

$$= \frac{x^2 + 3x + 2 - (x^2 - 7x + 12)}{(x-3)(x+1)}$$ Find products.

$$= \frac{x^2 + 3x + 2 - x^2 + 7x - 12}{(x-3)(x+1)}$$ Subtract.

$$= \frac{10x - 10}{(x-3)(x+1)}$$ Combine like terms.

$$= \frac{10(x-1)}{(x-3)(x+1)}$$ Factor numerator. ∎

group exploration

Performing operations with rational expressions

For Problems 1–3, a student tries to perform an operation and then simplify the result. If the result is correct, decide whether there is a more efficient way to do the problem. If the result is incorrect, describe any errors and do the problem correctly.

1. $\dfrac{7x}{x-3} - \dfrac{5x+9}{x-3} = \dfrac{7x - 5x + 9}{x - 3}$

$$= \dfrac{2x + 9}{x - 3}$$

2. $\dfrac{5}{4x} - \dfrac{3}{2x^2} = \dfrac{5}{4x} \cdot \dfrac{2x^2}{2x^2} - \dfrac{3}{2x^2} \cdot \dfrac{4x}{4x}$

$$= \dfrac{10x^2 - 12x}{8x^3}$$

$$= \dfrac{2x(5x - 6)}{8x^3}$$

$$= \dfrac{5x - 6}{4x^2}$$

3. $\dfrac{2}{x+3} \cdot \dfrac{5}{x} = \left(\dfrac{2}{x+3} \cdot \dfrac{x}{x} \right) \cdot \left(\dfrac{5}{x} \cdot \dfrac{x+3}{x+3} \right)$

$$= \dfrac{2x}{x(x+3)} \cdot \dfrac{5(x+3)}{x(x+3)}$$

$$= \dfrac{10x(x+3)}{x^2(x+3)^2}$$

$$= \dfrac{10}{x(x+3)}$$

TIPS FOR SUCCESS: Consider Various Uses of a Technique

At several points during the term, think about how a technique can be used in various contexts. For example, consider how we have used factoring in this course. In Sections 8.1–8.4, we learned how to factor polynomials. In Section 8.5, we used factoring to help us solve polynomial equations. In Section 8.6, we used factoring to help us solve quadratic equations that lead to estimates and predictions about authentic situations. In Section 9.2, we used factoring to help us simplify radical expressions, which also helped us simplify solutions of quadratic equations in Section 9.5. In Section 9.4, we used factoring to help us complete the square. In Sections 10.1–10.4, we used factoring to simplify rational expressions and to perform operations with rational expressions.

Considering the various ways we have used a technique will help you see the big picture. If you do that for all of the techniques which are used frequently in this course, you will be more likely to recognize what technique to use to solve a particular exercise and you will increase your understanding of how the topics in this course are connected. This will improve your performance on your upcoming final exam, as well as in the next mathematics course you may take.

HOMEWORK 10.4 FOR EXTRA HELP ▶

Student Solutions Manual · PH Math/Tutor Center · Math XL · MathXL® · MyMathLab · MyMathLab

Find the difference. Simplify the result.

1. $\dfrac{6}{x} - \dfrac{4}{x}$

2. $\dfrac{8}{x^3} - \dfrac{3}{x^3}$

3. $\dfrac{9x}{x-2} - \dfrac{2x}{x-2}$

4. $\dfrac{4x}{x+7} - \dfrac{6x}{x+7}$

5. $\dfrac{x}{x^2-9} - \dfrac{3}{x^2-9}$

6. $\dfrac{x}{x^2-25} - \dfrac{5}{x^2-25}$

7. $\dfrac{3r}{r+6} - \dfrac{7r-4}{r+6}$

8. $\dfrac{2t}{t-4} - \dfrac{3t-5}{t-4}$

9. $\dfrac{x^2}{x+1} - \dfrac{2x+3}{x+1}$

10. $\dfrac{x^2}{x-7} - \dfrac{3x+28}{x-7}$

11. $\dfrac{x^2+7x}{x^2-2x-8} - \dfrac{3x+32}{x^2-2x-8}$

12. $\dfrac{x^2-2x}{x^2+4x-5} - \dfrac{5x-6}{x^2+4x-5}$

13. $\dfrac{4a^2-5a-12}{a^2+8a+16} - \dfrac{3a^2-6a}{a^2+8a+16}$

14. $\dfrac{2k^2-4k}{k^2-10k+25} - \dfrac{k^2-2k+15}{k^2-10k+25}$

Find the difference. Simplify the result.

15. $\dfrac{3}{4x} - \dfrac{2}{x}$

16. $\dfrac{7}{x} - \dfrac{3}{5x}$

17. $\dfrac{5}{2b} - \dfrac{3}{8}$

18. $\dfrac{2}{3t} - \dfrac{5}{6}$

19. $\dfrac{5x}{8} - \dfrac{1}{6x}$

20. $\dfrac{5x}{9} - \dfrac{7}{12x}$

21. $\dfrac{5}{10x^4} - \dfrac{3}{15x^2}$

22. $\dfrac{1}{8x} - \dfrac{5}{4x^3}$

23. $\dfrac{a}{2b} - \dfrac{b}{3a}$

24. $\dfrac{a}{b} - \dfrac{b}{a}$

25. $\dfrac{3}{p-2} - \dfrac{5}{4p}$

26. $\dfrac{4}{w+6} - \dfrac{2}{3w}$

27. $\dfrac{7}{x-1} - \dfrac{3}{x+4}$

28. $\dfrac{3}{x+6} - \dfrac{4}{x-5}$

29. $\dfrac{3x}{x-2} - \dfrac{5}{x+3}$

30. $\dfrac{4}{x-4} - \dfrac{3x}{x+2}$

31. $\dfrac{1}{a+b} - \dfrac{1}{a-b}$

32. $\dfrac{2}{t+2r} - \dfrac{3}{t+3r}$

33. $\dfrac{c}{2c-8} - \dfrac{3}{c-4}$

34. $\dfrac{y}{4y-20} - \dfrac{6}{y-5}$

35. $\dfrac{4x}{6x-24} - \dfrac{7}{4x-16}$

36. $\dfrac{5x}{6x+42} - \dfrac{3}{8x+56}$

37. $\dfrac{x}{x-1} - \dfrac{2}{x^2-1}$

38. $\dfrac{3}{x-5} - \dfrac{5x-16}{x^2-25}$

39. $\dfrac{3x-1}{x^2+2x-15} - \dfrac{2}{x+5}$

40. $\dfrac{6x}{x^2-8x+15} - \dfrac{9}{x-3}$

41. $\dfrac{t}{3t-21} - \dfrac{4}{t^2-49}$

42. $\dfrac{3}{4a+4} - \dfrac{a}{a^2-1}$

43. $\dfrac{4x}{x^2-25} - \dfrac{2}{x^2+5x}$

44. $\dfrac{2x}{x^2-16} - \dfrac{3}{x^2-4x}$

45. $\dfrac{3}{(x-5)(x+2)} - \dfrac{4}{(x+2)(x+4)}$

46. $\dfrac{5}{(x-2)(x+3)} - \dfrac{2}{(x-6)(x-2)}$

47. $\dfrac{7}{x^2-5x+6} - \dfrac{2}{x^2-3x}$

48. $\dfrac{2}{x^2-4x-32} - \dfrac{6}{x^2+4x}$

49. $\dfrac{5b}{b^2+3b-10} - \dfrac{3}{b^2+4b-12}$

50. $\dfrac{4}{t^2-5t-24} - \dfrac{3t}{t^2+2t-3}$

51. $\dfrac{2x}{x^2+11x+18} - \dfrac{5}{x^2-5x-14}$

52. $\dfrac{3}{x^2-12x+35} - \dfrac{8}{x^2-3x-10}$

53. $\dfrac{x+3}{x-6} - 4$

54. $\dfrac{x-1}{x+5} - 2$

Find the difference. Simplify the result.

55. $\dfrac{x+2}{x-4} - \dfrac{x-3}{x+1}$

56. $\dfrac{x-6}{x-3} - \dfrac{x-4}{x+5}$

57. $\dfrac{x+2}{x-4} - \dfrac{4}{x^2-9x+20}$

58. $\dfrac{x-7}{x-2} - \dfrac{3}{x^2+3x-10}$

59. $\dfrac{5t}{t^2-10t+21} - \dfrac{t+4}{t-7}$

60. $\dfrac{4r}{r^2-9r+8} - \dfrac{r-3}{r-8}$

61. $\dfrac{x-4}{3x+3} - \dfrac{6}{x^2-1}$

62. $\dfrac{x-2}{3x-12} - \dfrac{5}{x^2-16}$

63. $\dfrac{3x}{x+y} - \dfrac{3xy}{2x^2+3xy+y^2}$

64. $\dfrac{2y}{x-y} - \dfrac{6xy}{3x^2-2xy-y^2}$

65. $\dfrac{2}{x^3-6x^2-3x+18} - \dfrac{3x}{5x^2-15}$

66. $\dfrac{4x}{x^3-2x^2+2x-4} - \dfrac{5}{4x^2+8}$

67. $\dfrac{3x}{x^3+1} - \dfrac{2}{x^2+2x+1}$

68. $\dfrac{2x}{x^3-64} - \dfrac{4}{x^2-8x+16}$

69. A student tries to find the difference $\dfrac{3x}{x+2} - \dfrac{5x+7}{x+2}$:

$$\dfrac{3x}{x+2} - \dfrac{5x+7}{x+2} = \dfrac{3x-5x+7}{x+2}$$

$$= \dfrac{-2x+7}{x+2}$$

Describe any errors. Then find the difference correctly.

70. A student tries to find the difference $\dfrac{5x}{x-7} - \dfrac{2x+4}{x-7}$:

$$\dfrac{5x}{x-7} - \dfrac{2x+4}{x-7} = \dfrac{5x-(2x+4)}{x-7}$$

$$= \dfrac{5x-2x+4}{x-7}$$

$$= \dfrac{3x+4}{x-7}$$

Describe any errors. Then find the difference correctly.

71. Describe how to subtract two rational expressions that have a common denominator.

72. Describe how to subtract two rational expressions that have different denominators.

Related Review

73. Find the quotient of $\dfrac{x+4}{x+3}$ and $\dfrac{x-7}{x+3}$.

74. Find the product of $\dfrac{x+4}{x+3}$ and $\dfrac{x-7}{x+3}$.

75. Find the difference of $\dfrac{x+4}{x+3}$ and $\dfrac{x-7}{x+3}$.

76. Find the sum of $\dfrac{x+4}{x+3}$ and $\dfrac{x-7}{x+3}$.

Perform the indicated operation. Simplify the result.

77. $\dfrac{x^2-9}{x^2+10x+25} \cdot \dfrac{x^2+5x}{x^2-4x-21}$

78. $\dfrac{x^2-2x-8}{x^2+10x+9} \div \dfrac{x^2-16}{x^2+18x+81}$

79. $\dfrac{2x}{x^2-4x+4} + \dfrac{4}{x^2-9x+14}$

80. $\dfrac{7}{x^2-11x+30} - \dfrac{3x}{x^2-7x+10}$

81. $\dfrac{x^2-10x+16}{5x^2-3x} \div \dfrac{x^2-3x-40}{25x^2-9}$

82. $\dfrac{4x^2-49}{4x-12} \cdot \dfrac{x^2-9}{2x^2+5x-7}$

83. $\dfrac{3}{x^2+4x-21} - \dfrac{5x}{2x-6}$

84. $\dfrac{2x}{5x-20} + \dfrac{4}{x^2-11x+28}$

85. Perform the indicated operation. Simplify the result.

a. $\dfrac{x}{2} + \dfrac{4}{x}$ b. $\dfrac{x}{2} - \dfrac{4}{x}$

c. $\dfrac{x}{2} \cdot \dfrac{4}{x}$ d. $\dfrac{x}{2} \div \dfrac{4}{x}$

86. Perform the indicated operation. Simplify the result.

a. $\dfrac{x+2}{x+3} + \dfrac{x+4}{x+5}$ b. $\dfrac{x+2}{x+3} - \dfrac{x+4}{x+5}$

c. $\dfrac{x+2}{x+3} \cdot \dfrac{x+4}{x+5}$ d. $\dfrac{x+2}{x+3} \div \dfrac{x+4}{x+5}$

Expressions, Equations, and Graphs

Perform the indicated instruction. Then use words such as linear, quadratic, cubic, power, radical, rational, polynomial, degree, one variable, *and* two variables *to describe the expression, equation, or system.*

87. Solve $3x^2 + 2x = 8$. **88.** Solve $5x^2 = 8$.

89. Factor $3x^2 + 2x - 8$.

90. Find the product $(4w-1)(3w^2+2w-5)$.

91. Find the difference $\dfrac{5}{3x^2+2x-8} - \dfrac{3x}{x^2-4}$.

92. Simplify $\sqrt{40ab^2}$. Assume that a and b are nonnegative.

10.5 SOLVING RATIONAL EQUATIONS

Objectives

▷ Solve *rational equations in one variable*.

▷ Compare solving rational equations with simplifying rational expressions.

▷ Use a rational model to make estimates and predictions.

In Sections 10.1–10.4, we worked mostly with rational expressions. In this section, we solve rational *equations* in one variable.

Solving Rational Equations in One Variable

Here are some examples of rational equations in one variable:

$$\frac{5}{x-2} = 7, \qquad \frac{2}{3x} + \frac{x}{4} = \frac{5}{6x}, \qquad \frac{2x}{x-3} - \frac{4}{x+5} = \frac{1}{x^2+2x-10}$$

A **rational equation in one variable** is an equation in one variable in which each side can be written as a rational expression. Recall from Section 4.4 that we can clear an equation of fractions by multiplying both sides of the equation by the LCD.

With rational equations, it is possible to take the usual steps for solving equations, yet arrive at x values that are excluded values for one or more of the fractions in the equation. These values of x are *not* solutions. We call these values **extraneous solutions.**

Example 1 Solving a Rational Equation

Solve $7 - \dfrac{4}{x} = 2 + \dfrac{6}{x}$.

Solution

We note that 0 is an excluded value. We clear the equation of fractions by multiplying both sides of the equation by x, which is the LCD of both of the fractions $\dfrac{4}{x}$ and $\dfrac{6}{x}$:

$$7 - \frac{4}{x} = 2 + \frac{6}{x} \qquad \text{Original equation}$$

$$x \cdot \left(7 - \frac{4}{x}\right) = x \cdot \left(2 + \frac{6}{x}\right) \qquad \text{Multiply both sides by LCD, } x.$$

$$x \cdot 7 - x \cdot \frac{4}{x} = x \cdot 2 + x \cdot \frac{6}{x} \qquad \text{Distributive law}$$

$$7x - 4 = 2x + 6 \qquad \text{Simplify.}$$

$$5x = 10 \qquad \text{Subtract 2x on both sides; add 4 to both sides.}$$

$$x = 2 \qquad \text{Divide both sides by 5.}$$

Since 2 is not an excluded value, we conclude that 2 is the solution of the equation. We check that 2 satisfies the original equation:

$$7 - \frac{4}{x} = 2 + \frac{6}{x}$$

$$7 - \frac{4}{2} \overset{?}{=} 2 + \frac{6}{2}$$

$$7 - 2 \overset{?}{=} 2 + 3$$

$$5 \overset{?}{=} 5$$

$$\text{true}$$

■

Example 2 Solving a Rational Equation

Solve $\dfrac{x}{4} - \dfrac{3}{2x} = \dfrac{5}{4}$.

Solution

We note that 0 is an excluded value. We clear the equation of fractions by multiplying both sides of the equation by $4x$, which is the LCD of all of the fractions:

$$\frac{x}{4} - \frac{3}{2x} = \frac{5}{4} \qquad \text{Original equation}$$

$$4x \cdot \left(\frac{x}{4} - \frac{3}{2x}\right) = 4x \cdot \frac{5}{4} \qquad \text{Multiply both sides by the LCD, 4x.}$$

$$4x \cdot \frac{x}{4} - 4x \cdot \frac{3}{2x} = 4x \cdot \frac{5}{4} \qquad \text{Distributive law}$$

$$x^2 - 6 = 5x \qquad \text{Simplify.}$$

$$x^2 - 5x - 6 = 0 \qquad \text{Write in } ax^2 + bx + c = 0 \text{ form.}$$

$$(x + 1)(x - 6) = 0 \qquad \text{Factor left side.}$$

$$x + 1 = 0 \quad \text{or} \quad x - 6 = 0 \qquad \text{Zero factor property}$$

$$x = -1 \quad \text{or} \quad x = 6$$

Since neither of our results, -1 and 6, is an excluded value, we conclude that the solutions are -1 and 6. We use "intersect" on a graphing calculator to check our work (see Fig. 12).

 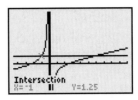 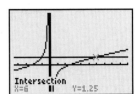

Figure 12 Verify the work

In Example 3, we will see the importance of keeping track of excluded values.

Example 3   Solving a Rational Equation

Solve $\dfrac{4}{w-3} + 2 = \dfrac{w+1}{w-3}$.

Solution

We note that 3 is an excluded value. We clear the equation of fractions by multiplying both sides of the equation by the LCD, $w - 3$:

$$\frac{4}{w-3} + 2 = \frac{w+1}{w-3} \qquad \text{Original equation}$$

$$(w-3)\cdot\left(\frac{4}{w-3}+2\right) = (w-3)\cdot\left(\frac{w+1}{w-3}\right) \qquad \text{Multiply both sides by LCD, } w-3.$$

$$(w-3)\cdot\frac{4}{w-3} + (w-3)\cdot 2 = (w-3)\cdot\frac{w+1}{w-3} \qquad \text{Distributive law}$$

$$4 + (w-3)\cdot 2 = w+1 \qquad \text{Simplify.}$$

$$4 + 2w - 6 = w + 1 \qquad \text{Distributive law}$$

$$2w - 2 = w + 1 \qquad \text{Combine like terms.}$$

$$w = 3 \qquad \text{Subtract } w \text{ from both sides; add 2 to both sides.}$$

Our result, 3, is *not* a solution, because 3 is an excluded value. Since the only possibility is not a solution of the original equation, we conclude that no number is a solution. We say that the solution set is the *empty set*. ∎

In Example 3, we multiplied both sides of the rational equation $\dfrac{4}{w-3}+2 = \dfrac{w+1}{w-3}$ by the LCD and simplified both sides. We got the equation $4 + (w-3)\cdot 2 = w+1$, which has the solution 3. However, 3 is *not* a solution of the original rational equation.

To see where we introduced the extraneous solution 3, notice that 3 does not satisfy the equation

$$(w-3)\cdot\frac{4}{w-3} + (w-3)\cdot 2 = (w-3)\cdot\frac{w+1}{w-3}$$

because 3 is an excluded value of the expression $(w-3)\cdot\dfrac{w+1}{w-3}$. However, 3 does satisfy the next equation, $4 + (w-3)\cdot 2 = w+1$:

$$4 + (w-3)\cdot 2 = w+1$$
$$4 + (3-3)\cdot 2 \overset{?}{=} 3+1$$
$$4 \overset{?}{=} 4$$
$$\text{true}$$

Since multiplying both sides of a rational equation by the LCD and then simplifying both sides may introduce extraneous solutions, we must always check that any proposed solution is not an excluded value.

Example 4 Solving a Rational Equation

Solve $\dfrac{x+4}{x-3} = \dfrac{x-6}{x+1}$.

Solution

We note that 3 and -1 are excluded values. We clear the equation of fractions by multiplying both sides of the equation by the LCD, $(x-3)(x+1)$:

$$\frac{x+4}{x-3} = \frac{x-6}{x+1} \qquad \text{Original equation}$$

$$(x-3)(x+1) \cdot \left(\frac{x+4}{x-3}\right) = (x-3)(x+1) \cdot \left(\frac{x-6}{x+1}\right) \qquad \begin{array}{l}\text{Multiply both sides by}\\ \text{LCD, } (x-3)(x+1).\end{array}$$

$$(x+1)(x+4) = (x-3)(x-6) \qquad \text{Simplify.}$$

$$x^2 + 5x + 4 = x^2 - 9x + 18 \qquad \text{Find the products.}$$

$$5x + 4 = -9x + 18 \qquad \text{Subtract } x^2 \text{ from both sides.}$$

$$14x = 14 \qquad \begin{array}{l}\text{Add } 9x \text{ to both sides; subtract 4}\\ \text{from both sides.}\end{array}$$

$$x = 1 \qquad \text{Divide both sides by 14.}$$

Figure 13 Verify the work

Since 1 is not an excluded value, the solution is 1. We use a graphing calculator table to check our work (see Fig. 13). ∎

To solve a rational equation, we factor the denominators of fractions to help us determine any excluded values, to find the LCD, and, later, to help us simplify rational expressions.

Example 5 Solving a Rational Equation

Solve $\dfrac{x}{x-5} + \dfrac{2}{x-6} = \dfrac{2}{x^2 - 11x + 30}$.

Solution

We begin by factoring the denominator of the expression on the right-hand side of the equation:

$$\frac{x}{x-5} + \frac{2}{x-6} = \frac{2}{x^2 - 11x + 30} \qquad \text{Original equation}$$

$$\frac{x}{x-5} + \frac{2}{x-6} = \frac{2}{(x-6)(x-5)} \qquad \text{Factor denominator.}$$

The excluded values are 5 and 6. Next, we clear the equation of fractions by multiplying both sides by the LCD, $(x-6)(x-5)$:

$$(x-6)(x-5) \cdot \left(\frac{x}{x-5} + \frac{2}{x-6}\right) = (x-6)(x-5) \cdot \frac{2}{(x-6)(x-5)}$$

On the left-hand side, we use the distributive law. On the right-hand side, we simplify:

$$(x-6)(x-5) \cdot \frac{x}{x-5} + (x-6)(x-5) \cdot \frac{2}{x-6} = 2$$

$$(x-6) \cdot x + (x-5) \cdot 2 = 2 \qquad \text{Simplify.}$$

$$x^2 - 6x + 2x - 10 = 2 \qquad \text{Distributive law}$$

$$x^2 - 4x - 10 = 2 \qquad \text{Combine like terms.}$$

$$x^2 - 4x - 12 = 0 \qquad \text{Write in } ax^2 + bx + c = 0 \text{ form.}$$

$$(x-6)(x+2) = 0 \qquad \text{Factor left-hand side.}$$

$$x - 6 = 0 \quad \text{or} \quad x + 2 = 0 \qquad \text{Zero factor property}$$

$$x = 6 \quad \text{or} \quad x = -2$$

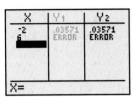

Figure 14 Verify the work

Since 6 is an excluded value, it is *not* a solution. The only solution is -2. We use a graphing calculator table to check our work (see Fig. 14). ∎

Solving a Rational Equation in One Variable

To solve a rational equation in one variable,

1. Factor the denominator(s) if possible.
2. Identify any excluded values.
3. Find the LCD of all of the fractions.
4. Multiply both sides of the equation by the LCD, which gives a simpler equation to solve.
5. Solve the simpler equation.
6. Discard any proposed solutions that are excluded values.

Solving Equations versus Simplifying Expressions

Throughout this course, we have solved equations and simplified expressions. In solving an equation, our objective is to find any *numbers* that satisfy the equation. In simplifying an expression, our objective is to find a simpler, yet equivalent, *expression*.

Solving a Rational Equation versus Simplifying a Rational Expression

To solve a rational equation, clear the fractions in it by multiplying both sides of the equation by the LCD. To simplify a rational expression, do *not* multiply it by the LCD. The only multiplication permissible is multiplication by 1, usually in the form $\dfrac{A}{A}$, where A is a nonzero polynomial.

Here, we compare solving a rational equation with simplifying a rational expression:

Solving the Equation $\dfrac{3}{2} = \dfrac{6}{x}$

The number 0 is an excluded value.

$$\dfrac{3}{2} = \dfrac{6}{x} \qquad \text{Original equation}$$

$$2x \cdot \dfrac{3}{2} = 2x \cdot \dfrac{6}{x} \qquad \begin{array}{l}\text{Multiply both}\\\text{sides by LCD, } 2x.\end{array}$$

$$3x = 12 \qquad \text{Simplify.}$$

$$x = 4 \qquad \begin{array}{l}\text{The result is a}\\\text{number.}\end{array}$$

Simplifying the Expression $\dfrac{3}{2} + \dfrac{6}{x}$

$$\dfrac{3}{2} + \dfrac{6}{x} = \dfrac{3}{2} \cdot \dfrac{x}{x} + \dfrac{6}{x} \cdot \dfrac{2}{2} \qquad \begin{array}{l}\text{Introduce missing}\\\text{factors.}\end{array}$$

$$= \dfrac{3x}{2x} + \dfrac{12}{2x} \qquad \text{Find products.}$$

$$= \dfrac{3x + 12}{2x} \qquad \begin{array}{l}\text{The result is an}\\\text{expression.}\end{array}$$

For the equation $\dfrac{3}{2} = \dfrac{6}{x}$, the solution is the *number* 4. By contrast, we simplify the expression $\dfrac{3}{2} + \dfrac{6}{x}$ by writing it as the *expression* $\dfrac{3x + 12}{2x}$. In general, **the result of solving a rational equation is the empty set or a set of one or more numbers. The result of simplifying a rational expression is an expression.**

Rational Models

When using a rational model to describe a situation, we can make predictions about the independent variable by substituting a value for the dependent variable and then solving the equation by using techniques discussed in this section.

Example 6   Using a Rational Model to Make a Prediction

The numbers of households in the United States that pay bills online are shown in Table 8 for various years.

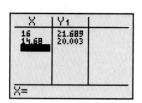

Figure 15 Verify the work

Table 8 Households That Pay Bills Online		
Year	Households That Pay Bills Online (millions)	Total Number of Households (millions)
1998	1.7	102.5
1999	4.1	103.9
2000	5.2	104.7
2001	7.3	108.2
2002	9.9	109.3

Sources: *Yankee Group; U.S. Census Bureau*

Let p be the percentage of households that pay their bills online at t years since 1995. A reasonable model is

$$p = \frac{196t - 416}{1.79t + 96.77}$$

1. Predict the percentage of households that will pay their bills online in 2011.
2. Predict when 20% of households will pay their bills online.

Solution

1. Since $2011 - 1995 = 16$, we substitute the input 16 for t in the equation $p = \dfrac{196t - 416}{1.79t + 96.77}$ and solve for p:

$$p = \frac{196(16) - 416}{1.79(16) + 96.77} \approx 21.69$$

So, 21.7% of households will pay their bills online in 2011, according to the model.

2. We substitute the output 20 for p in the equation $p = \dfrac{196t - 416}{1.79t + 96.77}$ and solve for t:

$$20 = \frac{196t - 416}{1.79t + 96.77}$$
Substitute 20 for p.

$$(1.79t + 96.77) \cdot 20 = (1.79t + 96.77) \cdot \frac{196t - 416}{1.79t + 96.77}$$
Multiply both sides by LCD, $1.79t + 96.77$.

$$35.8t + 1935.4 = 196t - 416$$
Simplify.

$$35.8t - 196t = -416 - 1935.4$$
Subtract $196t$ from both sides; subtract 1935.4 from both sides.

$$-160.2t = -2351.4$$
Combine like terms.

$$t \approx 14.68$$
Divide both sides by -160.2.

The value of t found represents the year $1995 + 15 = 2010$. So, 20% of households will pay their bills online in 2010, according to the model.

We use a graphing calculator table to verify our work in Problems 1 and 2 (see Fig. 15). ∎

group exploration

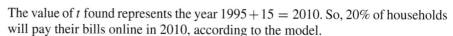

Looking ahead: Proportions

Recall from Section 2.5 that the ratio of a to b is the quotient $\dfrac{a}{b}$.

1. **a.** The price of a notebook is $3. For the given number of notebooks, find the ratio of the total cost to the number of notebooks. *Do not simplify the ratio.*
   i. 2 notebooks [**Hint:** First find the total cost of 2 notebooks. Then divide your result by 2. Include units in your result.]
   ii. 5 notebooks

b. Compare the results that you found in parts (i) and (ii). In particular, describe how the following equation is related to your results:

$$\frac{6}{2} = \frac{15}{5}$$

When an equation says that two ratios are equal, we call the equation a *proportion*.

c. For any number of notebooks, what is the ratio of the total cost to the number of notebooks? Explain.

2. a. A student earns \$8 per hour. For the given number of hours worked, find the ratio of the total income to the number of hours worked. *Do not simplify the ratio.*

   i. 3 hours

   ii. 5 hours

b. Write a proportion that is related to the results you found in parts (i) and (ii).

c. For any number of hours worked, what is the ratio of the total income to the number of hours worked? Explain how this ratio is related to the rate of change of income.

HOMEWORK 10.5 FOR EXTRA HELP ▶

 Student Solutions Manual   PH Math/Tutor Center *Math XP* MathXL® *MyMathLab* MyMathLab

---

Solve.

1. $\dfrac{3}{x} - 2 = \dfrac{7}{x}$

2. $\dfrac{9}{x} - 3 = \dfrac{3}{x}$

3. $5 - \dfrac{4}{x} = 3 + \dfrac{2}{x}$

4. $6 - \dfrac{8}{x} = 3 + \dfrac{4}{x}$

5. $\dfrac{5}{p-1} = \dfrac{2p+1}{p-1}$

6. $\dfrac{7}{y+3} = \dfrac{3y+4}{y+3}$

7. $\dfrac{8x+4}{x+2} = \dfrac{5x-2}{x+2}$

8. $\dfrac{3x-2}{x-5} = \dfrac{x+8}{x-5}$

9. $\dfrac{w+2}{w-4} + 3 = \dfrac{2}{w-4}$

10. $\dfrac{t-5}{t+1} + 2 = \dfrac{3}{t+1}$

11. $\dfrac{2}{x} + \dfrac{5}{4} = \dfrac{3}{x}$

12. $\dfrac{3}{x} - \dfrac{7}{2} = \dfrac{6}{x}$

13. $\dfrac{5}{6x} - \dfrac{1}{2} = \dfrac{3}{4x}$

14. $\dfrac{3}{8x} - \dfrac{3}{4} = \dfrac{1}{6}$

15. $\dfrac{4}{x-2} = \dfrac{2}{x+3}$

16. $\dfrac{3}{x+4} = \dfrac{4}{x-1}$

17. $\dfrac{2r+7}{4r} = \dfrac{5}{3}$

18. $\dfrac{5p-2}{3p} = \dfrac{6}{7}$

19. $\dfrac{5}{x+3} + \dfrac{3}{4} = 2$

20. $\dfrac{4}{x-3} + \dfrac{5}{2} = 3$

21. $\dfrac{2}{x-3} + \dfrac{1}{x+3} = \dfrac{5}{x^2-9}$

22. $\dfrac{3}{x+5} + \dfrac{2}{x-5} = \dfrac{1}{x^2-25}$

23. $\dfrac{4}{x+2} + \dfrac{3}{x+1} = \dfrac{3}{x^2+3x+2}$

24. $\dfrac{2}{x-3} + \dfrac{4}{x+5} = \dfrac{16}{x^2+2x-15}$

25. $\dfrac{5}{x^2-4} + \dfrac{2}{x+2} = \dfrac{4}{x-2}$

26. $\dfrac{3}{x^2-1} + \dfrac{7}{x-1} = \dfrac{4}{x+1}$

27. $\dfrac{3}{y-4} - \dfrac{4}{y-3} = \dfrac{3}{y^2-7y+12}$

28. $\dfrac{2}{r+2} - \dfrac{3}{r-6} = \dfrac{-16}{r^2-4r-12}$

29. $\dfrac{4}{x^2-3x} - \dfrac{5}{x} = \dfrac{7}{x-3}$

30. $\dfrac{2}{x^2+6x} - \dfrac{4}{x} = \dfrac{5}{x+6}$

31. $\dfrac{3}{2x-6} + \dfrac{5x}{6x-18} = \dfrac{2}{4x-12}$

32. $\dfrac{5}{4x+8} - \dfrac{3x}{2x+4} = \dfrac{2}{5x+10}$

33. $\dfrac{2}{x^2-x-6} - \dfrac{4}{x+2} = \dfrac{3}{2x-6}$

34. $\dfrac{2}{x^2+4x+3} - \dfrac{3}{x+1} = \dfrac{5}{3x+9}$

35. $\dfrac{3}{x} = \dfrac{4}{x^2} - 1$

36. $\dfrac{2}{x} = \dfrac{24}{x^2} - 2$

37. $1 = \dfrac{15}{t^2} - \dfrac{2}{t}$

38. $1 = \dfrac{-14}{w^2} + \dfrac{9}{w}$

39. $\dfrac{2}{x^2} + \dfrac{3}{x} = 5$

40. $\dfrac{4}{x^2} - \dfrac{1}{x} = 3$

41. $\dfrac{x-3}{x+2} = \dfrac{x+5}{x-1}$

42. $\dfrac{x+6}{x+2} = \dfrac{x-7}{x-3}$

43. $\dfrac{x}{x+1} = \dfrac{3}{x-1} + \dfrac{2}{x^2-1}$

44. $\dfrac{x}{x+3} = \dfrac{2}{x-3} - \dfrac{12}{x^2-9}$

45. $\dfrac{r}{r-3} - \dfrac{2}{r+5} = \dfrac{10}{r^2+2r-15}$

46. $\dfrac{4}{p+2} - \dfrac{p}{p+1} = \dfrac{5}{p^2 + 3p + 2}$

47. $\dfrac{1}{x-2} = \dfrac{2x}{x+1} - \dfrac{6}{x^2 - x - 2}$

48. $\dfrac{x}{x+5} = \dfrac{3}{x+4} + \dfrac{5}{x^2 + 9x + 20}$

49. $\dfrac{2}{x^2 - 9} = \dfrac{x}{x-3} - \dfrac{x-5}{x+3}$

50. $\dfrac{-1}{x^2 - 4} = \dfrac{x}{x-2} - \dfrac{x-4}{x+2}$

51. $\dfrac{3}{x-4} + \dfrac{7}{x-3} = \dfrac{x+4}{x-3}$

52. $\dfrac{5}{x+5} + \dfrac{4}{x-2} = \dfrac{x-5}{x+5}$

53. $\dfrac{2}{m-3} + \dfrac{5m}{m^2 - 9} = \dfrac{4}{m+3} - \dfrac{3}{m^2 - 9}$

54. $\dfrac{6w}{w^2 - 16} - \dfrac{3}{w+4} = \dfrac{5}{w^2 - 16} + \dfrac{2}{w-4}$

55. The consumer debt in the United States is shown in Table 9 for various years. Totals exclude loans secured by real estate.

Table 9 Consumer Debt

Year	Total Consumer Debt (trillions of dollars)	Number of Households (millions)
1999	1.5	103.6
2000	1.7	104.7
2001	1.8	108.2
2002	1.9	109.3
2003	2.0	111.6

Source: *Federal Reserve*

Let D be the average household debt (in *thousands* of dollars) at t years since 1990. A reasonable model of the situation is

$$D = \frac{120t + 460}{2.06t + 84.82}$$

a. Estimate the average household debt in 2003. Then compute the actual average household debt. Is your result from the model an underestimate or an overestimate?

b. Predict the average household debt in 2010.

c. Predict in what year the average household debt will be $24 thousand.

56. The numbers of households that have personal computers are shown in Table 10 for various years.

Table 10 Numbers of Households with Personal Computers

Year	Number of Households with Personal Computers (millions)	Total Number of Households (millions)
1995	31.36	99.0
1997	39.56	101.0
1999	50.00	103.6
2001	60.81	108.2
2003	68.78	111.3

Sources: *U.S. Department of Commerce; Media Matrix; U.S. Census Bureau*

Let p be the percentage of households with personal computers at t years since 1990. A reasonable model is

$$p = \frac{480t + 686}{1.59t + 90.31}$$

a. Use the model to estimate the percentage of households that had personal computers in 2003. Then compute the actual percentage. Is your result from the model an underestimate or an overestimate?

b. Predict the percentage of households that will have personal computers in 2011.

c. Predict when 84% of households will have personal computers.

57. In Exercise 75 of Homework 10.1, you worked with the model

$$p = \frac{-28t^2 + 2694t + 15{,}122}{50.8t + 275}$$

where p is the percentage of prisoners who are released in the year that is t years since 1980 (see Table 11).

Table 11 Prisoners and Releases from Prisons

Year	Number of Prisoners Released from Prison (thousands)	Year	Total Number of Prisoners (thousands)
1980	170	1980	316
1985	234	1985	481
1990	420	1990	740
1995	492	1995	1085
2000	585	2000	1331
2004	630	2003	1409

Source: *Bureau of Justice Statistics*

Predict in which year 38% of prisoners will be released.

58. In Exercise 76 of Homework 10.1, you worked with the model

$$p = \frac{-9.3t^2 + 428t + 1706}{1.26t + 80.51}$$

where p is the percentage of households that had cable television at t years since 1980 (see Table 12).

Table 12 Numbers of Households with Cable Television

Year	Number of Households with Cable Television (millions)	Total Number of Households (millions)
1980	17.5	80.8
1985	35.4	86.8
1990	50.5	93.3
1995	60.6	99.0
2000	66.3	104.7
2004	65.9	112.0

Sources: *Paul Kagan Associates, Inc; U.S. Census Bureau*

Estimate when exactly half of households had cable television.

59. A student tries to simplify $\dfrac{5}{x+2} + \dfrac{3}{x}$:

$$\frac{5}{x+2} + \frac{3}{x} = x(x+2) \cdot \left(\frac{5}{x+2} + \frac{3}{x} \right)$$

$$= x(x+2) \cdot \frac{5}{x+2} + x(x+2) \cdot \frac{3}{x}$$

$$= 5x + 3(x+2)$$

$$= 5x + 3x + 6$$

$$= 8x + 6$$

Describe any errors. Then simplify the expression correctly.

60. A student tries to solve the equation $\dfrac{x}{x-5} + \dfrac{1}{2} = \dfrac{2}{x-5}$:

$$\frac{x}{x-5} + \frac{1}{2} = \frac{2}{x-5}$$

$$\frac{2}{2} \cdot \frac{x}{x-5} + \frac{x-5}{x-5} \cdot \frac{1}{2} = \frac{2}{2} \cdot \frac{2}{x-5}$$

$$\frac{2x + x - 5}{2(x-5)} = \frac{4}{2(x-5)}$$

$$\frac{3x - 5}{2(x-5)} = \frac{4}{2(x-5)}$$

$$3x - 5 = 4$$

$$3x = 9$$

$$x = 3$$

Is the work correct? Is there a better way to solve the equation? Explain.

61. A student tries to solve $\dfrac{7}{x^2 + x - 20} = \dfrac{4}{x+5} - \dfrac{2}{x-4}$:

$$\frac{7}{x^2 + x - 20} = \frac{4}{x+5} - \frac{2}{x-4}$$

$$= \frac{4}{x+5} \cdot \frac{x-4}{x-4} - \frac{2}{x-4} \cdot \frac{x+5}{x+5}$$

$$= \frac{4(x-4) - 2(x+5)}{(x-4)(x+5)}$$

$$= \frac{4x - 16 - 2x - 10}{(x-4)(x+5)}$$

$$= \frac{2x - 26}{(x-4)(x+5)}$$

$$= \frac{2(x-13)}{(x-4)(x+5)}$$

Describe any errors. Then solve the equation correctly.

62. A student tries to solve a rational equation. The student's result is $\dfrac{x+1}{x-4}$. What would you tell the student?

63. When simplifying a rational expression, can we multiply it by the LCD? Explain. When solving a rational equation, can we multiply both sides of the equation by the LCD? Explain. (See page 9 for guidelines on writing a good response.)

64. Describe how to solve a rational equation. (See page 9 for guidelines on writing a good response.)

Related Review

Solve or simplify, whichever is appropriate.

65. $\dfrac{7}{x} + \dfrac{3}{x}$

66. $\dfrac{4}{x+5} - \dfrac{x+1}{x+5} = 2$

67. $\dfrac{7}{x} + \dfrac{3}{x} = 1$

68. $\dfrac{4}{x+5} - \dfrac{x+1}{x+5}$

69. $\dfrac{2}{x-3} + \dfrac{3x}{x+2} = \dfrac{2}{x^2 - x - 6}$

70. $\dfrac{5}{x-4} - \dfrac{2}{x^2 - 16}$

71. $\dfrac{2}{x-3} + \dfrac{3x}{x+2}$

72. $\dfrac{5}{x-4} - \dfrac{2}{x^2 - 16} = \dfrac{3}{x+4}$

Solve.

73. $3x^2 - 2x = 4$

74. $5(2x - 3) = 4x - 7$

75. $3x^3 + 8 = 2x^2 + 12x$

76. $x^2 = 4x + 12$

77. $2x^2 + 1 = 8$

78. $\dfrac{2x}{x-1} + \dfrac{5}{x+1} = \dfrac{4x-3}{x^2-1}$

Expressions, Equations, and Graphs

Perform the indicated instruction. Then use words such as linear, quadratic, cubic, power, radical, rational, polynomial, degree, one variable, *and* two variables *to describe the expression, equation, or system.*

79. Simplify $3(5m - 2n) - 4(m + 3n) + n$.

80. Find the quotient $\dfrac{x^2 - 9}{x^3 + 3x^2 - 28x} \div \dfrac{5x + 15}{x^3 - 16x}$.

81. Factor $2x^3 - 18x^2 + 28x$.

82. Graph $2x - 5y = 10$ by hand.

83. Solve $\dfrac{2}{x-4} - \dfrac{3}{x+4} = \dfrac{5}{x^2 - 16}$.

84. Solve $x(x - 2) = 48$.

10.6 PROPORTIONS; SIMILAR TRIANGLES

Objectives

▷ Use *proportions* to make estimates.

▷ Find the length of a side of a *similar triangle*.

In Section 2.5, we worked with ratios. In this section, we will discuss an equation called a *proportion*, which contains ratios. We will use proportions to make estimates of quantities and to find the lengths of the sides of special pairs of triangles called *similar triangles*.

Throughout this chapter, we have been working with rational expressions and equations. Recall from Section 10.1 that the word "rational" refers to a ratio. Recall from Section 2.5 that the ratio of a to b can be written as the fraction $\frac{a}{b}$ or as $a : b$. For example, if there are 9 women and 5 men on a coed softball team, then the ratio of the number of women to the number of men is

$$\frac{9 \text{ women}}{5 \text{ men}}$$

Proportions

A **proportion** is a statement of the equality of two ratios. For example,

$$\frac{4}{6} = \frac{2}{3}$$

is a proportion which says that the ratios $\frac{4}{6}$ and $\frac{2}{3}$ are equal.

There are many situations in which proportions can describe equal ratios. For example, suppose that a certain type of pen costs \$2. Then the ratio of the total cost of some pens to the number of pens is constant for any group of pens. Here is the ratio for a group of 3 pens:

$$\text{Total cost} \longrightarrow \frac{\$6}{3 \text{ pens}} = \$2 \text{ per pen} \longleftarrow \text{Number of pens}$$

Here is the ratio for a group of 5 pens:

$$\text{Total cost} \longrightarrow \frac{\$10}{5 \text{ pens}} = \$2 \text{ per pen} \longleftarrow \text{Number of pens}$$

The proportion

$$\frac{\$6}{3 \text{ pens}} = \frac{\$10}{5 \text{ pens}}$$

correctly states that the ratio of the total cost to the number of pens is the same for a group of 3 pens and a group of 5 pens. We say that the total cost and the number of pens are *proportional*.

In general, if the ratio of two related quantities is constant, we say that the two quantities are **proportional.**

When we work with a proportion of the form

$$\frac{a}{b} = \frac{c}{d}$$

if we know the values of three of the four variables, we can solve for the fourth variable.

Example 1 Using a Proportion to Make an Estimate

While commuting to work and running errands, a person travels 576 miles in a 3-week period.

1. Estimate over what period the person will travel 1000 miles while commuting to work and running errands.
2. Discuss any assumptions that you made in Problem 1. Describe a scenario in which the result in Problem 1 is an overestimate.

Solution

1. We let t be the period (in weeks) required for the person to travel 1000 miles while commuting to work and running errands. Assuming that the ratio of the total miles traveled to the number of weeks is constant, we set up a proportion:

$$\text{Total miles traveled} \longrightarrow \frac{576}{3} = \frac{1000}{t} \longleftarrow \text{Total miles traveled}$$
$$\text{Number of weeks} \longrightarrow \qquad\qquad \longleftarrow \text{Number of weeks}$$

Next, we multiply both sides of the equation by the LCD, $3t$:

$$\frac{576}{3} \cdot \frac{3t}{1} = \frac{1000}{t} \cdot \frac{3t}{1} \qquad \text{Multiply both sides by } 3t.$$

$$576t = 3000 \qquad \text{Simplify.}$$

$$t \approx 5.21 \qquad \text{Divide both sides by 576.}$$

So, the person will drive 1000 miles over about a 5.2-week period.

2. We assumed that the ratio of the total miles traveled to the number of weeks is constant. This will not be the case if the person's driving patterns over the course of the 1000 miles are significantly different from those for the 576 miles. For example, if the person is transferred to a work location that is closer to home and the person's driving patterns while running errands remained the same, our estimate in Problem 1 would be an overestimate. ■

In Example 1, the person traveled 576 miles in a 3-week period. For this period, the unit ratio of total miles traveled to number of weeks is

$$\frac{576 \text{ miles}}{3 \text{ weeks}} = \frac{192 \text{ miles}}{1 \text{ week}}$$

To estimate over what period the person would travel 1000 miles, we assumed that the ratio of total miles to number of weeks would remain $\frac{192 \text{ miles}}{1 \text{ week}}$. If that ratio were not close to $\frac{192 \text{ miles}}{1 \text{ week}}$, then our result would not be accurate.

If we use a proportion to estimate the value of one of two quantities, the accuracy of our result depends on how much the ratio of the two quantities varies. If the ratio is constant, the estimate will be accurate. If the ratio is approximately constant, the estimate will likely be reasonable. If the ratio varies a great deal, the estimate will likely be inaccurate.

Example 2 Using a Proportion to Make an Estimate

In a 2005 poll of 2209 adults, 1679 adults said that professional baseball players who use steroids should be banned from the National Baseball Hall of Fame (Source: *Harris Interactive, Inc.*).

1. Estimate how many of the 21,429 students at Grand Valley State University would say that professional baseball players who use steroids should be banned from the National Baseball Hall of Fame.
2. Discuss any assumptions that you made in Problem 1. Describe a scenario in which the result in Problem 1 is an underestimate.

Solution

1. We let n be the number of students at Grand Valley State University who would say that professional baseball players who use steroids should be banned from the National Baseball Hall of Fame. Assuming that the ratio of people who support the ban to the total number of people is the same at the university as in the poll, we set up a proportion:

$$\begin{array}{c} \text{Number of students in} \\ \text{favor of ban} \\ \text{Total number of students} \end{array} \longrightarrow \quad \frac{n}{21,429} = \frac{1679}{2209} \quad \longleftarrow \begin{array}{c} \text{Number of polled adults} \\ \text{in favor of ban} \\ \text{Total number of polled} \\ \text{adults} \end{array}$$

Next, we multiply both sides of the equation by 21,429:

$$\frac{n}{21,429} \cdot 21,429 = \frac{1679}{2209} \cdot 21,429 \qquad \text{Multiply both sides by 21,429.}$$

$$n \approx 16,288 \qquad \text{Simplify.}$$

So, about 16,288 of the 21,429 students would say that professional baseball players who use steroids should be banned from the National Baseball Hall of Fame.

2. We assumed that the ratio of people who support the ban to the total number of people is the same at the university as in the poll. This may not be the case. For example, if students at the university are more disapproving of steroid use than the adults in the survey, our estimate would be an underestimate. ■

Similar Triangles

Two triangles are called **similar triangles** if their corresponding angles are equal in measure. For example, the two triangles in Fig. 16 are similar triangles because the measures of angles A and X are equal, the measures of angles B and Y are equal, and the measures of angles C and Z are equal.

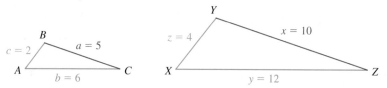

Figure 16 Two similar triangles

Here, we find the ratio of the lengths of corresponding sides of the two triangles in Fig. 16:

$$\frac{c}{z} = \frac{2}{4} = \frac{1}{2}, \qquad \frac{a}{x} = \frac{5}{10} = \frac{1}{2}, \qquad \frac{b}{y} = \frac{6}{12} = \frac{1}{2}$$

Notice that all three ratios are equal.

Side Lengths of Similar Triangles
The lengths of the corresponding sides of two similar triangles are proportional.

Similar triangles have the same shape, but not necessarily the same size.

Example 3 Finding the Length of the Side of a Triangle

The two triangles shown in Fig. 17 are similar. Find the length of the side labeled "x feet" on the larger triangle.

Figure 17 Two similar triangles

Solution

Since the triangles are similar, the lengths of corresponding sides are proportional:

$$\frac{2}{x} = \frac{7}{9} \qquad \text{Use a proportion.}$$

$$\frac{2}{x} \cdot 9x = \frac{7}{9} \cdot 9x \qquad \text{Multiply both sides by LCD, 9x.}$$

$$18 = 7x \qquad \text{Simplify.}$$

$$\frac{18}{7} = x \qquad \text{Divide both sides by 7.}$$

$$2.57 \approx x \qquad \text{Round to second decimal place.}$$

So, the length of the side labeled "x feet" is approximately 2.57 feet. ■

group exploration

Looking ahead: Inverse variation

1. Use ZDecimal on a graphing calculator to graph

$$y = \frac{1}{x}, \quad y = \frac{2}{x}, \quad y = \frac{3}{x}$$

in that order, and describe any patterns you observe.

2. By viewing any of the graphs on the calculator screen, determine what happens to the value of y as the value of x increases, for positive values of x.

3. Explain why the pattern you described in Problem 2 makes sense by referring to any of the three equations. [**Hint:** Are the values $\frac{1}{2}, \frac{1}{3}, \frac{1}{4}, \ldots$ getting larger or smaller?]

4. Let P be the pressure (in atmospheres) of air in a sealed syringe and V be the volume of air (in cubic centimeters). Here is a model that describes the relationship between P and V for a specific syringe:

$$P = \frac{7.4}{V}$$

Describe what happens to the pressure as the volume is increased. Explain in terms of the graph of the model and the model's equation.

5. For the situation described in Problem 4, describe what happens to the pressure as the volume is decreased. Explain in terms of the graph of the model and the model's equation.

6. In your opinion, is it possible to squeeze the syringe so hard (without breaking the seal) that the volume of air in the syringe becomes 0 cubic centimeters? Does the model support your theory? Explain. [**Hint:** In addition to substituting 0 for V in the model's equation, substitute a small value, such as 0.000001, for V.]

HOMEWORK 10.6

FOR EXTRA HELP ▶

Student Solutions Manual   PH Math/Tutor Center   MathXL®   MyMathLab

---

1. A person pays $95.60 for 4 months' use of AOL®. How much will the person pay for 6 months' use of AOL?

2. A household pays $146.25 for 3 months' use of cable TV. How much will the household pay for 5 months' use of cable TV?

3. If 0.75 cup of Post Grape-Nuts Flakes® cereal contains 4 grams of sugar, how many cups of that cereal contains 7 grams of sugar?

4. If 0.5 cup of Chef Boyardee Mini Ravioli® contains 4.5 grams of fat, how many cups of that ravioli would contain 11.25 grams of fat?

5. A 10-ounce solution contains 4 ounces of acid. How many ounces of acid are in 6 ounces of that solution?

6. A 7-ounce solution contains 3 ounces of lemon juice. How many ounces of lemon juice are in 5 ounces of the solution?

7. If 3 out of 5 students at a college are full-time students, what is the enrollment at the college if it has 21,720 full-time students?

8. If 2 out of 7 students at a college are business majors, what is the enrollment at the college if it has 4172 business majors?

9. If 1.25 inches on a map represent 50 miles, how many inches on the map represent 270 miles?

10. If 0.75 inch on a map represents 3 miles, how many inches on the map represent 10 miles?

11. If 5 U.S. dollars can be exchanged for 4.05 European euros, how many U.S. dollars can be exchanged for 260 euros?

12. If 15 U.S. dollars can be exchanged for 158.68 Mexican pesos, how many U.S. dollars can be exchanged for 3000 pesos?

13. The weight of an object on Earth and the weight of the same object on the Moon are proportional. An astronaut who weighs 180 pounds on Earth weighs 29.8 pounds on the Moon. What is the weight of a person on Earth if she weighs 28.5 pounds on the Moon?

14. The weight of an object on Earth and the weight of the same object on Jupiter are proportional. An astronaut who weighs 150 pounds on Earth would weigh 379.9 pounds on Jupiter.

What is the weight of a person on Earth if he would weigh 450 pounds on Jupiter?

15. A person pays $175.50 for 3 months of phone service.
   a. Estimate how much the person will pay for 7 months of phone service.
   b. Discuss any assumptions that you made in part (a). If the person begins to make a lot of long-distance calls during the 7-month period and the person's local-call patterns remain the same, is the result that you found in part (a) an underestimate or an overestimate? Explain.

16. A family spends $475 on groceries in a 5-month period.
   a. Estimate the family's total money for groceries during a 3-month period.
   b. Discuss any assumptions that you made in part (a). If hungry children were away during part of the 5-month period, but were home during the 3-month period, is the result that you found in part (a) an underestimate or an overestimate? Explain.

17. In a 2005 poll of 2022 adults in the United States, 1335 of the adults said that they had online access at home (Source: *Harris Interactive*).
   a. Estimate how many of the 13,590 adults in Deerfield, Illinois, have online access at home.
   b. Discuss any assumptions that you made in part (a). The 2000 median household income was $107,194 in Deerfield and $41,994 in the United States overall (Source: *U.S. Census Bureau*). Do you think that the result you found in part (a) is an underestimate or an overestimate? Explain.

18. In a 2004 poll of 2242 adults in the United States, 1143 of the adults believed that the Olympics should be restricted to amateur athletes (Source: *Harris Interactive*).
   a. Estimate how many of the 14,953 students at the Rockville Campus of Montgomery College believe that the Olympics should be restricted to amateur athletes.
   b. A much higher percentage of adults over age 40 believe in the restriction than do adults under age 40. Do you think that the result you found in part (a) is an underestimate or an overestimate? Explain.

19. When a 4000-pound car travels at 40 miles per hour and then hits a stationary object, such as a concrete wall, the force of impact is 26 tons. When the speed is 60 miles per hour instead, the force of impact is 55 tons. Are the speed and the force of impact proportional for a 4000-pound car? Explain.

20. If you park your car with the windows closed in a sunny location, the temperature inside the car will increase by 29°F in 20 minutes and by 34°F in 30 minutes (Source: *Jan Null*). Are the time elapsed and the increase in temperature proportional? Explain.

21. In fall 2003, the full-time equivalent (FTE) enrollment at Louisiana State University (LSU) was 27,764 students and the number of FTE faculty was 1294. The ratio of the FTE enrollment to the number of FTE faculty at 23 peer institutions is 17:1 (Source: *LSU*; U.S. News & World Report).
   a. Find the ratio of the FTE enrollment to the number of FTE faculty at LSU in fall 2003. Is the ratio greater than, equal to, or less than 17:1?
   b. The LSU deans and department chairs met to determine the cost of reducing their ratio of FTE enrollment to the number of FTE faculty to 17:1. How many FTE faculty would need to be hired to do that? The average cost (salary plus benefits) of adding an FTE faculty member at LSU in

2003 was $73,125. What would be the total cost of hiring enough faculty to reduce the ratio to 17:1?
   c. How much would the FTE enrollment at LSU have to be reduced to reduce the ratio of the FTE enrollment to the number of FTE faculty to 17:1? The revenue (tuition plus an academic excellence fee) from an FTE student was $3345 at LSU in 2003. What would be the total cost of reducing the FTE enrollment to reduce the ratio to 17:1?
   d. Which is the cheaper way to reduce the ratio, hiring FTE faculty or reducing the FTE enrollment?

22. Compare the meaning of *ratio* with the meaning of *proportion*.

For Exercises 23–28, the given figure shows two similar triangles. Find the length of the side labeled "x inches," "x feet," or "x meters." Round your result to the second decimal place.

23. See Fig. 18.

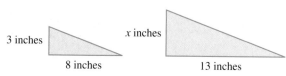

Figure 18 Exercise 23

24. See Fig. 19.

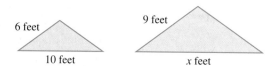

Figure 19 Exercise 24

25. See Fig. 20.

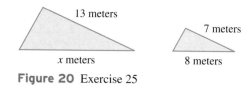

Figure 20 Exercise 25

26. See Fig. 21.

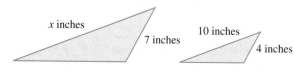

Figure 21 Exercise 26

27. See Fig. 22.

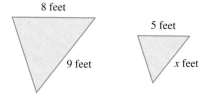

Figure 22 Exercise 27

28. See Fig. 23.

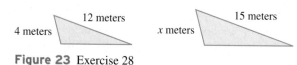

Figure 23 Exercise 28

Related Review

29. If x and y are proportional, then the ratio of x and y is constant. That is,

$$\frac{y}{x} = k$$

where k is a constant.

   a. For the equation $\frac{y}{x} = k$, get y alone on one side of the equation. What is the y-intercept of the graph of your result?

   b. Bicycle shop A charges \$15 per hour to rent a bicycle. Let C be the total cost (in dollars) of renting a bicycle for t hours. Find an equation to describe this situation. Refer to your result in part (a) to help you decide whether t and C are proportional. Explain.

   c. Bicycle shop B charges a flat fee of \$25, plus \$10 per hour, to rent a bicycle. Let C be the total cost (in dollars) of renting a bicycle for t hours. Find an equation to describe this situation. Refer to your result in part (a) to help you decide whether t and C are proportional. Explain.

   d. Find the total cost of renting a bicycle for 2 hours at bicycle shop B. Next, find the total cost of renting a bicycle for 3 hours. Finally, use your work to find two unit ratios to help you determine whether t and C are proportional. Is your conclusion the same as your conclusion in part (c)?

30. A student earns \$72 for working 8 hours at a clothing store.

   a. Use a proportion to find the student's total earnings for working 40 hours.

   b. What is the rate of change of the student's total earnings?

   c. Let E be the student's total earnings (in dollars) for working t hours. Find an equation to describe the situation.

   d. Use your linear model to find the student's total earnings for working 40 hours. Compare the result with the result that you found in part (a).

31. Consider the line shown in Fig. 24.

   a. Use the points A and B to find the slope of the line in terms of m and n.

   b. Use the points C and D to find the slope of the line in terms of p and q.

   c. Use the concept of similar triangles to help you explain why your results in parts (a) and (b) are equal.

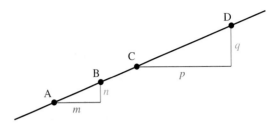

Figure 24 Exercise 31

   d. Explain why your work in parts (a), (b), and (c) shows that no matter which two distinct points on a line are used to find the slope of the line, the result will be the same.

32. Two similar right triangles are shown in Fig. 25. Find the length of the side labeled "x feet." [**Hint:** First use the Pythagorean theorem, and then set up a proportion.]

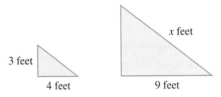

Figure 25 Exercise 32

Expressions, Equations, and Graphs

Perform the indicated instruction. Then use words such as linear, quadratic, cubic, power, radical, rational, polynomial, degree, one variable, *and* two variables *to describe the expression, equation, or system.*

33. Solve $2x^2 + 7x - 15 = 0$.

34. Find the sum $\dfrac{2x}{x-4} + \dfrac{5}{x+3}$.

35. Factor $2x^2 + 7x - 15$.

36. Find the difference $\dfrac{2x}{x-4} - \dfrac{5}{x+3}$.

37. Evaluate $2x^2 + 7x - 15$ for $x = -2$.

38. Find the quotient $\dfrac{2x}{x-4} \div \dfrac{5}{x+3}$.

10.7 VARIATION

The degree of one's emotions varies inversely with one's knowledge of the facts.
—Bertrand Russell

Objectives

▸ Know the meaning of *direct variation* and *inverse variation*.

▸ For direct variation and inverse variation, know how a change in the value of the independent variable affects the value of the dependent variable.

▸ Use a single point to find a direct variation equation or an inverse variation equation.

▸ Use direct variation models and inverse variation models to make estimates.

▸ Use a table of data to find a direct variation equation or an inverse variation equation.

In this section, we will discuss two simple, yet important, types of equations in two variables: *direct variation equations* and *inverse variation equations*.

Direct Variation

Recall from Section 10.6 that two related quantities are proportional if the ratio of the two quantities is constant. For example, x and y are proportional if

$$\frac{y}{x} = k$$

where k is a constant. If we multiply both sides of this equation by x, we have

$$y = kx$$

DEFINITION Direct variation

If $y = kx$ for some constant k, we say that **y varies directly as x** or that **y is proportional to x**. We call k the **variation constant** or the **constant of proportionality**. The equation $y = kx$ is called a **direct variation equation.**

For $y = 3x$, we say that y varies directly as x with variation constant 3. For $w = 8t$, we say that w varies directly as t with variation constant 8.

CHANGES IN VALUES FOR DIRECT VARIATION

A direct variation equation $y = kx$ is a linear equation whose graph has slope k and y-intercept $(0, 0)$. In Fig. 26, we sketch three such equations, with $k = \frac{1}{2}$, $k = 1$, and $k = 2$.

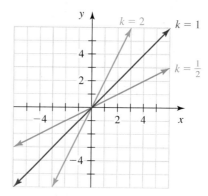

Figure 26 Graphs of $y = kx$ with $k = \frac{1}{2}$, $k = 1$, and $k = 2$

If k is positive (positive slope), then the graph of $y = kx$ is an increasing line. So, if the value of x increases, the value of y increases.

Changes in Values of Variables for Direct Variation

Assume that y varies directly as x for some positive variation constant k.

• If the value of x increases, then the value of y increases.
• If the value of x decreases, then the value of y decreases.

USING ONE POINT TO FIND A DIRECT VARIATION EQUATION

If one variable varies directly as another variable and we know one point that lies on the graph, we can find the variation constant k as well as the direct variation equation.

Example 1 Finding a Direct Variation Equation

The variable y varies directly as x with positive variation constant k.

1. What happens to the value of y as the value of x increases?
2. If $y = 7$ when $x = 2$, find an equation for x and y.

Solution

1. Since y varies directly as x with the positive variation constant k, the value of y increases as the value of x increases.
2. The equation is of the form $y = kx$. We find the constant k by substituting 2 for x and 7 for y:

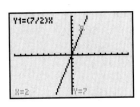

$$y = kx \qquad \text{y varies directly as x.}$$

$$7 = k(2) \qquad \text{Substitute 2 for x and 7 for y.}$$

$$\frac{7}{2} = k \qquad \text{Divide both sides by 2.}$$

Figure 27 Verify the work

The equation is $y = \dfrac{7}{2}x$. We use a graphing calculator graph to check that the curve $y = \dfrac{7}{2}x$ contains the point $(2, 7)$. See Fig. 27. ∎

Example 2 Using a Direct Variation Equation

The variable p varies directly as t. If $p = 6$ when $t = 2$, find the value of t when $p = 10$.

Solution

Since p varies directly as t, we can describe the relationship between t and p by the equation $p = kt$. We find the constant k by substituting 2 for t and 6 for p:

$$p = kt \qquad \text{p varies directly as t.}$$

$$6 = k(2) \qquad \text{Substitute 2 for t and 6 for p.}$$

$$3 = k \qquad \text{Divide both sides by 2.}$$

The equation is $p = 3t$. We find the value of t when $p = 10$ by substituting 10 for p in the equation $p = 3t$ and then solving for t:

$$10 = 3t \qquad \text{Substitute 10 for p.}$$

$$\frac{10}{3} = t \qquad \text{Divide both sides by 3.}$$

So, $t = \dfrac{10}{3}$ when $p = 10$. ∎

DIRECT VARIATION MODELS

Many authentic situations can be modeled well by direct variation equations. For example, when a person drives at a constant speed, the driving time and the distance traveled can be modeled by such an equation.

Example 3 Using a Direct Variation Model

For California residents, the cost of tuition at Solano Community College varies directly as the number of units (hours or credits) a student takes. For spring 2007, the cost of 6 units of classes is $120. Let C be the total cost (in dollars) of u units.

1. As the value of u increases, what happens to the value of C? What does that pattern mean in this situation?
2. Find an equation for u and C.
3. Find the total cost of 15 units of classes.
4. A student can afford to pay at most $275 for classes. How many units of classes can the student enroll in?

Solution

1. Since C varies directly as u, as u increases, the value of C increases. This means that the greater the number of units taken, the greater the total cost will be.

2. Since C varies directly as u, we can model the situation with the equation $C = ku$. We substitute 6 for u and 120 for C to find the constant k:

$$C = ku \qquad \text{C varies directly as u.}$$
$$120 = k(6) \qquad \text{Substitute 6 for u and 120 for C.}$$
$$20 = k \qquad \text{Divide both sides by 6.}$$

The equation is $C = 20u$.

3. We substitute 15 for u into the equation $C = 20u$ and solve for C:

$$C = 20(15) = 300$$

The cost of 15 units is \$300.

4. We substitute 275 for C in the equation $C = 20u$ and solve for u:

$$275 = 20u \qquad \text{Substitute 275 for C.}$$
$$\frac{275}{20} = u \qquad \text{Divide both sides by 20.}$$
$$13.75 = u \qquad \text{Compute.}$$

Even though 7, the first decimal place of 13.75, is more than 5, we still round 13.75 down to 13, because a student cannot enroll in a fraction of a class. The student can afford 13 units of classes. (The cost of 14 units is \$280 [try it], which is more than the student can afford.) We use a graphing calculator table to verify our work (see Fig. 28).

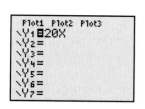

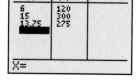

Figure 28 Verify the work

■

Finding and Using a Direct Variation Model

Assume that a quantity p varies directly as a quantity t. To make estimates about an authentic situation,

1. Substitute the values of a data point into the equation $p = kt$ and then solve for k.
2. Substitute the value of k into the equation $p = kt$.
3. Use the equation from step 2 to make estimates of quantity t or quantity p.

A VARIABLE VARYING DIRECTLY AS AN EXPRESSION

So far, we have described equations in which the dependent variable varies directly as the independent *variable*. Here, for constant k, we list some examples of equations in which the dependent variable varies directly as an *expression*:

$$y = kx^2 \qquad \text{y varies directly as } x^2.$$
$$w = kt^3 \qquad \text{w varies directly as } t^3.$$
$$H = k(p + 5) \qquad \text{H varies directly as } p + 5.$$

So, we use "varies directly" to mean that the dependent variable is equal to a constant, times an expression in terms of the independent variable.

Inverse Variation

In Sections 10.1 and 10.5, we worked with rational equations in two variables. Now we will focus on a simple type of rational equation in two variables: an inverse variation equation.

DEFINITION Inverse variation

If $y = \dfrac{k}{x}$ for some constant k, we say that **y varies inversely as x** or that **y is inversely proportional to x**. We call k the **variation constant** or the **constant of proportionality**. The equation $y = \dfrac{k}{x}$ is called an **inverse variation equation**.

For $y = \dfrac{7}{x}$, we say that y varies inversely as x with variation constant 7. For $r = \dfrac{4}{t}$, we say that r varies inversely as t with variation constant 4.

CHANGES IN VALUES FOR INVERSE VARIATION

In Fig. 29, we graph three equations of the form $y = \dfrac{k}{x}$, with $k = 1$, $k = 2$, and $k = 4$.

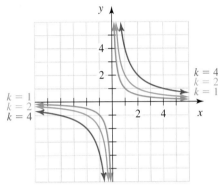

Figure 29 Graphs of $y = \dfrac{k}{x}$ with $k = 1$, $k = 2$, and $k = 4$

These graphs suggest that if k is positive, then the graph of $y = \dfrac{k}{x}$ is a decreasing curve for positive values of x, which is true. So, if the value of x increases, the value of y decreases.

Changes in Values of Variables for Inverse Variation

Assume that y varies inversely as x for some positive variation constant k. For positive values of x,

- If the value of x increases, then the value of y decreases.
- If the value of x decreases, then the value of y increases.

USE ONE POINT TO FIND AN INVERSE VARIATION EQUATION

If one variable varies inversely as another variable and we know one point that lies on the graph, we can find the variation constant k, as well as the inverse variation equation.

Example 4   Finding an Inverse Variation Equation

The variable y varies inversely as x with the positive variation constant k.

1. For positive values of x, what happens to the value of y as the value of x increases?
2. If $y = 3$ when $x = 5$, find an equation for x and y.

Solution

1. Since y varies inversely as x with positive variation constant k and the values of x are positive, the value of y decreases as the value of x increases.

2. The equation is of the form $y = \dfrac{k}{x}$. We find the constant k by substituting 5 for x and 3 for y in the equation $y = \dfrac{k}{x}$:

$$y = \frac{k}{x} \qquad \text{y varies inversely as x.}$$

$$3 = \frac{k}{5} \qquad \text{Substitute 5 for x and 3 for y.}$$

$$15 = k \qquad \text{Multiply both sides by the LCD, 5.}$$

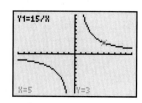

Y1=15/X

X=5 Y=3

Figure 30 Verify the work

The equation is $y = \dfrac{15}{x}$. We use a graphing calculator graph to check that the curve $y = \dfrac{15}{x}$ contains the point (5, 3). See Fig. 30. ∎

Example 5 Using an Inverse Variation Equation

The variable B varies inversely as P. If $B = 7$ when $P = 4$, find the value of P when $B = 5$.

Solution

Since B varies inversely as P, we can describe the relationship between P and B by the equation $B = \dfrac{k}{P}$. We find the constant k by substituting 4 for P and 7 for B:

$$B = \frac{k}{P} \qquad \text{B varies inversely as P.}$$

$$7 = \frac{k}{4} \qquad \text{Substitute 4 for P and 7 for B.}$$

$$28 = k \qquad \text{Multiply both sides by LCD, 4.}$$

The equation is $B = \dfrac{28}{P}$. We find the value of P when $B = 5$ by substituting 5 for B in the equation $B = \dfrac{28}{P}$ and then solving for P:

$$5 = \frac{28}{P} \qquad \text{Substitute 5 for B.}$$

$$5P = 28 \qquad \text{Multiply both sides by LCD, P.}$$

$$P = \frac{28}{5} \qquad \text{Divide both sides by 5.}$$

So, $P = \dfrac{28}{5}$ when $B = 5$. ∎

INVERSE VARIATION MODELS

Many authentic situations can be modeled well by inverse variation equations. For example, the length of a wrench handle and the force you must exert on the handle to loosen a bolt can be modeled by such an equation.

Example 6 Using an Inverse Variation Model

When a stone is tied to a string and whirled in a circle at constant speed, the tension T (in newtons) of the string varies inversely as the radius r (in centimeters) of the circle. If the radius is 50 centimeters, the tension is 96 newtons.

1. Find an equation for r and T.
2. For positive values of r, as the value of r increases, what happens to the value of T? What does that pattern mean in this situation?
3. Find the tension if the radius is 60 centimeters.

Solution

1. Since T varies inversely as r, we can model the situation with the equation $T = \dfrac{k}{r}$. We substitute 50 for r and 96 for T to find the constant k:

$$T = \frac{k}{r} \qquad \text{T varies inversely as r.}$$

$$96 = \frac{k}{50} \qquad \text{Substitute 50 for r and 96 for T.}$$

$$4800 = k \qquad \text{Multiply both sides by LCD, 50.}$$

The equation is $T = \dfrac{4800}{r}$.

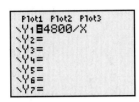

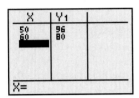

Figure 31 Verify the work

2. Since T varies inversely as r and k is positive, as the value of r increases, the value of T decreases. This means that the larger the radius, the smaller the tension will be.

3. We substitute 60 for r in the equation $T = \dfrac{4800}{r}$ and solve for T:

$$T = \frac{4800}{60} = 80$$

So, the tension is 80 newtons when the radius is 60 centimeters. We use a graphing calculator table to verify our work (see Fig. 31). ∎

Finding and Using an Inverse Variation Model

Assume that a quantity p varies inversely as a quantity t. To make estimates about an authentic situation,

1. Substitute the values of a data point into the equation $p = \dfrac{k}{t}$ and then solve for k.

2. Substitute the value of k into the equation $p = \dfrac{k}{t}$.

3. Use the equation from step 2 to make estimates of quantity t or quantity p.

A VARIABLE VARYING INVERSELY AS AN EXPRESSION

Here, for constant k, we list some examples of equations in which the dependent variable varies inversely as an *expression:*

$$y = \frac{k}{x^2} \qquad \text{y varies inversely as } x^2.$$

$$w = \frac{k}{r^3} \qquad \text{w varies inversely as } r^3.$$

$$p = \frac{k}{t - 2} \qquad \text{p varies inversely as } t - 2.$$

So, we use "varies inversely" to mean that the dependent variable is equal to a constant, divided by an expression in terms of the independent variable.

Example 7 Making an Estimate by Using an Inverse Variation Model

Radiation can be used to treat a tumor. The intensity of radiation varies inversely as the square of the distance from the machine that produces the radiation. If the intensity is 60 milliroentgens per hour (mr/hr) at a distance of 3 meters, at what distance is the intensity 40 mr/hr?

Solution

We let I be the intensity (in mr/hr) at distance d meters from the machine. Our desired model has the form

$$I = \frac{k}{d^2}$$

The denominator of the fraction on the right-hand side is d^2, because the intensity varies inversely as the *square* of the distance d.

Next, we substitute 3 for d and 60 for I and solve for k:

$$60 = \frac{k}{3^2} \qquad \text{Substitute 3 for d and 60 for I.}$$

$$60 = \frac{k}{9} \qquad 3^2 = 9$$

$$540 = k \qquad \text{Multiply both sides by LCD, 9.}$$

The model is $I = \dfrac{540}{d^2}$. Next, we substitute 40 for I and solve for d:

$$40 = \dfrac{540}{d^2} \qquad \text{Substitute 40 for } I.$$

$$40d^2 = 540 \qquad \text{Multiply both sides by } d^2.$$

$$d^2 = 13.5 \qquad \text{Divide both sides by 40.}$$

$$d = \sqrt{13.5} \qquad d \text{ is nonnegative, so disregard } -\sqrt{13.5}.$$

$$d \approx 3.67 \qquad \text{Approximate the square root.}$$

So, the intensity is 40 mr/hr at a distance of about 3.67 meters. ∎

Use a Table of Data to Find an Equation of a Model

If we assume that one quantity varies directly as another quantity, then we need only one point to find the direct variation equation. A similar idea applies to finding an inverse variation equation. However, if we do not make any assumptions about the relationship between two quantities, then we need to have *several* data points so that we can create a scattergram and determine which type of equation will best model the situation.

Table 13 The Lengths of the Telescopes and the Diameters of Vision

Length of Telescope (centimeters) L	Diameter of Vision (centimeters) D
10	70
15	46
20	36
25	29
30	24
35	21
40	19
45	16

Source: *J. Lehmann*

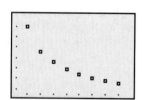

Figure 32 The telescope scattergram

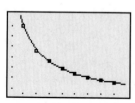

Figure 33 Check the fit

Example 8   Using a Table of Data to Find an Equation of a Model

A person creates "telescopes" by rolling and taping pieces of card stock. He makes eight telescopes that vary in length but have the same radius (4.8 centimeters). He then attaches a tape measure to a wall and extends the tape along the wall. While standing 1.5 meters away from the wall, he records how much of the tape measure he can see (the diameter of vision) through each telescope (see Table 13). Let D be the diameter of vision (in centimeters) at a distance of 1.5 meters for a telescope that is L centimeters long.

1. Create a scattergram of the data. Are the data modeled better by a direct variation equation or an inverse variation equation?
2. Find an equation of a model to describe the data.
3. What is the length of a telescope that has a diameter of vision equal to 26 centimeters (at a distance of 1.5 meters)?

Solution

1. A scattergram is shown in Fig. 32. It appears that the data are better modeled by an inverse variation model.
2. An inverse variation model is of the form $D = \dfrac{k}{L}$. We need only one point to find k. It seems that all of the points in the scattergram fit about the same pattern, so any of the points will likely generate a reasonable model. We choose the point $(30, 24)$ arbitrarily:

$$D = \dfrac{k}{L} \qquad \text{Inverse variation equation}$$

$$24 = \dfrac{k}{30} \qquad \text{Substitute 30 for } L \text{ and 24 for } D.$$

$$720 = k \qquad \text{Multiply both sides by 30.}$$

The equation is $D = \dfrac{720}{L}$. We see in Fig. 33 that the model fits the data quite well.

3. To estimate the length of a telescope with a diameter of vision equal to 26 centimeters (at a distance of 1.5 meters), we substitute 26 for D in the equation $D = \dfrac{720}{L}$ and solve for L:

$$26 = \dfrac{720}{L} \qquad \text{Substitute 26 for } D.$$

$$26L = 720 \qquad \text{Multiply both sides by LCD, } L.$$

$$L = \dfrac{720}{26} \qquad \text{Divide both sides by 26.}$$

$$L \approx 27.69 \qquad \text{Compute.}$$

So, the length of the telescope is about 27.7 centimeters. ■

group exploration

Comparing methods of finding a direct variation equation

In this exploration, you will use three different methods to find an equation for a direct variation model.

1. Suppose that an employee's total earnings E (in dollars) varies directly as the number t of hours worked. The employee earns \$72 for an 8-hour workday.
   a. Use the method discussed in this section to find an equation of a direct variation model to describe the data.
   b. What does the ordered pair $(0, 0)$ mean in this situation? Does that make sense? Use the slope formula $m = \dfrac{E_2 - E_1}{t_2 - t_1}$ with the ordered pairs $(0, 0)$ and $(8, 72)$ to help you find an equation of a model to describe the data.
   c. Use the information that the student earns \$72 for an 8-hour workday to compute the rate of change of earnings per hour. Then use the fact that the slope is a rate of change to help you find an equation of a model to describe the data.

2. Compare the three equations you found in Problem 1. Which method was easiest for you to use?

3. If a variable y varies directly as a variable x, does it follow that the rate of change of y with respect to x is constant? Explain.

TIPS FOR SUCCESS: Form a Study Group to Prepare for the Final Exam

To prepare for your final exam, it may be helpful to form a study group. Your group could list important concepts you all learned in the course and then discuss the meaning of these concepts and how to apply them. You could also list important techniques learned in the course and then practice them. Then set aside some solo study time after the group study session, to make sure that you can do the mathematics without the help of other members of the group.

HOMEWORK 10.7   FOR EXTRA HELP ▶

Student Solutions Manual   PH Math/Tutor Center   Math XL
MathXL®   MyMathLab
MyMathLab

Translate the sentence into an equation.

1. w varies directly as t.   **2.** p varies directly as r.

3. F varies inversely as r.   **4.** H varies inversely as x.

5. c varies directly as s squared.

6. y varies directly as the square root of w.

7. B varies inversely as t cubed.

8. z varies inversely as $d + 3$.

Translate the equation into a sentence by using the phrase "varies directly" or "varies inversely."

9. $H = ku$ **10.** $d = kt$ **11.** $P = \dfrac{k}{V}$

12. $w = \dfrac{k}{t}$ **13.** $E = kc^2$ **14.** $p = k\sqrt{x}$

15. $F = \dfrac{k}{r^2}$ **16.** $T = \dfrac{k}{d^3}$

Find an equation that meets the given conditions.

17. y varies directly as x, and $y = 8$ when $x = 4$.

18. y varies directly as x, and $y = 15$ when $x = 3$.

19. F varies inversely as r, and $F = 2$ when $r = 6$.

20. T varies inversely as d, and $T = 7$ when $d = 3$.

21. H varies directly as $\sqrt{u}$, and $H = 6$ when $u = 4$.

22. p varies directly as t^3, and $p = 40$ when $t = 2$.

23. C varies inversely as x^2, and $C = 2$ when $x = 3$.

24. M varies inversely as $\sqrt{r}$, and $M = 7$ when $r = 25$.

For Exercises 25–32, find the requested value of the variable.

25. If y varies directly as x, and $y = 20$ when $x = 5$, find y when $x = 3$.

26. If T varies directly as W, and $T = 28$ when $W = 4$, find T when $W = 6$.

27. If P varies inversely as V, and $P = 6$ when $V = 3$, find P when $V = 2$.

28. If H varies inversely as t, and $H = 8$ when $t = 4$, find H when $t = 16$.

29. If d varies directly as t^2, and $d = 12$ when $t = 2$, find d when $t = 4$.

30. If y varies directly as w^2, and $y = 32$ when $w = 4$, find w when $y = 32$.

31. If F varies inversely as r^2, and $F = 1$ when $r = 4$, find r when $F = 4$.

32. If I varies inversely as d^2, and $I = 9$ when $d = 2$, find I when $d = 3$.

33. The variable w varies directly as t with positive variation constant k. Describe what happens to the value of w as the value of t increases.

34. The variable p varies directly as r with positive variation constant k. Describe what happens to the value of p as the value of r decreases.

35. The variable z varies inversely as u with positive variation constant k. For positive values of u, what happens to the value of z as the value of u increases?

36. The variable B varies inversely as x with positive variation constant k. For positive values of x, what happens to the value of B as the value of x decreases?

37. The conduction of electricity in a strand of DNA varies directly with humidity for the positive variation constant k. If the humidity is increased, what happens to the conduction of electricity in a strand of DNA?

38. The intensity of the emission of longwave radiation from a cloud varies directly as the temperature for the positive variation constant k. If the temperature increases, what happens to the intensity of the emission of longwave radiation from a cloud?

39. The demand for refinancing home loans varies inversely with the rate of interest for the positive variation constant k. If the interest rate increases, what happens to the demand for refinancing home loans?

40. Relative humidity varies inversely with air temperature for the positive variation constant k. As the air warms, what happens to the relative humidity?

41. One version of "Murphy's law" is that if anything can go wrong, it will. A spinoff is "The number of doors left open varies inversely with the outside temperature." What does this spinoff mean? Why is it a version of Murphy's law?

42. The famous mathematician Bertrand Russell once said, "The degree of one's emotions varies inversely with one's knowledge of the facts." What does this quotation mean?

43. For an Indiana resident, the total cost of tuition at Ivy Tech State College varies directly as the number of credit hours (units) a student takes. For the academic year 2006–2007, the total cost of 12 credit hours is $1053. What is the total cost of 15 credit hours?

44. For a resident of Jackson County, Michigan, the total cost of tuition at Jackson Community College varies directly as the number of billing contact hours (units or credits) a student takes. For the academic year 2006–2007, the total cost of 9 billing contact hours is $706.50. What is the total cost of 12 billing contact hours?

45. The time it takes to fill up a car's gasoline tank varies directly as the total volume of gasoline being pumped. A Chevron® pump takes 81 seconds to pump 10 gallons of gasoline. How much gasoline can be pumped in 104 seconds?

46. A car is traveling at speed s (in mph) on a dry asphalt road, and the brakes are applied forcefully. The stopping distance d (in feet) varies directly as the square of the speed s. If a car traveling at 50 mph can stop in 83 feet, what is the stopping distance of a car traveling at 65 mph?

47. The weight of a pepperoni, mushroom, and garlic pizza varies directly as the square of the radius of the pizza. At Amici's East Coast Pizzeria in San Mateo, California, a pepperoni, mushroom, and garlic pizza with a diameter of 15 inches weighs 32 ounces. What is the weight of a 13-inch-diameter pizza with the same toppings?

48. The distance d that an object falls varies directly as the square of the time the object is in free fall. If an object falls freely for 2 seconds, it will fall 64.4 feet. To estimate the height of a cliff, a person drops a stone at the edge of the cliff and times how long it takes the stone to reach the base of the cliff. If it takes the stone 2.9 seconds to reach the base, what is the height of the cliff?

49. The force you must exert on a wrench handle to loosen a bolt on a bike varies inversely as the length of the handle.
   a. Must a greater force be exerted by a wrench with a short handle or a wrench with a long handle? Explain in terms of inverse variation.
   b. If a force of 40 pounds is needed when the handle is 6 inches long, what force is needed when the handle is 8 inches long?

50. A person plays an electric guitar outside. The sound level (in decibels) varies inversely as the square of the distance from the amplifier. If the sound level is 114 decibels 30 feet from the amplifier, what is the sound level at 40 feet?

51. The brightness of light, in milliwatts per square centimeter $\left(\text{mW/cm}^2\right)$, of a 25-watt light bulb varies inversely as the square of the distance, in centimeters, from the light bulb. If the brightness of light is 0.55 mW/cm^2 at a distance of 90 centimeters, at what distance is the brightness of light equal to 0.26 mW/cm^2?

52. The intensity of a television signal varies inversely as the square of the distance from the transmitter. If the intensity of a television signal is 30 watts per square meter $\left(\text{W/m}^2\right)$ at a distance of 2.6 km, at what distance is the intensity 20 W/m^2?

53. The more you stretch a spring with your hands, the more *force* the spring exerts on your hands. A person ran an experiment comparing the amount of stretch with the forces exerted by the spring (see Table 14).

Table 14 Stretches and Forces of a Spring

Stretch (meters)	Force (newtons)	Stretch (meters)	Force (newtons)
0.000	0.0	0.104	3.0
0.018	0.5	0.121	3.5
0.035	1.0	0.139	4.0
0.052	1.5	0.156	4.5
0.069	2.0	0.173	5.0
0.087	2.5		

Source: *Richard Taylor, The Hockaday School*

Let F be the force (in newtons) of the spring when the spring is stretched x meters.
   a. Create a scattergram of the data.
   b. Find an equation of a model to describe the data.
   c. In a sentence that uses the phrase "varies directly" or "varies inversely," describe how the stretch and force of the spring are related. (This law for springs is known as *Hooke's law.*)
   d. Use the model you found in part (b) to estimate at what stretch the force will be 4.2 newtons.

54. A person finds the total weights of various numbers of identical textbooks (see Table 15). Let w be the total weight (in pounds) of n textbooks.
   a. Draw a scattergram of the data.
   b. Find an equation of a model to describe the data.
   c. In a sentence that uses the phrase "varies directly" or "varies inversely," describe how the number of textbooks and the total weight are related.

Table 15 Total Weights of Identical Textbooks

Number of Textbooks	Total Weight (pounds)
0	0
1	3.5
2	7.1
3	10.7
4	14.1
5	17.6

Source: *J. Lehmann*

   d. Use the model you found in part (b) to estimate the total weight of 6 textbooks.
   e. Explain how you can tell that not all of the points of the scattergram lie on the graph of your model. What does this tell you about the books and/or the scale that the person used to weigh the books?

55. When a sealed syringe is filled with air, it gets harder and harder to squeeze it further the more you squeeze. Some volumes in cubic centimeters $\left(\text{cm}^3\right)$ and corresponding pressures in atmospheres (atm) for a sealed syringe are given in Table 16.

Table 16 Volumes and Pressures in a Syringe

Volume (cm^3)	Pressure (atm)	Volume (cm^3)	Pressure (atm)
3	2.23	13	0.56
5	1.46	15	0.48
7	1.05	17	0.42
9	0.83	19	0.37
11	0.67	20	0.35

Source: *J. Lehmann*

Let P be the pressure (in atm) in the syringe at volume V $\left(\text{in cm}^3\right)$.
   a. Create a scattergram of the data.
   b. Find an equation of a model to describe the data.
   c. In a sentence that uses the phrase "varies directly" or "varies inversely," describe how the volume and pressure are related.
   d. Use the model you found in part (b) to estimate at what volume the pressure will be 0.9 atm.

56. When a guitarist picks the thickest string on a six-string guitar (the "open E" string), the string vibrates at a frequency of 82.4 hertz, which means that it vibrates an average of 82.4 times per second. By using a finger to press the E string firmly against the fret board, the guitarist shortens the effective length of the string. As a result, the frequency increases (and so does the pitch of the note).

Effective String Length

We use some of the letters of the alphabet to refer to these notes. The effective lengths for the open E string and seven other notes for a Fender Squire® guitar and the corresponding frequencies are listed in Table 17.

Table 17 Effective Lengths and Frequencies of Eight Notes on the E String

Note	Effective Length of the E String (inches)	Frequency (in hertz)
E	25.50	82.4
F	24.07	87.3
G	21.44	98.0
A	19.10	110.0
B	17.02	123.5
C	16.06	130.8
D	14.31	146.8
E	12.75	164.8

Source: *J. Lehmann*

Let F be the frequency (in hertz) of the E string when the effective length is L inches.
a. Draw a scattergram of the data.
b. Find an equation of a model to describe the data.
c. In a sentence that uses the phrase "varies directly" or "varies inversely," describe how the effective length and the frequency are related.
d. When the E string is vibrating, what is its frequency if its effective length is 7.58 inches?

57. a. Give an example of an equation in which y varies inversely as x. Using your equation, find any outputs for the given input. State how many outputs there are for that single input.
   i. the input $x = 3$
   ii. the input $x = 5$
   iii. the input $x = -2$
   b. For your equation, how many outputs originate from a single input? Explain.
   c. For *any* equation in which y varies inversely as x, how many outputs originate from a single input? Explain.

58. Describe the meanings of *direct variation equation* and *inverse variation equation*. Compare these two types of equations. Compare some properties of these types of equations.

Related Review

59. a. If y varies directly as x, are x and y linearly related? Explain.
   b. If w and t are linearly related, does w vary directly as t? Explain.

60. a. If a quantity y varies directly as a quantity x, how many data points do you need to find an equation of a model to describe the situation?
   b. If quantities w and t are linearly related, how many data points do you need to find an equation of a model to describe the situation?

61. The total cost of 4 pillows is $50.
   a. Use a proportion to find the total cost of 6 pillows.
   b. Let c be the total cost (in dollars) of n pillows. Find a variation equation to describe this situation.
   c. Use the variation equation you found in part (b) to determine the total cost of 6 pillows. Is the result equal to your result in part (a)?

62. A person's earnings E (in dollars) vary directly as the amount of time t (in hours) that he has worked. The person earns $360 for working 40 hours. What is the slope of the model that describes the situation? What does the slope mean in this situation?

Expressions, Equations, and Graphs

63. Solve $2w^2 - 3w - 5 = 0$.

64. Solve $\dfrac{3}{x-1} + \dfrac{1}{x+2} = \dfrac{5}{x-1}$.

65. Evaluate $2w^2 - 3w - 5$ for $w = -1$.

66. Find the sum $\dfrac{3}{x-1} + \dfrac{1}{x+2}$.

67. Factor $2w^2 - 3w - 5$.

68. Find the difference $\dfrac{3}{x-1} - \dfrac{1}{x+2}$.

10.8 ## SIMPLIFYING COMPLEX RATIONAL EXPRESSIONS

Objectives

▸ Simplify *complex rational expressions*.

In this section, we will work with complex rational expressions. Here are some examples of such expressions:

$$\frac{\dfrac{8}{x}}{\dfrac{3}{x^2}} \qquad \frac{\dfrac{5x}{x+4}}{\dfrac{x^2}{x-1}} \qquad \frac{\dfrac{x}{7} - \dfrac{6}{x}}{\dfrac{x}{5} + \dfrac{3}{x}}$$

A **complex rational expression** is a rational expression whose numerator or denominator (or both) is a rational expression.

Here we find the values of two numerical complex rational expressions:

$$\dfrac{\dfrac{3}{3}}{3} = \dfrac{3}{1} = 3 \qquad \dfrac{\dfrac{3}{3}}{3} = \dfrac{1}{3}$$

From these two examples, we see that it is important to keep track of the main fraction bar (the longest one, in bold) of the complex fraction.

We will discuss two methods for simplifying complex rational expressions. Ask your instructor whether you are required to know method 1, method 2, or both methods. If the choice is yours, compare the use of method 1 in Examples 1 and 2 with the use of method 2 in Examples 3 and 4. The complex rational expressions in these examples are simplified by both methods, so you can get a sense of the advantages and disadvantages of each one.

Method 1: Writing a Complex Rational Expression as a Quotient of Two Rational Expressions

Recall from Section 2.2 that an expression in the form $\dfrac{R}{S}$, where R and S are expressions, can be written in the form $R \div S$. We use this idea to help simplify a complex rational expression:

$$\dfrac{\dfrac{2}{5}}{\dfrac{3}{7}} = \dfrac{2}{5} \div \dfrac{3}{7} \qquad \dfrac{R}{S} = R \div S$$

$$= \dfrac{2}{5} \cdot \dfrac{7}{3} \qquad \text{Multiply by the reciprocal of } \dfrac{3}{7}, \text{ which is } \dfrac{7}{3}.$$

$$= \dfrac{14}{15} \qquad \text{Multiply numerators; multiply denominators.}$$

We **simplify a complex rational expression** by writing it as a rational expression $\dfrac{P}{Q}$, with $\dfrac{P}{Q}$ in lowest terms.

Example 1   Simplifying Complex Rational Expressions by Method 1

Simplify by method 1.

1. $\dfrac{\dfrac{6}{x^3}}{\dfrac{10}{x}}$

2. $\dfrac{\dfrac{x}{x+5}}{\dfrac{3}{x^2-25}}$

Solution

1. $\dfrac{\dfrac{6}{x^3}}{\dfrac{10}{x}} = \dfrac{6}{x^3} \div \dfrac{10}{x} \qquad \dfrac{R}{S} = R \div S$

$$= \dfrac{6}{x^3} \cdot \dfrac{x}{10} \qquad \text{Multiply by reciprocal of } \dfrac{10}{x}, \text{ which is } \dfrac{x}{10}.$$

$$= \dfrac{2 \cdot 3}{x^3} \cdot \dfrac{x}{2 \cdot 5} \qquad \text{Factor numerator and denominator.}$$

$$= \dfrac{2 \cdot 3x}{2 \cdot 5x^3} \qquad \text{Multiply numerators; multiply denominators.}$$

$$= \dfrac{3}{5x^2} \qquad \text{Simplify: } \dfrac{2x}{2x} = 1$$

X	Y1	Y2
0	ERROR	ERROR
1	.6	.6
2	.15	.15
3	.06667	.06667
4	.0375	.0375
5	.024	.024
6	.01667	.01667

X=0

Figure 34 Verify the work

We use a graphing calculator table to verify our work (see Fig. 34). There are "ERROR" messages across from the value $x = 0$, because 0 is an excluded value of the numerator (and the denominator) of the original expression and 0 is an excluded value of the result.

2. $\dfrac{\dfrac{x}{x+5}}{\dfrac{3}{x^2-25}} = \dfrac{x}{x+5} \div \dfrac{3}{x^2-25}$ $\qquad \dfrac{R}{S} = R \div S$

$\qquad = \dfrac{x}{x+5} \cdot \dfrac{x^2-25}{3}$ $\qquad$ Multiply by reciprocal of $\dfrac{3}{x^2-25}$.

$\qquad = \dfrac{x}{x+5} \cdot \dfrac{(x-5)(x+5)}{3}$ $\qquad$ Factor numerator.

$\qquad = \dfrac{x(x-5)(x+5)}{3(x+5)}$ $\qquad$ Multiply numerators; multiply denominators.

$\qquad = \dfrac{x(x-5)}{3}$ $\qquad$ Simplify: $\dfrac{x+5}{x+5} = 1$ ∎

Next we use method 1 to simplify a complex rational expression that has two rational expressions in the numerator and two rational expressions in the denominator.

Example 2   Simplifying a Complex Rational Expression by Method 1

Simplify $\dfrac{\dfrac{1}{a} + \dfrac{2}{a^2}}{\dfrac{3}{4} - \dfrac{1}{a}}$ by method 1.

Solution

We write both the numerator and the denominator as a fraction and then simplify as before:

$\dfrac{\dfrac{1}{a} + \dfrac{2}{a^2}}{\dfrac{3}{4} - \dfrac{1}{a}} = \dfrac{\dfrac{1}{a} \cdot \dfrac{a}{a} + \dfrac{2}{a^2}}{\dfrac{3}{4} \cdot \dfrac{a}{a} - \dfrac{1}{a} \cdot \dfrac{4}{4}}$ $\qquad$ Introduce missing factors to get a common denominator, a^2. Introduce missing factors to get a common denominator, $4a$.

$\qquad = \dfrac{\dfrac{a}{a^2} + \dfrac{2}{a^2}}{\dfrac{3a}{4a} - \dfrac{4}{4a}}$ $\qquad$ Find products.

$\qquad = \dfrac{\dfrac{a+2}{a^2}}{\dfrac{3a-4}{4a}}$ $\qquad \dfrac{A}{B} + \dfrac{C}{B} = \dfrac{A+C}{B}$

$\qquad = \dfrac{a+2}{a^2} \div \dfrac{3a-4}{4a}$ $\qquad \dfrac{R}{S} = R \div S$

$\qquad = \dfrac{a+2}{a^2} \cdot \dfrac{4a}{3a-4}$ $\qquad$ Multiply by reciprocal of $\dfrac{3a-4}{4a}$.

$\qquad = \dfrac{4a(a+2)}{a^2(3a-4)}$ $\qquad$ Multiply numerators; multiply denominators.

$\qquad = \dfrac{4(a+2)}{a(3a-4)}$ $\qquad$ Simplify: $\dfrac{a}{a} = 1$

Since our result is in lowest terms, we are done. ∎

> **Using Method 1 to Simplify a Complex Rational Expression**
>
> To simplify a complex rational expression by method 1,
>
> 1. Write both the numerator and the denominator as fractions.
> 2. To write the complex rational expression as the quotient of two rational expressions, use the property
>
> $$\frac{\dfrac{A}{B}}{\dfrac{C}{D}} = \frac{A}{B} \div \frac{C}{D}, \text{ where } B, C, \text{ and } D \text{ are nonzero}$$
>
> 3. Divide the rational expressions.

Method 2: Multiplying by $\dfrac{\text{LCD}}{\text{LCD}}$

Instead of using method 1, we can simplify a complex rational expression by first finding the LCD of all of the fractions appearing in the numerator and denominator. Then, we multiply by 1 in the form $\dfrac{\text{LCD}}{\text{LCD}}$.

In Example 3, we simplify the same complex rational expressions that we simplified in Example 1, but now we use method 2.

Example 3　Simplifying Complex Rational Expressions by Method 2

Simplify by method 2.

1. $\dfrac{\dfrac{6}{x^3}}{\dfrac{10}{x}}$　　　　　　　　**2.** $\dfrac{\dfrac{x}{x+5}}{\dfrac{3}{x^2-25}}$

Solution

1. The LCD of $\dfrac{6}{x^3}$ and $\dfrac{10}{x}$ is x^3. So, we multiply the complex rational expression by 1 in the form $\dfrac{x^3}{x^3}$:

$$\frac{\dfrac{6}{x^3}}{\dfrac{10}{x}} = \frac{\dfrac{6}{x^3} \cdot \dfrac{x^3}{1}}{\dfrac{10}{x} \cdot \dfrac{x^3}{1}} \qquad \text{Multiply by } \frac{\text{LCD}}{\text{LCD}}, \frac{x^3}{x^3} = 1.$$

$$= \frac{\dfrac{6x^3}{x^3}}{\dfrac{10x^3}{x}} \qquad \text{Simplify.}$$

$$= \frac{6}{10x^2} \qquad \text{Simplify fractions in numerator and in denominator.}$$

$$= \frac{2 \cdot 3}{2 \cdot 5x^2} \qquad \text{Factor numerator and denominator.}$$

$$= \frac{3}{5x^2} \qquad \text{Simplify: } \frac{2}{2} = 1$$

The result is the same as our result in Problem 1 of Example 1.

2. $\dfrac{\dfrac{x}{x+5}}{\dfrac{3}{x^2-25}} = \dfrac{\dfrac{x}{x+5}}{\dfrac{3}{(x-5)(x+5)}}$ Factor numerators and denominators of fractions.

$$= \dfrac{\dfrac{x}{x+5}}{\dfrac{3}{(x-5)(x+5)}} \cdot \dfrac{\dfrac{(x-5)(x+5)}{1}}{\dfrac{(x-5)(x+5)}{1}}$$ Multiply by $\dfrac{\text{LCD}}{\text{LCD}}, \dfrac{(x-5)(x+5)}{(x-5)(x+5)}$.

$$= \dfrac{\dfrac{x(x-5)(x+5)}{x+5}}{\dfrac{3(x-5)(x+5)}{(x-5)(x+5)}}$$ $\dfrac{A}{B} \cdot \dfrac{C}{D} = \dfrac{AC}{BD}$

$$= \dfrac{x(x-5)}{3}$$ Simplify: $\dfrac{x+5}{x+5} = 1$, $\dfrac{(x-5)(x+5)}{(x-5)(x+5)} = 1$

The result is the same as our result in Problem 2 of Example 1.   ■

In comparing our work in Examples 1 and 3, we see that methods 1 and 2 required about the same number of steps. One advantage that method 1 has over method 2 for *these* complex rational expressions is that method 1 does not require us to find the LCD.

In Example 4, we simplify the same expression as in Example 2, but now we use method 2.

Example 4   Simplifying a Complex Rational Expression by Method 2

Simplify $\dfrac{\dfrac{1}{a} + \dfrac{2}{a^2}}{\dfrac{3}{4} - \dfrac{1}{a}}$ by method 2.

Solution

The LCD of the rational expressions in the numerator and the denominator is $4a^2$. To simplify, we multiply by $\dfrac{4a^2}{4a^2}$:

$$\dfrac{\dfrac{1}{a} + \dfrac{2}{a^2}}{\dfrac{3}{4} - \dfrac{1}{a}} = \dfrac{\dfrac{1}{a} + \dfrac{2}{a^2}}{\dfrac{3}{4} - \dfrac{1}{a}} \cdot \dfrac{4a^2}{4a^2}$$ Multiply by $\dfrac{\text{LCD}}{\text{LCD}}, \dfrac{4a^2}{4a^2}$.

$$= \dfrac{\dfrac{1}{a} \cdot \dfrac{4a^2}{1} + \dfrac{2}{a^2} \cdot \dfrac{4a^2}{1}}{\dfrac{3}{4} \cdot \dfrac{4a^2}{1} - \dfrac{1}{a} \cdot \dfrac{4a^2}{1}}$$ Distributive law

$$= \dfrac{4a + 8}{3a^2 - 4a}$$ Simplify.

$$= \dfrac{4(a+2)}{a(3a-4)}$$ Factor numerator and denominator.

Note that the result is the same as our result in Example 2.   ■

WARNING For the first step in Example 4, it would be incorrect to multiply by the fraction

$$\frac{\text{LCD of the numerator}}{\text{LCD of the denominator}} = \frac{a^2}{4a}:$$

$$\frac{\dfrac{1}{a} + \dfrac{2}{a^2}}{\dfrac{3}{4} - \dfrac{1}{a}} = \frac{\dfrac{1}{a} + \dfrac{2}{a^2}}{\dfrac{3}{4} - \dfrac{1}{a}} \cdot \frac{a^2}{4a} \qquad \text{Incorrect}$$

This is incorrect because the expression $\dfrac{a^2}{4a}$ is not equivalent to 1. It *is* correct to multiply by $\dfrac{4a^2}{4a^2} = 1$.

In comparing our work in Examples 2 and 4, we see that method 2 required fewer steps than method 1. In general, when the numerator, denominator, or both contain two rational expressions, method 2 is more efficient.

Using Method 2 to Simplify a Complex Rational Expression

To simplify a rational expression by method 2,

1. Find the LCD of all of the fractions appearing in the numerator and denominator.
2. Multiply by 1 in the form $\dfrac{\text{LCD}}{\text{LCD}}$.
3. Simplify the numerator and the denominator to polynomials.
4. Simplify the rational expression.

Example 5 Simplifying a Complex Rational Expression by Method 2

Simplify $\dfrac{3 - \dfrac{12}{x^2}}{2 - \dfrac{4}{x}}$ by method 2.

Solution

The LCD of the rational expressions in the numerator and the denominator is x^2. To simplify, we multiply by $\dfrac{x^2}{x^2}$:

$$\frac{3 - \dfrac{12}{x^2}}{2 - \dfrac{4}{x}} = \frac{3 - \dfrac{12}{x^2}}{2 - \dfrac{4}{x}} \cdot \frac{x^2}{x^2} \qquad \text{Multiply by } \frac{\text{LCD}}{\text{LCD}}, \frac{x^2}{x^2}.$$

$$= \frac{3x^2 - \dfrac{12}{x^2} \cdot \dfrac{x^2}{1}}{2x^2 - \dfrac{4}{x} \cdot \dfrac{x^2}{1}} \qquad \text{Distributive law}$$

$$= \frac{3x^2 - 12}{2x^2 - 4x} \qquad \text{Simplify.}$$

$$= \frac{3(x^2 - 4)}{2x(x - 2)} \qquad \text{Factor numerator and denominator.}$$

$$= \frac{3(x - 2)(x + 2)}{2x(x - 2)} \qquad \text{Factor numerator.}$$

$$= \frac{3(x + 2)}{2x} \qquad \text{Simplify: } \frac{x - 2}{x - 2} = 1$$

■

group exploration

Looking ahead: Combining like radicals

Recall from Section 4.2 that we can use the distributive law to combine like terms. For example, here we add the like terms $5x$ and $3x$: $5x + 3x = (5 + 3)x = 8x$. We can also use the distributive law to "combine" some radicals.

1. Use the distributive law to perform the operations.
   a. $5\sqrt{x} + 3\sqrt{x}$ **b.** $9\sqrt{2} - 6\sqrt{2}$
   c. $4\sqrt{7} + \sqrt{7} - 2\sqrt{7}$

2. Can you use the distributive law to "combine" $4\sqrt{3}$ and $6\sqrt{7}$ in the sum $4\sqrt{3} + 6\sqrt{7}$? If yes, show how. If no, explain why not.

3. Can you use the distributive law to "combine" $8\sqrt{2}$ and $3\sqrt{5}$ in the difference $8\sqrt{2} - 3\sqrt{5}$? If yes, show how. If no, explain why not.

4. Describe in general when you can use the distributive law to "combine" radicals in a sum or difference of two radicals.

5. Use the distributive law to perform the operations. [**Hint:** To begin, simplify the radicals.]
   a. $\sqrt{8} + \sqrt{18}$ **b.** $\sqrt{27} - \sqrt{12}$

HOMEWORK 10.8

FOR EXTRA HELP ▶

Student Solutions Manual   PH Math/Tutor Center   *Math XL* MathXL®   *MyMathLab* MyMathLab

Simplify.

1. $\dfrac{\frac{4}{5}}{\frac{8}{3}}$

2. $\dfrac{\frac{3}{4}}{\frac{5}{6}}$

3. $\dfrac{\frac{x}{4}}{\frac{x}{7}}$

4. $\dfrac{\frac{x}{5}}{x}$

5. $\dfrac{\frac{5x}{3}}{7x}$

6. $\dfrac{\frac{3x}{8}}{5x}$

7. $\dfrac{\frac{w}{6}}{\frac{w^2}{9}}$

8. $\dfrac{\frac{y}{12}}{y^2}$

9. $\dfrac{\frac{15x}{8}}{\frac{25x^3}{12}}$

10. $\dfrac{\frac{6x^3}{21}}{4x^2}$

11. $\dfrac{\frac{x^2-9}{21}}{x-3}$

12. $\dfrac{\frac{8}{x^2-16}}{\frac{20}{x+4}}$

13. $\dfrac{\frac{5a-10}{4}}{\frac{3a-6}{2}}$

14. $\dfrac{\frac{3t-12}{14}}{\frac{4t-16}{21}}$

15. $\dfrac{\frac{x^2-2x-15}{6x}}{\frac{x-5}{10}}$

16. $\dfrac{\frac{x^2+3x-28}{16x}}{\frac{x+7}{12}}$

17. $\dfrac{\frac{4}{x}+\frac{2}{x}}{\frac{9}{x}-\frac{7}{x}}$

18. $\dfrac{\frac{5}{x}-\frac{2}{x}}{\frac{3}{x}+\frac{4}{x}}$

19. $\dfrac{\frac{3}{8}+\frac{1}{4}}{\frac{1}{2}+\frac{5}{8}}$

20. $\dfrac{\frac{2}{3}+\frac{5}{6}}{\frac{7}{12}-\frac{1}{6}}$

21. $\dfrac{\frac{7}{4x}+\frac{1}{x}}{\frac{3}{2x}+\frac{5}{x}}$

22. $\dfrac{\frac{2}{x}+\frac{5}{3x}}{\frac{8}{5x}-\frac{1}{x}}$

23. $\dfrac{\frac{5}{3r}-\frac{3}{2r}}{\frac{1}{2r}-\frac{4}{3r}}$

24. $\dfrac{\frac{7}{2t}+\frac{3}{5t}}{\frac{3}{5t}+\frac{5}{2t}}$

25. $\dfrac{\frac{2}{3}-\frac{4}{3x}}{\frac{5}{6x}}$

26. $\dfrac{\frac{5}{4}-\frac{7}{4x}}{\frac{3}{8x}}$

27. $\dfrac{2+\frac{5}{x}}{4-\frac{1}{x}}$

28. $\dfrac{3-\frac{7}{x}}{5-\frac{6}{x}}$

29. $\dfrac{\frac{2}{b}-\frac{3}{b^2}}{\frac{4}{b}+\frac{5}{b^2}}$

30. $\dfrac{\frac{8}{w}+\frac{5}{w^2}}{\frac{6}{w}-\frac{3}{w^2}}$

31. $\dfrac{\frac{1}{x}-\frac{4}{x^3}}{\frac{1}{x}-\frac{2}{x^2}}$

32. $\dfrac{\frac{1}{x}-\frac{3}{x^2}}{\frac{1}{x}-\frac{9}{x^3}}$

33. $\dfrac{\frac{3}{4x}+\frac{1}{x^2}}{\frac{5}{2x}+\frac{1}{x^2}}$

34. $\dfrac{\frac{4}{3x}-\frac{1}{x^3}}{\frac{3}{2x}+\frac{2}{x^2}}$

35. $\dfrac{\frac{r}{3}-\frac{3}{r}}{\frac{r}{2}-\frac{2}{r}}$

36. $\dfrac{\frac{2}{a}-\frac{a}{2}}{\frac{a}{5}-\frac{5}{a}}$

37. $\dfrac{2-\frac{8}{x^2}}{1-\frac{2}{x}}$

38. $\dfrac{1-\frac{9}{x^2}}{2-\frac{6}{x}}$

39. A student tries to simplify a complex rational expression:

$$\frac{x}{\dfrac{1}{x}+\dfrac{1}{3}} = x \div \left(\frac{1}{x}+\frac{1}{3}\right)$$

$$= x \cdot \left(\frac{x}{1}+\frac{3}{1}\right)$$

$$= x \cdot (x+3)$$

$$= x^2 + 3x$$

Describe any errors. Then simplify the expression correctly.

40. A student tries to simplify a complex rational expression:

$$\frac{\dfrac{2}{x}-\dfrac{7}{x^2}}{\dfrac{4}{3}+\dfrac{5}{x}} = \frac{\dfrac{2}{x}-\dfrac{7}{x^2}}{\dfrac{4}{3}+\dfrac{5}{x}} \cdot \frac{x^2}{3x}$$

$$= \frac{\dfrac{2}{x}\cdot\dfrac{x^2}{1}-\dfrac{7}{x^2}\cdot\dfrac{x^2}{1}}{\dfrac{4}{3}\cdot\dfrac{3x}{1}+\dfrac{5}{x}\cdot\dfrac{3x}{1}}$$

$$= \frac{2x-7}{4x+15}$$

Describe any errors. Then simplify the expression correctly.

41. Describe a complex rational expression. Give an example. Then simplify your complex rational expression.

42. Describe how to simplify a complex rational expression.

Related Review

Simplify.

43. $\dfrac{x^2-9}{3x+15} \div \dfrac{x-3}{2x+10}$

44. $\dfrac{3x-6}{x^2+x-12} \div \dfrac{4x-8}{x+4}$

45. $\dfrac{\dfrac{x^2-9}{3x+15}}{\dfrac{x-3}{2x+10}}$

46. $\dfrac{\dfrac{3x-6}{x^2+x-12}}{\dfrac{4x-8}{x+4}}$

47. $\dfrac{2x^{-1}}{5\cdot3^{-1}}$

48. $\dfrac{4\cdot2^{-1}}{3x^{-1}}$

49. $\dfrac{2+x^{-1}}{5+3^{-1}}$ [**Hint:** First write as a complex rational expression.]

50. $\dfrac{4+2^{-1}}{3+x^{-1}}$ [**Hint:** First write as a complex rational expression.]

Expressions, Equations, and Graphs

Give an example of the following. Then solve, simplify, or graph, as appropriate.

51. rational equation in one variable

52. linear equation in one variable

53. system of two linear equations in two variables

54. quadratic equation in two variables

55. quadratic equation in one variable that can be solved by factoring

56. linear equation in two variables

57. quadratic equation in one variable that can't be solved by factoring

58. rational expression in one variable that is not in lowest terms

59. linear expression with four terms

60. quadratic expression with four terms

analyze this

GLOBAL WARMING LAB (continued from Chapter 9)

In previous Global Warming Labs, we found various models related to carbon emissions and human populations. We can use these models to find meaningful ratios and percentages of quantities.

As a warm-up activity, suppose that a student takes a quiz, correctly answering 15 out of 20 questions. To compute the percentage of questions answered correctly, we divide the number of questions answered correctly by the total number of questions and multiply the result by 100:

$$\text{Percentage of correct answers} = \frac{15}{20} \cdot 100 = 75\%$$

We will now take similar steps to find a model that describes the percentage of the world population that lives in the United States. Let u be the U.S. population and w be the world population (both in billions), both at t years since

1950. Here are some reasonable models:

$$u = 0.0024t + 0.16 \qquad \text{United States}$$

$$w = -0.000531t^2 + 0.136t + 0.63 \qquad \text{World}$$

Recall from the Global Warming Lab in Chapter 7 that the world population model works well only for the years 2000–2100.

To find a model of the percentage p of the world population that is in the United States at t years since 1950, we divide the right-hand side of the U.S. population model by the right-hand side of the world population model and multiply the result by 100:

$$p = \frac{0.0024t + 0.16}{-0.000531t^2 + 0.136t + 0.63} \cdot 100$$

$$= \frac{0.24t + 16}{-0.000531t^2 + 0.136t + 0.63}$$

We can this use this model to help us understand the situation better, which you will do in Problem 1 of this lab.

Analyzing the Situation

1. Use a graphing calculator to create a table of values of t and p for the percentage model

$$p = \frac{0.24t + 16}{-0.000531t^2 + 0.136t + 0.63}$$

Your table should contain the values 50, 60, 70, 80, 90, and 100 for t. What do your results tell you about this situation?

2. A model of U.S. carbon emissions c (in billions of metric tons) in the year that is t years since 1950 is $c = 0.017t + 0.68$. A model of the U.S. population u (in billions) at t years since 1950 is $u = 0.0024t + 0.16$. Find a model of the U.S. per-person carbon emissions y (in metric tons per person) in the year that is t years since 1950. [**Hint:** Find a quotient. You do *not* have to multiply the fraction by 100, because you are not trying to describe a percentage.] Perform a unit analysis of that model.

3. Use a graphing calculator to create a table of values of t and y for the model you found in Problem 2. Your table should contain the values 50, 60, 70, 80, 90, and 100 for t. What do your results tell you about this situation?

4. A student takes a quiz, answering 24 out of 30 questions correctly. What percentage of the questions did the student get right?

5. Let u be U.S. carbon emissions and w be world carbon emissions (both in billions of metric tons), both in the year that is t years since 1950. Here are some reasonable models:

$$u = 0.017t + 0.68 \quad \text{United States}$$
$$w = 0.11t + 1.68 \quad \text{World}$$

Find a model of the percentage p of worldwide carbon emissions emitted in the United States in the year that is t years since 1950.

6. Use a graphing calculator to create a table of values for t and p for the percentage model you found in Problem 5. Your table should contain the values 50, 60, 70, 80, 90, and 100 for t. What do your results tell you about this situation?

7. Explain how it is possible for the U.S. share of annual worldwide carbon emissions to be declining even though U.S. per-person annual carbon emissions is increasing. [**Hint:** There may be two reasons. Consider your result in Problem 1. Consider also what is happening in developing countries.]

ESTIMATING π LAB

One of the great moments in early mathematics was the discovery of π and its usefulness. Long before there were calculators, many great mathematicians tried to estimate π. In this lab, you will build on your work from the Volume Lab in Chapter 1 to find an estimate of π. If you haven't done that lab yet, do it now. Then respond to the questions that follow.

Analyzing the Situation

1. Recall from the Volume Lab in Chapter 1 that V is the volume of water (in ounces) in a cylinder when the height of the water in the cylinder is h centimeters (cm). In the Volume Lab, you likely sketched a linear model that contains the origin $(0, 0)$. If not, sketch a new linear model that does.

2. Find an equation of a model to describe the data.

3. What does the slope mean in this situation? Make sure that you include units in your description.

4. There are 29.57353 cm^3 in one ounce. Use this fact to help you find the slope of the model, in units of cm^2.

5. The formula for the volume of a cylinder is $V = \pi r^2 h$, where r is the cylinder's radius and h is the cylinder's height. In Problem 2, you found an equation of the form $V = mh$, where m is the slope. Explain why we can conclude that $m = \pi r^2$.

6. Substitute the slope of your model for m and the radius of the cylinder for r in the equation $m = \pi r^2$. Then treat π as if it is an unknown constant and solve for it. If you are using the data provided in the Volume Lab, the base radius of the cylinder is 4.45 cm. What is your estimate of π?

7. Use a calculator to help you find the error in your estimate of π.

Chapter Summary

Key Points
OF CHAPTER 10

Throughout these key points, assume that A, B, C, and D are polynomials.

Rational expression (Section 10.1)	If P and Q are polynomials with Q nonzero, we call the ratio $\dfrac{P}{Q}$ a **rational expression.**
Excluded value (Section 10.1)	A number is an **excluded value** of a rational expression if substituting the number into the expression leads to a division by 0.

Finding excluded values
(Section 10.1)

To find any excluded values of a rational expression $\dfrac{P}{Q}$,

1. Set the denominator Q equal to 0.

2. Solve the equation $Q = 0$.

Lowest terms and simplify a rational expression (Section 10.1)

A rational expression is in **lowest terms** if the numerator and denominator have no common factors other than 1 or -1. We **simplify a rational expression** by writing it in lowest terms.

Simplifying a rational expression
(Section 10.1)

To simplify a rational expression,

1. Factor the numerator and the denominator.

2. Use the property $\dfrac{AB}{AC} = \dfrac{A}{A} \cdot \dfrac{B}{C} = 1 \cdot \dfrac{B}{C} = \dfrac{B}{C}$, where A and C are nonzero, so that the expression is in lowest terms.

Rational equation in two variables
(Section 10.1)

A **rational equation in two variables** is an equation in two variables in which each side can be written as a rational expression.

Rational model (Section 10.1)

A **rational model** is a rational equation in two variables that describes the relationship between two quantities for an authentic situation.

Multiplying rational expressions
(Section 10.2)

If $\dfrac{A}{B}$ and $\dfrac{C}{D}$ are rational expressions and B and D are nonzero, then $\dfrac{A}{B} \cdot \dfrac{C}{D} = \dfrac{AC}{BD}$.

How to multiply rational expressions (Section 10.2)

To multiply two rational expressions,

1. Factor the numerators and the denominators.

2. Multiply by using the property $\dfrac{A}{B} \cdot \dfrac{C}{D} = \dfrac{AC}{BD}$, where B and D are nonzero.

3. Simplify the result.

Dividing rational expressions
(Section 10.2)

If $\dfrac{A}{B}$ and $\dfrac{C}{D}$ are rational expressions and B, C, and D are nonzero, then $\dfrac{A}{B} \div \dfrac{C}{D} = \dfrac{A}{B} \cdot \dfrac{D}{C}$.

How to divide rational expressions
(Section 10.2)

To divide two rational expressions,

1. Write the quotient as a product by using the property $\dfrac{A}{B} \div \dfrac{C}{D} = \dfrac{A}{B} \cdot \dfrac{D}{C}$, where B, C, and D are nonzero.

2. Find the product.

3. Simplify.

Converting units (Section 10.2)

To convert the units of a quantity,

1. Write the quantity in the original units.

2. Multiply by fractions equal to 1 so that the units you want to eliminate appear in one numerator and one denominator.

Adding rational expressions with a common denominator
(Section 10.3)

If $\dfrac{A}{B}$ and $\dfrac{C}{B}$ are rational expressions, where B is nonzero, then $\dfrac{A}{B} + \dfrac{C}{B} = \dfrac{A+C}{B}$.

Subtracting rational expressions with a common denominator
(Section 10.4)

If $\dfrac{A}{B}$ and $\dfrac{C}{B}$ are rational expressions, where B is nonzero, then $\dfrac{A}{B} - \dfrac{C}{B} = \dfrac{A-C}{B}$.

How to add or subtract two rational expressions with different denominators (Sections 10.3 and 10.4)

To add or subtract two rational expressions with different denominators,

1. Factor the denominators of the expressions if possible. Determine which factors are missing.

2. Use the property $\dfrac{A}{A} = 1$, where A is nonzero, to introduce missing factors.

3. Add the expressions by using the property $\dfrac{A}{B} + \dfrac{C}{B} = \dfrac{A+C}{B}$, where B is nonzero; or subtract the expressions by using the property $\dfrac{A}{B} - \dfrac{C}{B} = \dfrac{A-C}{B}$, where B is nonzero.

4. Simplify.

Subtract entire numerator
(Section 10.4)

When subtracting rational expressions, be sure to subtract the *entire* numerator.

Rational equation in one variable
(Section 10.5)

A **rational equation in one variable** is an equation in one variable in which each side can be written as a rational expression.

Solving a rational equation in one variable (Section 10.5)	To solve a rational equation in one variable,
	1. Factor the denominator(s) if possible.
	2. Identify any excluded values.
	3. Find the LCD of all of the fractions.
	4. Multiply both sides of the equation by the LCD, which gives a simpler equation to solve.
	5. Solve the simpler equation.
	6. Discard any proposed solutions that are excluded values.
Solving a rational equation versus simplifying a rational expression (Section 10.5)	To solve a rational equation, clear the fractions in it by multiplying both sides of the equation by the LCD. To simplify a rational expression, do *not* multiply it by the LCD. The only multiplication permissible is multiplication by 1, usually in the form $\dfrac{A}{A}$, where A is a nonzero polynomial.
Results of solving rational equations and simplifying rational expressions (Section 10.5)	The result of solving a rational equation is the empty set or a set of one or more numbers. The result of simplifying a rational expression is an expression.
Proportion (Section 10.6)	A **proportion** is a statement of the equality of two ratios.
Proportional (Section 10.6)	If the ratio of two related quantities is constant, we say that the two quantities are **proportional.**
Similar triangles (Section 10.6)	Two triangles are called **similar triangles** if their corresponding angles are equal in measure.
Side lengths of similar triangles (Section 10.6)	The lengths of the corresponding sides of two similar triangles are proportional.
Direct variation (Section 10.7)	If $y = kx$ for some constant k, we say that y **varies directly as x** or that y **is proportional to x.** We call k the **variation constant** or the **constant of proportionality.** The equation $y = kx$ is called a **direct variation equation.**
Changes in values of variables for direct variation (Section 10.7)	Assume that y varies directly as x for some positive variation constant k.
	• If the value of x increases, then the value of y increases.
	• If the value of x decreases, then the value of y decreases.
Finding and using a direct variation model (Section 10.7)	Assume that a quantity p varies directly as a quantity t. To make estimates about an authentic situation,
	1. Substitute the values of a data point into the equation $p = kt$ and then solve for k.
	2. Substitute the value of k into the equation $p = kt$.
	3. Use the equation from step 2 to make estimates of quantity t or quantity p.
Inverse variation (Section 10.7)	If $y = \dfrac{k}{x}$ for some constant k, we say that y **varies inversely as x** or that y **is inversely proportional to x.** We call k the **variation constant** or the **constant of proportionality.** The equation $y = \dfrac{k}{x}$ is called an **inverse variation equation.**
Changes in values of variables for inverse variation (Section 10.7)	Assume that y varies inversely as x for some positive variation constant k. For positive values of x,
	• If the value of x increases, then the value of y decreases.
	• If the value of x decreases, then the value of y increases.
Finding and using an inverse variation model (Section 10.7)	Assume that a quantity p varies inversely as a quantity t. To make estimates about an authentic situation,
	1. Substitute the values of a data point into the equation $p = \dfrac{k}{t}$ and then solve for k.
	2. Substitute the value of k into the equation $p = \dfrac{k}{t}$.
	3. Use the equation from step 2 to make estimates of quantity t or quantity p.
Complex rational expression (Section 10.8)	A **complex rational expression** is a rational expression whose numerator or denominator (or both) is a rational expression.
Using method 1 to simplify a complex rational expression (Section 10.8)	To simplify a complex rational expression by method 1,
	1. Write both the numerator and the denominator as fractions.

2. To write the complex rational expression as the quotient of two rational expressions, use the property

$$\frac{\dfrac{A}{B}}{\dfrac{C}{D}} = \frac{A}{B} \div \frac{C}{D}, \text{ where } B, C, \text{ and } D \text{ are nonzero}$$

3. Divide the rational expressions.

Using method 2 to simplify a complex rational expression (Section 10.8)

To simplify a complex rational expression by method 2,

1. Find the LCD of all of the fractions appearing in the numerator and denominator.
2. Multiply by 1 in the form $\dfrac{\text{LCD}}{\text{LCD}}$.
3. Simplify the numerator and the denominator to polynomials.
4. Simplify the rational expression.

CHAPTER 10 REVIEW EXERCISES

Find any excluded values.

1. $\dfrac{4}{x}$

2. $\dfrac{x}{7}$

3. $\dfrac{9}{3x-5}$

4. $\dfrac{x-5}{x^2-6x+8}$

5. $\dfrac{5t-6}{t^2+6t+9}$

6. $\dfrac{7w}{4w^2-9}$

7. $\dfrac{2x+3}{5x^2+18x-8}$

8. $\dfrac{x+9}{2x^3+5x^2-18x-45}$

Simplify.

9. $\dfrac{28x^3y^5}{35x^7y^2}$

10. $\dfrac{7x+21}{x^2+x-6}$

11. $\dfrac{w^2+7w+12}{w^2-16}$

12. $\dfrac{2a^2-2ab}{a^2-b^2}$

13. $\dfrac{x^2-8x+12}{3x^2-16x-12}$

14. $\dfrac{x^2+10x+25}{2x^3-3x^2-50x+75}$

Perform the indicated operation. Simplify the result.

15. $\dfrac{m-4}{m+3} \cdot \dfrac{m+2}{m-3}$

16. $\dfrac{25b^3}{b^2-b} \cdot \dfrac{b^2-1}{35b}$

17. $\dfrac{x^3+8x^2+16x}{x^2-10x+21} \cdot \dfrac{4x-12}{3x+12}$

18. $\dfrac{-3x+12}{x^2-1} \cdot \dfrac{-2x-2}{x^2-16}$

19. $\dfrac{5x-15}{6x^5} \div \dfrac{2x-6}{4x^2}$

20. $\dfrac{9x^2-36}{3x+3} \div \dfrac{2x^2-6x-20}{2x^2+7x+5}$

21. $\dfrac{7t+14}{t-7} \div (3t^2+2t-8)$

22. $\dfrac{a^2-9b^2}{4a^2-8ab} \div \dfrac{a^2-4ab-21b^2}{a^2-4ab+4b^2}$

23. $\dfrac{x^2}{x+7} + \dfrac{5x-14}{x+7}$

24. $\dfrac{3}{4x} + \dfrac{5}{6x^3}$

25. $\dfrac{5}{x^2+7x+6} + \dfrac{2x}{x^2-3x-4}$

26. $\dfrac{x+4}{x-2} + \dfrac{2}{x^2+4x-12}$

27. $\dfrac{p^2}{p^2-4} - \dfrac{4p+12}{p^2-4}$

28. $\dfrac{x-4}{x^2+2x-3} - \dfrac{x+2}{x^2-6x+5}$

29. $\dfrac{2xy}{x^2-10xy+25y^2} - \dfrac{2y}{x-5y}$

30. $\dfrac{y}{y^3+64} - \dfrac{3}{y^2+4y}$

31. Find the product of $\dfrac{8x^2+4x}{9x^4}$ and $\dfrac{12x^9}{10x+5}$.

32. Find the quotient of $\dfrac{x^2-36}{x^2+13x+36}$ and $\dfrac{x^2+12x+36}{9x+36}$.

33. Find the sum of $\dfrac{5m}{m^2+4m-45}$ and $\dfrac{3}{m^2+6m-27}$.

34. Find the difference of $\dfrac{c-5}{c+2}$ and $\dfrac{c+3}{c-1}$.

35. A student tries to find the difference $\dfrac{2x}{x+4} - \dfrac{7x-3}{x+4}$:

$$\dfrac{2x}{x+4} - \dfrac{7x-3}{x+4} = \dfrac{2x-7x-3}{x+4}$$
$$= \dfrac{-5x-3}{x+4}$$

Describe any errors. Then find the difference correctly.

For Exercises 36 and 37, round approximate results to two decimal places. Refer to the list of equivalent units in the margin of p. 510 as needed.

36. A TI-84 graphing calculator weighs 0.62 pound. What is the calculator's weight in ounces?

37. Americans consume an average of 121 gallons of water per year (Source: *Wirthlin Worldwide*). What is this average in cups per day?

Solve.

38. $\dfrac{3x-2}{x+4} = \dfrac{9}{x+4}$

39. $\dfrac{5}{2x} - \dfrac{4}{x} = \dfrac{3}{2}$

40. $\dfrac{3}{r-1} + \dfrac{2}{4r-4} = \dfrac{7}{4}$

41. $\dfrac{3}{t-2} + \dfrac{2}{t+2} = \dfrac{4}{t^2-4}$

42. $\dfrac{5}{x-4} - \dfrac{2}{x-7} = \dfrac{4}{x^2 - 11x + 28}$

43. $\dfrac{3}{x+3} - \dfrac{2}{x-3} = \dfrac{-12}{x^2 - 9}$

44. $\dfrac{4}{x} = \dfrac{3}{x^2} - 2$

45. $\dfrac{-3}{x+6} + \dfrac{2}{x+1} = \dfrac{x-2}{x+6}$

46. Textbook sales and college enrollments are shown in Table 18 for various years.

Table 18 Textbook Sales and College Enrollments

Year	Textbook Sales (millions of dollars)	Year	College Enrollment (millions)
2000	4265	1985	12.5
2001	4571	1988	13.1
2002	4899	1992	14.0
2003	5086	1996	15.2
2004	5479	2000	15.3
2005	5703	2003	16.6

Sources: *Book Industry Study Group, Inc.; U.S. Census Bureau*

Let M be the average amount of money spent on textbooks per college student (in dollars per student) during the year that is t years since 1980. A reasonable model is

$$M = \dfrac{288.6t - 1493}{0.22t + 11.4}$$

a. Use the model to estimate the average amount of money that was spent on textbooks per student in 2003. Then find the actual average. Is your result from using the model an underestimate or an overestimate?

b. Predict the average amount of money that will be spent on textbooks per student in 2011.

c. During which year will the average amount of money spent on textbooks per student equal $450?

47. If a recipe for 4 servings of chicken cacciatore calls for 14 ounces of diced tomatoes, how many ounces of diced tomatoes should be used to make 7 servings?

48. Two similar triangles are shown in Fig. 35. Find the length of the side labeled "x yards." Round your result to the second decimal place.

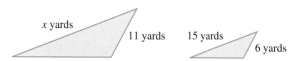

Figure 35 Exercise 48

Translate the sentence into an equation.

49. A varies directly as w.

50. F varies inversely as r squared.

Translate the equation into a sentence by using the phrase "varies directly" or "varies inversely."

51. $V = kr^3$

52. $I = \dfrac{k}{d^2}$

For Exercises 53 and 54, find the requested value of the variable.

53. If w varies directly as t^2, and $w = 18$ when $t = 3$, find t when $w = 50$.

54. If H varies inversely as d, and $H = 4$ when $d = 3$, find H when $d = 6$.

55. The variable w varies directly as p with positive variation constant k. What happens to the value of w as the value of p decreases?

56. The variable C varies inversely as t with positive variation constant k. For positive values of t, what happens to the value of C as the value of t increases?

57. An author's total royalties vary directly as the number of books sold. If the total royalties are $47,500 from selling 5000 books, what would be the total royalties from selling 7500 books?

58. The time it takes a commuter to drive to work varies inversely as the average driving speed. If it takes the commuter 40 minutes when traveling at an average speed of 45 mph, how long will the commute take if the average speed is 50 mph?

59. As you move away from an object, it appears to decrease in size. To describe this relationship, a person stood 4 feet away from a painting, held a yardstick at arm's length (2 feet), and measured the image of the painting. The image had a height of 14 inches. For this exercise, we call the height of such an image the *apparent height* of the painting. Apparent heights of the painting were collected at various distances from it (see Table 19).

Table 19 Apparent Heights of a Painting

Distance away from Painting (feet)	Apparent Height (inches)
4	14.0
5	10.3
6	8.8
7	7.8
8	7.0
9	6.0
10	5.5
11	5.0

Source: *J. Lehmann*

Let A be the apparent height (in inches) of the painting when the person is D feet away.

a. Create a scattergram of the data.

b. Find an equation of a model to describe the data.

c. In a sentence that uses the phrase "varies directly" or "varies inversely," describe how the distance from the painting and the apparent height are related.

d. Find the apparent height of the painting when the person is standing 12 feet away from it.

e. Estimate the actual height of the painting. [**Hint:** Take into account the fact that the person held the yardstick at arm's length (2 feet).]

Simplify.

60. $\dfrac{\frac{12}{x^2}}{\frac{9}{x^3}}$

61. $\dfrac{\frac{x^2 - 6x - 16}{25x}}{\frac{x - 8}{35}}$

62. $\dfrac{5 - \frac{2}{w}}{1 - \frac{3}{w}}$

63. $\dfrac{\frac{3}{2b} + \frac{1}{b^2}}{\frac{1}{3b} - \frac{2}{b^2}}$

CHAPTER 10 TEST

For Exercises 1–3, find any excluded values.

1. $\dfrac{x}{2}$

2. $\dfrac{w-2}{w+9}$

3. $\dfrac{x-5}{x^2+3x-54}$

4. Give three examples of rational expressions, each of whose only excluded value is −4.

Simplify.

5. $\dfrac{p^2-16}{p^2-9p+20}$

6. $\dfrac{5m^2+10m}{3m^3-2m^2-12m+8}$

Perform the indicated operation. Simplify the result.

7. $\dfrac{x^2+5x+6}{x^2-10x+25} \cdot \dfrac{x^2-25}{x^2-2x-8}$

8. $\dfrac{x^2-2xy}{20x^6y} \div \dfrac{xy-2y^2}{45x^4y^2}$

9. $\dfrac{5a}{a^2-a-20} + \dfrac{2}{a^2-3a-10}$

10. $\dfrac{t}{t^2-49} - \dfrac{5}{2t-14}$

11. Find the quotient of $\dfrac{9x^2-4}{-2x-8}$ and $\dfrac{6x^2-x-2}{-3x-12}$.

12. Find the difference of $\dfrac{x-1}{x+4}$ and $\dfrac{x+2}{x-7}$.

For Exercises 13 and 14, solve.

13. $\dfrac{4}{p-2} + \dfrac{3}{5} = 1$

14. $\dfrac{3}{w^2-4w-21} - \dfrac{5}{w-7} = \dfrac{2}{3w+9}$

15. The speed limit is 130 kilometers per hour on motorways in France. Describe this speed limit in miles per hour. Round your result to two decimal places. Refer to the list of equivalent units in the margin of p. 510 as needed.

16. The numbers of U.S. households that have been burglarized are shown in Table 20 for various years. Let p be the percentage of households that have been burglarized in the year that is t years since 1980. A reasonable model is

$$p = \dfrac{-22t+804}{1.28t+80.45}$$

a. Use the model to estimate the percentage of households that were burglarized in 2000. Then find the actual percentage. Is your result from using the model an underestimate or an overestimate?

Table 20 Numbers of Households Burglarized

Year	Number of Households Burglarized (millions)	Total Number of Households (millions)
1980	8.19	80.8
1985	6.53	86.8
1990	6.02	93.3
1995	4.88	99.0
2000	3.33	104.7
2003	2.99	111.3

Sources: *Bureau of Justice Statistics; U.S. Census Bureau*

b. Predict the percentage of households that will be burglarized in 2010.

c. Predict in what year 1% of households will be burglarized.

17. A 2004 Toyota Prius® uses 3 gallons of gasoline to travel 153 miles on highways. Estimate how many gallons of gasoline are required to travel 400 miles on highways.

18. Two similar triangles are shown in Fig. 36. Find the length of the side labeled "x meters." Round your result to the second decimal place.

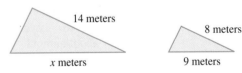

14 meters

x meters

8 meters

9 meters

Figure 36 Exercise 18

19. If H varies directly as t, and $H = 12$ when $t = 3$, find H when $t = 6$.

20. If p varies inversely as w^2, and $p = 9$ when $w = 2$, find w when $p = 4$.

21. The variable c varies inversely as u for the positive variation constant k. For positive values of u, describe what happens to the value of c as the value of u increases.

22. The total in-district tuition at Kalamazoo Valley Community College varies directly as the number of credit hours (units). For winter term 2007, the total tuition is $732 for 12 credit hours. If a student can afford $960 for tuition, in how many credit hours can she enroll?

23. Simplify $\dfrac{\dfrac{4}{3x} - \dfrac{1}{x^2}}{\dfrac{1}{2x} - \dfrac{5}{x^2}}$.

11

More Radical Expressions and Equations

Move out of your comfort zone. You can only grow if you are willing to feel awkward and uncomfortable when you try something new.

—Brian Tracy

Table 1 Numbers of McDonald's Restaurants

Year	Number of Restaurants (thousands)
1999	26
2000	29
2001	30
2002	31
2003	31
2004	32
2005	32

Source: *McDonald's*

Has a McDonald's® restaurant opened in your neighborhood in the last year? From 1999 to 2003, the number of McDonald's restaurants in the world increased by about 6 thousand restaurants (see Table 1). In Exercise 55 of Homework 11.3, you will predict when there will be 36 thousand McDonald's restaurants in the world.

In Sections 9.1 and 9.2, we simplified radical expressions. In this chapter, we will discuss how to add, subtract, and multiply radical expressions. We will also discuss how to solve equations that have radicals, and we will use this skill to make predictions about authentic situations such as the McDonald's restaurant data.

11.1 ADDING AND SUBTRACTING RADICAL EXPRESSIONS

Objectives

▷ Know the meaning of *like radicals*.

▷ Add and subtract radical expressions.

Recall from Section 4.2 that we can use the distributive law to add like terms, such as $3x$ and $6x$:

$$3x + 6x = (3 + 6)x = 9x$$

How do we add (or subtract) radical expressions? We can again use the distributive law if the radicals are like radicals. We say that $3\sqrt{x}$ and $6\sqrt{x}$ are like radicals, because they have the same radicand. In general, square root radicals that have the same radicand are called **like radicals.**

We add the like radicals $3\sqrt{x}$ and $6\sqrt{x}$ as follows:

$$3\sqrt{x} + 6\sqrt{x} = (3 + 6)\sqrt{x} = 9\sqrt{x}$$

To add or subtract like radicals, we use the distributive law. When we add or subtract like radicals, we say that we *combine like radicals.*

Example 1 Adding and Subtracting Radical Expressions

Perform the indicated operation.

1. $3\sqrt{2} + 5\sqrt{2}$ **2.** $7\sqrt{x} + \sqrt{x}$ **3.** $9\sqrt{5x} - 2\sqrt{5x}$

4. $2\sqrt{3} + 6\sqrt{7}$ **5.** $5t\sqrt{2} + 3t\sqrt{2}$

Solution

1. $3\sqrt{2} + 5\sqrt{2} = (3 + 5)\sqrt{2} = 8\sqrt{2}$
2. $7\sqrt{x} + \sqrt{x} = 7\sqrt{x} + 1\sqrt{x} = (7 + 1)\sqrt{x} = 8\sqrt{x}$
3. $9\sqrt{5x} - 2\sqrt{5x} = (9 - 2)\sqrt{5x} = 7\sqrt{5x}$
4. Since the radicals $2\sqrt{3}$ and $6\sqrt{7}$ have different radicands, they are not like radicals and we cannot use the distributive law. The expression $2\sqrt{3} + 6\sqrt{7}$ is already in simplified form.
5. $5t\sqrt{2} + 3t\sqrt{2} = (5t + 3t)\sqrt{2} = 8t\sqrt{2}$ ■

Example 2   Adding and Subtracting Radical Expressions

Perform the indicated operations: $5\sqrt{x} + 3\sqrt{2} - 7\sqrt{x}$.

Solution

$$
\begin{aligned}
5\sqrt{x} + 3\sqrt{2} - 7\sqrt{x} &= 5\sqrt{x} - 7\sqrt{x} + 3\sqrt{2} && \text{Rearrange terms.} \\
&= (5 - 7)\sqrt{x} + 3\sqrt{2} && \text{Distributive law} \\
&= -2\sqrt{x} + 3\sqrt{2} && \text{Subtract.}
\end{aligned}
$$ ■

Sometimes, simplifying radicals will allow us to combine like radicals.

Example 3   Adding and Subtracting Radical Expressions

Perform the indicated operations.

1. $\sqrt{8} + 5\sqrt{2}$ 2. $5\sqrt{27} - 2\sqrt{75}$ 3. $6\sqrt{4w} - 7\sqrt{9w} + 5\sqrt{w}$

Solution

1.
$$
\begin{aligned}
\sqrt{8} + 5\sqrt{2} &= \sqrt{4 \cdot 2} + 5\sqrt{2} && \text{4 is a perfect square.} \\
&= \sqrt{4}\sqrt{2} + 5\sqrt{2} && \sqrt{ab} = \sqrt{a}\sqrt{b} \\
&= 2\sqrt{2} + 5\sqrt{2} && \sqrt{4} = 2 \\
&= 7\sqrt{2} && \text{Combine like radicals.}
\end{aligned}
$$

2.
$$
\begin{aligned}
5\sqrt{27} - 2\sqrt{75} &= 5\sqrt{9 \cdot 3} - 2\sqrt{25 \cdot 3} && \text{9 and 25 are perfect squares.} \\
&= 5\sqrt{9}\sqrt{3} - 2\sqrt{25}\sqrt{3} && \sqrt{ab} = \sqrt{a}\sqrt{b} \\
&= 5 \cdot 3 \cdot \sqrt{3} - 2 \cdot 5 \cdot \sqrt{3} && \sqrt{9} = 3; \sqrt{25} = 5 \\
&= 15\sqrt{3} - 10\sqrt{3} && \text{Multiply.} \\
&= 5\sqrt{3} && \text{Combine like radicals.}
\end{aligned}
$$

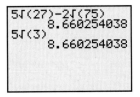

Figure 1 Verify the work

We use a graphing calculator to verify our work (see Fig. 1).

3.
$$
\begin{aligned}
6\sqrt{4w} - 7\sqrt{9w} + 5\sqrt{w} &= 6\sqrt{4}\sqrt{w} - 7\sqrt{9}\sqrt{w} + 5\sqrt{w} && \sqrt{ab} = \sqrt{a}\sqrt{b} \\
&= 6 \cdot 2 \cdot \sqrt{w} - 7 \cdot 3 \cdot \sqrt{w} + 5\sqrt{w} && \sqrt{4} = 2; \sqrt{9} = 3 \\
&= 12\sqrt{w} - 21\sqrt{w} + 5\sqrt{w} && \text{Multiply.} \\
&= -4\sqrt{w} && \text{Combine like radicals.}
\end{aligned}
$$

We use a graphing calculator table to verify our work (see Fig. 2).   ■

Recall from Section 9.1 that if x is nonnegative, then $\sqrt{x^2} = x$.

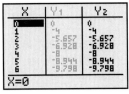

Figure 2 Verify the work

Example 4   Adding and Subtracting Radical Expressions

Perform the indicated operations. Assume that x is nonnegative.

1. $5\sqrt{3x^2} - x\sqrt{12}$
2. $x\sqrt{20} - 6\sqrt{2x} + 4\sqrt{45x^2}$

Solution

1. $5\sqrt{3x^2} - x\sqrt{12} = 5\sqrt{x^2 \cdot 3} - x\sqrt{4 \cdot 3}$ Rearrange factors; x^2 and 4 are perfect squares.

$\qquad\qquad\qquad = 5\sqrt{x^2}\sqrt{3} - x\sqrt{4}\sqrt{3}$ $\sqrt{ab} = \sqrt{a}\sqrt{b}$

$\qquad\qquad\qquad = 5x\sqrt{3} - 2x\sqrt{3}$ $\sqrt{x^2} = x$ for nonnegative x; $\sqrt{4} = 2$

$\qquad\qquad\qquad = 3x\sqrt{3}$ Combine like radicals.

2. $x\sqrt{20} - 6\sqrt{2x} + 4\sqrt{45x^2}$

$\qquad = x\sqrt{4 \cdot 5} - 6\sqrt{2x} + 4\sqrt{9x^2 \cdot 5}$ 4, 9, and x^2 are perfect squares.

$\qquad = x\sqrt{4}\sqrt{5} - 6\sqrt{2x} + 4\sqrt{9}\sqrt{x^2}\sqrt{5}$ $\sqrt{ab} = \sqrt{a}\sqrt{b}$

$\qquad = x \cdot 2 \cdot \sqrt{5} - 6\sqrt{2x} + 4 \cdot 3 \cdot x \cdot \sqrt{5}$ $\sqrt{4} = 2$; $\sqrt{9} = 3$; $\sqrt{x^2} = x$ for nonnegative x

$\qquad = 2x\sqrt{5} - 6\sqrt{2x} + 12x\sqrt{5}$ Multiply.

$\qquad = 2x\sqrt{5} + 12x\sqrt{5} - 6\sqrt{2x}$ Rearrange terms.

$\qquad = 14x\sqrt{5} - 6\sqrt{2x}$ Combine like radicals.   ∎

group exploration

Looking ahead: Multiplying two radical expressions

In Section 9.1, we discussed the product property for square roots:

$$\sqrt{ab} = \sqrt{a}\sqrt{b} \qquad \text{where } a \text{ and } b \text{ are nonnegative numbers}$$

We can use this property to multiply two radical expressions:

$$\sqrt{2}\sqrt{7} = \sqrt{2 \cdot 7} = \sqrt{14}$$

Perform the indicated operation.

1. $\sqrt{3} \cdot \sqrt{5}$ **2.** $6\sqrt{5} \cdot \sqrt{2}$ **3.** $\sqrt{3}(\sqrt{2} - \sqrt{5})$

4. $(x + \sqrt{5})(x + \sqrt{7})$ **5.** $(x - \sqrt{2})(x + \sqrt{3})$ **6.** $(x + \sqrt{3})(x - \sqrt{3})$

7. $(x - \sqrt{5})^2$

HOMEWORK 11.1

FOR EXTRA HELP ▶

Student Solutions Manual   PH Math/Tutor Center   MathXL®   MyMathLab

Perform the indicated operations. Use a graphing calculator table to verify your work.

1. $4\sqrt{3} + 5\sqrt{3}$ **2.** $2\sqrt{7} + 6\sqrt{7}$

3. $-3\sqrt{5} + 9\sqrt{5}$ **4.** $-5\sqrt{6} + 3\sqrt{6}$

5. $4\sqrt{x} + \sqrt{x}$ **6.** $\sqrt{x} + 7\sqrt{x}$

7. $\sqrt{7} + \sqrt{7}$ **8.** $\sqrt{3} + \sqrt{3}$

9. $2\sqrt{6} - 9\sqrt{6}$ **10.** $5\sqrt{10} - 8\sqrt{10}$

11. $-4\sqrt{5t} - 7\sqrt{5t}$ **12.** $-8\sqrt{3a} - 5\sqrt{3a}$

13. $7\sqrt{5} - 3\sqrt{7}$ **14.** $8\sqrt{2} - 5\sqrt{6}$

15. $5\sqrt{2} - \sqrt{2} + 2\sqrt{2}$ **16.** $8\sqrt{3} - \sqrt{3} + 5\sqrt{3}$

17. $-4\sqrt{7} + 8\sqrt{7} - 3\sqrt{5}$ **18.** $-3\sqrt{2} + 5\sqrt{3} - 7\sqrt{2}$

19. $7\sqrt{p} - 4\sqrt{p} - 3\sqrt{p}$ **20.** $9\sqrt{b} - 5\sqrt{b} - 4\sqrt{b}$

21. $\sqrt{18} + 4\sqrt{2}$ **22.** $\sqrt{12} + 6\sqrt{3}$

23. $5\sqrt{6} - \sqrt{24}$ **24.** $7\sqrt{5} - \sqrt{45}$

25. $-4\sqrt{5} - \sqrt{20}$ **26.** $8\sqrt{3} - \sqrt{27}$

27. $\sqrt{28} + \sqrt{63}$ **28.** $\sqrt{32} + \sqrt{8}$

29. $2\sqrt{27} + 5\sqrt{12}$ **30.** $3\sqrt{20} + 4\sqrt{45}$

31. $5\sqrt{8} - 5\sqrt{18}$ **32.** $2\sqrt{32} - 3\sqrt{50}$

33. $\sqrt{20} - 4\sqrt{5} + \sqrt{45}$ **34.** $\sqrt{32} - 3\sqrt{2} + \sqrt{18}$

35. $5\sqrt{12} + 4\sqrt{75} - 2\sqrt{3}$ **36.** $3\sqrt{8} + 2\sqrt{18} - 8\sqrt{2}$

37. $\sqrt{4x} + \sqrt{81x}$ **38.** $\sqrt{64x} + \sqrt{16x}$

39. $2\sqrt{9x} - 3\sqrt{4x}$ **40.** $4\sqrt{25x} - 5\sqrt{9x}$

41. $\sqrt{50x} - \sqrt{20} - \sqrt{32x}$ **42.** $\sqrt{98x} - \sqrt{28} - \sqrt{72x}$

43. $3\sqrt{12t} - 2\sqrt{44} + 5\sqrt{27t}$ **44.** $4\sqrt{5a} - 5\sqrt{63} + 3\sqrt{45a}$

Perform the indicated operations. Assume that any variables are nonnegative. Use a graphing calculator table to verify your work.

45. $x\sqrt{20} + x\sqrt{45}$ **46.** $x\sqrt{18} + x\sqrt{50}$

47. $x\sqrt{40} - 5\sqrt{10x^2}$ **48.** $x\sqrt{32} - 3\sqrt{2x^2}$

49. $\sqrt{5w^2} - \sqrt{45w^2}$ **50.** $\sqrt{7b^2} - \sqrt{28b^2}$

51. $2\sqrt{75x^2} + 3\sqrt{12x^2}$ 52. $4\sqrt{27x^2} + 8\sqrt{300x^2}$
53. $\sqrt{3x^2} - 2x\sqrt{3} + 5\sqrt{4x}$ 54. $\sqrt{8x^2} - 5x\sqrt{2} + 6\sqrt{3x}$
55. $5\sqrt{48y^2} - 3\sqrt{20y} - 2y\sqrt{12}$
56. $\sqrt{54t^2} - 9\sqrt{24t} - t\sqrt{600}$

57. The time it takes a planet to make one revolution around the Sun is called the planet's *period*. The period T (in years) of a planet whose average distance from the Sun is d million kilometers is modeled by the equation $T = 0.0005443d\sqrt{d}$.
   a. What is the period of Saturn, whose average distance from the Sun is 1427.0 million kilometers?
   b. If a person is 59 years old in "Earth years," approximately how old is the person in "Saturn years"? [**Hint:** 1 "Earth year" is the amount of time it takes Earth to revolve around the Sun once.]

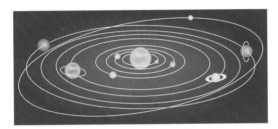

58. A person builds a *pendulum* by tying one end of some thread to a washer (a metal ring) and attaching the other end to the bottom of a table so that the washer is suspended and can swing freely. The *period* of the pendulum is the amount of time it takes the washer to swing forward and backward once. Let T be the period (in seconds) of the pendulum, where L is the length (in feet) of the thread. A model of the situation is $T = \dfrac{\pi}{4}\sqrt{2L}$. Estimate the period of the pendulum if its length is 2 feet.

59. The number of border control agents in the Tucson, Arizona, sector of the U.S. Border Patrol has been increased since 1998 due to efforts to curb the flow of illegal immigrants and, more recently, to apprehend terrorists (see Table 2).

Table 2 Number of Border Control Agents in the Tucson Sector

Year	Number of Tucson Sector Agents
1998	1050
1999	1380
2000	1530
2001	1590
2002	1700
2003	1750
2004	1750

Source: *U.S. Customs and Border Protection*

Let n be the number of border control agents in the Tucson sector at t years since 1998. A model of the situation is $n = 308\sqrt{t} + 1050$.
   a. Use a graphing calculator to draw the graph of the model and, in the same viewing window, the scattergram of the data. Does the model fit the data well?
   b. Substitute a value for one of the variables in the model's equation to predict the number of agents in the Tucson sector in 2010.

c. Use TRACE on a graphing calculator to predict when there will be 2000 agents in the Tucson sector. [**Hint:** You may need to Zoom Out.]

60. The sales of the pain reliever Celebrex® are shown in Table 3 for various years.

Table 3 Sales of Celebrex

Year	Sales (billions of dollars)
1999	1.4
2000	2.1
2001	2.5
2002	2.6
2003	2.6
2004	2.7

Source: *IMS Health*

Let s be Celebrex sales (in billions of dollars) in the year that is t years since 1999. A model of the situation is $s = 0.65\sqrt{t} + 1.4$.
   a. Use a graphing calculator to draw the graph of the model and, in the same viewing window, the scattergram of the data. Does the model fit the data well?
   b. Substitute a value for one of the variables in the model's equation to predict the sales in 2011.
   c. Use TRACE on a graphing calculator to predict when the sales will be \$3.5 billion. [**Hint:** You may need to Zoom Out.]

61. A student tries to simplify $2\sqrt{5} + 4\sqrt{5}$:
$$2\sqrt{5} + 4\sqrt{5} = 6\sqrt{10}$$
Describe any errors. Then simplify the expression correctly.

62. A student tries to simplify $6\sqrt{2} + 7\sqrt{3}$:
$$6\sqrt{2} + 7\sqrt{3} = 13\sqrt{5}$$
Describe any errors. Then simplify the expression correctly.

63. Explain how the distributive law can be used to show that the statement $4\sqrt{7} + 5\sqrt{7} = 9\sqrt{7}$ is true.

64. Explain how the distributive law can be used to show that the statement $9\sqrt{2} - 6\sqrt{2} = 3\sqrt{2}$ is true.

Related Review

Simplify.

65. $5x - 7x$ 66. $2x + 4x$
67. $5\sqrt{x} - 7\sqrt{x}$ 68. $2\sqrt{x} + 4\sqrt{x}$
69. $\sqrt{12x^2} + \sqrt{3x^2}$ 70. $\sqrt{2x^2} - \sqrt{18x^2}$
71. $12x^2 + 3x^2$ 72. $2x^2 - 18x^2$

Expressions, Equations, and Graphs

Perform the indicated instruction. Then use words such as linear, quadratic, cubic, power, radical, rational, polynomial, degree, one variable, *and* two variables *to describe the expression, equation, or system.*

73. Solve $t(t - 2) = 7$.
74. Factor $10p^2 - 13pq - 3q^2$.
75. Find the difference $\dfrac{3w}{w^2 - 9} - \dfrac{1}{4w + 12}$.
76. Solve $t^2 - 3t - 18 = 0$.   77. Factor $x^3 - 10x^2 + 25x$.
78. Graph $y = x^2 - 1$ by hand.

11.2 MULTIPLYING RADICAL EXPRESSIONS

Objectives

▷ Multiply two radical expressions.

▷ Find the product of two *radical conjugates*.

▷ Simplify the square of a radical expression with two terms.

In Section 11.1, we added and subtracted radical expressions. How do we multiply two radical expressions?

Multiplication of Radical Expressions

In Section 9.1, we discussed the product property for square roots, namely, $\sqrt{ab} = \sqrt{a}\sqrt{b}$, where a and b are nonnegative. We can use this property to help us multiply radical expressions. Here, we multiply $\sqrt{2}$ and $\sqrt{3}$:

$$\sqrt{2} \cdot \sqrt{3} = \sqrt{2 \cdot 3} = \sqrt{6}$$

Next, we find the power $\left(\sqrt{x}\right)^2$, where x is nonnegative:

$$\left(\sqrt{x}\right)^2 = \sqrt{x}\sqrt{x} = \sqrt{x^2} = x$$

Power Property for Square Roots

If x is nonnegative, then

$$\left(\sqrt{x}\right)^2 = \sqrt{x}\sqrt{x} = x$$

It is good practice to check whether the product of radical expressions can be simplified.

Example 1 Finding the Product of Radical Expressions

Find the product.

1. $\sqrt{5} \cdot \sqrt{7}$ **2.** $5\sqrt{3} \cdot \sqrt{6}$

3. $2\sqrt{x} \cdot 3\sqrt{x}$ **4.** $\sqrt{3t} \cdot \sqrt{5t}$

Solution

1. $\sqrt{5} \cdot \sqrt{7} = \sqrt{5 \cdot 7}$ $\sqrt{a}\sqrt{b} = \sqrt{ab}$

           $= \sqrt{35}$ Multiply.

2. $5\sqrt{3} \cdot \sqrt{6} = 5\sqrt{18}$ $\sqrt{a}\sqrt{b} = \sqrt{ab}$

           $= 5\sqrt{9 \cdot 2}$ 9 is a perfect square.

           $= 5\sqrt{9}\sqrt{2}$ $\sqrt{ab} = \sqrt{a}\sqrt{b}$

           $= 5 \cdot 3\sqrt{2}$ $\sqrt{9} = 3$

           $= 15\sqrt{2}$ Multiply.

3. $2\sqrt{x} \cdot 3\sqrt{x} = 2 \cdot 3 \cdot \sqrt{x} \cdot \sqrt{x}$ Rearrange factors.

             $= 6x$ Multiply; $\sqrt{x}\sqrt{x} = x$ for $x \geq 0$

4. $\sqrt{3t} \cdot \sqrt{5t} = \sqrt{15t^2}$ $\sqrt{a}\sqrt{b} = \sqrt{ab}$

           $= \sqrt{15}\sqrt{t^2}$ $\sqrt{ab} = \sqrt{a}\sqrt{b}$

           $= t\sqrt{15}$ $\sqrt{x^2} = x$ for $x \geq 0$ ■

Now we use the distributive law to help us find products of radical expressions.

Example 2   Finding the Product of Radical Expressions

Find the product.

1. $\sqrt{5}(3 - \sqrt{2})$
2. $\sqrt{x}(\sqrt{x} - \sqrt{5})$
3. $2\sqrt{3}(4\sqrt{3} + 6\sqrt{7})$

Solution

1. $\sqrt{5}(3 - \sqrt{2}) = \sqrt{5} \cdot 3 - \sqrt{5} \cdot \sqrt{2}$ Distributive law

   $\qquad\qquad = 3\sqrt{5} - \sqrt{10}$ Rearrange factors; $\sqrt{a}\sqrt{b} = \sqrt{ab}$

2. $\sqrt{x}(\sqrt{x} - \sqrt{5}) = \sqrt{x} \cdot \sqrt{x} - \sqrt{x} \cdot \sqrt{5}$ Distributive law

   $\qquad\qquad = x - \sqrt{5x}$ $\sqrt{x}\sqrt{x} = x$ for $x \geq 0$; $\sqrt{a}\sqrt{b} = \sqrt{ab}$

3. $2\sqrt{3}(4\sqrt{3} + 6\sqrt{7}) = 2\sqrt{3} \cdot (4\sqrt{3}) + 2\sqrt{3} \cdot (6\sqrt{7})$ Distributive law

   $\qquad\qquad = 2 \cdot 4\sqrt{3}\sqrt{3} + 2 \cdot 6\sqrt{3}\sqrt{7}$ Rearrange factors.

   $\qquad\qquad = 8 \cdot 3 + 12\sqrt{21}$ $\sqrt{x}\sqrt{x} = x$ for $x \geq 0$; $\sqrt{a}\sqrt{b} = \sqrt{ab}$

   $\qquad\qquad = 24 + 12\sqrt{21}$ Multiply.   ■

Next, we find the product of two radical expressions for which each factor has two terms.

Example 3   Finding the Product of Radical Expressions

Find the product.

1. $(5 - \sqrt{3})(4 + \sqrt{3})$
2. $(\sqrt{x} + \sqrt{3})(\sqrt{x} + \sqrt{5})$

Solution

1. We begin by multiplying each term in the first radical expression by each term in the second radical expression:

   $(5 - \sqrt{3})(4 + \sqrt{3}) = 20 + 5\sqrt{3} - 4\sqrt{3} - \sqrt{3}\sqrt{3}$ Multiply pairs of terms.

   $\qquad\qquad = 20 + \sqrt{3} - 3$ Combine like radicals; $\sqrt{x}\sqrt{x} = x$ for $x \geq 0$

   $\qquad\qquad = 17 + \sqrt{3}$ Simplify.

2. Again, we begin by multiplying each term in the first radical expression by each term in the second radical expression:

   $(\sqrt{x} + \sqrt{3})(\sqrt{x} + \sqrt{5}) = \sqrt{x}\sqrt{x} + \sqrt{x}\sqrt{5} + \sqrt{3}\sqrt{x} + \sqrt{3}\sqrt{5}$ Multiply pairs of terms.

   $\qquad\qquad = x + \sqrt{5x} + \sqrt{3x} + \sqrt{15}$ $\sqrt{x}\sqrt{x} = x$ for $x \geq 0$; $\sqrt{a}\sqrt{b} = \sqrt{ab}$

   We cannot combine the terms $\sqrt{5x}$ and $\sqrt{3x}$, because they are not like radicals. So, our result is simplified. We use a graphing calculator table to verify our work (see Fig. 3).   ■

Figure 3 Verify the work

Product of Two Radical Expressions, Each with Two Terms

To find the product of two radical expressions in which each factor has two terms,

1. Multiply each term in the first radical expression by each term in the second radical expression.
2. Combine like radicals.

Example 4 Finding the Product of Radical Expressions

Find the product $(5 - 2\sqrt{7})(3 - 4\sqrt{7})$.

Solution

We begin by multiplying each term in the first radical expression by each term in the second radical expression:

$$
\begin{aligned}
(5 - 2\sqrt{7})(3 - 4\sqrt{7}) &= 15 - 5(4\sqrt{7}) - (2\sqrt{7})3 + (2\sqrt{7})(4\sqrt{7}) && \text{Multiply pairs of terms.} \\
&= 15 - 5 \cdot 4 \cdot \sqrt{7} - 2 \cdot 3 \cdot \sqrt{7} + 2 \cdot 4 \cdot \sqrt{7}\sqrt{7} && \text{Rearrange factors.} \\
&= 15 - 20\sqrt{7} - 6\sqrt{7} + 8 \cdot 7 && \sqrt{x}\sqrt{x} = x \text{ for } x \geq 0 \\
&= 15 - 26\sqrt{7} + 56 && \text{Combine like radicals;} \\
& && \text{multiply.} \\
&= 71 - 26\sqrt{7} && \text{Simplify.} \quad \blacksquare
\end{aligned}
$$

Multiplication of Two Radical Conjugates

Recall from Section 7.5 that we call binomials such as $3x + 5$ and $3x - 5$ conjugates of each other. Similarly, we call the radical expressions $3 + 5\sqrt{x}$ and $3 - 5\sqrt{x}$ radical conjugates of each other. To find the **radical conjugate** of a radical expression with two terms, we change the addition symbol to a subtraction symbol or vice versa.

Product of Two Radical Conjugates

To find the product of two radical conjugates, use the property

$$(A + B)(A - B) = A^2 - B^2$$

In Example 5, we find products of two radical conjugates.

Example 5 Finding the Product of Two Radical Conjugates

Find the product.

1. $(\sqrt{7} + \sqrt{2})(\sqrt{7} - \sqrt{2})$
2. $(5\sqrt{a} - 4\sqrt{b})(5\sqrt{a} + 4\sqrt{b})$

Solution

1. We substitute $\sqrt{7}$ for A and $\sqrt{2}$ for B in the property for the product of two radical conjugates:

$$
\begin{aligned}
(A + B)(A - B) &= A^2 - B^2 \\
(\sqrt{7} + \sqrt{2})(\sqrt{7} - \sqrt{2}) &= (\sqrt{7})^2 - (\sqrt{2})^2 && \text{Substitute.} \\
&= 7 - 2 && (\sqrt{x})^2 = x \text{ for } x \geq 0 \\
&= 5 && \text{Subtract.}
\end{aligned}
$$

2. We substitute $5\sqrt{a}$ for A and $4\sqrt{b}$ for B in the property for the product of two radical conjugates:

$$
\begin{aligned}
(A - B)(A + B) &= A^2 - B^2 \\
(5\sqrt{a} - 4\sqrt{b})(5\sqrt{a} + 4\sqrt{b}) &= (5\sqrt{a})^2 - (4\sqrt{b})^2 && \text{Substitute.} \\
&= 5^2(\sqrt{a})^2 - 4^2(\sqrt{b})^2 && \text{Raise each factor to} \\
& && \text{second power: } (xy)^n = x^n y^n \\
&= 25a - 16b && (\sqrt{x})^2 = x \text{ for } x \geq 0 \quad \blacksquare
\end{aligned}
$$

Simplifying the Square of a Binomial

Recall from Section 7.5 the properties for the square of a binomial:

$$(A + B)^2 = A^2 + 2AB + B^2 \qquad \text{\textit{Square of a sum}}$$
$$(A - B)^2 = A^2 - 2AB + B^2 \qquad \text{\textit{Square of a difference}}$$

For each problem in Example 6, we will first simplify a radical expression without using these properties. We will then simplify the expression again, using one of the properties.

Example 6 Simplifying the Square of a Radical Expression with Two Terms

Simplify.

1. $\left(3 + \sqrt{5}\right)^2$
2. $\left(\sqrt{x} - \sqrt{3}\right)^2$

Solution

1. We begin by using the fact that $C^2 = CC$ and then multiply pairs of terms:

$$\begin{aligned}
\left(3 + \sqrt{5}\right)^2 &= \left(3 + \sqrt{5}\right)\left(3 + \sqrt{5}\right) & & \textit{C}^2 = \textit{CC} \\
&= 9 + 3\sqrt{5} + 3\sqrt{5} + \sqrt{5}\sqrt{5} & & \textit{Multiply pairs of terms.} \\
&= 9 + 6\sqrt{5} + 5 & & \textit{Combine like radicals; } \sqrt{x}\sqrt{x} = x \text{ for } x \geq 0 \\
&= 14 + 6\sqrt{5} & & \textit{Simplify.}
\end{aligned}$$

Another way to simplify $\left(3 + \sqrt{5}\right)^2$ is to substitute 3 for A and $\sqrt{5}$ for B in the property for the square of a sum:

$$(A + B)^2 = A^2 + 2AB + B^2$$
$$\left(3 + \sqrt{5}\right)^2 = 3^2 + 2(3)\sqrt{5} + \left(\sqrt{5}\right)^2 \qquad \textit{Substitute.}$$
$$= 9 + 6\sqrt{5} + 5 \qquad (\sqrt{x})^2 = x \text{ for } x \geq 0$$
$$= 14 + 6\sqrt{5} \qquad \textit{Simplify.}$$

2. We begin by using the fact that $C^2 = CC$ and then multiply pairs of terms:

$$\begin{aligned}
\left(\sqrt{x} - \sqrt{3}\right)^2 &= \left(\sqrt{x} - \sqrt{3}\right)\left(\sqrt{x} - \sqrt{3}\right) & & \textit{C}^2 = \textit{CC} \\
&= \sqrt{x}\sqrt{x} - \sqrt{3}\sqrt{x} - \sqrt{3}\sqrt{x} + \sqrt{3}\sqrt{3} & & \textit{Multiply pairs of terms.} \\
&= x - \sqrt{3x} - \sqrt{3x} + 3 & & \begin{array}{l}\sqrt{x}\sqrt{x} = x \text{ for } x \geq 0; \\ \sqrt{a}\sqrt{b} = \sqrt{ab}\end{array} \\
&= x - 2\sqrt{3x} + 3 & & \textit{Combine like radicals.}
\end{aligned}$$

Another way to simplify $\left(\sqrt{x} - \sqrt{3}\right)^2$ is to substitute $\sqrt{x}$ for A and $\sqrt{3}$ for B in the property for the square of a difference:

$$(A - B)^2 = A^2 - 2AB + B^2$$
$$\left(\sqrt{x} - \sqrt{3}\right)^2 = \left(\sqrt{x}\right)^2 - 2\sqrt{x}\sqrt{3} + \left(\sqrt{3}\right)^2 \qquad \textit{Substitute.}$$
$$= x - 2\sqrt{3x} + 3 \qquad (\sqrt{x})^2 = x \text{ for } x \geq 0; \sqrt{a}\sqrt{b} = \sqrt{ab}$$

∎

In Example 6, we showed two ways to simplify the square of a radical expression with two terms: (1) using the definition of "square" and (2) using the property for the square of a sum or difference. When simplifying similar expressions in the Homework, experiment with both methods to help you decide which one works best for you. Many students are more successful with writing the square as a product of two identical expressions and then multiplying pairs of terms.

Simplifying the Square of a Radical Expression

To simplify the square of a radical expression with two terms,

- Use $C^2 = CC$ to write the square as a product of two identical expressions, and then multiply each term in the first radical expression by each term in the second radical expression,

or

- Use the property $(A + B)^2 = A^2 + 2AB + B^2$ or the property $(A - B)^2 = A^2 - 2AB + B^2$.

WARNING   In simplifying $(x + k)^2$, it is important to remember the middle term of $x^2 + 2kx + k^2$. Likewise, in simplifying $(x - k)^2$, it is important to remember the middle term of $x^2 - 2kx + k^2$. Do not make the following typical error in simplifying $\left(\sqrt{x} + \sqrt{7}\right)^2$:

$$\left(\sqrt{x} + \sqrt{7}\right)^2 = \left(\sqrt{x}\right)^2 + \left(\sqrt{7}\right)^2 = x + 7 \qquad \text{Incorrect}$$

$$\left(\sqrt{x} + \sqrt{7}\right)^2 = \left(\sqrt{x}\right)^2 + 2\sqrt{7}\sqrt{x} + \left(\sqrt{7}\right)^2 = x + 2\sqrt{7x} + 7 \qquad \text{Correct}$$

group exploration

Rationalizing the denominator

In Section 9.2, we "rationalized the denominator" of fractions of the form $\dfrac{1}{\sqrt{a}}$ by finding an equivalent expression that does not have a radical in any denominator. Here, you will explore how to rationalize the denominator of a fraction when that denominator is a sum or a difference involving radicals.

1. Find the product.

     a. $\left(x + \sqrt{3}\right)\left(x - \sqrt{3}\right)$ **b.** $\left(x - \sqrt{7}\right)\left(x + \sqrt{7}\right)$

     c. $\left(2 + \sqrt{x}\right)\left(2 - \sqrt{x}\right)$ **d.** $\left(6 - \sqrt{x}\right)\left(6 + \sqrt{x}\right)$

2. What patterns do you notice from your work in Problem 1?

3. Rationalize the denominator of the expression $\dfrac{1}{5 - \sqrt{x}}$ by performing the multiplication

$$\frac{1}{5 - \sqrt{x}} \cdot \frac{5 + \sqrt{x}}{5 + \sqrt{x}}$$

Use a graphing calculator table to verify your work.

4. Rationalize the denominator of the expression $\dfrac{1}{8 + \sqrt{x}}$.

5. Describe how to rationalize the denominator of a radical expression.

HOMEWORK 11.2   FOR EXTRA HELP ▶

Student Solutions Manual   PH Math/Tutor Center   Math XL
MathXL®   MyMathLab
MyMathLab

Perform the indicated operation. Assume that any variables are nonnegative. Use a graphing calculator to verify your work when possible.

1. $\sqrt{2} \cdot \sqrt{5}$

2. $\sqrt{7} \cdot \sqrt{3}$

3. $-\sqrt{6} \cdot \sqrt{2}$

4. $-\sqrt{2} \cdot \sqrt{10}$

5. $\sqrt{17}\sqrt{17}$

6. $\sqrt{19}\sqrt{19}$

7. $2\sqrt{5} \cdot \sqrt{10}$

8. $2\sqrt{15} \cdot \sqrt{3}$

9. $7\sqrt{2}\left(-3\sqrt{14}\right)$

10. $4\sqrt{10}\left(-2\sqrt{5}\right)$

11. $5\sqrt{t} \cdot 7\sqrt{t}$

12. $2\sqrt{a} \cdot 8\sqrt{a}$

13. $\sqrt{7t}\sqrt{3t}$

14. $\sqrt{5b}\sqrt{2b}$

15. $\left(8\sqrt{x}\right)^2$

16. $\left(6\sqrt{x}\right)^2$

17. $\sqrt{ab}\sqrt{bc}$

18. $\sqrt{rw}\sqrt{rtw}$

19. $\sqrt{5}(1 + \sqrt{7})$

20. $\sqrt{2}(4 + \sqrt{3})$

21. $-\sqrt{7}(\sqrt{2} + \sqrt{5})$

22. $-\sqrt{3}(\sqrt{5} + \sqrt{7})$

23. $5\sqrt{c}(9 - \sqrt{c})$

24. $3\sqrt{w}(8 - \sqrt{w})$

25. $-4\sqrt{5}(3\sqrt{2} - \sqrt{5})$

26. $-3\sqrt{2}(6\sqrt{7} - 5)$

Perform the indicated operation. Use a graphing calculator table to verify your work when possible.

27. $(4 + \sqrt{5})(2 - \sqrt{5})$

28. $(6 + \sqrt{2})(3 - \sqrt{2})$

29. $(\sqrt{3} - \sqrt{5})(\sqrt{3} + \sqrt{2})$

30. $(\sqrt{5} - \sqrt{2})(\sqrt{5} + \sqrt{3})$

31. $(\sqrt{x} - \sqrt{7})(\sqrt{x} - \sqrt{2})$

32. $(\sqrt{x} - \sqrt{5})(\sqrt{x} - \sqrt{3})$

33. $(y + \sqrt{2})(y - \sqrt{11})$

34. $(a + \sqrt{5})(a - \sqrt{2})$

35. $(4 - 2\sqrt{5})(2 - 3\sqrt{5})$

36. $(1 - 3\sqrt{2})(3 - 4\sqrt{2})$

37. $(\sqrt{3} + \sqrt{7})(\sqrt{3} - \sqrt{7})$

38. $(\sqrt{2} + \sqrt{5})(\sqrt{2} - \sqrt{5})$

39. $(r + \sqrt{3})(r - \sqrt{3})$

40. $(t + \sqrt{5})(t - \sqrt{5})$

41. $(5\sqrt{2} - 2\sqrt{3})(5\sqrt{2} + 2\sqrt{3})$

42. $(4\sqrt{3} - 3\sqrt{5})(4\sqrt{3} + 3\sqrt{5})$

43. $(3\sqrt{a} - 5\sqrt{b})(3\sqrt{a} + 5\sqrt{b})$

44. $(4\sqrt{m} - 7\sqrt{n})(4\sqrt{m} + 7\sqrt{n})$

45. $(4 + \sqrt{7})^2$

46. $(6 + \sqrt{2})^2$

47. $(\sqrt{3} - \sqrt{5})^2$

48. $(\sqrt{2} - \sqrt{7})^2$

49. $(b + \sqrt{2})^2$

50. $(w + \sqrt{6})^2$

51. $(\sqrt{x} - \sqrt{6})^2$

52. $(\sqrt{x} - \sqrt{5})^2$

53. $(2\sqrt{5} - 3\sqrt{2})^2$

54. $(3\sqrt{3} - 4\sqrt{5})^2$

55. A student tries to simplify $(x + \sqrt{3})^2$:

$$(x + \sqrt{3})^2 = x^2 + (\sqrt{3})^2 = x^2 + 3$$

Describe any errors. Then simplify the expression correctly.

56. A student tries to simplify $(x - \sqrt{5})^2$:

$$(x - \sqrt{5})^2 = x^2 - (\sqrt{5})^2 = x^2 - 5$$

Describe any errors. Then simplify the expression correctly.

57. A student tries to find the product $3(2\sqrt{5})$:

$$3(2\sqrt{5}) = 6\sqrt{15}$$

Describe any errors. Then find the product correctly.

58. A student tries to find the product $(4\sqrt{7})(5\sqrt{7})$:

$$(4\sqrt{7})(5\sqrt{7}) = 20\sqrt{7}$$

Describe any errors. Then find the product correctly.

59. a. Decide whether each of the following is true or false:

  i. $\sqrt{ab} = \sqrt{a}\sqrt{b}$

  ii. $\sqrt{a + b} = \sqrt{a} + \sqrt{b}$

  iii. $(ab)^2 = a^2 b^2$

  iv. $(a + b)^2 = a^2 + b^2$

  v. $\dfrac{1}{ab} = a^{-1}b^{-1}$

  vi. $\dfrac{1}{a + b} = a^{-1} + b^{-1}$

 b. Compare the types of equations that are true and the types of equations that are false in part (a). What patterns do you notice?

60. Describe how to simplify a radical expression. (See page 9 for guidelines on writing a good response.)

Related Review

Perform the indicated operation.

61. $3\sqrt{x} + 2\sqrt{x}$

62. $5\sqrt{x} - 7\sqrt{x}$

63. $(3\sqrt{x})(2\sqrt{x})$

64. $(5\sqrt{x})(-7\sqrt{x})$

65. $(3 + \sqrt{x})(2 + \sqrt{x})$

66. $(5 + \sqrt{x})(7 - \sqrt{x})$

67. $(3 + x)(2 + x)$

68. $(5 + x)(7 - x)$

69. $(3\sqrt{x})^2$

70. $(-5\sqrt{x})^2$

71. $(3 + \sqrt{x})^2$

72. $(5 - \sqrt{x})^2$

73. $(3 + x)^2$

74. $(5 - x)^2$

Expressions, Equations, and Graphs

Perform the indicated instruction. Then use words such as linear, quadratic, cubic, power, radical, rational, polynomial, degree, one variable, *and* two variables *to describe the expression, equation, or system.*

75. Graph $y = -\dfrac{2}{5}x - 1$ by hand.

76. Solve $-3x = 7x - 4(2x - 1)$.

77. Solve $\dfrac{p}{p + 4} - \dfrac{4}{p - 4} = \dfrac{p^2 + 16}{p^2 - 16}$.

78. Simplify $(\sqrt{a} + 2)^2$.

79. Find the quotient $\dfrac{x^2 - 1}{x^2 + 5x + 6} \div \dfrac{x^2 - 3x - 4}{x^2 + 3x}$.

80. Simplify $3\sqrt{75} - 5\sqrt{12}$.

11.3 SOLVING SQUARE ROOT EQUATIONS

Objectives

▷ Solve *square root equations* in one variable.

▷ Graph a square root equation in two variables.

▷ Use a *square root model* to make predictions.

In Sections 9.1, 9.2, 11.1, and 11.2, we worked with radical *expressions*—specifically, square root expressions. In this section, we solve square root *equations*.

Solving Square Root Equations in One Variable

A **square root equation** is an equation that contains at least one square root expression. Here are some examples of a square root equation in one variable:

$$\sqrt{x} = 7 \qquad 2\sqrt{4x - 7} = 12 \qquad \sqrt{x - 5} = x - 8$$

Consider the square root equation

$$\sqrt{x} = 5$$

If x is negative, then $\sqrt{x}$ cannot be 5, because $\sqrt{x}$ is not a real number. If x is nonnegative, then $\left(\sqrt{x}\right)^2 = x$. To get the left side of the equation $\sqrt{x} = 5$ to be x, we square both sides:

$$\left(\sqrt{x}\right)^2 = 5^2$$
$$x = 25$$

So, the solution is 25. This checks out, because $\sqrt{25} = 5$.

This work suggests using the **squaring property of equality** to help us solve square root equations.

Squaring Property of Equality

If A and B are expressions, then all solutions of the equation $A = B$ are *among* the solutions of the equation $A^2 = B^2$. That is, the solutions of an equation are among the solutions of the equation obtained by squaring both sides.

Recall from Section 10.5 that if we clear a rational equation of fractions and arrive at a value of x that is an excluded value, then we call that result an extraneous solution. In general, if a proposed solution of any type of equation is *not* a solution, we call it an **extraneous solution.**

Squaring both sides of an equation can introduce extraneous solutions. Consider the simple equation $x = 3$, whose only solution is 3. Here we square both sides of $x = 3$ and then solve by using the square root property (Section 9.3):

$$
\begin{array}{ll}
x = 3 & \text{The only solution is 3.} \\
x^2 = 3^2 & \text{Square both sides.} \\
x^2 = 9 & 3^2 = 9 \\
x = \pm\sqrt{9} & x^2 = k \text{ is equivalent to } x = \pm\sqrt{k} \\
& \text{for nonnegative constant } k. \\
x = \pm 3 & \sqrt{9} = 3
\end{array}
$$

Squaring both sides of the equation $x = 3$ introduced the extraneous solution -3 (which is *not* a solution of the original equation).

Checking Proposed Solutions

Since squaring both sides of a square root equation may introduce extraneous solutions, it is essential to check that each proposed solution satisfies the original equation.

Example 1 Solving a Square Root Equation

Solve $\sqrt{6x - 2} = 4$.

Solution

$$\sqrt{6x - 2} = 4 \qquad \text{Original equation}$$

$$\left(\sqrt{6x - 2}\right)^2 = 4^2 \qquad \text{Square both sides.}$$

$$6x - 2 = 16 \qquad (\sqrt{x})^2 = x \text{ for } x \geq 0$$

$$6x = 18 \qquad \text{Add 2 to both sides.}$$

$$x = 3 \qquad \text{Divide both sides by 6.}$$

We check that 3 satisfies the original equation:

$$\sqrt{6x - 2} = 4 \qquad \text{Original equation}$$

$$\sqrt{6(3) - 2} \stackrel{?}{=} 4 \qquad \text{Substitute 3 for x.}$$

$$\sqrt{16} \stackrel{?}{=} 4 \qquad \text{Simplify.}$$

$$4 \stackrel{?}{=} 4 \qquad \sqrt{16} = 4$$

$$\text{true}$$

So, the solution is 3. ∎

Example 2 Solving a Square Root Equation

Solve $\sqrt{x + 5} = -3$.

Solution

$$\sqrt{x + 5} = -3 \qquad \text{Original equation}$$

$$\left(\sqrt{x + 5}\right)^2 = (-3)^2 \qquad \text{Square both sides.}$$

$$x + 5 = 9 \qquad (\sqrt{x})^2 = x \text{ for } x \geq 0$$

$$x = 4 \qquad \text{Subtract 5 from both sides.}$$

We check that 4 satisfies the original equation:

$$\sqrt{x + 5} = -3 \qquad \text{Original equation}$$

$$\sqrt{4 + 5} \stackrel{?}{=} -3 \qquad \text{Substitute 4 for x.}$$

$$\sqrt{9} \stackrel{?}{=} -3 \qquad \text{Add.}$$

$$3 \stackrel{?}{=} -3 \qquad \sqrt{9} = 3$$

$$\text{false}$$

The result, 4, is an extraneous solution—it is *not* a solution. The solution set is the empty set. ∎

In Example 2, we introduced an extraneous solution by squaring both sides of an equation. Remember that if you square both sides of an equation, you must check that each proposed solution satisfies the original equation.

Example 3 Solving a Square Root Equation

Solve $\sqrt{6x - 5} = \sqrt{4x + 2}$.

Solution

$$\sqrt{6x - 5} = \sqrt{4x + 2} \qquad \text{Original equation}$$

$$\left(\sqrt{6x - 5}\right)^2 = \left(\sqrt{4x + 2}\right)^2 \qquad \text{Square both sides.}$$

$$6x - 5 = 4x + 2 \qquad (\sqrt{x})^2 = x \text{ for } x \geq 0$$

$$2x = 7 \qquad \text{Subtract 4x from both sides; add 5 to both sides.}$$

$$x = \frac{7}{2} \qquad \text{Divide both sides by 2.}$$

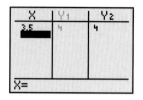

Figure 4 Verify the work

We check that $\frac{7}{2}$ satisfies the original equation:

$$\sqrt{6x - 5} = \sqrt{4x + 2} \qquad \text{Original equation}$$

$$\sqrt{6\left(\frac{7}{2}\right) - 5} \stackrel{?}{=} \sqrt{4\left(\frac{7}{2}\right) + 2} \qquad \text{Substitute } \frac{7}{2} \text{ for } x.$$

$$\sqrt{21 - 5} \stackrel{?}{=} \sqrt{14 + 2} \qquad \text{Simplify.}$$

$$\sqrt{16} \stackrel{?}{=} \sqrt{16} \qquad \text{Simplify.}$$

$$\text{true}$$

The solution is $\frac{7}{2}$. We can also use a graphing calculator table to verify our work (see Fig. 4). ∎

Example 4 Isolating a Square Root to One Side of the Equation

Solve $2 + 3\sqrt{y} = 23$.

Solution

We isolate $\sqrt{y}$ to one side of the equation and then square both sides:

$$2 + 3\sqrt{y} = 23 \qquad \text{Original equation}$$

$$3\sqrt{y} = 21 \qquad \text{Subtract 2 from both sides.}$$

$$\sqrt{y} = 7 \qquad \text{Divide both sides by 3.}$$

$$\left(\sqrt{y}\right)^2 = 7^2 \qquad \text{Square both sides.}$$

$$y = 49 \qquad (\sqrt{x})^2 = x \text{ for } x \geq 0$$

We check that 49 satisfies the original equation:

$$2 + 3\sqrt{y} = 23 \qquad \text{Original equation}$$

$$2 + 3\sqrt{49} \stackrel{?}{=} 23 \qquad \text{Substitute 49 for } y.$$

$$2 + 3(7) \stackrel{?}{=} 23 \qquad \sqrt{49} = 7$$

$$23 \stackrel{?}{=} 23 \qquad \text{Simplify.}$$

$$\text{true}$$

The solution is 49. ∎

We see from Example 4 that to solve a square root equation, we isolate a square root term to one side of the equation before squaring both sides.

Solving a Square Root Equation in One Variable

We solve a square root equation in one variable by following these steps:

1. Isolate a square root term to one side of the equation.

2. Square both sides.

3. Solve the new equation.

4. Check that each proposed solution satisfies the original equation.

Example 5 Isolating a Square Root to One Side of the Equation

Solve $\sqrt{3x - 5} + 4 = x + 1$.

Solution

We isolate $\sqrt{3x-5}$ to one side of the equation and then square both sides:

$$\sqrt{3x-5}+4 = x+1 \qquad \text{Original equation}$$
$$\sqrt{3x-5} = x-3 \qquad \text{Subtract 4 from both sides.}$$
$$\left(\sqrt{3x-5}\right)^2 = (x-3)^2 \qquad \text{Square both sides.}$$
$$3x-5 = x^2-6x+9 \qquad (\sqrt{x})^2 = x \text{ for } x \geq 0;\ (A-B)^2 = A^2-2AB+B^2$$
$$0 = x^2-9x+14 \qquad \text{Write in } 0 = ax^2+bx+c \text{ form.}$$
$$0 = (x-2)(x-7) \qquad \text{Factor right-hand side.}$$
$$x-2 = 0 \quad \text{or} \quad x-7 = 0 \qquad \text{Zero factor property}$$
$$x = 2 \quad \text{or} \qquad x = 7$$

We check that both 2 and 7 satisfy the original equation:

Check $x = 2$

$$\sqrt{3x-5}+4 = x+1$$
$$\sqrt{3(2)-5}+4 \overset{?}{=} 2+1$$
$$\sqrt{1}+4 \overset{?}{=} 3$$
$$5 \overset{?}{=} 3$$
false

Check $x = 7$

$$\sqrt{3x-5}+4 = x+1$$
$$\sqrt{3(7)-5}+4 \overset{?}{=} 7+1$$
$$\sqrt{16}+4 \overset{?}{=} 8$$
$$8 \overset{?}{=} 8$$
true

So, the only solution is 7.

In Example 5, we simplified the right-hand side of the equation

$$\left(\sqrt{3x-5}\right)^2 = (x-3)^2$$

by using the property for the square of a difference, namely, $(A-B)^2 = A^2-2AB+B^2$:

$$(x-3)^2 = x^2-6x+9 \qquad \text{Correct}$$

WARNING   Remember that, in general, $(A-B)^2$ is *not* equal to A^2-B^2, so it is incorrect to say

$$(x-3)^2 = x^2-9 \qquad \text{Incorrect}$$

If an equation contains two square root terms, we may need to use the squaring property of equality twice.

Example 6   Using the Squaring Property of Equality Twice

Solve $\sqrt{x+21} = \sqrt{x}+3$.

Solution

$$\sqrt{x+21} = \sqrt{x}+3 \qquad \text{Original equation}$$
$$\left(\sqrt{x+21}\right)^2 = \left(\sqrt{x}+3\right)^2 \qquad \text{Square both sides.}$$
$$x+21 = \left(\sqrt{x}\right)^2 + 2\left(\sqrt{x}\right)3 + 3^2 \qquad (\sqrt{x})^2 = x \text{ for } x \geq 0;\ (A+B)^2 = A^2+2AB+B^2$$
$$x+21 = x+6\sqrt{x}+9 \qquad (\sqrt{x})^2 = x \text{ for } x \geq 0; \text{ simplify.}$$
$$12 = 6\sqrt{x} \qquad \text{Subtract } x \text{ from both sides; subtract 9 from both sides.}$$
$$2 = \sqrt{x} \qquad \text{Divide both sides by 6.}$$
$$2^2 = \left(\sqrt{x}\right)^2 \qquad \text{Square both sides.}$$
$$4 = x \qquad \text{Simplify.}$$

We check that 4 satisfies the original equation:

$$\sqrt{x+21} = \sqrt{x}+3 \qquad \text{Original equation}$$
$$\sqrt{4+21} \overset{?}{=} \sqrt{4}+3 \qquad \text{Substitute 4 for } x.$$
$$5 \overset{?}{=} 5 \qquad \text{Simplify.}$$
true

The solution is 4.

Graphs of Square Root Equations in Two Variables

So far in this section, we have solved square root equations in *one* variable. We can describe the solutions of square root equations in *two* variables with a graph. Here are some examples of square root equations in two variables:

$$y = \sqrt{x} \qquad y = \sqrt{x+5} \qquad y = 2\sqrt{x}+3$$

Example 7   Graphing a Square Root Equation

Sketch the graph of $y = \sqrt{x}$.

Solution

We list some solutions in Table 4. We choose perfect squares as values of x, because we can find their principal square roots mentally. Since the radicand of $\sqrt{x}$ must be nonnegative, we cannot choose any negative numbers as values of x. Then we plot the points that correspond to the solutions we found and sketch a curve that contains these points (see Fig. 5).

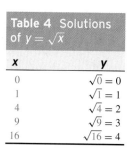

Table 4 Solutions of $y = \sqrt{x}$

x	y
0	$\sqrt{0} = 0$
1	$\sqrt{1} = 1$
4	$\sqrt{4} = 2$
9	$\sqrt{9} = 3$
16	$\sqrt{16} = 4$

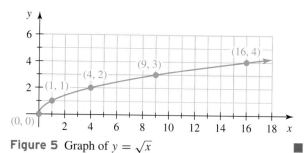

Figure 5 Graph of $y = \sqrt{x}$

In Fig. 6, we show the graphs of $y = \sqrt{x+5}$ and $y = 2\sqrt{x}+3$.

In Example 5, we found that 7 is the solution of the equation $\sqrt{3x-5}+4 = x+1$. We can verify our work by using "intersect" on a graphing calculator (see Fig. 7).

Figure 6 Graphs of two square root equations in two variables

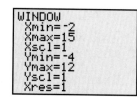 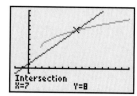 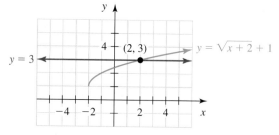

Figure 7 Verify the work

Example 8   Solving a Radical Equation in One Variable by Graphing

The graphs of $y = \sqrt{x+2}+1$ and $y = 3$ are shown in Fig. 8. Use these graphs to solve the equation $\sqrt{x+2}+1 = 3$.

Figure 8 Solving a radical equation in one variable

Solution

The graphs of $y = \sqrt{x+2}+1$ and $y = 3$ intersect only at the point $(2, 3)$, which has x-coordinate 2. So, 2 is the solution of $\sqrt{x+2}+1 = 3$. ∎

Square Root Models

Just as we have used linear equations, quadratic equations, and rational equations to model authentic situations, we can use square root equations in two variables to model such situations. A **square root model** is a square root equation in two variables that is used to describe an authentic situation.

Example 9   Using a Square Root Model

The percentages of American adults who watch cable television are shown in Table 5 for various income groups.

Table 5 Percentages of American Adults Who Watch Cable Television		
Thousands of Dollars		
Annual Income Group	**Income Used to Represent Income Group**	**Percent**
0–9.999	5.0	55.9
10–19.999	15.0	62.3
20–29.999	25.0	67.8
30–34.999	32.5	72.2
35–39.999	37.5	74.2
40–49.999	45.0	77.2
50+	70.0	84.8

Source: *Mediamark Research, Inc.*

Let p be the percentage of American adults with an annual income of I thousand dollars who watch cable television. A square root model of the situation is

$$p = 4.7\sqrt{I} + 45$$

1. Verify that the model fits the data well.
2. Find the p-intercept. What does it mean in this situation?
3. Estimate the percentage of adults with an income of $18 thousand who watch cable television.
4. Estimate the income at which 73% of adults watch cable television.

Solution

1. We draw the graph of the model and, in the same viewing window, the scattergram of the data (see Fig. 9). It appears that the model fits the data well.
2. To find the p-intercept, we substitute 0 for I in the equation $p = 4.7\sqrt{I} + 45$ and solve for p:

$$p = 4.7\sqrt{0} + 45 = 45$$

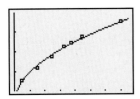

Figure 9 Check how well the model fits the data

The p-intercept is $(0, 45)$. According to the model, 45% of adults who have no income watch cable television. Model breakdown has likely occurred, because it is hard to imagine that as many as 45% of American adults with no income could afford to subscribe to cable television.

3. We substitute 18 for I in the equation $p = 4.7\sqrt{I} + 45$ and solve for p:

$$p = 4.7\sqrt{18} + 45 \approx 64.94$$

So, 64.9% of adults with an income of $18 thousand watch cable television.

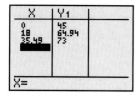

Figure 10 Verify the work

4. We substitute 73 for p in the equation $p = 4.7\sqrt{I} + 45$ and solve for I:

$$73 = 4.7\sqrt{I} + 45 \qquad \text{Substitute 73 for } p.$$
$$28 = 4.7\sqrt{I} \qquad \text{Subtract 45 from both sides.}$$
$$\frac{28}{4.7} = \sqrt{I} \qquad \text{Divide both sides by 4.7.}$$
$$\left(\frac{28}{4.7}\right)^2 = (\sqrt{I})^2 \qquad \text{Square both sides.}$$
$$35.49 \approx I \qquad \text{Calculate; } (\sqrt{x})^2 = x \text{ for } x \geq 0$$

Our result 35.49 does approximately satisfy $73 = 4.7\sqrt{I} + 45$. (Try it.) So, 73% of Americans whose income is \$35.5 thousand watch cable television.

We use a graphing calculator table to verify our work in Problems 2–4 (see Fig. 10). ∎

group exploration

Solving square root equations

1. A student tries to solve $\sqrt{x} + 3 = 5$:

$$\sqrt{x} + 3 = 5$$
$$(\sqrt{x})^2 + 3^2 = 5^2$$
$$x + 9 = 25$$
$$x = 16$$

Describe any errors. Then solve the equation correctly.

2. A student tries to solve $\sqrt{2x - 5} = -3$:

$$\sqrt{2x - 5} = -3$$
$$(\sqrt{2x - 5})^2 = (-3)^2$$
$$2x - 5 = 9$$
$$2x = 14$$
$$x = 7$$

Describe any errors. Then solve the equation correctly.

3. A student tries to solve $3\sqrt{x} - 2 = 5$:

$$3\sqrt{x} - 2 = 5$$
$$3\sqrt{x} = 7$$
$$(3\sqrt{x})^2 = 7^2$$
$$3x = 49$$
$$x = \frac{49}{3}$$

Describe any errors. Then solve the equation correctly.

TIPS FOR SUCCESS: Retake Quizzes and Exams to Prepare for the Final Exam

To study for your final exam, consider retaking your quizzes and other exams. These quizzes and exams can reveal your weak areas. If you have difficulty with a certain concept, you can refer to Homework exercises that address that concept. Reflect on *why* you are having such difficulty, rather than just doing more Homework exercises that address the concept.

HOMEWORK 11.3 FOR EXTRA HELP ▶

 Student Solutions Manual

 PH Math/Tutor Center

 Math XL MathXL®

 MyMathLab MyMathLab

Solve.

1. $\sqrt{x} = 6$ **2.** $\sqrt{x} = 2$ **3.** $\sqrt{5x+1} = 4$

4. $\sqrt{2x-6} = 2$ **5.** $\sqrt{x} = -7$ **6.** $\sqrt{x} = -3$

7. $\sqrt{3t-2} = 2$ **8.** $\sqrt{4w-3} = 5$ **9.** $\sqrt{x+4} = -2$

10. $\sqrt{x-8} = -5$ **11.** $\sqrt{5x-2} = \sqrt{3x+8}$

12. $\sqrt{7x-1} = \sqrt{4x+17}$ **13.** $\sqrt{25p^2+4p-8} = 5p$

14. $\sqrt{9r^2-7r+21} = 3r$ **15.** $\sqrt{x^2-6x} = 4$

16. $\sqrt{x^2+5x} = 6$ **17.** $\sqrt{x}+3 = 7$

18. $\sqrt{x}-5 = 1$ **19.** $3\sqrt{x} = 6$

20. $4\sqrt{x} = 20$ **21.** $5\sqrt{w}+3 = 23$

22. $4\sqrt{m}-1 = 11$ **23.** $-3\sqrt{x}+7 = 2$

24. $-2\sqrt{x}+10 = 3$ **25.** $4\sqrt{x}-3 = -5$

26. $5\sqrt{x}-2 = -12$ **27.** $5+\sqrt{a-3} = 8$

28. $7+\sqrt{b+1} = 10$ **29.** $\sqrt{x-2} = x-2$

30. $\sqrt{x+4} = x-2$ **31.** $\sqrt{x+3} = x+1$

32. $\sqrt{x-5} = x-7$ **33.** $\sqrt{4r-3}-r = -2$

34. $\sqrt{5t-1}-t = 1$ **35.** $\sqrt{(x+5)(x+2)} = x+3$

36. $\sqrt{(x-1)(x+4)} = x+1$

37. $\sqrt{x+5} = \sqrt{x}+1$

38. $\sqrt{x-4} = \sqrt{x}-2$ **39.** $\sqrt{m-5}+2 = \sqrt{m+3}$

40. $\sqrt{p+2}-3 = \sqrt{p-7}$

Graph the equation by hand. Then use a graphing calculator to verify your work.

41. $y = 1+\sqrt{x}$ **42.** $y = 2\sqrt{x}$

For Exercises 43–46, use the graph of $y = \sqrt{x+4}-1$ shown in Fig. 11 to solve the given equation.

43. $\sqrt{x+4}-1 = -1$ **44.** $\sqrt{x+4}-1 = 0$

45. $\sqrt{x+4}-1 = -2$ **46.** $\sqrt{x+4}-1 = 1$

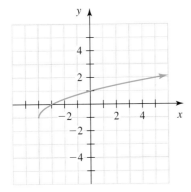

Figure 11 Exercises 43–46

Use "intersect" on a graphing calculator to solve the equation. Round the solution(s) to the second decimal place.

47. $\sqrt{x}-4 = 3-x$ **48.** $\sqrt{x}+2 = 2x-5$

49. $\sqrt{x}+3 = x^2-5$ **50.** $\sqrt{x}-4 = 3-x^2$

51. A student tries to solve $\sqrt{x^2+x+6} = x+4$:

$$\sqrt{x^2+x+6} = x+4$$
$$\left(\sqrt{x^2+x+6}\right)^2 = (x+4)^2$$
$$x^2+x+6 = x^2+16$$
$$x = 10$$

Describe any errors. Then solve the equation correctly.

52. A student tries to solve $\sqrt{x-1} = x-3$:

$$\sqrt{x-1} = x-3$$
$$\left(\sqrt{x-1}\right)^2 = (x-3)^2$$
$$x-1 = x^2-6x+9$$
$$0 = x^2-7x+10$$
$$0 = (x-2)(x-5)$$
$$x-2 = 0 \quad \text{or} \quad x-5 = 0$$
$$x = 2 \quad \text{or} \quad x = 5$$

The student says that the solutions are 2 and 5. Is the student correct? Explain.

53. The number of acres devoted to genetically modified crops has increased greatly in the past few years (see Table 6).

Table 6 Genetically Modified Crops

Year	Genetically Modified Crops (millions of acres)
1997	28
1998	69
1999	99
2000	109
2001	139

Sources: *Clive James; Department of Agriculture*

Let c be the amount of land used for genetically modified crops (in millions of acres) at t years since 1997. A model of the situation is $c = 50\sqrt{t}+28$.
a. Use a graphing calculator to draw the graph of the model and, in the same viewing window, the scattergram of the data. Does the model fit the data well?
b. What is the c-intercept? What does it mean in this situation?
c. Predict the number of acres of genetically modified crops in 2010. Is your result larger than the land area of Texas, which is 167.6 million acres?
d. Predict when there will be 225 million acres of genetically modified crops.

I've been growing genetically modified oranges for years and they're the most natural thing you can imagine.

!!!

54. The most popular video game console in the United States is the Sony PlayStation®. The numbers of households in the United States that have this console are listed in Table 7 for various years.

Table 7 Numbers of Households That Have a Sony PlayStation

Year	Number of Households That Have a Sony PlayStation (millions)
1998	16.0
1999	21.9
2000	25.0
2001	27.4
2002	28.8

Source: *International Development Group*

Let n be the number of households (in millions) that have a Sony PlayStation at t years since 1998. A model of the situation is $n = 6.4\sqrt{t} + 16$.
a. Use a graphing calculator to draw the graph of the model and, in the same viewing window, the scattergram of the data. Does the model fit the data well?
b. What is the n-intercept? What does it mean in this situation?
c. Predict when 38 million households will have a Sony PlayStation.

55. The numbers of McDonald's restaurants around the world are shown in Table 8 for various years. Let n be the number of McDonald's restaurants (in thousands) at t years since 1999. A model of the situation is $n = 2.65\sqrt{t} + 26$.
a. Use a graphing calculator to draw the graph of the model and, in the same viewing window, the scattergram of the data. Does the model fit the data well?
b. Predict the number of McDonald's restaurants in 2011.
c. There are 192 countries in the world. What will be the average number of McDonald's restaurants per country in 2011?
d. Predict when there will be 36 thousand McDonald's restaurants.

Table 8 Numbers of McDonald's Restaurants

Year	Number of Restaurants (thousands)
1999	26
2000	29
2001	30
2002	31
2003	31
2004	32
2005	32

Source: *McDonald's*

56. The percentages of convicts released from a state prison who have subsequently been arrested for a new crime are shown in Table 9 for various numbers of years since their release.

Table 9 Released Convicts Who Have Been Arrested for a New Crime

Number of Years Since Release	Percent
0	0
0.5	30
1	44
2	59
3	68

Source: *U.S. Department of Justice*

Let p be the percentage of convicts released from a state prison who have been arrested for a new crime after being out of prison for t years. A model of the situation is $p = 41\sqrt{t}$.
a. Use a graphing calculator to draw the graph of the model and, in the same viewing window, the scattergram of the data. Does the model fit the data well?
b. Estimate the percentage of released convicts who have been arrested for a new crime after being out of prison for 5 years.
c. Estimate the number of years since their release when all convicts will have been arrested for a new crime.

57. Tire dumps are unsightly and sometimes catch fire, releasing hazardous chemicals. In 2003 alone, Americans threw out 300 million tires, one per person in the United States. Table 10 shows the percentages of tires that were reused or recycled for various years.

Table 10 Reused or Recycled Tires

Year	Percent
1990	11
1992	27
1994	55
1996	62
1998	67
2001	78
2003	83

Source: *Rubber Manufacturers Association*

Let p be the percentage of tires that have been reused or recycled at t years since 1990. A model of the situation is $p = 20\sqrt{t} + 11$.
a. Use a graphing calculator to draw the graph of the model and, in the same viewing window, the scattergram of the data. Does the model fit the data well?
b. Predict when all tires will be reused or recycled.

c. Currently, there are two U.S. plants that burn tires to create energy. A third plant has been proposed that would burn 10 million tires per year and produce enough energy for up to 20 thousand homes. However, there are concerns about the release of harmful chemicals from the plant. Should the plant be built? To decide, make the following assumptions:

- Construction would have begun in 2006, and it would take two years for the plant to be built.

- It is better to reuse or to recycle a tire than to burn it for energy.

- If a tire can't be reused or recycled, it is better to burn it for energy than to dump it in a landfill or tire pile.

[**Hint:** Refer to your result in part (b).]

58. Describe how to solve a square root equation. Is checking your answers for extraneous solutions optional?

Related Review

Solve.

59. $2x^2 + 4x = 7$

60. $2(3x - 5) - 4(2x + 1) = 8$

61. $\sqrt{x + 2} = x - 4$

62. $x^2 - 5x = 24$

63. $\dfrac{3}{x - 2} - \dfrac{5x}{x^2 - 4} = \dfrac{7}{x + 2}$

64. $(x + 3)^2 = 40$

Expressions, Equations, and Graphs

Give an example of each. Then solve, simplify, or graph, as appropriate.

65. sum of two rational expressions

66. radical equation in one variable

67. linear equation in two variables

68. system of two linear equations in two variables

69. quadratic equation in two variables

70. linear equation in one variable

71. rational equation in one variable

72. quadratic polynomial with four terms

73. radical expression with four terms

74. quadratic equation in one variable

Chapter Summary

Key Points
OF CHAPTER 11

Like radicals (Section 11.1)	Square root radicals that have the same radicand are called **like radicals.**
Adding or subtracting like radicals (Section 11.1)	To add or subtract like radicals, we use the distributive law.
Power property for square roots (Section 11.2)	If x is nonnegative, then $(\sqrt{x})^2 = \sqrt{x}\sqrt{x} = x$.
Product of two radical expressions, each with two terms (Section 11.2)	To find the product of two radical expressions in which each factor has two terms, 1. Multiply each term in the first radical expression by each term in the second radical expression. 2. Combine like radicals.
Radical conjugate (Section 11.2)	To find the **radical conjugate** of a radical expression with two terms, we change the addition symbol to a subtraction symbol or vice versa.
Product of two radical conjugates (Section 11.2)	To find the product of two radical conjugates, use the property $(A + B)(A - B) = A^2 - B^2$.
Simplifying the square of a radical expression (Section 11.2)	To simplify the square of a radical expression with two terms, • Use $C^2 = CC$ to write the square as a product of two identical expressions, and then multiply each term in the first radical expression by each term in the second radical expression, or • Use the property $(A+B)^2 = A^2 + 2AB + B^2$ or the property $(A - B)^2 = A^2 - 2AB + B^2$.
Square root equation (Section 11.3)	A **square root equation** is an equation that contains at least one square root expression.
Squaring property of equality (Section 11.3)	If A and B are expressions, then all solutions of the equation $A = B$ are *among* the solutions of the equation $A^2 = B^2$. That is, the solutions of an equation are among the solutions of the equation obtained by squaring both sides.
Checking proposed solutions (Section 11.3)	Since squaring both sides of a square root equation may introduce extraneous solutions, it is essential to check that each proposed solution satisfies the original equation.

Solving a square root equation in one variable (Section 11.3)	We solve a square root equation in one variable by following these steps:
	1. Isolate a square root term to one side of the equation.
	2. Square both sides.
	3. Solve the new equation.
	4. Check that each proposed solution satisfies the original equation.
Square root model (Section 11.3)	A **square root model** is a square root equation in two variables that is used to describe an authentic situation.

CHAPTER 11 REVIEW EXERCISES

Perform the indicated operation(s). Assume that any variables are nonnegative. Use a graphing calculator to verify your work when possible.

1. $6\sqrt{3} + 9\sqrt{3}$

2. $-4\sqrt{2x} + 3\sqrt{2x}$

3. $7\sqrt{5x} - 9\sqrt{5x}$

4. $8\sqrt{3} + 4\sqrt{6}$

5. $2\sqrt{5} - 3\sqrt{7} - 8\sqrt{5}$

6. $\sqrt{27} + 8\sqrt{3}$

7. $-3\sqrt{6} - \sqrt{24}$

8. $\sqrt{32} + \sqrt{18}$

9. $\sqrt{45} - \sqrt{20}$

10. $5\sqrt{8} + 3\sqrt{50}$

11. $2\sqrt{40} - 5\sqrt{90}$

12. $\sqrt{32} - 5\sqrt{3} - \sqrt{72}$

13. $7\sqrt{5} + 4\sqrt{20} - 8\sqrt{24}$

14. $\sqrt{64x} + \sqrt{81x}$

15. $2\sqrt{25x} - 3\sqrt{49x}$

16. $\sqrt{12x} + \sqrt{8} - \sqrt{75x}$

17. $w\sqrt{28} - w\sqrt{63}$

18. $\sqrt{7a^2} + \sqrt{28a^2}$

19. $5\sqrt{3x^2} - 3x\sqrt{48}$

20. $2\sqrt{8x} - 4\sqrt{45x^2} + x\sqrt{20}$

21. $\sqrt{7a}\sqrt{5b}$

22. $4\sqrt{3}\left(-2\sqrt{6}\right)$

23. $\left(5\sqrt{x}\right)^2$

24. $\sqrt{2}\left(\sqrt{3} + \sqrt{7}\right)$

25. $-\sqrt{3}\left(\sqrt{7} - \sqrt{5}\right)$

26. $\sqrt{x}\left(\sqrt{x} - \sqrt{2}\right)$

27. $2\sqrt{7}\left(5\sqrt{3} + \sqrt{7}\right)$

28. $\left(\sqrt{5} - 3\right)\left(\sqrt{5} + 6\right)$

29. $\left(\sqrt{2} - 9\right)\left(\sqrt{2} - 8\right)$

30. $\left(\sqrt{2} - \sqrt{7}\right)\left(\sqrt{3} - \sqrt{5}\right)$

31. $\left(\sqrt{b} + 8\right)\left(\sqrt{b} - 1\right)$

32. $\left(t + \sqrt{3}\right)\left(t + \sqrt{5}\right)$

33. $\left(3\sqrt{7} - 1\right)\left(2\sqrt{7} - 2\right)$

34. $\left(\sqrt{5} - \sqrt{7}\right)\left(\sqrt{5} + \sqrt{7}\right)$

35. $\left(4\sqrt{a} + 3\sqrt{b}\right)\left(4\sqrt{a} - 3\sqrt{b}\right)$

36. $\left(b - \sqrt{3}\right)\left(b + \sqrt{3}\right)$

37. $\left(4 - \sqrt{5}\right)^2$

38. $\left(\sqrt{x} + 6\right)^2$

39. $\left(x + \sqrt{3}\right)^2$

40. $\left(3\sqrt{5} - 4\sqrt{2}\right)^2$

Solve.

41. $\sqrt{9x^2 - 5x + 20} = 3x$

42. $\sqrt{2x - 8} = 6$

43. $\sqrt{3r + 8} = -7$

44. $\sqrt{4p + 3} = \sqrt{6p - 2}$

45. $\sqrt{x^2 - 3x} = 2$

46. $\sqrt{x} + 2 = 9$

47. $2\sqrt{x} - 5 = 11$

48. $-3\sqrt{x} + 4 = 16$

49. $3 + \sqrt{y - 2} = 5$

50. $\sqrt{p + 5} = x + 3$

51. $\sqrt{(x + 3)(x - 4)} = x - 2$

52. $1 + \sqrt{2x - 3} = x$

53. $\sqrt{5w + 1} - w = -1$

54. $\sqrt{t + 5} = \sqrt{t} + 1$

55. Use "intersect" on a graphing calculator to solve the equation $-2\sqrt{x} + 3 = \frac{1}{2}x^2 - 6$. Round the solution(s) to the second decimal place.

56. In 1992, the pharmaceutical industry pledged to give the Food and Drug Administration (FDA) millions of dollars each year, provided that a specified amount of money would be spent on new-drug approvals. Due to waning support from Congress, the FDA has cut spending in all areas—including checking the safety of drugs already in use—except new-drug approvals (see Table 11).

Table 11 Percentages of the FDA Budget Spent on New-Drug Approvals and on Drug Safety and Other Areas

	Percent	
Year	New-Drug Approvals	Drug Safety and Other Areas
1993	52.8	47.2
1995	63.2	36.8
1997	69.6	30.4
1999	76.8	23.2
2001	77.6	22.4
2003	79.0	21.0

Source: *Food and Drug Administration*

Let p be the percentage of the FDA budget that went to new-drug approvals for the year that is t years since 1993. A model of the situation is $p = 9\sqrt{t} + 52.3$.

a. Use a graphing calculator to draw the graph of the model and, in the same viewing window, the scattergram of the data. Does the model fit the data well?

b. Predict what percentage of the FDA budget will go to new-drug approvals in 2011.

c. Predict when all of the FDA budget will go to new-drug approvals.

CHAPTER 11 TEST

Perform the indicated operation(s). Assume that any variables are nonnegative.

1. $5\sqrt{3} - 3\sqrt{2} - 7\sqrt{3}$ 2. $-3\sqrt{20x} - 2\sqrt{45x}$

3. $5\sqrt{45} - 5\sqrt{18x} - 2\sqrt{32x}$

4. $3\sqrt{24b^2} - 7\sqrt{6b^2}$

5. $-8\sqrt{14} \cdot 5\sqrt{2}$ 6. $\sqrt{x}(\sqrt{x} - 3)$

7. $(\sqrt{5} - 2)(\sqrt{5} + 4)$ 8. $(\sqrt{2} - \sqrt{5})(\sqrt{2} + \sqrt{7})$

9. $(5\sqrt{a} - 2\sqrt{b})(5\sqrt{a} + 2\sqrt{b})$

10. $(4 + \sqrt{3})^2$ 11. $(\sqrt{x} - \sqrt{5})^2$ 12. $(3\sqrt{2} + 2\sqrt{3})^2$

Solve.

13. $\sqrt{3x - 5} = 4$ 14. $\sqrt{2x + 7} = \sqrt{5x - 8}$

15. $4\sqrt{t} - 3 = 5$ 16. $3 + \sqrt{w + 5} = 9$

17. $\sqrt{x - 3} = x - 5$ 18. $\sqrt{x + 8} = \sqrt{x} + 2$

For Exercises 19–22, use the graph of $y = \sqrt{x + 5} - 3$ shown in Fig. 12 to solve the given equation.

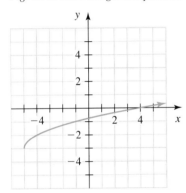

Figure 12
Exercises 19–22

19. $\sqrt{x + 5} - 3 = 0$ 20. $\sqrt{x + 5} - 3 = -1$

21. $\sqrt{x + 5} - 3 = -3$ 22. $\sqrt{x + 5} - 3 = -4$

23. The percentage of high school students taking foreign-language instruction has more than doubled since 1982 (see Table 12).

Table 12 High School Students Taking Foreign-Language Instruction	
Year	**Percent**
1982	21
1985	31
1990	37
1994	41
2000	43

Source: *American Council on the Teaching of Foreign Languages*

Let p be the percentage of high school students who are taking foreign-language instruction at t years since 1982. A model of the situation is $p = 5.5\sqrt{t} + 21$.

a. Use a graphing calculator to draw the graph of the model and, in the same viewing window, the scattergram of the data. Does the model fit the data well?

b. Predict the percentage of high school students who will take foreign-language instruction in 2010.

c. Predict when half of high school students will take foreign-language instruction.

CUMULATIVE REVIEW OF CHAPTERS 1-11

Perform the indicated operation(s) and simplify the result. State whether the expression is a linear polynomial, a quadratic polynomial, a cubic polynomial, a fourth-degree polynomial, a rational expression, or a radical expression.

1. $-3ab^2(5a - 2b)$

2. $7 - 2(4p + 3) - 5p$

3. $(3x^2 - 5x + 1) + (6x^2 - 2x - 4)$

4. $(4x^3 - x^2 + x) - (5x^3 - 2x + 1)$

5. $(2m - 5n)(6m + 7n)$

6. $(3x + 2)(x^2 - 4x + 3)$

7. $(2p + 3q)^2$

8. $\dfrac{5m - 10}{m^3 - 2m^2 - 15m} \cdot \dfrac{3m^2 + 9m}{m^2 - 4}$

9. $\dfrac{2w^2 + 5w - 3}{w^2 - 10w + 25} \div \dfrac{4w^2 - 1}{2w - 10}$

10. $\dfrac{2x}{x^2 - 36} + \dfrac{4}{5x - 30}$

11. $\dfrac{3}{x^2 - x - 6} - \dfrac{5x}{x^2 + 6x + 8}$

12. $\sqrt{\dfrac{3}{7}}$

13. $2\sqrt{27} - 5\sqrt{12}$

14. $(\sqrt{x} - \sqrt{2})(\sqrt{x} + \sqrt{5})$

15. $(2\sqrt{b} - 5\sqrt{c})(2\sqrt{b} + 5\sqrt{c})$

16. $(\sqrt{3} - \sqrt{7})^2$

17. Write $y = -3(x + 2)^2 + 1$ in standard form.

18. Simplify $(3x^4y^8)^3(2xy^4)$.

Simplify. Assume that any variables are nonnegative.

19. $\sqrt{50x^2y}$ 20. $\sqrt{\dfrac{t}{7}}$

Evaluate the expression for $a = -2$, $b = 8$, and $c = -4$.

21. $ac^2 - \dfrac{b}{c}$ 22. $\dfrac{a + bc}{ab - c}$

Factor the expression.

23. $16p^2 - 81q^2$ 24. $a^2 - 6a - 27$

25. $w^2 + 14w + 49$

26. $6m^2n + 13mn^2 + 6n^3$

27. $2x^2 - 16x + 32$

28. $x^3 + 5x^2 - 9x - 45$

29. $12x^4 + 4x^3 - 40x^2$

30. $x^3 + 27$

31. Solve the inequality $-4(x - 2) > 20$. Describe the solution set as an inequality, in interval notation, and in a graph.

32. Graph the inequality $3x - 2y \geq 2$.

33. Graph the solution set of the system

$$y > \frac{1}{4}x - 3$$
$$y \geq -2x$$

Solve. State whether the equation is linear, quadratic, rational, or radical.

34. $2x + 5 = 7x - 3$

35. $6(2x - 3) = 4(3x + 1) - 3(2x - 5)$

36. $w^2 = 5w + 24$ **37.** $(2r - 1)(3r - 2) = 1$

38. $3x^2 - 7 = 13$ **39.** $(x + 4)^2 = 60$

40. $3x^2 - 5x - 4 = 0$ **41.** $2x(x + 3) = -3$

42. $\dfrac{6}{p - 2} = \dfrac{5}{p - 3}$

43. $\dfrac{5}{m - 3} = \dfrac{m}{m - 2} + \dfrac{m}{m^2 - 5m + 6}$

44. $5\sqrt{x} - 2 = 1$ **45.** $\sqrt{x} = x - 2$

46. Solve the formula $a = \dfrac{v - v_0}{t}$ for t.

47. Solve by completing the square: $3x^2 - 12x = 42$.

48. Give examples of three equations that are equivalent to $2(x - 3) = 10$.

Solve the system.

49. $2x - 7y = 3$ **50.** $3x - 4y = 35$
     $-5x + 3y = 7$ $y = 2x - 5$

Graph the equation.

51. $4x - 3y = 6$ **52.** $y = 2x^2 - 5$

Find all x-intercepts.

53. $2x - 5y = 20$

54. $y = x^2 + 2x - 48$

55. Ski run A declines steadily for 130 yards over a horizontal distance of 610 yards. Ski run B declines steadily for 165 yards over a horizontal distance of 700 yards. Which run is steeper? Explain.

56. Find the slope of the line that contains the points $(-5, -4)$ and $(-1, -2)$. State whether the line is increasing, decreasing, horizontal, or vertical.

57. Find an equation of the line that has slope $m = -\dfrac{2}{5}$ and contains the point $(-3, 4)$.

58. Find an equation of the line that passes through the points $(-2, 4)$ and $(6, -3)$.

59. Find an equation of the line sketched in Fig. 13.

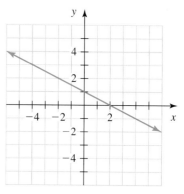

Figure 13 Exercise 59

60. The lengths of two sides of a right triangle are given (see Fig. 14). Find the length of the third side. (The triangle is not drawn to scale.)

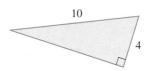

Figure 14 Exercise 60

61. Find any excluded values of the expression $\dfrac{5}{2x^2 - 7x + 5}$.

62. Simplify $\dfrac{-7x + 14}{x^2 - 7x + 10}$.

63. Two hours ago the temperature was $6°$F. Now the temperature is $-3°$F. What is the change in temperature over the past two hours?

64. A recipe for 8 servings of bread pudding calls for 3 tablespoons of sugar. How many tablespoons of sugar are required for 6 servings?

65. If p varies directly as w^2, and $p = 12$ when $w = 2$, find p when $w = 3$.

66. If c varies inversely as r, and $c = 10$ when $r = 2$, find r when $c = 4$.

67. Simplify $\dfrac{\dfrac{1}{4x} + \dfrac{3}{x^2}}{\dfrac{1}{2x} - \dfrac{5}{x^2}}$.

68. A person plans to invest a total of $12,000. She will invest in both an account at 4% annual interest and an account at 7% annual interest. How much should she invest in each account so that the total interest in one year will be $615?

69. Just before beginning its descent, an airplane is at an altitude of 27,500 feet. The airplane then descends steadily at a rate of 1350 feet per minute. Let A be the airplane's altitude (in feet) at t minutes after the plane has begun its descent.
   a. Find an equation of a model to describe the data.
   b. Estimate the airplane's altitude 7 minutes after the plane has begun its descent.
   c. Estimate when the airplane reached an altitude of 1200 feet.
   d. What is the slope? What does it mean in this situation?
   e. What is the t-intercept? What does it mean in this situation?

70. Table 13 shows the percentages of American adults of various age groups who say that they are interested in soccer.

Table 13 American Adults Who Say That They Are Interested in Soccer

Age (in years)	Age Used to Represent Age Group (in years)	Percent
18–24	21.0	32.8
25–34	29.5	26.9
35–44	39.5	25.1
45–54	49.5	23.1
55–64	59.5	19.0
over 64	70	16.2

Source: *ESPN Sports Poll*

Let p be the percentage of American adults at age a years who say that they are interested in soccer.

a. Use a graphing calculator to draw a scattergram of the data. Can the data be modeled better by a linear equation or a quadratic equation? Explain.

b. Find an equation of a model to describe the data.

c. What is the slope? What does it mean in this situation?

d. Estimate what percentage of 25-year-old Americans say that they are interested in soccer.

e. Estimate at what age 21% of American adults say that they are interested in soccer.

f. Find the a-intercept. What does it mean in this situation?

71. Average daily oil production, both onshore and offshore, in the United States is shown in Table 14 for various years.

Table 14 Onshore and Offshore Oil Production

Year	Average Daily Oil Production (millions of barrels a day) Onshore	Offshore
1990	6.3	1.1
1992	6.0	1.2
1994	5.3	1.3
1996	4.9	1.6
1998	4.6	1.7
2000	4.0	1.8
2002	3.7	2.0
2003	3.7	2.0

Source: *Energy Information Administration*

Let p be the average daily oil production (in millions of barrels a day) in the year that is t years since 1990. Reasonable models for onshore and offshore production are

$$p = -0.21t + 6.27 \quad \text{onshore}$$
$$p = 0.074t + 1.08 \quad \text{offshore}$$

a. Find the sum of the expressions $-0.21t + 6.27$ and $0.074t + 1.08$. What does the result represent?

b. Evaluate the result in part (a) for $t = 21$. What does your result mean in this situation?

c. Estimate the rate of change of total average daily oil production (both onshore and offshore).

d. Use substitution or elimination to predict when onshore and offshore average daily oil production will be equal. Find that average daily oil production.

72. A student drives 12 miles south from home and then 3 miles east to go to school. How many miles would the trip take if she could drive along a straight line from home to school?

73. The revenue from selling Dr. Grip® pens varies directly as the number of pens sold. The revenue from selling 1275 pens is $8861.25. If the revenue is $10,591.80, how many pens were sold?

74. The percentages of male drivers who were speeding when they became involved in fatal crashes are shown in Table 15 for various age groups.

Table 15 Speeding Male Drivers in Fatal Crashes

Age Group (years)	Age Used to Represent Age Group (years)	Percent
15–24	21.0	38
25–34	29.5	28
35–44	39.5	20
45–54	49.5	15
55–64	59.5	13
65–74	69.5	10
75+	80.0	7

Source: *National Center for Statistics & Analysis*

Let p be the percentage of male drivers at age a years who were speeding when they became involved in a fatal crash.

a. Use a graphing calculator to draw a scattergram of the data. Can the data be modeled better by a linear equation or a quadratic equation? Explain.

b. A model of the situation is $p = 0.0086a^2 - 1.35a + 61.3$. Use a graphing calculator to draw the graph of the model and, in the same viewing window, the scattergram of the data. Does the model fit the data well?

c. Estimate the percentage of 23-year-old male drivers who were speeding when they became involved in a fatal crash.

d. Estimate at what age 25% of male drivers were speeding when they became involved in a fatal crash.

75. The average percentages of Americans who regularly attend church in a typical week are shown in Table 16 for various age groups.

Table 16 Percentages of Americans Who Regularly Attend Church in a Typical Week

Age Group (years)	Age Used to Represent Age Group (years)	Percent
20s	25	31
30s	35	42
40s	45	47
50s	55	48
60 or more	65	53

Source: *Barna Research Group*

Let p be the average percentage of Americans at the age that is a years more than 25 years who regularly attend church in a typical week. (So, $a = 5$ represents the age $25 + 5 = 30$ years.) A model of the situation is $p = 3.4\sqrt{a} + 31$.

a. Use a graphing calculator to draw the graph of the model and, in the same viewing window, the scattergram of the data. Does the model fit the data well?

b. Estimate what percentage of 30-year-old Americans go to church in a typical week. [**Hint:** Recall that a is the age that is a years more than 25 years.]

c. Estimate at what age half of all Americans go to church in a typical week.

Using a TI-83 or TI-84 Graphing Calculator

The more you experiment with a graphing calculator, the more comfortable and efficient you will become with it.

A TI graphing calculator can detect several types of errors and display an error message. When this occurs, refer to Section A.23 for explanations of some common error messages and how to fix these types of mistakes. Errors do not hurt the calculator. In fact, you can't hurt the calculator regardless of the order in which you press its keys. So, the more you experiment with the calculator, the better off you will be.

To access a TI-83 command written in yellow above a key, first press $\boxed{\text{2nd}}$, then the key. Whenever a key must follow the $\boxed{\text{2nd}}$ key, this appendix will use brackets for the key. For example, "Press $\boxed{\text{2nd}}$ [OFF]" means to press $\boxed{\text{2nd}}$ and then press $\boxed{\text{ON}}$ (because "OFF" is written in yellow above the $\boxed{\text{ON}}$ key). The same applies for TI-84 commands written in blue above a key.

Aside from different-colored keys, the TI-83 and TI-84 are similar calculators: Virtually all of the key combinations for a TI-83 and a TI-84 are the same.

Instructions for using a TI-85 and TI-86 (as well as a TI-82 and TI-83) graphing calculator are available at the website www.prenhall.com/divisions/esm/app/calc_v2/. This site also can serve as a cross-reference for TI-83 graphing calculator instructions. Since the TI-83 and TI-84 are so similar, TI-84 users will find this site helpful even though the TI-84 is not mentioned.

A.1 TURNING A GRAPHING CALCULATOR ON OR OFF

To turn a graphing calculator on, press $\boxed{\text{ON}}$. To turn it off, first press $\boxed{\text{2nd}}$. Then press [OFF].

A.2 MAKING THE SCREEN LIGHTER OR DARKER

To make the screen darker, first press $\boxed{\text{2nd}}$ (then release it); then hold the $\boxed{\triangle}$ key down for a while. To make the screen lighter, first press $\boxed{\text{2nd}}$ (then release it); then hold the $\boxed{\triangledown}$ key down for a while.

A.3 ENTERING AN EQUATION

To enter the equation $y = 2x + 1$,

1. Press $\boxed{\text{Y=}}$.
2. If necessary, press $\boxed{\text{CLEAR}}$ to erase a previously entered equation.

When we show two or more buttons in a row, press them one at a time and in order.

Figure 1 Entering an equation

3. Press **2** $\boxed{\text{X,T,}\Theta\text{,}n}$ $\boxed{+}$ **1**. The screen will look like the one displayed in Fig. 1.

4. If you want to enter another equation, press $\boxed{\text{ENTER}}$. Then type in the next equation.

5. Use the $\boxed{\triangle}$ or $\boxed{\triangledown}$ key to get from one equation to another.

A.4 GRAPHING AN EQUATION

To graph the equation $y = 2x + 1$,

1. Enter the equation $y = 2x + 1$; see Section A.3.

2. Press $\boxed{\text{ZOOM}}$ **6** to draw a graph of your equation between the values of -10 and 10 for both x and y.

3. See Section A.6 if you want to zoom in or zoom out to get another part of the graph to appear on the calculator screen. Or see Section A.7 to change the window format manually; then press $\boxed{\text{GRAPH}}$.

A.5 TRACING A CURVE WITHOUT A SCATTERGRAM

To *trace a curve*, we find coordinates of points on the curve. To trace the line $y = 2x + 1$,

1. Graph $y = 2x + 1$ (see Section A.4).

2. Press $\boxed{\text{TRACE}}$.

3. If you see a flashing "×" on the curve, the coordinates of that point will be listed at the bottom of the screen. If you don't see the flashing "×," press $\boxed{\text{ENTER}}$, and your calculator will adjust the viewing window so that you can see it.

4. To find coordinates of points on your curve that are off to the right, press $\boxed{\triangleright}$.

5. To find coordinates of points on your curve that are off to the left, press $\boxed{\triangleleft}$.

6. Find the y-coordinate of a point by entering the x-coordinate. For example, to find the y-coordinate of the point that has x-coordinate 3, press **3** $\boxed{\text{ENTER}}$. The screen will look like the one displayed in Fig. 2. This feature works for values of x between Xmin and Xmax, inclusive (see Section A.7).

7. If more than one equation has been graphed, press $\boxed{\triangledown}$ to trace the second equation. Continue pressing $\boxed{\triangledown}$ to trace the third equation, and so on. Press $\boxed{\triangle}$ to return to the previous equation. Notice that the equation of the curve being traced is listed in the upper left corner of the screen.

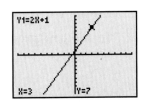

Figure 2 Tracing a curve

A.6 ZOOMING

The $\boxed{\text{ZOOM}}$ menu has several features that allow you to adjust the viewing window. Some of the features adjust the values of x that are used in tracing.

- **Zoom In** magnifies the graph around the cursor location. The following instructions are for zooming in on the graph of $y = 2x + 1$:
  1. Graph $y = 2x + 1$ (see Section A.4).
  2. Press $\boxed{\text{ZOOM}}$ **2**.
  3. Use $\boxed{\triangleleft}$, $\boxed{\triangleright}$, $\boxed{\triangle}$, and $\boxed{\triangledown}$ to position the cursor on the portion of the line that you want to zoom in on.
  4. To zoom in, press $\boxed{\text{ENTER}}$.

If you lose sight of the line, you can always press $\boxed{\text{TRACE}}$ $\boxed{\text{ENTER}}$.

5. To zoom in on the graph again, you have two options:
   a. To zoom in at the same point, press ENTER .
   b. To zoom in at a new point, move the cursor to the new point; then press ENTER .
6. To return to your original graph, press ZOOM 6. Or zoom out (see the next instruction) the same number of times you zoomed in.

- **Zoom Out** does the reverse of Zoom In: It allows you to see *more* of a graph. To zoom out, follow the preceding instructions, but press ZOOM 3 instead of ZOOM 2 in step 2.
- **ZStandard** will change your viewing screen so that both x and y will go from -10 to 10. To use ZStandard, press ZOOM 6.
- **ZDecimal** lets you trace a curve by using the numbers $0, \pm 0.1, \pm 0.2, \pm 0.3, \ldots$ for x. ZDecimal will change your viewing screen so that x will go from -4.7 to 4.7 and y will go from -3.1 to 3.1. To use ZDecimal, press ZOOM 4.
- **ZInteger** allows you to trace a curve by using the numbers $0, \pm 1, \pm 2, \pm 3, \ldots$ for x. ZInteger can be used for any viewing window, although it will change the view slightly. To use ZInteger, press ZOOM 8 ENTER .
- **ZSquare** will change your viewing window so that the spacing of the tick marks on the x-axis is the same as that on the y-axis. To use ZSquare, press ZOOM 5.
- **ZoomStat** will change your viewing window so that you can see a scattergram of points that you have entered in the statistics editor. To use ZoomStat, press ZOOM 9.
- **ZoomFit** will adjust the dimensions of the y-axis to display as much of a curve as possible. The dimensions of the x-axis will remain unchanged. To use ZoomFit, press ZOOM 0.

When zooming out, you will return to the original graph only if you did not move the cursor while zooming in.

A.7 SETTING THE WINDOW FORMAT

To graph the equation $y = 2x + 1$ between the values of -2 and 3 for x and between the values of -5 and 7 for y,

1. Enter the equation $y = 2x + 1$ (see Section A.3).
2. Press WINDOW . Then change the window settings so that the window looks like the one displayed in Fig. 3 after you have used steps 3–8.
3. Press (-) 2 ENTER to set the smallest value of x to -2.
4. Press 3 ENTER to set the largest value of x to 3.
5. Press 1 ENTER to set the scaling for the x-axis to increments of 1.
6. Press (-) 5 ENTER to set the smallest value of y to -5.
7. Press 7 ENTER to set the largest value of y to 7.
8. Press 1 ENTER to set the scaling for the y-axis to increments of 1.
9. Press GRAPH to view the graph of $y = 2x + 1$. The screen will look like the graph drawn in Fig. 4.

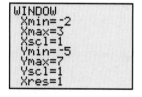

Figure 3 Window settings

If you press ZOOM 6 or ZOOM 9 or zoom in or zoom out, your window settings will change accordingly.

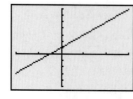

Figure 4 Graph of $y = 2x + 1$

Table 1 Creating a Scattergram

x	y
2	4
3	7
4	10
5	11

Make sure that you press CLEAR rather than DEL . If you press DEL , the column will vanish. If you ever do this by mistake, press STAT 5 ENTER to get back the missing column.

If Plot 1 is off, your points will be saved in columns L_1 and L_2, but they will not be plotted.

A.8 PLOTTING POINTS IN A SCATTERGRAM

To create a scattergram of the data displayed in Table 1,

1. To enter the data, press STAT 1 .

2. If there are numbers listed in the first column (list L_1), clear the column by pressing ◁ as many times as necessary to get to column L_1. Next, press △ once to get to the top of column L_1. Then press CLEAR ENTER .

3. If there are numbers listed in the second column (list L_2), clear the column by pressing ▷ to move the cursor to column L_2. Then press △ CLEAR ENTER .

4. To return to the first entry position of list L_1, press ◁ .

5. Press 2 ENTER 3 ENTER 4 ENTER 5 ENTER to enter the data in column L_1. (If you make a mistake, you can delete an entry by pressing DEL ; then insert an entry by pressing 2nd [DEL].)

6. Press ▷ to move to the first entry position of list L_2.

7. Press 4 ENTER 7 ENTER 10 ENTER 11 ENTER to enter the elements of L_2.

8. Press 2nd [STAT PLOT].

9. Press **1** to select Plot 1.

10. Press ENTER to turn Plot 1 on.

11. Press ▽ ENTER to choose the scattergram mode.

12. Press ▽ so that the cursor is at "Xlist." Then press 2nd [L_1].

13. Press ▽ so that the cursor is at "Ylist." Then press 2nd [L_2].

14. Use squares, plus signs, or dots to represent the points plotted on the scattergram. These three symbols are called "Marks." Press ▽ once so that the cursor is on one of the three Mark symbols. Next, press ▷ and/or ◁ to select a symbol. Then press ENTER . The screen will look like the one displayed in Fig. 5.

15. Press ZOOM **9**. The screen will look like the one displayed in Fig. 6.

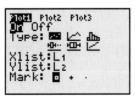

Figure 5 Setting up Plot 1

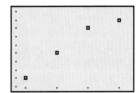

Figure 6 Creating a scattergram

A.9 TRACING A SCATTERGRAM

To see the coordinates of a point in a scattergram,

1. Draw a scattergram (see Section A.8).

2. Press TRACE .

3. Notice the flashing "×" on one of the points of the scattergram. The coordinates of this point are listed at the bottom of the screen.

4. To find the coordinates of the next point to the right, press ▷ .

5. To find the coordinates of the next point to the left, press ◁ .

Table 2 Creating a Scattergram

x	y
2	4
3	7
4	10
5	11

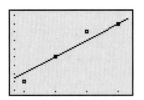

Figure 7 Graphing an equation and a scattergram

Recall from Section A.5 that if you do not see the flashing "×," press ENTER and the calculator will adjust the viewing window so that you can see it.

A.10 GRAPHING EQUATIONS WITH A SCATTERGRAM

To graph the equation $y = 2x + 1$ with a scattergram of the data displayed in Table 2,

1. Enter the equation $y = 2x + 1$ (see Section A.3).
2. Follow the instructions in Section A.8 to draw the scattergram. (The graph of the equation will also be drawn, because you turned the equation on.) The screen will look like the one displayed in Fig. 7.

A.11 TRACING A CURVE WITH A SCATTERGRAM

To trace a curve with a scattergram,

1. Graph an equation with a scattergram (see Section A.10).
2. Press TRACE to trace points that make up the scattergram. Press TRACE ▽ to trace points that lie on the curve. If other equations are graphed, continue pressing ▽ to trace the second equation, and so on. Press △ to begin to return to the scattergram. Notice that the label "P1:L_1, L_2" is in the upper left corner of the screen when Plot 1's points are being traced and that the equation entered in the Y= mode is listed in the upper left corner of the screen when the curve is being traced.

A.12 TURNING A PLOTTER ON OR OFF

To change the on/off status of the plotter,

1. Press Y=.
2. Press △. A flashing rectangle will be on "Plot 1."
3. Press ▷, if necessary, to move the flashing rectangle to the plotter you wish to turn on or off.
4. Press ENTER to turn your plotter on or off. The plotter is on if the plotter icon is highlighted.

A.13 CREATING A TABLE

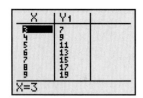

Figure 8 Table of ordered pairs for $y = 2x + 1$

To create a table of ordered pairs for the equation $y = 2x + 1$, where the values of x are 3, 4, 5, ... (see Fig. 8),

1. Enter the equation $y = 2x + 1$ for Y_1 (see Section A.3).
2. Press 2nd [TBLSET].
3. Press 3 ENTER to tell the calculator that the first x value in your table is 3.
4. Press 1 ENTER to tell the calculator that the x values in your table increase by 1.
5. Press ENTER ▽ ENTER to highlight "Auto" for both "Indpnt" and "Depend." The screen will now look like the one displayed in Fig. 9.
6. Press 2nd [TABLE] to create the table shown in Fig. 8.

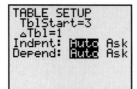

Figure 9 Table setup

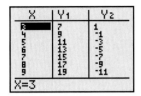

Figure 10 Table for two
equations

A.14 CREATING A TABLE FOR TWO EQUATIONS

To create a table of ordered pairs for the equations $y = 2x + 1$ and $y = -2x + 7$,
where the values of x are $3, 4, 5, \ldots$ (see Fig. 10),

1. Enter the equation $y = 2x + 1$ for Y_1, and enter the equation $y = -2x + 7$ for Y_2
(see Section A.3).

2. Follow steps 2–5 of Section A.13.

A.15 USING "ASK" IN A TABLE

To use the Ask option in the Table Setup mode to complete Table 3 for $y = 2x + 1$,

1. Enter the equation $y = 2x + 1$ for Y_1 (see Section A.3).

2. Press $\boxed{\text{2nd}}$ [TBLSET].

3. Press $\boxed{\text{ENTER}}$ twice. Next, press $\boxed{\triangleright}$. Then press $\boxed{\text{ENTER}}$. The Ask option for
"Indpnt" will now be highlighted. Make sure that the Auto option for "Depend" is
highlighted.

4. Press $\boxed{\text{2nd}}$ [TABLE].

5. Press **2** $\boxed{\text{ENTER}}$ **2.9** $\boxed{\text{ENTER}}$ **5.354** $\boxed{\text{ENTER}}$ **7** $\boxed{\text{ENTER}}$ **100** $\boxed{\text{ENTER}}$. The
screen will now look like the one displayed in Fig. 11.

Table 3 Using "Ask" in
a Table with $y = 2x + 1$

x	y
2	
2.9	
5.354	
7	
100	

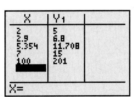

Figure 11 Using "Ask" for a table
with $y = 2x + 1$

A.16 PLOTTING POINTS IN TWO SCATTERGRAMS

It is possible to draw two scattergrams on the same calculator screen and use different
markings for the two sets of points. To begin, follow the instructions in Section A.8 to
create a scattergram of the data values in Table 4.

These data are stored in columns L_1 and L_2. The points are plotted by the plotter
called "Plot 1."

You will now create a scattergram of the data values in Table 5.

These data will be stored in columns L_3 and L_4. The points will be plotted by the
plotter called "Plot 2." To do this,

1. To enter the data, press $\boxed{\text{STAT}}$ **1**.

2. To clear list L_3, press $\boxed{\triangleright}$ and/or $\boxed{\triangleleft}$ to move the cursor to column L_3. Then press
$\boxed{\triangle}$ $\boxed{\text{CLEAR}}$ $\boxed{\text{ENTER}}$.

3. To clear list L_4, press $\boxed{\triangleright}$ to move the cursor to column L_4. Then press
$\boxed{\triangle}$ $\boxed{\text{CLEAR}}$ $\boxed{\text{ENTER}}$.

4. To return to the first entry position of list L_3, press $\boxed{\triangleleft}$.

5. Press **2** $\boxed{\text{ENTER}}$ **2** $\boxed{\text{ENTER}}$ **3** $\boxed{\text{ENTER}}$ **5** $\boxed{\text{ENTER}}$ to enter the elements of L_3.

6. Press $\boxed{\triangleright}$ to move to the first entry position of list L_4.

7. Press **11** $\boxed{\text{ENTER}}$ **9** $\boxed{\text{ENTER}}$ **6** $\boxed{\text{ENTER}}$ **4** $\boxed{\text{ENTER}}$ to enter the elements
of L_4.

8. Press $\boxed{\text{2nd}}$ [STAT PLOT].

9. Press **2** to select "Plot 2."

Table 4 Creating the First
of Two Scattergrams

x	y
2	4
3	7
4	10
5	11

Make sure that you press $\boxed{\text{CLEAR}}$
rather than $\boxed{\text{DEL}}$. If you press
$\boxed{\text{DEL}}$, the column will vanish. If
you ever do this by mistake, press
$\boxed{\text{STAT}}$ **5** $\boxed{\text{ENTER}}$ to get back the
missing column.

Table 5 Creating the
Second of Two Scattergrams

x	y
2	11
2	9
3	6
5	4

Figure 12 Setting up Plot 2

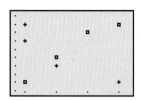

Figure 13 Creating two scattergrams

10. Press ENTER to turn Plot 2 on.

11. Press ▽ twice so that the cursor is at "Xlist." Then press 2nd [L_3].

12. Press ENTER so that the cursor is at "Ylist." Then press 2nd [L_4].

13. Press ▽ once so that the cursor is on one of the three choices for "Mark." Next, press ▷ and/or ◁ to select a symbol different from the one you used for the first scattergram. Then press ENTER. The screen will look like the one in Fig. 12.

14. Press ZOOM 9 to obtain the two scattergrams with different symbols. The screen will look like the one displayed in Fig. 13.

A.17 FINDING THE INTERSECTION POINT(S) OF TWO CURVES

To find the intersection point of the lines $y = 2x + 1$ and $y = -2x + 7$,

1. Enter the equation $y = 2x + 1$ for Y_1, and enter the equation $y = -2x + 7$ for Y_2 (see Section A.3).

2. By zooming in or out or by changing the window settings, draw a graph of both curves so that you can see an intersection point. For our example, press ZOOM 6.

3. Press 2nd [CALC]. The screen will look like the one displayed in Fig. 14.

4. Press **5** to select "intersect."

5. You will now see a flashing cursor on your first curve. If there is more than one intersection point on your display screen, move the cursor by pressing ▷ or ◁ so that it is much closer to the intersection point you want to find. The screen will look something like the one displayed in Fig. 15.

6. Press ENTER to put the cursor on the second curve. Press ENTER again to display "Guess?" Press ENTER once more. The screen will look like the one displayed in Fig. 16. The intersection point is (1.5, 4).

Figure 14 Menu of choices

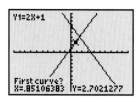

Figure 15 Put cursor near intersection point

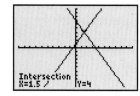

Figure 16 Location of intersection point

A.18 FINDING THE MINIMUM OR MAXIMUM OF A CURVE

To find the minimum point of the curve $y = x^2 - 3x + 1$,

1. Enter the equation $y = x^2 - 3x + 1$ (see Section A.3).

2. Use ZDecimal to draw a graph of the equation (see Section A.6).

3. Press 2nd [CALC].

4. Press **3** to select "minimum."

5. Move the flashing cursor by pressing ◁ or ▷ so that it is to the left of the minimum point, and press ENTER.

6. Move the flashing cursor by pressing ◁ or ▷ so that it is to the right of the minimum point, and press ENTER.

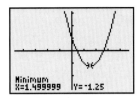

Figure 17 Finding the minimum point of $y = x^2 - 3x + 1$

7. Press $\boxed{\text{ENTER}}$. The calculator will display the coordinates of the minimum point—about $(1.50, -1.25)$. See Fig. 17.

You can find the maximum point of a curve in a similar fashion, but press **4** to select the "maximum" option, rather than the "minimum" option, in step 4.

A.19 FINDING ANY x-INTERCEPTS OF A CURVE

To find the x-intercept of the line $y = x - 2$,

1. Enter the equation $y = x - 2$ (see Section A.3).
2. Use ZDecimal to draw a graph of the equation (see Section A.6).
3. Press $\boxed{\text{2nd}}$ [CALC].
4. Press **2** to choose the "zero" option.
5. Move the flashing cursor by pressing $\boxed{\triangleleft}$ or $\boxed{\triangleright}$ so that it is to the left of the x-intercept, and press $\boxed{\text{ENTER}}$. Or type a number between Xmin and the x-coordinate of the x-intercept, and press $\boxed{\text{ENTER}}$.
6. Move the flashing cursor by pressing $\boxed{\triangleleft}$ or $\boxed{\triangleright}$ so that it is to the right of the x-intercept, and press $\boxed{\text{ENTER}}$. Or type a number between the x-coordinate of the x-intercept and Xmax, and press $\boxed{\text{ENTER}}$.
7. Press $\boxed{\text{ENTER}}$. The screen will look like the one displayed in Fig. 18. The x-intercept is $(2, 0)$.

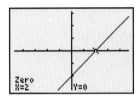

Figure 18 Finding the x-intercept of $y = x - 2$

A.20 TURNING AN EQUATION ON OR OFF

You can graph an equation only if its equals sign is highlighted (the equation is then "on"). Up to 10 equations can be graphed at one time. To change the on–off status of an equation,

1. Press $\boxed{\text{Y=}}$.
2. Move the cursor to the equation whose status you want to change.
3. Use $\boxed{\triangleleft}$ to place the cursor over the "=" sign of the equation.
4. Press $\boxed{\text{ENTER}}$ to change the status.

A.21 FINDING COORDINATES OF POINTS

To find the coordinates of particular points,

1. Press $\boxed{\text{GRAPH}}$ to get into graphing mode.
2. Press $\boxed{\triangleright}$ to get a cursor to appear on the screen. (If you cannot see it, it is probably on one or both of the axes. If it is on an axis, you should still be able to see a small flashing dot.) Notice that the coordinates of the point where the cursor is currently positioned are at the bottom of the screen.
3. Use $\boxed{\triangleleft}$, $\boxed{\triangleright}$, $\boxed{\triangle}$, or $\boxed{\triangledown}$ to move the cursor left, right, up, or down, respectively.

A.22 ENTERING AN EQUATION BY USING Y_n REFERENCES

To enter the complicated equation $y = \dfrac{x+1}{x-3} \div \dfrac{x-2}{x+5}$ by using Y_n references,

1. Enter $Y_1 = \dfrac{x+1}{x-3}$ and $Y_2 = \dfrac{x-2}{x+5}$ (see Section A.3).
2. Turn both equations off (see Section A.20).
3. Move the flashing cursor to the right of "$Y_3 = .$"
4. Press $\boxed{\text{VARS}}$ $\boxed{\triangleright}$ $\boxed{\text{ENTER}}$.
5. Move the cursor to "1:Y_1" and press $\boxed{\text{ENTER}}$. "Y_1" will now appear to the right of "$Y_3 =$" in the $\boxed{Y=}$ window.
6. Press $\boxed{\div}$.
7. Press $\boxed{\text{VARS}}$ $\boxed{\triangleright}$ $\boxed{\text{ENTER}}$.
8. Move the cursor to "2:Y_2" and press $\boxed{\text{ENTER}}$. "Y_1/Y_2" will now appear to the right of "$Y_3 =$" in the $\boxed{Y=}$ window.

A.23 RESPONDING TO ERROR MESSAGES

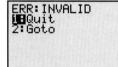

Figure 19 "Syntax" error message

Here are several common error messages and how to respond to them:

- The **Syntax** error (see Fig. 19) means that you have misplaced one or more parentheses, operations, or commas. The calculator will find this type of error if you choose "Goto" by pressing $\boxed{\triangledown}$, then $\boxed{\text{ENTER}}$. Your error will be highlighted by a flashing black rectangle.

  The most common "Syntax" error is pressing $\boxed{(\text{-})}$ when you should have pressed $\boxed{-}$, or vice versa:

  1. Press the $\boxed{(\text{-})}$ key when you want to take the opposite of a number or are working with negative numbers. To compute $-5(-2)$, press $\boxed{(\text{-})}$ **5** $\boxed{(}$ $\boxed{(\text{-})}$ **2** $\boxed{)}$.
  2. Press the $\boxed{-}$ key when you want to subtract two numbers. To compute $5-2$, press **5** $\boxed{-}$ **2**.

- The **Invalid** error (see Fig. 20) means that you have tried to enter an inappropriate number, expression, or command. The most common "Invalid" error is to try to enter a number that is not between Xmin and Xmax, inclusive, when you use a command such as $\boxed{\text{TRACE}}$, "minimum," or "maximum."

Figure 20 "Invalid" error message

- The **Invalid dimension** error (see Fig. 21) means that you have the plotter turned on (see Fig. 22) but have not entered any data points in the STAT list editor (see Fig. 23). In this case, first press $\boxed{\text{ENTER}}$ to exit the error message display; then either turn the plotter off or enter data in the STAT list editor.

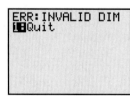

Figure 21 "Invalid dimension" error message

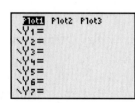

Figure 22 Plotter is on

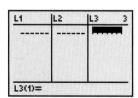

Figure 23 STAT list editor's columns are empty

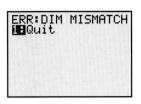

Figure 24 "Dimension mismatch" error message

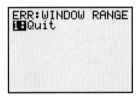

Figure 26 "Window range" error message

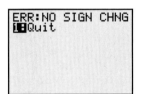

Figure 27 "No sign change" error message

- The **Dimension mismatch** error (see Fig. 24) is fixed in two ways:
  1. In the STAT list editor, one column that you are using to plot has more numbers than the other column has (see Fig. 25). In this case, first press ENTER to exit the error message display; then add or delete numbers so that the two columns have the same length.

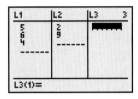

Figure 25 Columns of unequal length in STAT list editor

  2. In the STAT list editor, one column that you are using to plot has more numbers than the other column has, but you didn't notice the difference in length because you deleted one or both of the columns by mistake. You can find the missing column(s) by pressing STAT 5 ENTER.
- The **Window range** error (see Fig. 26) means one of two things:
  1. You made an error in setting up your window. This usually means that you entered a larger number for *Xmin* than for *Xmax* or that you entered a larger number for *Ymin* than for *Ymax*. In this case, first press ENTER to exit the error message display; then change your window settings accordingly (see Section A.7).
  2. You pressed ZOOM 9 when only one data-point pair was entered in the STAT list editor. (For some TI graphing calculators, the command ZoomStat works only when you have two or more pairs of data points in the STAT list editor.) In this case, first press ENTER to exit the error message display; then either add more points to the STAT list editor or avoid pressing ZOOM 9 and set up your window settings manually (see Section A.7).
- The **No sign change** error (see Fig. 27) means one of two things:
  1. You are trying to locate a point that does not appear on the screen. For example, you may be trying to find an intersection point of two curves that intersect offscreen. Or you may be trying to find a zero of an equation that does not appear on the screen. In this case, press ENTER and change your window settings so that the point you are trying to locate is on the screen.
  2. You are trying to locate a point that does not exist. For example, you may be trying to find an intersection point of two parallel lines. Or you may be trying to find a zero of an equation that does not have one. In this case, press ENTER and stop looking for the point that doesn't exist!
- The **Nonreal answer** error (see Fig. 28) means that your computation did not yield a real number. For example, $\sqrt{-4}$ is not a real number. The calculator will locate this computation if you choose "Goto" by pressing ▽, then ENTER.

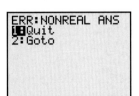

Figure 28 "Nonreal answer" error message

- The **Divide by 0** error (see Fig. 29) means that you asked the calculator to perform a calculation that involves a division by zero. For example, $3 \div (5 - 5)$ will yield the error message shown in Fig. 29.

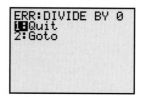

Figure 29 "Divide by zero" error message

The calculator will locate the division by zero if you choose "Goto" by pressing ▽, then ENTER .

Answers to Odd-Numbered Exercises

Answers to most discussion exercises and to exercises in which answers may vary have been omitted.

Chapter 1

Homework 1.1   **1.** 25,000 fans attended the concert.   **3.** In 2003, 159 million Americans used cell phones.   **5.** In 2004, 4.4 million iPods were sold.   **7.** The company lost $45 thousand last year.   **9.** The statement $t = 9$ represents the year 2009.   **11.** The statement $t = -3$ represents the year 2002.   **13.** h; 67, 72; -5, 0; answers may vary.   **15.** p; 12.99, 17.99; -2, -8; answers may vary.   **17.** T; 15, 40; 240, -10; answers may vary.   **19.** s; 25, 32; -15, -9; answers may vary.

21. a. The rectangles are not drawn to scale. Answers may vary.   4 inches☐ 3 inches☐ 1 inch▭   6 inches 8 inches 24 inches   **b.** W, L **c.** A

23. a. The rectangles are not drawn to scale. Answers may vary.   5 feet ☐ 3 feet ☐ 1 foot ▭   5 feet 7 feet 9 feet   **b.** W, L **c.** P

25. a. The rectangles are not drawn to scale. Answers may vary.   1 inch ▭ 2 inches ▭ 3 inches ☐   4 inches 5 inches 6 inches   **b.** W, L, A **c.** None

27. a. The rectangles are not drawn to scale. Answers may vary.   2 yards☐ 2 yards☐ 2 yards☐   2 yards 3 yards 4 yards   **b.** L, P **c.** W

29. number line from -3 to 5   **31.** $-\frac{5}{3}$ $-\frac{2}{3}$ $\frac{7}{3}$ number line from -2 to 2   **33.** 2.8; 0.4 1.1 1.8 2.5 3.6 number line from 0 to 4

35. -1.8 0.5 1.2 3.1 number line from -3 to 3   **37.** number line from 3 to 9   **39.** number line from -3 to 3

41. number line from -3 to 5   **43.** number line from -4 to 0   **45.** 3, 356   **47.** -4 **49.** $\sqrt{7}, \pi$ **51.** $-2, -5, -7$; answers may vary.

53. $-8, -9, -27$; answers may vary.   **55.** $-2, -5, -40$; answers may vary.   **57.** $\frac{5}{4}, \frac{3}{2}, \frac{7}{4}$; answers may vary.   **59.** $-2.1, -2.3, -2.8$; answers may vary.   **61.** Average: 11 units; 9 15; number line 4–16, Number of units u **63.** Average: 12.8%; 10 14 16; number line 9–17, Percent p **65.** 22 24 30 32; number line 21–33, Hours L

67. 5; number line -6 to 6, Degrees Fahrenheit F **69. a.** Average: 18.4 products; 2 9 19 62; number line 0–80, Number of products n **b.** increase   **c.** increase

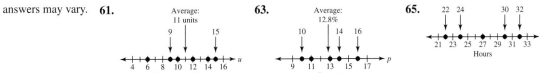

71. a. Average: 177.5 thousand complaints; 86 162 215 247; number line 0–250, Thousands of complaints n **b.** increase   **c.** decrease   **73. a.** -5 **b.** no; answers may vary.

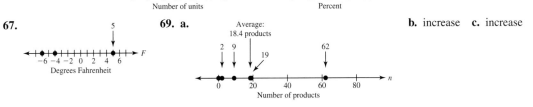

75. a. i. 8 number line 6–10   **ii.** 3 number line 0–6   **iii.** 5 number line 1–9   **b.** Answers may vary.
c. infinitely many   **77.** Answers may vary.

Homework 1.2   **1–15 odd.** $(-5, 4)$, $(-3, 0)$, $(0, 2)$, $(5, 1)$, $(4, -2)$, $(-1.3, -3.9)$, $(2.5, -4.5)$, $(-3, -6)$ **17.** 2   **19.** independent: n; dependent: s **21.** independent: a; dependent: h

23. independent: c; dependent: T **25.** independent: A; dependent: n **27.** independent: t; dependent: h **29.** A telemarketer who works 32 hours per week will sell an average of 43 magazine subscriptions per week.   **31.** 38% of Americans at age 21 years say that they volunteer.   **33.** $328 billion was spent on defense in 2002.   **35.** 42 million travelers booked trips online in 2003.

37.
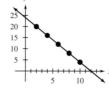

39. A$(-4, -3)$, B$(-5, 0)$, C$(-2, 4)$, D$(1, 3)$, E$(0, -2)$, F$(5, -4)$

41. a.   **b.** The fifth book   **c.** The third book to the fourth book

43. a.   **b.** 1995; 26 hours   **c.** 1965; 9 hours   **45. a.**   **b.** increase   **c.** increase

47. a.   **b.** 60–69-year-old drivers   **c.** 16-year-old drivers   **d.** 16 years and 17 years; answers may vary.   **e.** Answers may vary.

49. a.   **b.** It has taken less time for recent inventions to reach mass use; answers may vary.   **c.** No; it took longer for the microwave to reach mass use than it did for several other earlier inventions.   **d.** Answers may vary.   **e.** It took longer for the automobile to reach mass use than it did for the earlier inventions of electricity and the telephone; answers may vary.

51. a. computer science; $53 thousand   **b.** social science; $32 thousand   **c.** $44 thousand   **53.** Answers may vary; answers may vary; the points lie on the same vertical line; answers may vary.   **55.** Three possibilities; one possibility: (6, 1) and (6, 5); another possibility: $(-2, 1)$ and $(-2, 5)$ **57. a.** x-coordinate: positive; y-coordinate: positive   **b.** x-coordinate: negative; y-coordinate: positive   **c.** x-coordinate: negative; y-coordinate: negative   **d.** x-coordinate: positive; y-coordinate: negative   **59.** Answers may vary.

Homework 1.3   **1.** 2   **3.** 6   **5.** (2, 0)   **7.** -2 **9.** -6 **11.** $(0, -1)$ **13. a–b.**   **c.** 18   **d.** 9   **e.** (0, 24)   **f.** (12, 0)

15. a. 18 thousand gallons   **b.** 4.2 hours   **c.** 30 thousand gallons   **d.** 5 hours   **17.** No; answers may vary.

19. a.
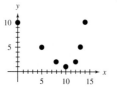
b. No   **21. a.**
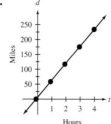
b. 150 miles   **c.** 3.5 hours

23. a. 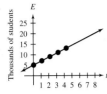 **b.** 17 thousand students **c.** 7 years **25. a.** 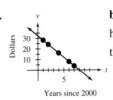 **b.** 2005 **c.** (8, 0); the stock will have no value in 2008. **d.** (0, 32); the value of the stock was $32 in 2000.

27. a. 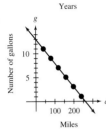 **b.** 6 gallons **c.** 220 miles **d.** (260, 0); the gasoline tank will be empty after 260 miles of driving (if no refueling takes place). **e.** (0, 13); the car has a 13-gallon gasoline tank.

29. a. 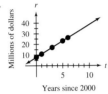 **b.** $38 million **c.** 2002 **d.** (0, 8); the revenue was $8 million in 2000.

31. a. 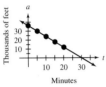 **b.** 22 thousand feet **c.** 30 minutes **d.** underestimate

33. a. 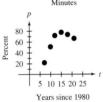 **b.** no **35.** no; the y-coordinate (not the y-intercept) of (2, 5) is 5.

37. no; the y-coordinate of an x-intercept must be 0. **39.** No; an x-intercept is a point that corresponds to an ordered pair with two coordinates (not a single number). **41.** Answers may vary. **43. a. i.** Answers may vary. **ii.** Answers may vary. **iii.** Answers may vary. **b.** one; answers may vary. **c.** one; answers may vary. **45.** Answers may vary.

Homework 1.4

Throughout this section, answers may vary.

1. a. and c. 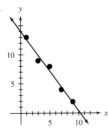 **b.** approximately linearly related **d.** (8, 3.2) **e.** (5.9, 6) **f.** (0, 14) **g.** (10.3, 0)

3. a. and b. 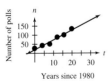 **c.** 2004 **d.** 171 polls; 6 polls per week

5. a. and c. 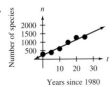 **b.** approximately linearly related **d.** 1997 **e.** 1620 species

7. a. and b.

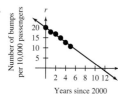

c. (0, 20); the voluntary bumping rate in 2000 was 20 bumps per 10,000 passengers.   **d.** 2009
e. (11, 0); there will be no voluntary bumping in 2011; it is highly likely that model breakdown has occurred.

9. a. and b.

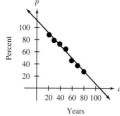

c. 92%   **d.** 59 years old   **e.** (107, 0); no 107-year-old people go to the movies; model breakdown has occurred.

11. a. and b.

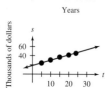

c. (0, 19); the average salary was $19 thousand in 1980.   **d.** $53 thousand   **e.** 2007

13. a. and b.

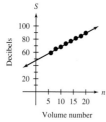

c. 88 decibels   **d.** Volume number 15

15. a. 99 infections per 1000 PCs a month   **b.** 113 infections per 1000 PCs a month   **c.** underestimate; the line is under the data point; -14 infections per 1000 PCs a month   **17.** overestimate   **19.** Answers may vary.   **21.** Answers may vary.   **23.** Not necessarily; model breakdown may occur for some points on the line.

Chapter 1 Review

1. The DVD revenue was $17.5 billion in 2003.   **2.** 2011   **3.** p; 60, 70 (Answers may vary.); -12, 107 (Answers may vary.)
4. a. The rectangles are not drawn to scale. Answers may vary.   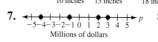   **b.** W, L **c.** P

5. 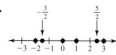   **6.**   **7.**   **8.**

9. -6 **10.** -4 **11.** independent: a; dependent: p **12.** independent: t; dependent: a **13.** There were 313 U.S. billionaires in 2004.
14. In 2000, there were 10.6 thousand injuries.   **15.**   **16. a.**   **b.** 2000, 2003   **c.** 1970

17. a. France; 78%   **b.** United States; 20%   **c.** 50%   **18.** -1 **19.** -5 **20.** 4   **21.** -6 **22.** (0, -2)   **23.** (-4, 0)
24. a. and b.   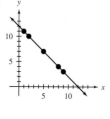   **c.** 1   **d.** 7   **e.** (12, 0)   **f.** (0, 12)   **25. a.**   **b.** linearly related

26. a.

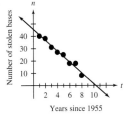

b. $2 million   **c.** 2002   **d.** (0, 22); in 2000, the profit was $22 million.   **e.** (11, 0); in 2011, the profit will be 0 dollars.

27. 0   **28. a. and c.**

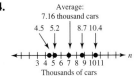

b. approximately linearly related   **d.** (5, 13.5)   **e.** (2, 20)   **f.** (0, 24.3)   **g.** (11.2, 0)

29. a. and b.

c. 26%   **d.** 2010   **30. a. and b.**

c. (0, 45.2); Mays stole 45 bases in 1955, according to the model.   **d.** (10.4, 0); Mays did not steal any bases in 1965, according to the model.   **e.** overestimate; yes; answers may vary.   **f.** underestimate; yes; answers may vary.

Chapter 1 Test

1. a. The rectangles are not drawn to scale. Answers may vary.   6 feet ☐   4 feet ☐   2 feet ☐   **b.** W, L **c.** A
6 feet   9 feet   18 feet

2.

3.

4.

5. c; answers may vary.   **6.** In 2003, 27% of ATMs were privately owned.

7. a.

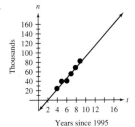

b. (21, 50); Americans in the age group 18–24 are the most likely to have been without health insurance.
c. (59.5, 17); Americans in the age group 55–64 are the least likely to have been without health insurance.

8. −3   **9.** 4   **10.** (0, −1)   **11.** (2, 0)   **12. a.**

b. $29 thousand   **c.** 7 years   **d.** (0, 21); when the person was hired, her salary was $21 thousand.

13. Answers may vary.   **14. a. and b.**

c. 145 thousand   **d.** 2006

15. overestimate; answers may vary.

Chapter 2

Homework 2.1   **1.** 8   **3.** 3   **5.** 42   **7.** 2   **9.** 12   **11.** 36   **13.** 52; the total cost of 4 CDs is $52.   **15.** 88 points per test
17. a.

Number of Shares	Total Value (dollars)
1	$5 \cdot 1$
2	$5 \cdot 2$
3	$5 \cdot 3$
4	$5 \cdot 4$
n	$5n$

$5n$ **b.** 35; the total value of 7 shares is $35.   **19. a.**

Tuition (dollars)	Total Cost (dollars)
400	$400 + 12$
401	$401 + 12$
402	$402 + 12$
403	$403 + 12$
t	$t + 12$

$t + 12$

b. 429; if the tuition is $417, then the total cost is $429.

21. a.

Number of Hours of Courses	Total Cost (dollars)
1	$87 \cdot 1$
2	$87 \cdot 2$
3	$87 \cdot 3$
4	$87 \cdot 4$
n	$87n$

$87n$ **b.** 1305; the total cost for 15 hours of classes is $1305.   **23.** $x + 4$; 12   **25.** $x \div 2$; 4

27. $x - 5$; 3   **29.** $7x$; 56   **31.** $16 \div x$; 2   **33.** The number divided by 2   **35.** 7 minus the number   **37.** The number plus 5   **39.** The product of 9 and the number   **41.** The difference of the number and 7   **43.** The number times 2   **45.** 9   **47.** 3   **49.** 18   **51.** xy; 27
53. $x - y$; 6   **55.** 186; the car traveled 186 miles when driven for 3 hours at 62 mph.   **57.** 20; if a car can travel 240 miles on 12 gallons of gasoline, the car's gas mileage is 20 miles per gallon.   **59.** $170 thousand   **61. a.** 5, 10, 15, 20; the student earns $5, $10, $15, and $20 for working 1, 2, 3, and 4 hours, respectively.   **b.** $5 per hour   **c.** Answers may vary.   **63. a.** 50, 100, 150, 200; the person drives 50, 100, 150, and 200 miles in 1, 2, 3, and 4 hours, respectively.   **b.** 50 mph   **c.** Answers may vary.   **65.** Answers may vary.
67. Answers may vary.

Homework 2.2   **1.** 7   **3.** $2 \cdot 2 \cdot 5$ **5.** $2 \cdot 2 \cdot 3 \cdot 3$ **7.** $3 \cdot 3 \cdot 5$ **9.** $2 \cdot 3 \cdot 13$ **11.** $\dfrac{3}{4}$ **13.** $\dfrac{1}{4}$ **15.** $\dfrac{3}{5}$ **17.** $\dfrac{2}{5}$ **19.** $\dfrac{1}{5}$ **21.** $\dfrac{5}{6}$
23. $\dfrac{2}{15}$ **25.** $\dfrac{3}{10}$ **27.** $\dfrac{5}{3}$ **29.** $\dfrac{5}{6}$ **31.** $\dfrac{2}{3}$ **33.** $\dfrac{2}{15}$ **35.** $\dfrac{5}{7}$ **37.** $\dfrac{3}{4}$ **39.** $\dfrac{1}{5}$ **41.** $\dfrac{1}{3}$ **43.** $\dfrac{3}{4}$ **45.** $\dfrac{19}{12}$ **47.** $\dfrac{14}{3}$ **49.** $\dfrac{1}{9}$ **51.** $\dfrac{17}{63}$
53. $\dfrac{11}{5}$ **55.** 1   **57.** 599   **59.** Undefined   **61.** 0   **63.** 1   **65.** 0   **67.** $\dfrac{1}{3}$ **69.** $\dfrac{9}{5}$ **71.** $\dfrac{1}{3}$ **73.** 0.17   **75.** 1.33   **77.** 0.43
79. Answers may vary.   **81.** $\dfrac{1}{10}$ square mile   **83.** $\dfrac{1}{4}$ of the course points

85. The quotient of the number and 3   **87.**

Number of People	Cost per Person (dollars per person)
2	$\dfrac{19}{2}$
3	$\dfrac{19}{3}$
4	$\dfrac{19}{4}$
5	$\dfrac{19}{5}$
n	$\dfrac{19}{n}$

$\dfrac{19}{n}$ **89. a. i.** $\dfrac{5}{9}$ **ii.** $\dfrac{5}{4}$ **iii.** $\dfrac{3}{2}$ **iv.** $\dfrac{1}{6}$
b. Answers may vary.

91. Answers may vary.   **93.** Answers may vary.

95. a. and b.

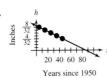

Years since 1950

c. $\dfrac{3}{32}$ inch **d.** 2015 **e.** $(85, 0)$; there will be no grass on putting surfaces in 2035. This prediction is highly unlikely.

97. a. i. -6 **ii.** -8 **iii.** -7 **b.** The results are all negative. **c.** -9 **d.** Answers may vary.

Homework 2.3 **1.** 4 **3.** -7 **5.** 3 **7.** 8 **9.** -4 **11.** -7 **13.** -5 **15.** -5 **17.** 2 **19.** -3 **21.** -10 **23.** -3 **25.** 0 **27.** 0 **29.** -13 **31.** -22 **33.** -1145 **35.** 0 **37.** -6.7 **39.** -4.8 **41.** -97.3 **43.** $\dfrac{2}{7}$ **45.** $-\dfrac{1}{4}$ **47.** $-\dfrac{3}{4}$ **49.** $\dfrac{7}{12}$ **51.** 6221.4 **53.** $-97,571.14$ **55.** -0.11 **57.** -1 **59.** -6 **61.** $x+2$; -4 **63.** $-4+x$; -10 **65.** \$175

67.

Check No.	Date	Description of Transaction	Payment	Deposit	Balance
					-89.00
	7/18	Transfer		300.00	211.00
3021	7/22	State Farm	91.22		119.78
3022	7/22	MCI	44.26		75.52
	7/31	Paycheck		870.00	945.52

69. -2871 dollars **71.** -1633 dollars **73.** $4°F$

75. a.

Weight before Diet (pounds)	Weight after Diet (pounds)
160	$160 + (-20)$
165	$165 + (-20)$
170	$170 + (-20)$
175	$175 + (-20)$
B	$B + (-20)$

$B + (-20)$ **b.** 149; the person's current weight is 149 pounds.

77. a.

Deposit (dollars)	New Balance (dollars)
50	$-80 + 50$
100	$-80 + 100$
150	$-80 + 150$
200	$-80 + 200$
d	$-80 + d$

$-80 + d$ **b.** 45; the new balance is \$45.

79. negative **81.** The numbers are equal in absolute value and opposite in sign. **83. a.** 3 **b.** 4 **c.** 6 **d.** no; answers may vary.

Homework 2.4 **1.** -2 **3.** -6 **5.** 9 **7.** -1 **9.** -3 **11.** 11 **13.** -6 **15.** -79 **17.** 420 **19.** -5.4 **21.** -11.3 **23.** 5.7 **25.** 15.98 **27.** -1 **29.** $\dfrac{1}{2}$ **31.** $\dfrac{3}{4}$ **33.** $-\dfrac{13}{24}$ **35.** 2 **37.** -2 **39.** $-\dfrac{1}{4}$ **41.** -2.7 **43.** -7 **45.** -2 **47.** -3128.17 **49.** 112,927.91 **51.** -0.95 **53.** $-12°F$ **55.** $11°F$ **57. a.** $-12°F$ **b.** $-6°F$ **c.** Answers may vary. **59.** 20,602 feet **61. a.** 0.7 percentage point, -2.5 percentage points, 4.5 percentage points, -7.7 percentage points, 0.5 percentage point, 6.0 percentage points **b.** 6.0 percentage points **c.** The absolute value of the change in percent turnout between 1992 and 1996 is greater than the others; the large decrease in percent turnout suggests that not many of those 11 million people voted; answers may vary. **63. a.** \$2.1 billion **b.** From 2001 to 2003 **c.** From 1997 to 2000

65. a.

Score on the Second Exam (points)	Change in Score (points)
80	80 − 87
85	85 − 87
90	90 − 87
95	95 − 87
p	$p − 87$

$p − 87$ **b.** −6; the score decreased by 6 points.

67. a.

Change in Enrollment	Current Enrollment
100	100 + 24,500
200	200 + 24,500
300	300 + 24,500
400	400 + 24,500
c	$c + 24,500$

$c + 24,500$

b. 23,800; the current enrollment is 23,800 students due to a decrease in enrollment of 700 students in the past year.   **69.** Answers may vary.   **71. a. i.** 2   **ii.** 8   **iii.** 5   **b.** Answers may vary.   **73.** −3   **75.** −7   **77.** 9   **79.** −3 − x; 2   **81.** x − 8; −13   **83.** x − (−2); −3   **85. a. i.** −10   **ii.** −24   **iii.** −63   **b.** Answers may vary.   **c.** −21   **d.** Answers may vary.   **87. a.** 3   **b.** −3   **c.** They are equal in absolute value and opposite in sign.   **d.** −6, 6; they are equal in absolute value and opposite in sign.   **e.** Answers may vary.   **f.** They are equal in absolute value and opposite in sign.

Homework 2.5   **1.** 0.63   **3.** 0.09   **5.** 8%   **7.** 0.073   **9.** 5.2%   **11.** $2.80   **13.** 125 students   **15.** 175 cars   **17.** −12   **19.** 18   **21.** −1   **23.** −8   **25.** −5   **27.** 8   **29.** 555   **31.** −39   **33.** 0.08   **35.** −0.975   **37.** −0.3   **39.** −9   **41.** 4   **43.** $-\dfrac{1}{10}$ **45.** $\dfrac{1}{15}$ **47.** $-\dfrac{9}{14}$ **49.** $\dfrac{15}{14}$ **51.** −3   **53.** 13   **55.** 6   **57.** −100   **59.** $-\dfrac{1}{4}$ **61.** $\dfrac{4}{3}$ **63.** $-\dfrac{11}{12}$ **65.** $-\dfrac{9}{20}$ **67.** $-\dfrac{4}{5}$ **69.** $\dfrac{3}{4}$ **71.** $-\dfrac{1}{2}$ **73.** 1   **75.** $-\dfrac{1}{24}$ **77.** 10,252.84   **79.** −6.78   **81.** 0.48   **83.** −8.07   **85.** −24   **87.** $-\dfrac{3}{2}$ **89.** −48   **91.** $\dfrac{1}{2}$ **93.** $\dfrac{w}{2}$; −4   **95.** $w(−5)$; 40   **97.** $\dfrac{3}{4}$ **99.** $\dfrac{2.25}{1}$; the proposed Freedom Tower would be 2.25 times taller than the John Hancock Tower.   **101.** $\dfrac{1.37}{1}$; the number of U.S. billionaires in 2004 is 1.37 times the number of U.S. billionaires in 2002.

103. a. $\dfrac{0.8\text{ red bell pepper}}{1\text{ black olive}}$; for each black olive used, 0.8 bell pepper is needed.   **b.** $\dfrac{1.25\text{ black olives}}{1\text{ red bell pepper}}$; for each red bell pepper used, 1.25 olives are required.   **105. a.** $\dfrac{12.77}{1}$; the FTE enrollment at Texas A&M University is 12.8 times larger than that at St. Olaf College.   **b.** $\dfrac{2.93}{1}$; the number of FTE faculty at University of Massachusetts–Amherst is 2.9 times greater than that at Butler University.   **c.** Butler University: $\dfrac{12.44}{1}$, St. Olaf College: $\dfrac{11.87}{1}$, Stonehill College: $\dfrac{13.21}{1}$, University of Massachusetts: $\dfrac{17.32}{1}$, Texas A&M University: $\dfrac{21.1}{1}$ **d.** Texas A&M University; St. Olaf College   **e.** Answers may vary.   **107. a.** $\dfrac{2.39}{1}$ **b.** For each $1 the person pays to her MasterCard account, she should pay about $2.39 to her Discover account.   **109.** −3162 dollars   **111.** −29.52 dollars   **113. a.** −6   **b.** 8   **c.** A negative number times a negative number is equal to a positive number.   **d.** Answers may vary.   **115.** $\dfrac{a}{b} = \dfrac{-a}{-b}$, $\dfrac{-a}{b} = \dfrac{a}{-b} = -\dfrac{a}{b} = -\dfrac{-a}{-b}$ **117.** Answers may vary.   **119.** One number is positive and one number is negative.   **121.** a or b is zero.   **123. a.** 16   **b.** 1   **c.** yes; answers may vary.

Homework 2.6   **1.** 64   **3.** 32   **5.** −64   **7.** 64   **9.** $\dfrac{36}{49}$ **11.** 12   **13.** −18   **15.** 10   **17.** −3   **19.** $-\dfrac{5}{2}$ **21.** $-\dfrac{2}{3}$ **23.** 10   **25.** −17   **27.** −50   **29.** −10   **31.** 20   **33.** 15   **35.** $\dfrac{1}{2}$ **37.** 27   **39.** −48   **41.** 1   **43.** −17   **45.** 5   **47.** 2   **49.** −41   **51.** 3   **53.** $-\dfrac{9}{7}$ **55.** −9   **57.** 48   **59.** −613.37   **61.** −1.54   **63.** −1.33   **65.** −14   **67.** −8   **69.** −5   **71.** 40   **73.** $\dfrac{5}{4}$ **75.** $-\dfrac{13}{7}$

77. $-\dfrac{5}{2}$ **79.** $\dfrac{3}{5}$ **81.** -27 **83.** -12 **85.** 32 **87. a.**

Years since 1975	Congressional Pay (thousands of dollars)
0	$4 \cdot 0 + 44.6$
1	$4 \cdot 1 + 44.6$
2	$4 \cdot 2 + 44.6$
3	$4 \cdot 3 + 44.6$
4	$4 \cdot 4 + 44.6$
t	$4t + 44.6$

$4t + 44.6$

b. 184.6; congressional pay will be about $184.6 thousand in 2010. **89. a.**

Years since 1980	Population (thousands)
0	$-2 \cdot 0 + 145$
1	$-2 \cdot 1 + 145$
2	$-2 \cdot 2 + 145$
3	$-2 \cdot 3 + 145$
4	$-2 \cdot 4 + 145$
t	$-2 \cdot t + 145$

$-2t + 145$

b. 83; Gary's population will be 83 thousand in 2011. **91.** $5 + (-6)x$; 29 **93.** $\dfrac{x}{-2} - 3$; -1 **95.** 8 cubic feet **97.** Answers may vary.
99. no; answers may vary. **101. a.** -2 **b.** 3 **c.** Answers may vary. **103. a.** 24 **b.** 24 **c.** The results are equal. **d.** -40; -40; the results are equal. **e.** Answers may vary. **f.** yes **g.** Answers may vary. **105. a.** 4; 1; 0; 1; 4 **b.** nonnegative **c.** nonnegative

Chapter 2 Review

1. 6 **2.** -12 **3.** -3 **4.** 10 **5.** -16 **6.** -4 **7.** -3 **8.** 12 **9.** -1 **10.** $-\dfrac{3}{2}$ **11.** -11 **12.** -13 **13.** -16 **14.** -4 **15.** 8
16. -20 **17.** -12 **18.** 4 **19.** 14 **20.** 0.06 **21.** 10.9 **22.** $-\dfrac{2}{15}$ **23.** $\dfrac{5}{6}$ **24.** $\dfrac{7}{9}$ **25.** $\dfrac{1}{24}$ **26.** $-\dfrac{1}{6}$ **27.** 64 **28.** -64
29. 16 **30.** $\dfrac{27}{64}$ **31.** -54 **32.** 3 **33.** 0 **34.** $\dfrac{2}{3}$ **35.** $-\dfrac{8}{11}$ **36.** -19 **37.** -3 **38.** 58 **39.** $\dfrac{3}{4}$ **40.** $-\dfrac{4}{5}$ **41.** -8.68
42. 0.62 **43.** $2\dfrac{1}{6}$ yards **44.** $4095.49 **45.** -4700 feet **46. a.** $-12°$F **b.** $-4°$F **c.** Answers may vary. **47. a.** $26 million
b. -5 million dollars **c.** from 1996 to 2000; $26 million **d.** from 2000 to 2004; $42 million **48.** 2.29; the number of ring tones on cell phones sold in 2004 is 2.29 times larger than the number of ring tones on phones sold in 2003. **49.** 0.75 **50.** 0.029 **51.** $37.41
52. 74 students **53.** -4394.4 dollars **54.** -10 **55.** 57 **56.** -2 **57.** $-\dfrac{11}{4}$ **58.** 55 **59.** $-\dfrac{1}{2}$ **60.** $x + 5$; 2 **61.** $-7 - x$; -4
62. $2 - x(4)$; 14 **63.** $1 + \dfrac{-24}{x}$; 9 **64.** 50; if there are 13 players on the team, the cost is $50 per player.
65. a.

Time (hours)	Volume of Water (cubic feet)
0	$-50 \cdot 0 + 400$
1	$-50 \cdot 1 + 400$
2	$-50 \cdot 2 + 400$
3	$-50 \cdot 3 + 400$
4	$-50 \cdot 4 + 400$
t	$-50t + 400$

$-50t + 400$ **b.** 50; there will be 50 cubic feet of water in the basement after 7 hours of pumping.

Chapter 2 Test

1. -13 **2.** 63 **3.** -6 **4.** -8 **5.** $\dfrac{1}{2}$ **6.** 3 **7.** -25 **8.** -0.08 **9.** $-\dfrac{45}{4}$ **10.** $\dfrac{13}{40}$ **11.** 81 **12.** -16 **13.** 6 **14.** -17
15. $-\dfrac{21}{4}$ **16.** $-4°$F **17. a.** 0.7 audit per 1000 tax returns **b.** -3.2 audits per 1000 returns **c.** Answers may vary. **18.** 2.17; the

average ticket price in 2004 is 2.17 times larger than the average ticket price in 1991.   **19.** -33 **20.** $-\dfrac{2}{3}$ **21.** 11   **22.** 124

23. $2x - 3x$; 5   **24.** $\dfrac{-10}{x} - 6$; -4 **25. a.**

Years since 1999	Number of Books Published
0	$7(0) + 25$
1	$7(1) + 25$
2	$7(2) + 25$
3	$7(3) + 25$
4	$7(4) + 25$
t	$7t + 25$

$7t + 25$

b. 102; there will be 102 books about obesity published in 2010.

Cumulative Review of Chapters 1 and 2

1. a. The rectangles are not drawn to scale. Answers may vary.      **b.** W, L **c.** P

2. [number line from -3 to 4]   **3.** [number line from -3 to 4, labeled c, Dollars]   **4.** -5 **5.** independent variable: t; dependent variable: V

6. a. [graph, p vs t, Dollars, Years since 1995]   **b.** 1997   **c.** 2002   **d.** from 1997 to 1998; -200 dollars   **e.** from 2001 to 2002; -30 dollars   **7.** -3 **8.** 4   **9.** $(0, -1)$ **10.** $(2,0)$

11. a. [graph B, Thousands of dollars vs Months]   **b.** $12 thousand   **c.** 7 months after the person was laid off   **d.** $(0, 20)$; when the person was laid off, the balance was $20 thousand.   **e.** $(10, 0)$; there was no money in the account 10 months after the person was laid off.

12. a. and b. [graph c, Dollars per household vs Years since 1995]   **c.** $(0, 22)$; the average monthly spending on cable TV per household in 1995 was $22.   **d.** 2008   **e.** $46

13. $\dfrac{3}{4}$ **14.** -4 **15.** $\dfrac{18}{25}$ **16.** $-\dfrac{11}{24}$ **17.** -17 **18.** $-\dfrac{2}{9}$ **19.** $-8°$F   **20.** $1865   **21.** $-\dfrac{1}{2}$ **22.** 49   **23.** $x - \dfrac{-12}{x}$; -7

24. $7 + (-2)x$; 15   **25.** 7.14; the percent growth of the investment was 7.14%.

26. a.

Years since 2000	Sales (thousands of cameras)
0	$4 \cdot 0 + 15$
1	$4 \cdot 1 + 15$
2	$4 \cdot 2 + 15$
3	$4 \cdot 3 + 15$
4	$4 \cdot 4 + 15$
t	$4t + 15$

$4t + 15$ **b.** 59; the sales will be 59 thousand cameras in 2011.

Chapter 3

Homework 3.1   **1.** $(-3, -10)$, $(2, 0)$ **3.** $(0, 7)$, $(4, -5)$

5. $(0, 2)$ **7.** $(0, -4)$ **9.** $(0, 0)$ **11.** $(0, 0)$ **13.** $(0, 0)$ **15.** $(0, 0)$ **17.** $(0, 0)$

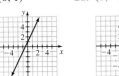

19. $(0, 1)$ **21.** $(0, -3)$ **23.** $(0, 5)$ **25.** $(0, -3)$ **27.** $(0, -3)$ **29.** $(0, 1)$

31. a.   Answers may vary.   **b.**   **c.** For each solution, the y-coordinate is 3 less than twice the x-coordinate.

x	y
0	-3
1	-1
2	1

33. a. i. 7; one   **ii.** 13; one   **iii.** -5; one   **b.** one; answers may vary.   **c. i.** Answers may vary; one   **ii.** Answers may vary; one
iii. Answers may vary; one   **d.** one; answers may vary.   **e.** one; answers may vary.

35. a. i.   **ii.**   **iii.**   **b.** x-intercept: $(0, 0)$; y-intercept: $(0, 0)$

x-intercept: $(0, 0)$; x-intercept: $(0, 0)$; x-intercept: $(0, 0)$;
y-intercept: $(0, 0)$ y-intercept: $(0, 0)$ y-intercept: $(0, 0)$

37. Answers may vary.   **39.** 3   **41.** 0   **43.** 4   **45.** -2 **47.** C, D, E   **49.** Answers may vary; infinitely many   **51.** $y = x + 3$
53. $y = x$ **55. a.** Answers may vary.   **b.** $y = 3x$ **57.**   **59.** Answers may vary.   **61.** Answers may vary.

Homework 3.2

1. a.

Drink Cost (dollars)	Total Cost (dollars)
n	T
2	$2 + 3$
3	$3 + 3$
4	$4 + 3$
5	$5 + 3$
d	$d + 3$

$T = d + 3$ **b.** The units for both of the expressions T and $d + 3$ are dollars.

c.

d. $(0, 3)$; if the person does not buy any drinks, then the total cost is $3. **e.** $13

3. a.

Number of Credits	Total Cost (in dollars)
c	T
3	$56 \cdot 3$
6	$56 \cdot 6$
9	$56 \cdot 9$
12	$56 \cdot 12$
c	$56 \cdot c$

$T = 56c$ **b.** The units for both of the expressions T and $56c$ are dollars.

c.   **d.** $840

5. a.

Time at Company (years)	Salary (thousands of dollars)
t	s
0	$3 \cdot 0 + 24$
1	$3 \cdot 1 + 24$
2	$3 \cdot 2 + 24$
3	$3 \cdot 3 + 24$
4	$3 \cdot 4 + 24$
t	$3 \cdot t + 24$

$s = 3t + 24$ **b.** The units for both of the expressions $3t + 24$ and s are thousands of dollars. **c.**

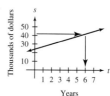

d. $(0, 24)$; the starting salary is $24 thousand. **e.** 6 years

7. a.

Time (weeks)	Value (dollars)
t	v
0	$65 - 5 \cdot 0$
1	$65 - 5 \cdot 1$
2	$65 - 5 \cdot 2$
3	$65 - 5 \cdot 3$
4	$65 - 5 \cdot 4$
t	$65 - 5 \cdot t$

$v = 65 - 5t$ **b.** The units for both of the expressions $65 - 5t$ and v are dollars.

c.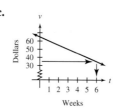

d. 6 weeks after the publicity was released

9. a. $a = r - 5$ **b.** The units for both of the expressions a and $r - 5$ are minutes. **c.**

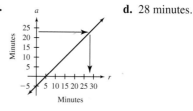

d. 28 minutes.

11. a. $d = 60t$ **b.** The units for both of the expressions d and $60t$ are miles.

c.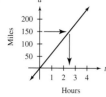

d. $(0, 0)$; the person will not travel any distance in 0 hours of driving. **e.** 2.5 hours

13. a. $T = 26u + 31$ **b.** The units for both of the expressions T and $26u + 31$ are dollars. **c.** $421

15. a. $g = 11 - 2t$ **b.** The units for both of the expressions g and $11 - 2t$ are gallons.

c.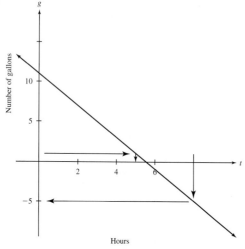

d. 5 hours **e.** -5 gallons; model breakdown has occurred.

17. a.

Answers may vary. **b.** $s = 2.5t + 4.3$ **c.**

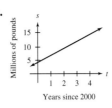

Years Since 2000	Sales (millions of pounds)
t	s
0	4.3
1	6.8
2	9.3
3	11.8
4	14.3

19. $n = 45$;

21.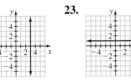

23.

25.   **27.**   **29.**   **31.**   **33.**

$x = 0$

35.   **37.**   **39.**   **41.**   **43.**

45. $x = -3$ **47. a. i.** 4   **ii.** 0   **b. i.** -4 **ii.** 0   **49. a. i.** 5   **ii.** 3   **b. i.** -5 **ii.** -3

51. a.

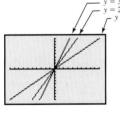

$y = 3x$
$y = 2x$
$y = x$

b. $y = x$, $y = 2x$, $y = 3x$ **c.** Answers may vary.   **d.** Answers may vary.

53. Answers may vary.

Homework 3.3

1. road A   **3.** ski run A   **5.**   run: 2; rise: 3; slope: $\dfrac{3}{2}$ **7.**   run: 2; rise: -4; slope: -2

9.   run: 6; rise: 4; slope: $\dfrac{2}{3}$ **11.** 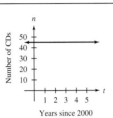   run: 2; rise: -4; slope: -2 **13.** 2; increasing   **15.** -4; decreasing

17. $\dfrac{1}{2}$; increasing   **19.** $-\dfrac{1}{3}$; decreasing   **21.** -1; decreasing   **23.** $-\dfrac{1}{2}$; decreasing   **25.** 2; increasing   **27.** $\dfrac{3}{2}$; increasing

29. $-\dfrac{4}{5}$; decreasing   **31.** $\dfrac{2}{3}$; increasing   **33.** $-\dfrac{1}{2}$; decreasing   **35.** 0; horizontal   **37.** undefined slope; vertical   **39.** -9.25; decreasing

41. 1.14; increasing   **43.** −0.21; decreasing   **45.** 0.71; increasing   **47.** $\frac{2}{3}$ **49.** −3   **51. a.** negative   **b.** positive   **c.** undefined
d. zero   **53.** Answers may vary.   **55.** Answers may vary.   **57.** Answers may vary.   **59.** Answers may vary.   **61.** Answers may vary; $\frac{4}{3}$
63. Answers may vary; $\frac{13}{4}$ **65.** Answers may vary; yes   **67. a.** Answers may vary. **b.** no **c.** yes **d.** Answers may vary.
69. Answers may vary.   **71.** Answers may vary.   **73.** Answers may vary.   **75.** not necessarily; answers may vary.

Homework 3.4

1.   **3.**   **5.**   **7.**   **9.**   **11.**

13.   **15.**   **17.** slope: $\frac{2}{3}$; y-intercept: (0, −1)   **19.** slope: $-\frac{1}{3}$; y-intercept: (0, 4)   **21.** slope: $\frac{4}{3}$; y-intercept: (0, 2)

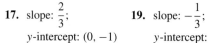

23. slope: $-\frac{4}{5}$; y-intercept: (0, −1)   **25.** slope: $\frac{1}{2}$; y-intercept: (0, 0)   **27.** slope: $-\frac{5}{3}$; y-intercept: (0, 0)   **29.** slope: 4; y-intercept: (0, −2)   **31.** slope: −2; y-intercept: (0, 4)

33. slope: −4; y-intercept: (0, −1)   **35.** slope: 1; y-intercept: (0, 1)   **37.** slope: −1; y-intercept: (0, 3)   **39.** slope: −3; y-intercept: (0, 0)   **41.** slope: 1; y-intercept: (0, 0)

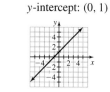 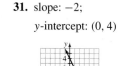 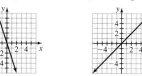

43. slope: 0; y-intercept: (0, −3)   **45.** slope: 0; y-intercept: (0, 0)

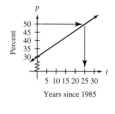 $y = 0$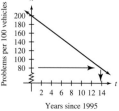

47. a.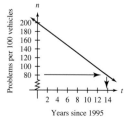
Percent — Years since 1985   **b.** 2010   **49. a.** Problems per 100 vehicles — Years since 1995   **b.** 2009

51. a. m is positive; b is negative **b.** m is zero; b is negative **c.** m is negative; b is positive **d.** m is positive; b is positive
53. Answers may vary. **55.** Answers may vary. **57.** Answers may vary. **59.** $y = 3x - 4$ **61.** $y = -\dfrac{6}{5}x + 3$ **63.** $y = -\dfrac{2}{7}x$
65. $y = 2x + 3$ **67.** perpendicular **69.** neither **71.** parallel **73.** neither **75.** parallel **77.** perpendicular **79.** no; answers may
vary. **81.** no; answers may vary. **83.** Answers may vary. **85. a.** 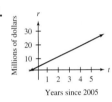 **b.**

x	y
-2	1
0	2
2	3

Answers may vary.

c. For each solution, the y-coordinate is two more than half the x-coordinate. **87.** Answers may vary;   **89.** -2 **91.** 3

93. $\dfrac{1}{3}$ **95. a.**   **b.** $y = 2x + 3$ **97. a.** The slope of each line is zero. **b.** 0

99. a. k; answers may vary. **b.** b; answers may vary.

Homework 3.5
1. \$1550 per year **3.** -1650 feet per minute **5.** 26.4 employers per year **7.** 3.4 percentage points per year
9. -2.55 thousand Steller's sea lions per year **11.** -115 million barrels per year **13.** \$4950 per person
15. a. yes; 4; the revenue increases by \$4 million per year. **b. i.** $r = 4t + 3$

ii.

Years Since 2005	Revenue (millions of dollars)
t	r
0	3
1	7
2	11
3	15
4	19

Answers may vary. **iii.**

17. a. yes; -21.9; the number of sites is decreasing by about 21.9 sites per year. **b.** $(0, 402)$; in 2004, there were 402 sites.
c. $n = -21.9t + 402$ **d.** 249 sites **e.** -36 sites; model breakdown has occurred. **19. a.** -650; the balance declines by \$650 per
month. **b.** $(0, 4700)$; the balance is \$4700 on September 1. **c.** $B = -650t + 4700$ **d.** The units for both of the expressions B and
$-650t + 4700$ are dollars. **e.** \$800 **21. a.** 385; the tuition increases by \$385 per credit. **b.** $T = 385c + 10$ **c.** The units for both of
the expressions T and $385c + 10$ are dollars. **d.** \$3475 **23. a.** -0.02; the car uses 0.02 gallon of gasoline per mile. **b.** $(0, 11.9)$; there
were 11.9 gallons of gasoline in the tank at the start of the trip. **c.** $G = -0.02d + 11.9$ **d.** The units for both of the expressions G and
$-0.02d + 11.9$ are gallons. **e.** 10.5 gallons **25. a.** -3.65; the number of refineries is decreasing by about 3.65 refineries per year.
b. $(0, 157.31)$; there were about 157 refineries in 2000. **c.** 117 refineries

27. a. yes **b.** 20.7; the number of users is increasing by 20.7 million users per year. **c.** 21, 24, 21, 17, 21, 23,
17, 24 (all in millions of users per year); answers may vary. **d.** $(0, 19.6)$; there were 19.6 million users in
1995. **e.** 330.1 million users; model breakdown has occurred.

29. a. yes **b.** 3.56; the result overestimates the qualifying core GPA by 0.01. **c.** -0.00254; the qualifying
core GPA decreases by 0.00254 for an increase of 1 point on the SAT. **d.** $(0, 4.58)$; the qualifying core
GPA is 4.58 for an SAT score of 0; model breakdown has occurred, because the highest possible GPA is
4.0 and the lowest possible SAT score is 400 points. **31. a.** yes; answers may vary. **b.** 60 miles per
hour **33. a.** yes; answers may vary. **b.** -1.5 gallons per hour **35.** yes; 30; the total charge increases
by \$30 per hour.

37. equations 1 and 3 **39.**

Equation 1		Equation 2		Equation 3		Equation 4	
x	y	x	y	x	y	x	y
0	3	0	99	21	16	43	17
1	8	1	92	22	14	44	20
2	13	2	85	23	12	45	23
3	18	3	78	24	10	46	26
4	23	4	71	25	8	47	29

41. Set 1: $y = 2x + 5$; Set 2: $y = -3x + 20$; Set 3: $y = 8x + 21$; Set 4: $y = -5x + 9$ **43.** y increases by 7. **45.** Answers may vary.
47. a. The slope 3 is the number multiplied times x. **b.** If the run is 1, the rise is 3. **c.** As the value of x increases by 1, the value of y increases by 3. **d.** Answers may vary. **49.** Answers may vary.

Chapter 3 Review

1. $(-3, 9), (4, -5)$ **2.** -1 **3.** -3 **4.** -2 **5.** -2 **6.** -4 **7.** 4 **8.** airplane A **9.** 2; increasing **10.** -1; decreasing **11.** $\frac{1}{2}$; increasing **12.** $-\frac{1}{2}$; decreasing **13.** -1; decreasing **14.** $-\frac{1}{3}$; decreasing **15.** $-\frac{9}{8}$; decreasing **16.** $-\frac{2}{3}$; decreasing **17.** undefined slope; vertical **18.** 0; horizontal **19.** -3.62; decreasing **20.** 0.94; increasing **21.** Answers may vary.

22. **23.** **24.** **25.** slope: $\frac{3}{4}$; y-intercept: $(0, -1)$ **26.** slope: $-\frac{1}{2}$; y-intercept: $(0, 3)$

27. slope: $-\frac{2}{5}$; y-intercept: $(0, -1)$ **28.** slope: $\frac{2}{3}$; y-intercept: $(0, 0)$ **29.** slope: -4; y-intercept: $(0, 0)$ **30.** slope: 2; y-intercept: $(0, -4)$ **31.** slope: -3; y-intercept: $(0, 1)$

32. slope: 1; y-intercept: $(0, 2)$ **33.** slope: 0; y-intercept: $(0, -5)$ **34.** **35.**

36. a.

x	y
-1	3
0	1
1	-1

Answers may vary. **b.**   **c.** For each solution, the y-coordinate is one more than -2 times the x-coordinate.

37. a. $B = -3t + 19$ **b.** 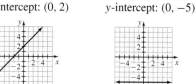 **c.** $(0, 19)$; when the person first lost his job, the balance was $19 thousand. **d.** 5 months

38. $x = 5$ **39.** $y = -\frac{2}{3}x + 4$ **40.** neither **41.** perpendicular **42.** perpendicular **43.** parallel **44.** $-1.5°$F per hour
45. $75 million per year **46. a.** $c = 2d + 2.5$ **b.** The units for both of the expressions c and $2d + 2.5$ are dollars. **c.** $36.50

47. a. -4; the person loses 4 pounds per month.   **b.** $w = -4t + 195$ **c.** 171 pounds   **48. a.** 17.75; the average monthly cost increases by \$17.75 per year.   **b.** $c = 17.75t + 516$ **c.** \$693.50   **49.** yes; 130; the cost is \$130 per calculator.   **50.** equations 1, 3, and 4

51.

Equation 1		Equation 2		Equation 3		Equation 4	
x	y	x	y	x	y	x	y
0	50	0	12	61	25	26	-4
1	41	1	16	62	23	27	-1
2	32	2	20	63	21	28	2
3	23	3	24	64	19	29	5
4	14	4	28	65	17	30	8

52. The value of y decreases by 6 when the value of x is increased by 1.

Chapter 3 Test

1. 3   **2.** 3   **3.** $(0, 1)$ **4.** $(1.5, 0)$ **5.** ski run B   **6.** 3; increasing   **7.** $-\frac{1}{2}$; decreasing   **8.** 0; horizontal   **9.** undefined slope; vertical
10. 1.29; increasing   **11.**   **12.** slope: $-\frac{3}{2}$; y-intercept: $(0, 2)$ **13.** slope: $\frac{5}{6}$; y-intercept: $(0, 0)$ **14.** slope: 3; y-intercept: $(0, -4)$

15. slope: 0; y-intercept: $(0, 2)$ **16.** slope: -2; y-intercept: $(0, 3)$ **17.** $y = \frac{1}{2}x + 1$ **18. a.** $v = -2t + 17$ **b.**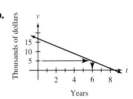

c. $(0, 17)$; the used car is currently worth \$17 thousand.   **d.** 6 years from now   **19. a.** m is positive; b is positive   **b.** m is zero; b is negative   **c.** m is negative; b is positive   **d.** m is negative; b is negative   **20.** neither   **21.** parallel   **22.** -525 dollars per month
23. 5.53 million doses per year   **24. a.** 0.17; total sales are increasing by \$0.17 billion per year.   **b.** $s = 0.17t + 1.6$ **c.** \$2.79 billion
25. a. 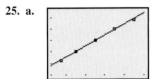   yes   **b.** 0.24; the cooking time increases by 0.24 hour per pound of turkey.   **c.** $(0, 1.64)$; the cooking time of a 0-pound turkey is 1.64 hours; model breakdown has occurred.   **d.** 6.2 hours   **26. a.** yes   **b.** 8 miles per hour   **27.** Set 1: $y = -3x + 25$; Set 2: $y = 4x + 2$; Set 3: $y = -5x + 12$; Set 4: $y = 6x + 47$ **28.** The value of y increases by 3 as the value of x is increased by 1.

Chapter 4

Homework 4.1   **1.** $10x$ **3.** $36x$ **5.** $-4x$ **7.** $\frac{7x}{4}$ **9.** $3x + 27$ **11.** $2x - 10$ **13.** $-2t - 10$ **15.** $10x - 30$ **17.** $-24x - 42$
19. $6x - 10y$ **21.** $-20x - 15y + 40$ **23.** $2x + 5$ **25.** $-0.3x - 0.06$ **27.** $4a + 19$ **29.** $-12x + 9$ **31.** $-9a + 19$
33. $10.5x - 38.56$ **35.** $-t - 2$ **37.** $-8x + 9y$ **39.** $-5x - 8y + 1$ **41.** $-3x + 3$ **43.** $-x + 5$ **45.** $-2t + 3$ **47.** $2x + 4$
49. $8x$; 16; answers may vary.   **51.** $5x - 35$; -25; answers may vary.   **53.** $-3x - 7$; -13; answers may vary.   **55.** $3(x + 2)$; $3x + 6$
57. $-4(2x - 5)$; $-8x + 20$ **59.** $5 - 2(x - 4)$; $-2x + 13$ **61.** $-2 + 7(2x + 1)$; $14x + 5$ **63. a.** $2x - 2$ **b.** 8   **c.** 8   **d.** The results are equal; the result in part (a) might be correct.   **65.** 18; 10; the work is incorrect.   **67. a.** 24; 48   **b.** Answers may vary.   **c.** $(ab)c$
69. $y = 2x - 6$ **71.** $y = -3x - 3$ **73.** $y = -6x$ **75.** Answers may vary.   **77.** Answers may vary.

    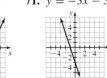

79. $3 = 1$; answers may vary. **81.** commutative law for addition; associative law for addition; commutative law for addition; associative law for addition **83.** commutative law for multiplication; distributive law; commutative law for multiplication **85.** $-a = -1 \cdot a$; associative law for multiplication; the product of two real numbers with different signs is negative; $-1 \cdot a = -a$; $-(-a) = a$

Homework 4.2 **1.** $7x$ **3.** $5x$ **5.** $-13w$ **7.** $4t$ **9.** $-0.5x$ **11.** $\dfrac{7}{3}x$ **13.** $-3x - 3$ **15.** $-2p - 7$ **17.** $3x + y + 1$
19. $-9.9x + 1.1y + 2.1$ **21.** $-a + 15$ **23.** $43.16x + 23.08$ **25.** $7a - 8b$ **27.** $-x + 2$ **29.** $2x - 2y$ **31.** $-13t - 5$
33. $-2x - 4y + 6$ **35.** $-8x - 14$ **37.** $-24x - 38y$ **39.** 0 **41.** $-4x + 7y - 21$ **43.** $7x - 13y - 2$ **45.** $-\dfrac{2}{7}a + \dfrac{6}{7}$ **47.** $3x - 3$
49. $x + 5x$; $6x$ **51.** $4(x - 2)$; $4x - 8$ **53.** $x + 3(x - 7)$; $4x - 21$ **55.** $2x - 4(x + 6)$; $-2x - 24$ **57.** Twice the number plus 6 times
the number (answers may vary); $8x$ **59.** 7 times the difference of the number and 5 (answers may vary); $7x - 35$ **61.** The number, plus 5
times the sum of the number and 1 (answers may vary); $6x + 5$ **63.** Twice the number minus 3 times the difference of the number and 9
(answers may vary); $-x + 27$ **65.** $8x - 5$ **67.** $-3x + 9$ **69.** $5x + 2$; 22; answers may vary. **71.** $3x + 11$; 23; answers may vary.
73. a. $8x + 12$ **b.** 28 **c.** 28 **d.** The results are equal; answers may vary. **75.** $-2(x - 3)$, $2(3 - x)$, $-3(x - 2) + x$, $-2x + 6$ **77.** 5,
14, 14, 14; $9x - 13$ **79.** Answers may vary. **81.** $y = -2x$ **83.** $y = 2x - 4$ **85.** $y = 3x - 4$ **87.** Answers may vary.

Homework 4.3 **1.** yes **3.** no **5.** yes **7.** 5 **9.** -14 **11.** 24 **13.** -28 **15.** -3 **17.** 3 **19.** 3 **21.** -4 **23.** 5 **25.** $\dfrac{4}{3}$
27. $\dfrac{6}{5}$ **29.** 0 **31.** 15 **33.** $\dfrac{21}{2}$ **35.** -12 **37.** $-\dfrac{10}{3}$ **39.** 6 **41.** -3 **43.** $\dfrac{1}{2}$ **45.** -11.1 **47.** 41.76 **49.** 2.3 **51.** -4 **53.** 4
55. 5 **57.** 4 **59.** 3 **61.** -3 **63.** 3 **65.** -2 **67.** 2 **69.** 2 **71.** Answers may vary; 5 **73.** $4(3)$ is equal to 12; $\dfrac{4(3)}{4} = 3$ is equal
to $\dfrac{12}{4} = 3$; 3 is equal to 3. **75.** yes; answers may vary. **77.** Answers may vary. **79.** Answers may vary. **81.** no **83.** Answers may
vary. **85.** Answers may vary. **87.** yes **89. a.** 5 **b.** 4 **c.** $k - b$ **91.** Answers may vary.

Homework 4.4 **1.** 5 **3.** -5 **5.** $-\dfrac{4}{3}$ **7.** 12 **9.** 2 **11.** 6 **13.** -5 **15.** 2 **17.** 3 **19.** $\dfrac{3}{4}$ **21.** -1 **23.** $\dfrac{5}{4}$ **25.** $\dfrac{5}{2}$
27. $\dfrac{8}{5}$ **29.** 6 **31.** $-\dfrac{41}{3}$ **33.** $-\dfrac{27}{10}$ **35.** $\dfrac{12}{13}$ **37.** -7 **39.** $\dfrac{11}{3}$ **41.** 1.67 **43.** 6.34 **45.** 21.85 **47. a.** $a = -5.6t + 483$
b. 421.4 students per counselor **c.** 2042 **49. a.** $v = 3.36t + 8.16$ **b.** \$18.24 **c.** 6 months after Stewart was sentenced.

51. a. 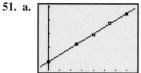 yes **b.** 1.30; the percentage of freshmen in college whose average grade in high school was an A is
increasing by 1.30 percentage points per year. **c.** 2006 **d.** 55.5% **e.** Answers may vary.

53. 3 **55.** -4 **57.** -4 **59.** -7 **61.** Twice the number, minus 3, is 7 (answers may vary); 5 **63.** Six times the number, minus 3, is
equal to 8 times the number, minus 4 (answers may vary); $\dfrac{1}{2}$ **65.** Twice the difference of the number and 4 is 10 (answers may vary); 9
67. Four, minus 7 times the sum of the number and 1 is 2 (answers may vary); $-\dfrac{5}{7}$ **69.** -1 **71.** -2 **73.** 2.83 **75.** 3.45 **77.** -5.54
79. 2 **81.** -2 **83.** 4 **85.** -1 **87.** 2 **89.** -3 **91.** set of all real numbers; identity **93.** empty set; inconsistent equation **95.** 2;
conditional equation **97.** set of all real numbers; identity **99.** empty set; inconsistent equation **101.** Answers may vary; answers may
vary; 7 **103.** Answers may vary; 7 **105.** Answers may vary. **107.** 3; answers may vary. **109.** 3; 3; answers may vary.

Homework 4.5 **1.** linear equation **3.** linear expression **5.** linear expression **7.** linear equation **9.** 2 **11.** $7x$ **13.** $-4b + 5$
15. $\dfrac{5}{4}$ **17.** $\dfrac{7}{19}$ **19.** $19x - 7$ **21.** $-\dfrac{3}{4}$ **23.** $-16x - 12$ **25.** $5.9p - 4.5$ **27.** 2 **29.** 3.5 **31.** $-6.1x + 35.28$ **33.** $-\dfrac{3w}{4}$
35. -2 **37.** $\dfrac{x}{12} + \dfrac{1}{2}$ **39.** -6 **41.** $\dfrac{14}{33}$ **43.** $\dfrac{17}{4}x - \dfrac{7}{6}$ **45.** Answers may vary. **47.** Answers may vary. **49.** no; answers may vary.
51. Answers may vary; $\dfrac{7x}{12}$ **53.** Answers may vary. **55.** Answers may vary. **57.** Answers may vary. **59.** $3 + 2x = -10$; $-\dfrac{13}{2}$
61. $4 - 6(x - 2)$; $-6x + 16$ **63.** $-9x = x - 5$; $\dfrac{1}{2}$ **65.** $\dfrac{x}{2} = 3(x - 5)$; 6 **67.** $x + x(6)$; $7x$ **69.** $x + \dfrac{x}{2}$; $\dfrac{3}{2}x$ **71.** Answers may vary.

Homework 4.6 **1.** $P = 4S$ **3.** $P = 2H + 2S + B$ **5.** $P = 2A + 2B + 2C + 2D$ **7.** $V = 5$ **9.** $B = 4$ **11.** $t \approx 1.86$
13. $L = 3$ **15.** $b = 7$ **17.** 14.5 feet **19.** 16 inches **21.** 59 feet **23. a.** $A = 3x$ **b.** $P = 2x + 6$ **c.** The units for both of the

expressions P and $2x + 6$ are inches.   **25. a.** Answers may vary.   **b.** Answers may vary.   **c.** Answers may vary.   **27. a.** 30 cents   **b.** 40 cents   **c.** $T = 10d$ **d.** The units for both of the expressions T and $10d$ are cents.   **29.** $T = 12.95n$ **31. a.** $T = 32.50x$ **b.** 18,522; if 18,522 people buy tickets, the total sales will be \$601,965.   **33. a.** $C = 10k$ **b.** $E = 95n$ **c.** $T = 10k + 95n$ **d.** 2000;   for 8000 \$10 tickets and 2000 \$95 tickets, the total cost is \$270,000.   **35. a.** $V = LWH$ **b.** 8 feet   **37. a.** $I = Prt$ **b.** \$600   **c.** $B = P + Prt$ **d.** \$2400   **39.** 82   **41.** $W = \dfrac{A}{L}$ **43.** $T = \dfrac{PV}{nR}$ **45.** $M = -\dfrac{Ur}{Gm}$ **47.** $B = \dfrac{2A}{H}$ **49.** $t = \dfrac{v - v_0}{g}$ **51.** $r = \dfrac{A - P}{Pt}$ **53.** $b = 3A - a - c$ **55.** $x = \dfrac{y - k + mh}{m}$ **57.** $y = a - x$ **59.** $y = -\dfrac{3}{4}x + 4$ **61.** $y = -\dfrac{1}{2}x + 2$ **63.** $y = \dfrac{5}{2}x - 3$ **65.** $y = -\dfrac{3}{7}x - \dfrac{5}{7}$ **67. a.** 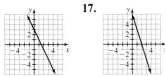 yes   **b.** $t = \dfrac{c - 135.2}{44.2}$ **c.** 2005   **d.** 2006, 2007, 2008,   2010, 2011

69. a. $s = 28.25t + 155$ **b.** $t = \dfrac{s - 155}{28.25}$ **c.** No; 2012   **d.** 2009, 2010, 2012, 2014, and 2016   **71. a.** $n = 1.3t + 17.7$ **b.** 2011   **c.** $t = \dfrac{n - 17.7}{1.3}$ **d.** 2011   **e.** The results are the same; $t = \dfrac{n - 17.7}{1.3}$; answers may vary.   **f.** $n = 1.3t + 17.7$; answers may vary; 30.7   thousand visas   **73. a.** The entries in the third column are 200, 210, 130, 275, st, all in miles; $d = st$ **b.** The units of the expressions on   both sides of the formula are miles.   **c.** $t = \dfrac{d}{s}$ **d.** 4.5; the person will travel 315 miles if he travels at 70 miles per hour for 4.5 hours.   **e.** 6.44 hours   **75.** The units of s are miles per hour and the units of dt are miles-hours. Since the units are different, the formula is   incorrect.   **77. a.** It doubles the perimeter.   **b.** The area is multiplied by 4.   **79.** Answers may vary.

Chapter 4 Review

1. $-20x$ **2.** $-24x - 12$ **3.** $12y - 28$ **4.** $-3x + 6y + 8$ **5.** $\dfrac{7}{9}x$ **6.** $4a - 9b - 7$ **7.** $-18x - 8y$ **8.** $-8.06x - 20.2$ **9.** $-5m - 4$ **10.** $-3a - 40b$ **11.** $-4(x - 7); -4x + 28$ **12.** $-7 + 2(x + 8); 2x + 9$ **13.** Answers may vary; $a(b + c) = ab + ac$ **14.** Answers   may vary.   **15.** $-5(x - 4), 5(4 - x), -2(x - 10) - 3x, -5x + 20$ **16.** 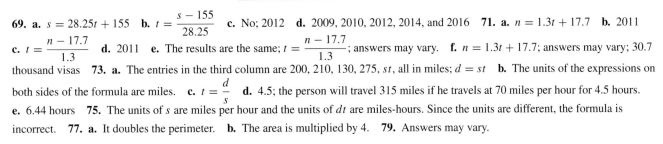   **17.**   **18.** no   **19.** 7   **20.** -5

21. 3   **22.** -6 **23.** 2.7   **24.** $-\dfrac{7}{2}$ **25.** $\dfrac{5}{11}$ **26.** 1   **27.** $\dfrac{15}{8}$ **28.** $\dfrac{38}{3}$ **29.** $\dfrac{44}{3}$ **30.** Answers may vary; 7   **31.** Answers may vary.   **32.** -8.31 **33. a.** $n = 0.06t + 1.9$ **b.** 2.5 million prisoners   **c.** 2012   **34.** $\dfrac{39}{4}$ **35.** -39 **36.** -1.64 **37.** 3   **38.** 1   **39.** 2   **40.** 0   **41.** the set of all real numbers; identity   **42.** $\dfrac{22}{15}$; conditional equation   **43.** empty set; inconsistent equation   **44.** Answers may   vary; 12   **45.** linear expression   **46.** linear equation   **47.** $-2t$ **48.** 3   **49.** -5 **50.** $0.2a + 0.1$ **51.** $-3p - 54$ **52.** $-\dfrac{11}{32}$ **53.** -18 **54.** $\dfrac{13}{3}r - \dfrac{11}{12}$ **55.** no; answers may vary.   **56.** Answers may vary; $\dfrac{2}{3}x + \dfrac{7}{5}$ **57.** $4(6 - x) = 17; \dfrac{7}{4}$ **58.** $x - \dfrac{x}{2}; \dfrac{x}{2}$ **59.** $P = 2A + 2B + 2C + 2D + 2E$ **60. a.** $T = 15n + 25w$ **b.** 220; for 370 \$15 tickets and 220 \$25 tickets, the total cost is \$11,050.   **61.** $r = \dfrac{C}{2\pi}$ **62.** $c = P - a - b$ **63.** $y = \dfrac{1}{2}x - 3$ **64.** $T = \dfrac{2A - HB}{H}$ **65. a.** $v = 0.04t + 1.17$ **b.** $t = \dfrac{v - 1.17}{0.04}$ **c.** 2011   **d.** 2011   **e.** The results are the same; $t = \dfrac{v - 1.17}{0.04}$; answers may vary.   **f.** $v = 0.04t + 1.17$; answers may vary; 1.57 billion visits

Chapter 4 Test

1. $-4x + 6$ **2.** $-4.08x + 0.96$ **3.** $-22w + 53$ **4.** $-11a - 3b - 2$ **5.** Answers may vary; $-2x - 10$ **6.**   **7.** $\dfrac{11}{3}$

8. 10   **9.** 7   **10.** $\dfrac{10}{13}$ **11.** $-\dfrac{41}{4}$ **12.** $-\dfrac{32}{25}$ **13.** 1.18   **14.** -32 **15.** $23x + 24$ **16.** $-\dfrac{12}{11}$ **17.** no; answers may vary.   **18.** Answers may vary.   **19.** $5(x - 2) = 29; \dfrac{39}{5}$ **20.** $2 + 4(3 + x); 4x + 14$ **21.** 2   **22.** 4   **23.** -2 **24.** 4   **25. a.** $n = 30t + 303$ **b.** 603 thousand   **c.** 2012   **26.** 18 feet   **27.** $a = 2A - b$

Cumulative Review of Chapters 1-4

1. n; 275, 300; 0, -150; answers may vary. **2.**

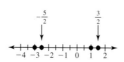

3. In 2006, there were 72.1 million unique visitors.

4. -2 **5.** 4 **6.** $-\dfrac{3}{10}$ **7.** $-\dfrac{47}{40}$ **8.** $-\dfrac{3}{5}$ **9.** $\dfrac{19}{6}$ feet **10.** -7 points **11.** -27 **12.** $-\dfrac{1}{2}$; decreasing **13.** undefined slope; vertical

14. road A **15.**   **16.** **17.** **18.** **19.** $y = 3x - 2$

20. -932 dollars per month **21.** 39.5 thousand reports per year **22. a.** 16.7; sales of convertibles increase by 16.7 thousand cars per year. **b.** $s = 16.7t + 242$ **c.** 425.7 thousand convertibles **d.** 2012 **23.** equations 2, 3, and 4

24. a. 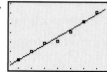 yes **b.** 4.01; the number of live births from fertility treatments is increasing by 4.01 thousand births per year. **c.** (0, 16.69); in 1995, there were 16.7 thousand live births from fertility treatments. **d.** 76.8 thousand live births **e.** 2012

25. 3 **26.** -8 **27.** $-3a - 7b + 6$ **28.** $14p + 7$ **29.** 2 **30.** $-6a - 4$ **31.** -5.86 **32.** $x + 9\left(\dfrac{x}{3}\right); 4x$ **33.** $2(7 - 2x) = 87; -\dfrac{73}{4}$

34. $h = \dfrac{A}{2\pi r}$ **35.** $y = \dfrac{2}{3}x - 2$

Chapter 5

Homework 5.1

1. slope: 2; y-intercept: (0, -3)

3. slope: -3; y-intercept: (0, 5)

5. slope: $-\dfrac{3}{5}$; y-intercept: (0, -2)

7. slope: $\dfrac{1}{2}$; y-intercept: (0, 2)

9. slope: -3; y-intercept: (0, 0)

11. slope: 0; y-intercept: (0, -4)

13. slope: -1; y-intercept: (0, 3)

15. slope: -2; y-intercept: (0, 4)

17. slope: 2; y-intercept: (0, -1)

19. slope: $\dfrac{2}{3}$; y-intercept: (0, 0)

21. slope: $\dfrac{5}{2}$; y-intercept: (0, -3)

23. slope: $\dfrac{4}{5}$; y-intercept: (0, -3)

25. slope: $\dfrac{3}{4}$; y-intercept: (0, -2)

27. slope: $\dfrac{2}{5}$; y-intercept: (0, -2)

29. slope: $-\dfrac{1}{4}$; y-intercept: (0, 1)

31. slope: $-\dfrac{1}{2}$; y-intercept: (0, -2)

33. slope: -4; y-intercept: (0, -2)

35. slope: $\dfrac{3}{2}$; y-intercept: (0, 2)

37. slope: $-\dfrac{5}{3}$; y-intercept: (0, 0)

39. slope: 0;
y-intercept: (0, 3)

41. x-intercept: (6, 0);
y-intercept: (0, −2)

43. x-intercept: (5, 0);
y-intercept: (0, 3)

45. x-intercept: (−6, 0);
y-intercept: (0, 4)

47. x-intercept: (2, 0);
y-intercept: (0, 6)

49. x-intercept: (4, 0);
y-intercept: (0, 6)

51. x-intercept: (3, 0);
y-intercept: (0, 5)

53. x-intercept: (1.21, 0);
y-intercept: (0, 2.68)

55. x-intercept: (−2.05, 0);
y-intercept: (0, 3.47)

57. x-intercept: (2.07, 0); y-intercept: (0, 9.32)   **59.** x-intercept: (−14.94, 0); y-intercept: (0, −37.21)   **61.**

63.

65.

67.

69. a. 1 **b.** 2 **c.**

71. no; $-\dfrac{2}{3}$

73. a.

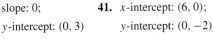

b. Answers may vary.   **c.** For each solution, the difference of three times the x-coordinate
and five times the y-coordinate is equal to 10.

75. a. x-intercept: (5, 0); y-intercept: (0, 7)   **b.** x-intercept: (4, 0); y-intercept: (0, 6)   **c.** x-intercept: (a, 0); y-intercept: (0, b)

d. $\dfrac{x}{2} + \dfrac{y}{5} = 1$ **77.** Answers may vary.   **79.** Answers may vary.   **81.** 1   **83.** 2   **85.** (4, 0)   **87.** $\dfrac{1}{2}$ **89.** $-6x + 26$; linear expression
in one variable   **91.** 3; linear equation in one variable

Homework 5.2   **1.** $y = 2x - 1$ **3.** $y = -3x + 1$ **5.** $y = 2x + 2$ **7.** $y = -6x - 15$ **9.** $y = \dfrac{2}{5}x - \dfrac{1}{5}$ **11.** $y = -\dfrac{3}{4}x - \dfrac{13}{2}$

13. $y = 3$ **15.** $x = -2$ **17.** $y = 2.1x - 13.67$ **19.** $y = -5.6x - 22.4$ **21.** $y = -6.59x - 17.95$ **23.** $y = 2x - 4$

25. $y = -2x + 8$ **27.** $y = 5x - 2$ **29.** $y = x - 2$ **31.** $y = x$ **33.** $y = -4x + 9$ **35.** $y = 2$ **37.** $x = -4$ **39.** $y = \dfrac{1}{2}x + 1$

41. $y = \dfrac{2}{3}x - \dfrac{1}{3}$ **43.** $y = -\dfrac{1}{6}x + \dfrac{3}{2}$ **45.** $y = -\dfrac{2}{7}x + \dfrac{3}{7}$ **47.** $y = \dfrac{3}{5}x + \dfrac{2}{5}$ **49.** $y = \dfrac{3}{2}x - 2$ **51.** $y = -1.23x + 8.62$

53. $y = -1.70x - 5.46$ **55.** $y = -1.01x + 0.65$ **57.** $y = 0.48x - 6.11$ **59.** $y = 2x - 1$ **61.** $y = -4x - 7$ **63.** $y = -2x + 10$

65. $y = 4x + 3$ **67.** $y = -\dfrac{4}{3}x + \dfrac{23}{3}$ **69. a.** $y = \dfrac{3}{2}x - 3$ **b.**

c. Answers may vary.

71. a. possible; answers may vary.   **b.** possible; answers may vary.   **c.** not possible; answers may vary.   **d.** possible; $y = 0$

73. $E = 2t + 5$ **75. a. i.** $y = 3x - 5$ **ii.** $y = 3x - 5$ **b.** The results are the same.   **77. a.** Answers may vary.   **b.** Answers may
vary.   **c.** Answers may vary.   **d.** no such line; answers may vary.   **79.** Set 1: $y = -2x + 25$; Set 2: $y = 4x + 12$; Set 3: $y = -5x + 77$;
Set 4: $y = 3$ **81.** Answers may vary.   **83. a.**

b. $y = 2x - 3$; it is the same equation.

85. linear equation in two variables **87.** 2; expression in two variables

Homework 5.3

For many exercises in this section, answers may vary.

1. $n = 14.65t + 179$ **3.** $n = -220.25t + 3367.25$ **5. a.** $p = 0.51t + 8.08$ **b.** 0.51; the percentage of sexual harassment charges filed by men is increasing by 0.51 percentage point per year. **c.** (0, 8.08); in 1990, 8.08% of sexual harassment charges were filed by men.
7. a. $n = -3.67t + 17.67$ **b.** −3.67; the number of U.S. flag desecrations is decreasing by 3.67 desecrations per year. **c.** (0, 17.67); there were about 18 U.S. flag desecrations in 2000. **9.**  increase m and decrease b

11. $y = 2.5x - 2.2$; answers may vary. **13.** $y = -1.11x + 20.83$; answers may vary. **15.** student C

17. a. 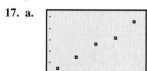 **b.** $p = 1.98t + 68.98$; answers may vary. **c.**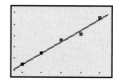

19. a. 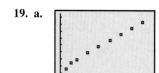 **b.** $H = 4.5d + 15.3$; answers may vary. **c.**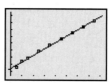

21. a. 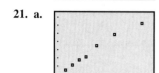 **b.** $p = 1.08d + 8.81$; answers may vary. **c.**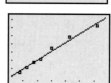

23. a. 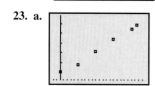 **b.** $p = 0.47t + 6.99$; answers may vary. **c.** 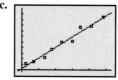 **d.** 0.47; the percentage of Americans who have a college degree increases by 0.47 percentage point per year.
e. (0, 6.99); in 1960, 6.99% of Americans earned a college degree.

25. a. 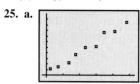 **b.** $r = 7.54t + 23.99$; answers may vary. **c.** 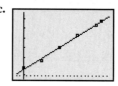 **d.** 7.54; the land speed record is increasing by 7.54 mph each year. **e.** (0, 23.99); in 1900, the land speed record was 23.99 mph.

27. a. 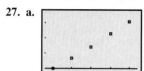 **b.** $p = 7.8r - 31.8$; answers may vary. **c.** 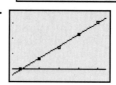 **29.** Answers may vary.

31. a. $V = 2t + 10$; the units for both of the expressions $2t + 10$ and V are dollars. **b.** 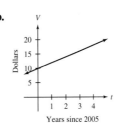 **c.** Answers may vary.

33. a. $n = -298t + 50{,}575$ **b.** -298; the average attendance is decreasing by 298 spectators per year. **c.** The units for both of the expressions $-298t + 50{,}575$ and n are number of spectators. **35.** $\frac{5}{2}$; linear equation in one variable

37.

linear equation in two variables

Homework 5.4
1. a. 90.8% **b.** 2016; model breakdown has occurred. **c.** $(0, 68.98)$; 69.0% of all seats were filled in May of 2000 for the six largest airlines. **d.** 1.98; each year, the percentage of all seats that are filled in May increases by 1.98 percentage points for the six largest airlines. **3. a.** $38.1 thousand **b.** 1.08; the percentage of households that own a computer increases by 1.08 percentage points for each $1 thousand increase in income. **c.** 12%; the error is -10 percentage points; answers may vary. **5. a.** 14 dog years **b.** 146 human years **c.** 23 dog years **d.** 4.5; a dog aging 1 year is equivalent to a human aging 4.5 years. **e.** $H = 7d$ **f.** 7; a dog aging 1 year is equivalent to a human aging 7 years; the slope of the graph of the equation in part (e) is greater than the slope found in part (d). **7. a.**  **b.** $p = -0.56t + 53.17$ **c.** 2010 **d.** $(0, 53.17)$; in 1970, 53.2% of children lived in "very happy" marriages. **e.** $(94.95, 0)$; in 2065, no children will live in "very happy" marriages; model breakdown has likely occurred.

9. a. 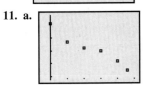 **b.** $w = -1.27t + 158.93$ **c.** -1.27; the math professor lost 1.27 pounds per week. **d.** 11 weeks **e.** $(125.14, 0)$; the math professor will be weightless in about 125 weeks; model breakdown has occurred.

11. a. **b.** $p = -0.14t + 11.68$ **c.** 2010 **d.** $(0, 11.68)$; in 1980, 11.7% of high school students dropped out of school. **e.** $(83.43, 0)$; in 2063, no high school students will drop out of school; model breakdown has likely occurred.

13. a. 7.8; the percent increase in patients who would die increases by 7.8 percentage points if the patient-to-nurse ratio is increased by 1. **b.** 46.2% **c.** 7.2%; overestimate; answers may vary. **d.** Answers may vary. **15.** 2010 **17.** 9.1% **19.** 2014 **21.** 2012 **23. a.** $n = 0.55t + 27.2$ **b.** $(0, 27.2)$; in 2000, there were 27.2 million Americans who lived alone. **c.** The units for both of the expressions $0.55t + 27.2$ and n are millions of people. **d.** 2026 **25. a.** $s = -20.67t + 152$ **b.** $(0, 152)$; in 2004, the annual echinacea sales were $152 million. **c.** The units for both of the expressions $-20.67t + 152$ and s are millions of dollars. **d.** 2006 **27.** $\frac{17}{18}$; linear equation in one variable **29.** $-18x + 17$; linear expression in one variable

Homework 5.5
1. true **3.** true **5.**  **7.** **9.** **11.**

13.

In Words	Inequality	Graph	Interval Notation
numbers greater than or equal to 4	$x \geq 4$	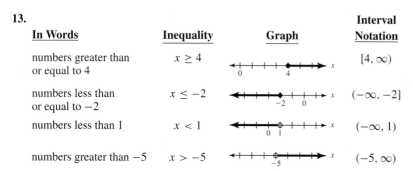	$[4, \infty)$
numbers less than or equal to -2	$x \leq -2$		$(-\infty, -2]$
numbers less than 1	$x < 1$		$(-\infty, 1)$
numbers greater than -5	$x > -5$		$(-5, \infty)$

15. $3, 6$ **17.** -4 **19.** $x > 1$; $(1, \infty)$; **21.** $x < -3$; $(-\infty, -3)$; **23.** $x \leq 3$; $(-\infty, 3]$;

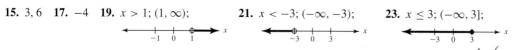

25. $x \geq -2$; $[-2, \infty)$; **27.** $t \leq -2$; $(-\infty, -2]$; **29.** $x > -4$; $(-4, \infty)$; **31.** $x < -\dfrac{1}{2}$; $\left(-\infty, -\dfrac{1}{2}\right)$;

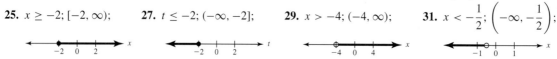

33. $x \leq 0$; $(-\infty, 0]$; **35.** $x > -2$; $(-2, \infty)$; **37.** $a < 6$; $(-\infty, 6)$; **39.** $x \leq -3$; $(-\infty, -3]$;

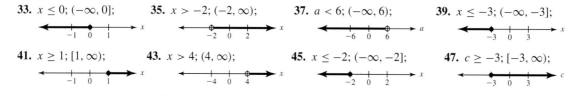

41. $x \geq 1$; $[1, \infty)$; **43.** $x > 4$; $(4, \infty)$; **45.** $x \leq -2$; $(-\infty, -2]$; **47.** $c \geq -3$; $[-3, \infty)$;

49. $x \geq -3$; $[-3, \infty)$; **51.** $x < 2.5$; $(-\infty, 2.5)$; **53.** $b < 2$; $(-\infty, 2)$; **55.** $x > -5$; $(-5, \infty)$;

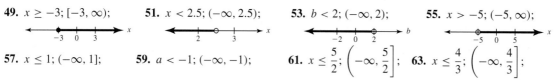

57. $x \leq 1$; $(-\infty, 1]$; **59.** $a < -1$; $(-\infty, -1)$; **61.** $x \leq \dfrac{5}{2}$; $\left(-\infty, \dfrac{5}{2}\right]$; **63.** $x \leq \dfrac{4}{3}$; $\left(-\infty, \dfrac{4}{3}\right]$;

65. $x \leq -1.7$; $(-\infty, -1.7]$; **67.** $y \geq \dfrac{5}{3}$; $\left[\dfrac{5}{3}, \infty\right)$; **69.** $x > 7$; $(7, \infty)$; **71.** $x > -\dfrac{3}{5}$; $\left(-\dfrac{3}{5}, \infty\right)$;

73. $r \leq 3$; $(-\infty, 3]$;

75. a. 0.83; the percentage of teachers who are "very satisfied" with teaching is increasing by 0.83 percentage point per year. **b.** after 2007 **77. a.** $p = -0.58t + 74.85$ **b.** before 1998 **c.** The number of nonmarried households is growing at a greater rate than the number of married households. **79. a.** $r = -1.80t + 64.93$ **b.** 28.9 births per 1000 women; 300,502 births **c.** after 2021
81. Answers may vary; $x > -5$ **83. a.** Answers may vary. **b.** Answers may vary. **85.** Answers may vary. **87.** 4
89. $x < 4$; $(-\infty, 4)$; **91.** $x + 5 > 2$; $x > -3$; $(-3, \infty)$; **93.** $2x \leq 5x - 6$; $x \geq 2$; $[2, \infty)$; **95.** Answers may vary.
97. Answers may vary.

Chapter 5 Review

1. slope: 4;
y-intercept: $(0, -5)$

2. slope: -3;
y-intercept: $(0, 4)$

3. slope: $\dfrac{1}{2}$;
y-intercept: $(0, 1)$

4. slope: $-\dfrac{2}{3}$;
y-intercept: $(0, -2)$

5. slope: $\dfrac{5}{3}$;
y-intercept: $(0, 0)$

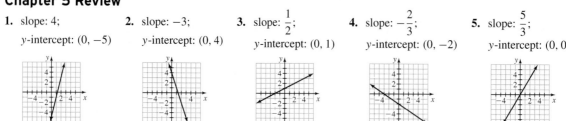

6. slope: 0;
y-intercept: (0, −5)

7. slope: 2;
y-intercept: (0, −5)

8. slope: $\frac{3}{2}$;
y-intercept: (0, 3)

9. slope: $-\frac{4}{5}$;
y-intercept: (0, 2)

10. slope: $-\frac{1}{3}$;
y-intercept: (0, 2)

11. slope: $-\frac{2}{5}$;
y-intercept: (0, 4)

12. slope: 0;
y-intercept: (0, 4)

13. x-intercept: (5, 0);
y-intercept: (0, −4)

14. x-intercept: (2, 0);
y-intercept: (0, 3)

15. x-intercept: (−4, 0);
y-intercept: (0, −3)

16. x-intercept: (2, 0);
y-intercept: (0, −4)

17. x-intercept: (3, 0);
y-intercept: (0, 3)

18. x-intercept: (3, 0);
y-intercept: (0, −2)

19. x-intercept: (9.48, 0); y-intercept: (0, −22.95)

20. x-intercept: (−37.99, 0); y-intercept: (0, 97.25)   **21. a.** x-intercept: $\left(-\frac{7}{3}, 0\right)$; y-intercept: (0, 7)   **b.** x-intercept: $\left(-\frac{9}{2}, 0\right)$;
y-intercept: (0, 9)   **c.** x-intercept: $\left(-\frac{b}{m}, 0\right)$; y-intercept: (0, b)   **22. a.** slope: $\frac{1}{2}$;
y-intercept: (0, −2)   **b.** x-intercept: (4, 0);
y-intercept: (0, −2)

c. The graphs are the same; answers may vary.   **23.** $y = -4x + 7$ **24.** $y = 2x + 15$ **25.** $y = -3x - 7$ **26.** $y = -\frac{2}{5}x + \frac{24}{5}$
27. $y = \frac{3}{7}x + \frac{57}{7}$ **28.** $y = -\frac{2}{3}x - 8$ **29.** $x = 2$ **30.** $y = -4$ **31.** $y = -5.29x - 17.26$ **32.** $y = 1.45x - 3.08$ **33.** $y = 2x - 3$
34. $y = -2x + 5$ **35.** $y = 3x - 1$ **36.** $y = 5x - 15$ **37.** $y = -\frac{3}{2}x + \frac{1}{2}$ **38.** $y = \frac{1}{4}x - 3$ **39.** $y = -\frac{5}{3}x + 4$ **40.** $y = \frac{3}{2}x - 4$
41. $x = 5$ **42.** $y = -3$ **43.** $y = -0.85x + 12.16$ **44.** $y = -0.92t - 2.31$ **45.** $y = -\frac{1}{5}x - \frac{17}{5}$ **46.** $y = -2.13x + 30.38$
47. a.

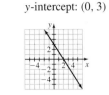

b. $p = -0.27t + 17.36$ **c.** 2011   **d.** (64.30, 0); in 2044, no restaurant patrons will order dessert; this prediction seems highly unlikely.   **e.** (0, 17.36); in 1980, 17.4% of restaurant patrons ordered dessert.

48. a.

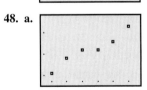

b. $s = 0.19t + 1.86$ **c.** 0.19; sales are increasing by 190 million units per year.   **d.** 5.5 billion units
e. 2012

49. a. $n = 0.3t + 9.6$ **b.** 0.3; the number of living Americans who have been diagnosed with cancer increases by 0.3 million Americans per year.   **c.** 12.6 million Americans   **d.** 2015   **50.** 2012   **51.** $x > 3$; (3, ∞);   **52.** $x \geq -1$; [−1, ∞);

53. $w \leq -5$; (−∞, −5];   **54.** $p > 6$; (6, ∞);   **55.** $x > -2$; (−2, ∞);   **56.** $x > -3$; (−3, ∞);

57. $a \le -3$; $(-\infty, -3]$;   **58.** $b \ge -4$; $[-4, \infty)$;

59. a. $r = -3.18t + 61.60$ **b.** -3.18; the violent crime rate is decreasing by 3.18 violent crimes per 1000 people age 12 or older per year.
c. after 2006

Chapter 5 Test

1. slope: -3;
y-intercept: $(0, -1)$

2. slope: $\dfrac{2}{5}$;
y-intercept: $(0, -2)$

3. slope: 0;
y-intercept: $(0, 5)$

4. x-intercept: $(3, 0)$;
y-intercept: $(0, -6)$

5. x-intercept: $(7.38, 0)$; y-intercept: $(0, -9.10)$ **6.** x-intercept: $(2, 0)$; y-intercept: $(0, 7)$ **7.** no; $\dfrac{5}{4}$ **8.** $y = 7x + 10$

9. $y = -\dfrac{2}{3}x + 3$ **10.** $x = 2$ **11.** $y = -\dfrac{1}{2}x + 4$ **12.** $y = -1.92t - 3.64$ **13.** $y = -\dfrac{1}{3}x - \dfrac{7}{3}$

14.
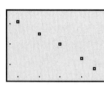
decrease m and increase b **15.** $y = 1.15x + 6.88$

16. a.

b. $p = -1.25t + 97.70$ **c.** -1.25; the percentage of employers who offer pensions is decreasing by 1.25 percentage points per year.   **d.** $(0, 97.70)$; in 1980, 97.7% of employers offered pensions.
e. $(78.16, 0)$; in 2058, no company will offer a pension, according to the model.   **f.** 1978; a little research would show that this is false.   **g.** after 2018

17. $x \le 2$; $(-\infty, 2]$;   **18.** $x < 2$; $(-\infty, 2)$;

Chapter 6

Homework 6.1   **1.** $(1, -1)$ **3.** $(3, -2)$ **5.** $(2, 1)$ **7.** $(-2, 1)$ **9.** $(0, 0)$ **11.** $(2, -4)$ **13.** $(4, -3)$ **15.** $(3, -2)$ **17.** $(2, -3)$
19. $(3, -2)$ **21.** $(1, 3)$ **23.** $(1.16, -2.81)$ **25.** $(-4.67, -3.83)$ **27.** $(4.14, -4.90)$ **29.** $(-4.83, 1.07)$ **31.** all solutions of the line
$y = -2x + 4$; dependent system   **33.** $(1, -3)$ **35.** empty set solution; inconsistent system   **37.** 1983; 478 thousand graduates of each
gender   **39. a.** $r = -0.064t + 27.00$ **b.** $r = -0.029t + 22.10$ **c.** 2040; 18.04 seconds   **41. a.** B, D   **b.** B, C   **c.** B
d. A, E, F   **43.** $(2.8, -2.4)$ **45.** $(14, 1)$ **47.** $(3, 9)$ **49.** Answers may vary.   **51.** no; answers may vary.   **53. a.** The lines are
parallel.   **b.** no   **c.** empty set solution   **55.** Answers may vary.   **57.** $(-2, 1)$; system of two linear equations in two variables
59. -2; linear equation in one variable

Homework 6.2   **1.** $(2, 4)$ **3.** $(1, 1)$ **5.** $(-2, 3)$ **7.** $(-3, -1)$ **9.** $(4, 1)$ **11.** $(-2, -4)$ **13.** $(-2, 3)$ **15.** $(0, 0)$
17. $(1, -6)$ **19.** $(-2.07, 1.76)$ **21.** $(3.35, -1.71)$ **23.** $(-0.22, 4.34)$ **25.** $(-2, -5)$ **27.** $(2, -1)$ **29.** $(-1, 3)$ **31.** $(-2, -5)$
33. $(1, -3)$ **35.** infinite number of solutions of the equation $x = 4 - 3y$; dependent system   **37.** $(2, -4)$ **39.** empty set solution;
inconsistent system   **41.** infinite number of solutions of the equation $x = 3y - 1$; dependent system   **43.** empty set solution; inconsistent
system   **45.** $(17, 56)$ **47.** A: $(0, 0)$; B: $(0, 8)$; C: $(5, 3)$; D: $\left(\dfrac{7}{2}, 0\right)$ **49. a.** $(1, 2)$ **b.** $(1, 2)$ **c.** The results are the same.

51. Answers may vary.   **53. a.** $(2, 1)$ **b.** $(2, 1)$ **c.** 2   **d.** They are the same.   **e.** 3; 3; they are the same.   **f.** Answers may vary.

55.

linear equation in two variables   **57.** $(-3, 14)$; system of two linear equations in two variables

Homework 6.3   **1.** (2, 1)   **3.** (2, 4)   **5.** (2, −3)   **7.** (1, −2)   **9.** (−1, −2)   **11.** (1, −3)   **13.** (−2, −5)   **15.** (1, −3)
17. (−4, 5)   **19.** (3, 10)   **21.** the infinite number of solutions of the equation $4x − 7y = 3$; dependent system   **23.** empty set solution;
inconsistent system   **25.** (−4, 1)   **27.** empty set solution; inconsistent system   **29.** the infinite number of solutions of the equation
$3x − 9y = 12$; dependent system   **31.** (3.36, 5.51)   **33.** (1.52, 5.11)   **35.** (3.03, −0.71)   **37.** (−2, 4)   **39.** (5, 1)   **41.** (−4, 3)
43. (−4, −1)   **45.** (−2, 1)   **47.** (−1, 3)   **49.** (3, 2)   **51.** (2, 1); answers may vary.   **53. a.** (1, 2)   **b.** (1, 2)   **c.** The results are the
same.   **55.** (23, 24)   **57.** A: (0, 0); B: (0, 4); C: (3, 2); D: $\left(\dfrac{9}{5}, 0\right)$ **59.** Answers may vary.   **61. a.** $5 = 2m + b$ **b.** $9 = 4m + b$
c. $m = 2; b = 1$ **d.** $y = 2x + 1$ **e.** Answers may vary.   **63.** $−8x − 4$; linear expression in one variable   **65.** $−\dfrac{1}{2}$; linear equation in
one variable

Homework 6.4   **1.** 1983; 478 thousand graduates of each gender   **3.** 2040; 18.04 seconds   **5. a.** $p = 6.32t − 23.71$
b. $p = 4.26t + 7.79$ **c.** 2005; 72.9%   **7. a.** $s = −2.17t + 97.74$ **b.** $s = 2.02t − 4.84$ **c.** 2004; 44.6%   **d.** 11.8%   **9. a.** 2070
b. 19 women; from the 102nd Congress to the 103rd Congress; no   **c.** 2093   **d.** Answers may vary.   **11. a.** $d = 1.6t + 56.9$
b. $d = 2.4t + 36.2$ **c.** 2026; 98.3 thousand degrees   **13. a.** $s = 2.5t + 4.6$ **b.** $s = −2.6t + 20$ **c.** 2003; $12.1 million
15. a. $V = −1903t + 18{,}249$ **b.** $V = −1225t + 14{,}564$ **c.** 2007; $7906   **17. a.** $s = −1.2t + 29.5$ **b.** $s = 1.7t + 12.1$
c. 2011; $22.3 million   **19. a.** $c = 850t + 5425$ **b.** $c = 570t + 6557$ **c.** 2009, $8861   **21.** 2079; 3280 calories per day
23. (1,4); system of two linear equations in two variables   **25.** 1; linear equation in one variable

Homework 6.5   **1.** width: 114.50 feet, length: 185.50 feet   **3.** width: 8 feet, length: 13 feet   **5.** width: 36 feet, length: 78 feet
7. width: 19 inches, length: 35 inches   **9.** width: 18 yards, length: 55 yards   **11.** 1500 $15 tickets, 500 $22 tickets   **13.** 39 *Amnesiac*
CDs, 214 *Hail to the Thief* CDs   **15.** 1300 books and 200 CDs   **17.** main level: $26, balcony: $14   **19.** general: $37, reserved: $62
21. a. $200   **b.** $280   **c.** 0.08*d*   **23. a.** $480   **b.** $420   **c.** $0.03x + 0.06y$ **25.** First Funds TN Tax-Free I account: $14,000, W & R
International Growth C account: $6000   **27.** Middlesex Savings Bank CD account: $2500, First Funds Growth & Income I account: $6000
29. Limited Term NY Municipal X account: $7500, Calvert Income A account: $2500   **31.** $5000 should be invested in each account.
33. a. 1.3 ounces   **b.** 1.95 ounces   **c.** 0.65*x* ounces   **35. a.** 1.7 ounces   **b.** 35% solution: 6 ounces, 10% solution: 9 ounces
37. 20% solution: 4 quarts, 30% solution: 1 quart   **39.** 10% solution: 1 gallon, 25% solution: 2 gallons   **41.** 15% solution: 1 quart, 35%
solution: 3 quarts   **43.** 12% solution: 2 gallons, 24% solution: 6 gallons   **45.** 20% solution: 3 ounces, water: 2 ounces   **47. a.** $3000
b. $4500   **c. i.** no; answers may vary.   **ii.** no; answers may vary.   **iii.** yes; 250 $10 tickets, 50 $15 tickets   **49.** Answers may vary.
51. Answers may vary.   **53.** $L = \dfrac{P − 2W}{2}$ **55.** $r = \dfrac{A − P}{Pt}$ **57.** $−28x + 12$; linear expression in one variable   **59.** $\dfrac{10}{33}$; linear
equation in one variable

Homework 6.6   **1.** (4, 1)   **3.** (−1, −4), (5, 0)   **5.** 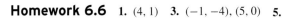   **7.**   **9.**   **11.**

13. 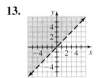   **15.**   **17.**   **19.**   **21.**   **23.**

25.   **27.**   **29.**   **31.**   **33.**   **35.**

37.   **39.**   **41.**   **43.**   **45.**   **47.**

49. **51.** **53.** **55.** **57.**

59. a. $w \le 3.50h - 80.97$, $w \ge 3.08h - 64.14$, $h \ge 63$, $h \le 78$ **b.** 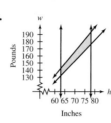 **c.** The ideal weights are between 145 pounds and 157 pounds, inclusive.

61. $y \le -\dfrac{1}{2}x + 2$ **63.** no; answers may vary. **65. a.** A, B, C, D **b.** A, B, C, F, G, H **c.** A, B, C **d.** E **67.** Answers may vary.

69. Answers may vary. **71.** **73.** **75.** $x \le 2$; $[-\infty, 2)$; 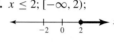 **77.**

79. $(2, 3)$ **81.** Answers may vary. **83.** Answers may vary. **85.** Answers may vary.

Chapter 6 Review

1. $(2, 1)$ **2.** $(-4, -2)$ **3.** $(0, 0)$ **4.** $(1, -1)$ **5.** $(-3, 1)$ **6.** $(-3, -2)$ **7.** $(3, -1)$ **8.** $(2, 1)$ **9.** $(0, 0)$ **10.** $(2, -12)$
11. $(1, -2)$ **12.** $(-3, 4)$ **13.** $(2.61, -2.20)$ **14.** $(-1.19, 4.76)$ **15.** $(3, 2)$ **16.** $(-4, -1)$ **17.** $(-2, 1)$ **18.** $(-3, -2)$
19. $(4, -2)$ **20.** $(2, 1)$ **21.** $(-1, 2)$ **22.** $(-2, 1)$ **23.** $(0.77, 4.79)$ **24.** $(2.14, -1.62)$ **25.** $(-3, 1)$ **26.** $(-5, 4)$ **27.** $(-2, 0)$
28. $(4, 3)$ **29.** $(-4, 2)$ **30.** $(5, 3)$ **31.** $(-3, -2)$ **32.** $(1, -2)$ **33.** empty set solution; inconsistent system **34.** all solutions of the
equation $y = -4x + 3$; dependent system **35. a.** Answers may vary. **b.** Answers may vary. **c.** Answers may vary. **d.** $(2, 3)$
36. $(-1.4, 2.8)$ **37.** $(14, 47)$ **38.** $(1, 3)$; answers may vary. **39. a.** $p = 4.4t + 7.4$ **b.** $p = -4.7t + 86.8$ **c.** 2009; 45.8%
40. a. $p = -1.25t + 58$ **b.** $p = 1.25t + 42$ **c.** 2001 **d.** 2001 **e.** car model's slope: -1.25, light-truck model's slope: 1.25; the
slopes are equal in absolute value and opposite in sign. **41.** width: 5 feet, length: 17 feet **42.** 6500 $22 tickets, 1500 $39 tickets

43. **44.** **45.** **46.** **47.** **48.**

49. 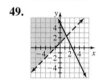 **50.** **51.** **52.**

53. a. $w \le 3h - 54$, $w \ge 3h - 68$, $h \ge 58$, $h \le 74$ **b.** 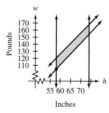 **c.** The ideal weight range is between 142 pounds and 156 pounds, inclusive.

54. Answers may vary.

Chapter 6 Test

1. $(5, -3)$ **2.** $(-3.18, 1.88)$ **3.** $(-2, -7)$ **4.** $(-1, 3)$ **5.** $(2, -3)$ **6.** $(-4, 2)$ **7.** $(-2, 3)$ **8.** all solutions of the equation $2x - 3y = 4$; dependent system **9.** empty set solution; inconsistent system **10.** $(5, 2)$ **11.** $(3.45, 1.93)$ **12. a.** C, D **b.** D, F **c.** D **d.** A, B, E **13.** $(24, 5)$ **14.** Answers may vary. **15.** A: $(0, 0)$; B: $(0, 11)$; C: $\left(\frac{9}{2}, 5\right)$; D: $(2, 0)$ **16. a.** $p = -8.42t + 62.63$ **b.** $p = 6.96t + 7.51$ **c.** 2004; 32.5% **17. a.** $n = 4.9t + 230$ **b.** $n = 2.9t + 280$ **c.** 2025; no; 352.5 million guns **18.** 3% account: $2000, 7% account: $5000 **19.**

 20. **21.** **22.**

Cumulative Review of Chapters 1-6

1. Fahrenheit degrees **2.** A cricket chirps 129 times per minute when the temperature is 70°F. **3.** $-\frac{8}{9}$ **4.** $\frac{46}{35}$ **5.** 21

6. $-\frac{3}{4}$; decreasing **7. a.** m is negative; b is positive. **b.** m is positive; b is negative. **c.** m is zero; b is positive. **d.** m is negative; b is negative. **8.** equation 1: $y = -8x + 49$; equation 2: $y = 4x + 11$; equation 3: $y = -2x + 45$; equation 4: $y = 3x + 8$ **9.** $-\frac{10}{13}$ **10.** $-13p + w + 5$ **11.** $4w + 6y - 10$ **12.** $\frac{22}{9}$ **13.** $y = \frac{c - ax}{b}$ **14.** $6 + 3(4 + x)$; $3x + 18$ **15.** $4 - \frac{x}{3} = 2$; 6

16. slope: 2; y-intercept: $(0, -4)$ **17.** slope: $\frac{1}{2}$; y-intercept: $(0, -3)$ **18.** slope: $-\frac{5}{2}$; y-intercept: $(0, 6)$ **19.** slope: 0; y-intercept: $(0, -3)$ **20.** x-intercept: $(5, 0)$; y-intercept: $(0, -2)$

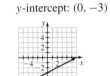

21. x-intercept: $(2, 0)$; y-intercept: $(0, 4)$ **22.** **23.** not parallel **24.** $y = -\frac{2}{5}x - \frac{4}{5}$ **25.** $y = -\frac{4}{3}x - \frac{17}{3}$ **26.** $x = -2$ **27.** 2

28. 6 **29.** $(3, 0)$ **30.** $y = -\frac{1}{3}x + 1$ **31.** $x < -1$; $(-\infty, -1)$ **32.** $(2, 1)$ **33.** $(3.31, -1.80)$ **34.** $(3, -2)$ **35.** $(2, -1)$

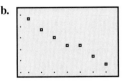

36. $(23.90, 28.49)$ **37.** $(2, -5)$ **38.** $(15, 37)$ **39.** **40.**

41. a. Answers may vary. **b.** **c.** $-0.0186t + 1.72$ **d.** $(0, 1.72)$; there were 1.72 deaths per 100 million miles traveled in 1990. **e.** 2029; yes; answers may vary. **f.** 1.39 deaths per 100 million miles traveled

42. a. $r = 0.080t + 1.30$ **b.** $r = -0.064t + 4.79$ **c.** 2014; 3.2 **43. a.** -375; the enrollment is decreasing by 375 students per year. **b.** $E = -375t + 25{,}700$ **c.** 2011 **d.** after 2015 **44. a.** $n = -6.8t + 181.3$ **b.** $n = 6.8t + 9.4$ **c.** When one of the Bell companies loses control of a line, one of the other companies gains control of that line. Apparently, the total number of lines has been approximately constant (190.7 million lines) since 2000. **d.** 2013 **e.** after 2013 **45.** 16% solution: 8 quarts, 28% solution: 4 quarts

Chapter 7

Homework 7.1

1. **3.** **5.** **7.** **9.** **11.**

13. **15.** **17.** -5 **19.** $-2, 0$ **21.** -1 **23.** no such value **25.** $(-3, 0), (1, 0)$ **27.** $(-1, 4)$ **29.** 3

31. $1, 3$ **33.** no such value **35.** $(0, 3)$ **37.** $(2, -1)$ **39. a.** **b.** Answers may vary. **c.** For each solution, the y-coordinate is 3 less than the square of the x-coordinate.

41. 3 **43.** 1, 3 **45. a.** 0 **b.** 3 **c.** $(0, 3)$; answers may vary. **47. a. i.** 5; one **ii.** 17; one **iii.** 10; one **b.** one; answers may vary. **c. i.** Answers may vary; one **ii.** Answers may vary; one **iii.** Answers may vary; one **d.** one; answers may vary. **e.** one; answers may vary. **49.** **51.** **53. a.** **b.** **c.** The y-intercept is $(0, 0)$ for both equations. **d.** $y = x^2$; answers may vary.

55. $-\dfrac{1}{5}$; linear equation in one variable **57.** $20x + 4$; linear expression in one variable

Homework 7.2 **1.** quadratic model **3.** no; answers may vary.

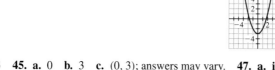

5. a. 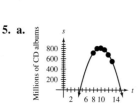 yes **b.** 93 million CD albums **c.** 1996, 2004 **d.** $(9.9, 811)$; in 2000, sales were 811 million CD albums—the largest sales ever, according to the model. Answers may vary. **e.** $(4.4, 0), (15.3, 0)$; there were no sales in 1994 and 2005; model breakdown has occurred for both results.

7. a. 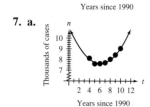 yes **b.** 9.9 thousand cases; 41.3 cases per workday **c.** 1992, 2000 **d.** 1996; in both 1995 and 1996, there were 7.6 thousand cases—the least number of cases for years between 1994 and 2000, inclusive.

9. a. 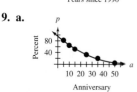 yes **b.** $(0, 95)$; 95% of married people are still married when they first get married; model breakdown has occurred. **c.** 43% **d.** 17th anniversary **e.** no; answers may vary.

11. a. $157 billion **b.** $143 billion **c.** overestimate; the data point for 1998 is below the model's graph; $14 billion

13. a. 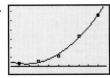 yes   **b.** 206.1 thousand electric cars   **c.** 2009   **d.** no; answers may vary.

15. a. 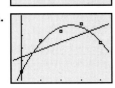 quadratic model   **b.** linear model; answers may vary.   **c.** 2002–2003 season; 59 wins   **d.** 1999–2000 season and 2004–2005 season

17. overestimate; answers may vary.   **19.** Answers may vary.   **21.** Answers may vary.

23. a. 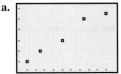 **b.** $M = 0.52t + 107.20$ **c.** 122.8 male births per 100 female births   **d.** 1966   **25.** Answers may vary.

27. $(2, 5)$; system of two linear equations in two variables   **29.** 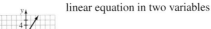 linear equation in two variables

Homework 7.3
1. quadratic (or second-degree) polynomial in one variable   **3.** cubic (or third-degree) polynomial in one variable
5. seventh-degree polynomial in two variables   **7.** $6x$ **9.** $-13x$ **11.** $8t^2$ **13.** $-11a^4b^3$ **15.** $5x^2$ **17.** The terms $7x^2$ and $3x$ cannot be combined.   **19.** $-3b^3$ **21.** $6x^6$ **23.** The terms $2t^3w^5$ and $4t^5w^3$ cannot be combined.   **25.** $7.4p^4$ **27.** $5x^2 - 3x$
29. $-x^2 + 14x - 8$ **31.** $11x^3 - x^2 + 7x$ **33.** $-45.1t^3 + 114.1t^2 - 9t$ **35.** $-4x^2 - 7x + 6$ **37.** 0 **39.** $-3x^3 - 10x^2 + 7x - 11$
41. $13a^3 - 3a^2 - a - 1$ **43.** $a^2 - 9ab + 9b^2$ **45.** $5.9x^3 + 20.9x^2 - 14.3x + 64.1$ **47.** $-7x + 11$ **49.** $-2x^2 - 4x + 6$
51. $y^3 + 6y^2 + y + 2$ **53.** 0 **55.** $-2x^2 + 11xy + y^2$ **57.** $6.75x^2 + 14.74x - 17.28$ **59.** $2x^2 + 3x + 14$ **61.** $5x^3 - 8x^2 - 2x + 11$
63. $-x^2 + 11x - 5$ **65.** $9x^3 - 5x^2 + 5x - 2$ **67. a.** $16.70t + 908.23$; the expression represents the total number (in thousands) of women and men who have earned bachelor's degrees in the year that is t years since 1980.   **b.** 1425.93; there will be a total of about 1.43 million women and men who will earn a bachelor's degree in 2011.   **c.** $9.86t - 28.05$; the expression represents the difference in the number of bachelor's degrees earned by women and men since 1980.   **d.** 277.61; women will earn 277.61 thousand more bachelor's degrees than men in 2011.

69. a.    **b.** $0.017t^2 + 1.5t + 3.9$; the expression represents the percentage of vehicles that are either crossover utility vehicles or SUVs.   **c.** 36.41; the percentage of U.S. vehicle sales that are either crossover utility vehicles or SUVs will be 36.4% in 2008.   **d.** $0.293t^2 - 4.94t + 6.66$; the expression represents the difference of the percentages of crossover utility vehicles and SUVs at t years since 1990.   **e.** 12.67; the percentage of U.S. vehicle sales that are crossover utility vehicles will be 12.7 percentage points more than the percentage of U.S. vehicle sales that are SUVs in 2008.   **71.** Answers may vary; $2x^2 + x + 6$ **73.** Answers may vary.   **75.** Answers may vary.
77. $2x^3 + 5x^3 = (2 + 5)x^3 = 7x^3$ **79.** $x^2 - 3x + 5x^2$; $6x^2 - 3x$ **81.** $4x^2 - (x^2 + x)$; $3x^2 - x$

83. a. 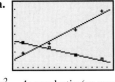   **b.** $0.18t + 14.99$; the expression represents the total petroleum consumed (in millions of barrels per day) in the United States at t years since 1980.   **c.** 20.57; the United States will consume 20.57 million barrels of petroleum per day in 2011.   **d.** 1992

85. $x^2 - 1$; quadratic (or second-degree) polynomial in one variable   **87.** quadratic equation in two variables

Homework 7.4
1. x^7 **3.** w^9 **5.** $30x^7$ **7.** $-36p^5t^3$ **9.** $-\dfrac{14}{5}x^5$ **11.** $3w^2 - 6w$ **13.** $26.32x^2 - 20.44x$ **15.** $-8x^3 - 12x$
17. $6m^3n^2 + 10mn^3$ **19.** $6x^3 - 4x^2 + 14x$ **21.** $-6t^4 - 12t^3 + 6t^2$ **23.** $6x^3y - 8x^2y^2 + 10xy^3$ **25.** $x^2 + 6x + 8$
27. $x^2 + 3x - 10$ **29.** $a^2 - 5a + 6$ **31.** $x^2 - 36$ **33.** $x^2 - 14.5x + 48.76$ **35.** $15y^2 + 14y - 8$ **37.** $4x^2 + 16x + 16$

39. $9x^2 - 6x + 1$ **41.** $12x^2 - 17xy - 5y^2$ **43.** $6a^2 - 13ab + 6b^2$ **45.** $9x^2 - 16$ **47.** $81x^2 - 16y^2$ **49.** $11.5x^2 + 22.61x - 70.07$
51. $-0.629x^2 - 15.969x + 1038.86$ **53.** $x^3 + 6x^2 - 3x - 18$ **55.** $6t^3 - 4t^2 - 15t + 10$ **57.** $6a^4 + a^2b^2 - 15b^4$
59. $x^3 + 5x^2 + 11x + 10$ **61.** $x^3 + 8$ **63.** $2b^3 - 11b^2 + 14b - 8$ **65.** $6x^4 - 2x^3 + 17x^2 - 3x + 12$
67. $12w^4 - 6w^3 - 5w^2 + 4w - 2$ **69.** $6x^4 + 10x^3 - 3x^2 + 9x - 2$ **71.** $a^2 + b^2 - c^2 + 2ab$ **73. a.** millions of dollars
b. $-1.2t^2 + 75.28t + 2018.56$; the expression represents the total monthly cost (in millions of dollars) of cable TV in the United States
at t years since 2000. **c.** 2701.44; the total monthly cost of cable TV in the United States in 2011 will be about \$2701 million, or about
\$2.7 billion. **75. a.** billions of dollars **b.** $0.05t^2 + 3.10t + 37.43$; the expression represents the total money (in billions of dollars) paid
for teacher salaries in the year that is t years since 1980. **c.** 175.43; the total money paid for teacher salaries will be \$175.43 billion in
2010. **77.** Answers may vary; $-24x^2$ **79. a. i.** $8x^2 + 22x + 15$; quadratic **ii.** $15x^2 - 29x - 14$; quadratic **b.** Answers may vary;
quadratic polynomial **c.** quadratic polynomial; answers may vary. **81. a.** $x^2 + 11x + 28$ **b.** $x^2 + 11x + 28$ **c.** Answers may vary.
83. $(x - 5)(x + 2)$, $x(x - 3) - 10$, $x^2 - 3x - 10$, $(x + 2)(x - 5)$ **85.** $6x^3 - 22x^2 + 26x - 10$ **87.** $-2x^2 + 7x - 7$ **89.** $y = 3x^2 - 6x$;
quadratic; parabola **91.** $y = x^2 - 7x + 12$; quadratic; parabola **93.** $y = -2x - 3$; linear; line **95.** $y = 10x^2 + x - 2$; quadratic;
parabola **97.** -2; linear equation in one variable **99.** $14w^2 - 25w - 25$; quadratic (or second-degree) polynomial in one variable

Homework 7.5 **1.** $x^8 y^8$ **3.** $36x^2$ **5.** $64x^3$ **7.** $64x^2$ **9.** $-27x^3$ **11.** $-a^5$ **13.** $x^2 + 10x + 25$ **15.** $x^2 - 8x + 16$
17. $4x^2 + 12x + 9$ **19.** $25y^2 - 20y + 4$ **21.** $9x^2 + 36x + 36$ **23.** $81x^2 - 36x + 4$ **25.** $4a^2 + 20ab + 25b^2$
27. $64x^2 - 48xy + 9y^2$ **29.** $y = x^2 + 12x + 36$ **31.** $y = x^2 - 6x + 10$ **33.** $y = 2x^2 + 16x + 29$ **35.** $-3x^2 + 6x - 5$ **37.** $x^2 - 16$
39. $t^2 - 49$ **41.** $x^2 - 1$ **43.** $9a^2 - 49$ **45.** $4x^2 - 9$ **47.** $9x^2 - 36y^2$ **49.** $9t^2 - 49w^2$ **51.** $x^2 - 121$ **53.** $-2x^3 + 4x^2 + 3x$
55. $36m^2 - 24mn + 4n^2$ **57.** $-10t^3$ **59.** $3x^3 + x^2 + 2x + 8$ **61.** $4t^2 - 9p^2$ **63.** $8x^3y^2 - 16x^2y^2 - 10xy^2$ **65.** $-8x^2 - x - 3$
67. $x^2 - 4x - 21$ **69.** $16w^2 - 64w + 64$ **71.** $x^3 - x^2 - 11x + 15$ **73.** $6x^2 - 5xy - 21y^2$ **75.** $36x^2 - 49$ **77.** $4t^2 + 28t + 49$
79. Answers may vary; $16x^2$ **81.** Answers may vary; $x^2 + 14x + 49$ **83.** Answers may vary; $x^2 - 10x + 25$; answers may vary.
85. a. Answers may vary. **b.** $x^2 + 8x + 16$ **c.** Answers may vary. **87.** $25 = 13$; answers may vary; $(A + B)^2 = A^2 + 2AB + B^2$
89. $(x - 2)^2$, $x(x - 4) + 4$, $x^2 - 4x + 4$ **91.** Answers may vary. **93. a. i.** 32, 16, 8, 4, 2 **ii.** Answers may vary. **iii.** 1 **b.** 81, 27, 9,
3; 1 **c.** 1 **95.** $y = 10x^2 - 6x$; quadratic; parabola **97.** $y = 24x^2 - 38x + 15$; quadratic; parabola **99.** $y = 6x - 9$; linear; line
101. $(5, -1)$; system of two linear equations in two variables **103.** linear equation in two variables

Homework 7.6 **1.** x^8 **3.** r^6 **5.** $15x^9$ **7.** $32b^8$ **9.** $54a^6b^8$ **11.** r^7t^7 **13.** $64x^2$ **15.** $32x^5y^5$ **17.** $16a^4$ **19.** 1 **21.** a^3
23. $2x^4$ **25.** $\dfrac{5x^3y^7}{4}$ **27.** $\dfrac{t^7}{w^7}$ **29.** $\dfrac{27}{t^3}$ **31.** 1 **33.** r^8 **35.** x^{36} **37.** $36x^6$ **39.** t^{12} **41.** $8a^{27}$ **43.** $x^{13}y^{20}$ **45.** $45x^{16}$
47. $-3c^{26}$ **49.** x^9y^{19} **51.** $\dfrac{5t^8}{4}$ **53.** $\dfrac{3}{4}$ **55.** $\dfrac{y^3}{8x^3}$ **57.** $\dfrac{x^8}{y^{20}}$ **59.** $\dfrac{r^{12}}{36}$ **61.** $\dfrac{8a^{12}}{27b^6}$ **63.** 1 **65.** $\dfrac{8a^{18}b^3}{27c^{15}}$ **67.** $x^{11}y^4$ **69.** $\dfrac{w^7}{32}$
71. $2x^2y^7$ **73.** 400 feet **75.** 507.66 watts **77.** Answers may vary; x^8 **79.** Answers may vary; $25x^6$ **81.** Answers may vary; $16x^8$
83. x^5 **85.** $x^3 + x^2$ **87.** $5x^4$ **89.** $6x^8$ **91.** $9x^2$ **93.** $x^2 + 6x + 9$ **95.** 5; linear equation in one variable **97.** $\dfrac{1}{6}x - \dfrac{5}{6}$; linear (or
first-degree) polynomial in one variable

Homework 7.7 **1.** $\dfrac{1}{36}$ **3.** $\dfrac{1}{x^4}$ **5.** $\dfrac{1}{b}$ **7.** 16 **9.** w^2 **11.** $\dfrac{1}{x^3y^5}$ **13.** $\dfrac{b^2}{a^4}$ **15.** $\dfrac{2a^3c^8}{5b^5}$ **17.** $-\dfrac{2w}{3x^9y^4}$ **19.** $\dfrac{1}{x^{14}}$ **21.** t^{12}
23. $\dfrac{30}{t^2}$ **25.** $-\dfrac{12}{x^9}$ **27.** $\dfrac{4}{x^2y^3}$ **29.** $\dfrac{1}{x^4}$ **31.** $\dfrac{1}{a^8}$ **33.** a^5 **35.** $\dfrac{7x^6}{4}$ **37.** $\dfrac{1}{32}$ **39.** $\dfrac{1}{25}$ **41.** $\dfrac{9}{w^5}$ **43.** $\dfrac{64}{c^6}$ **45.** $\dfrac{x^5}{32}$ **47.** $\dfrac{x^{12}}{y^{30}}$
49. $\dfrac{a^{18}}{8b^3}$ **51.** $\dfrac{b^6}{a^3}$ **53.** x^3y^3 **55.** $\dfrac{1}{x^{12}y^8}$ **57.** $\dfrac{1}{8c^{21}}$ **59.** $\dfrac{27}{r^{15}t^{27}}$ **61.** $\dfrac{a^6d^8}{64b^2c^{10}}$ **63.** $\dfrac{a^6d^6}{72c^2b^3}$ **65.** $\dfrac{3c^7}{4b^5}$ **67. a.** $s = \dfrac{d}{t}$ **b.** 62; an
object that travels 186 miles in 3 hours at a constant speed is traveling at a speed of 62 miles per hour. **69. a.** $L = \dfrac{5760}{d^2}$ **b.** 90; the
sound level is 90 decibels at a distance of 8 yards from the amplifier. **71.** 49,000 **73.** 0.00859 **75.** 0.000295 **77.** $-451,200,000$
79. 4.57×10^7 **81.** 6.59×10^{-5} **83.** -5.987×10^{12} **85.** 1×10^{-6} **87.** from top to bottom of the second column: 0.000000048;
0.0000017; 35,800,000; 1,280,000,000 **89.** 400,000,000,000 **91.** 46,600,000,000 **93.** 0.00063 **95.** 3.72×10^6 **97.** 2.7×10^7
99. 7.5×10^{-6} **101.** Answers may vary; $\dfrac{5y^2}{x^3w}$ **103.** Answers may vary; x^{10} **105. a.** 1 **b.** 1 **c.** 1 **d.** 1 **107.** $-\dfrac{6}{x^3}$ **109.** $\dfrac{9}{x^6}$
111. $-\dfrac{27}{x^9}$ **113.** $-\dfrac{1}{27x^9}$ **115. a. i.** x^7 **ii.** x^7 **b.** They are the same. **117.** Answers may vary. **119.** Answers may vary.
121. Answers may vary.

Chapter 7 Review

1.   **2.**   **3.**   **4.** 2 **5.** 1 **6.** 1, 3 **7.** 2 **8. a.** 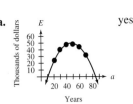   yes

b. \$21 thousand **c.** 27 years, 67 years **d.** (47, 50); the average annual expenditure for 47-year-old Americans is \$50 thousand, the largest average annual expenditure for any age. **e.** (10, 0), (84, 0); both 10-year-old and 84-year-old Americans do not spend any money; model breakdown has occurred. **9. a.** 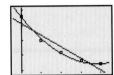   quadratic model **b.** quadratic model **c.** (3.83, 12.03); in 2004, 12.0% of Americans had confidence in executives at major corporations, the lowest percentage in any year. **d.** 17.1% **e.** 2008

10. $-9x^3 - 2x^2 - x + 3$ **11.** $-2x^3 + 2x^2 - 3x + 5$ **12.** $-7x^3 - 8x^2 - 10x + 2$ **13. a.** $-0.3t + 94.2$; the expression represents the percentage of airplanes that are regional jets, turboprops, or large jets. **b.** 91.2; in 2010, 91.2% of airplanes will be regional jets, turboprops, or large jets. **c.** $9.1t - 79.4$; the expression represents the difference of the percentage of airplanes that are regional jets and the percentage of airplanes that are turboprops or large jets. **d.** 11.6; in 2010, the percentage of airplanes that are regional jets will be 11.6 percentage points more than the percentage of airplanes that are turboprops or large jets. **14.** $-21x^7$ **15.** $10x^5 - 35x^4 + 20x^3$
16. $w^2 - 12w + 27$ **17.** $6a^2 - ab - 40b^2$ **18.** $15x^3 + 18x^2 - 20x - 24$ **19.** $x^3 + x^2 - 7x + 20$ **20.** $8b^3 - 30b^2 + 13b - 21$
21. $-8t^3$ **22.** $x^2 + 14x + 49$ **23.** $x^2 - 8x + 16$ **24.** $4p^2 + 20p + 25$ **25.** $-5c^2 - 20c - 20$ **26.** $x^2 - 36$ **27.** $16m^2 - 49n^2$
28. $9p^2 - 12pt + 4t^2$ **29.** $2a^4 + 3a^3 - a^2 + 7a - 3$ **30.** cubic polynomial; answers may vary. **31.** $y = x^2 - 10x + 28$
32. $y = -2x^2 - 12x - 24$ **33.** x^{10} **34.** $12x^7$ **35.** $-40a^6b^{12}$ **36.** $\dfrac{xy^3}{2}$ **37.** $\dfrac{x^3}{8}$ **38.** $32x^{45}y^{15}$ **39.** $75x^{14}$ **40.** $\dfrac{3c^5}{2}$ **41.** $\dfrac{a^8}{81}$
42. 1 **43.** $\dfrac{3x^3y^5}{2}$ **44.** $\dfrac{27x^9}{64}$ **45.** $\dfrac{1}{64}$ **46.** x^5 **47.** $\dfrac{1}{r^{10}}$ **48.** $\dfrac{1}{t^3}$ **49.** $\dfrac{y^2w^7}{x^5}$ **50.** $\dfrac{1}{c^{24}}$ **51.** $-\dfrac{20}{x^6}$ **52.** $\dfrac{x^{15}}{32}$ **53.** $\dfrac{x^{12}}{27y^{18}}$
54. $\dfrac{3a^3}{4b^3}$ **55.** $\dfrac{9b^{11}}{8a^7}$ **56.** $\dfrac{y^{14}}{x^{21}w^{28}}$ **57.** Answers may vary; x^{15} **58.** 583,200,000 **59.** 0.000317 **60.** 7.42×10^7 **61.** 8×10^{-5}
62. 1.426×10^9

Chapter 7 Test

1.   **2.** 5 **3.** $-3, -1$ **4.** -2 **5.** no such value **6.** $(-4, 0), (0, 0)$ **7.** $(-2, -4)$

8. a.   yes **b.** 19 million pairs **c.** 1988, 1998 **d.** (13, 11); sales were 11 million pairs in 1993, the lowest sales ever.

9. $-4x^3 - x^2 - 2x$ **10.** $-15x^4 + 40x^3 - 10x^2$ **11.** $12p^2 + 10p - 12$ **12.** $6k^3 - 23k^2 + 16k + 10$ **13.** $x^2 - 12x + 36$
14. $16a^2 + 56ab + 49b^2$ **15.** $3w^4 + w^3 - 5w^2 + 7w - 6$ **16. a.** $-0.15t + 92.90$; the expression represents the total viewing share for broadcast and ad-supported cable television at t years since 1980. **b.** 88.25; the total viewing share for broadcast and ad-supported cable television will be 88.25% in 2011. **c.** $-4.19t + 102.58$; the expression represents the difference in viewing share for broadcast television and ad-supported cable television at t years since 1980. **d.** -27.31; broadcast television will have 27.31 less percentage points of viewing share than ad-supported cable television in 2011. **17.** $y = -3x^2 + 6x + 2$ **18.** Answers may vary; $x^2 + 8x + 16$ **19.** $\dfrac{3x^4}{4y^5}$

20. $64a^{15}b^{16}$ **21.** $\dfrac{x^{18}}{y^{24}}$ **22.** $\dfrac{x^6}{49}$ **23.** $\dfrac{x^8w^{12}}{y^{24}}$ **24.** $\dfrac{t^5}{2p^3}$ **25.** 4.68×10^{-4} **26.** 16,500,000

Chapter 8

Homework 8.1 **1.** $(x+2)(x+3)$ **3.** $(t+4)(t+5)$ **5.** $(x+4)^2$ **7.** $(x-4)(x+2)$ **9.** $(a-8)(a+2)$ **11.** $(x+8)(x-3)$
13. prime **15.** $(t-4)(t+7)$ **17.** $(x-8)(x-2)$ **19.** $(x-8)(x-3)$ **21.** prime **23.** $(r-5)^2$ **25.** $(x-6)^2$
27. $(x+y)(x+9y)$ **29.** $(m-3n)(m+2n)$ **31.** $(a-6b)(a-b)$ **33.** $(x-5)(x+5)$ **35.** $(x-9)(x+9)$ **37.** $(t-1)(t+1)$
39. prime **41.** $(2x-5)(2x+5)$ **43.** $(9r-1)(9r+1)$ **45.** prime **47.** $(7p-10q)(7p+10q)$ **49.** $(8m-3n)(8m+3n)$
51. $(x-6)(x+3)$ **53.** $(x+7)^2$ **55.** $(a-2)(a+2)$ **57.** prime **59.** $(x-6)(x-2)$ **61.** $(w-8)(w+6)$ **63.** $(x-4)^2$
65. prime **67.** $(m-9n)(m+3n)$ **69.** $(x-16)(x-2)$ **71.** $(10p-3t)(10p+3t)$ **73.** $(p+6)^2$ **75.** Answers may vary; the
polynomial is prime. **77.** $(x-3)(x+7)$, $x^2+4x-21$, $(x+7)(x-3)$ **79.** $(x-8)(x+3)$; $x^2-5x-24$; answers may vary.
81. Answers may vary. **83.** $-13, -8, -7, 7, 8, 13$ **85.** Answers may vary. **87.** $x^2-7x-18$ **89.** $(x-5)(x-10)$ **91.** $9x^2-49$
93. $(5x-6)(5x+6)$ **95.** $3p+11w$; linear (or first-degree) polynomial in two variables **97.** $10p^2-6pw-28w^2$; quadratic (or
second-degree) polynomial in two variables **99.** $(p-9w)(p-2w)$; quadratic (or second-degree) polynomial in two variables

Homework 8.2 **1.** $2(3x+4)$ **3.** $5w(4w+7)$ **5.** $6x(2x-5)$ **7.** $3ab(2a-3)$ **9.** $4x^2y^2(2x+3y)$ **11.** $5(3x^3-2x-6)$
13. $4t(3t^3+2t^2-4)$ **15.** $5ab(2a^3-3a^2+5)$ **17.** $2(x-3)(x+3)$ **19.** $3(m+2)(m+5)$ **21.** $2(x-6)(x-3)$
23. $3x(x-3)(x+3)$ **25.** $4r(r-5)(r+1)$ **27.** $6x^2(x-2)(x+2)$ **29.** $2m^2n(2m-3)(2m+3)$ **31.** $5x^2(x-4)(x+6)$
33. $-2x(x-2)(x+2)$ **35.** $4t(t+1)(t+8)$ **37.** $-3x(2x-3)(2x+3)$ **39.** $-3x(x-2)(x+8)$ **41.** $6a^2b(a+3)^2$
43. $(x-3)(5x^2+2)$ **45.** $(2x+5)(6x^2-7)$ **47.** $(p+3)(2p^2+5)$ **49.** $(3x-1)(2x^2+7)$ **51.** $(3w+1)(5w^2-2)$
53. $(2x-3)(2x+3)(4x-3)$ **55.** $(b-3)(b+3)(2b-5)$ **57.** $(x-1)^2(x+1)$ **59.** $(3+a)(x+y)$ **61.** $(2x+3)(y-4)$
63. $(9x-5)(9x+5)$ **65.** $(w-8)(w-2)$ **67.** $(x-6)(x-4)$ **69.** $5ab(4a-3b^2)$ **71.** $3(r+5)^2$ **73.** $x(8x-7)(8x+7)$
75. $-(m-3)^2$ **77.** $(x-2)(x+2)(x+9)$ **79.** $2mn(m-2n)(m-3n)$ **81.** Answers may vary; $(2x^2+5)(3x+4)$ **83.** Answers
may vary; $4x(x+2)(x+5)$ **85.** Answers may vary. **87.** Answers may vary. **89.** Answers may vary. **91.** $2x^3+2x^2-24x$
93. $5(x-4)^2$ **95.** $(2x-3)(3x^2-2)$ **97.** $x^3-3x^2+5x-15$ **99.**

linear equation in two variables

101. 1; linear equation in one variable **103.** $(1, -3)$; system of two linear equations in two variables

Homework 8.3 **1.** $(x+3)(2x+1)$ **3.** $(x+2)(5x+1)$ **5.** $(x+2)(3x+2)$ **7.** $(t+2)(2t-3)$ **9.** $(2x-3)(3x-2)$
11. $(2x+5)^2$ **13.** $(x-6)(6x-1)$ **15.** prime **17.** $(3x+4)(6x-1)$ **19.** $(m-6)(3m-4)$ **21.** $(x-8)(2x-5)$
23. $(2a+3b)(a+b)$ **25.** $(5x-2y)(x+4y)$ **27.** $3(2b-c)(b-2c)$ **29.** $2(x+5)(2x+3)$ **31.** $5(2a-3)(2a-1)$
33. $3(x+1)(8x-3)$ **35.** $-2(2x-3)(5x+2)$ **37.** $2w^2(2w^2-3w-6)$ **39.** $5x^2(x+2)(2x-5)$ **41.** $2(a-6b)(3a+b)$
43. $4r(r+2w)(3r+4w)$ **45.** $(x-9)(x+3)$ **47.** $-8x(6x-5)$ **49.** $(a^2-3)(5a+2)$ **51.** prime **53.** $(2x-3)^2$
55. $-17(p+1)(p-1)$ **57.** $(x+4)(x+6)$ **59.** $(b-7c)(b+4c)$ **61.** $(4t-3)(2t-1)$ **63.** $7x^2(x-2)(x+2)$
65. $(2p+3)(2p-3)(3p-1)$ **67.** prime **69.** $3x^2(x-5)(x-2)$ **71.** $2x^2(8x-5)(x-2)$ **73.** $(6a+7b)(6a-7b)$
75. $-2y(x-6)(x+2)$ **77.** $y(4y+3)(y-3)$ **79.** no; answers may vary. **81.** Answers may vary. **83.** Answers may vary.
85. $(x+6)(3x-2)$ **87.** $12x^2-25x+7$ **89.** $2x^3-3x^2-14x+15$ **91.** $2x(3x-1)(x+2)$
93.

quadratic equation in two variables

95. $(x-3)(x+1)$; quadratic (or second-degree) polynomial in one variable **97.** 32; quadratic (or second-degree) polynomial in one
variable

Homework 8.4 **1.** $(x+3)(x^2-3x+9)$ **3.** $(x+5)(x^2-5x+25)$ **5.** $(x-2)(x^2+2x+4)$ **7.** $(x-1)(x^2+x+1)$
9. $(2t+3)(4t^2-6t+9)$ **11.** $(3x-2)(9x^2+6x+4)$ **13.** $5(x+2)(x^2-2x+4)$ **15.** $2(x-3)(x^2+3x+9)$
17. $(2x+3y)(4x^2-6xy+9y^2)$ **19.** $(4a-3b)(16a^2+12ab+9b^2)$ **21.** $(x-8)(x+8)$ **23.** $(m+4)(m+7)$ **25.** $2(t-3)(t+4)$
27. prime **29.** $-3(x-5)(x-3)$ **31.** $(3p-1)(5p-1)$ **33.** $(x-1)^2$ **35.** $4(2r+1)(3r-1)$ **37.** $-2ab^2(2b-3a)$
39. $(a-5b)(a+4b)$ **41.** $(2x-5)(2x-1)(2x+1)$ **43.** $-4x^3(3x+1)$ **45.** $5a^2(a+1)(3a+2)$ **47.** $(x-2)(x-12)$
49. $2w^2(w^2+2w-4)$ **51.** $3x^2(2x-3)(2x+3)$ **53.** $(x+5)^2$ **55.** $(x-3)(2x+7)$ **57.** $m(m-9n)(m-4n)$ **59.** prime

61. $(10x - 3y)(10x + 3y)$ **63.** $(2x + 3)^2$ **65.** $(2x + 3)(3x - 2)(3x + 2)$ **67.** $a(a - 2b)(3a - 4b)$ **69.** prime
71. $(x - 10)(x^2 + 10x + 100)$ **73.** $(4a + 3)(16a^2 - 12a + 9)$ **75.** $3(x + 2)(x^2 - 2x + 4)$ **77.** Answers may vary. **79.** Answers may
vary. **81.** Answers may vary; $x^3 - 27 = (x - 3)(x^2 + 3x + 9)$ **83.** no; answers may vary. **85.** Answers may vary. **87.** $25x^2 - 49$
89. $(9x - 4)(9x + 4)$ **91.** $3x(x - 1)(x + 4)$ **93.** $-14x^3 + 10x^2 - 2x$ **95.**   linear equation in two variables

97. 7; linear equation in one variable **99.** $12x^2 - 7x - 10$; quadratic (or second-degree) polynomial in one variable

Homework 8.5 **1.** 2, 7 **3.** $-5, -2$ **5.** $-7, 4$ **7.** 7 **9.** $-4, 1$ **11.** 2, 3 **13.** $-8, 8$ **15.** $-\dfrac{7}{6}, \dfrac{7}{6}$ **17.** $-7, 7$ **19.** 0, 1
21. $-\dfrac{5}{2}, \dfrac{7}{3}$ **23.** $\dfrac{3}{2}, \dfrac{1}{2}$ **25.** $-\dfrac{4}{5}, \dfrac{1}{2}$ **27.** $-\dfrac{2}{5}$ **29.** $-6, 2$ **31.** $2, \dfrac{4}{3}$ **33.** 1, 6 **35.** $-\dfrac{5}{2}, \dfrac{3}{2}$ **37.** $\dfrac{1}{2}, 3$ **39.** 2, 5 **41.** $-6, -2$
43. $-4, \dfrac{4}{3}$ **45.** $-3, 4$ **47.** 2, 4 **49.** 3 **51.** $-3, -1$ **53.** no real-number solutions **55.** $-0.79, 3.79$ **57.** $-3.54, 2.54$ **59.** 1, 3
61. 2 **63.** $(-4, 0), (7, 0)$ **65.** $(4, 0)$ **67.** $(-1, 0), \left(\dfrac{3}{2}, 0\right)$ **69.** $-3, 3, \dfrac{5}{2}$ **71.** $-2, 2, \dfrac{4}{3}$ **73.** $-\dfrac{2}{3}, \dfrac{2}{3}, \dfrac{3}{2}$ **75.** $-4, 2$ **77.** $-5, 0, 5$
79. $-3, -2, 0$ **81.** $-2, 0, 4$ **83.** Answers may vary; 0, 2 **85.** Answers may vary; 5, 8 **87.** Answers may vary.
89. a. 1, 5 **b.** 3 **c.** 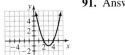 **91.** Answers may vary. **93.** $-2, 6$ **95.** $-\dfrac{3}{2}$ **97.** $P = \dfrac{A}{1 + RT}$

99. $(x - 5)(x + 1)$; quadratic (or second-degree) polynomial in one variable **101.** $-1, 5$; quadratic equation in one variable
103. 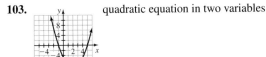 quadratic equation in two variables

Homework 8.6
1. a.  yes **b.** 307 thousand **c.** 1988, 1998; a little research would show that model breakdown has occurred for
the 1988 estimate.

3. a. 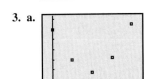 quadratic equation **b.** 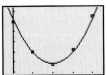 yes **c.** 1999, 2005 **d.** (2, 538); total spending in 2002 was
$538 billion, the lowest in any year.

5. a. 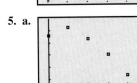 quadratic equation **b.** 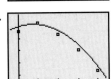 yes **c.** $3.2 billion **d.** $(-4.5, 0), (6, 0)$; the revenue was
zero dollars in 1994 and 2005; model breakdown has occurred.

7. a. 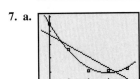 quadratic model **b.** linear model **c.** quadratic model **d.** 2000, 2004 **9.** 2002, 2011 **11.** 2007, 2010

13. a. 0 seconds, 4 seconds; answers may vary. **b.** 1 second, 3 seconds; answers may vary. **c.** 2 seconds; answers may vary.
15. width: 5 feet, length: 12 feet **17.** width: 7 feet, length: 14 feet **19.** Answers may vary.

21. a.   linear equation   **b.** $T = 121.07t + 422.71$ **c.** 2010

23. a. $c = 0.14t + 10.3$ **b.** 2011   **25.** \$7064   **27.** Answers may vary.   **29.** Answers may vary.   **31.** Answers may vary.
33. Answers may vary.

Chapter 8 Review

1. $(x + 4)(x + 5)$ **2.** $2(3x - 4)(x + 1)$ **3.** $(9x - 7)(9x + 7)$ **4.** $(x + 7)^2$ **5.** $-3t^2(2t + 5)(3t - 2)$ **6.** $(p - 9q)(p + 6q)$
7. $(x - 8)(x - 4)$ **8.** $-(3x - 2)(3x + 2)$ **9.** prime   **10.** $5mn(4m - 9n^2)$ **11.** $2x(x + 7)(x + 1)$ **12.** $16x(x - 1)^2$
13. $8x^2(3x - 4)$ **14.** $5x^2y(x - 3)(x - 4)$ **15.** $-(m - 5)(m + 7)$ **16.** $(2r^2 + 3)(2r - 5)$ **17.** $(2p + 1)(p + 3)$ **18.** $(x - 5)(x - 4)$
19. $(3t - 2y)(2t + 5y)$ **20.** $2x(x - 5)(x + 5)$ **21.** $(x - 5)^2$ **22.** $(p - 9)(p + 9)$ **23.** prime   **24.** prime
25. $(x - 1)(2x - 3)(2x + 3)$ **26.** $(2x + 5)^2$ **27.** $2w(6w - 1)(w - 4)$ **28.** $(7a - 3b)(7a + 3b)$ **29.** prime
30. $(x - 2)(x + 2)(x + 3)$ **31.** $(r + 2)(r^2 - 2r + 4)$ **32.** $(2t - 3)(4t^2 + 6t + 9)$ **33.** Answers may vary; the polynomial is prime.
34. Answers may vary; $5x(x + 3)(x + 4)$ **35.** $-25, -14, -11, -10, 10, 11, 14, 25$ **36.** $-7, -2$ **37.** $-6, -2, 0$ **38.** $-1, 2$ **39.** 3
40. $3, 5$ **41.** $-\dfrac{9}{5}, \dfrac{9}{5}$ **42.** $-1, 1, \dfrac{7}{2}$ **43.** $-\dfrac{2}{3}, \dfrac{1}{2}$ **44.** $0, 5$ **45.** $\dfrac{3}{2}, \dfrac{3}{4}$ **46.** $-5, 7$ **47.** $-2, 6$ **48.** $-5, 6$ **49.** $-3, 3, \dfrac{2}{3}$ **50.** $-2, 2$
51. $-6, 2$ **52.** $-2.56, 1.56$ **53.** 1   **54.** $-2, 4$ **55.** no real-number solutions   **56.** $0, 2$ **57.** $(-7, 0), (7, 0)$ **58.** $\left(-\dfrac{3}{4}, 0\right),$
$\left(\dfrac{5}{2}, 0\right)$ **59.** Answers may vary.   **60.** Answers may vary.

61. a. 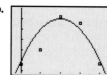   quadratic equation   **b.** 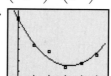   yes   **c.** 34 episodes per 1000 people   **d.** 1997, 2005
e. $(2, 43)$; in 2001, the annual episode rate was 43 episodes per 1000 people—the largest rate ever, according to the model.

62. width: 3 feet, length: 10 feet

Chapter 8 Test

1. $(x - 8)(x + 5)$ **2.** $(x - 6)(x - 4)$ **3.** $2m^2n(4n^2 - 5m)$ **4.** $(p - 10q)(p - 4q)$ **5.** $(5p - 6y)(5p + 6y)$ **6.** $3x^2y(x - 4)(x - 3)$
7. $(2x - 3)(2x + 3)(2x + 5)$ **8.** $(4x - 3)(2x - 5)$ **9.** $(4x - 1)(16x^2 + 4x + 1)$ **10.** $x^2 - 3x - 10, (x - 5)(x + 2), (x + 2)(x - 5)$
11. Answers may vary; $(x - 2)(x + 2)(5x + 3)$ **12.** $4, 9$ **13.** $-\dfrac{3}{7}, \dfrac{3}{7}$ **14.** -6 **15.** $-4, 6$ **16.** $-2, -\dfrac{2}{3}, 2$ **17.** $-1, 0, 5$
18. $-5, -1$ **19.** $-4, -2$ **20.** -3 **21.** no real-number solution   **22.** $\left(-\dfrac{2}{5}, 0\right), \left(\dfrac{3}{2}, 0\right)$ **23.** Answers may vary.

24. a. 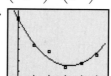   quadratic equation; answers may vary.   **b.**   yes   **c.** 76%   **d.** 2001, 2006

e. $(3.25, 47.88)$; in 2003, 48% of Americans thought that the environment should have been given top priority, the lowest percentage in any year, according to the model (the actual percentage in 2003 was 47%).   **25.** 2001, 2013

Cumulative Review of Chapters 1-8

1. independent: a, dependent: p **2.** underestimate   **3.** $-6°\text{F}$ **4.** $\dfrac{1}{2}$; increasing   **5.** The value of y decreases by 4 as the value of x is
increased by 1.   **6.** $y = \dfrac{2}{3}x - \dfrac{14}{3}$ **7.** $y = -\dfrac{5}{3}x + \dfrac{1}{3}$ **8.** $y = \dfrac{1}{2}x + 1$ **9.** $3(x + 5) = 9; -2$ **10.** $7 - 4\left(\dfrac{x}{2}\right); 7 - 2x$ **11.** $(2, -1)$
12. $(4, -2)$ **13.** $(-3, 2)$ **14.** $(9, 0)$ **15.** -42 **16.** 15   **17.** 84   **18.** $-\dfrac{21}{8}$ **19.** -5 **20.** 1   **21.** $0, 2$ **22.** no such value
23. $(-1, 0), (3, 0)$ **24.** $(1, 4)$ **25.**   **26.**   quadratic equation in two variables

27. $(2m - 7n)(2m + 7n)$; quadratic (or second-degree) polynomial in two variables   **28.** $9p^2 - 42pq + 49q^2$; quadratic (or second-degree) polynomial in two variables   **29.** $(w + 7)(w - 2)$; quadratic (or second-degree) polynomial in one variable   **30.** $-4, \dfrac{2}{5}$; quadratic equation in one variable   **31.** $-5, 7$; quadratic equation in one variable   **32.** $xy(2x - 3)(3x + 5)$; fourth-degree polynomial in two variables   **33.** $16p^2 + 18p - 9$; quadratic (or second-degree) polynomial in one variable   **34.** $-20x^3y - 8x^2y^2 + 12xy^3$; fourth-degree polynomial in two variables   **35.** $(a - 8b)(a + 5b)$; quadratic (or second-degree) polynomial in two variables   **36.** $\dfrac{7}{6}$; linear equation in one variable   **37.** $-4x^2 - 7x + 9$; quadratic (or second-degree) polynomial in one variable   **38.** $3(x - 9)(x - 2)$; quadratic (or second-degree) polynomial in one variable   **39.**   linear equation in two variables

40. $6m^3 - 19m^2 + 18m - 20$; cubic (or third-degree) polynomial in one variable   **41.** $-2x^3 + 5x^2 - 6x$; cubic (or third-degree) polynomial in one variable   **42.** $(r + 2)(2r - 3)(2r + 3)$; cubic (or third-degree) polynomial in one variable
43. $(2x - 3)\left(4x^2 + 6x + 9\right)$; cubic (or third-degree) polynomial in one variable   **44.** $h = \dfrac{S - 2\pi r^2}{r}$ **45.** $x \geq -4$; $[-4, \infty)$

46.   **47.**   **48.** $(x - 2)(x + 4)$, $x^2 + 2x - 8$, $(x + 4)(x - 2)$, $x(x + 2) - 8$ **49.** $y = -3x^2 + 12x - 12$

50. $(-7, 0), (3, 0)$ **51.** $(10, 0)$ **52.** $-8a^3b^6$ **53.** $\dfrac{r^5}{8}$ **54.** $\dfrac{8y^3}{x^5}$ **55.** $\dfrac{x^6w^{12}}{y^{18}}$ **56.** 4200 \$20 tickets, 1800 \$35 tickets

57. a. 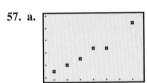   **b.** $B = 0.063t + 10.13$ **c.** 0.063; Jones's best time in the 100-meter run increases by 0.063 second each year.   **d.** 10.95 seconds   **e.** 0 seconds; 2001 to 2002   **f.** Answers may vary.

58. a. $L = 0.10t + 77.2$ **b.** $L = 0.20t + 76.7$ **c.** 2000; 77.7 years   **59. a.** 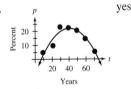   yes

b. $(7.0, 0), (74.3, 0)$; no 7-year-old children visit online trading sites. Also, no 74-year-old adults visit online trading sites; model breakdown has occurred.   **c.** $(40.7, 22.0)$; this means that 22% of 41-year-old adults visit online trading sites, the highest percentage for any age group, according to the model. In reality, 22.9% of Americans between the ages of 25 and 34 years, inclusive, visit online trading sites.   **d.** 12.9%   **e.** 26-year-old and 55-year-old Americans   **60.** 2001, 2011   **61.** width: 6 feet, length: 14 feet

Chapter 9

Homework 9.1   **1.** 2   **3.** 9   **5.** 11   **7.** 12   **9.** -4 **11.** -9 **13.** not a real number   **15.** not a real number   **17.** irrational; 5.48   **19.** irrational; 8.83   **21.** rational; 14   **23.** $2\sqrt{5}$ **25.** $3\sqrt{5}$ **27.** $3\sqrt{3}$ **29.** $5\sqrt{2}$ **31.** $10\sqrt{3}$ **33.** $-7\sqrt{2}$ **35.** $24\sqrt{2}$ **37.** $6\sqrt{30}$ **39.** $3\sqrt{x}$ **41.** $8\sqrt{t}$ **43.** $9x$ **45.** $15tw$ **47.** $x\sqrt{5}$ **49.** $7x\sqrt{39y}$ **51.** $2\sqrt{3p}$ **53.** $6\sqrt{7x}$ **55.** $2ab\sqrt{15}$ **57.** $15y\sqrt{5x}$ **59.** 68.7 miles per hour   **61.** no; answers may vary.   **63.** 4 and 5   **65.** 8 and 9   **67. a. i.** 2   **ii.** 5   **iii.** 8   **b.** Each of the numbers 2, 5, and 8 is larger than its principal square root.   **c. i.** $\sqrt{0.2}$ **ii.** $\sqrt{0.5}$ **iii.** $\sqrt{0.8}$ **d.** Each of the numbers 0.2, 0.5, and 0.8 is smaller than its principal square root.   **e.** a number greater than 1; a positive number less than 1   **69.** no; answers may vary.   **71.**

73. $49x^2$ **75.** $7x$ **77.**

quadratic equation in two variables **79.** 2, 4; quadratic equation in one variable

81. $2\sqrt{17x}$; radical expression in one variable

Homework 9.2

1. $\dfrac{5}{6}$ **3.** $\dfrac{11}{x}$ **5.** $\dfrac{\sqrt{7x}}{5}$ **7.** $\dfrac{\sqrt{19}}{a}$ **9.** $\dfrac{\sqrt{5}}{xy}$ **11.** $-\dfrac{2\sqrt{2}}{7}$ **13.** $\dfrac{2\sqrt{5}}{9}$ **15.** $\dfrac{5\sqrt{3w}}{6}$ **17.** $\dfrac{2\sqrt{a}}{b}$ **19.** $\dfrac{4\sqrt{5}}{x}$
21. $\dfrac{r\sqrt{7t}}{9}$ **23.** $\dfrac{2\sqrt{3}}{3}$ **25.** $\dfrac{6\sqrt{5}}{5}$ **27.** $\dfrac{a\sqrt{13}}{13}$ **29.** $\dfrac{7\sqrt{x}}{x}$ **31.** $\dfrac{\sqrt{14}}{7}$ **33.** $\dfrac{\sqrt{22}}{2}$ **35.** $\dfrac{\sqrt{7p}}{p}$ **37.** $\dfrac{\sqrt{6}}{4}$ **39.** $\dfrac{\sqrt{6x}}{10}$ **41.** $\dfrac{w\sqrt{15}}{6}$
43. $\dfrac{x\sqrt{35y}}{5}$ **45.** $\dfrac{3+\sqrt{2}}{2}$ **47.** $2-\sqrt{7}$ **49.** $\dfrac{4+6\sqrt{13}}{3}$ **51.** $\dfrac{2+\sqrt{3}}{4}$ **53.** $\dfrac{2-\sqrt{2}}{4}$ **55.** $\dfrac{3-\sqrt{5}}{2}$ **57. a.** $T=\dfrac{\sqrt{h}}{4}$ **b.** 9.5
seconds **59.** Answers may vary; $\dfrac{5\sqrt{3}}{3}$ **61.** yes; answers may vary. **63.** $4(2x+3)$ **65.** $4\left(3+2\sqrt{3}\right)$ **67.** $-7(3x-2)$
69. $7(2-3\sqrt{7})$ **71.** 0, 4; quadratic equation in one variable **73.** $x(x-4)$; quadratic (or second-degree) polynomial in one variable
75.

quadratic equation in two variables

Homework 9.3

1. ± 2 **3.** ± 14 **5.** 0 **7.** $\pm\sqrt{15}$ **9.** $\pm 2\sqrt{5}$ **11.** $\pm 3\sqrt{3}$ **13.** no real-number solutions **15.** $\pm 2\sqrt{7}$
17. no real-number solutions **19.** $\pm\dfrac{\sqrt{5}}{2}$ **21.** $\pm\dfrac{\sqrt{35}}{5}$ **23.** $\pm\dfrac{\sqrt{10}}{4}$ **25.** $\pm\dfrac{\sqrt{6}}{2}$ **27.** $\pm\dfrac{\sqrt{70}}{5}$ **29.** $-6, 2$ **31.** $-3, 9$ **33.** $7\pm\sqrt{13}$
35. no real-number solutions **37.** $-2\pm 3\sqrt{2}$ **39.** $6\pm 2\sqrt{6}$ **41.** 5 **43.** $-2, 6$ **45.** ± 9 **47.** $2\pm 2\sqrt{6}$ **49.** $\pm\dfrac{\sqrt{33}}{3}$ **51.** $-5, \dfrac{3}{2}$
53. 0, 2 **55.** $\sqrt{41}$ **57.** $2\sqrt{14}$ **59.** $2\sqrt{11}$ **61.** 10 **63.** $\sqrt{13}$ **65.** $4\sqrt{5}$ **67.** 8 **69.** $\sqrt{21}$ **71.** $3\sqrt{5}$ **73.** 18.8 miles **75.** yes
77. 14.7 inches **79.** 28.2 inches **81.** 3.2 miles **83.** 340 feet (113 yards and 1 foot) **85.** 439.8 miles **87. a.** ± 6 **b.** ± 6
c. They are the same results. **d.** Answers may vary. **89. a.** yes; $2\pm\sqrt{7}$ **b.** no **c.** no; answers may vary. **91.** Answers may vary.
93. a. ± 6 **b.** 6 **c.** Answers may vary. **95. a.** Answers may vary. **b.** Answers may vary. **c.** the square **d.** W, L, D **e.** A
97. $12x^4+3x^3-23x^2-2x+10$; fourth-degree polynomial in one variable **99.** $3x^2(x-7)(x+3)$; fourth-degree polynomial in one
variable **101.**

linear equation in two variables

Homework 9.4

1. 9; $(x+3)^2$ **3.** 49; $(x+7)^2$ **5.** 1; $(x+1)^2$ **7.** 16; $(x-4)^2$ **9.** 25; $(x-5)^2$ **11.** 100; $(x-10)^2$
13. $-3\pm\sqrt{14}$ **15.** $-7\pm\sqrt{29}$ **17.** $5\pm\sqrt{21}$ **19.** $1\pm\sqrt{6}$ **21.** $-6\pm 4\sqrt{2}$ **23.** $2\pm 3\sqrt{2}$ **25.** $4\pm 2\sqrt{6}$ **27.** no real-number
solutions **29.** $-10\pm 2\sqrt{15}$ **31.** $-3\pm 2\sqrt{2}$ **33.** $2\pm\sqrt{3}$ **35.** $5\pm 4\sqrt{2}$ **37.** no real-number solutions **39.** $-4\pm 2\sqrt{3}$
41. $4\pm\sqrt{19}$ **43.** $-1\pm 2\sqrt{2}$ **45.** $5\pm 3\sqrt{2}$ **47.** $-1\pm\sqrt{2}$ **49.** $3\pm 2\sqrt{2}$ **51.** $-2\pm 2\sqrt{2}$ **53.** ± 3 **55.** 5, 6 **57.** $5\pm 4\sqrt{2}$
59. $-3, \dfrac{4}{3}$ **61.** $\pm\sqrt{13}$ **63.** $3\pm 2\sqrt{3}$ **65. a.** 2, 4 **b.** 2, 4 **c.** Answers may vary. **67. a.** yes; $-2\pm\sqrt{11}$ **b.** no **c.** Answers
may vary. **69.** Answers may vary. **71.** $x^2+14x+49$ **73.** $(x-8)^2$ **75.** $x^2-8x+16$; quadratic (or second-degree) polynomial in
one variable **77.** $4\pm\sqrt{3}$; quadratic equation in one variable **79.** $(p-4q)^2$; quadratic (or second-degree) polynomial in two variables

Homework 9.5

1. $-1, -\dfrac{3}{2}$ **3.** $\dfrac{-7\pm\sqrt{17}}{8}$ **5.** $\dfrac{-3\pm\sqrt{29}}{2}$ **7.** $\dfrac{5\pm\sqrt{61}}{6}$ **9.** $\dfrac{-1\pm\sqrt{6}}{5}$ **11.** $\dfrac{3\pm\sqrt{3}}{3}$ **13.** no real-number
solutions **15.** $\dfrac{-5\pm\sqrt{57}}{4}$ **17.** $\dfrac{-1\pm\sqrt{5}}{2}$ **19.** no real-number solutions **21.** $\dfrac{7\pm\sqrt{13}}{6}$ **23.** $\dfrac{1\pm\sqrt{5}}{6}$ **25.** $-1\pm 2\sqrt{3}$ **27.** -1.17,
1.92 **29.** $-4.90, 1.66$ **31.** $-4.65, 2.31$ **33.** $-6.32, 14.82$ **35.** $(0.22, 0), (2.28, 0)$ **37.** $(-0.64, 0), (1.24, 0)$ **39.** $(-2.29, 0)$,
$(0.89, 0)$ **41.** $(-2.06, 0), (1.41, 0)$ **43.** $-9, -2$ **45.** $-4\pm\sqrt{13}$ **47.** $\dfrac{-5\pm\sqrt{37}}{6}$ **49.** $\pm\dfrac{7}{6}$ **51.** $0, \dfrac{3}{2}$ **53.** $\pm\dfrac{\sqrt{35}}{7}$ **55.** $\dfrac{1}{2}, \dfrac{2}{3}$
57. no real-number solutions **59.** $\dfrac{1\pm\sqrt{22}}{7}$ **61.** $-2\pm\sqrt{2}$ **63.** $-1, 3$ **65.** 1 **67.** $-5, -1$ **69.** no real-number solutions
71. $-1.24, 3.24$ **73.** $-0.23, 2.90$ **75.** $-2, 4$ **77.** 1 **79.** Answers may vary; $\dfrac{-5\pm\sqrt{37}}{2}$ **81. a.** $-3, 1$ **b.** $-3, 1$ **c.** $-3, 1$
d. The results are the same. **e.** Answers may vary. **83.** Answers may vary. **85.** $\dfrac{7}{2}$ **87.** $\pm\dfrac{\sqrt{14}}{2}$ **89.** $\dfrac{3\pm\sqrt{11}}{2}$ **91.** $(1.54, 0)$,

$(-0.87, 0)$ **93.** $(0.67, 0)$ **95.** $\dfrac{-3 \pm \sqrt{29}}{2}$; quadratic equation in one variable **97.** $(3x + 1)(2x - 1)$; quadratic (or second-degree) polynomial in one variable **99.** $-25x + 23$; linear (or first-degree) polynomial in one variable

Homework 9.6

1. a. 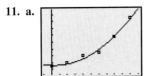 yes **b.** 1991, 1999 **c.** 1995; 13; actually, the rank was better in 1996 when the rank was 11.

d. 86; answers may vary.

3. a. 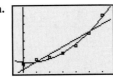 yes **b.** 2000; 155 bank robberies **c.** 1996, 2004 **5. a.** 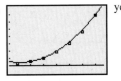 yes **b.** 76.6%

c. $36.4 thousand, $204.8 thousand; model breakdown has occurred for the $204.8 thousand estimate. **7. a.** 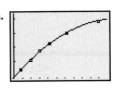 yes

b. $162.1 billion **c.** 2009 **9. a.** 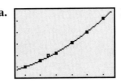 The quadratic model describes the situation better. **b.** 44.24%; 43.50%; the quadratic model's estimate is better. **c.** $(0, 41.54)$; $(0, 42.4)$; the quadratic model's p-intercept describes the situation better. **d.** 2004

11. a. 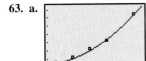 yes **b.** $10.97 billion; 5.77% **c.** 2010 **13. a.**  yes **b.** 19 years

c. i. $62,800 **ii.** $17,124 **iii.** 5.2 years **iv.** $424,976 **v.** underestimate **15.** Answers may vary. **17. a.** $n = 1.96t + 6.51$
b. 2007 **c.** 47.7 thousand students **19. a.** $p = 4t + 54$ **b.** 82% **c.** 2015 **21.** 6.9% **23.** Answers may vary. **25.** Answers may vary. **27.** Answers may vary. **29.** Answers may vary. **31.** Answers may vary.

Chapter 9 Review

1. 14 **2.** -8 **3.** not a real number **4.** not a real number **5.** 9.75 **6.** -45.15 **7.** $2\sqrt{21}$ **8.** $6\sqrt{x}$ **9.** $3b\sqrt{2a}$ **10.** $7mn\sqrt{2}$
11. 6 and 7 **12.** $\dfrac{\sqrt{5}}{3}$ **13.** $\dfrac{5\sqrt{2y}}{x}$ **14.** $\dfrac{4\sqrt{x}}{x}$ **15.** $\dfrac{b\sqrt{10}}{8}$ **16.** Answers may vary; $\dfrac{3\sqrt{7}}{7}$ **17.** $\pm 3\sqrt{5}$ **18.** $\pm\dfrac{\sqrt{35}}{5}$ **19.** $\pm\dfrac{2\sqrt{10}}{5}$
20. $-4 \pm 3\sqrt{3}$ **21.** no real-number solutions **22.** 8 **23.** $4\sqrt{5}$ **24.** $3\sqrt{3}$ **25.** $\sqrt{58}$ **26.** $2\sqrt{14}$ **27.** yes **28.** $-3 \pm \sqrt{11}$
29. $-1 \pm 3\sqrt{2}$ **30.** $4 \pm 2\sqrt{5}$ **31.** no real-number solutions **32.** $-2 \pm \sqrt{10}$ **33.** $3 \pm 3\sqrt{2}$ **34.** $\dfrac{-7 \pm \sqrt{37}}{6}$ **35.** $\dfrac{5 \pm \sqrt{57}}{4}$ **36.** no
real-number solutions **37.** $\dfrac{3 \pm \sqrt{29}}{2}$ **38.** $\dfrac{1 \pm \sqrt{13}}{4}$ **39.** $\dfrac{1 \pm \sqrt{11}}{2}$ **40.** $-0.46, 1.80$ **41.** $-4.47, 1.63$ **42.** Answers may vary;
$\dfrac{-3 \pm \sqrt{65}}{4}$ **43.** $-2, 7$ **44.** $\dfrac{2 \pm \sqrt{7}}{3}$ **45.** no real-number solutions **46.** $4 \pm 2\sqrt{6}$ **47.** $-5, 7$ **48.** $\pm\dfrac{2\sqrt{15}}{5}$ **49.** $\dfrac{7 \pm \sqrt{37}}{2}$
50. $-2, \dfrac{4}{3}$ **51.** $-7 \pm 3\sqrt{3}$ **52.** $\dfrac{-7 \pm \sqrt{73}}{4}$ **53.** $\dfrac{2 \pm \sqrt{14}}{2}$ **54.** $\dfrac{1}{2}, \dfrac{3}{2}$ **55.** 2 **56.** no real-number solutions **57.** 0, 4 **58.** $-2, 6$
59. $-2.08, 2.41$ **60.** $(0.28, 0), (2.39, 0)$ **61.** $(-2.95, 0), (4.77, 0)$ **62. a.** $-3, 5$ **b.** $-3, 5$ **c.** $-3, 5$ **d.** The results are the same.
63. a. 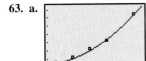 yes **b.** 1.4% **c.** 87 years **64. a.**  yes **b.** 47.5% **c.** 2012

Chapter 9 Test

1. -11 **2.** not a real number   **3.** $4\sqrt{3}$ **4.** $8\sqrt{x}$ **5.** $3a\sqrt{5b}$ **6.** $\dfrac{3\sqrt{7}}{7}$ **7.** $\dfrac{m\sqrt{2n}}{n}$ **8.** $\dfrac{\sqrt{15}}{10}$ **9.** $4\sqrt{10}$ **10.** 12.96 inches

11. $\pm\dfrac{\sqrt{15}}{3}$ **12.** $\pm\dfrac{4\sqrt{10}}{5}$ **13.** $5\pm3\sqrt{2}$ **14.** $\dfrac{-3\pm\sqrt{57}}{4}$ **15.** $-5,8$ **16.** $\dfrac{1\pm\sqrt{13}}{3}$ **17.** $2\pm\sqrt{11}$ **18.** $-4.96, 6.61$

19. $(1.39, 0), (30.19, 0)$ **20.** $2, 4$ **21.** $1, 5$ **22.** 3 **23.** no real-number solutions   **24.** Answers may vary; $\dfrac{-2\pm\sqrt{10}}{2}$

25. a. 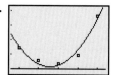   yes   **b.** 2000; \$14.8 million; underestimate   **c.** \$679 million   **d.** 2004

Chapter 10

Homework 10.1   **1.** 0 **3.** no excluded values   **5.** 4 **7.** -9 **9.** 4 **11.** $-\dfrac{4}{3}$ **13.** $-3, -2$ **15.** $-5, 7$ **17.** 5 **19.** ±4

21. $\pm\dfrac{7}{5}$ **23.** $-5, -\dfrac{3}{2}$ **25.** $\dfrac{2}{3}, \dfrac{3}{2}$ **27.** ±2 **29.** $\dfrac{2x}{3}$ **31.** $\dfrac{4t^2}{5}$ **33.** $\dfrac{2x}{3y^3}$ **35.** $\dfrac{3}{5}$ **37.** $\dfrac{2}{3}$ **39.** $\dfrac{a}{7}$ **41.** $\dfrac{x-y}{3}$ **43.** $\dfrac{4}{x+5}$

45. $\dfrac{5}{x-2}$ **47.** $\dfrac{t+1}{t+5}$ **49.** $\dfrac{x-2}{x-1}$ **51.** $\dfrac{x-2}{x-5}$ **53.** $\dfrac{x+4}{x-4}$ **55.** $\dfrac{2}{3x+8}$ **57.** $\dfrac{-4}{w+4}$ **59.** $\dfrac{2x+5}{x+3}$ **61.** $\dfrac{3(x+2)}{6x-1}$ **63.** $\dfrac{a+b}{a-b}$

65. $\dfrac{5}{x^2-3}$ **67.** $\dfrac{1}{t-1}$ **69.** $\dfrac{x^2-2x+4}{x+5}$ **71.** $\dfrac{x^2+4x+16}{x+4}$ **73. a.** $p = \dfrac{60}{n}$

Number of Students	Cost per Student (dollars per student)
n	p
10	$\dfrac{60}{10}$
20	$\dfrac{60}{20}$
30	$\dfrac{60}{30}$
40	$\dfrac{60}{40}$
n	$\dfrac{60}{n}$

b. The units on both sides of the equation are dollars per student.   **c.** 2.4; the per-person cost is \$2.40 if 25 students go to the party.

d. The value of p decreases; the greater the number of students at the party, the lower is the per-person cost.   **75. a.** 44.8%   **b.** 44.0%

c. overestimate   **d.** 39.3%   **77.** no; answers may vary.   **79.** $\dfrac{5}{2}$ **81.** Answers may vary; the expression is already simplified.

83. Answers may vary.   **85.** Answers may vary.   **87.** Answers may vary.   **89.** $\dfrac{y^9 w^2}{5x^4}$ **91.** $\dfrac{4}{x-4}$ **93.** $\dfrac{3-2\sqrt{2}}{4}$

95. $(p-3)(p+3)(2p-3)$; cubic (or third-degree) polynomial in one variable   **97.**   quadratic equation in two variables

99. $\dfrac{1\pm\sqrt{13}}{3}$; quadratic equation in one variable

Homework 10.2   **1.** $\dfrac{15}{x^2}$ **3.** $\dfrac{x^2}{9}$ **5.** $\dfrac{18}{5a^6}$ **7.** $\dfrac{2(x-4)}{(x-3)(x+5)}$ **9.** $\dfrac{(k-2)(k+4)}{(k-6)(k+6)}$ **11.** $\dfrac{10}{21}$ **13.** 6 **15.** $\dfrac{2w^4(w+5)}{w+3}$

17. $\dfrac{5(x+1)}{3(x-7)}$ **19.** $\dfrac{30(x-4)}{7(x+5)}$ **21.** $\dfrac{5t^4}{6}$ **23.** $\dfrac{6}{x^3(x-7)}$ **25.** $\dfrac{3a^3}{10b^7}$ **27.** $\dfrac{x+4}{x-5}$ **29.** $\dfrac{t-6}{t+9}$ **31.** $\dfrac{4(x+4)}{3x}$ **33.** $\dfrac{(x+1)(x+2)}{(x-6)(x-4)}$

35. $\dfrac{(a+3)(a+5)}{4a(a-3)}$ **37.** $\dfrac{(x-3)(x+3)}{(x-5)(x+1)}$ **39.** $\dfrac{(x^2+16)(x-4)}{4x(x+4)}$ **41.** $\dfrac{x(3x+5)}{(x-4)(x+1)}$ **43.** $\dfrac{p^2(p+1)}{2}$ **45.** $\dfrac{x+5}{(x-5)(x+1)}$

47. $\dfrac{2(x-3)}{x+7}$ **49.** $\dfrac{2b(b+5)}{b-1}$ **51.** $\dfrac{x}{(x-4)(x+8)}$ **53.** $\dfrac{a-b}{2a+3b}$ **55.** $\dfrac{x(x-y)}{3(x+y)}$ **57.** $\dfrac{3(x^2+3)}{x(x+2)}$ **59.** $\dfrac{x(x+3)}{6}$

61. a. $\dfrac{(x+1)^2}{(x-2)^2}$ **b.** $\dfrac{(x-1)^2}{(x+2)^2}$ **63.** 36 inches   **65.** 6.21 miles   **67.** 0.71 ounce per day   **69.** 14.78 miles per gallon   **71.** 0.76 gram

73. $\dfrac{x^2}{4}$ **75.** $\dfrac{(x-2)(x+1)}{(x-6)(x+4)}$ **77.** Answers may vary; $\dfrac{x+2}{x+6}$ **79.** Answers may vary; $\dfrac{x(x+4)}{3(x-7)}$ **81.** Answers may vary.

83. Answers may vary. **85.** $2x^3 - 20x^2 + 66x - 72$ **87.** $\dfrac{2}{x-4}$ **89.** $-3x^2 + 30x - 75$; quadratic (or second-degree) polynomial in one variable **91.** $5 \pm 2\sqrt{2}$; quadratic equation in one variable **93.** -27; quadratic (or second-degree) polynomial in one variable

Homework 10.3 **1.** $\dfrac{9}{x}$ **3.** $\dfrac{8x}{x-1}$ **5.** $\dfrac{8x+2}{x+3}$ **7.** $t+2$ **9.** $\dfrac{1}{x-2}$ **11.** $\dfrac{x-4}{x+2}$ **13.** $\dfrac{r-7}{r+3}$ **15.** $\dfrac{11}{2x}$ **17.** $\dfrac{5w+9}{6w}$

19. $\dfrac{10x^2+9}{12x}$ **21.** $\dfrac{12x^2+25}{40x^3}$ **23.** $\dfrac{a^2+b^2}{ab}$ **25.** $\dfrac{13x+15}{4x(x+3)}$ **27.** $\dfrac{7(r-1)}{(r-5)(r+2)}$ **29.** $\dfrac{3(11x-1)}{5x(x-1)}$ **31.** $\dfrac{5x^2+12x-8}{(x-4)(x+2)}$

33. $\dfrac{2a}{(a+b)(a-b)}$ **35.** $\dfrac{5x+2}{5(x+4)}$ **37.** $\dfrac{1}{3}$ **39.** $\dfrac{2(3x+10)}{15(x-5)}$ **41.** $\dfrac{3}{x-1}$ **43.** $\dfrac{t+4}{t+9}$ **45.** $\dfrac{x^2+3x+6}{2(x-3)(x+3)}$ **47.** $\dfrac{x-1}{x(x-2)}$

49. $\dfrac{2(3a+5)}{(a-3)(a+1)(a+4)}$ **51.** $\dfrac{8x-17}{(x-4)(x+1)(x+4)}$ **53.** $\dfrac{2(2x^2+9x-3)}{(x-3)(x+4)(x+6)}$ **55.** $\dfrac{x^2-x-16}{(x+3)(x+4)(x+5)}$ **57.** $\dfrac{4w+13}{w+5}$

59. $\dfrac{2(x^2+11)}{(x-4)(x+5)}$ **61.** $\dfrac{2y^2-13}{(y-3)(y+2)}$ **63.** $\dfrac{x^2-x-18}{(x-5)(x+3)}$ **65.** $\dfrac{b^2-b-1}{(b-4)(b-3)}$ **67.** $\dfrac{x^2-5x+26}{4(x-3)(x+3)}$ **69.** $\dfrac{2x^2}{(x-y)^2}$

71. $\dfrac{11x-20}{3(x-4)(x^2+2)}$ **73.** $\dfrac{9x^2+18x+16}{(x-2)(x+2)(x^2+2x+4)}$ **75. a.** $\dfrac{18}{d^2} + \dfrac{18}{(2d)^2}$ **b.** $\dfrac{45}{2d^2}$ **c.** 15.63; the total illumination is

15.63 W/m^2 when the person is 1.2 meters away from the closer light and 2.4 meters away from the other light.

77. Answers may vary; $\dfrac{8x+19}{(x+2)(x+3)}$ **79.** yes; yes **81.** Answers may vary. **83.** $x^2 - 5x + 6$ **85.** $x-2$ **87.** $(-3, -11)$; system of two linear equations in two variables **89.**   linear equation in two variables **91.** -3; linear equation in one variable

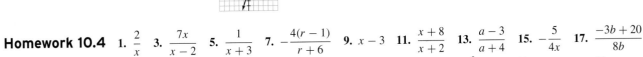

Homework 10.4 **1.** $\dfrac{2}{x}$ **3.** $\dfrac{7x}{x-2}$ **5.** $\dfrac{1}{x+3}$ **7.** $-\dfrac{4(r-1)}{r+6}$ **9.** $x-3$ **11.** $\dfrac{x+8}{x+2}$ **13.** $\dfrac{a-3}{a+4}$ **15.** $-\dfrac{5}{4x}$ **17.** $\dfrac{-3b+20}{8b}$

19. $\dfrac{15x^2-4}{24x}$ **21.** $\dfrac{-2x^2+5}{10x^4}$ **23.** $\dfrac{3a^2-2b^2}{6ab}$ **25.** $\dfrac{7p+10}{4p(p-2)}$ **27.** $\dfrac{4x+31}{(x-1)(x+4)}$ **29.** $\dfrac{3x^2+4x+10}{(x-2)(x+3)}$ **31.** $\dfrac{-2b}{(a+b)(a-b)}$

33. $\dfrac{c-6}{2(c-4)}$ **35.** $\dfrac{8x-21}{12(x-4)}$ **37.** $\dfrac{x+2}{x+1}$ **39.** $\dfrac{1}{x-3}$ **41.** $\dfrac{t^2+7t-12}{3(t-7)(t+7)}$ **43.** $\dfrac{2(2x^2-x+5)}{x(x-5)(x+5)}$ **45.** $\dfrac{-x+32}{(x-5)(x+2)(x+4)}$

47. $\dfrac{5x+4}{x(x-3)(x-2)}$ **49.** $\dfrac{5b^2+27b-15}{(b-2)(b+5)(b+6)}$ **51.** $\dfrac{2x^2-19x-45}{(x-7)(x+2)(x+9)}$ **53.** $-\dfrac{3(x-9)}{x-6}$ **55.** $\dfrac{10(x-1)}{(x-4)(x+1)}$

57. $\dfrac{x^2-3x-14}{(x-5)(x-4)}$ **59.** $\dfrac{(t-6)(t+2)}{(t-7)(t-3)}$ **61.** $\dfrac{(x-7)(x+2)}{3(x-1)(x+1)}$ **63.** $\dfrac{6x^2}{(2x+y)(x+y)}$ **65.** $\dfrac{-3x^2+18x+10}{5(x-6)(x^2-3)}$

67. $\dfrac{x^2+5x-2}{(x+1)^2(x^2-x+1)}$ **69.** Answers may vary; $\dfrac{-2x-7}{x+2}$ **71.** Answers may vary. **73.** $\dfrac{x+4}{x-7}$ **75.** $\dfrac{11}{x+3}$ **77.** $\dfrac{x(x-3)}{(x-7)(x+5)}$

79. $\dfrac{2(x^2-5x-4)}{(x-7)(x-2)^2}$ **81.** $\dfrac{(x-2)(5x+3)}{x(x+5)}$ **83.** $\dfrac{-5x^2-35x+6}{2(x-3)(x+7)}$ **85. a.** $\dfrac{x^2+8}{2x}$ **b.** $\dfrac{x^2-8}{2x}$ **c.** 2 **d.** $\dfrac{x^2}{8}$ **87.** $-2, \dfrac{4}{3}$; quadratic

equation in one variable **89.** $(3x-4)(x+2)$; quadratic (or second-degree) polynomial in one variable **91.** $\dfrac{-9x^2+17x-10}{(x-2)(x+2)(3x-4)}$; rational expression in one variable

Homework 10.5 **1.** -2 **3.** 3 **5.** 2 **7.** empty set solution **9.** 3 **11.** $\dfrac{4}{5}$ **13.** $\dfrac{1}{6}$ **15.** -8 **17.** $\dfrac{3}{2}$ **19.** 1 **21.** $\dfrac{2}{3}$

23. empty set solution **25.** $-\dfrac{7}{2}$ **27.** empty set solution **29.** $\dfrac{19}{12}$ **31.** $-\dfrac{6}{5}$ **33.** 2 **35.** $-4, 1$ **37.** $-5, 3$ **39.** $-\dfrac{2}{5}, 1$ **41.** $-\dfrac{7}{11}$

43. 5 **45.** $-4, 1$ **47.** $\dfrac{7}{2}$ **49.** $\dfrac{17}{11}$ **51.** 7 **53.** -7 **55. a.** \$18.1 thousand; \$17.9 thousand; overestimate **b.** \$22.7 thousand

c. 2012 **57.** 2012 **59.** Answers may vary; $\dfrac{2(4x+3)}{x(x+2)}$ **61.** Answers may vary; $\dfrac{33}{2}$ **63.** Answers may vary. **65.** $\dfrac{10}{x}$ **67.** 10

69. $2, \dfrac{1}{3}$ **71.** $\dfrac{(x-1)(3x-4)}{(x-3)(x+2)}$ **73.** $\dfrac{1 \pm \sqrt{13}}{3}$ **75.** $\pm 2, \dfrac{2}{3}$ **77.** $\pm \dfrac{\sqrt{14}}{2}$ **79.** $11m - 17n$; linear (or first-degree) polynomial in two

variables **81.** $2x(x-7)(x-2)$; cubic (or third-degree) polynomial in one variable **83.** 15; rational equation in one variable

Homework 10.6 **1.** \$143.40 **3.** 1.31 cups **5.** 2.4 ounces **7.** 36,200 students **9.** 6.75 inches **11.** 320.99 U.S. dollars

13. 172 pounds **15. a.** \$409.50 **b.** Answers may vary; underestimate **17. a.** 8973 adults **b.** Answers may vary; underestimate

19. no **21. a.** 21.5:1; greater **b.** 339.2 FTE faculty; \$24.8 million **c.** 5766 students; \$19.3 million **d.** reduce FTE enrollment

23. 4.88 inches **25.** 14.86 meters **27.** 5.63 feet **29. a.** $y = kx$; $(0, 0)$ **b.** $C = 15t$; the variables t and C are proportional; answers may vary.   **c.** $C = 10t + 25$; the variables t and C are not proportional; answers may vary.   **d.** $45; $55; the variables t and C are not proportional; yes   **31. a.** $\dfrac{n}{m}$ **b.** $\dfrac{q}{p}$ **c.** Answers may vary.   **d.** Answers may vary.   **33.** $-5, \dfrac{3}{2}$; quadratic equation in one variable   **35.** $(2x - 3)(x + 5)$; quadratic (or second-degree) polynomial in one variable   **37.** -21; quadratic (or second-degree) polynomial in one variable

Homework 10.7   **1.** $w = kt$ **3.** $F = \dfrac{k}{r}$ **5.** $c = ks^2$ **7.** $B = \dfrac{k}{t^3}$ **9.** H varies directly as u.   **11.** P varies inversely as V.   **13.** E varies directly as c^2.   **15.** F varies inversely as r^2.   **17.** $y = 2x$ **19.** $F = \dfrac{12}{r}$ **21.** $H = 3\sqrt{u}$ **23.** $C = \dfrac{18}{x^2}$ **25.** 12   **27.** 9   **29.** 48   **31.** ± 2 **33.** increases   **35.** decreases   **37.** increases   **39.** decreases   **41.** Answers may vary.   **43.** $1316.25   **45.** 12.8 gallons   **47.** 24 ounces   **49. a.** short handle   **b.** 30 pounds   **51.** 130.9 centimeters

53. a. 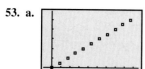   **b.** $F = 28.92x$ **c.** The force varies directly as the stretch.   **d.** 0.145 meter

55. a. 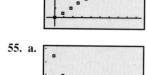   **b.** $P = \dfrac{7.18}{V}$ **c.** The pressure varies inversely as the volume.   **d.** 8.0 cm^3

57. a. i. Answers may vary; one   **ii.** Answers may vary; one   **iii.** Answers may vary; one   **b.** one; answers may vary.   **c.** one; answers may vary.   **59. a.** yes; answers may vary.   **b.** not necessarily; answers may vary.   **61. a.** $75   **b.** $c = 12.5n$ **c.** $75; yes   **63.** $\dfrac{5}{2}, -1$; quadratic equation in one variable   **65.** 0; quadratic (or second-degree) polynomial in one variable   **67.** $(2w - 5)(w + 1)$; quadratic (or second-degree) polynomial in one variable

Homework 10.8   **1.** $\dfrac{3}{10}$ **3.** $\dfrac{7}{4}$ **5.** $\dfrac{14}{5}$ **7.** $\dfrac{3}{2w}$ **9.** $\dfrac{9}{10x^2}$ **11.** $\dfrac{2}{3(x + 3)}$ **13.** $\dfrac{5}{6}$ **15.** $\dfrac{5(x + 3)}{3x}$ **17.** 3   **19.** $\dfrac{5}{9}$ **21.** $\dfrac{11}{26}$ **23.** $-\dfrac{1}{5}$ **25.** $\dfrac{4(x - 2)}{5}$ **27.** $\dfrac{2x + 5}{4x - 1}$ **29.** $\dfrac{2b - 3}{4b + 5}$ **31.** $\dfrac{x + 2}{x}$ **33.** $\dfrac{3x + 4}{2(5x + 2)}$ **35.** $\dfrac{2(r - 3)(r + 3)}{3(r - 2)(r + 2)}$ **37.** $\dfrac{2(x + 2)}{x}$ **39.** Answers may vary; $\dfrac{3x^2}{x + 3}$ **41.** Answers may vary.   **43.** $\dfrac{2(x + 3)}{3}$ **45.** $\dfrac{2(x + 3)}{3}$ **47.** $\dfrac{6}{5x}$ **49.** $\dfrac{6x + 3}{16x}$ **51.** Answers may vary.   **53.** Answers may vary.   **55.** Answers may vary.   **57.** Answers may vary.   **59.** Answers may vary.

Chapter 10 Review

1. 0   **2.** no excluded values   **3.** $\dfrac{5}{3}$ **4.** 2, 4   **5.** -3 **6.** $\pm\dfrac{3}{2}$ **7.** $-4, \dfrac{2}{5}$ **8.** $-\dfrac{5}{2}, \pm 3$ **9.** $\dfrac{4y^3}{5x^4}$ **10.** $\dfrac{7}{x - 2}$ **11.** $\dfrac{w + 3}{w - 4}$ **12.** $\dfrac{2a}{a + b}$ **13.** $\dfrac{x - 2}{3x + 2}$ **14.** $\dfrac{x + 5}{(x - 5)(2x - 3)}$ **15.** $\dfrac{(m - 4)(m + 2)}{(m - 3)(m + 3)}$ **16.** $\dfrac{5b(b + 1)}{7}$ **17.** $\dfrac{4x(x + 4)}{3(x - 7)}$ **18.** $\dfrac{6}{(x - 1)(x + 4)}$ **19.** $\dfrac{5}{3x^3}$ **20.** $\dfrac{3(x - 2)(2x + 5)}{2(x - 5)}$ **21.** $\dfrac{7}{(t - 7)(3t - 4)}$ **22.** $\dfrac{(a - 3b)(a - 2b)}{4a(a - 7b)}$ **23.** $x - 2$ **24.** $\dfrac{9x^2 + 10}{12x^3}$ **25.** $\dfrac{2x^2 + 17x - 20}{(x - 4)(x + 1)(x + 6)}$ **26.** $\dfrac{x^2 + 10x + 26}{(x - 2)(x + 6)}$ **27.** $\dfrac{p - 6}{p - 2}$ **28.** $-\dfrac{14}{(x - 5)(x + 3)}$ **29.** $\dfrac{10y^2}{(x - 5y)^2}$ **30.** $-\dfrac{2(y^2 - 6y + 24)}{y(y + 4)(y^2 - 4y + 16)}$ **31.** $\dfrac{16x^6}{15}$ **32.** $\dfrac{9(x - 6)}{(x + 6)(x + 9)}$ **33.** $\dfrac{5m^2 - 12m - 15}{(m - 5)(m - 3)(m + 9)}$ **34.** $\dfrac{-11c - 1}{(c - 1)(c + 2)}$ **35.** Answers may vary; $\dfrac{-5x + 3}{x + 4}$ **36.** 9.92 ounces   **37.** 5.30 cups per day   **38.** $\dfrac{11}{3}$ **39.** -1 **40.** 3   **41.** $\dfrac{2}{5}$ **42.** $\dfrac{31}{3}$ **43.** empty set solution   **44.** $\dfrac{-2 \pm \sqrt{10}}{2}$ **45.** $\pm\sqrt{11}$ **46. a.** $313; $306; overestimate   **b.** $409   **c.** 2015   **47.** 24.5 ounces   **48.** 27.5 yards   **49.** $A = kw$ **50.** $F = \dfrac{k}{r^2}$ **51.** V varies directly as r^3.   **52.** I varies inversely as d^2.   **53.** ± 5 **54.** 2   **55.** decreases   **56.** decreases   **57.** $71,250   **58.** 36 minutes   **59. a.** [graph]   **b.** $A = \dfrac{54.36}{D}$ **c.** The apparent height varies inversely as the distance.   **d.** 4.5 inches   **e.** 27.2 inches   **60.** $\dfrac{4x}{3}$ **61.** $\dfrac{7(x + 2)}{5x}$ **62.** $\dfrac{5w - 2}{w - 3}$ **63.** $\dfrac{3(3b + 2)}{2(b - 6)}$

Chapter 10 Test

1. no excluded values **2.** -9 **3.** $-9, 6$ **4.** Answers may vary. **5.** $\dfrac{p+4}{p-5}$ **6.** $\dfrac{5m}{(m-2)(3m-2)}$ **7.** $\dfrac{(x+3)(x+5)}{(x-5)(x-4)}$ **8.** $\dfrac{9}{4x}$

9. $\dfrac{5a^2+12a+8}{(a-5)(a+2)(a+4)}$ **10.** $\dfrac{-3t-35}{2(t-7)(t+7)}$ **11.** $\dfrac{3(3x+2)}{2(2x+1)}$ **12.** $\dfrac{-14x-1}{(x-7)(x+4)}$ **13.** 12 **14.** $-\dfrac{22}{17}$ **15.** 80.75 miles per hour

16. a. 3.4%; 3.2%; overestimate **b.** 1.2% **c.** 2011 **17.** 7.84 gallons **18.** 15.75 meters **19.** 24 **20.** ±3 **21.** decreases

22. 15 credit hours **23.** $\dfrac{2(4x-3)}{3(x-10)}$

Chapter 11

Homework 11.1
1. $9\sqrt{3}$ **3.** $6\sqrt{5}$ **5.** $5\sqrt{x}$ **7.** $2\sqrt{7}$ **9.** $-7\sqrt{6}$ **11.** $-11\sqrt{5t}$ **13.** $7\sqrt{5}-3\sqrt{7}$ **15.** $6\sqrt{2}$ **17.** $4\sqrt{7}-3\sqrt{5}$
19. 0 **21.** $7\sqrt{2}$ **23.** $3\sqrt{6}$ **25.** $-6\sqrt{5}$ **27.** $5\sqrt{7}$ **29.** $16\sqrt{3}$ **31.** $-5\sqrt{2}$ **33.** $\sqrt{5}$ **35.** $28\sqrt{3}$ **37.** $11\sqrt{x}$ **39.** 0
41. $\sqrt{2x}-2\sqrt{5}$ **43.** $21\sqrt{3t}-4\sqrt{11}$ **45.** $5x\sqrt{5}$ **47.** $-3x\sqrt{10}$ **49.** $-2w\sqrt{5}$ **51.** $16x\sqrt{3}$ **53.** $-x\sqrt{3}+10\sqrt{x}$
55. $16y\sqrt{3}-6\sqrt{5y}$ **57. a.** 29.3 years **b.** 2 "Saturn years" **59. a.**   yes **b.** 2117 agents **c.** 2008

61. Answers may vary; $6\sqrt{5}$ **63.** $4\sqrt{7}+5\sqrt{7}=(4+5)\sqrt{7}=9\sqrt{7}$ **65.** $-2x$ **67.** $-2\sqrt{x}$ **69.** $3x\sqrt{3}$ **71.** $15x^2$ **73.** $1\pm2\sqrt{2}$;
quadratic equation in one variable **75.** $\dfrac{11w+3}{4(w-3)(w+3)}$; rational expression in one variable **77.** $x(x-5)^2$; cubic (or third-degree)
polynomial in one variable

Homework 11.2
1. $\sqrt{10}$ **3.** $-2\sqrt{3}$ **5.** 17 **7.** $10\sqrt{2}$ **9.** $-42\sqrt{7}$ **11.** $35t$ **13.** $t\sqrt{21}$ **15.** $64x$ **17.** $b\sqrt{ac}$ **19.** $\sqrt{5}+\sqrt{35}$
21. $-\sqrt{14}-\sqrt{35}$ **23.** $45\sqrt{c}-5c$ **25.** $-12\sqrt{10}+20$ **27.** $3-2\sqrt{5}$ **29.** $3+\sqrt{6}-\sqrt{15}-\sqrt{10}$ **31.** $x-\sqrt{7x}-\sqrt{2x}+\sqrt{14}$
33. $y^2-y\sqrt{11}+y\sqrt{2}-\sqrt{22}$ **35.** $38-16\sqrt{5}$ **37.** -4 **39.** r^2-3 **41.** 38 **43.** $9a-25b$ **45.** $23+8\sqrt{7}$ **47.** $8-2\sqrt{15}$
49. $b^2+2b\sqrt{2}+2$ **51.** $x-2\sqrt{6x}+6$ **53.** $-12\sqrt{10}+38$ **55.** Answers may vary; $x^2+2x\sqrt{3}+3$ **57.** Answers may vary; $6\sqrt{5}$
59. a. i. true **ii.** false **iii.** true **iv.** false **v.** true **vi.** false **b.** Answers may vary. **61.** $5\sqrt{x}$ **63.** $6x$ **65.** $x+5\sqrt{x}+6$
67. x^2+5x+6 **69.** $9x$ **71.** $x+6\sqrt{x}+9$ **73.** x^2+6x+9 **75.**   linear equation in two variables

77. empty set solution; rational equation in one variable **79.** $\dfrac{x(x-1)}{(x-4)(x+2)}$; rational expression in one variable

Homework 11.3
1. 36 **3.** 3 **5.** empty set solution **7.** 2 **9.** empty set solution **11.** 5 **13.** 2 **15.** $-2, 8$ **17.** 16 **19.** 4
21. 16 **23.** $\dfrac{25}{9}$ **25.** empty set solution **27.** 12 **29.** $2, 3$ **31.** 1 **33.** 7 **35.** -1 **37.** 4 **39.** 6 **41.** 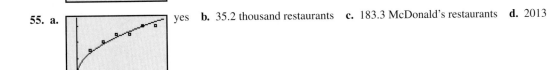 **43.** -4

45. empty set solution **47.** 4.81 **49.** 3.13 **51.** Answers may vary; $-\dfrac{10}{7}$
53. a. yes **b.** $(0, 28)$; there were 28 million acres of genetically modified crops in 1997. **c.** 208.3 million acres;
yes **d.** 2013

55. a. yes **b.** 35.2 thousand restaurants **c.** 183.3 McDonald's restaurants **d.** 2013

57. a. 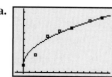 yes **b.** 2010 **c.** Since all tires could be reused or recycled by 2010, the plant would need to be in operation for only two years. This would not be enough time to warrant building such a plant.

59. $\dfrac{-2 \pm 3\sqrt{2}}{2}$ **61.** 7 **63.** $\dfrac{20}{9}$ **65.** Answers may vary. **67.** Answers may vary. **69.** Answers may vary. **71.** Answers may vary.
73. Answers may vary.

Chapter 11 Review

1. $15\sqrt{3}$ **2.** $-\sqrt{2x}$ **3.** $-2\sqrt{5x}$ **4.** $8\sqrt{3}+4\sqrt{6}$ **5.** $-6\sqrt{5}-3\sqrt{7}$ **6.** $11\sqrt{3}$ **7.** $-5\sqrt{6}$ **8.** $7\sqrt{2}$ **9.** $\sqrt{5}$ **10.** $25\sqrt{2}$
11. $-11\sqrt{10}$ **12.** $-2\sqrt{2}-5\sqrt{3}$ **13.** $15\sqrt{5}-16\sqrt{6}$ **14.** $17\sqrt{x}$ **15.** $-11\sqrt{x}$ **16.** $-3\sqrt{3x}+2\sqrt{2}$ **17.** $-w\sqrt{7}$ **18.** $3a\sqrt{7}$
19. $-7x\sqrt{3}$ **20.** $4\sqrt{2x}-10x\sqrt{5}$ **21.** $\sqrt{35ab}$ **22.** $-24\sqrt{2}$ **23.** $25x$ **24.** $\sqrt{6}+\sqrt{14}$ **25.** $-\sqrt{21}+\sqrt{15}$ **26.** $x-\sqrt{2x}$
27. $10\sqrt{21}+14$ **28.** $3\sqrt{5}-13$ **29.** $-17\sqrt{2}+74$ **30.** $\sqrt{6}-\sqrt{10}-\sqrt{21}+\sqrt{35}$ **31.** $b+7\sqrt{b}-8$ **32.** $t^2+t\sqrt{5}+t\sqrt{3}+\sqrt{15}$
33. $-8\sqrt{7}+44$ **34.** -2 **35.** $16a-9b$ **36.** b^2-3 **37.** $-8\sqrt{5}+21$ **38.** $x+12\sqrt{x}+36$ **39.** $x^2+2x\sqrt{3}+3$ **40.** $-24\sqrt{10}+77$
41. 4 **42.** 22 **43.** empty set solution **44.** $\dfrac{5}{2}$ **45.** $-1, 4$ **46.** 49 **47.** 64 **48.** empty set solution **49.** 6 **50.** -1 **51.** $\dfrac{16}{3}$
52. 2 **53.** 7 **54.** 4 **55.** 3.28 **56. a.** 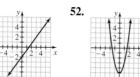 yes **b.** 90.5% **c.** 2021; model breakdown has likely occurred.

Chapter 11 Test

1. $-2\sqrt{3}-3\sqrt{2}$ **2.** $-12\sqrt{5x}$ **3.** $-23\sqrt{2x}+15\sqrt{5}$ **4.** $-b\sqrt{6}$ **5.** $-80\sqrt{7}$ **6.** $x-3\sqrt{x}$ **7.** $2\sqrt{5}-3$ **8.** $2+\sqrt{14}-\sqrt{10}-\sqrt{35}$
9. $25a-4b$ **10.** $8\sqrt{3}+19$ **11.** $x-2\sqrt{5x}+5$ **12.** $12\sqrt{6}+30$ **13.** 7 **14.** 5 **15.** 4 **16.** 31 **17.** 7 **18.** 1 **19.** 4 **20.** -1
21. -5 **22.** empty set solution **23. a.** yes **b.** 50.1% **c.** 2010

Cumulative Review of Chapters 1–11

1. $-15a^2b^2+6ab^3$; fourth-degree polynomial **2.** $-13p+1$; linear polynomial **3.** $9x^2-7x-3$; quadratic polynomial
4. $-x^3-x^2+3x-1$; cubic polynomial **5.** $12m^2-16mn-35n^2$; quadratic polynomial **6.** $3x^3-10x^2+x+6$; cubic polynomial
7. $4p^2+12pq+9q^2$; quadratic polynomial **8.** $\dfrac{15}{(m-5)(m+2)}$; rational expression **9.** $\dfrac{2(w+3)}{(w-5)(2w+1)}$; rational expression
10. $\dfrac{2(7x+12)}{5(x-6)(x+6)}$; rational expression **11.** $\dfrac{-5x^2+18x+12}{(x-3)(x+2)(x+4)}$; rational expression **12.** $\dfrac{\sqrt{21}}{7}$; radical expression **13.** $-4\sqrt{3}$;
radical expression **14.** $x-\sqrt{2x}+\sqrt{5x}-\sqrt{10}$; radical expression **15.** $4b-25c$; radical expression **16.** $10-2\sqrt{21}$; radical
expression **17.** $y=-3x^2-12x-11$ **18.** $54x^{13}y^{28}$ **19.** $5x\sqrt{2y}$ **20.** $\dfrac{\sqrt{7t}}{7}$ **21.** -30 **22.** $\dfrac{17}{6}$ **23.** $(4p-9q)(4p+9q)$
24. $(a-9)(a+3)$ **25.** $(w+7)^2$ **26.** $n(2m+3n)(3m+2n)$ **27.** $2(x-4)^2$ **28.** $(x-3)(x+3)(x+5)$ **29.** $4x^2(x+2)(3x-5)$
30. $(x+3)(x^2-3x+9)$ **31.** $x<-3; (-\infty, -3)$ **32.** **33.** **34.** $\dfrac{8}{5}$; linear equation

35. $\dfrac{37}{6}$; linear equation **36.** $-3, 8$; quadratic equation **37.** $\dfrac{1}{6}, 1$; quadratic equation **38.** $\pm\dfrac{2\sqrt{15}}{3}$; quadratic equation
39. $-4\pm2\sqrt{15}$; quadratic equation **40.** $\dfrac{5\pm\sqrt{73}}{6}$; quadratic equation **41.** $\dfrac{-3\pm\sqrt{3}}{2}$; quadratic equation **42.** 8; rational equation
43. 5; rational equation **44.** $\dfrac{9}{25}$; radical equation **45.** 4; radical equation **46.** $t=\dfrac{v-v_0}{a}$ **47.** $2\pm3\sqrt{2}$ **48.** Answers may vary.
49. $(-2, -1)$ **50.** $(-3, -11)$ **51.** **52.** **53.** $(10, 0)$ **54.** $(-8, 0), (6, 0)$ **55.** Ski run B

56. $\frac{1}{2}$; increasing   **57.** $y = -\frac{2}{5}x + \frac{14}{5}$ **58.** $y = -\frac{7}{8}x + \frac{9}{4}$ **59.** $y = -\frac{1}{2}x + 1$ **60.** $2\sqrt{21}$ **61.** $1, \frac{5}{2}$ **62.** $-\frac{7}{x-5}$ **63.** $-9°F$

64. 2.25 tablespoons   **65.** 27   **66.** 5   **67.** $\frac{x+12}{2(x-10)}$ **68.** 4% account: $7500, 7% account: $4500   **69. a.** $A = -1350t + 27,500$

b. 18,050 feet   **c.** 19.48 minutes   **d.** -1350; the airplane descends by 1350 feet per minute.   **e.** (20.37, 0); the airplane will land

20.4 minutes after beginning its descent.   **70. a.** 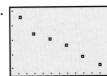 linear equation   **b.** $p = -0.31a + 37.92$ **c.** -0.31; the

percentage of American adults who say that they are interested in soccer decreases by 0.31 percentage point per year of age.

d. 30.2%   **e.** 55 years   **f.** (122.32, 0); no 122-year-old

Americans are interested in soccer. A little research would show that the oldest American to date was 113 years old.

71. a. $-0.136t + 7.35$; the expression represents the total average daily oil production in the year that is t years since 1990.   **b.** 4.5; in

2011, total average daily oil production will be 4.5 million barrels per day.   **c.** Each year, total average daily oil production declines by

136 thousand barrels.   **d.** 2008; 2.43 million barrels per day   **72.** 12.4 miles   **73.** 1524 pens

74. a. 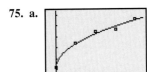 quadratic equation; answers may vary.   **b.** 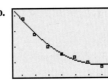 yes   **c.** 35%   **d.** 34 years

75. a. 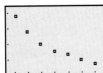 yes   **b.** 38.6%   **c.** 56 years

Index